Cannabis

Cannabis

EVOLUTION AND ETHNOBOTANY

ROBERT C. CLARKE

AND

MARK D. MERLIN

UNIVERSITY OF CALIFORNIA PRESS

Berkeley   Los Angeles   London

University of California Press, one of the most distinguished
university presses in the United States, enriches lives around
the world by advancing scholarship in the humanities, social
sciences, and natural sciences. Its activities are supported by the
UC Press Foundation and by philanthropic contributions from
individuals and institutions. For more information, visit
www.ucpress.edu.

University of California Press
Berkeley and Los Angeles, California
University of California Press, Ltd.
London, England

Limited hard cover edition, published 2013.

First Paperback Printing 2016

Library of Congress Cataloging-in-Publication Data
Clarke, Robert Connell, 1953–
    Cannabis : evolution and ethnobotany/Robert C. Clarke,
Mark D. Merlin.
        p. cm.
    Includes bibliographical references and index.
    ISBN 978-0-520-29248-2   (pbk.: alk. paper)
    1. Cannabis. 2. Cannabis—Evolution. 3. Cannabis—
Utilization. 4. Human-plant relationships. I. Merlin, Mark
David. II. Title.
SB295.C35C54 2013
362.29'5—dc23   2012036385

Manufactured in China
22   21
10   9   8   7   6   5   4

The paper used in this publication meets the minimum requirements of ANSI/NISO Z39.48-1992 (R 2002) (*Permanence of
Paper*). ⊗

DEDICATION

Richard Evans Schultes (1915–2001) was born, raised, and worked most of his life in and around his beloved Harvard University, where he was an undergraduate and graduate student and then a longtime professor of botany and director of the Botanical Museum. However, Dick, as he liked us to call him, is best known for his 13 years of field research in the upper Amazon Basin, principally in Colombia. One of his several well-known and very accomplished graduate students, Wade Davis, wrote a well-received book (*One River*, Simon & Schuster, 1996) that features the challenging and engaging experiences of Professor Schultes among numerous indigenous groups in remote tropical rainforest regions of northern South America. Readers are encouraged to read Davis's book and appreciate the zeal and fortitude of Schultes's botanical and cultural explorations but, more important, to fathom the profound human-plant relationships of the native people whom he befriended and deeply respected. Schultes was especially sensitive to the ecological, spiritual, and medicinal plant relationships that indigenous peoples developed throughout the world. His profound research and mentorship have become almost legendary. The authors of this book are among the many students, scientists, and scholars who have been deeply inspired by the scientific goals and methodologies of our dear friend, Dick Schultes. We dedicate this book to his memory and celebrate his many contributions to ethnobotany, of which he is certainly one of our modern founding fathers.

CONTENTS

PREFACE

Cannabis is one of the world's most useful plant groups. It has been a part of human culture for thousands of years beginning in Eurasia, and today it is associated with people in almost all parts of the world. Although Cannabis is most often thought of as a "drug plant," its use for a huge number of other purposes including fiber, food, paper, medicine, and so on is almost unparalleled, ranking it with the coconut palm and bamboos. Cannabis is truly a remarkable genus of multipurpose plants with extensive and complicated histories. A fully comprehensive, documented history of Cannabis's evolution and its widespread, diverse use by humans has never been published. This book is an attempt to accomplish that task. The evolution of Cannabis and the great variety of human-Cannabis relationships are presented here in greater depth than ever before. How this project developed and progressed is an interesting story in itself.

The coauthors have worked earnestly over the past 15 years or more to produce this book; however, they began to focus their scholarly and scientific interests on Cannabis well before their collaboration started in 1996. Both Mark Merlin and Robert Clarke first initiated their research on Cannabis while enrolled in undergraduate programs in the University of California system at their respective campuses decades ago. Their independent and joint field work involving the genus necessitated extensive travel across several continents to countless libraries and museums, complemented by innumerable interviews in regions where Cannabis has either ancient roots or only a relatively modern history of cultivation and use. This volume represents the better part of two scholarly careers spent following the historic trail of both the evolutionary biology and ethnobotanical heritage of Cannabis.

It is quite fitting that University of California Press has decided to publish this extensive body of research. Mark Merlin's father graduated from UC Berkeley in 1925 with a bachelor's degree in philosophy and Mark earned his undergraduate degree in history at UC Santa Barbara in 1967. Robert Clarke's mother was a budget analyst for UC San Diego and he received his degree in biology from UC Santa Cruz in 1976.

In the fall of 1967, as an incoming graduate student at the University of Hawai'i at Mānoa (UHM), Mark began to think about what topic he would choose to study for his master's thesis in geography. He had been exposed to the exponential rise in the use of psychoactive Cannabis in California in the second half of the 1960s. Out of almost nowhere the illicit smoking and eating of Cannabis began to spread at a phenomenal rate—at UC Santa Barbara, across the state of California, in several other areas in North America, and beyond. Mark wondered, what is it about this "drug plant" that allowed it to become such an attractive item of recreational and spiritual interest over what seemed to him to be a very short time? Where did these plants come from, what were their geographical origins and what was their history of relationships with humans over time? And indeed why was Cannabis being so actively suppressed by governments across the globe?

Mark decided that the ancient history of human-Cannabis relationships would be an interesting topic and consulted with his graduate advisors at UHM. He informed them that he was inspired by the "man-land" (human-environment) tradition in geography and most specifically with the work of Carl O. Sauer, the UC Berkeley geographer who explored the origins of agricultural crops and how their relationships with humans changed landscapes far and wide over much of the planet. His advisors eventually gave Mark permission to pursue his academic goals. Mark spent the next two years carrying out research and writing a thesis that was accepted in 1969. Then with the support of two professors at UHM, one in the English department and another in the history department, Fairleigh Dickinson University Press published his thesis as a hardback book, Man and Marijuana: Some Aspects of Their Ancient Relationships, in 1972, and A. S. Barnes and Company published it as a paperback in 1973. The book was well received and is still a key resource for the early history of Cannabis.

Richard Evans Schultes, the highly respected former director of the Botanical Museum of Harvard University and longtime professor at that institution, was much impressed by Mark's thesis on Cannabis, as well as his later published PhD dissertation on the evolution and early human use of the opium poppy (On the Trail of the Ancient Opium Poppy, Associated University Presses, 1984). As a result, he invited Mark to spend part of his first sabbatical leave from UHM with him at Harvard. During that time in 1985, the "seeds" of this book were sown in Mark's mind. Subsequently, Schultes and Professor Harold St. John (UHM) nominated Mark for membership in the Linnaean Society of London into which he was inducted in 1987 for his contributions to ethnobotany. Some years later, Mark's coauthored a well-received book on kava (Lebot, Merlin, and Lindstrom, *Kava: The Pacific Drug*, Yale University Press, 1992) which was published with the scholarly support of Professor Schultes. Later, Michael Pollan referred to Mark's earlier book on marijuana as an important historical reference in his best-selling book *The Botany of Desire* (Random House, 2001).

Early in the 1970s, Rob began his scholarly involvement with *Cannabis*. While attending UC Santa Cruz as an undergraduate biology major, he asked his advisor Jean Langenheim, who dedicated her career to the study of plant terpenes (volatile unsaturated hydrocarbons found in the essential oils of plants) and the distribution of fossil amber–producing plants, if he could focus on *Cannabis* for his senior thesis. He argued that the book Mark Merlin had published on marijuana was proof that a scholarly investigation of the biological and cultural origins of *Cannabis* was a valid line of academic inquiry. Rob's advisor was impressed with the references Rob had already amassed and granted him permission to pursue his research on the botany and ecology of *Cannabis*.

The written product of Rob's extensive botanical research not only satisfied his undergraduate requirement and earned him honors but also was self-published in 1977 as *Botany and Ecology of Cannabis*. That publication was expanded and took on a form that has been in print for more than three decades, now in three languages (*Marijuana Botany: The Propagation and Breeding of Distinguished Cannabis*, And/Or Press, 1981). Rob spent much of his career based in Amsterdam working as a plant breeder with HortaPharm BV and serving as projects manager for the International Hemp Association. He also wrote another lengthy book on *Cannabis* (*Hashish!*, Red Eye Press, 1998). In addition, Rob collaborated with John McPartland and David Watson on *Hemp Diseases and Pests: Management and Biological Control* (CABI Press, 2000) and has written many peer-reviewed articles on various aspects of *Cannabis* botany, hemp cultivation and processing, hempen textiles, and various other manifestations of human-*Cannabis* relationships. Michael Pollan interviewed Rob in Amsterdam while writing *The Botany of Desire* and included a number of relevant comments by Rob in his best-selling book.

Rob and Mark communicated by mail back in the 1970s when Rob wrote Mark to tell him that his master's thesis cum book had inspired his own academic research. Mark congratulated him on his research and initial publication, indicating that it significantly expanded the literature available on the botany of this important genus. For the next two decades, the two were out of touch.

Then in 1996, while they were, unbeknownst to one another, both attending the joint annual meetings of the Society for Economic Botany and the Society for Ethnopharmacology at Imperial College in London, they met in person for the first time. After lengthy discussions concerning their respective careers and scholarly involvement with *Cannabis*, Mark and Rob came to the joint conclusion that there was a real need for a comprehensive study of *Cannabis*'s evolutionary biology and ethnobotanical history and that they should write it together.

The two scholars left those meetings in London with a commitment to work in tandem and produce a definitive history of *Cannabis* through the ages. This project became a saga that included many trips to each other's home cities, Honolulu and Amsterdam. Extensive correspondence over the years and occasional brainstorming as their paths crossed in such places as California, China, Thailand, and Vietnam drove them forward while many other independent projects also had to be attended to by each. Nevertheless, they pushed on, sometimes in great spurts of research and writing, other times somewhat bogged down with their own respective research and instructional duties. Slowly but surely the large volume that follows took shape and was refined and edited with their critical feedback to each other's work. A final flurry of work over the past few years led to completion of their labor of love, fascinated as they both remain with this intriguing genus of plants.

The authors are now confident that their synthesis is well worth the large portion of their lives' work that went into its production. Many people helped them along the way, some with extraordinary efforts to improve their manuscript or inform them about important sources of information of which they were not yet aware. Many of those fine folks are listed in the Acknowledgments.

Human-*Cannabis* relationships are more important today than ever and have evolved to a greater extent during the duration of this book project than in many preceding centuries. For students, lawyers, physicians, activists, law enforcement personnel, legislators, elected leaders, farmers, and an enormous range of scholars and aficionados, there is something, perhaps a lot, that will be of interest. If you know of additional sources of accurate, documented information pertinent to our discussion of one of humanity's more ancient plant resources, please let the authors know for future editions that we hope will be inspired by this publication.

ACKNOWLEDGMENTS

Our acknowledgments were written last, which in theory gave us more time to think about them than any other aspect of the 15-year academic odyssey that this book represents. In practice, this time period amounts to a quarter of the authors' lives. As a result, it is impossible to recall each of the countless persons who offered their helpful support, ranging from critical scholarly input to providing comfortable and thought-provoking sets and settings. Regrettably, we have omitted many of you who may simply take to heart that we very much appreciate your friendship.

Although we accept all responsibility for editing oversights, we relied gratefully on the editorial experience and exuberant participation of Annie Riecken, Michael Backes, Craig Fleetham, and others. We were also very much encouraged by the initial reviews by Richard Tucker, David Burney, Ethan Russo, and Norman Grinspoon.

David Watson pioneered many aspects of the modern *Cannabis* realm and since our early friendship has consistently shared both insights and inanities surrounding the human-*Cannabis* relationship. Karl Hillig's investigation of *Cannabis's* evolution using modern molecular techniques provided the basic taxonomic system we adopted for this book. *Cannabis* breeder Etienne de Meijer and colleagues published many core papers concerning *Cannabis's* genetic diversity and the inheritance of the genus. John McPartland and colleagues explored the cutting edge of human-*Cannabis* coevolution, and his meta-analytical analyses provided us with much food for thought. Academic inspiration and support flowed freely from fellow members of the Society for Economic Botany, and we wish to thank in particular Michael Balick, Paul Cox, Trish Flaster, Peter Matthews, Will McClatchey, John Rashford, and Doug Yen. Sid Kahn provided wizened views into Central Asian *Cannabis* history.

Unique images were graciously supplied by Shelly Benoit, Jiang Hong-en, Li Xiao, Rohit Markande, Todd McCormick, Nick van der Merwe, Marc Richardson, Mojave Richmond, Neil Schultes, Judy Sky, David Watson, and the University of Kentucky Archives. All figures were adeptly manipulated in the virtual realm by Nancy Hulbirt. Matt Barbee drafted our wonderful maps.

Thanks to the following for their generous academic support and expertise: Hannu Ahokas, Michael Aldrich, John Allen, Elizabeth Barber, Janos Berenji, Jean Black, Ivan Bócsa, Jeff Boutain, Otto Brinkhammer, Teresa Brugh, Ricardo Gonzalez, Gu Wen-feng, Dorian Fuller, Allan Hall, Ho Huu Nhi, Stefanie Jacomets, Hongen Jiang, S. K. Jain, Petr Kuneš, Nicolai Lemeshev, Laurel Kendall, Marsha Levine, Erica Longtin, Ilze Loze, Marco Madella, Jesse Merlin, Nguyen Van Viet, Sandra Olsen, Pham Thi Minh Loan, Claire Shimabukuro, Andrew Sherratt, Sergei Shuvalov, Claudia Sureda, Sim Yeon-ok, Joseph Streeter, Siim Veski, Ben Wadman, Goro Yamada, Yan Xia, Yang Jongsung, and perhaps most of all, the University of Hawai'i at Mānoa and the International Hemp Association.

We would also like to acknowledge the profound inspiration received from the works of Carl O. Sauer (professor of geography at the University of California at Berkeley), Charles Heiser (professor of botany at Indiana University), Edgar Anderson (professor of botany at Washington University in St. Louis), Jack Harlan (professor of plant genetics at the University of Illinois Urbana-Champaign), Nikolai Vavilov (former director of the Russian All-Union Institute of Agricultural Sciences at Leningrad), and Richard Evans Schultes (professor of botany at Harvard University). Their scientific and scholarly ideas regarding human-environment and plant-people relationships permeate our discussions of the evolutionary biology and ethnobotanical history of *Cannabis*. We stand on the shoulders of giants.

Ultimately, we respect our patient yet ever-questioning parents, who encouraged us to pursue that which we truly enjoyed in their hopes that we should do well.

NOTES TO READERS

During the long course of writing this book, we have adopted several conventions that may differ from other scientific publications, and they are explained here.

Acronyms

Throughout this book we follow the migration routes and evolutionary interactions of various *Cannabis* gene pools and their representative taxa. For clarity and readability we represent these gene pools with acronyms such as BLH, rather than broad-leaf hemp or *Cannabis indica* ssp. *chinensis*. Full names are used when they first appear in each chapter, and readers should refer often to Table 1, which lists the acronyms used in this book.

Place Names

The time frame of this book stretches from prehistory into the future. Geographical and political names have changed repeatedly during the historical period and are likely to do so again. We have chosen to use the names and political boundaries of present-day countries and regions, even though they may not have been used in the past since this seems easier for the reader than switching names and boundaries during each time period. Geographical terms are the most recently accepted Roman character versions of foreign terms, many of which have only recently come into common Western usage, such as the "Hengduan Mountains" and "Yungui Plateau" of southwestern China. We also choose to capitalize geographical features such as Yangzi River but not to capitalize political designations such as Sichuan province.

Spelling

We use Roman character versions of foreign words and phrases familiar to those who predominately read English. Most Chinese words in the text are presented in modern-day Pinyin without diacritical marks, as well as translated to English, and simplified Chinese characters and Pinyin versions with diacritical marks are included for particularly important terms or where linguistic connections would be less clear without them. We capitalize Chinese and Korean dynasty names as well as historical periods and geological eras and epochs.

Dates

In a modern cross-cultural treatise for readers from diverse religious and historic backgrounds, we feel it is appropriate to use BCE (before the Common Era) rather than BC (before Christ), and CE (of the Common Era) rather than AD (*anno Domini*, "in the year of the Lord"). Dates BCE are noted as such, as are early dates CE (generally before 1000 CE), and after this time the epithet is dropped as in 1492 or 2010. For geological time spans and most archeological data, we choose to use BP (before the present) rather than the date (BCE or CE) so the reader will understand how long ago things happened. We make an exception when referring to historical dynastic eras, where we use dates.

Photographs

All photos in this book were taken by Robert C. Clarke unless otherwise noted.

TABLE 1

Acronyms used in this book to represent differing *Cannabis* gene pools, based largely on Hillig (2005a/b).
Biotypes and their corresponding acronyms, Latin names (scientific binomials),
traditional geographical ranges, population status, and primary uses are listed here.

Acronym	Biotype	Binomial	Early range	Population status	Uses
PA	Putative ancestor	*Cannabis ruderalis*	Northern Central Asia	Putative *C. sativa* and *C. indica* ancestor; either wild or ancient feral escapes	Possible ancient use for seed and crude fiber
PHA	Putative hemp ancestor	Either extant and unrecognized or extinct	Balkan Peninsula and Caucasus Mountains during last ice age	Hypothetical *C. sativa* ancestor	Possible ancient use for seed and crude fiber
NLHA	Narrow-leaf hemp ancestor	*Cannabis sativa* ssp. *spontanea*	Eastern Europe and Central Asia	Putative NLH ancestor; more likely feral NLH	Seed and crude fiber
NLH	Narrow-leaf hemp	*Cannabis sativa* ssp. *sativa*	Europe	Cultivated and feral	Seed and textile fiber
PDA	Putative drug ancestor	Either extant and unrecognized or extinct	Hengduan Mountains and Yungui Plateau during last ice age	Hypothetical *C. indica* ancestor	Possible ancient use for ritual and medicinal drugs
BLHA	Broad-leaf hemp ancestor	Either extant and unrecognized or extinct	Eastern Asia	Hypothetical BLH ancestor	Possible ancient use for seed and crude fiber
BLH	Broad-leaf hemp	*Cannabis indica* ssp. *chinensis*	China, Korea, Japan, and Southeast Asia	Cultivated and feral	Seed and textile fiber
NLDA	Narrow-leaf drug ancestor	*Cannabis indica* ssp. *kafiristanica*	Himalayan Foothills—Kashmir to Myanmar	Putative NLD ancestor; more likely feral NLD	Drugs—marijuana and hashish
NLD	Narrow-leaf drug	*Cannabis indica* ssp. *indica*	South and Southeast Asia, Middle East	Cultivated and feral	Drugs—marijuana and hashish; also fiber and seed
BLD	Broad-leaf drug	*Cannabis indica* ssp. *afghanica*	Northern Afghanistan and Pakistan	Cultivated and possibly feral	Drugs—hashish

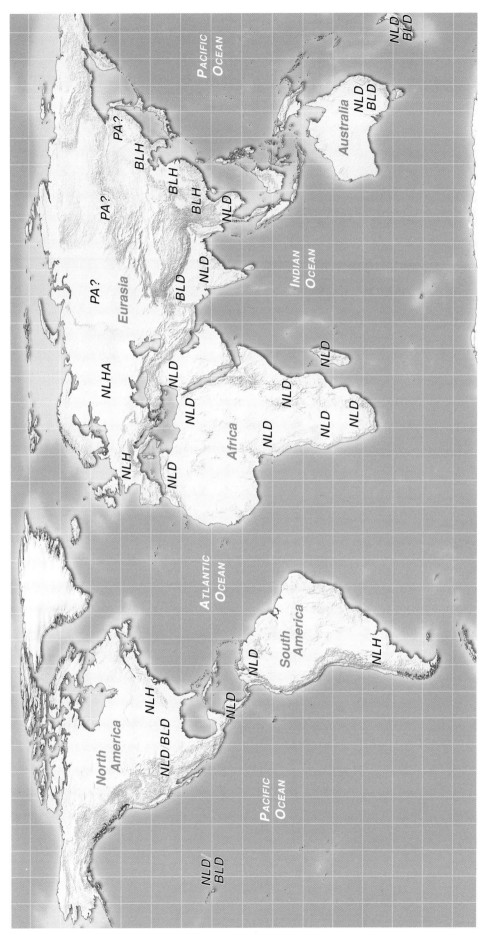

MAP 1. Present-day ranges of *Cannabis* gene pools indicated by their biotype acronyms (cartography by Matt Barbee).

CHAPTER ONE

Introduction to the Multipurpose Plant *Cannabis*

Throughout the ages, [*Cannabis*] has been extolled as one of man's greatest benefactors—
and cursed as one of his greatest scourges. [*Cannabis*] is undoubtedly an herb that has
been many things to many people. Armies and navies have used it to make war, men
and women to make love. Hunters and fishermen have snared the most ferocious creatures, from the tiger to the shark, in its Herculean weave. Fashion designers have dressed
the most elegant women in its supple knit. Hangmen have snapped the necks of thieves
and murderers with its fiber. Obstetricians have eased the pain of childbirth with its
leaves [female flowers]. Farmers have crushed its seeds and used the oil within to light
their lamps. Mourners have thrown its seeds [inflorescences] into blazing fires and have
had their sorrow transformed into blissful ecstasy by the fumes that filled the air.

(ABEL 1980)

IN THE BEGINNING: CIRCUMSTANCES OF EARLY HUMAN
CONTACT WITH *CANNABIS*

A BRIEF SUMMARY OF THE LONG AND DIVERSE HISTORY OF
RELATIONSHIPS BETWEEN *CANNABIS* AND HUMANS

WHAT SHALL WE CALL THESE PLANTS?

SHOULD WE PRAISE OR CONDEMN THIS MULTIPURPOSE PLANT?

WHAT WE DISCUSS IN THIS BOOK

In the Beginning: Circumstances of Early Human Contact with *Cannabis*

Over the vast time span within which humans have known
and used *Cannabis* for many purposes, it has been heralded as
one of humankind's supreme resources and cursed as one of
our utmost burdens. As an introduction to this controversial
plant, we have constructed a possible scenario for the origins
of *Cannabis* use by humans, utilizing botanical, ecological,
and archeological evidence. Hypothetical early human contact with *Cannabis* and the subsequent discovery and application of its useful resources took place during the distant
past in one of the more temperate and well-watered areas of
ancient Central Asia.

It was springtime many thousands of years ago. A long ice
age had recently ended, and a small group of nomadic people
was on the move, venturing far from their ancestral territory.
Finding a suitable clearing near the bend of a meandering
river, they stopped to camp. They had migrated into this
remote location under pressure from other more powerful
and aggressive human groups.

In their new open environment, they constructed simple
thatch shelters in which to sleep, store their few belongings,
and protect their families from the elements. At this time,
humans had not yet developed techniques for cultivating
plants and domesticating animals. Like all other peoples
during this ancient era, this group depended completely
on hunting and gathering their food and other required
resources.

Women spent much of the day searching for and collecting
seasonal wild edible fruits, roots, grains, vegetables, grubs, and
nuts, as well as cordage fibers and fuel wood. Meanwhile, men
tracked and stalked deer, pigs, goats, horses, certain birds, and
other land animals in nearby forests and grasslands, as well as
assisting with seasonal gathering. The river adjacent to their

new settlement supplied water and promised other important
natural resources critical for survival. Fish were also potentially useful if they could figure out how to catch them.

As time passed, they increasingly disturbed the clearing
surrounding their settlement and in the process, inadvertently created nitrogen-rich soil environments by depositing
organic waste materials in dump heaps. By trampling and
cutting back much of the original vegetation, the immigrants
unintentionally favored several sun-loving plants that were
preadapted to the new, human-made open scars with waste-
enriched soil.

One plant that often colonizes dump heaps or waste areas
in open environments is *Cannabis*, a tall herb that is naturally adapted to disturbed or sunny habitats. Toward the
end of the short, warm summer, women gathering seasonal
fruits and nuts discovered stands of wild hemp full of ripe
seeds along the river near their settlement. They teased out
and tasted a few seeds and decided they were worthwhile
food. Unable to remove the myriad of seeds easily, they cut
whole plants with seeds still attached and dragged them
back to camp. Thus seeds of this conspicuous herb were
brought into the group's clearing during their search for
food. Here, *Cannabis* found a favorable niche in the sunny,
moist, and well-drained soil, nutrient-enriched by human
activities.

Women experimented with these plants, letting them dry
and flailing them against cleared ground. As they whipped
the dry plants against the open earth, seeds flew into the air.
Most landed near the threshing where they were swept up,
but a few strayed farther and were not retrieved. Others seeds
were left behind in threshed plants that were discarded onto
dump heaps.

By the end of the next cold season, new spring show-
ers gave the forgotten hemp seeds the necessary moisture required for germination and growth, and the plants

FIGURE 1. *Cannabis* naturally colonizes open streamside habitats with ample sunlight, water, nutrients, and air movement. The plants shown here are *Cannabis* spontaneously along a watercourse (A) in the northeastern State of Megalaya in India. Feral *Cannabis* is highly adaptable and can grow and reproduce in a wide variety of temperate habitats, even under extreme conditions such as in a concrete culvert (B) along a highway in rural southwestern China. *Cannabis* is opportunistic and thrives in nutrient-rich waste heaps resulting from human activities (C) as illustrated by this population in Arunachal Pradesh, India. *Cannabis* is also a formidable weed in field crops as seedlings (D) grow rapidly in competition for sunlight.

(*continued*)

FIGURE 1 (*continued*). Adult plants commonly appear in temperatefield crops *Cannabis* on open fertile land, as in Yunnan province, China (E). Feral plants are often favored by humans and encouraged to survive, either because they are potentially useful for seed and drugs, as in western Nepal (F), or simply because they are strikingly attractive, as along a main street in Kunming in Yunnan province, China (G).

FIGURE 2. Humans living across a wide geographical range have found diverse uses for *Cannabis*. Examples of this remarkable variety include the cloth of Hmong tribal weavers (A) in rural villages of southwestern China, snack seed food sold (B) in eastern Chinese cities, ritual clothing worn in traditional Korean Confucian funerary worship (C) and contemporary Western recreational and medicinal drug use (D).

FIGURE 3. *Cannabis* is a versatile crop plant grown with various strategies for different uses. Hemp fiber fields are sown close together to promote stalk elongation and fiber yield, as in the Netherlands (A); hemp seed fields are sown farther apart to allow branching to maximize seed yield as in Gansu province, China (B); and uniform crops of potent seedless female flowers for medicinal and recreational uses are produced by transplanting genetically identical vegetative cuttings as in Switzerland (C).

flourished through the summer, thriving on available water, sunlight, and nitrogen-rich piles of organic waste. Soon the women began to harvest the hemp seeds around their nearby rubbish piles, making fewer trips farther from home in search of wild hemp. Within a few years, trips to collect wild hemp ceased, and then the seed was harvested only from self-sown *Cannabis* plants in disturbed environments near their settlement. Human and plant interactions such as these were the bridge between hunting and gathering and agriculture; these were the incipient moments of early settled farming.

Like all traditional peoples, past and present, these early humans knew their immediate environment intimately through their own experiences and information passed on orally from their ancestors. As a key element of survival, they were quite familiar with local plants, animals, and inorganic materials, and most of their hunting and gathering equipment was fabricated from local plant and animal sources.

The group's store of knowledge developed slowly, and when challenged by a new living situation, they were eager to develop new techniques to utilize unfamiliar animal and plant resources. As the newly introduced *Cannabis* populations grew larger around the settlement, they became increasingly conspicuous. Could *Cannabis* offer other benefits for survival? Their curiosity grew, and through a process of trial and error, they experimented with its uses.

They knew initially that edible *Cannabis* seeds borne in clusters on the female plants contained a nutritious oily substance. Soon they discovered that they could also be used as a source of oil for cooking, fuel, or even as a base material for crude soap. They already knew about the uses of fibers and eventually recognized the extraordinary fibrous qualities of *Cannabis*. They wore animal skins and furs held together with thongs and were always searching for new plants and animals that could provide durable fibers. However, they had yet to learn the crafts of spinning and weaving.

These early settlers eventually learned they could peel bark from the hollow *Cannabis* stalk and extract long fibers that were easily utilized. They also learned that hemp fibers were very strong, long lasting, and water resistant. As they experimented with methods for fiber extraction, the group saw that by soaking long *Cannabis* stalks in pools along the river and letting them partly decompose, the process now known as retting took place. After sufficient time, most of the adhesive layers of the stalk decomposed into water-soluble juices, and the insoluble, water-resistant materials (the long fiber cells) were left to be more easily collected and dried. They experimented with the fibers, creating strong, durable, waterproof cords and later discovered how to spin yarn and weave cloth with hemp fiber.

Fish in the retting pools were stunned by a lack of oxygen and/or the water-soluble plant juices and floated to the surface in a senseless state. They were in no way rendered inedible; however, in a stupefied state they were easily gathered. Relatively easy access to an important food resource stimulated early humans to experiment with the construction of fish lines and nets made from water-resistant hemp fibers.

But was the need for fiber or food the only reason for their interest in *Cannabis*? Perhaps it was first used for its spiritual or euphoric value and thus initially employed for entertainment or ceremonial purposes. In their ceaseless quest for food, they could have first realized *Cannabis*'s psychoactive potential while eating its seeds. The small, resin-covered bracts surrounding the seeds are potentially psychoactive and could have been ingested along with the seeds; however, the potent smoke breathed in when *Cannabis* plants were burned would have induced a more rapid onset of mind-altering experiences. At first unintentionally, early humans ventured into new realms of cognitive experience and soon favored *Cannabis* as a spiritual, recreational, or medicinal ally.

Psychoactive *Cannabis* resin (complex mixture of aromatic compounds and cannabinoids) can induce rapturous and joyous sensations, ranging from mild reverie and a general sense of well-being to ecstasy and hallucination. In our ancient past, these experiences probably generated a deeper interest in the plant as they do for some today. If only temporarily, the mind-altering resin could have opened new "doors of perception" for early peoples. Use of the psychoactive resin may have become a key mental and physical refuge from frequently monotonous and strenuous patterns of life.

Consuming *Cannabis* also could have had an explosive effect on early people's world view and ideology. Early hunting and gathering groups guarded and handed down "mysteries" or cosmological explanations that served as their interpretations of reality, and these spiritual explanations helped them understand life and death in their own cultural contexts. The ecstatic, visionary effects of *Cannabis* ingestion may have morphed these mysteries into a new system of beliefs and symbols, psychologically precipitating the invention and interpretation of invisible spirits, both malevolent and benevolent. If so, these early people came to regard the plant as a gift from their ancestors and their gods to be used as a vehicle for transcending to higher planes of consciousness. Essentially, *Cannabis* would have provided a means by which they could communicate with their deities—an early "Plant of the Gods" (Schultes and Hofmann 1992).

Regardless of their initial motivation for using *Cannabis*, the group soon realized its many possibilities. They used the plant as a food supplement, an important source of fiber, fuel, and medicine, and they revered its psychoactive properties as a mental elixir for relaxation, recreation, and spiritual communication. Most importantly, by consciously or inadvertently carrying seeds as they migrated, *Cannabis* became part of their transported entourage. Humans and *Cannabis* became linked in a number of ways very early on and have remained so until modern times.

The scenario presented earlier involves a series of hypothetical yet plausible ancient Holocene events in the lives of a Mesolithic hunting and gathering group that was just beginning to experiment with fishing, farming, weaving, and ritual plant use. This succession of events probably recurred often in several regions during the recession of the last glacial age that began the Holocene Epoch about 12,000 years ago and possibly much earlier in the Pleistocene Epoch. This hypothetical group's experiences symbolize some of the possible circumstances behind early human experimentation with *Cannabis*, which evolved into an important and long-lasting multipurpose relationship affecting the evolution of both

human culture and *Cannabis* as a crop plant. The antiquity and depth of this relationship forms the basis of this book.

A Brief Summary of the Long and Diverse History of Relationships between *Cannabis* and Humans

Cannabis has played a profound role on the stage of human history. The development of agriculture, which began approximately 10,000 years ago, has had monumental consequences for humans and our planet, allowing us to exert more control over our food supply and vastly increase our populations and success as a species. In this book, we argue that in some areas of Eurasia, *Cannabis* was a major, if not crucial, player in this transformational change in human ecology. The so-called agricultural revolution in fact took millennia to unfold and is still progressing with new scientific breakthroughs in genetic engineering and environmental manipulation. These modern innovations also affect the role and impact of *Cannabis* in our lives. Through artificial selection of desirable qualities and for a variety of purposes, humans have been manipulating *Cannabis* plants for many thousands of years.

The saga of human-*Cannabis* relationships has been a long, drawn-out affair, an epic association of people and a plant that has influenced history on many fronts in various regions of the world. For instance, hemp was a significant and possibly crucial source of rope used to trap, harness, and command the power and versatility of horses, beginning thousands of years ago in the Eurasian steppes. In this huge region horses have long been used in transportation, hunting, farm work, recreation, and war. Hemp also provided rigging and sails that allowed sailing vessels of the great fleets of Europe and Asia to navigate the oceans for exploration, exploitation, battle, commerce, and travel. *Cannabis*'s function as a vital, nutritious food and source of vegetable oil was significant in the past. Its use for drug purposes, medicinal and mind-altering, licit and illicit, has been widespread not only in our time but also throughout history.

A review of the ancient biogeography, history, breeding, genetics, and multiple uses of *Cannabis* provides us with an enlightened perspective on this age-old natural resource. Before we roll back the clock and consider how our ancient roots intertwine with *Cannabis*, let us review some basics about the genus as it grows naturally in the wild and as a crop plant under cultivation.

What Shall We Call These Plants?

There are many names for the plant in question. You say "weed," and I say "hemp." You say "marijuana," and I say "*Cannabis*." Are we talking about the same plant?

If you call this plant a weed, you may be right depending on your definition of a weedy plant. Some define a weed as a plant growing where it is unwanted. Others refer to a weed as a plant that has escaped cultivation. It is true that in some regions, such as Central Asia, which is probably its original homeland, or in other areas that have similar ecological conditions, such as the American Midwest, *Cannabis* escaped from hemp fields and thrives as a feral plant or naturalized alien weed.

On the other hand, if you define a weed as a plant considered troublesome or useless, you may or may not be right. *Cannabis* plants are troublesome to some, especially farmers

FIGURE 4. *Cannabis* provides many natural resources for humans. Hemp stalks provide fiber used to make cordage (A) and weave cloth. Female flowers (B) provide medicinal and recreational drugs, edible seed (C), hemp seed oil and essential oils (D) are used to make packaged foods and beverages (E).
(*continued*)

FIGURE 4 (*continued*). Whole plants serve as fodder (F), provide educators with a compelling example of a traditional, multipurpose plant (G) and add ornamental beauty to our lives (H).

as well as officials enforcing laws prohibiting cultivation, possession, and use. On the other hand, for many centuries *Cannabis* has provided us with valuable resources, including fiber, food, medicine, and religious sacrament, and so it can hardly be considered useless. Use of the term "weed" is also a colloquialism, being one of many English language nicknames for drug type *Cannabis*.

What about the term "hemp"? The word hemp originally, and still formally, refers to *Cannabis sativa*, a tall Eurasian herb that is widely cultivated for its tough bast (bark) fiber. However, in more recent times the word "hemp" has been applied as a collective noun representing many additional fiber-bearing plants. Today, "true hemp" or "common hemp" refers to *Cannabis*, or more specifically European

Cannabis sativa or narrow-leaf hemp (NLH). The complex history of *Cannabis* as a fiber source in ancient East and South Asia, somewhat later in Western Europe, and during more recent times in North America, is described in detail in Chapter 5.

"Marijuana" is a common, often notorious nickname for our ancient cultivated and weedy ally and is probably of Hispanic derivation. The name "marijuana" refers to both the plant and the dried leaves and flowers that are smoked for mind-altering purposes. The term "sinsemilla" is derived from the Spanish phrase meaning "without seed" and is the name most commonly used for seedless marijuana.

There are many other colloquial or ethnic names that refer to the plant, or more specifically its mind-altering products,

FIGURE 5. The range of variation within each *Cannabis* species and their respective biotypes can be quite pronounced, yet each taxon is characterized by certain common phenotypic traits. Narrow-leaf hemp ancestor (NLHA), *C. sativa* ssp. *spontanea* plants are most often of medium height with medium length internodes and less developed branching, small light to medium green leaves with narrow leaflets, and small, leafy inflorescences and seeds, as in this Central Asian population (A). Narrow-leaf hemp (NLH), *C. sativa* ssp. *sativa* plants are taller with long internodes and more branch development, medium green leaves with large narrow leaflets and larger inflorescences with medium size seeds, such as this French fiber and seed cultivar growing in the Netherlands (B). Broad-leaf hemp (BLH), *C. indica* ssp. *chinensis* plants have more robust stalks and medium to long internodes, more and longer branches, larger deep green leaves with broad leaflets and much larger inflorescences with large seeds, as seen in this fiber and seed crop from Yunnan province, China (C). *(continued)*

including grass, pot, hashish, *kif, ganja*, and so on. Biological taxonomists (scientists who systematically classify and name plants and animals) have placed all marijuana, or true hemp plants, in genus *Cannabis*. Today, many professional and nonprofessional people in countries around the world refer to these plants as "cannabis". Although both common and scientific names will be used in this study, we will predominantly refer to them as *Cannabis*.

It is particularly important to understand the names used throughout this book for various kinds of *Cannabis*. A modern taxonomic treatment of *Cannabis* is presented in Chapter 11. The use of acronyms (e.g., PA, BLD, NLH) to represent different *Cannabis* gene pools, in addition to their scientific Latin binomials with subspecies designations (indicating biotypes or groups of organisms with similar phenotypes or observable characteristics) should encourage readers to think in terms of gene pools while tracing the natural and human-directed evolution of *Cannabis*. Even readers familiar with the history of the taxonomic study of *Cannabis* will find the modern system presented in this book to be quite different from what was previously proposed and of considerable use in resolving previous taxonomic discrepancies.

Should We Praise or Condemn This Multipurpose Plant?

People have cultivated many different kinds of plants since the dawn of agriculture thousands of years ago. Yet perhaps none have been both praised and condemned so much as *Cannabis*. As noted in our hypothetical scenario, *Cannabis*, the notorious provider of mind-altering Δ⁹-tetrahydrocannabinol (Δ⁹-THC, or simply THC) and other psychoactive substances, has been utilized for many purposes in addition to its common use as a religious and recreational drug plant. Its stalks can be used as a fuel source. As we noted earlier, the durable tissue derived from the elongated cells found in the outer bark of *Cannabis* provides fiber for clothing, furnishings, sails, rope, canvas, and other woven materials, as well as for paper. Its nutrient-rich seeds feed humans and domesticated animals; the seed oil is useful for cooking and burning as lamp fuel. And of course, *Cannabis* has many medicinal applications, most underrealized today but which proved helpful in the past and may become even more valuable in the future.

Although people have enjoyed an extraordinarily long and diverse association with *Cannabis* over large parts of our planet, it is now almost universally illegal to grow and sell. There are many books describing various generally illicit aspects of this relationship, but few offer a satisfactory look into the vast and interesting shared history of humans and *Cannabis*, especially our early biological and ethnobotanical experiences with this useful, hardy plant. Hopefully this book sheds more light on a most controversial subject and will help readers better understand why *Cannabis* has been referred to as both the "devil's weed" (from the film *Marijuana* 1936) and the "giver of delight" (from ancient Vedic texts).

FIGURE 5 (*continued*). Putative narrow-leaf ancestor, *C. indica* ssp. *kafiristanica* plants are short to medium in height, with short internodes and many branches, small medium green leaves with narrow leaflets, and small inflorescences with small seeds, such as this example from Bihar state, India (D). Narrow-leaf drug (NLD), *C. indica* ssp. *indica* plants are often tall with medium to long internodes and well developed branches, medium to large medium green leaves with narrow leaflets, and large, long inflorescences and medium size seeds, such as this Mexican variety growing in California (E). Broad-leaf drug (BLD), *C. indica* ssp. *afghanica* plants are often short to medium in height with short internodes, well developed branches, large dark green leaves with broad leaflets, and large leafy inflorescences with medium to large seeds, as seen in this Afghan hashish variety growing in California (F).

What We Discuss in This Book

Where, when, why, and how did early humans use *Cannabis*? What were the environmental conditions in which our relationships with these plants actually began? How and why did the many uses of *Cannabis* spread from one society to another during ancient times? How have humans and *Cannabis* interacted and affected each other, even perhaps directing their coevolution? And what are the lessons to be learned for our time and for the future? This is the story of the evolution of *Cannabis* and its connections to our cultural evolution.

As the "father of modern ethnobotany," Richard Evans Schultes (1969b), pointed out more than 40 years ago, progress in the study of psychoactive drug plants during the second half of the twentieth century owes its "spectacular success to interdisciplinary studies and consequent integration of data gleaned from many seemingly unrelated fields of investigation: anthropology, botany, ethnobotany, chemistry, history, linguistics, medicine, pharmacognosy, pharmacology and psychology." This investigation of *Cannabis* brings together information from all these disciplines and more.

Chapter 2 presents an overview of the natural origins and early evolution of *Cannabis* and proposes a hypothetical model for the evolution of *Cannabis* up to the time when early humans first encountered it. Is the scenario presented in this introduction a realistic description of the earliest human contact and use of *Cannabis*? A coherent, convincing hypothesis regarding the origin of human relationships with any plant must be based on a fundamental understanding of the botany of that species, and especially important is a thorough ecological understanding including environmental adaptations to such variables as climate, soil, landform, and other organisms such as humans. Thus a brief botanical and ecological description of *Cannabis* is also presented in Chapter 2, including an attempt to identify the probable centers of species formation—the regions where the early evolution of *Cannabis* occurred.

Chapter 3 provides a critical, in-depth discussion of the cultural circumstances within which people began to use *Cannabis*, offering an overview of how the early collection of wild plants led to cultivation and in due course to artificial selection of desirable traits and eventually domestication. Human selection has been the most important determinant of *Cannabis*'s evolution and has radically influenced the geographical range of divergent *Cannabis* taxa (reflected in phenotypically differing genotypes or genetic inheritance). Many botanical aspects of *Cannabis* relevant to its natural evolution are discussed, along with a detailed discussion of evolutionary changes in *Cannabis* imposed through artificial selection, using the present-day distribution of chemical variants as a case in point.

Chapter 4 presents methodology for interpreting material evidence for *Cannabis*'s antiquity, its biogeographical spread, and general aspects of its multipurpose use over time and space. This includes reference to pertinent paleobotanical, archaeobotanical, archeological, and historical evidence. Chapter 4 also presents our hypotheses for the ancient use of *Cannabis* seed and fiber in fishing and the early relationships among humans, hemp, and horses tracing a series of phases of cultural dispersal of *Cannabis* from its Eurasian homeland to various regions of the Old World. In the prehistoric period this involved migrating nomads, who perhaps first spread it into China, India, Southwest Asia, Europe, the Mediterranean, and Africa. A six-phase model is presented for the historical dispersal of *Cannabis* by humans.

Chapters 5 and 6 present detailed histories of different traditional uses of *Cannabis* fiber and seed. Thus we learn how *Cannabis* fiber has been utilized for cordage, cloth, and paper in Chapter 5, and then how it provides seed oil, food, and fodder in Chapter 6. In these two chapters, the ancient as well as modern cultural dispersal of *Cannabis* for fiber and seed use throughout much of Eurasia, Africa, and the Western Hemisphere are described and explained.

Chapter 7 presents a detailed discussion of ancient through more modern *Cannabis* use for psychoactive purposes in ritual and recreational contexts. Special consideration is given here to traditional ritualistic applications in order to address the question of whether or not *Cannabis* was originally a "Plant of the Gods" (see Schultes and Hofmann 1992; Merlin 1972, 2003). Ethnobotanical traditions of *Cannabis* use for medicinal purposes are explored in Chapter 8. *Cannabis* also has a long, widespread, and continuing history as a therapeutic agent, and we review medical usage in East and South Asia, the Middle East, Europe, Africa, and the New World.

Some major aspects of *Cannabis*'s use in modern medicine are also discussed.

Chapter 9 examines the large body of evidence from ancient texts, historical accounts, ethnographic research, and archeological sites concerning the ritual use of nonpsychoactive *Cannabis* in various parts of Europe and East Asia. Here we focus attention on symbolic, ceremonial, and spiritual aspects of hemp in traditional cultures where it has long been cultivated.

Chapter 10 continues our investigation of the influence of humans on the evolution of *Cannabis* by following the work of plant breeders during the last century. Through the use of Mendelian selection and breeding techniques, plant breeders developed industrial hemp varieties—seed-propagated cultivars (cultivated varieties) suitable for fiber and/or seed production. Other breeders, both professionals and amateurs, have developed a myriad of seed and vegetatively propagated cultivars for recreational and medical purposes.

Chapter 11 reviews the taxonomic history of *Cannabis*, explores the realms of modern chemotaxonomy and molecular taxonomy, and assesses their bearing on both the *Cannabis* "species question" and evolutionary studies. Recent phylogenetic research is also described and analyzed in terms of its relevance to our understanding of the evolutionary and ethnobotanical history of *Cannabis*.

Chapter 12 expands upon ideas in Chapter 2 and proposes a model for the early evolution of the hemp and hop family (Cannabaceae) and the migration of *Cannabis* as the Holocene epoch began 12,000 years ago.

Chapter 13 reviews the long and complex history of *Cannabis* and its use by humans, continues to examine the depth of our relationship with *Cannabis*, and critically evaluates new evidence for biological coevolution of humans and *Cannabis* at the DNA level. Its past, present, and future multipurpose uses are reviewed with special focus on the present position and future potential of recreational, ritualistic, medical, agricultural, and industrial applications of this resource-rich genus. Here we also present new evidence to support our model for the evolution of *Cannabis* and offer food for thought for future researchers. The comprehensive references section at the end of the book includes citations of all authors cited in the text and should serve as a valuable bibliographic source for further study.

CHAPTER TWO

Natural Origins and Early Evolution of *Cannabis*

What seest thou else
In the dark backward and abysm of time?

(WILLIAM SHAKESPEARE, *The Tempest*)

Introduction

Where and when did humans first come into contact with *Cannabis*? And how and why did people begin to employ these extremely useful plants? In order to support our hypothetical scenario of early human interactions with *Cannabis* presented in Chapter 1, we need to identify when and where the species originated. What were our planet's environmental and biotic conditions during *Cannabis*'s early evolution? What are the environmental conditions in which it grows naturally without human help? Can we realistically understand how *Cannabis* evolved? And if so, where and how did it evolve? To answer these questions we must investigate the basic life cycle and ecological requirements of *Cannabis*. After understanding the botany and ecology of *Cannabis*, including an identification of its closest botanical relatives, an analysis of the processes through which it reproduces, and an application of these parameters to ancient vegetation and climate reconstructions, then we can begin to comprehend its geographical and evolutionary origins (see Chapter 12 for discussions of climate reconstruction, refugia, and species formation).

Many avenues must be explored during a thorough study of the origin and evolution of any cultivated plant. We have utilized diverse subdisciplines of botany, along with archeology, paleontology, history, linguistics, and geography to provide useful insights into the origin and dispersal of *Cannabis* (also see Chapter 4). The data gleaned from each approach have been filtered and then assembled carefully in order to gain the clearest possible view of the plant's antiquity. Several great plant geographers studied *Cannabis*, and we investigate their theories of origin in Central and South Asia. Even so,

the information is often incomplete for a particular region or era, and it is important to resist the temptation to indulge in idle conjecture without sufficient facts to back up hypotheses. With this in mind, we have attempted to assemble enough data to support the hypotheses for the natural origin and early evolution of *Cannabis* presented here. However, we have constructed only the skeleton of our model, and an additional, more rigorous, biosystematic analysis is still needed to flesh out more detail.

Basic Life Cycle of *Cannabis*

Typically, *Cannabis* is a medium to tall, erect, annual herb, but environmental influences strongly affect the growth habits of individual plants throughout its range. Provided with an open, sunny environment; light, well-drained soil; and sufficient nutrients and water, *Cannabis* can grow to a height of 5 meters (16 feet) in a four- to six-month growing season. Exposed riverbanks, lakesides, and agricultural lands are ideal habitats for *Cannabis* since they normally offer good sunlight, moist and well-drained soil, and ample nutrients. When growing in arid locations with negligible soil nutrients, *Cannabis* develops minimal foliage and may mature and bear seed when only 20 centimeters (8 inches) tall. When planted in close stands on fertile soil, as in fiber hemp cultivation, plants do not branch but grow as tall, slender, and straight stalks. If a plant is not crowded (e.g., when cultivated for seed or drug production), limbs bearing flowers will grow from small axial meristems (growing points) located at the nodes (intersections of the petioles or leaf stalks) along the main stalk.

FIGURE 6. Overwintered or sown *Cannabis* seeds sprout (A) in spring as the soil warms, and young seedlings (B) quickly develop into pre-floral juvenile plants (C). Once flowering commences, plants form either female (D center left) or male (D top right) flowers. Individual female flowers emerge in clusters (E), each containing a single ovary (not visible) surrounded by a resin-covered bract from which a pair of receptive stigmas emerges. (*continued*)

FIGURE 6 (*continued*). Clusters of male flowers (F) open to expose stamens that release pollen grains carried by breezes (G), and once exhausted the male flowers fall to the ground (H). A pollen grain carried by the wind lands on a receptive stigma; and if the ovule is fertilized it swells within the closely adhering bract until the seed reaches maturity (I). Seeds fall from the flowers and eventually the whole plant falls to the ground, leaving them dispersed and awaiting the next spring to germinate and grow again (J).

In temperate environments, seeds are sown outdoors in the springtime and usually germinate in three to seven days. The first true leaves arise about 10 centimeters (4 inches) or less above the cotyledons (seed leaves) as a pair of oppositely oriented single leaflets. Subsequent leaves arise in opposing pairs, and a variously shaped leaf sequence develops with the second pair of leaves having 3 leaflets, the third pair having 5, and so on up to 9, 11, and even 13 leaflets. In some warm, sunny climates with favorable soil conditions, *Cannabis* can grow taller by as much as 10 centimeters (4 inches) a day. The rapidly elongating stalks produce a strong bast (bark) fiber used for cordage and woven textiles.

Cannabis exhibits a dual response to day length. During the first two or three months, it responds to increasing day length with more vigorous vegetative growth, but later in the season *Cannabis* requires shorter autumn days to flower and complete its life cycle. Cultivated *Cannabis* produces flowers when it is exposed to a day length (photoperiod) of 12 to 14 hours, which varies with the strain, and all varieties have an absolute requirement of a minimum number of short day lengths (or, more accurately, long nights) that will induce fertile flowering. Dark (night) cycles of 10 to 12 hours must be uninterrupted by light periods in order to induce flowering.

Cannabis is normally dioecious, which means that unisexual male or female flowers develop on separate plants, although cosexual monoecious or hermaphrodite examples with both sexes produced on one plant occasionally do occur (see Chapter 10). *Cannabis* is anemophilous (wind-pollinated) and relies on air currents to carry pollen grains from male plants to female plants.

The first sign of flowering is the appearance of undifferentiated floral primordia along the main stalk at the nodes, one primordium behind each of the paired stipules (leaf spurs), located one on each side of the base of each leaf's petiole. Before flowering, the sexes of *Cannabis* are indistinguishable except for general trends in growth habit—in less crowded conditions female plants tend to be shorter and produce more branches than male plants. When flowering is initiated, the male flower primordium can be identified by its curved, crab's claw shape. This is soon followed by the differentiation of dense clusters of round, pointed flower buds, each having five radial segments. The female primordium can be identified by the enlargement of a tapered curved tubular bract (floral sheath). In both sexes, when flowering begins, the pattern of increasing numbers of leaflets reverses, and as flowering progresses, the number of leaflets per leaf decreases until only a small single leaflet appears below each pair of flowers. The phyllotaxy (leaf arrangement along the main stalk) also changes from opposite to alternate (and remains alternate) throughout flowering regardless of sexual type.

Development of branches bearing flowering organs varies greatly between males and females. Female plants are leafy to the top with many small leaflets subtending the flowers tightly crowded within erect compact clusters, while male plants have only a few small leaves growing sparsely along the elongated flowering limbs. Male flowers hang from long, multibranched, loose clusters formed of small (approximately five millimeters, or 1/5 inch long) individual flower buds along an axis up to 30 centimeters (12 inches) long. Tightly clustered female flowers have two long white, yellowish, or pinkish stigmas (female sexual organs receptive to pollen) protruding from each bract. The bract measures two to eight millimeters (1/12–1/3 inch) in length and adheres closely to the single ovary, completely surrounding it. The bract is covered with hundreds of glandular trichomes (plant hairs). These glands and their resinous secretion may protect the reproductive organs from excessive transpiration and may also repel pests (Clarke 1981). It is this aromatic resin that contains the psychoactive properties that have attracted human attention for millennia (Pollan 2001; also see Merlin 1972; Clarke 1981).

The differences in flowering patterns of male and female plants are expressed in many ways. Soon after pollen is shed, the male plant dies. The female plant may mature for up to five months after viable flowers are formed if little or no fertilization occurs and if it is not killed by frost, pests, or disease. Compared with female plants, male plants show a more rapid increase in height as well as a more rapid decrease in leaf size and leaflet number approaching the single leaflets that accompany the flower clusters.

Many factors contribute to sex determination in flowering *Cannabis* plants. Under average conditions, with a normal day length of 12 to 14 hours, a *Cannabis* population will flower and produce approximately equal numbers of male and female plants, their sex determined by X and Y sexual inheritance (see the following and Chapter 12). Monoecy (male and female flowers on the same plant) is an aberration used by breeders to create relatively stable monoecious hemp cultivars. Artificial selection under cultivation for sex-related characters such as female flower form and seed size may also result in abnormal sex ratios (see Chapters 3 and 10). However, under conditions of extreme stress, such as nutrient excess or deficiency, mutilation, extreme cold, or radically altered light cycles, populations have been shown to depart greatly from the expected one-to-one male-to-female sex ratio, and cosexual individuals of many different phenotypes (observable traits) may arise. Environmental influences and genetic mechanisms affecting monoecious sexual differentiation in *Cannabis* are poorly understood.

Pollination of the female flower results in the browning, shriveling, and eventual loss of the paired stigmas as well as a swelling of the tubular bract inside which the fertilized ovule is enlarging. After approximately three to six weeks, the seed matures and after some time is harvested and dispersed by humans or simply drops to the ground. This completes the normally four- to six-month life cycle in as little as two months or as long as ten months, varying according to its biotype (group of organisms sharing the same genotype) or ecotype (group within a species sharing similar ecological adaptations) as well as ambient environmental conditions. Fresh and fully mature seeds approach 100 percent viability, but this decreases with age. For example, usually at least 50 percent of seeds will germinate after three to five years of storage at room temperature, but without refrigeration, viability of seeds rarely exceeds 10 years. On the other hand, uninterrupted freezing can preserve seeds for decades. We would expect viability over time to be much lower under most natural conditions.

The mature, achene fruit (seed) is partially surrounded by the bract. The calyx is reduced to a seed coat variously patterned in gray, brown, or black. The seed is slightly elongated and compressed, measuring two to six millimeters (1/12–1/4 inch) in length and one to four millimeters (1/24–1/6 inch) in maximum diameter. Seed weights vary from 600 seeds per gram (16,800 seeds per ounce) in wild varieties to very large seeds comprising only 15 seeds per gram (420 seeds per ounce) in cultivated varieties. Larger seeds have long been used as edible grains. *Cannabis* seeds provide an excellent nutritional source of easily digestible protein and essential fatty acids (EFAs; Deferne and Pate 1996; see Chapter 6). Now our focus turns to some of the more significant environmental aspects of *Cannabis* adaptation, growth, and development.

Ecological Requirements of *Cannabis*: Sunlight, Temperature, Water, and Soil

Relationships between an individual plant and its environment are complex and, for most plants, poorly understood. Environmental conditions, in association with the genotype, determine what the phenotype will be—that is, the actual individual organism. We know that a number of different genetic and environmental variables affect the morphology and physiology of a *Cannabis* plant. Key environmental factors influencing growth and development of *Cannabis* plants include sunlight, temperature, moisture, and soil condition.

Although *Cannabis* plants are thermophilic (warmth-loving) and heliotropic (sun-loving), they are more tolerant of shade than many crop plants and may survive in shaded areas, but their biomass and production of pollen and seed will be greatly reduced. *Cannabis* thrives best in exposed places where it does not have to compete with taller plants for available sunlight. Therefore, *Cannabis* plants find a suitable habitat for their energy needs in open environments, such as scars in vegetation created by stream erosion, landslides, and various forms of human landscape alteration. Disturbed soils are vital for proper establishment of feral populations, which are derived from escapes from cultivation and are composed of self-sown, naturalized individuals. If seeds cannot find a crevice in the soil in which to sprout safely, they will be eaten by birds or small mammals; if they do germinate atop the soil, they will dry out and die if their roots cannot find moist soil to penetrate.

Cannabis can become acclimated to high temperatures if sufficient water and nutrients are available, but it does not tolerate extreme cold. Seedlings and young plants are more frost resistant than plants nearing maturity. At higher latitudes, hemp is traditionally planted in late spring and harvested at the end of the short summer, avoiding the cold temperatures and short day length of the low sun in autumn. This climatic adaptation indicates that *Cannabis* is native to a northern temperate region where it can successfully complete its life cycle between spring and autumn without experiencing lethal frosts.

Impact of temperature variation is regulated by the transpiration rate of the plant, or how fast it loses moisture. In hot, dry climates, *Cannabis*'s high transpiration rate makes it very susceptible to wilting. However, the pubescence of glandular trichomes concentrated around the inflorescences, especially of female flowers, helps protect vital reproductive tissues from drying out, slowing water loss by producing a lower surface temperature and slowing transpiration.

Cannabis needs relatively small amounts of water to merely survive, except during germination and establishment. It flourishes on well-drained soils where ample supplies of water are available; on the other hand, stressful arid conditions or waterlogged soil can cause severe stunting and death. *Cannabis* matures and reproduces under a wide range of moisture regimes, especially in subhumid to moderately arid conditions. However, water deficiencies negatively affect root proliferation, branch and leaf development, flower formation, seed production, and resin secretion.

The *Farmer's Cyclopedia* (United States Department of Agriculture 1914) briefly describes the moisture requirements for fiber *Cannabis* cultivation in the temperate continental United States, providing us with insight into its natural adaptation: "Hemp requires about 110 days for its growth. It should have a rainfall of at least 10 inches [25 centimeters] during this period. If the level of free water in the soil is within 8 to 10 feet [2.5–3.0 meters] from the surface, as is often the case in alluvial river-bottom lands, and the character of the soil is such that there is good capillary action to bring the water up, hemp will not suffer from drought, even should there be very little rainfall." *Cannabis* plants display prominent adaptations to a variety of moisture conditions that relate closely to its differing uses. For example, the long fibrous cells in the stalk are much more durable and flexible when grown under mild humid conditions (Klages 1942). Where moisture stress is high, as in hot and dry environments, these same cells are less well developed and more brittle. This difference in stalk cell formation is an important consideration when the plant is cultivated specifically for strong and flexible fiber. Seed yield and quality are also lower in crops starved for water. *Cannabis* must have evolved in at least a seasonally moist and temperate region with warm, wet summers, such as areas with continental temperate or subtropical climates.

Cannabis plants also need well-drained soils. This is an important ecological requirement, as the roots are attacked by various fungi and cannot tolerate standing water. *Cannabis* is generally a tall plant growing in open environments, and the extensive root system needs a friable but nutrient-rich soil to allow proper root growth, adequate drainage, and efficient uptake of vital soil minerals. Under natural conditions *Cannabis* grows best in sandy and loamy alluvial (river valley) soils, and these edaphic limitations help us determine its original geographical origin.

Cannabis Origin and Evolution Studies

Humans have been attracted to *Cannabis* for a very long time, resulting in its wide distribution and multiple uses. Generally, we assume that the longer people use a plant, the greater the number of applications they will find for it. *Cannabis* has been used for millennia as a fiber, food, and drug plant and ranks among the very oldest of economic plants. Many varieties of *Cannabis* have evolved through the pressures of natural selection within the diverse environments into which humans have introduced it, compounded by varying human selective pressures to provide hemp fiber, seed, or resin. We should point out here that a controversy surrounds the taxonomy of *Cannabis*, which has been classified either as a monotypic genus (i.e., containing only a single species), *Cannabis sativa*, or a polytypic genus (i.e., including up to three species), *Cannabis sativa* (NLH and NLHA), *Cannabis indica* (NLD, NLDA, BLD, and BLH), and possibly *Cannabis ruderalis* (PA). We support the latter taxonomy (see Table 1 and also Chapter 11 for a more detailed discussion of *Cannabis* taxonomy). In any case, we suggest that there are three population types for *Cannabis* plants based on their natural origins and associations with humans: (1) those that are truly wild, (2) those that are cultivated, and (3) those that grow spontaneously in areas associated with (and often disturbed by) humans, either derived from truly wild populations or from feral escapes from cultivation. We rely on the ecological requirements and reproductive strategies of *Cannabis* to offer clues as to which regions it inhabited prior to human contact. The present geographical distribution of truly wild and feral

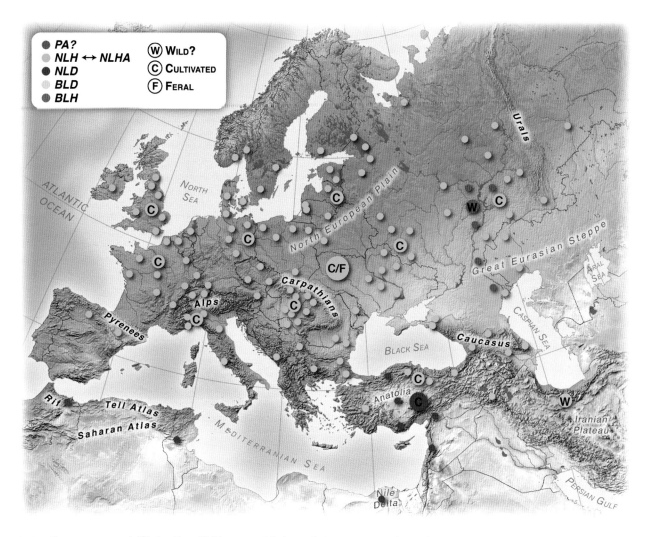

MAP 2. European ranges of differing *Cannabis* biotypes and their population status. Data derived from the written and oral reports of twentieth-century botanists, agronomists, and travelers. Larger circles indicate increased density of populations (cartography by Matt Barbee). (See TABLE 1 at the beginning of this book for explanations of *Cannabis* gene pool acronyms.)

populations should also provide us with a good indication of the geographical region, or at least the ecological conditions, within which *Cannabis* evolved.

The first criterion when searching for the geographical origin of a cultivated plant is to determine the range of its truly wild growth (de Candolle 1967). At first this may seem straightforward, with relatively easy solutions resulting from a survey of herbarium specimens, biodiversity surveys, and guidebooks to native floras. *Cannabis*, however, is particularly difficult to study in this respect as it was among the very early plants to be cultivated and spread by humans. Consequently, it has escaped from cultivation repeatedly and has become naturalized (feral) in a wide range of environments throughout Eurasia and North America. Early *Cannabis* has been characterized as a weedy camp follower, livingon nutrient-rich dump heaps associated with human occupation,and as such was preadapted to cultivation (Merlin 1972; see also Anderson 1967) was preadapted to cultivation (Anderson 1967; Merlin 1972). Consequently, it is difficult for observers to accurately determine if a self-sown population of *Cannabis* is truly wild, and therefore indigenous to a region, or if it is growing spontaneously as a feral escape from ancient or recent cultivation.

Cannabis is particularly adept at naturalizing to a range of temperate and subtropical climates. The contemporary geographical range of *Cannabis* in all its biotypes (ecotypes, subspecies, or varieties) is immense, and it grows spontaneously or cultivated, or both, in many regions. If a plant was recorded in a region and at a later date vanished, it may be assumed that it was either not indigenous to that region and only introduced for a time, or it was native but became extinct in that part of its original, truly wild range. Conversely, because a plant maintains its spontaneous growth in an area does not necessarily mean that it is indigenous to that region; as an introduced species, it might, in fact, have found a niche favorable for its continued proliferation and become naturalized and even invasive. For example, *Cannabis sativa* or narrow-leaf hemp (NLH) is found today growing as a weed along streams, drainage ditches, and in farm fields across temperate continental areas of North America where it was introduced from Europe in the seventeenth century.

The diversity of *Cannabis* populations, both in terms of morphology and economic usage, varies from region to region. Areas of rich diversity are often interpreted as probable places of origin, or at least areas with lengthy periods of naturalization, since increased diversity can be a product of increased

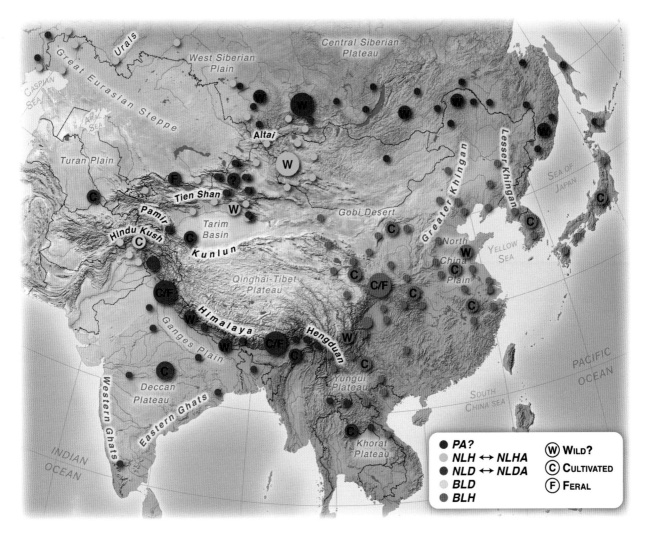

MAP 3. Asian ranges of differing *Cannabis* biotypes and their population status. Data derived from the written and oral reports of twentieth-century botanists, agronomists, and travelers. Larger circles indicate increased density of populations (cartography by Matt Barbee). (See TABLE 1 at the beginning of this book for explanations of *Cannabis* gene pool acronyms.)

time during which to diversify. However, great diversity within a region is not always a sign of antiquity. Alien plants often evolve quite rapidly under a new set of natural and/or human cultural (artificial) selection pressures encountered in new habitats and can diversify extensively in a relatively short time.

The famous Swiss botanist Alphonse de Candolle (1967, originally published in 1882) postulated that agricultural crops in particular are subject to sudden and often radical evolutionary pressures of human selection for a particular plant product such as fiber, food, or drug. A cultivated plant varies from its wild ancestor primarily in those economically or culturally valuable characteristics for which it is grown and selected. Feral escapes from cultivation will often vary in these same characteristics. Other characteristics tend to vary much less, as they are of less importance to the farmer and thus are not as affected by careful scrutiny and selection. De Candolle's basic principles accounting for morphological changes in crop plants during domestication still hold true. However, physiological changes are probably of greater importance as plants adapt to new ecological conditions but are harder to recognize as they leave no direct fossil evidence.

As we have noted, *Cannabis* grows in a wide variety of areas across distant regions of the world and thrives in temperate continental climates. But because its distribution is often so closely associated with human settlements or trade routes, the original native range is obscured. Today it is widely believed that *Cannabis* is indigenous to some area in the broad region referred to as Central Asia (e.g., Vavilov 1931; Schultes 1969a/b; Merlin 1972; Damania 1998).

CENTRAL ASIA: VAVILOV AND THE ORIGINS OF *CANNABIS*

Parts of Central Asia (from the Caucasus to the Altai Mountains), South Asia (through the foothills of the Himalayan and Hindu Kush Mountains), and East Asia (in the mountainous Hengduan-Yungui region or along the Yangzi River and Huang He [Yellow River] of present-day China) have all been proposed as possible locations for the area of natural origin and/or primary domestication of *Cannabis*, and all these regions likely played a role in *Cannabis* evolution at one time or another. Exact geographical origin is unclear today because *Cannabis*'s range shifted repeatedly during glacial-interglacial cycles covering hundreds of thousands of years. Perhaps soon after Holocene warming began about 12,000 years ago, or later during the advent of agriculture, it was spread across Eurasia by humans. In any case, we believe Central Asia offers by far the most plausible location for the primary origin and early evolution of *Cannabis*.

De Candolle (1967) stated that *Cannabis* occurs "wild" only south of the Caspian Sea, in Siberia near the Irtysch River, and in the Khirgiz Desert beyond Lake Baikal; he also suggested that it was first cultivated in southern Siberia. The *Indian Hemp Drugs Commission Report* (1893–94; see Kaplan 1969) identified a broad area encompassing the southern Himalayan foothills from Kashmir through Nepal and northeastern India as the region of spontaneous growth. Currently, the range of self-sown growth also extends throughout Eastern Europe into the western and central regions of the former Soviet Union and across northern South and Southeast Asia. *Cannabis* also grows spontaneously in its introduced ranges in parts of Africa south of the Sahara Desert (NLD) and in parts of temperate, central North America (NLH) (Hulten 1970) and as a weed in farm fields and disturbed niches across temperate China (BLH) where it has escaped from cultivation (International Association of Agricultural Economists 1973).

Fieldwork and theories of famous Russian botanist Nicolai Ivanovich Vavilov (1931) added considerably to our understanding of crop plant origins. Vavilov studied phenotypic diversity (variation of observable traits) within Central Asian *Cannabis*, made many firsthand observations of the genus, and used it as an example of how to differentiate between a genuinely wild plant and a more recent escape from cultivation. Some of these characteristics of domestication are also found in wild cereals. His four criteria for identifying wild *Cannabis* were as follows:

1. Germination of seed is slow and irregular,
2. Seed coat [reduced perianth (calyx and corolla)] persists as an outer husk around the seed, developing a camouflaging pattern,
3. Seeds have oil glands, and these attract various insects that remove and distribute them,
4. Inflorescence shatters and distributes the seeds.

Vavilov and Bukinich (1929) reported that weedy *Cannabis* occurred commonly in irrigated parts of Afghanistan. More important, Vavilov (1931) also wrote about his 1929 visit to Chinese Turkestan to look for evidence of proposed origins of several wild and cultivated plants. Chinese Turkestan, in present-day Xinjiang province of China, lies north and northwest of the Himalayan Mountains and Qinghai-Xizang Plateau, southwest of the Tian Shan Mountains, and northeast of the Pamir Plateau; it is separated from the whole of China by the Taklimakan Desert to the east. Vavilov reported numerous thick stands of cultivated *Cannabis* in valleys of Chinese Turkestan and along the slopes south of the Tian Shan Mountains as well as its occurrence as a common weed throughout the Russian provinces of Irkutsk and Omsk and east of the Amur River. He concluded that the majority of the cultivated plants of the region were predominantly imports from China to the east or Afghanistan and Pakistan to the southwest. However, Vavilov considered *Cannabis* to be a native crop that originated in Central Asia.

Vavilov characterized wild and weedy *Cannabis* populations from Chinese Turkestan and northern Central Asia (1931) as "shattering forms with a horseshoe at the base of the fruit, with seeds of different size, up to the dimensions of the cultivated large-seeded forms." Wild hemp, according to

Vavilov, was utilized only occasionally by local people for the manufacture of cordage, and he commented that the people of Central Asia extracted hemp fibers in a most primitive way and without retting—merely pulling the fibers from the dry stalks. Its utilization, however, was especially extensive in the Altai Mountains, and he surmised that it was there that wild hemp could have been a likely candidate for cultivation near settled populations. Vavilov observed what may have been a vestige of ancient hunter-gatherer (by then pastoralist) use of *Cannabis*, collecting wild hemp fiber in the mountains at the beginning of autumn before moving into lower valleys to avoid the cold winter. Similarly, the Nu, an ethnic minority of part-time pastoralists living in Yunnan province, China, will sow hemp seeds along ridges in early summer while grazing their livestock and leave the crop unattended until they return in autumn to collect winter fodder, when they thresh the hemp seeds; strip off the bark; and haul it back to town for processing, spinning, and weaving (Clarke 1996, personal observations).

Vavilov and Bukinich (1929) generally characterized Afghan *Cannabis* as short in stature with short internodes and profuse branching from the first node. Eastern Afghan varieties were described as having small leaves with egg-shaped leaflets with their narrow ends toward the base and extremely small dark-colored seeds that shattered and dispersed easily—characteristic of wild plants. Vavilov termed this wild *Cannabis* of eastern Afghanistan, which commonly had dark gray seeds with a marbled seed coat pattern, *C. indica* var. *kafiristanica*. Hillig and Mahlberg (2004) preserved this name and assigned it to wild and feral narrow-leaf drug ancestor (NLDA) biotypes. A second variety was described with a colorless seed coat and named *C. indica* var. *afghanica*. Hillig and Mahlberg (2004) also preserved this name to represent broad-leaf drug (BLD) biotypes, which they called wide-leaf drug biotypes (see also Chapter 11). Vavilov concluded that Afghan *Cannabis* varieties were entirely different from both wild and cultivated European and Asiatic *Cannabis* and therefore must be considered as varieties of *C. indica* Lam. Vavilov also pointed out that "*C. sativa* L." of the European type was cultivated for hashish in northern Afghanistan. We recognize this as the *C. indica* ssp. *indica* narrow-leaf drug (NLD) biotype based on its resemblance to European hemp (tall with narrow leaves) and its high THC content (Hillig and Mahlberg 2004).

Earlier, Russian botanist D. E. Janischevsky (1924) described and published descriptions of a new species, *Cannabis ruderalis*, growing wild in the Volga River region, Western Siberia, and Central Asia. Hillig (2005a/b) recognized *C. ruderalis* as the putative ancestor (PA) of *C. sativa* (see Chapter 11). However, present-day narrow-leaf hemp (NLH), narrow-leaf ancestor (NLHA), and PA populations overlap in range and traits and there is no clear differentiation between the taxa (Hillig and Mahlberg 2004). Janischevsky's work was part of a large-scale Soviet agricultural research program carried out under the direction of Vavilov during the 1920s and 1930s. Vavilov, with help from a team of experts, conducted an extensive series of expeditions to many continents, collecting information that contributed to identifying and understanding regions of species diversity, which Vavilov argued were the areas of species formation. De Candolle (1967) first used this criterion, although he did not rely so heavily on it and took a more comprehensive approach in his attempt to determine *Cannabis* origins, integrating a greater variety of sources than Vavilov.

Based on the work of Janischevsky and others, Vavilov (1949–51) classified *Cannabis* as indigenous in three major

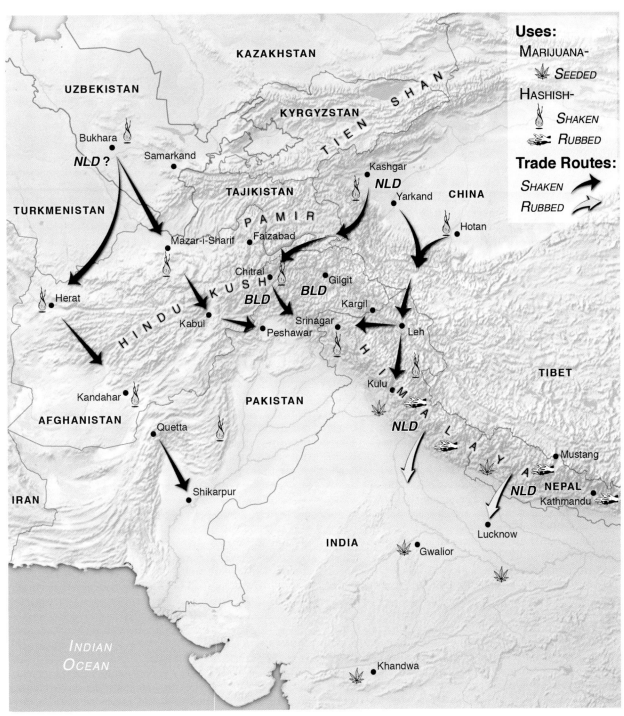

MAP 4. Eighteenth- and nineteenth-century observers described several different *Cannabis* biotypes within a mountainous region known as the "Pamir Knot" where Central Asia and South Asia meet. Within this relatively small region are found narrow-leaf drug ancestor (NLDA) and narrow-leaf drug (NLD) biotypes used for *ganja* (marijuana) production in lowland India and *charas* (rubbed hashish) production along the Himalayan foothills, as well as broad-leaf drug (BLD), biotypes from Afghanistan and northern Pakistan used during the late twentieth century for making sieved hashish. In the late 1920s, Russian botanists reported that NLD cultivars were grown in Turkestan for sieved hashish production (cartography by Matt Barbee). (See TABLE 1 at the beginning of this book for explanations of *Cannabis* gene pool acronyms.)

"centers" of species formation described later in this chapter. These "centers" we may categorize with hindsight as areas of hybridization induced by relatively intense, directional, artificial selection for desired crop characteristics as well as regions of trade and exchange rather than areas of evolutionary origin. Each of these factors can promote variation within a cultivated species. Under the category of fiber plants, Vavilov placed varieties of *Cannabis sativa* that produce large seeds, which we now consider broad-leaf hemp (BLH) biotypes (after Hillig 2005a/b), in his "Chinese Center" of cultivated plants, which includes the mountainous regions of central and western China and adjacent lowlands. Under the category of spice plants and stimulants, Vavilov listed *Cannabis indica,* which we now consider to be a NLD biotype (Hillig 2005a/b), as originating in the "Indian Center," which includes all the Indian subcontinent except northwestern India, Punjab, and the Northwest Frontier (now a part of Pakistan). Finally, under the category of grain crops in his "Central Asiatic Center," Vavilov again listed *Cannabis indica,* which we recognize as BLD biotypes, once again after Hillig (2005a/b). This comparatively small area includes northern Pakistan, all of Afghanistan, the Central Asian Republics of Tajikistan, and Uzbekistan as well as the western Tian Shan Mountains. We propose that these three centers were more likely areas of early agriculture and selection for specific uses (food, drug, and fiber) following the early Holocene dispersal of *Cannabis* throughout Eurasia.

Although Vavilov's evidence for centers of agricultural crop diversity is convincing, his interpretations of the evidence are not as well accepted (e.g., see Merlin 1972). The use of phenotypic diversity as a key to plant origins has been thoroughly challenged in recent decades. The idea of a "center of origin" might be intellectually satisfying, but it does not always follow as a logical conclusion from an analysis of the data. Although patterns of variation can supply valuable information about the genome of a crop, the question of agricultural and species origins "is much too complex to be solved by such a simple device, and every scrap of evidence is needed from any source that might be even inferentially pertinent" (Harlan 1971). Phenotypic change upon dispersal away from the area of origin and into an introduced environment is common and has obviously occurred in the case of *Cannabis.* In addition, evolution, and thus variation, has been greatly accelerated by human selection during domestication.

David Harris (1967) emphasized the significance of crop hybridization with weeds in the evolution of species diversity and argued that a large number of weedy plants "are derivative from, rather than ancestral to, their associated crops, and consequently Vavilov's centers of maximum diversity are not necessarily centers of primary domestication." Edgar Anderson (1967) in his informative book exploring the antiquity of the ongoing important relationship between humans and plants suggested that Vavilov's areas of greater variability are places where flora previously separated came together and hybridized. Indeed the continued existence of primitive varieties of cultivated plants among traditional peoples, often found in remote areas (such as regions of temperate Eurasia where spontaneously growing *Cannabis* is found today), is probably a result of the basic conservatism of these isolated peoples. Thus Vavilov's ancient centers of species formation may very well be centers of early human cultural and agricultural survival rather than centers of origin. Actually, Vavilov (1931) suggested this possibility when he discussed the origin and distribution of *Cannabis* in Central Asia. At first, he asserted that *Cannabis* was most likely one of the few indigenous crops of the area:

> The autochthonic [indigenous] crops of Central Asia are few, but still such ones may be found. Of the field crops the first to be mentioned is hemp. All over [the] northern Tian Shan [Mountains], on its slopes, in the valleys to the north of it, wild growing hemp is of common occurrence. The waste lots of the town of Yarkand in Xinjiang province China are covered with thick stands of hemp. It grows on the ridges of fields, not infrequently forming broad borders along the roads. In ravines, on forest skirts, on marshy ground, on waste land near the villages—weed hemp is the commonest of plants.

But then Vavilov reevaluated this assumption in the very same paper and noted that there is also good reason to believe that hemp is not endemic to Central Asia: "We admit that the introduction of hemp, as of a wild growing plant characterized by a vast area stretching from the southeast of European USSR to the Pacific, has taken place simultaneously, as well as at different times, in different regions. It may as well have taken place in the agricultural districts of Central Asia." The former Harvard University economic botanist Oakes Ames (1939) referred to scholars of his time who generally believed *Cannabis* "to be indigenous to the temperate parts of Asia near the Caspian sea, southern Siberia, the Kirghiz Desert and Persia." Ames' student, Richard Evans Schultes, who eventually took Ames's position at Harvard, stated that *Cannabis* is "one of the most ancient of cultivated plants [and] is native probably to Central Asia" (Schultes 1969a/b).

Even though the arguments for hemp being endemic to Central Asia are not conclusive and, in fact, the origin and first use of *C. sativa* and *C. indica* may have occurred elsewhere, we suggest, with the same cautious reserve as Vavilov, that the natural origin of *Cannabis* took place in Central Asia, possibly in the upland valleys of the Tian Shan or Altai Mountains and that very early, if not the first, cultural applications of *Cannabis* took place in this same general area during the early Pleistocene. If *Cannabis* originated in Central Asia, it would have been ideally situated for diffusion southeastward into eastern Asia and southwestward into Europe as Pleistocene ice sheets advanced. The Central Asian biotypes observed by Vavilov and later researchers are the result of later introductions accompanying human dispersal during the early Holocene (see Chapter 12).

CANNABIS AND VITIS

Here it is relevant to briefly review the biological evolution and domestication of the common grapevine (*Vitis vinifera*), which was derived from the wild grapevine (*Vitis vinifera* subsp. *silvestris*). Evolution of this useful plant, under the processes of domestication, provides us with a model that has strong parallels to that of *Cannabis*. *Vitis* is similar to *Cannabis* in many respects, including its biology, reproduction, geographical origins, and human influences that include long-distance seed dispersal followed by localized distribution of select individuals via asexual propagation. The grapevine is much farther down the vegetative domestication path than *Cannabis*, and therefore an understanding of grapevine history and its interaction with humans may help us predict the future of *Cannabis* evolution.

Vitis vinifera is the only member of the grape genus indigenous to Eurasia, possibly originating in the Near East (e.g., see Myles et al. 2010), with evolutionary origins dating back to around 65 million years ago (This et al. 2006). Following the last glacial maximum (LGM) about 18,000 years ago, grape populations began spreading northward from the Italian Peninsula and also westward from the Caucasus, resulting in some admixture in central Europe (Grassi et al. 2008). Presently, the truly wild form, *V. vinifera* subsp. *silvestris*, is relatively rare. It is occasionally found in environments from sea level up to 1,000 meters (about 3,300 feet) in elevation all the way from the southern Atlantic coast of Europe to the western Himalayas, from Portugal in the west to Turkmenistan in the east, and from the Rhine River valley in the north to Tunisia in the south, where it grows as a vine on the surrounding tree canopy (This et al. 2006). The common, domesticated grapevine is one of the oldest fruit crops; it is cultivated extensively worldwide and is of great economic importance in its use for table fruit, raisins, sweet preserves, juice, and wine. Its domestication, occurring between 9,000 to 7,500 years ago in the Near East, with the earliest archeological evidence for this in northern Iran, Georgia, and Turkey, was coincident with the discovery of wine (This et al. 2006; also see McGovern 2003). Domestication brought many changes to grape's agronomic traits including greater fruit yield and sugar content. Truly wild grapevines are dioecious and wind-pollinated with bird-mediated dispersal (similar to *Cannabis*) while domesticated grapevines are self-pollinating hermaphrodites (Grassi et al. 2008). How did this change occur?

Selection for higher yield, more sugar content, and determinant maturation resulted in changes in berry color and berry and bunch size, as well as a crucial change from dioecious to hermaphrodite sexuality; this eliminated the need to maintain male plants as pollinators and allowed the self-fertilizing of mutant phenotypes. By 5,500 to 5,000 years ago, early domesticates were spread by humans to Egypt and Lower Mesopotamia, followed by dispersal into several Mediterranean cultural realms, especially the Roman Empire, eventually reaching as far as China and Japan by 200 CE. In the process, humans shaped the diversity of grape cultivars extant today. Long-range transport was facilitated by seeds, new cultivars arose from sexual crosses between seedlings, and unique offspring with favorable characteristics were multiplied asexually, producing populations of identical clones (This et al. 2006). The domestication process presently followed in indoor drug *Cannabis* production is much the same—sexual crossing of pollen and seed parents to produce genetically diverse seeds, which are transported to new environments, sown, and grown with only a few select female plants reproduced asexually through rooted cuttings, thereby fixing the selected traits and allowing no further evolution.

As a result of centuries of exchange of genetic material (seeds and cuttings), it is difficult to determine the original home of widespread domesticated plants such as grape and hemp. Wild-growing grape populations have been documented from many regions of Europe, but it is often unclear if they are truly undomesticated *silvestris* rather than *vinifera* cultivars growing as feral escapes from cultivation or possibly hybrids resulting from crosses between wild and cultivated plants (This et al. 2006). This situation must also be rectified in *Cannabis* before its evolutionary pathways can be deciphered. Are there any truly wild progenitor populations of *Cannabis* extant today?

THEORIES FOR SOUTH ASIAN ORIGIN OF DOMESTICATED *CANNABIS*

South Asia also presents another possible location for the origin and/or early domestication of *Cannabis*. Used in preparation of a ritual drink known as *bhang*, *Cannabis* was referred to in the ancient Indian *Atharva Veda* or "Science of Charms" (written sometime between 4,000 and 3,400 BP) as one of the "five kingdoms of herbs. . . . which release us from anxiety" (Abel 1980; see also Booth 2003 and Chapter 7). Carolus Linnaeus (Carl von Linné), the "Father of Taxonomy," believed *Cannabis sativa* was native to India, although he never collected or categorized specimens from the area. The great diversity of *Cannabis* varieties and usages in northern India and Nepal along the foothills of the Himalayas may indicate that this region was one of the first areas where *Cannabis* was extensively utilized, most likely for mind-altering purposes.

Sharma (1979, 1980) used phenotypic diversity as a major criterion in his conclusion that *Cannabis* originated in the valleys along the southern slope of the Himalaya Mountains from Kashmir through Nepal and Bhutan to Burma. He noted that wild (or nearly wild) populations occur in relatively unpopulated areas throughout the Himalayan region and that significant variation can be measured between glandular trichome characteristics and epidermal (leaf surface) patterns of populations from differing climates. He did not, however, offer evidence that leaf surface traits are sufficient taxonomic criteria to determine races of *Cannabis*. Furthermore, like Vavilov, Sharma assumed that an area with the greatest diversity within a species is also the area in which the natural origin of the species occurred, rather than recognizing that such variation may be derivative instead of ancestral. In other words, if *Cannabis* was introduced to the southern slopes of the Himalaya Mountain range and then intensively cultivated, the evolution of many varieties through artificial selection and hybridization by humans, in conjunction with substantial ecological variation along steep elevation gradients, may have occurred subsequent to its introduction. In addition, it should be remembered that perceptions of significant variation are often subjective. Whether it was Vavilov or Sharma who observed the "most" variation in spontaneously growing populations they investigated (in Central Asia and the Himalayan foothills respectively) is impossible to determine by studying their reports, and neither traveled in the study area of the other.

Although we have argued that *Cannabis* evolved naturally in Central Asia, if *Cannabis* did originate in northern South Asia, it most likely would have evolved along or relatively near streams in the Himalayan foothills. According to our Holocene dispersal scenario, *Cannabis* arrived in this region early on as it expanded westward from the Hengduan Mountains and Yungui Plateau in southwestern China. Much later, traders could have carried *Cannabis* west to the Middle East. NLD varieties eventually spread westward by sea traders to the east coast of Africa and eastward through Burma into Southeast Asia. Following this scenario further, NLH would have evolved at higher latitudes from South Asian NLD varieties and spread farther north into southern Russia and then west into Europe. Some varieties could have migrated so far north that the summer season was too short to produce psychoactive levels of THC and evolved into fiber or seed varieties under human selection. Under this scenario, the PA, *C. ruderalis*, collected during the twentieth century, in turn, would likely have evolved from *C. sativa* as this species spread farther north into the Central Asian region formerly known as Turkestan. However, there is little evidence to support this scenario and several reasons to challenge it.

The massive Himalaya and Hindu Kush Mountains, which have proven such a mighty barrier to plant and animal dispersals (including humans), lie between the origin regions proposed by Vavilov and Sharma. No examples of crop plant co-origin both north and south of the Himalaya and Hindu Kush Mountains have yet been reported (Simmonds 1976; Smartt and Simmonds 1995). In addition, genetic data does not reveal any links between South Asian NLD and European NLH populations except those resulting from more recent hybridization influenced by cultivation and breeding (Hillig 2005a/b). Although *Cannabis* now grows spontaneously throughout Eurasia, but not necessarily as a native plant, it seems unlikely to us that the genus originated both north and south of these mountain ranges. However, human migrations spread *Cannabis* throughout the Himalayan and Hindu Kush Mountains early in prehistory, probably starting sometime after the beginning of the Holocene, but possibly much earlier as anatomically modern humans (AMHs) first began their advance across Eurasia.

Models of early use and domestication (see Chapter 3), archeological data (see Chapter 4), and historical records (see Chapters 5 through 9), in conjunction with evolutionary studies involving reproductive strategies and geography (see Chapter 12), lead us to conclude that *Cannabis* originated somewhere in Central Asia rather than South or East Asia, although these regions may have served as glacial refugia where speciation occurred. China and India were both regions of early *Cannabis* evolution under domestication and foci for later diffusion, resulting in the broad diversity of phenotypes selected for various uses appearing across both East and South Asia. In the following discussion we evaluate the, at times, seemingly contradictory data and opinions in a temporal framework, then rectify many of the discrepancies and propose a hypothetical model for the early evolution of *Cannabis*.

Model for the Early Evolution of *Cannabis*

How long ago did *Cannabis* originate, and when did AMHs begin their association with these useful plants? We know that the earliest angiosperms (flowering plants) probably evolved more than 140 million years ago (e.g., see Soltis et al. 2008); early humans appear to have evolved into *Homo sapiens* in Africa about 200,000 years ago (e.g., University of Utah 2005; McDougall et al. 2005); and AMHs began extensively colonizing the Middle East during the Upper Paleolithic about 45,000 to 40,000 years ago, reaching the steppes of Central Asia and highland southern East Asia by about 35,000 years ago (Wells 2002; Finlayson 2005). Dispersals then radiated outward, reaching Europe and South Asia by about 30,000 years ago and northeastern Asia by around 5,000 to 20,000 years ago (Meltzer 2009; Kunzig 2004; also see Wells 2002). The ice age of the LGM reached its peak about 21,000 to 18,000 years ago (Soffer and Gamble 1990; Otto-Bliesner et al. 2006), and the warming Holocene epoch began about 12,000 years ago (Roberts 1998). Early farming commenced relatively soon after the end of the Pleistocene and spread widely from a series of centers in the Old and New

Worlds (Bellwood 2005). The timing and location of the earliest cultivation of *Cannabis*, as with most plants, may never be completely ascertained, and although we do not have archeological evidence for very early cultivation of *Cannabis* in Central Asia (probably due to the lack of sufficient research in that general region), we do know that hemp was planted quite early on in East Asia and most likely much later in Europe (see Chapter 4 for a full discussion of cultural spread and early farming of *Cannabis* and Chapter 12 for a detailed look at climate change and glacial refugia).

When and where along this continuum did (1) family Cannabaceae appear, (2) *Cannabis* and *Humulus* (the hop genus) diverge, and (3) *Cannabis*'s species evolve? When was the natural evolution of *Cannabis* first affected by human contact? How did the various subspecies, biotypes, and ecotypes evolve? Answering these questions will allow us to advance our hypothesis for the early evolution of *Cannabis*. In the absence of pre-Holocene *Cannabis* seeds, only limited ancient pollen (which may be hard to identify with certainty; see Chapter 4), and without fossils of a clearly identified *Cannabis* progenitor, it is difficult to determine with any accuracy when *Cannabis* evolved into the biotypes we see today. The survey of reproductive strategies presented earlier indicates that *Cannabis*, an herbaceous, sun-loving, short-day flowering annual, most likely evolved somewhere in temperate latitudes of the northern hemisphere, and data from published research favors Eurasia, especially Central Asia, as its region of origin. Future DNA research and additional forms of molecular genetic investigation may help to more accurately determine the original home of *Cannabis*.

In the meantime, a review of evidence for the origin and prehistoric dispersal of *Cannabis* offered by the disciplines of paleoclimatology, archeology, and taxonomy supports our model for the evolution of *Cannabis*. During the last interglacial period (approximately 135,000 to 110,000 years ago), the northern hemisphere, including the vast region of Eurasia, was relatively warm and humid; it is somewhere within this huge area that the ancestors of modern *Cannabis* and *Humulus* would have found environmental niches suitable for their evolution and proliferation. Around 50,000 years ago, AMHs began dispersing northeastward out of Africa and by 35,000 years ago into middle Eurasia where their populations thrived and multiplied, eventually spreading both west and east to occupy vast areas of the earth's landmass (Kunzig 2004). This middle Eurasian cauldron of human evolutionary and cultural change lay within the natural range of *Cannabis*, and Pleistocene early humans would have been attracted to its readily apparent attributes.

Our assertion that Central Asia was both the original Pleistocene home and center for evolution and dispersal of *Cannabis* within the past 50,000 years is supported by our reconstruction of the climatic conditions across Eurasia during past geological periods. In order to further explore the human-*Cannabis* relationship, it is also important to determine in which regions early people may have lived nearby *Cannabis* populations. This can be ascertained by present-day human genome analysis combined with paleoclimate reconstructions. Adams and Faure (1998), in their survey of plant and animal remains from various time periods and geographical locations, made correlations with the environmental requirements of related extant species and produced a map series of reconstructed world vegetation. Gepts (2004) listed *Cannabis* as originating in the temperate steppe biome. Possible habitats conducive to the growth and spread of *Cannabis* during the early Holocene are based on ranges of vegetation zones supporting feral growth today. These coincide with three paleoenvironmental classifications: (1) cool, temperate, deciduous broadleaf and coniferous forests with a fairly open canopy; (2) semiarid temperate woodland or scrub; and (3) herbaceous forest steppe with clumps of trees in favorable locations. These vegetation zones existed at each time period reconstructed by Adams and Faure, but their ranges shifted between different time periods, and they occurred in different geographical regions than today.

Archeological sites provide physical evidence that bands of hunter-gatherers were living in these regions during the Upper Paleolithic (50,000 to 10,000 BP) many millennia before the LGM (e.g., see Madeyska 1990). As climate cooled, leading up to the LGM 18,000 years ago, and early humans moved southward, they could have taken *Cannabis* seeds with them. After PA populations dispersed southward and diverged geographically, two populations may have survived in two separate isolated locations and evolved into two new species—in temperate foothills of southern and southeastern European mountain ranges, (1) the putative hemp ancestor (PHA) and progenitor of modern *C. sativa,* and in temperate mountain valleys of southern East Asia, (2) the putative drug ancestor (PDA), the progenitor of modern *C. indica.* After several millennia, as northern latitudes began to warm and the Holocene commenced, early humans could have returned northward, carrying the progenitors of modern *Cannabis* taxa across much of Eurasia from their twin origins. Following the LGM, and throughout the early Holocene, the Magdalenian cultural complex of central Europe expanded to the northeast onto the northern European plains and into the steppe regions of Eastern Europe, while the Solutrean cultural complex spread across Mediterranean southern Europe to the Black Sea (Bar-Yosef 1990). By this time, Paleolithic cultures were well distributed across modern-day China, Korea, and Japan, and early Huang He and then Yangzi River farming cultures soon began to radiate across East Asia (Chen and Olsen 1990; also see Xue et al. 2006).

However, the divergence of *C. sativa* and *C. indica* likely occurred during much earlier glaciations, and AMHs encountered *Cannabis* much later as it began to spread from its most recent refugia following the LGM. Speciation occurred during an earlier glacial period when advancing ice sheets pushed ancestral populations of plants and animals into more southerly refugia, *C. sativa* evolving in refugia in southeastern Europe and *C. indica* in southern East Asia, where their respective ranges were likely reduced in subsequent glacial periods leading up to the LGM. If the PA (*C. ruderalis*) exists today, it must have survived at low population density in cryptic refugia at more northern latitudes than *C. sativa* or *C. indica.* During interglacial warming, *Cannabis* populations evolved naturally as they expanded northward, recolonizing niches for which they were preadapted, only to be restricted to temperate refugia during a subsequent glacial cold period. During the LGM, European *C. sativa* NLH populations likely found refuge in the foothills of the Caucasus Mountains and on the Balkan Peninsula, while Asian *C. indica* ssp. *chinensis* (broad-leaf hemp, or BLH) populations survived within the Hengduan Mountain-Yungui Plateau region of present-day southwestern China and possibly also in coastal northeastern China, Korea, and Japan. *C. indica* ssp. *indica* NLD populations may have survived in the Hengduan Mountain-Yungui Plateau region or along the Himalayan foothills, while *C. indica* ssp. *afghanica* BLD populations evolved later in the foothills of the Hindu Kush Mountains. It is unlikely that any

Cannabis populations survived the LGM outside of refugia, which may have been more in number than we can presently identify. After the LGM, *Cannabis* populations expanded once again, and their dispersal and introduction into newly disturbed niches was often aided by humans dispersing from their temperate refugia. During rapid diffusion into new niches, the *Cannabis* genome narrowed from founder effects resulting from inheritance of a limited subset of genes from only one or a few pioneering parents. It then diversified by ecological adaptation to each new niche. Meanwhile, populations remaining within upland refugia with varying topography likely remained genetically diverse due to individual adaption to differing microclimates within a small geographical range. Variation extant today at the subspecies and biotype levels results from relatively recent post-LGM expansion with human assistance and builds upon a much more ancient evolutionary foundation. Human-imposed geographical isolation and selection have proved sufficient to preserve species integrity while increasing biotype diversity.

Putative progenitor populations (PA, PHA, and PDA) are the "missing links" in our model of early evolution and are very likely extinct. Due to the high probability of intercrossing with neighboring feral or cultivated populations in more recent times, it is unlikely that any genetically pure ancestral populations survive today even in remote regions of Central Asia. It is even more unlikely that there would be any relict populations remaining in the regions where the hypothetical progenitor populations of hemp and drug *Cannabis* (PHA and PDA) originated, as *Cannabis* has been cultivated for at least two millennia across Europe and much longer in East Asia (see Chapters 4 through 9). If specimens tentatively identified by Hillig (2005a/b) and others as *C. ruderalis* do not represent relict populations of the original PA, then how did they arise and how do they differ genetically from other extant taxa? It seems likely to us that Central Asian populations studied during the twentieth century and perceived as putative ancestors were products of mixed heritage (PA introgressed with NLHA and NLH) combined with lack of human selection and ecological adaptation to marginal environments. Without human selection, *Cannabis* has a tendency to revert to atavistic (ancient ancestral) genetic combinations quite rapidly and atavistic traits would be expressed frequently, especially when populations are genetically isolated and subjected to increased inbreeding. Naturally growing and seemingly wild populations that could be interpreted as descendants of putative ancestors have also been observed in Kashmir (Watson personal communications 1978–2007), as well as Shandong and Yunnan provinces in China, lowland Nepal, and northeastern India (Clarke personal observations 1993, 1995, 2006, 2009, respectively).

By 8,000 years ago, large tracts of northern Eurasia had a suitable temperate climate for supporting climax broadleaf and coniferous woodland vegetation cover and allowing *Cannabis* to proliferate. We assume that humans spread *Cannabis* easily via their hunting and gathering activities and eventually introduced it into their new agricultural settlements where and when these became established during the Holocene. Responding to a constantly changing natural environment and early unconscious human selective pressures, the NLH ancestor (NLHA) slowly evolved through intermediate populations into the *C. sativa* interbreeding complex (NLHA-NLH) extant today in Europe and Western Asia. Uniformity of surrounding climate and vegetation and restricted latitudinal spread within a relatively homogenous cultural setting may account for lack of genetic diversity within *C. sativa* (Hillig 2005a/b). The present-day range of *C. sativa* NLH includes Europe and North America yet is relatively small in comparison to the worldwide ranges of *C. indica* biotypes BLH, BLD, and NLD (see Table 1).

In response to entirely different sets of natural and human selective pressures, PDAs also adapted and evolved as they migrated into regions that were both climatically and culturally diverse, became isolated, and were exposed to a far wider range of selective pressures than PHAs. In response, *C. indica* evolved into three biotypes or subspecies. BLH landraces (local varieties arising from unconscious human selection in concert with natural selection) likely evolved in China very early on, in close association with the expansion of Chinese agriculture, and relatively soon spread to Korea and Japan where additional BLH populations may already have been growing if they survived escaped the LGM; feral populations can presently be found in several regions across China and Korea and on Hokkaido Island, Japan. Although the closest relatives of BLH are the highly psychoactive BLD and NLD biotypes, East Asian hemp varieties are relatively low in THC. Since there was little traditional psychoactive use following the rise of Confucianism, BLH landraces were only rarely selected for drug content in the past two millennia. BLD varieties evolved under extremely arid conditions in an isolated mountain range within present-day Afghanistan, were eventually used for producing hashish, and are the most morphologically distinct of the *Cannabis* taxa (see Chapter 11). NLD biotypes are also high in drug content. Along the Himalayan foothills in northern South Asia, NLDA populations introgress with NLD cultivars to form an interbreeding NLDA-NLD complex similar to that of the NLHA-NLH complex of Europe and western Asia. According to our taxonomy (following Hillig 2004a/b, 2005a/b), *C. indica* cultivars are the most geographically widespread and most widely utilized biotypes today, growing on all continents and used for recreational and medicinal drugs as well as fiber and seed production, while *C. sativa* cultivars are presently grown only for fiber and seed on limited acreage in Europe and North America.

Summary and Conclusions

Glacial ice sheets advanced and retreated many times during the earth's history, and species have either moved or perished as they advanced, survivors recolonizing their previous homelands as the climate warmed and glaciers retreated. During Quaternary glaciations *Cannabis*'s range would have been highly restricted to two or more isolated refugia (located in distant parts of southern Eurasia or possibly within smaller cryptic refugia at more northern latitudes) with climatic conditions similar to those favored by *Cannabis* today. Isolation of populations during times of glacial advance could have led to speciation within genus *Cannabis*. There were several series of Quaternary glaciations during the past two million years as well as many Tertiary glaciations before them, and *Cannabis* would have moved southward during times of cold and back northward during warm periods several times during its evolution; adaptive radiation during the Holocene is only the most recent cycle of expansion. Today, possible refugia are represented by favorable microclimates where *Cannabis* survived to later disperse and re-enlarge its range. It is more difficult to determine both areas of origin or endemism and potential refugia in organisms such as *Cannabis* with widespread ecological ranges and partial fossil records and even more difficult to determine in plants with ancient human relationships.

Feral *Cannabis* populations are found today growing in temperate climates at northern latitudes. These are usually characterized as warm continental regions with spring and early summer rains, followed by a dry cool autumn and accompanied by the widely fluctuating day length (photoperiod) afforded by more northern (and more southern) latitudes; indeed, feral *Cannabis* only flourishes in this narrow climate niche. The vegetation cover most favorable for *Cannabis* is temperate-climate upland open woodland growing in valleys with alluvial soil deposits and slopes for drainage with sufficient sunlight and summer rainfall. Suitable regions would have had moist temperate conditions during glaciations without being so near the equator as to lose short-day flowering response and cold hardiness. Humans created many favorable open habitats, but *Cannabis* thrived in more or less these same conditions long before we entered the scene.

Equally important in determining *Cannabis*'s prehistoric range are the conditions it does not tolerate such as extreme heat, cold, aridity, or humidity; heavy or waterlogged soils; and permafrost. Many presently warm and humid tropical equatorial regions were arid deserts during the LGM. In addition, *Cannabis* could not survive too much humidity; today *Cannabis* does not become feral in subtropical monsoon regions. During glacial periods *Cannabis*'s range would not have included semitropical and tropical regions as this is not where natural wild or feral *Cannabis* populations flourish today. Mediterranean climates with cool, wet winters and hot, dry summers are also not conducive to the natural growth of *Cannabis* because it requires summer rain. Some *Cannabis* populations became extinct while some survived in suitable microclimates, providing additional chances for isolated populations to evolve independently within their refugial ranges. Topography within large southern montane refugia is complex and local microclimates abound. Each river valley offered isolation from neighboring populations and a unique suite of selective pressures, an ideal setting for genetic divergence and speciation. Several regions of ancient Eurasia presented likely locations for Pleistocene *Cannabis* refugia. We propose that such favorable LGM refugia for *C. sativa* could have existed within the Caucasus Mountains with another for *C. indica* in the Hengduan Mountains and Yungui Plateau and possibly also along the Himalayan foothills as well as on the Shandong and Korean Peninsulas and the Japan Archipelago.

Most plant species have very limited distributions, so why and how has *Cannabis* become so widespread and abundant? Animals including humans are more mobile than plants and can move away from advancing ice sheets. Plant populations are much more sedentary, moving no farther spatially than their propagules. During glacial advances plant populations do not move so much as just die off as the climate becomes less favorable and their range becomes more restricted. During times of glacial advance, ice sheets would expand southward, encroaching upon the expanded range of *Cannabis*. In populations adapting to changes nearest the ice sheets, female plants would drop their seeds nearby at the end of the season, but male plants could spread their adaptive success via windblown pollen deep into the extant population. It is only during interglacial warming that *Cannabis* would have expanded from its reduced refugial range.

Today *Cannabis* is widely distributed around the world largely as a consequence of the human-*Cannabis* relationship, but it may also have been endemic in several regions of Eurasia prior to human contact. Some plants are aided in the long-distance transportation of their seed by migrating birds and hoofed mammals, which also could have played a part in *Cannabis*'s early distributional changes, although humans have certainly had the greatest effect since the Holocene began. However, because *Cannabis* seed is not regularly disseminated by animal, water, or wind vectors and most seeds remain near the seed plant, the postglacial range of *Cannabis* would have expanded much more slowly without the assistance of humans.

Cannabis likely originated millions of years ago in northern Eurasia and moved ahead of climate changes, dispersing (likely without human assistance) southward during glaciations to escape unfavorable conditions. During a glacial maximum, *Cannabis* populations were forced into refugia in southern Europe and southwestern East Asia, possibly leading to speciation events giving rise to European *C. sativa* and Asian *C. indica*. During this time, *C. indica* evolved an enhanced biosynthetic capacity to produce THC. Early humans utilized both *C. sativa* and *C. indica* for fiber and seed, but only *C. indica* has a history of drug use. *Cannabis* thrived during the early Holocene as the earth warmed, and with human assistance its range expanded around the world. Range expansion continues, although genetic diversity has decreased, also as a result of human influence. Self-sowing feral *Cannabis* presently occupies a restricted ecological belt extending around the world.

Cannabis's annual life cycle and its ecological requirements for open environments, ample water, and well-drained soils favor origin in moist riverside environments. Studies of the reproductive strategies of *Cannabis* indicate probable evolution in northern temperate latitudes. Early researchers such as de Candolle (1967, originally published in 1882) and Vavilov (1931) favored Central Asia as the likely region of origin, in which case, *Cannabis* was advantageously positioned for dispersal throughout Europe, southern Asia, and the Far East. More recent studies indicate that if primordial *Cannabis* naturally evolved in Central Asia prior to contact with humans, it must have moved to warmer, more southern latitudes several times before, and again during, the LGM, possibly carried by early humans and then redistributed throughout Eurasia by humans moving northward as climate warmed during the Holocene. *Cannabis* was pre-adapted for successful growth upon its return from southern refugia as it originated farther north many millennia earlier. Present-day *C. ruderalis*, the putative ancestor of extant *Cannabis* taxa, grows throughout Central Asia and most likely represents a degenerate, inbred, and unselected hybrid blend of various *Cannabis* gene pools that survived as feral escapes, rather than direct descendants of the now long-extinct ancestral population in its original home.

CHAPTER THREE

Ethnobotanical Origins, Early Cultivation, and Evolution through Human Selection

The origins of agriculture involve both human intentionality and a
set of underlying ecological and evolutionary principles.

(FLANNERY 1986)

Introduction

When and where did humans first encounter *Cannabis*? How
and why did people first use the plant? Since our evidence is
still indirect, there can be no firm answers to questions about
the origins of people's initial associations with *Cannabis*. However, we can formulate logical hypotheses with the aid of various sets of information. Speculating about the cultural roots of
ancient economic plants like *Cannabis* can be both interesting
and revealing because our associations with the natural world
rely so much on our relationships with domesticated plants

and animals. Human history has been strongly affected by our
relationships with cultivated plants, even those that have provided valuable resources no longer utilized.

In this chapter we begin with botanical, environmental,
and anthropological evidence to formulate our theoretical
reconstruction of the earliest uses of *Cannabis*. These hypotheses raise significant questions about its long-term importance to human culture. Could *Cannabis* have been one of
the earliest cultivated plants and therefore have played a key
role in the crucial transition from hunting and gathering to
incipient agriculture, at least in some regions of Eurasia?

Once humans began to use and cultivate *Cannabis* for varying purposes, they at first unwittingly, and then intentionally, began to sow seeds from those plants they deemed superior; these included those few plants that significantly deviated from the norm in a valuable and easily measurable trait such as yield or chemical composition. The majority of this chapter explores the various uses humans have found for *Cannabis* and how we have directed its evolution in several directions to satisfy our needs.

As nomadic groups settled in fixed locations, natural *Cannabis* populations were likely depleted by persistent annual collection. Consequently, ancient humans found it more convenient to cultivate the plant near their home rather than travel farther and farther each year in search of wild plants. Early farmers realized the superiority of cultivated plants over their wild relatives—in this case, increased resources they extracted from cultivated *Cannabis*. As a consequence, early farmers eventually favored landraces and domesticated cultivars over wild types. Today, there are few cultures that utilize wild or feral *Cannabis* in preference to cultivated varieties for any usage, unless cultivated *Cannabis* is unavailable.

During its early association with incipient farmers, cultivated *Cannabis* was almost certainly in frequent contact with its truly wild as well as recently feral relatives. Although continuing introgressive hybridization between wild/feral and cultivated populations might be expected, we have found little supporting evidence for its evolutionary importance. Those characteristics of *Cannabis* cultivars that distinguish themselves from wild plants probably evolved in the remote past and remain essentially the same today—simply recombined through selection and breeding. Whatever the case, independently operating sets of selective pressures have resulted in evolution of populations with differing phenotypic norms (median sets of observable characteristics of individuals within a population). This occurs under the influence of disruptive selection, which comes about when natural ecological conditions favor selections toward one norm, while humans artificially favor selections toward another set of characteristics. Self-sown plants exhibit traits that favor survival as natural (wild) or naturalized (feral) populations, while cultivated varieties commonly exhibit traits that humans prefer.

First Contacts: Origins of "Human-*Cannabis*" Relationships

People have a very long and curious association with *Cannabis*. In different places, and often for different reasons, humans have developed important uses for this unusual genus of plants. As a diverse ancient resource, *Cannabis* played an important, sometimes crucial role in fundamental cultural changes that occurred after the glacial Pleistocene Epoch. Furthermore, over time humans were probably instrumental in some if not most of the dispersal and increased geographical range of *Cannabis*.

Did humans first use *Cannabis* as a source of fiber to make cordage and clothing or as a source of seed to provide food? Or could *Cannabis* have been used initially as a medicinal plant? More challenging is the possibility that it was originally utilized as a psychoactive substance for spiritual purposes to communicate with ancestors and other supernatural forces. There is the intriguing possibility that early cultivation of plants such as *Cannabis* may have been inspired by religious or ceremonial motives. In a letter to one of the authors of this book (Merlin in 1968), the

influential geographer Carl O. Sauer, a longtime professor at the University of California at Berkeley, posed the following rhetorical question: "Did *Homo religiosus* precede *Homo economicus*?" Sauer's seminal if speculative thoughts about motives involved in the profound cultural innovation of agriculture are combined with the insightful ideas of Edgar Anderson and Richard Evans Schultes in the following discussion concerning the ethnobotanical origins and early cultivation of *Cannabis*.

Most of early human existence was spent roving around in small nomadic groups, hunting animals and gathering wild foods in varied environments. Then near or shortly after the end of the Pleistocene glaciations, human settlement and subsistence patterns began to change in profound ways. Global temperature increased, and as a result, continental ice sheets and alpine glaciers receded. Higher latitudes, formerly covered by massive ice fields, opened up to colonization. Meltwater from receding glaciers swelled existing rivers and created new ones while also raising sea levels. Sauer (1967) suggested that river valleys in moderate environments spawned human resourcefulness: "It was above all a rarely favorable time for humans to test out the possibilities of water-side life and especially of living along fresh water." As a consequence, some human groups began to take up more permanent residence in relatively comfortable and safe river valleys: "In large measure primitive man is a riparian creature anywhere" (Sauer 1958). With stationary settlement patterns, the pace of human cultural evolution accelerated rapidly. Our distant ancestors domesticated dogs from wolves, possibly at first for sentries, food, or pets. They also produced such cultural innovations as newly styled chisel-edged stone axes and adzes, useful pottery, bows and arrows for hunting, and lines and nets for fishing.

In some areas, fishing played a very important role in cultural development. Lacking agriculture and utilizing rather simple technologies, hunting and gathering groups led the mobile lives required of a foraging economy following seasonal cycles of food availability. But the development of methods to hook, stun, net, trap, and otherwise collect fish and waterfowl provided a relatively stable and stationary, high-protein food supply. These fixed and abundant food resources allowed some early fishers to become more sedentary. A more settled life allowed for additional free time to experiment with a great diversity of plants and animals, and this stimulated an elaboration of new arts and crafts (e.g., see Mithen 2004; Sauer 1967, 1969).

In some cases, settlers inspired by the success of fishing began experimenting with plant cultivation. Through a long, slow, and important process known as the "Neolithic Revolution" (Childe 1936), farming was invented; humans essentially became active producers of their own food supply. This gradually changed subsistence patterns from hunting, fishing, and gathering to a new mode of life based on the economy of farming. *Cannabis* provided a variety of resources in abundant and easily cultivated plants, and it may therefore have played an important role in the development of agriculture in areas such as Central Asia and northern China (see Chapters 4, 5, and 6).

Innovative early humans who adopted more sedentary modes of life—gathering, hunting, and fishing along rivers and streams, were also some of the earliest cultivators of plants. There is good reason to assume that among the oldest crops were versatile species grown for nutritious starch-type food, tough water-resistant fibers, euphoric and medicinal drugs, and fish poisons (Sauer 1952; Merlin 1972). Some of our oldest cultivated plants have multiple uses, such as flax (*Linum usitatissimum*) for seed and fiber and opium poppy (*Papaver somniferum*) for seed and medicine. Diverse usage of a plant is a strong indication of how long humans have been associated with that species (Ames 1939) and perhaps why they were initially attracted to it. *Cannabis* provides multipurpose and accessible plants, especially for well-situated, folk living in relatively mild temperate climates along fresh waters. Ecological relationships between these early fishers and *Cannabis*, at least in parts of Eurasia, support the notion that *Cannabis* was one of our oldest important plant resources and why it may have been among the earliest cultivated plants.

Cannabis can be found growing wild, or at least self-sown, in many parts of Eurasia, especially in Central Asia where it is adapted to slopes of undulating foothills and mountains up to several thousand meters above sea level depending on latitude. It also grows spontaneously on rich, fertile alluvial ground of river flood plains and valley bottoms and frequently in recently disturbed open environments along streams. Environments of this last type are quite often created by the independent or complementary work of streams and humans. Erosion and depositional actions of stream flow produce open environments by scouring the terrain and dumping mud, sand, and gravel in new places. People, who disturb the mantle of vegetation, affect waterways by increasing erosion and deposit nutrient-rich waste in dump heaps near their streamside residences.

When early hunter-gatherer-fishers settled down along stream-scoured banks, they cleared some of the surrounding land and constructed shelters. In removing vegetation they helped create one of nature's relatively rare environments— the "open habitat." Newly cleared habitats are quickly invaded by a series of pioneer plants. Sun-loving *Cannabis* thrives in open environments with relatively well-drained soils rich in nitrogen compounds such as those found in and around dump heaps. Camp-following *Cannabis* was among the first plants to colonize newly opened habitats, often brought in by humans who collected seeds elsewhere and dropped them inadvertently near settlements. Preadaptation of *Cannabis* to open nitrogen-rich soil environments was one important reason early humans developed significant ecological relationships with this genus.

Humans provided *Cannabis* with a suitable habitat and soon learned to utilize plants growing on or near their waste piles rather than traveling to collect them. Different plant parts were used as sources of fiber, food, seed oil, medicine, and mind-altering drugs. As new uses were discovered, this increased our dependence upon *Cannabis* for meeting specific needs and interests. Over time, humans more directly contributed to the success of *Cannabis* by protecting or consciously cultivating it near their homes; during migrations, they transported it with them purposely or inadvertently. These ecological, biogeographic, and cultural aspects enlighten the theme of our study—human-*Cannabis* relationships.

Transitions to Cultivation and Civilization

Cannabis, as we have argued, has a very long relationship with humans and may have been one of our first cultivated plants in some regions of Eurasia. Referring to evidence provided to him by a friend who had spent time with forest "Pygmies" in Africa, Carl Sagan (1977), a well-known professor of astronomy, tells us that when these people stalk and hunt

FIGURE 7. Mature female inflorescences containing relatively large ripe seeds are covered with thousands of glandular trichomes, each of which reflects sunlight, attracting foraging humans to the extremely fragrant and sticky resin. These easily observed characteristics made *Cannabis*'s economic value readily apparent to early humans.

mammals and fish, sometimes poised patiently for an hour with fishing spear aloft, "they prepare themselves through marijuana intoxication, which helps to make the long waits, boring to anyone further evolved than a Komodo dragon, at least moderately tolerable." Sagan referred to *Cannabis* as the Pygmies' "only cultivated crop" and suggested that "it would be wryly interesting if, in human history, the cultivation of marijuana led generally to the invention of agriculture and thereby to civilization." Although *Cannabis* was introduced to Africa by humans only 2,000 years ago, the Pygmy example and Sagan's comments regarding *Cannabis*'s relationship to hunting, fishing, and early agriculture are insightful. The shift from food procurement to food production was one of the most important developments in human history (e.g., Damania 1998; Bellwood 2005). This slow but momentous development increased people's food supply significantly and thereby stimulated substantial population growth (Bocquet-Appel 2011), and in some regions it led to overexploitation of natural resources.

Systematic gardening and farming spread slowly during the Neolithic stage, beginning sometime about 10,000 years ago, and in the process of contact and acculturation, these new subsistence activities replaced those of the Mesolithic hunting, gathering, and fishing cultures. Eventually the development of cultivation and domestication techniques further stimulated farming culture, forcing the decline of hunting and fishing economies to a secondary role except in marginal areas of the inhabited world.

During the early Holocene in Southwest Asia, and perhaps in parts of China such as the Huang He (Yellow River) and Yangzi River watersheds—where farming may have originated independently and megalithic civilizations evolved—climatic conditions were favorable for incipient agriculture. However, alluvial lowlands in the river basin areas of the Tigris, Euphrates, Nile, and Indus River valleys were still quite arid in this period of prehistory. Although the annual flooding of alluvial plains brought new soil, fresh nutrients, and moisture to certain lowland valleys, the magnitude and seasonality of the moisture deficit necessitated a certain degree of social organization and perhaps subdivision of labor for flood control, irrigation construction, and canal maintenance (Hawkes 1964; Bellwood 2005).

Therefore, it was in the hills or uplands of Southwest Asia (or other regions of Eurasia such as China) that systematic farming most likely began, especially in places where rainfall was adequate for "dry farming," and wild, potentially cultivable, annual plants already grew. Over time, some combination of social stratification, population pressure, and perhaps innate initiative precipitated the diffusion of peoples and culture from small upland settlements into the semiarid riverside lowlands of Southwest Asia, or other areas such as the lower Huang He and Yangzi River basins, where large-scale civilizations began. Between 8000 and 7000 BP, early farmers moved into the lowlands of the Tigris and Euphrates and eventually into the Huang He, Nile, and Indus River valleys; it was in these regions that large-scale "hydraulic" civilizations and associated monumental architecture first appeared. In the section that follows, we elaborate on our hypothesis that the first use of *Cannabis* took place within its native range in Central Asia, and it was subsequently grown in many regions of early civilization, such as China, India, and other areas of Eurasia.

FIGURE 8. *Cannabis*'s fibers would have been obvious to early humans. Fresh stalks are easily broken and the fibrous bark peeled by hand as in this demonstration in Shandong province, China.

Earliest Uses of *Cannabis*: Useful Traits for Ancient People

People have long been attracted to *Cannabis* for its various useful products—fiber, food, and drugs—some for economic reasons and others perhaps for spiritual purposes. As noted earlier, it is presently impossible to say with certainty which products were utilized first. However, some simple hypotheses can be advanced that also bear on the later evolution of *Cannabis* under human selection.

When ancient humans encountered *Cannabis* growing naturally in their cultural landscape, they must have soon discovered its usefulness. Its vigorous growth in open areas makes it stand out, and its unique appearance makes it readily distinguishable. The economic attributes of *Cannabis* are also easily noticed. Its fibers persist on the ground after the stem rots; the seeds are relatively large, prevalent, and eaten by birds; and sticky resin glands glisten attractively in sunlight.

Considering the nature of the primary economic attributes of *Cannabis* (fiber, seed, and psychoactive resin) and the possible means by which ancient humans discovered them, one might argue that fibers were the first product of apparent value. *Cannabis* drops its leaves as it dies back in autumn, leaving stalks that are eventually blown to the ground by winter winds. The stalks rot and decompose, exposing long fibers that are more resistant to rotting than softer stalk tissues. These tough and flexible fibers remain on the ground until the following spring. At this time humans would have noticed them rather easily, and if needed, they could have gathered the fibers for making cordage, nets, or other fabricated items. In several regions of the Himalayan foothills, giant nettle (*Girardinia diversifolia*) fiber is found lying on top of the soil under the plants and used to weave cloth. Observations of natural stem rotting may have led to the technique of hemp "retting" (intentional, controlled rotting to free the fiber from the woody core of the stalk). In addition, examination of naturally dense-growing stands probably encouraged the first hemp farmers to sow seeds closely together to suppress branching as the plants matured and thereby improve fiber length and quality (see Chapter 5). It should be pointed out that hemp fiber, which is suitable for spinning yarn and weaving fine cloth, can only be derived from cultivated hemp grown close together, not from self-sown feral plants or plants with branches, although these latter growth forms are used for seed and psychoactive resin production. The production of fiber suitable for weaving clothing must have been one of the earliest incentives to begin cultivating *Cannabis*. Other fiber-bearing plants native to temperate and subtropical regions of Eurasia, such as elm (*Ulmus*), lime (*Tilia*), and willow (*Salix*), provide long fibers that can be collected from wild populations and are suitable for making cordage, while cultivated European nettle (*Urtica dioica*) and flax (*Linum usitatissimum*) fiber, like cultivated hemp fiber, are also suitable for spinning yarn and weaving. It is likely that wild sources were extensively utilized when suitable natural habitats were abundant. But as human populations grew, environmental alteration increased, and wild fiber sources dwindled, it became necessary to cultivate hemp (and other fiber crops) to meet increasing demand for woven fabrics. In sum, *Cannabis* use for fiber is an ancient human trait. Indeed, it has been referred to as the oldest known cultivated fiber plant (Small, Beckstead, and Chan 1975; also see Small and Marcus 2002). According to Damania (1998), "*Cannabis sativa*" was first used for its fiber and is "one of the oldest cultivated non-food crops."

In northern temperate regions of Eurasia where spontaneous (self-sown) fiber *Cannabis* grows today, there are few other native fiber plants. Thus it is quite likely that *Cannabis* played a significant role in early cordage and textile manufacture in these areas. Flax was the only other fiber plant widely utilized during the early history of Europe. Damania (1998) tells us that flax was probably domesticated about the "same time (ca. 8200 BP) as wheat and barley in the Near East from where it spread to other parts of the Old World along trade routes." Although flax was distributed over a large area in ancient times, mainly through human agency, *Cannabis* was even more widely distributed across Eurasia. Genera in the nettle family (Urticaceae) such as *Urtica* (nettle) in Europe and *Boehmeria* (ramie) in East Asia may also have been utilized for fiber early on.

In South Asia, *Cannabis* was rarely utilized to make cordage because several other fiber plants were more commonly grown, for example, kudzu (*Pueraria thunbergiana*), banana or abaca (*Musa textilis*), jute (*Corchorus* spp.), and ramie (*Boehmeria nivea*). Consequently, no fiber varieties of *Cannabis* originate from South Asia. In India and Nepal, narrow-leaf drug (NLD) *Cannabis* has occasionally been used to make cordage or yarn for weaving, but the fiber is extracted from landrace varieties primarily grown for drug production, which contain much higher levels of THC than European narrow-leaf hemp (NLH) varieties (see Chapters 5 and 7).

It may also be argued that early humans gathered *Cannabis* seeds for food prior to (or along with) their use of these plants for fiber or drugs. In regions where hemp seeds were consumed during the historical period, they were often considered a famine food, eaten only during times of shortage of more desirable foods. However, in ancient northern China, *Cannabis* seeds were formerly one of the four most important "cereals" (see Chapter 6), along with two millet species (*Setaria italica* and *Panicum miliaceum*) and buckwheat (*Fagopyrum esculentum*). This changed after soy beans (*Glycine max*) were domesticated and upland races of rice (*Oryza sativa*) were introduced and spread into northern China, both by about 3000 BP (Chang 1976; Harlan 1995; Lee et al. 2007); subsequently, *Cannabis* seeds were decreasingly utilized by the the majority of Chinese and presently are enjoyed only as snack foods. Although *Cannabis* seeds were not a high-priority food among most ancient societies and are relatively rarely consumed today, they must have been eaten more extensively at an earlier date before introduction of the presently dominant cereal crops (e.g., rice, maize, and wheat), at least in northern China, if not elsewhere too.

Burning *Cannabis* plants, ignited by a natural or man-made fire, may have alerted humans to the psychoactivity of its resinous flowers. If chaff left after threshing seeds was added to a fire burning within an enclosed structure, breathing the smoke could have proved quite psychoactive. The Scythians, according to the Greek historian Herodotus, used this technique during funerary rituals well before the time of Christ (Herodotus 1921; Brunner 1977; see also Chapters 7 and 9). We argue that discovery of the mind-altering capability of *Cannabis* was more likely an accident resulting from the search for food.

Cannabis seeds are hidden within the sticky, resin-covered female flowers. Wild *Cannabis* seeds shatter readily from mature, drying inflorescences and drop to the ground. Therefore, early humans either had to search through fresh inflorescences to pick out the seeds or harvest the mature seedy inflorescences before the seeds fell out. The inflorescences would have been first harvested and dried and then the seeds would have been threshed from the floral parts. At first, humans probably tried to eat the foliage or entire inflorescence, but this would not have been an effective way to feed. Picking through fresh flowers in search of edible seeds would certainly result in sticky resins adhering to gatherers' fingertips. *Cannabis* resins are known as *hashish* or *charas* and are one of the most potent forms of *Cannabis* drugs, largely because psychoactive THC is a major component of *Cannabis* resin (see Chapter 7). The sticky accumulation of resin on the fingers would eventually interfere with seed collection. One obvious way to remove the accumulation would be to scrape it off with one's teeth. If early *Cannabis* plants contained even small amounts of mind-altering THC, this would a likely means of unwitting ingestion through which the plant's effects could have been noticed.

Oral ingestion was the most common method of consuming *Cannabis* drugs prior to the spread of smoking from the New World after the fifteenth century, along with tobacco (*Nicotiana tabaccum* or *N. rustica*; Clarke 1998a; also see Merlin 2003). If early humans cooked seeds with any resinous bracts adhering to them, THC-acid (THC as it is found in fresh plant material) would have become more potent through the process of decarboxylation (loss of carbon dioxide, i.e., elimination of a carboxyl, COOH, group). Possibly they tossed threshed female flowering branches or balls of sticky resin into a fire to dispose of them, accidentally breathing the smoke. In any event, if early humans ate or breathed enough THC, they would have experienced mind-altering effects.

If foraging led to the discovery of *Cannabis*'s psychoactive properties, then one could assume that the first, and certainly the most widely used, concentrated *Cannabis* drug was hand-rubbed resin (*charas*) collected in the same way as it is today in much of South Asia. *Cannabis* resin forms into a dense, compact lump that is easily transported. Nomadic peoples could have regularly collected hand-rubbed resin during their autumn migrations and carried it for use throughout the year.

As the valuable products of *Cannabis* are quite useful in their wild forms, we doubt that they came under much artificial human selective pressure within the plant's original, wild range. Artificial selection more likely began as *Cannabis* moved into new centers of agriculture such as eastern China, India, and Europe and encountered new habitats and human requirements. Cultivated *Cannabis* likely originated from spontaneously growing ancestors that were transported away from, and introduced outside, their original natural range, probably soon after initial human contact. In

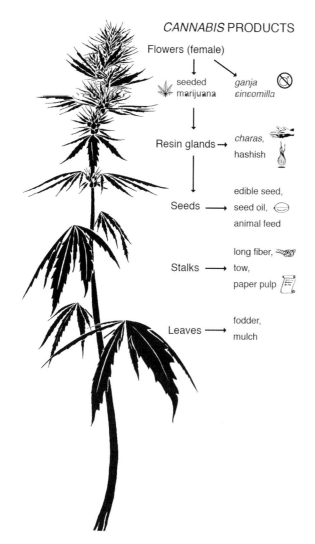

CANNABIS PRODUCTS

Flowers (female)

seeded → ganja
marijuana   sinsemilla

Resin glands → charas, hashish

Seeds → edible seed, seed oil, animal feed

Stalks → long fiber, tow, paper pulp

Leaves → fodder, mulch

FIGURE 9. Humans make use of each part of the *Cannabis* plant.

other words, *Cannabis* was dispersed by people, both consciously and unconsciously, and escaped widely before it was selected intensively for individual economic traits desirable to specific cultures (see Chapter 4).

Initial human selections from wild populations probably came from one or a few plants that produced fiber, seed, or resin and happened to appeal (consciously or not) to the person collecting seeds for sowing. As a result, each of these collections formed a relatively narrow genetic base (or founder effect) for subsequent generations. This led to new populations that were genetically composed of closely related individuals. Even in a dioecious, obligate outcrosser such as *Cannabis*, with separate male and female plants, the founder effect results from population inbreeding between very closely related siblings (Pickersgill and Heiser 1976; see also Hey et al. 2005).

As we noted previously, *Cannabis* has been characterized as a "camp follower," which means it was moved around by people either consciously or accidentally, and found favorable niches, such as refuse heaps or other areas disturbed by human occupation, where it flourished and was repeatedly utilized, selected and hybridized by ancient people. Anderson (1956) pointed out that plants follow humans not out of preference for but rather tolerance to human disturbances.

He also argued that crop plant evolution proceeds through bursts of domestication, in various times and in different places. Evolutionary activity involving human input, according to Anderson, is concentrated in disturbed habitats and, in the case of crop plants, is directed by human needs for improved plant products.

In order to accurately determine origins of domesticated traits in crop plants it is necessary to understand the selective pressures that humans and nature have imposed on plants during their early contact. Keep in mind that cultivated varieties (and feral escapes) do not arise solely from human selection but rather by a combination of human and natural selective pressures. However, cultivated plants are more or less protected from the forces of natural selection, and this isolation, accompanied by fluctuations in population size, leads to evolution of variability within the crop (Pickersgill and Heiser 1976; also see Bellwood 2005). Specific selective pressures leading to the evolution of individual varieties undoubtedly differ somewhat in each case. However, generalizations can be made that pertain directly to the evolution of *Cannabis* as a whole.

Spontaneously growing *Cannabis* populations worldwide share some characteristics of gross morphology and development, such as strongly erect growth, early flowering, and uneven maturation—a consequence of adaptation to open growth conditions common to truly wild as well as feral or weedy *Cannabis* everywhere—and throughout much of Eurasia, it is difficult to determine if noncultivated feral populations are truly wild. However, no uniform weedy phenotype (individual response to the environment) is found throughout the range of self-sown *Cannabis*, and spontaneous varieties vary widely from region to region. It is possible, therefore, that *Cannabis* was naturally widespread throughout Eurasia very early on and had differentiated through natural selection to the species or subspecies level prior to human contact. However, when humans first encountered *Cannabis*, we believe they realized its value rather quickly, collected the seeds, and spread them far and wide, trading them to other early agricultural groups. Whatever the case, early, self-sowing populations of *Cannabis* soon became established in introduced environments, and artificial selection continued for fiber, seed, and/or drug use, leading eventually to a multitude of new landraces and cultivars. Self-sown feral populations may be descendants from ancient naturalized populations that escaped cultivation early on and would be easily confused with truly wild populations. Plants that escaped cultivation more recently exhibit relatively similar phenotypes when compared to neighboring cultivated populations and are easier to distinguish than earlier escapes. Despite its great antiquity as an economic plant, *Cannabis* may not have been among the very first cultivated food plants, even within its native range. However, *Cannabis* was certainly among the very first cultivated fiber and drug plants and thus provided a primary source of fiber and seed oil to northern temperate cultures for millennia. The use of wild *Cannabis* most likely predates the dawn of agriculture. Although the present archeological record does not yet support this assertion in any great detail, it is logical to assume that the use of "wild" plants always predates the development and use of "cultivated" plants (for detailed discussions of the early history of cultivation and use of *Cannabis*, see Chapters 4 through 9, especially the section in Chapter 4 that focuses on the early association of *Cannabis* in ancient Japan).

TABLE 2

Plant parts used	Use category	Material type or other benefits
Stem bark	Cordage	Long cellulose fibers
Stem fiber	Cordage and woven textiles, building materials	Long cellulose fibers, concrete reinforcement
Stems (wood and bark)	Paper	Long and short cellulose fibers
Stem wood w/o bark	Building materials, animal bedding	Chip board, concrete matrix
All parts: Primarily female flowers and seeds	Medicinal	Herbal remedies, pharmaceuticals, nutraceuticals
Female flowers and associated resin glands	Recreational drugs	Marijuana (*ganja*), hashish (*charas*)
Seeds, seed oil	Human food	Proteins and essential fatty acids, essential fatty acids (*omega*-3 and *omega*-6)
Seeds, seed cake, foliage	Animal feed	Proteins and essential fatty acids, proteins and trace fatty acids, vegetable mass
Seed oil	Industrial feedstock	Oil used in paint and plastic manufacture
Stem wood w/o bark, seed oil	Fuel	Heat, light
All parts: Primarily bark, seeds, and female flowers	Ritual and social	Social activities employing various plant parts such as healing and life cycle rituals and inebriation
Populations	Environmental	Erosion control and CO_2 fixation
The plant, people, and their interplay	Aesthetic	Intrinsic beauty of the plant
The genus	Educational	Iconic example of an economic plant and its ancient human relationships

NOTE: Stalks provide fiber used to twist cordage and yarn, weave textiles, and make paper. Medicinal remedies are made from all parts of the plant, especially the female flowers and seeds; the female flowers also provide psychoactive drugs. Seeds and seed oil provide vital dietary requirements for humans and livestock as well as biofuels and industrial oils, and all plant parts are used in various ritual contexts. *Cannabis* is an iconic example of an economic plant with a long history of human use and serves as an excellent educational tool. Populations perform valuable ecological functions; most importantly humans find satisfaction in aesthetic appreciation of its attributes.

Evolution of *Cannabis* through Human Selection

The actual use of *Cannabis* by past societies is not surprising as it is a source of three useful products; hemp fibres [*sic*] and seeds as well as the resin itself and it can also grow wild requiring very little attention. (DERHAM 2004)

A combination of natural and human selective pressures molded *Cannabis* into a phenotypically diverse genus. Selection of traits that favor fiber, seed, or drug production has led to diverse *Cannabis* cultivars. The following section explores several basic principles of crop plant evolution and how each applies to the evolution of *Cannabis* in particular. First we investigate the basic evolutionary effects of human selections for various plant products and identify the specific phenotypic changes that accompanied the domestication of *Cannabis*; this discussion focuses on directional evolutionary changes in two of the most significant parameters of *Cannabis*'s evolution—cannabinoid profile and date of maturation. We then turn our attention to the evolution of divergent cannabinoid phenotypes and their worldwide distribution in the context of historically verifiable dispersal events.

Disruptive Selection

The largest single influence that humans imposed on the evolution of *Cannabis* crop plants was through disruptive selection. This type of selection favors extreme traits (e.g., long internodes and suppressed branching vs. short internodes and many branches) over intermediate traits (e.g., medium length internodes and moderate branching). Diversity in heritable variation of phenotypes results from selection for different plant products, in this case, fibers versus flowers. Natural disruptive selection also plays an important role in speciation but is more difficult to detect.

Monopodal (single stalk) annuals such as *Cannabis* suppress lateral branching when grown in thick stands and branch freely when grown in the open (Iltis 1983). As a result, *Cannabis* is naturally adapted to cultivation, either in dense stands for long fiber production or in open conditions to promote branching and flower formation for seed or drug production. In order to mold this basic phenotypic plasticity, humans have selected the *Cannabis* plants with specific mutations, leading ultimately to the development of cultivars suitable for particular uses. Fiber varieties have longer internodes and are less branched, even when grown in open conditions. On the other hand, seed and drug varieties have shorter internodes and are more highly branched, even when grown in crowded conditions.

Narrow-leaf drug (NLD) and broadleaf drug (BLD) cultivars produce much greater quantities of psychoactive THC than narrow-leaf hemp (NLH) and broad-leaf hemp (BLH) fiber and seed varieties. In drug cultivars, THC content in the dried female flowers can reach 15 to 20 percent dry weight, while spontaneous NLD and BLD populations rarely exceed 5 percent dry weight THC. NLH and BLH landrace varieties rarely exceed 1 percent THC, while legally registered European NLH cultivars must produce less than 0.3 percent THC, the legislated maximum allowable content for industrial hemp in the European Union, Canada, Australia, and New Zealand (e.g., see Small and Marcus 2002). Great variation in cannabinoid content evolved as a result of three forms of selection: natural selection (where THC may or may not provide a selective advantage), human selection for both fiber and seed use (where THC content is less important or undesirable), or human selection for psychoactive use (where high THC content is of the utmost importance).

Cannabis evolution has resulted in a large amount of phenotypic variation through isolation of cultivated populations from wild hemp plants, as well as repeated disruptive selection accompanied by limited introgression (discussed later). Ancient as well as modern *Cannabis* biotypes have been altered by disruptive human selection, and their genetic integrity has been preserved through geographical isolation.

Origin from Weedy Populations

Weed is a word with many meanings. Some people even casually refer to *Cannabis* as "weed." A particular plant may or may not be a weed depending on one's definition of the term. Here we refer to weeds as self-sown plants occurring in areas disturbed by humans and often deemed undesirable.

Although humans transport many weedy plant species as they move about the earth, sometimes intentionally but most often by accident, the history of this complicated dispersal process is largely unstudied. Ironically, science knows relatively little about many of the most common plants, and *Cannabis* is a striking example. In fact, the history of human association with *Cannabis* is just one aspect of a more challenging and larger problem: understanding the complex history of weeds. *Cannabis* is an extremely aggressive weed. As long as it finds a favorable habitat, spontaneous populations will grow near and within cultivated areas throughout much of its range. *Cannabis* has significant economic impact as a persistent weed throughout much of temperate Eurasia and North America, necessitating costly eradication when deemed necessary.

On the other hand, weeds serve as reservoirs of natural variation that can inject variability into more genetically limited cultivated crops through intermediate hybrids; in some cases, weeds were progenitors of important crops such as rice and wheat (Simmonds 1976; Hawkes 1983; see Abbo et al. 2005 for a recent critical review of this subject). However, since weeds are more or less poorly adapted to undisturbed preagricultural conditions (Harlan 1965), it is difficult to imagine weedy populations giving rise to modern *Cannabis*.

Truly wild plants are defined as native plants occurring in natural sites undisturbed by purposeful or accidental human activity. However, it is often difficult to determine reliably if a site has remained undisturbed, particularly in ancient times, and if it was disturbed, if that disturbance occurred naturally or was the work of humans or perhaps by human and natural forces combined. Both truly wild and naturalized populations are self-sowing and exist either as feral plants or may be remnants of truly wild populations. Although *Cannabis* is well adapted to naturally disturbed environments as well as weedy habitats or other areas disturbed by humans, it probably did not evolve as a weed in cultivated fields. Its attributes are too obvious and its discovery was too early for *Cannabis* to have been accidentally or purposefully selected from weeds growing within other crops.

Feral escapes from cultivation, however, may be important in the evolution of cultivated varieties. Weedy feral *Cannabis* plants can invade a crop field and interbreed with *Cannabis* cultivars. Although this initial introduction of genes from weedy populations is very dilute, if subsequent natural selection favors weed-type hybrids, their genes could proliferate and eventually constitute a significant portion of the cultivated variety's genotype.

Whenever any organism is introduced to a new geographical region as a result of purposeful or accidental human activity, it becomes an alien species (as opposed to native species) in that new area. Alien species often undergo accelerated evolution when introduced to new environments as they are exposed to new natural selective pressures (e.g., Lavergne and Molofsky 2007); they may also become invasive and proliferate widely in their introduced habitat, competing with and often displacing native vegetation. Human selective pressures during domestication further accelerate evolution (Cox 2004).

As alien *Cannabis* was dispersed by humans into a wide range of environments where it was native or had previously been introduced, conditions for hybridization between established and introduced populations occurred frequently. Ellstrand and Schierenbeck (2000) demonstrated that hybridization between previously isolated taxa can produce a stimulus for evolution of invasiveness, the ability to spread beyond the site of introduction and become established in new locations. A plant may simply be introduced to a new habitat and escape into the environment where it proliferates, or if the taxa previously existed in this habitat, the new introduction may hybridize with the local population (wild, feral, or cultivated) and their offspring will have a higher chance of becoming well established as feral populations.

Through the agency of human dispersal, feral *Cannabis* populations flourished across northern temperate latitudes of Eurasia. This brought members of the primary *Cannabis* biotypes (e.g., NLH, BLH, and NLD) into contact with one another. As a result, hybridization likely occurred. Invasive weedy populations in the United States are descendants of hybrids between European NLH and East Asian BLH gene pools created during early twentieth-century industrial hemp breeding experiments (e.g., 'Kentucky' and 'Minnesota' hemp; see Chapter 10). Increased vigor observed in these new hybrid hemp varieties could also have conferred evolutionary advantages as feral escapes or weeds. However, it appears unlikely that weedy populations played a role in the evolution of *Cannabis* cultivars by providing new traits through this process of introgression, which is discussed in the following section.

Natural Hybridization:
Introgression versus Isolation

Among anemophilous (wind-pollinated) and dioecious (male and female flowers occurring on separate individuals) plants like *Cannabis*, outcrossing is obligatory because male and female flowers are almost always borne on separate plants thus favoring hybridization. There are no sterility barriers between *Cannabis* taxa, and all hybrid crosses produce fertile seed. In some cases, weedy populations are known to naturally hybridize with cultivated populations (Chrtek 1981). Hybridization is the cross-fertilization of plants of different taxa and usually occurs at the genus level or below (e.g., between genera, species, subspecies, varieties, and forms); it is essentially the recombination of separate gene pools through sexual reproduction. The more specific case of introgressive (or natural) hybridization is the transfer of genes from one self-sown population to another through an intermediate hybrid population. Effective introgression relies on repeated backcrossing of the hybrid population with one or both parental populations to facilitate gene transfer. If evidence for hybrid populations and backcrossing are weak, then introgression is less likely (Heiser 1973). Artificial hybrids are created by plant breeders who transfer genes from one parental population to the other through carefully controlled, repeated backcrossing. This is an example of artificial introgression through intentional hybridization. The important difference between natural and artificial introgression is that natural selection takes place in the first case, while artificial selection is exercised in the second case. Natural crosses often occur between self-sown and cultivated varieties, and hybrid populations can be selected naturally or artificially, depending on where the hybrids grow in relation to the self-sown and cultivated varieties (for more on introgression, hybridization, and evolutionary processes, see Harrison 1993; for an excellent example of introgression between the world's oldest fruit crop, the grape, and its wild ancestor in the genus *Vitis*, see Myles et al. 2011).

Introgression is to be expected in crop plants because disturbed areas in close proximity to cultivated populations offer perfect niches for the proliferation of hybrid swarms (self-sown hybrid populations), and in such swarms, the taxa involved come into contact within sympatric (overlapping) areas of their ranges (Anderson 1948; Ellstrand et al. 1999). One might also expect that domesticated plants have little adaptive significance to offer self-sown populations because they share much of the same genetic complement, with exception of some mutations selected during domestication that may not be favored in self-sown populations (Heiser 1973). Since self-sown and cultivated populations have reached different adaptive norms under varying conditions, hybrids between them would tend to be less fit and less able to compete with either parental population. However, in the case of agricultural weeds, humans have created a series of environmentally disturbed niches well suited for hybrids. On the other hand, although hybridization and backcrossing do occur, in most cases of introgression, gene flow remains highly localized (Heiser 1973).

The question most pertinent to the identification of truly wild *Cannabis* populations concerns the genetic integrity of the putative ancestor (PA) *Cannabis ruderalis*; if hybrid swarms appear, do they survive and eventually lead to actual introgression that results in the exchange of genes between wild and feral or cultivated populations, or are hybrid swarms merely transitory? The answer bears on several important issues. If introgression is frequent, it is doubtful that truly wild populations (i.e., *C. ruderalis*) still exist; if this is the case, all self-sown populations are descended from, or have introgressed with, cultivated populations. However, *Cannabis* is a genus composed of six subspecies and many varieties, and these distinct taxa must have evolved in genetic isolation without significant introgression. If introgression is common, then we should rethink some of our taxonomic notions about the number of *Cannabis* species. The multispecies view relies heavily on geographical and genetic isolation of taxa as a vehicle for their rapid evolution.

Frequency of hybrid fertilization and effectiveness of hybridization in creating a unique population must be considered when assessing the possibility of introgression. More important, the effectiveness of introgression in allowing gene exchange at a frequency that produces change in either population must be analyzed. According to Harlan (1965), "The typical pattern is a localized irruption of a hybrid swarm that quickly subsides in a few generations. There are usually rather formidable barriers to gene exchange, and selection pressure for either weed or crop is apparently strong." Introgression, therefore, would occur only in situations where two *Cannabis* populations, either self-sown or cultivated, occur sympatrically (occupying the same or overlapping geographic areas), and hybridization between them is frequent. The collective genotype of a hybrid swarm is continually changing, balanced between natural selection and the indirect selective forces of humans via new escapes from cultivation. In general, introgression is highly likely under three simultaneous conditions: (1) the hybrid swarm persists, (2) backcrosses to one or both parental populations are frequent, and (3) both natural and artificial selection are not too restrictive.

Escaped and weedy *Cannabis* grows along roadsides, in ditches, and on fallow land or fields sown in *Cannabis* or other crops. If hybrid swarms persist, they may act as a bridge for gene exchange between populations. Some fields usually lie fallow each year, and if escaped populations can reestablish themselves quickly, or if new escapes occur, these feral plants offer many chances for hybridization and possible gene transfer. However, under these conditions, the effects of introgression would be minimal because genomes of recent feral escapes are very similar to the crops from which they are derived.

In the cultivation of drug *Cannabis,* much time may be spent transplanting, irrigating, and feeding plants on an individual basis. Males are often removed to prevent pollination of females, and weedy or otherwise undesirable types are culled. The chance of a weedy plant of either sex surviving until reproductive age is very low. In fiber hemp fields, and more rarely in fields of drug *Cannabis* (e.g., hashish fields in Lebanon and Morocco), plants are grown very closely spaced, usually without furrows, so little room is available for competing weeds. If particularly vigorous weedy female *Cannabis* plants survived until harvest, they could possibly make a genetic contribution if the plants are allowed to go to seed and the seed is saved in bulk for sowing the following year. However, so many plants are grown in such a broad area that the genetic contribution of the weedy intruder would be diluted to the point of ineffectiveness in relation to the remainder of the cultivated gene pool. Weedy *Cannabis* introductions would have to be frequent and become established long enough to evolve under natural selection before enough differing genes could be transferred from the weedy population to the cultivated one; under such conditions, introgression could affect a shift in genotype leading to observable phenotypic variation.

Gene transfer from self-sown to cultivated populations is more likely in a fiber field than a seed or drug field. In hemp fiber fields, the plants are grown too close together for farmers to detect weedy individuals or to readily remove them even if they could be seen. European narrow-leaf hemp ancestor (NLHA) × narrow-leaf hemp (NLH) populations are common and may lead to introgression of cultivated genes to wild populations; this may account, in part, for the genetic similarities between NLHA and NLH ecotypes. Hashish varieties are sown close together in a broadcast fashion in Morocco and Lebanon, but no feral populations have been reported from these regions. In a situation where seeds are collected in bulk for sowing the following year, seed from a weedy plant (or seeds resulting from fertilization by a weedy plant) will end up in the next generation. In general, cultivated drug varieties, and therefore their escapes as well, tend to mature later than acclimated, self-sown populations. Escaped male plants in particular might have a good chance to make a genetic contribution to a spontaneous population by fertilizing many self-sown female plants, although escaped males would likely flower later than hybrid spontaneous plants. Escaped males would then be in direct competition with males of the hybrid swarm to fertilize the females; only later maturing, self-sown females are likely to be fertilized by an escaped male. It is less likely that an escaped, later maturing female plant would flower early enough to produce viable seeds from fertilization by males of the hybrid swarm since the males would probably mature earlier than the escaped female. However, even if hybrids appear, it is doubtful that any domesticated traits would offer an adaptive advantage to a self-sown wild population because domesticated traits that reduce fitness in the feral population are unlikely to be proliferated in hybrid swarms and transferred to the wild population through backcrossing. It is therefore unlikely that genes of drug populations will be transferred to wild populations by introgression via hybrid swarms, although narrow-leaf drug ancestor (NLDA) × narrow-leaf drug (NLD) hybrid populations certainly exist in South Asia.

Sinskaja (1925) referred to cultivated *Cannabis* varieties in the Altai Mountains of Central Asia that produced lighter colored seeds than "wild" types, which had a marbled mosaic fruit coat and a horseshoe-shaped base. Sinskaja assumed that the cultivated variety was imported from another region, possibly China, because it matured much later than wild populations. Wild and cultivated types were usually found in allopatric (nonoverlapping) populations. Local inhabitants used wild seed for cultivation in times of seed shortages, and intermediate hybrid phenotypes were observed. However, despite possible introgression, perhaps for many years, weedy and cultivated varieties were easily distinguished. Differences in maturation would have prevented some cross-pollination between cultivated fields and wild-type populations. The farmers Sinskaja observed may have considered the light-seeded variety a preferable cultivar and continued selecting it, even though it matured later. In any case, disruptive selection

seems to have resulted in evolution and maintenance of two distinct varieties. The situation described by Sinskaja could serve as an example of introgression, but this is impossible to verify as the effect of gene transfer on the phenotypes of the original parental populations cannot be determined.

How do we explain the continued existence of traditional Hungarian landrace varieties (e.g., 'Tiborszalasi') grown today on a small scale in the Carpathian Basin of southeastern Europe? Certainly many plants from cultivated landrace fields have escaped, forming self-sown populations that grow nearby and persist even though the landraces are now rarely cultivated. Hemp farmers have rarely bothered to uproot neighboring feral populations, allowing free gene flow via pollen exchange between landrace and feral populations and vice versa. In fact, farmers have been more likely to have accidentally or even intentionally included feral female plants in their seed harvest and possibly sowed these seeds the following year; this would allow limited gene flow from the feral population to the cultivated landrace via the feral female's seed as well. In this case, isolation is ineffective and opportunities for introgression are frequent; indeed the escaped feral population and the traditional landrace variety are very similar in appearance. Although landrace varieties may no longer be cultivated, self-sowing feral landrace populations thrive and are large enough to maintain their genetic diversity.

Hybrid swarms are usually short-lived because they are subject to strong selection as either a weed or a crop (Harlan 1965). Selective pressures, both natural and human, are constantly changing. It seems logical that introgression with weedy populations could provide natural variation in the cultivated varieties, thus allowing flexibility under changing selection (Pickersgill and Heiser 1976). However, we have not found any objectively verified examples of introgression between weedy or feral self-sown *Cannabis* and cultivated *Cannabis*. Apparently, introgression has not played a significant part in the evolution of biotypes and cultivars yet has been quite significant in the blending of feral escapes and original wild populations. Although natural hybridization does occur and gene pools may introgress, the combining of genomes more usually occurs through intentional artificial hybridization and should be considered as another element of human selection and breeding rather than natural introgression.

Artificial Hybridization

Rapid evolution is likely to occur in crop plants when previously isolated varieties are allowed to form a hybrid population, and, as a result, a wider variety of possible genetic combinations are brought together. *Cannabis* populations became geographically isolated as they spread across Eurasia into diverse environments. Subsequently, they became genetically isolated through artificial selection for a variety of traits. In most of these introduced regions there were no indigenous wild types and no hybridization with spontaneous populations could occur, at least within early generations before escapes from cultivation became established feral populations. Known as the "founder effect" (Mayr 1942), this extreme isolation in various regions of Eurasia favored evolution of localized, variant phenotypes based on a narrow gene pool. Later on, hybridization may have occurred in these regions between evolving cultivated varieties and their feral relatives rather than between cultivars and truly wild varieties. The most rapid evolution has occurred when local landraces have been hybridized with landraces imported from distant origins. In such a hybrid population, two genetically differing genomes are combined resulting in more genotypic variation and consequently more chances for change through evolution.

Evolution under domestication progresses rapidly when hybrid offspring are artificially selected. This can lead to new, desirable combinations. For example, during the eighteenth and nineteenth centuries, Chinese BLH fiber cultivars were introduced into Europe and the Americas. The introduced BLH varieties were then crossed with the local NLH varieties, and the recombinant offspring provided improved parents for future crosses (see Chapter 5 for a detailed discussion of hemp fiber history).

This situation also developed in North America and Europe more recently during the late 1970s and early 1980s but with a distinct difference in the traits that were artificially selected. In this case, many high-THC NLD varieties from diverse regions of the world (where they had evolved in relative isolation) were brought together and intentionally hybridized; generally, this was in an effort to breed marijuana cultivars that were more potent and productive when grown outdoors at higher latitudes or elevations with short growing seasons. Profound recombination of gene pools isolated for hundreds or thousands of years resulted in an exponential proliferation of hybrids and potential cultivars (see Chapter 10 for a history of modern *Cannabis* breeding).

Evolution of *Cannabis* accelerated significantly under cultivation, and cultivated varieties may have been modified to some degree by hybridization with escaped *Cannabis*. Schultes (1970) suggested that no truly wild populations of *Cannabis* still exist. He felt that reciprocal, introgressive hybridization could easily occur between escaped or truly wild forms and cultivated varieties. Only a very isolated region with no *Cannabis* agriculture could have a truly primitive, unhybridized wild population of *Cannabis*. Following Schultes's line of reasoning, all self-sowing *Cannabis* populations that exhibit wild traits are simply in a more advanced stage of escaping and becoming feral.

If, however, the effects of introgression are not as widespread as Schultes assumed, then the possibility of discovering truly wild or natural populations is increased. Throughout the history of *Cannabis*, dispersal has generally been directed away from Central Asia, and the centers of intense *Cannabis* agriculture and evolution have been and still are far removed from the original prehistoric home of *Cannabis*. It is possible that truly wild varieties spread across Eurasia as the Holocene commenced and before agriculture began; if so, relics of these wild populations could still survive today beyond the influence of humans. It is also possible that truly "wild" populations such as those described by Vavilov and his colleagues might still be found within some area(s) of the vast Central Asian or European Steppe regions, far removed from modern agriculture. Securing seeds of self-sown Central Asian populations is a high priority for future taxonomic studies (see Chapter 11).

Although we suspect natural introgressive hybridization may at times have played a role in the evolution of *Cannabis*, we have not been able to find any clear examples. It is apparent that artificial hybridization has played a much greater role in the evolution of landrace varieties and modern cultivars. In addition, appearances of some wild traits in domesticated populations, and vice versa, could also be explained by a process called atavism.

Atavism

Atavism is the reappearance of a primitive characteristic after its absence for several generations, usually through chance recombination of genes. Most examples in cultivated plants appear as the recurrence of wild traits within domesticated populations. In fact, atavism occurs in several of the domesticated traits of *Cannabis*, including seed shattering, floral anatomy, and maturation date. One or more atavistic traits may occur at any time, and when members of the population escape cultivation and slip from the reins of human selection, those atavistic traits providing increased fitness for survival may once again become common. Atavistic chance recombinations may also appear in complex hybrids. Webbed leaves and nested bracts (reduced floral leaflets covering each seed), both traits resembling those of the hop vine (*Humulus*), as well as fused bracts and monoecy (both male and female flowers occurring on the same plant) are rare characteristics that may also arise atavistically.

Atavistic recovery of wild traits is an important consideration when characterizing the wild-domesticated complex in *Cannabis*, and indeed it can confuse cases of possible introgression between spontaneous and cultivated populations. Sympatric *Cannabis* populations may interbreed and introgression can occur, but the appearance of wild-type traits such as freely shattering seeds, lax floral clusters, and early maturation in some members of cultivated populations does not necessarily mean that introgression has taken place. These traits occasionally appear atavistically in intercrossing, cultivated populations without introgression occurring with spontaneous populations. In addition, the appearance of domesticated traits in spontaneous populations does not necessarily indicate introgression of genes from cultivated plants to a hybrid swarm, as these individuals exhibiting cultivated traits could more recently have escaped from cultivation or could be expressing atavistic traits similar in appearance to the early domesticates initially selected by prehistoric humans.

Dewey (1913) referred to a possibly atavistic purple-leafed mutation in an isolated and highly inbred 'Kentucky' fiber variety. Derived from the seeds of one selected female plant, the variety was grown for seven years in complete isolation. Theoretically, this should have resulted in a strong founder effect and consequently a population with a narrowed genome. Following two years of cultivation in Kentucky the variety was planted out at the United States Department of Agriculture (USDA) experimental station in Washington, DC, where two purple-leaved mutants appeared in a population of 100 plants and were among the most vigorous. This mutation had never been noticed in 'Kentucky' or other varieties, prior to this time. Dewey's most interesting observation, however, concerns seed collected from the best purple-leaved mutant: "It is of brownish color, easily distinguished from the normal gray of the other seeds, and it is larger in size than the others." Since this population was produced in isolation, there was no chance of introgression, and atavistic recovery of ancient traits through inbreeding is a logical explanation. Atavistic traits occasionally appear and make introgression difficult to verify in *Cannabis*. However, plants exhibiting atavistic traits (both domesticated and wild) can provide valuable breeding material for the creation of artificial hybrids in plant improvement programs.

Isolation of Populations

Isolation has played an important role in the evolution of *Cannabis* varieties. Dispersal of a plant via human migration and trade to remote locations where the plant never existed, or no longer exists, results in significant and at times extreme geographical isolation. Several other factors also encourage the isolation of populations on a localized level.

Cannabis is a wind-pollinated genus in which the male plants release clouds of pollen grains that are readily carried by air currents. Therefore, it might appear that effective isolation of *Cannabis* populations would rarely be achieved. Whitehead (1983) considered many factors important in successful wind pollinators. *Cannabis* exhibits many of these characteristics in addition to dioecy that make it well adapted to wind pollination. For example, the pollen grains are small (25 to 30 microns in diameter), and the stamens (male sexual organs producing pollen) are borne in pendulous clusters at the ends of branches near the tops of male plants where the male gametes are well positioned for wind dispersal. The paired stigmas (female sexual organs receptive to pollen and connected to the ovule) are erect and feathery, borne in clusters on female plants that are often shorter, leafier, and more highly branched than male plants. Male plants compose approximately 50 percent of the population, and dehiscence (shedding) of pollen occurs over a three- to six-week period. Sparse foliage, especially in the upper part of the male plant, prevents it from trapping pollen as it is released; denser female foliage, as well as the female inflorescences, serve as a sieve helping to slow and intercept passing pollen. *Cannabis* frequently grows in open slope environments and along exposed waterways that are characteristically breezy. This obviously favors effective wind pollination, and because it is wind pollinated, those are the places it thrives.

Simultaneously, several characteristics of *Cannabis* prevent wind pollination from being completely successful. Although dense female foliage may slow wind velocity and allow pollen to settle on receptive stigmas, it also provides a multitude of nonreceptive surfaces upon which pollen may fall. A pollen grain must land on a stigma for it to have a chance to germinate and fertilize an ovule. The myriad glandular and eglandular trichomes on the foliage and inflorescence easily trap most of the pollen grains, and this will also prevent fertilization. Furthermore, *Cannabis* blooms in mid to late summer when neighboring vegetation is at its peak of leafiness and the surrounding plants can trap large amounts of pollen, thereby limiting effective dispersal. If *Cannabis* evolved along riverine flood plains or open stream banks, there may not have been much taller vegetation around when it was flowering.

It seems at first that due to wind pollination, neighboring *Cannabis* populations would cross freely and introgression would be favored. In actuality, although individual pollen grains can be carried for great distances, the vast majority of the total pollen released travels no more than a few meters before it lands. Pollen density decreases exponentially as one gathers samples farther away from the source (Whitehead 1983). When elite hemp seed is produced in breeding projects and accidental hybridization must be prevented, fields are spaced at least five kilometers (about three miles) apart (Small and Antle 2003).

Selection of drug and fiber varieties often results in populations that mature somewhat later than their weedy or feral counterparts and thus produces temporal isolation. Under such conditions, self-sown, early maturing male plants would release their pollen long before a late-maturing drug variety female could produce many receptive stigmas. However, hybridization could be achieved if pollen from the earliest maturing males within the generally later maturing cultivated population is carried to the later maturing female plants in the spontaneous population, which would still be producing many receptive stigmas. Male plants are removed systematically by cultivators of drug *Cannabis* who want the resin-producing females to survive and mature. Under cultivation regimes that aim to produce seedless *Cannabis* and maximize drug production, isolation would be even more complete because there would be far fewer, if any, male plants remaining in the cultivated population. Therefore, based on pollination mechanisms alone, we feel that hybridization between sympatric populations is not favored, and *Cannabis*, once again, is not a likely candidate for introgressive evolution, at least between populations with differing maturation dates. In fact, since *Cannabis* is normally dioecious, and outcrossing between nearby individuals within populations is almost always obligatory, sufficient heterogeneity may be encouraged through recombination between members of the same small population; in this case, gene introductions from neighboring populations are not required to increase variation and thereby confer an evolutionary advantage.

Biological reproductive barriers such as self-sterility are also unnecessary for the survival of *Cannabis* as it is usually dioecious. Geographical isolation and disruptive selection have fostered and preserved great genetic variation in *Cannabis*. Even when hybridization does occur, disruptive selection on the part of humans appears to reinforce genetic isolation sufficiently and effectively limit introgression.

Sympatric (or at least contiguous) feral and cultivated *Cannabis* populations have been reported within a limited area of Central Asia as well as Afghanistan (Vavilov and Bukinich 1929; Vavilov 1931; Schultes et al. 1974) and South Asia (Sharma 1980). In both fiber and seed fields, many genes could be transferred from the cultivated crop to a nearby feral population through windblown pollen (i.e., pollen of early maturing males in cultivated populations could fertilize late maturing females in the feral population). In this case, gene flow would be unidirectional from the cultivated to the feral population. However, according to personal communications from travelers to these regions during the 1970s, populations grew sympatrically without a noticeable blending or loss of the distinctive traits of either spontaneous or cultivated populations. Today, isolation of varieties is maintained by both artificial and natural selection. Growers of *Cannabis* for sieved hashish production traditionally select plants of short stature with tight, leafy resinous inflorescences (Clarke 1998a). If pollen blows into a hashish field from a nearby spontaneous or cultivated *Cannabis* population and forms hybrid seed, the farmer is likely to weed out taller and thinner feral-like hybrid offspring by selecting seeds for the following year from shorter, bushier, earlier maturing and more resinous plants that strongly favor hashish production. This significantly reduces the frequency through which new genetic information might enter a cultivated landrace population from feral *Cannabis* plants. On the other hand, if pollen from hashish fields blows into spontaneous *Cannabis* populations, hybrid offspring will most likely be less able to compete with preadapted, purely feral or truly wild plants. Inflorescences of hashish varieties have fewer flowers and more leaves, and the smaller number of seeds produced is so tightly clustered within broad leaflets that they cannot readily disperse. Since hybrids produced by crossing cultivated and feral *Cannabis* plants would normally be less adapted to compete, their survival under natural conditions is reduced, and consequently the frequency of cultivated genes in self-sown *Cannabis* populations remains low. Therefore, populations of cultivars and self-sown *Cannabis* generally retain separate and exclusive gene pools, even when growing in close proximity. This is true, for instance, with fiber cultivars and feral populations of escaped traditional landraces in the Carpathian Basin of Hungary, Romania, and Yugoslavia (Clarke personal observation 1993, also see previous text). In that region, escapes from cultivated landrace varieties grown since antiquity (and occasionally still cultivated) are found, self-sown, in wooded or brushy areas close by cultivated fields of modern cultivars. These highly adapted, feral populations can be found growing within the uncultivated strips of land between farm fields, and their seeds are spread by rodents, birds, water, and farmers' ploughs onto the open sunny edges of hemp fields up to several meters from their shady refuge. In these cases, they are usually unable to compete with the densely sown fiber hemp crop and perish before they reproduce. In seed hemp fields, these self-sown plants are removed by farmers because they grow more slowly and have a different growth form than the cultivated plants. In both cases, no exchange of genes from feral to cultivated populations occurs because no male or female feral plants survive to reproductive age within the cultivated field. Self-sown populations located near a hemp seed breeder's field are always searched out and uprooted prior to pollen dehiscence to prevent introgression of genes from the feral population to the cultivated one because this would spoil the genetic purity of the seed breeder's crop.

Even in situations where phenotypically different cultivated and feral populations occur sympatrically, disruptive selection and temporal reproductive isolation are often sufficient to ensure that the integrity of each variety is preserved. However, due to years of sympatry and largely natural selection, populations of *Cannabis* that might be considered truly wild are, in reality, more likely semiwild, or even non-native feral escapes. These are important considerations when making taxonomic decisions concerning *Cannabis*.

Population Size and Changes in Variability

Cultivated crops are somewhat protected from natural selective pressures because they are tended by humans who remove competing weeds and provide water and nutrients. During early domestication, soon after collection in the wild and initial seed dispersal, field crops such as fiber and seed *Cannabis* would have undergone a phase of expansion in their new environments during which plants are harvested in mass, and their seeds blended together, giving rise to many surviving progeny. Since human experience with the plant was limited during early stages of domestication, intentional selective pressures probably were not so keen. When cultivation of a new crop first begins, rare mutants have their highest chance of both survival and giving rise to additional

mutant offspring. Following an initial increase in variability, human selection becomes more rigorous. As a result, only the most favorable phenotypes (with their respective genotypes) are actively protected and encouraged, and these become the forerunners of modern cultivars (Pickersgill and Heiser 1976).

Cannabis probably went through its initial geographic range expansion and genetic diversification phases very early on, owing to its long association with humans. Subsequently, additional expansions occurred repeatedly as *Cannabis* was disseminated to new regions. Although early diversification has not been documented, more recent historical expansion phases in *Cannabis* can be cited as examples. In North America, drug *Cannabis* varieties were first cultivated on a frequent basis, albeit illegally, beginning in the mid-1960s. At this time relatively scarce seeds collected from potent imported drug varieties were grown, and since each plant was cared for individually, almost all offspring had a chance to survive. Seedless *Cannabis* cultivation was not yet popular, and many plants produced seeds that were saved for sowing the following season. These seeds were traded from grower to grower and often gave rise to offspring that once again lived as a small founder population in relative isolation. *Cannabis* variability increased because of expansion in total population size, while localized populations were started from limited genetic bases, often just a single pair of founding parents. This gave rise to proliferation of unique genotypes in isolation. Variability soon increased even more as farmers began to grow plants from more than one seed source simultaneously, and as a consequence hybrid seeds were created. Subsequent generations of crosses between these sources (F_2, F_3, etc., hybrids) showed additional variability and were later selected as parents for further hybridization. The advent of illicit *sinsemilla* (Spanish colloquialism for seedless marijuana) cultivation, both outdoors from seed and indoors using clones reproduced vegetatively, was accompanied by much more intensive selection and breeding, combining a great variety of imported landraces and hybrid individuals from a myriad of isolated locations. The mixture of genes from geographically isolated gene pools led to a broad range of new drug cultivars as is evidenced by the Dutch marijuana seed catalogs of the 1980s and 1990s (Clarke 2001).

When crops are grown on a very small scale, the rate of evolution under domestication may be more limited, as chances for survival of a rare mutant are less likely (Pickersgill and Heiser 1976); this presents a contrasting situation to that of the larger populations of expansion phases with increased survival of their early mutants. Slower evolutionary change can occur in *sinsemilla* gardens where the total number of female plants is low, and only a very few or no male plants reach maturity. However, genetic narrowing in small populations is balanced by intense artificial selection for favorable traits and multiplication of selected varieties. A unique phenotype is more likely to be noticed in a small population, and if it is regarded favorably, its genotype is much more likely to be reproduced and passed on to the next generation. Small population sizes and recurring founder effects may account for the radical proliferation of divergent gross phenotypes now found in drug *Cannabis*.

Evolutionary Effects of Dioecy

Most agricultural crops that reproduce sexually (including the majority of cereal grains and most vegetable crops) are self-pollinating and therefore inbreeding can occur. Maize

(*Zea mays*), rye (*Secale cereale*), buckwheat (*Fagopyrum esculentum*), and broad beans (*Vicia faba*) are exceptions that cross-pollinate through self-incompatibility. Many important tropical crops such as taro (*Colocasia esculenta*), sweet potato (*Ipomoea batatas*), yam (*Dioscorea spp.*), and banana (*Musa spp.*) are reproduced vegetatively. Self-pollinated species usually consist of clusters of inbred lines, while cross-pollinated species such as those of *Cannabis* are usually made up of distinct cultivated varieties (Zohary 1984).

Separation of sexes on individual male and female plants prevents self-fertilization of an individual plant and makes outcrossing obligatory. Dioecy promotes more rapid evolution through recombination, even in situations where sibling matings are commonplace, because self-fertilization of individual plants is prevented. As we previously noted, *Cannabis* populations are strongly dioecious with male and female flowers borne on separate plants with only rare monoecious plants bearing both sexes. Dioecy sets the stage for increased heterozygosity and variability even within geographically isolated populations by forcing obligate outcrossing, but it does not prevent sibling crosses (fertilization between two individuals descendant from the same "mother" plant), promoting homozygosity and uniformity. Since there are few, if any, self-sown populations of monoecious *Cannabis*, it appears that dioecy, and therefore obligate outcrossing, are positively selected. In other plant taxa, more efficient methods of reproduction ensuring obligate outcrossing have evolved, such as self-incompatibility (inability to self-fertilize) and temporal isolation (male flowers releasing pollen and female flowers becoming receptive to pollen at different times).

Because it is dioecious, and therefore the male vs. female relationship is very pronounced, *Cannabis* may well have been one of the first plants to be consciously hybridized. Few other common annual crop plants (e.g., New World papaya, *Carica papaya*, and Old World spinach, *Spinacea oleracea*) are as obviously dioecious. It must have been recognized very early in the history of human-*Cannabis* relationships that male and female plants differed in their economic importance. Fibers of varying quality are extracted from each sex and seeds are produced only in female plants. Psychoactive resins are produced in much greater amounts by female plants. The pollen from male plants causes females to produce seeds. If male plants are removed prior to pollen shedding, female plants will be seedless and more psychoactive. This natural dichotomy led early farmers into an intimate understanding of the relationship between male and female plants in production of seed for subsequent crops. Seeds from select female plants often produce improved offspring, even if there is no selection of the male pollen parent. Planning eventually replaced accident, and artificial hybrids were intentionally created between select parental lines (see Chapter 10 for a discussion of *Cannabis* breeding).

In an outcrosser such as *Cannabis*, it is much more likely that a dominant mutation will be expressed than a recessive mutation; on the other hand, a recessive mutation is much more likely to become fixed in an inbreeding population of self-fertilizing plants. Even if a desirable recessive mutation in an outcrosser is expressed as a homozygote (a trait expressed by two identical alleles) and then recognized as desirable by the farmer and thus saved as the seed parent for the following generation, it is most likely that the recessive mutant plant would be randomly fertilized by unselected pollen carrying the dominant nonmutant allele. Therefore the mutation would probably not appear in the following heterozygous

generation. A dominant mutation in an outcrosser will produce offspring of which at least half will be heterozygous (a trait expressed by two differing alleles) but will exhibit the mutation. Further selections in subsequent generations will rapidly increase the prevalence of the dominant mutant allele (Pickersgill and Heiser 1976).

Early humans, making fairly simple selections based completely on phenotype, would tend to favor recessive mutants in inbreeding crops and dominant mutants in outcrossers such as *Cannabis*. Outcrossers would also be expected to resist domestication more effectively than inbreeders and to continue producing wild-type plants for a number of generations following initial selections (Pickersgill and Heiser 1976). This may also encourage the appearance of atavistic traits as wild-type characteristics reappear in domesticated varieties.

Possible deleterious effects of dioecy, such as two plants required for fertilization and seed being produced by only half of the population, are overcome by virtue of differences in resource allocation. A spontaneous stand of *Cannabis* often begins growth as many overcrowded seedlings. Soon, taller and more vigorous plants begin to overshadow less vigorous ones, which become stunted. Male plants are usually taller and thinner than females with longer internodes and reduced foliage near the growing shoot. They overgrow the more robust female plants, but they shade them very little as their terminal foliage is sparse. Male "tallness" was also naturally selected, as noted earlier, because it allows pollen to be blown away from nearby plants (which are the most likely to be sister plants) and therefore lowers the chance of sibling crosses and inbreeding. Male plants mature and begin to drop pollen as the first receptive stigmas begin to emerge on female plants. Pollen shedding continues for three to six weeks during which time the females continue to produce fresh stigmas that trap pollen and allow fertilization. After three to four weeks male plants begin to senesce but may produce pollen for up to two more weeks. In this example of natural adaptation, males shed pollen over several weeks, and female inflorescences mature over the same period, producing some early, some medium, and some later appearing flowers. Thus plants mature over a range of time, an adaptation to variable weather patterns that allows for the continual presence of viable receptive stigmas. As males die off further competition with the seed-bearing females is eliminated, leaving more space for the females to receive additional sunlight, nutrients, and water for the maturation of seeds. In contrast, male organs on monoecious plants remain in competition for resources much longer; therefore, the same plant that devoted energy to producing pollen must then devote additional resources to seed production. This increases the load on a single plant and lowers its overall ability to set and mature seeds.

Dioecy also assists in herbivory and parasite control. Male *Cannabis* plants are attacked by pests and pathogens as they begin to senescence and lose their natural defenses long before females. Thus pests and pathogens are attracted to the more susceptible male plants and feed on them in preference to the female plants, which in turn protects the female seed-bearing plant from predation (e.g., see McPartland et al. 2000). Since male plants become expendable after shedding their pollen, female plants are partially protected from predation and competition for resources, which is vital because they require additional time in sufficiently good health to mature their seeds.

In sum, dioecy in *Cannabis* has resulted in outcrossing and the favoring of dominant alleles, leading to rapid evolution of selected cultivars. Dioecy is also important to the survival of *Cannabis* through resource allocation and control of predation.

Effects of Human Selection on Sexual Expression for Different Products

Dioecious, self-sown *Cannabis* has a one-to-one sex ratio of approximately 50 percent male and 50 percent female individuals with few, if any, monoecious plants. Dioecy appears to be the natural sexual condition of *Cannabis*, and monoecy is an artificially derived state. However, offspring of some drug varieties can range from 50 to 100 percent female individuals, and human selection for either seed, fiber, or drug production can not only affect the basic sex ratio but also may increase the frequency of naturally rare monoecious individuals.

SEEDS

Cannabis has a very long history of cultivation for its vegetal oil and protein. Hemp remains a traditional source of edible seeds and seed oil in Eastern Europe and China, while hemp seed cultivars were more recently introduced to Western Europe and Canada. Hemp seeds are also grown for poultry feed, birdseed, and even fishing bait in some areas (see Chapter 6). Maximum seed yields of *Cannabis* are achieved by allowing males to fertilize females throughout the fourth through sixth (or later) weeks of their flowering cycle. Fresh stigmas are produced daily for several weeks, and a constant rain of pollen will ensure heavy seed set.

Since male plants are relatively short-lived and their individual periods of pollen shedding vary, it is advantageous to have many pollen-producing male parents, as this ensures consistent pollination over a longer period. Yet at the same time, for maximum seed yield, a population must have as many seed-producing female parents as possible. This process results in reinforcement of the 50 percent male to 50 percent female wild-type sex ratio since, in actuality, both sexes are of economic importance in seed production. Simple Mendelian X and Y sexual inheritance also supports a one-to-one sex ratio.

Monoecy has not been selected in seed varieties because monoecious populations usually have lower seed yields than dioecious populations. However, monoecious fiber varieties are also used to produce sowing seed for fiber crops (see later in this chapter). Stands of hybrid 'Uniko-B' bred by Ivan Bócsa at Kompolt in Hungary, produce more than 90 percent female plants and less than 10 percent monoecious plants and yield up to 1,200 kilograms of seed per hectare (ca. 1075 pounds per acre) as compared to dioecious Hungarian landrace cultivars (e.g., 'Kompolti') that yield 800 to 900 kilograms per hectare (ca. 700 to 800 pounds per acre) and French monoecious cultivars such as 'Fibrimon' and 'Fedora' that yield only 600 to 800 kilograms per hectare (ca. 550 to 700 pounds per acre; Bócsa personal communication 1993).

FIBERS

The majority of contemporary European fiber hemp cultivars (with the exception of those grown in France) are dioecious.

Both male and female plants produce valuable bast fibers in the phloem tissues of the bark surrounding the stalk, and both are utilized in hemp fiber production (see Chapter 5 for a detailed discussion of hemp fiber history). However, male plants produce a higher quality, finer primary fiber because their internodes are longer and they die off earlier, thereby producing a lower percentage of coarse, lignified secondary fiber. In fact, traditional Chinese, Korean, Japanese, Nepali, and Hungarian hemp spinners identify and prefer a special, high quality hemp fiber harvested only from male plants. Because fiber quality is also largely dependent on the maturity of the stalk, the timing of hemp harvests is very important. Hemp plants usually begin to flower abundantly just before the fibers have reached maturity. Since males and females flower and mature at different times during the season, it is difficult to time the harvest to achieve the maximum average quality for both. The fibers of males mature earlier than females; however, since cultivated hemp plants are normally spaced close together (over 100 plants per square meter or over 9 plants per square foot), it is difficult to harvest only male plants without damaging unripe female plants. Chinese, Korean, and Japanese hemp farmers harvest the entire field just before flowering commences as the primary fiber from juvenile plants is much more uniform, even though they yield less.

In order to overcome staggered maturation, French industrial fiber hemp breeders developed monoecious varieties with high quality, evenly maturing fibers. As both male and female flowers are borne on the same plant, the discrepancy in maturation between the sexes is eliminated. Inbreeding of monoecious varieties has also allowed French breeders to lower the THC level to less than other European industrial hemp varieties. However, breeders have found it difficult to combine the monoecious character with high fiber quality and yield (Bócsa personal communication 1993; also see Bócsa 1998). Perhaps varying male and female reproductive strategies compete against each other when combined in a monoecious line (see the previous section on the evolutionary effects of dioecy).

MARIJUANA AND HASHISH

"Marijuana" and "hashish" are common names for differing psychoactive *Cannabis* products. Hashish is a mechanically extracted, concentrated resin gland preparation. Seeded marijuana is merely the dried inflorescences from THC-bearing varieties of *Cannabis*. If nature is allowed to take its course, male plants will thoroughly fertilize the females producing many seeds. If all fit males are allowed to reach reproductive age, as in the wild, then there is no selective pressure on sex ratio. Seedless marijuana (*ganja* or *sinsemilla*), on the other hand, is produced by removing all male plants from a field prior to pollen shedding, thereby preventing fertilization and seed set. This technique favors flower and resin production instead of seeds and results in particularly potent, seedless marijuana, which has a higher market value than seeded marijuana. The more seeds present in flowers, the lower the price the farmer receives. This has been the economic situation since at least the early nineteenth century in India (Kaplan 1969).

A few seeds (resulting from accidental pollinations) are found in many *Cannabis* crops even when male plants are intentionally removed. Since it was most likely the intention

TABLE 3

Human impact on the evolution of *Cannabis* through selection of favorable traits for different crop uses.

Wild vs. selected character states of economically important traits of *Cannabis* fruits, stalks, and flowers are shown along with type of inheritance (qualitative vs. quantitative), type of phenotypic change (morphological vs. physiological), cropping system (F = fiber; S = seed; D = drug), and type of human selection (I = intentional; U = unintentional).

Plant part	Wild trait	Selected trait	Inheritance	Change	Crop	Selection
Fruits	Freely shattering	Persistent	Qualitative	Morphological	F, S, D	U, U, U
	Horseshoe base	Reduced horseshoe	Qualitative	Morphological	F, S, D	U, U, U
	Fruit size: small	Fruit size: large	Qualitative	Morphological	F, S, D	U, I, U
	Fruit color: dark	Fruit color: light	Qualitative	Morphological	F, S	U, U
	Fruit coat: mottled	Fruit coat: less mottled	Qualitative	Morphological	F, S	U, U
	Protein content: lower	Protein content: higher	Quantitative	Physiological	S	I
	Oil content: lower	Oil content: higher	Quantitative	Physiological	S	I
	Germination: delayed	Germination: rapid	Qualitative	Physiological	F, S, D	U, U, U
	Germination: staggered	Germination: uniform	Qualitative	Physiological	F, S, D	U, U, U
Stalks	Fibers: brittle	Fibers: supple	Qualitative	Morphological	F	I
	Fibers: brittle	Fibers: very brittle	Qualitative	Morphological	D	U
	Fibers: short	Fibers: long	Qualitative	Morphological	F	I
	Fiber content: low	Fiber content: high	Quantitative	Morphological	F	I
	Branching: moderate	Branching: sparse	Qualitative	Morphological	F	I
	Branching: moderate	Branching: profuse	Qualitative	Morphological	S, D	I, I
	Internodes: medium	Internodes: longer	Qualitative	Morphological	F	I
	Internodes: medium	Internodes: shorter	Qualitative	Morphological	S, D	I, I
Flowers	Fruits: few	Fruits: many	Quantitative	Morphological	S	I
	Bracts: few	Bracts: many	Quantitative	Morphological	S, D	I, I
	Inflorescences: few	Inflorescences: many	Quantitative	Morphological	S, D	I, I
	Resin glands: few	Resin glands: many	Quantitative	Morphological	D	I
	THC level: low	THC level: high	Quantitative	Physiological	D	I
	THC level: low	THC level: very low	Quantitative	Physiological	F, S	I, I
	Terpenoid profile: simple	Terpenoid profile: complex	Qualitative	Physiological	D	U
	Maturation: early	Maturation: late	Qualitative	Physiological	D	I
	Maturation: early	Maturation: very early	Qualitative	Physiological	F, S	I, I

of the grower to produce completely seedless marijuana, these few seeds usually result from some overlooked, ignored, and therefore unselected male plant. Pollen can come from a variety of sources such as a neighboring cultivated field, feral plants, early flowering males removed after they dropped pollen, late-flowering males that avoided detection during male removal, or intersex plants with only a few male flowers that are not detected (an intersex plant is an individual that has a mixture of male and female reproductive structures, i.e., is monoecious). Depending on the pollen source, the sex ratios of following generations can be affected.

If pollen comes from a neighboring cultivated field, selective pressures exercised in that field will influence the genetic makeup of the pollen. If no selections have been made that would favor certain male plants, there will be no effect on the sex ratio of the subsequent generation.

Likewise, if pollen is introduced from a wild or weedy population that has undergone natural selection for a 50 percent male to 50 percent female sex ratio, there will be no effect on the sex ratio of the subsequent generation. Even if the cultivated population has undergone extensive selection influencing sex ratio, the introduced, naturally selected pollen will tend to restore the wild-type one-to-one sex ratio.

If pollen comes from early flowering males that are either ignored or removed after pollen dehiscence, the pollen should carry the genes for early maturation, and plants in the subsequent generation will tend to flower earlier than previously. On the other hand, if pollen comes from late-flowering males that escaped detection, it should carry genes for late male flowering, and males in the subsequent generation will flower later (Clarke personal observations). Of course, in seedless *Cannabis* cultivation, growers attempt to remove all male plants, the majority of which would mature near the mean (average) of maturation times. The very early maturing or very late maturing exceptions that escape removal are the only pollen plants with a possibility to fertilize seed plants. These two factors, working together, result in offspring that either tend to mature particularly early or particularly late, with fewer plants of intermediate maturation. This type of disruptive selection does not significantly alter the sex ratio, but it does influence the maturation characteristics of the population.

Pollen introduced from intersex plants can also change the sexual expression of subsequent generations. Intersex plants can range from being fully monoecious, with approximately equal numbers of male and female flowers produced contemporaneously throughout the floral phase, to those that are only slightly intersex and produce only a few flowers of the opposite sex, either male or female appearing at different, but often overlapping, times during floral maturation. True hermaphrodites, with both male and female organs borne together within the same flower, are very rare and not often fully fertile. In general, intersex plants tend to give rise to intersex offspring (Clarke personal observation). Indeed, this was the basis for specialized breeding of French monoecious industrial hemp fiber cultivars (see earlier in this chapter and Chapter 10). However, monoecy must be selected for constantly, or else freely pollinating populations will drift back to dioecy, as the number of dioecious individuals increases in each generation.

Intersex male plants that give rise to only a few female flowers are of little evolutionary consequence as they are usually culled along with the males before they can spread pollen or set seed. If they should survive, develop a few mature seeds, and transmit their intersex genes, they would be negatively selected in the next generation. However, female plants will occasionally give rise to male flowers late in floral maturation. If these male flowers pollinate receptive stigmas and sufficient time remains for the seeds to mature, then the female intersex characteristic can be carried to subsequent generations. In fact, female plants that give rise to few male flowers are often overlooked since they express this trait after the removal of male plants is complete. As a result, although rare, intersex plants that are largely female with only a few male flowers are much more common in seedless drug varieties than predominately male intersex plants with a few female flowers. In crosses between a female seed plant and an intersex female plant with only a few male flowers, the offspring population will tend to be predominately female with a minority of intersex females (Clarke personal observations). This also skews the sex ratio of the next generation in favor of female individuals. Male flowers on predominately female intersex plants are often the source of pollen that produces undesired seeds in a *sinsemilla* crop, and female intersex plants will appear again at low frequency in the next crop cycle. Female intersex plants were responsible for so much late-season seed set in the *ganja*-growing regions of ninteenth-century India that specialists (*poddars*) walked through the fields toward the end of the growth season, removing any late-maturing intersex plants in order to prevent unwanted seed set (Prain 1904, Kaplan 1969). Male plants in populations cultivated for drug production are sometimes nearly as large and vigorous as females. Among *sinsemilla* growers, a well branched male plant with dense inflorescences resembling the female floral form and growth habit is considered an obvious choice for breeding.

Female plants usually begin flowering when males do but have produced only a few primordial flowers along the main

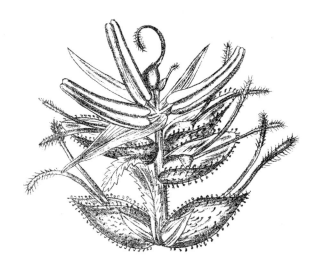

FIGURE 10. Pistillate (female) inflorescences with staminate (male) organs arising at the terminal meristem late in development were relatively common in Bengal *ganja* fields, requiring field workers to destroy the potential pollen sources before they fertilized the seedless females (from Prain 1904).

stalk at the time pollen dehiscence begins, and thus early female flowers are the only ones to be fertilized. Often, formation of female inflorescences that bear most of the fertile ovules commences after many of the earlier flowering male plants have died. This is possibly a mechanism to prevent earlier and earlier maturation of subsequent generations.

In sum, the cultivation and breeding of *Cannabis* for its varying products led to changes in flowering and maturation times and has produced sex ratios differing from the wild-type one to one.

Sexual Dimorphism and Selection

In terms of size and longevity, natural selection has favored the female plant as the hardy seed bearer and slighted the male plant as the more ephemeral pollen bearer. In general, through natural and/or artificial selection, the female plant has evolved a more robust stature needed to support the weight of seeded inflorescences.

The majority of *Cannabis* products are derived from female plants. Fiber is extracted from both sexes, but seeds and psychoactive resin are produced by females alone. Male leaves do have some glandular trichomes, normally only a few, that produce small amounts of THC. Since economically valuable traits recognized and selected by humans are primarily expressed in female plants, selection has largely been limited to them. Even when males are intentionally selected, it is much more difficult to recognize their potential as parents because they lack the economic traits. This has had a marked effect on the evolution of *Cannabis* under domestication. Since there is less artificial selection imposed on male plants by humans, there may, in turn, be more natural selection affecting male plants than female.

In a freely pollinating situation such as fiber or seed cultivation, fertilization (and therefore gene transfer) is random in each generation. Even if female plants are carefully cultivated and selected for economically valuable characteristics, the male genetic contribution remains relatively random and unselected. Breeding progresses slowly under these conditions, since males are not affected by direct artificial selection. Breeding advances more quickly if both female and male plants are selected for favorable traits. Rare, self-pollinated intersex plants can confer fixed recessive characters in their offspring; however, they usually transmit the intersex characteristic as well and can suffer an undesirable loss of vigor due to inbreeding.

When breeding for increased fiber content, samples from several male plants are collected by removing the upper half of the central stalk before they shed pollen, and these samples are used to determine fiber percentage. Inferior males with lower fiber content are destroyed, and only the males with the highest fiber content are allowed to fertilize the females. Female stalks are tested after the seed is harvested, and only seeds collected from the females of high fiber content are sown during the following crop cycle. This basic sampling, analysis, and selection protocol is also applied to raising or lowering cannabinoid levels in medical and recreational *Cannabis* cultivars.

The selection of male drug plants, for instance, is more often intuitive rather than empirical because potency is harder to recognize and quantify in males. Even if it were easier to select potentially potent males as pollen parents, other practical reasons would make growing male plants unprofitable. It is difficult for *sinsemilla* farmers, generally working illegally (since the mid-twentieth century), to devote time, space, and energy cultivating and selecting favorable male plants since they provide no direct economic benefit. In order to facilitate *sinsemilla* production, worthless male plants are not grown. Plant breeders in some areas apply analytical laboratory methods to assist in selection of appropriate parents. In fiber variety development the healthiest male plants are used for selections and the remainder destroyed. All the selected males are analyzed for fiber and THC percentages, and only high fiber with low drug content males are used as pollen sources. Very rapid breeding progress can be made in this way (see Chapter 10).

The selection of female plants also affects the phenotypes of their male offspring. All males have a female parent that may (or may not) have been artificially selected for its economically valuable traits. Therefore, although selection for males is often unintentional, males are heavily influenced by the selective pressures exerted on their female parent. Like female plants, male plants from drug and seed varieties are more robust and more highly branched than male plants from fiber varieties, only to a lesser degree than the females (Clarke personal observations). Male flowering branches are often less elongated and more crowded in drug than fiber varieties. Changes in male morphology parallel changes in female morphology, although artificial selection has been exercised almost entirely on female plants.

Only a few males are needed to produce sufficient pollen to thoroughly fertilize all nearby females and thereby produce enough seed to ensure survival. During "seedless" marijuana production, if seeds are desired for sowing future crops, the majority of males are removed, leaving only a few select ones. If seeds from all the female plants are pooled, and a random selection is sown during the following crop cycle, there will be no directed selection imposed on female plants. In this case, selection is exercised more on males than females. The situation is reversed if most of the males are allowed to flower, as in the production of hemp seed, and individual females are selected at harvest time. Selective pressure is

then predominately on females. In the case of drug *Cannabis* cultivation in South and Southeast Asia, it is difficult to determine if selecting males was the tradition within the geographical range of *ganja* production stretching from Kerala in southern India to Bengal in northeastern India and eastward through Thailand. If fertilization has traditionally been strictly avoided whenever possible according to traditional cultivation practices in these regions, then seeds were only produced by accidental pollinations, and intentional selection of male pollen plants was precluded.

In any case, most marijuana production worldwide involves little if any artificial human selection. Male plants are normally removed without any thought of preserving a select plant for breeding. Female flowers are dried, stripped from their stems, and mixed together without making any individual selections. In fact, volunteer seedlings that appear spontaneously are often preserved and used for marijuana production (Clarke personal observation). This situation hardly differs from hashish production where female plants are pollinated by a wide range of males, and the females are processed in bulk. There is no individual plant selection in these cases; any crop improvement usually results from the importation of seeds providing new germplasm. Natural selection is the primary agent of evolution within freely cross-pollinating populations, and human-selected traits, such as potency or high fiber content, steadily disappear under this natural process. The role of humans in both the unconscious and conscious selection of differing economically valuable traits during domestication has greatly altered the phenotypic characteristics of *Cannabis* landrace varieties. In the next section we will investigate some of the more important phenotypic changes in *Cannabis* morphology and physiology during the domestication process.

Phenotypic Changes during Domestication

Naturally growing stands of *Cannabis* provide many useful products without any cultivation, selection, or domestication. No clear indications of sequential stages in domestication have, as yet, been found in the archeological record for *Cannabis*. Increase in seed size, changes in seed dehiscence characteristics (i.e., presence or absence of a horseshoe-shaped fruit base or "caruncle"), and possibly changes in fiber quality are the only archeological indications of domesticating trends in *Cannabis*, and not enough samples have been characterized to draw any solid conclusions (Fleming and Clarke 1998; also see Chapter 4). In fact, until the advent of asexual propagation for growing *sinsemilla*, *Cannabis* had not been domesticated to the point of relying solely upon humans for its survival, and therefore, by strict definition, we do not consider *Cannabis* to be fully domesticated.

Changes in phenotype can be induced under cultivation in a number of ways. For example, phenotypic alteration can be brought on by moving plants from their native environment into new regions and introducing novel methods of cultivation; this can occur by changing the maintenance system for the cultivated variety (e.g., sexual vs. asexual multiplication) and through selection pressures associated with varying types of human use of the plant (Zohary 1984). Domestication of a plant often effects changes in both its morphology (form) and its physiology (function). Indeed, these changes

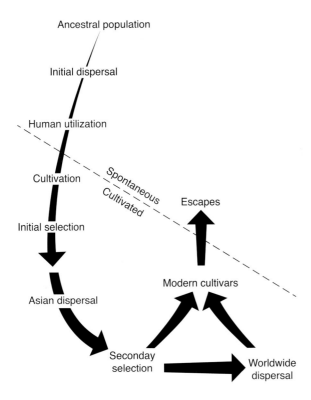

FIGURE 11. Stages in the evolution and dispersal of *Cannabis* during cultivation and human selection.

SEEDS

The most apparent physical changes in *Cannabis* during domestication have been in seed characteristics. Whenever seeds are collected for food or sowing, humans unconsciously select for nonshattering inflorescences. Normally, seeds that are persistent and remain on the plants are harvested and carried away rather than naturally dispersing and falling on the ground. In fact, one of the primary characteristics distinguishing many domesticated plants from their wild ancestors is loss of the seed dehiscence dispersal mechanism.

Hemp fiber production requires sowing thousands of seeds. Selection for nonshattering types is strongest in fiber and oil seed varieties selected for production of the seeds themselves. However, drug cultivars also have very persistent seeds. In a self-sowing spontaneous population growing in the same location year after year, there is no human selection for nonshattering phenotypes. *Cannabis* landrace populations utilized by humans exercising little intentional selection often serve as both producers of fiber and/or drug, as well as next year's sowing seed, collected as a by-product during primary product extraction. Under such conditions a nonshattering inflorescence type is favored, since the seeds remain in the inflorescence where they can be retrieved and later sown.

Seeds of *Cannabis* cultivars are often larger and lighter in color than those of spontaneously growing varieties. In addition, seeds produced by cultivars lack the horseshoe-shaped base and a mosaic perianth pattern associated with freely shattering, camouflaged, wild-type seeds (Vavilov 1931). Seed length varies from very small (< 1 millimeter, or 1/24 inch) to quite large (6 millimeters, or ¼ inch). In general, self-sown spontaneous varieties are considered to have the smallest seeds and varieties cultivated for seed the largest. European industrial hemp cultivars usually produce seeds ranging in size from about 40 to 60 per gram (approximately 1,100 to 1,700 per ounce). Tiny feral seeds from China can range from 100 to 500 per gram (approximately 2,800 to 14,000 per ounce), while large seeds eaten as snack foods in China might only include as few as 15 per gram (approximately 420 per ounce; Clarke personal observations). This represents more than a 20-fold range in seed size. Thus China appears to be a center of hemp seed variation and selection during domestication, and this is probably explained by the lengthy history of cultivation and use of hemp for human food, domesticated animal feed, and vegetal oil in East Asia (see Chapter 6).

However, seed size can be deceiving, especially when making selections during breeding or when using seeds in taxonomic identification. If a sample is more or less fully seeded, then seed size seems to be an accurate taxonomic marker. A problem arises if a sample is only partially seeded, as is often the case in *sinsemilla* marijuana varieties. A few seeds are able to concentrate more energy and resources required for growth without competition from many other developing seeds. Consequently, the resulting seeds are disproportionately larger, but subsequent generations, if fully seeded, as in the spontaneous state, will produce more average-sized seeds. Thus seeds from populations where males were removed are artificially large and therefore should not be used as a natural taxonomic character for classification. Seed size is also strongly influenced by climate and soil conditions. Under highly favorable conditions seeds will form that are as large as the variety can produce. Under marginal conditions viable seeds are still produced, but they are often much smaller. Spontaneous populations often grow in marginal areas and

are interpreted as indications that domestication has at least partially occurred or is taking place. Both conscious and unconscious human selection can produce many phenotypic differences between a cultivated plant and its wild ancestor. The most apparent changes occur in the plant parts most intensively utilized because farmers normally select for these desired traits. Obvious morphological changes in size, shape, and color of seeds and loss of seed dispersal mechanisms are often accompanied by more subtle physiological changes such as variations in oil and protein content and secondary metabolite synthesis. Physiological changes are more difficult to detect but are likely to be just as common as morphological changes since domesticated plants have become physiologically adapted to changes in climatic and soil conditions. The ultimate source of all variation is mutation. New mutations are subjected to natural selection, and in the case of crop plants, they are also often subjected to additional human selection (Pickersgill and Heiser 1976).

Domesticated plants can differ from their wild progenitors both qualitatively and quantitatively. Morphological differences are most often qualitative, as they involve anatomical changes frequently controlled by only one or a few genes, and are inherited by relatively simple genetic mechanisms. Physiological changes more often have a quantitative effect, such as increased vigor and yield, or variations in the amount of secondary metabolites. Physiological changes are usually controlled by a group of genes, and consequently their inheritance is often complex. Hybrids between varieties varying in a quantitative characteristic usually show values intermediate between the two parents (Pickersgill and Heiser 1976). However, hybrid vigor often masks the blending effect and hybrids may express quantitative traits at levels higher than either parent.

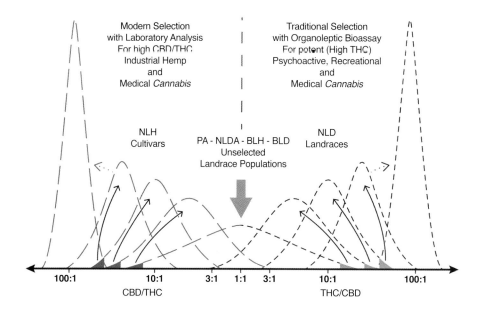

Modern *vs.* Traditional phenotype selection

FIGURE 12. Naturally growing and unselected *Cannabis* populations (PA, NLDA, BLH, and BLD) usually express a bell curve of individual plant cannabinoid ratios ranging from high THC and low CBD to low THC and high CBD, with a population average of around 1:1. Beginning with narrow-leaf drug (NLD) landrace populations, farmer's traditional individual plant selection for potency altered the cannabinoid ratio of these landrace populations to greatly favor THC (right side of diagram). Laboratory analysis was required in twentieth-century narrow-leaf hemp (NLH) breeding programs to lower THC content and raise CBD content (left side of diagram).

consequently their normally small seeds may be larger when grown under more favorable ecological conditions.

As noted earlier, physiological changes also accompany domestication. Mechanisms for perennation (inhibition of seed germination that allows seeds to survive through the winter before germination) are often lost under domestication. The seeds of wild varieties usually germinate slowly and unevenly, while the seeds of domesticated varieties germinate quickly and uniformly. This germination timing of the spontaneous varieties is, of course, a naturally selected adaptation to climatic fluctuations from year to year (e.g., the distribution of rainfall and sunny days may vary somewhat). Under cultivation, hemp seeds are collected by farmers and stored in a situation inhibiting germination until the next sowing season, thus artificially ensuring survival through the winter. The wild characteristic of perennation is no longer of advantage in a domesticated variety. Hemp farmers unconsciously select for uniform and rapid germination by providing a crowded field situation where smaller late-germinating seedlings have less ability to compete with well-established early seedlings. Modern plant breeders have also consciously selected for rapid and even germination as this facilitates mechanized field agriculture by allowing more uniform treatment of the crop.

Cultivated *Cannabis* varieties have lost all specialized dehiscence mechanisms as well as physiological means of delaying germination. In extreme cases, mature seeds will germinate within the female inflorescence, as when they are wetted by warm autumn rains. Usually, germination causes the seed to crack, and on occasion seedlings will grow from the female inflorescence (Clarke personal observation 1977). Human intervention is required to effectively perpetuate

these varieties, which is a strong sign of domestication. Premature germination has not been reported in any self-sown populations of *Cannabis*.

Cannabis seeds vary somewhat in oil content (Mölleken and Theimer 1997). Seeds from fiber, seed, and drug varieties have average oil content of approximately 30 percent. Variation in oil content and quality arise largely from natural selection accompanied by some unconscious human selection. However, a former Soviet oil seed cultivar 'Oleifera' reportedly contained approximately 40 percent oil (Small et al. 1975).

Oil seed varieties have usually been selected for high seed yield rather than high seed oil and protein content. This is manifested by the dearth of scientific articles pertaining to breeding for oil quality and a lack of correlation between oil content and cultivar status (Mölleken and Theimer 1997). A plant with many seeds makes an obvious choice for selection and breeding, but it is difficult to detect changes in oil content without sophisticated analysis. Therefore, with traditional breeding techniques it is more straightforward to increase total oil production by selecting for higher seed yield rather than by breeding for higher oil content in the individual seeds.

FIBERS

Hemp bast (bark) fibers are extracted from the primary and secondary phloem layers of the main stalk. Fiber length and quality have been strongly selected during domestication and breeding of hemp varieties. Changes in fiber characteristics are usually exhibited in plants of both sexes. Varieties selected for fiber production have longer and more flexible fibers, while seed and drug varieties have shorter and more brittle fibers. Unbranched, slender stalks with long internodes are preferred for fiber production. In order to maximize flower and glandular trichome production, profusely branched stalks with very short internodes are preferred for *Cannabis* seed and drug production. Long, supple fibers are associated with herbaceous, unbranched stalks and long internodes. Short, brittle fibers are associated with woody, well-branched stalks and short internodes. Fiber cultivars have also been selected for total fiber yield, and exceptional plants of Hungarian fiber cultivars (e.g., 'Kompolti TC') can produce over 40 percent dry weight of bast fiber (Bócsa 1994).

INFLORESCENCES

The floral characteristics of *Cannabis* have been modified primarily by human selection of seed and drug varieties. Increased seed and resin production both depend on increased floral production. Seed for sowing is also a major product of fiber varieties and high seed yield is important. Consequently, both seed and drug varieties usually have larger inflorescences containing many more flowers than fiber varieties, which have larger inflorescences than wild and feral types. Increase in the number of flowers is most apparent in modern seedless drug varieties. As we have previously pointed out, seeds are precluded by removing males, which eliminates pollen; consequently, seed set is normally prevented in all but a few inflorescences. At most, only a few thousand seeds are needed to produce the following year's crop; therefore, seed productivity is not so important to the survival of seedless drug varieties. Vegetatively reproduced cultivars require no seed production as propagation is by rooted vegetative cuttings and is entirely asexual. This clonal method of reproduction was developed in modern times by drug *Cannabis* cultivators to ensure the perpetuation of high THC-yielding plants. Seedless varieties develop many more flowers crowded within the inflorescence because there is no need to allow for the swelling of maturing seeds. Drug varieties have been selected for larger inflorescences in order to provide more surface area for glandular trichomes, which increases potency. Consequently, when modern drug cultivars with the potential to produce seeds are grown in full sun to a large size, approximately two meters (6.5 feet) or more in height and covering two to three square meters (20 to 30 square feet) of field area, they can yield up to one kilogram (2.2 pounds of seed on each plant; Clarke personal observations).

Directional Evolutionary Changes

Many evolutionary changes in *Cannabis* resulting from domestication are unidirectional. "One-way" evolution starts with the wild condition of a trait as the norm, and through artificial selection the general evolutionary trend is directed away from the wild characteristic toward a desirable form or function that we refer to as a domesticated characteristic (e.g., from small to large seed). However, this does not mean that traits cannot revert atavistically to a more primitive condition or that such traits do not introgress between the wild and domesticated condition (see earlier sections of this chapter). Quantitative traits, such as fiber or seed yields, regularly

occur as a suite of related characteristics controlled by several to many genes, while qualitative traits (e.g., changes in seed morphology) are controlled by fewer or a single gene.

Other evolutionary changes induced through artificial selection are bidirectional. This sort of "two-way" evolution proceeds in two conceptually opposite directions toward polar extremes. The result of bidirectional evolution is a continuum with the wild condition somewhere between two domesticated phenotypic extremes. Unidirectional evolution can also result in a phenotypic continuum, but the wild condition lies at one end of the continuum opposite to the other, domesticated extreme. Bidirectional evolution is a strong indication that disruptive selection has taken place. Variations in both cannabinoid content and timing of maturation have resulted from disruptive selection and bidirectional evolutionary changes.

CANNABINOID PROFILE

Variations in CBD (cannabidiol) and THC levels evolved bidirectionally. Present phytochemical knowledge suggests that cannabinoid synthesis is a unique characteristic of *Cannabis* since no other genus is known to synthesize cannabinoids (see Toyota et al. 2002 for an alternative view). Primitive populations of *Cannabis* probably possessed the B_T allele controlling THC biosynthesis and produced at least some psychoactive THC (e.g., see Hillig and Mahlberg 2004). However, it is readily apparent that domestication of *Cannabis* for drug use, especially during the latter part of the twentieth century, significantly raised THC levels in dried floral clusters (up to 25 percent dry weight in extreme cases; Watson 1998, personal communication). All cultivated drug varieties are higher in THC than fiber varieties or wild or feral spontaneous populations. The self-sown NLHA and BLH normally produce low levels of THC (approximately 0.5 to 2.0 percent dry weight). NLH fiber varieties produce even lower levels of THC (< 0.1 to 1 percent dry weight). BLH populations exhibit a typical bell curve of cannabinoid content. CBD to THC ratios range from mostly CBD with almost no THC to mostly THC with almost no CBD; CBD or THC contents in female inflorescences can reach 4 percent dry weight. However, the vast majority of plants contain approximately equal amounts of CBD and THC and a total cannabinoid content of 2 to 3 percent dry weight. This indicates that the primitive condition of *Cannabis*, which we feel is most closely represented by present-day spontaneous populations, fell somewhere between the two extremes of very low and very high THC content.

For example, *C. indica* ssp. *chinensis* (BLH) landraces grown for fiber and seed by the Hmong ethnic minority in southwestern China and northern Laos, Thailand, and Vietnam exhibit a bell curve of cannabinoid ratios, and if seed selections are made from the few highest CBD/THC ratio and highest THC/CBD ratio plants and sown in isolated gardens, the cannabinoid ratios of the offspring will be skewed in favor of the parental selections. If this process is repeated for only a few generations, two distinct varieties will result, one with a very high average and narrower range of CBD/THC ratios and a second with a very high average and narrower range of THC/CBD ratios. Since THC is psychoactive and easily detected by ingestion, early *Cannabis* farmers readily selected higher potency plants, sowed their seeds, and developed potent landraces. However, CBD, now recognized as a physiologically active, medically useful chemical compound, is not psychoactive and therefore is only detected in landrace populations by sophisticated laboratory analysis, increased by controlled breeding, and preserved through vegetative propagation.

Artificial selection for psychoactive potency has raised THC levels radically to reach extremely high concentrations in modern times. Simultaneously, artificial selection to decrease potency in NLH cultivars has resulted in levels lower than those found in spontaneous NLHA populations. For example, twentieth-century European industrial hemp breeders developed many cultivars below 0.3 percent THC (the European Union maximum allowable THC content). CBD levels also vary considerably as a result of human selection. Spontaneous *Cannabis* populations express levels of CBD approximately equal to those of THC (0.5 to 2.0 percent dry weight). However, many NLD landraces and the majority of modern high-THC drug cultivars are nearly devoid of CBD, while European industrial hemp laws stipulate that the CBD to THC ratio must exceed three to one. CBD levels have also been increased in medical varieties produced by GW Pharmaceuticals in the United Kingdom (Russo 2006; also see Chapter 8).

TIMING OF FLORAL MATURATION

The timing of floral maturation offers another example of bidirectional evolution. *Cannabis* likely originated in Central Asia probably before the Würm glaciation lasting from 110,000 to 10,000 years ago. As the ice receded from this glaciation, we believe that *Cannabis* evolved into two separate geographically isolated populations: on one hand, the putative hemp ancestor (PHA), which became *C. sativa* and radiated across Europe, and on the other hand, the putative drug ancestor (PDA), which became *C. indica* and radiated from southeastern Asia across South and East Asia (see Chapters 2 and 12). Therefore, extremely early flowering northern temperate hemp seed varieties (e.g., 'Finola') and late flowering tropical drug landraces (e.g., from India and Thailand) each evolved from two separate ancestral populations sharing common ancient roots in Central Asia. Division of the primordial *Cannabis* gene pool likely happened more than 20,000 years ago resulting from disruptive natural selection caused by ice sheets advancing into new, differing environments (e.g., southern Europe and southeastern Asia) and was possibly assisted by early humans. The later dispersal of early *C. sativa* and *C. indica* resulted from human migration, and varietal differences were established over time through geographical isolation and disruptive selection. Later, bidirectional evolution most likely accompanied widespread Holocene dispersal of *Cannabis* into ecologically divergent environments such as those in higher and lower latitudes of northern and southern hemispheres respectively. The main stalk characteristics influencing fiber quality, such as internode length and branching pattern, also appear to have evolved bidirectionally from a median condition.

Natural evolutionary determinants were important before humans started using and spreading *Cannabis*, and natural selection will always play a role in its evolution. Nevertheless, human artificial selection during the process of domestication has had by far the greater influence on changes in *Cannabis* phenotypes, both morphological and physiological. This process accelerated during the last half of the twentieth century as industrial hemp and marijuana breeders developed new cultivars through vigorous artificial selection (see Chapter 10).

Evolution of Cannabinoid Phenotypes

Cannabinoids are non-nitrogen-containing terpenophenols (compounds that combine a terpene molecule with a phenol structure), and they are only found in *Cannabis*. Levels of the major pentyl (five carbon side chain) cannabinoids (THC, CBC, CBD, and CBG) and their propyl (three carbon side chain) homologs (THCV, CBCV, CBDV, and CBGV) vary widely. Each of these compounds can be absent or present in trace to relatively large amounts in *Cannabis*, especially in the glandular trichomes found on the female flowers and associated leaves. When ancient humans first encountered wild *Cannabis*, it almost certainly contained cannabinoids, including some psychoactive THC. The use of *Cannabis* for drugs is ancient. Vedic and Chinese texts first discussed *Cannabis* in terms of its psychoactive effects thousands of years ago; however, this was long after initial contact with *Cannabis* was made and its use for drugs had already become widespread (Merlin 1972, 2003). Extensive human-driven dispersals and early artificial selections had already taken place 5,000 to 10,000 years ago or possibly even much earlier, before these early historical accounts (see Chapters 4 through 9 for detailed archaeobotanical and historical discussions).

When did cannabinoid biosynthesis evolve? How and why have such diverse cannabinoid chemotypes (chemical profile phenotypes) evolved? How were the varying chemical profiles proliferated, and how are they maintained? The cannabinoid profiles of *Cannabis* varieties have changed markedly during their evolutions. Modern *Cannabis* landraces from northern temperate latitudes are generally much lower in THC than landraces from tropical and subtropical latitudes (Clarke 1977). While this has resulted largely from human selection for various uses in different regions, some marijuana growers in North America have recognized that high-potency, semitropical and tropical drug varieties may be fairly potent the first year they are cultivated at northern latitudes but that subsequent generations rapidly lose potency. Others have reported that European hemp cultivars gain potency over generations when grown in warmer latitudes closer to the equator. In both cases, acclimatization (adaptation to changes in environment) resulted from genetic adaptation of populations to change in climate, and in particular photoperiod, which is based on latitude and determines when and how plants mature. Alterations in THC production associated with introduction into a new environment have often been attributed to climatic acclimatization alone, and this led to the misconception that cannabinoid production is controlled entirely by environment without any genetic control. Given that any population consists of a variety of individuals not equally fit for survival, cultivation of fiber types at tropical latitudes, for instance, allows a few late-maturing types, which might perish at temperate latitudes before reproducing, to reach their full biosynthetic potential and produce small amounts of THC in a longer growing season. However, this only results from immediate environmental influence on THC synthesis. If seeds of these plants are sown the following year, an increase in average THC level could occur through a change in frequency of THC-producing plants resulting from natural selection for late maturity and consequently increased THC levels. On the other hand, in most regions of North America many subtropical and equatorial NLD varieties could not mature outdoors before frost and perished or were harvested before seeds ripened. Those that did mature and bear seed, however, became the genetic stock from which development of modern marijuana cultivars occurred (Clarke 1981).

During the late 1970s and early 1980s, early maturing landrace varieties from Afghanistan (BLD) and South Africa (NLD) were also introduced into North America. A few potent and early maturing individuals were used to breed relatively strong NLD × BLD hybrids adapted to temperate habitats. This is a striking example of the founder effect, where germplasm from only one or just a few individual plants establish a new population. Some of the world's most potent hybrid cultivars are now grown outdoors in North America, indicating that northern temperate latitudes are suitable for production of high-THC cultivars as long as they are strongly selected for high THC content and early maturation. Tropical NLD varieties lose potency in northern latitudes primarily as a result of inadequate selection (both natural and human) for high THC content and secondarily because of ecological conditions.

Cannabis probably evolved its resin-secreting glandular trichomes primarily as a defense against herbivory (attack by plant-eating organisms), although glands may also prevent desiccation by lowering airflow and reflecting sunlight. *Cannabis* resins are made up almost entirely of cannabinoids with a lesser amount of terpenoids (10 or 15 carbon, ring-shaped aromatic compounds). Cannabinoid production, as noted earlier, is restricted to *Cannabis* and does not even occur in the closely related hop genus *Humulus*. Terpenoids and other aromatic volatile compounds are well represented in the resinous secretions of a wide variety of plant taxa (Langenheim 2003). In general, terpenoids are considered to be feeding deterrents, but it has not been proven that cannabinoids have a similar effect (McPartland et al. 2000). In fact, pest populations can become rampant indoors under lights and in glasshouses, and the highest THC clones are no more resistant to pathogens and pest infestations than lower THC clones (Clarke, personal observation). The majority of supposed terpenoid defense systems operate effectively with a terpenoid complement similar to that of *Cannabis*. It is possible that cannabinoids are not required to prevent herbivory but rather to supplement the chemical arsenal for another adaptive reason (e.g., attraction of humans or other animals).

Assuming terpenoids present an adequate defense against pests, then cannabinoids are not required to protect from herbivory. Terpenophenolic cannabinoids are produced as an alternate shunt to the terpenoid pathway leading to mono-and sesquiterpenoid compounds; therefore, cannabinoid synthesis may be in direct biochemical competition for substrate molecules with terpenoid synthesis. If terpenoids are the primary repellent agents in *Cannabis* defense systems, then production of large amounts of cannabinoids in artificially selected drug cultivars might very well diminish the efficiency of terpenoid defense systems by using up terpenoid/cannabinoid precursor molecules for cannabinoid rather than terpenoid biosynthesis. In this scenario, natural selection would not favor extremely high cannabinoid content, a condition reflected in the modest cannabinoid levels of self-sown and unselected drug populations (1 to 4 percent dry weight). If *Cannabis* terpenoids evolved to prevent herbivory, it is possible that human selection for increased mind-altering THC has attracted rather than repelled pests, as the terpenoid defense system may have been compromised in favor of THC production.

Escaped and unselected drug populations lose potency rapidly due to both natural environmental selection and genetic drift (random change in gene frequencies), and consequently maintenance of high-potency varieties requires constant, rigorous selection. Open pollinations in feral situations do not impose selective pressure favoring high THC content. Rapid atavistic loss of THC synthesis may also be another indication that THC content is not of great natural adaptive significance. From evolutionary and ethnobotanical points of view, THC is to a large extent a human artifact, a reflection of ancient relationships between humans and *Cannabis*, a "Plant of the Gods" helping them communicate with their ancestors and other spirits (Schultes and Hofmann [1979] 1992; Merlin 2003; also see Chapter 7).

Sexual dimorphism in cannabinoid content is well documented (Clarke 1981). In high-potency drug varieties, female plants are much higher in THC (up to 10 to 20 times) than males, and normally only females enter into the marijuana trade as recreational and medicinal products. In low-potency NLH fiber and seed cultivars, sexual dimorphism in THC content is less pronounced, and male and female plants express similarly low levels of THC production. Medium potency BLH landraces grown in southwestern China for fiber or seed exhibit more of a tendency toward cannabinoid sexual dimorphism than artificially selected European NLH varieties, but not nearly so much as hybrid drug varieties (Clarke, personal observation 2003).

Small et al. (1975) working at 45 degrees north latitude in their outdoor research gardens in Ottawa, Canada, were unable to observe exaggerated sexual dimorphism in drug varieties. The majority of landrace drug varieties they investigated originated south of 30 degrees north and were unable to produce any mature female inflorescences at such a northerly garden. They reported that fiber types, which were very low in THC to begin with, showed the most pronounced sexual dimorphism in cannabinoid content since early maturing fiber types were the only ones that could mature enough to flower and exhibit sexual dimorphism. Modern *sinsemilla* cultivars were adapted through human selection to flower early enough to mature outdoors in northern temperate latitudes.

Although there has been more rigorous artificial selection of females than males in drug varieties, males in these varieties are still higher in THC than most fiber hemp males or females. This results from extreme selection for THC content in female plants, which in turn give rise to male offspring, some of which may have inherited increased potential for THC synthesis, although they possess far fewer of the secretory glandular trichomes required to express the trait. THC production is not strictly sex-linked to female plants. Female plants, however, produce more THC than male plants by virtue of more vigorous growth, a greater number of glandular trichomes, and an extended growing season. Hybrids between high and low potency varieties almost always result in an F_1 generation of intermediate potency skewed toward the more potent partner (although a few individuals may exceed the THC content of either parent due to the effects of hybrid vigor), regardless of whether the male plant is the high or low potency parent (Beutler and der Marderosian 1978). Male plants carry the genes for THC synthesis, but it is difficult to directly determine their psychoactive potential.

As noted, the primary psychoactive constituent of *Cannabis* is THC. A THC content of 1 percent was chosen as the cutoff point between psychoactive and nonpsychoactive varieties by Small and Beckstead (1973a/b), based on their statistical groupings of cannabinoid profiles used to establish chemotypes for *Cannabis* taxa. US government seizures of illegally imported NLD and BLD *Cannabis* usually range from 1 percent THC up to 20 percent THC or more, with average values much closer to 4 or 5 percent (Clarke 1998a). Dried female inflorescences of highly selected NLD and BLD varieties from Europe and North America may reach THC levels up to 25 percent but average closer to 15 percent. NLH and BLH fiber cultivars are typically well below 1 percent THC. However, escaped fiber varieties in the midwestern United States were found to have THC content up to 3 percent, without any apparent hybridization with drug populations (Haney and Kutscheid 1973). Since these feral plants originated as fiber varieties from Europe and China that probably had low to moderate THC content, at least some individuals may have regained increased THC synthesis through atavism and adaptation. On the other hand, both traditional Chinese BLH and European NLH landraces were higher in THC than modern hemp cultivars descendant from them, as they were not selected for abnormally low THC content until the 1970s. Unselected BLH landrace populations appear to be intermediate in THC content, although few average higher than 2 percent THC. These observations indicate that ancient populations of *Cannabis* may have had an average THC content around one to 2 percent and that some plants may have had levels reaching 3 percent or more. Therefore, ancient humans probably encountered wild *Cannabis* plants with sufficient THC content to be at least mildly psychoactive. This is an important point when considering the origins of religious, recreational, and medicinal uses (see Chapters 7 and 8).

How did cannabinoid biosynthesis evolve to levels of THC production as high as 25 percent dry weight of the inflorescences? Early gatherers, as we have suggested, probably encountered *Cannabis* populations containing individuals with sufficient psychoactivity to warrant interest. One or more traits associated with potency, such as later maturation, larger inflorescences, or increased resin production, provided clues to early humans as to which plants were most psychoactive. The quality and quantity of resin secreted by glandular trichomes is directly correlated to potency. It is readily apparent which plant's inflorescences sparkle most in sunlight and feel the stickiest, indicating high resin content. Perhaps early people found an individual plant with exceptionally high THC content (maybe 3 to 5 percent or more) and collected seeds for sowing. Small (1978) states that in *Humulus* (hop) there has been "disruptive selection of resin content, the domesticated being much more productive," resulting from selection for increased humulone (a bitter terpenoid desirable in beer brewing) levels. Similar disruptive selection must have occurred throughout the evolution of drug varieties of *Cannabis*, differentiating them from fiber and seed varieties, as well as wild and feral populations.

As early humans' awareness of drug plants increased, wild phenotypes that had proven to be the most potent were selected as a source of seed for early cultivation and domestication, or perhaps selections for drug varieties were made from plants already brought into cultivation for fiber or seed. In either case, continued artificial selections concentrating on increased potency led to relatively high THC content in NLD and BLD landraces. More recently, hybridization between potent NLD and BLD landraces resulted in modern *sinsemilla* cultivars with up to 25 percent THC content.

These highly selected modern *sinsemilla* varieties are the most potent *Cannabis* in the world. Varieties from Thailand, India, and South Africa are all very potent direct descendants of the South Asian *ganja* NLD gene pool. The *sinsemilla* NLD varieties usually have a relatively high THC content with little CBD or other cannabinoids. *Sinsemilla* cultivation has provided

a rigorous selective regimen, which, as we pointed out, has resulted in cultivars with extremely high THC levels. Since *Cannabis* plants are often harvested and dried individually, it is possible for a farmer to smoke flowers of individual plants and in a qualitative and subjective procedure select seeds from the most potent ones with the highest THC (and consequently the lowest CBD) content. Hashish, on the other hand, is collected from plants either by hand-rubbing the resin (*charas*) from a number of plants as in Nepal and India or by harvesting and drying a seeded population and sieving the resin from the bulked plants as in Afghanistan, Morocco, and the Middle East. As a result, individual plant selection is impossible, and the cannabinoid phenotypes of individual hashish plants vary widely from nearly all THC to nearly all CBD with a population bell curve in between focused on a median ratio of 1 to 1.

Why are fiber varieties so low in THC? Disruptive selection has favored increasing THC production and lowering CBD production in drug varieties while diminishing THC production and increasing CBD production in fiber varieties, resulting in bidirectional evolution of cannabinoid phenotype. This divergence may have first occurred many thousands of years ago when the putative hemp ancestors (PHAs) of the NLH gene pool and the putative drug ancestors (PDAs) of the BLH, NLD, and BLD gene pools were driven from Central Asia by advancing ice sheets. Likely beginning more than 5,000 years ago, fiber cultivars were selected to mature early and produce tall, unbranched stalks. As noted, seed production, and hence floral development, is relatively less important for a hemp fiber crop, which is sometimes harvested before seeds ripen. In modern hemp-growing regions, farmers sow some seed of their select fiber varieties in widely spaced rows so they will branch and bear more flowers and seeds for production of sowing seed for the following year. Since there is no selection for abnormally high THC levels, natural selection acts to maintain THC at a relatively low level around 1 to 3 percent dry weight. Modern European selections for extremely low THC content (formerly below 0.3 percent and now below 0.2 percent dry weight) cultivars were developed in reaction to perceived increases in *Cannabis* smoking by Europeans, based on an unreasonable fear that fiber hemp can and will be smoked for mind-altering purposes.

It is also possible that THC production was not naturally selected for, or favored by, the short summer growing season and long photoperiod at northern latitudes where fiber cultivars evolved. THC levels drop off rapidly if plants are not constantly selected for high potency. If the potential for THC biosynthesis was not realized because northern temperate seasons did not allow complete maturation, and therefore artificial selection could not be exercised, varieties would lose their ability to produce significant levels of THC.

In contemporary North America, biosynthesis of small amounts of THC in escaped hemp populations has lowered their chances of survival. If escaped hemp produced no THC at all, its survival would not be threatened by law enforcement officers and naive drug users who confuse it with potent drug *Cannabis*.

Short growing seasons in northern temperate latitudes are favorable for some high-THC cultivars that have been bred to mature under longer day and shorter night conditions. However, THC production is favored by the longer season of more tropical latitudes. Even North American drug varieties selected for early maturation tend to be lower in THC than those selected for later maturation.

In sum, extremely low-THC with high-CBD hemp varieties and extremely high-THC with low-CBD drug varieties were derived by disruptive human selection for either industrial hemp or recreational and/or medicinal drugs from a median THC level, likely around 1 to 3 percent dry weight.

Geographical Distribution of Cannabinoid Phenotypes

Researchers have repeatedly attempted to classify and group *Cannabis* biotypes from diverse geographic origins by determining cannabinoid chemotypes, forming groups and circumscribing chemotaxonomic races. Much of this work was prompted by forensic scientists attempting to "fingerprint" the geographical origins of illicit *Cannabis* seizures made by government agencies and was largely funded by the United Nations. Davis et al. (1963) measured THC, CBN, and CBD levels in *Cannabis* from several origins and plotted CBD percentage versus THC percentage + CBN percentage (THC percentage and CBN percentage were grouped together because CBN is a degradation product of THC, and combining them gives a more accurate measure of the original THC content in old and degraded samples). Davis et al. (1963) graphed these data against the geographical origins of the samples and concluded that climatic factors such as mean annual cloudiness and mean temperature influenced cannabinoid chemotype—sunny and warm regions favoring high-THC biosynthesis and cloudy and cool regions favoring high CBD biosynthesis; these researchers thereby attributed variations in cannabinoid production directly to climate instead of genetic background determined by environmental and human selection.

Grlic (1968) classified various samples from *Cannabis* seizures into five "ripening" groups using the ratio of THC percentage + CBN percentage to CBD-acid percentage + CBD percentage. Jenkins and Patterson (1973) also analyzed samples of various origins and plotted CBD percentage, THC percentage, and CBN percentage on triangular coordinate graphs. They concluded from the clusters on the graphs that there was a "reasonable similarity in the results of analysis of samples from any one country, and that the analytical pattern of results varies from one country to another." However, geographical origins of samples overlapped and it was not possible to differentiate each sample from all the others for forensic purposes. The spread of individual clusters of data points in this study was attributed to the breakdown of THC to CBN over time, since THC percentage and CBN percentage were plotted on separate coordinates and samples were stored for varying lengths of time.

Moving away from analyzing drug seizures, Fetterman et al. (1971) grew seeds from seven different geographical origins. Cannabinoid content was determined by gas-liquid chromatography (GLC) and the samples were divided into two chemotypes—drug-type and fiber-type. They proposed a ratio similar to Grlic (1968) with the addition of the acid fractions of THC, CBN, and CBD as follows: Δ^9-THC percentage + THC-acid percentage + CBN percentage + CBN-acid percentage to CBDA percentage + CBD percentage. In this chemotype ratio, Δ^9-THC percentage + THC-acid percentage + CBN percentage + CBN-acid percentage was considered to approximate the original total THC content of fresh samples since CBN was thought to be the sole degradation product of THC (Levine 1944). During gas chromatography

(GC) analysis, cannabinoid acids are decarboxylated to their neutral forms as the samples are vaporized. Only low temperature GLC can distinguish between the carboxylic acid and decarboxylated forms. Samples with chemotype ratios exceeding one to one were classified as Type I (drug-type), and samples with chemotype ratios less than one to one were classified as Type II (fiber-type). The researchers concluded that "the phenotype of one variety of marijuana remains the same regardless of the plant part, sex, age, year, or place of growth" and that the cannabinoid chemotype "is a relatively simple and useful means of distinguishing between the drug and fiber types of marijuana." In other words, they concluded that genetic background holds sway over ecological conditions in the control of cannabinoid biosynthesis.

Utilizing GC, Small and Beckstead (1973b) analyzed 350 samples of *Cannabis* from police seizures and germplasm collections representing approximately 50 countries that were grown outdoors during the summer of 1971 at Ottawa, Canada. Graphs of Δ^9-THC percentage versus CBD percentage produced "three large modally distinct groups with respect to cannabinoid content." Type I samples contained larger amounts of THC (> 0.3 percent) and limited amounts of CBD (< 0.5 percent). Type I samples typically originated from south of 30 degrees north latitude. Type II varieties contained relatively high levels of both THC (> 0.3 percent) and CBD (> 0.5 percent), and Type III varieties contained very little THC (< 0.3 percent). Types II and III usually originated north of 30 degrees north latitude. Small and Beckstead (1973b) also recognized Type IV varieties from northeastern Asia containing small amounts of CBGM (cannabigerol monomethyl ether) at ~0.05 percent. Based on the ratio of THC percentage to CBD percentage, Type II varieties could not consistently be distinguished from Type III, and they felt that Type II varieties could have originated from hybridization of Type I with Type III. Type I is considered to be drug *Cannabis* and Types II and III are considered to be either fiber or seed *Cannabis*. THC content in Type II varieties was usually higher in female than male plants. They attributed the difference in varieties to origin north or south of 30 degrees north latitude and to selection for fiber and seed varieties at northern latitudes and drug varieties at southern latitudes.

Clarke (1977, 1981) plotted graphs of latitude versus chemotype (total THC percentage + CBN percentage to total CBD percentage) using data collected from published scientific literature. Once again, a correlation between latitude and chemotype was reported, substantiating the observation of Small and Beckstead that fiber and seed types occur north of approximately 30 degrees north and drug types occur south of 30 degrees. However, these results did not take into account historical parameters and cultural preferences influencing cannabinoid profile selection.

According to Beutler and der Marderosian (1978), controlled crosses between high and low THC varieties produce offspring of intermediate THC content, supporting the theory of Small and Beckstead (1973b) that intermediate Type II chemotypes could have resulted from hybridization between Type I and Type III chemotypes. De Meijer et al. (2003) corroborated these findings and proposed that cannabinoid biosynthesis is controlled by a pair of alleles, B_D coding for CBD synthesis and B_T coding for THC synthesis.

Turner et al. (1979) challenged the use of cannabinoid ratios to determine chemotaxonomic groupings in *Cannabis*. Using an Indian variety, they demonstrated that variables of sex, age, and plant part determined whether the chemotype fell within either of the two groups reported by Fetterman et al. (1971) or any of the four groups of Small and Beckstead (1973b). However, their Indian variety had an intermediate THC percentage to CBD percentage ratio of close to one to one, and therefore relatively slight changes in cannabinoid content were interpreted as a shift from fiber chemotype to drug chemotype. We feel that the cannabinoid ratios of varieties that are not of intermediate type (close to one to one) are not significantly affected by small fluctuations in cannabinoid content and that their cannabinoid ratios can still be considered characteristic of their selective breeding and possibly their origin. Turner et al. (1979) also pointed out that in all previous analyses, CBC had been erroneously identified as CBD. They proposed the following ratio for determining cannabinoid chemotype: Δ^9-THC + Δ^9-THCV + CBN + Δ^8-THC + CBL to CBC + CBD + CBDV + CBG + CBGM.

This ratio separates the cannabinoids by their ring structure, placing those of closed central ring structure (THC, CBN, and CBL) in the numerator and those of open central ring structure (CBC, CBD, and CBG) in the denominator. This natural division occurs at the crucial step in the biosynthetic pathway of conversion of CBG (cannabigerol) to THC (Taura et al. 1995, 1996) and thus reflects the two allelic forms of the THC/CBD synthase gene—B_T coding for THC biosynthesis and B_D coding for CBD biosynthesis. It also corrects for the confusion in previous analyses between CBC and CBD by grouping them together and thereby allowing comparison with previous data sets.

Pate (1983) published an evolutionary correlation between cannabinoid chemotype and ultraviolet (UV) light levels. He concluded that THC production evolved as a protective mechanism against damage caused by UV light. Although his correlations were statistically significant and showed that varieties originating from lower latitudes were higher in THC, his choice of THC as the most likely candidate for UV protection did not take into account that CBD may offer UV protection as well. Terpenoid compounds may also protect the plant from UV radiation (e.g., see Lydon et al. 1987).

In the following section, we provide a list of geographical regions, characterized by their cannabinoid chemotypes, including brief geographical and historical explanations.

NORTH AMERICA

North American self-sown populations have the lowest chemotype ratios (THC etc. to CBD etc.) since varieties were originally introduced for fiber use. NLH was introduced from Western Europe in the early seventeenth century and BLH was introduced from China in the late nineteenth century. This accounts for the similarity between cannabinoid ratios for North America and those for Western Europe and China and reflects the primary uses for *Cannabis* in these regions for fiber and seed.

WESTERN EUROPE

After North America, Western Europe has the second lowest chemotype ratios with proportions that never exceed one to one. European hemp gene pools include NLHA, which may either be native wild or escaped feral fiber or seed varieties; NLH cultivars; BLH varieties imported to a lesser extent from eastern Asia during the nineteenth century; and various hybrids between these taxa, all of which were used for fiber and seed rather than drugs.

EASTERN EUROPE

Eastern Europe has slightly higher average chemotype ratios than Western Europe, yet they rarely exceed one to one. Traditional use in this region is also for fiber and seed, although its use for ritualistic, psychoactive purposes was important in some areas until the rise and spread of monotheism. Chemotypes for Western and Eastern Europe overlap slightly at their southern latitudes. Slightly lower ratios for Western Europe than Eastern Europe may reflect recent selection for extremely low THC content as legislated by Western European nations, which set their legally acceptable THC content to below 0.3 percent (now 0.2 percent) and chemotype ratio less than one to three. Historically, Eastern European hemp breeders used native NLHA populations in their selection of useful varieties, while in some areas of Western and Central Europe, Far Eastern BLH varieties were hybridized with NLH varieties early on to make improved industrial hemp cultivars. As a result, some European hybrid varieties (e.g., the Hungarian 'Kompolti TC' and 'Uniko-B') exhibit traits from Far Eastern varieties, such as very thin and tall stalks with long internodes and large broad leaves.

CENTRAL AMERICA AND THE CARIBBEAN

All cannabinoid ratios from this region exceed one to one. Although the earliest introductions were NLH fiber varieties from Western Europe (middle sixteenth century, late eighteenth century, and early nineteenth century), later introductions were NLD varieties from India (middle nineteenth century). Hybridization is suspected here also and is indicated by intermediate cannabinoid ratios. Currently, the largest use in this region is for marijuana production.

SOUTH AMERICA

South America exhibits a wide range of cannabinoid ratios, most of which exceed one to one. Early introductions of NLH from Western Europe for fiber use (middle sixteenth and early nineteenth centuries), and later dispersal of NLD seeds from the Caribbean coast (following 1834), have resulted in a wide range of chemotypes. Ratios for this region, although averaging above one to one, are lower than most other drug-producing regions. These intermediate values could have resulted from hybridization, or they could reflect a lack of intense selection for drug types. The only country in South America where *Cannabis* has been consistently produced for marijuana on a relatively large scale is Colombia, although large quantities are currently produced in Paraguay (e.g., see United Nations Office on Drugs and Crime 2011). Ratios for Colombian samples are well above one to one and are the highest in South America.

MIDDLE EAST

The Middle East provides an even wider range of ratios from one to ten to ten to one; however, most average values still exceed one to one. Although some NLH fiber varieties were introduced into Syria during the nineteenth century, traditional use throughout the predominantly Islamic nations of this region has been for the production of *Cannabis* drugs, and more specifically, in recent times, for sieved hashish made from varieties of NLD (and possibly BLD). Fiber varieties were probably introduced much earlier, for paper, cordage, and possibly cloth production (Zaimeche 2005). Higher levels of CBD and lower THC to CBD ratios are quite common in hashish varieties because bulk processing during sieved hashish production does not allow selection of high-THC individuals, resulting in an average cannabinoid ratio of close to one to one. Also, the wide range of ratios within this region could once again have resulted from hybridization of drug (hashish) varieties with fiber varieties derived from separate historical introductions from various geographical regions for different uses. Wide-ranging intermediate ratios have probably developed in this region because the Middle East is a crossroads where both European NLH and Asian NLD varieties could have been introduced.

EAST ASIA

East Asia also exhibits a wide range of cannabinoid chemotype ratios. Chinese ratios fall well into the fiber range, while ratios from Korea and Japan fall well into the drug range. This creates an apparent anomaly, since neither Japan nor Korea has a recorded history of *Cannabis* drug use prior to World War II. Northern Chinese varieties may have descended more directly from wild Central Asian PA populations, which could explain their relatively low cannabinoid ratios. However, a recent genetic (and chemotaxonomic) analysis indicates that Chinese BLH varieties are more closely related to *C. indica* than to *C. sativa* and therefore should be classified as *C. indica ssp. chinensis* (see Hillig 2005a/b and Hillig and Mahlberg 2004). Chinese BLH varieties were considered among the finest fiber cultivars in the world; during the late nineteenth and early twentieth centuries they were widely disseminated to Western Europe and the New World and thus contributed more to modern European hemp cultivars than any other imported gene pool. Drug types from Korea and Japan are native landraces that were likely not selected for either high or low THC content as there were no known indigenous *Cannabis* drug cultures, especially after the rise and spread of Confucianism and Buddhism. On the northern Japanese island of Hokkaido feral populations that escaped from World War II fiber hemp production are common and illicitly harvested for marijuana use. Feral Korean *Cannabis* is also harvested for drug use and a few accessions were used during the late twentieth century to breed *sinsemilla* cultivars. These observations indicate that both Korean and Japanese landraces have the potential to produce psychoactive levels of THC. During the 1970s, Japanese breeders improving a local landrace developed 'Tochigi Shiro' a low-THC cultivar grown for fiber production. Small and Beckstead (1973a) classified Japanese varieties as Type IV since they contain CBGM, which is not found in other varieties. This may indicate a somewhat different, possibly independent evolution for Japanese varieties, or at least a pronounced founder effect. It could also have resulted from a very ancient presence and use of *Cannabis* in Japan where the oldest known macrofossils (seeds) have been found in association with pottery shards

and dated to about 10,000 BP (Kudo et al. 2009; Okazaki et al. 2011; see Chapter 4 for detail regarding these remarkable remains in ancient Japan).

INDIAN SUBCONTINENT

Samples from the Indian Subcontinent exhibit very high cannabinoid ratios, and most countries in this region provide several examples of nearly CBD-less varieties. Sri Lanka is the only exception, where the British introduced hemp varieties in the late nineteenth century. In addition, Sri Lanka does not have the long history of *Cannabis* drug use associated with mainland India, where its use originated over 3,500 years ago and continued selection led to the formation of the NLD gene pool. THC synthesis has been so strongly selected in South Asia that varieties in this region are often nearly devoid of CBD. Over the centuries, NLD varieties were disseminated around the world, and all trace their origins to the original South Asian drug-type gene pool (Hillig 2005a/b; Hillig and Mahlberg 2004).

SOUTHEAST ASIA

Neighboring Southeast Asian NLD varieties also have very high chemotype ratios, with values always exceeding one to one, and the traditional use in this region is also primarily for marijuana. Many apparently CBD-less samples are found here also, as well as samples containing traces of tetrahydrocannabiverol (THCV), the propyl homolog of THC. Seedless *ganja* cultivation is common to both India and Southeast Asia, and the majority cultures of Southeast Asia and India have no tradition of hemp fiber or seed use.

EQUATORIAL AFRICA

Varieties in Western and Central equatorial Africa have high cannabinoid ratios, with moderately high THC level and relatively low CBD level. NLD varieties were extensively introduced into this region following World War II and have been used for marijuana production since that time. This region also has no history of *Cannabis* fiber or seed use.

SOUTH AND EAST AFRICA

South and East African varieties have extremely high cannabinoid ratios and many examples of apparently CBD-less varieties also appear. The occurrence of this chemotype in both India and South Africa supports the diffusion of Indian varieties to South Africa ca. 2,000 BP (Burney et al. 1994; Merlin 2003). The traditional use in South and East Africa is for drugs, as in India. Separate clusters of data for Equatorial vs. South and East Africa likely result from introductions of differing founder NLD populations. Equatorial West Africa did not receive drug varieties of *Cannabis* until after World War II, centuries later than South and East Africa and likely from different sources.

Although THC to CBD ratios below one to one are limited to higher latitudes far from the equator in the northern hemisphere, and ratios exceeding one to one are limited to more equatorial latitudes, this is not the case in the southern hemisphere. Indeed the highest cannabinoid ratios, and a preponderance of apparently CBD-less varieties, occur in South Africa far from the equator, possibly a founder effect resulting from limited NLD introductions from India. Ratios appear to be correlated with northern vs. southern origins or vice versa but not with distance from the equator. This indicates that latitudinal constraints on cannabinoid production may be of little significance in determining cannabinoid profile compared to human influences such as seed dispersal, cultural preference, and selective breeding.

It should be noted that, as a group, NLH varieties cluster more tightly than NLD and BLD varieties and may have evolved from a narrower genetic base. This could result from the relatively recent breeding of European and New World NLH fiber cultivars, which promoted blending and homogenization of gene pools rather than repeated dispersals and subsequent isolation accompanied by localized selection as in the evolution of NLD and BLD landrace biotypes. During the middle nineteenth century, fiber varieties were collected from China, Europe, Russia, and the United States, then hybridized by Western plant breeders and eventually distributed throughout hemp growing regions of the New World, Europe, and Russia. During the 1970s, many very low-THC NLH varieties were developed in Europe from traditional landraces, and introduced varieties were shared by many hemp breeding programs (see Chapter 10). This could also account for similarities between chemotypes for NLH varieties as a whole. On the other hand, the illegality of drug *Cannabis* in the twentieth century limited the international exchange of *Cannabis* germplasm. The dispersal of drug varieties began earlier on, and these varieties, with adaptations to differing natural and human selective pressures, were eventually spread over a vast and diverse cultural and geographical landscape, resulting in more diverse chemotypes. Until recent years, drug varieties remained more isolated in small regions within much larger and more diverse geographical and climatic ranges, and therefore they have had more time to evolve characteristic traits and fewer chances to hybridize and homogenize their genes. Thus a wide variety of high-THC landraces evolved.

Repeated human selection and crossing of plants with favorable fiber, seed, or drug traits has amplified the natural genetic diversity of *Cannabis*. The role of humans in initially selecting wild *Cannabis*, developing cultivars for varying uses, and isolating them from introgression with wild and escaped populations through continued cultivation and selection has most certainly had the largest impact on *Cannabis*'s evolution. Natural selection has also favored the segregation of characters between wild and cultivated populations and further promoted the isolation of *Cannabis* gene pools.

The overlapping ranges of chemotype ratios for each geographical region are too extensive to provide accurate identification for taxonomic or forensic purposes. It is our opinion that cannabinoids support only general taxonomic differentiation of taxa (e.g., *C. sativa* vs. *C. indica*) and therefore have only limited value as taxonomic tools. However, cannabinoid data does reflect specific dispersals and varying uses of *Cannabis* accounted for in the literature and are consequently of ethnobotanical and evolutionary interest. Cannabinoid ratios can also be used to indicate evolutionarily significant hybridization between distinct gene pools following their dispersal into new regions.

Methods of utilizing *Cannabis* vary profoundly among different cultures. European, northern Asiatic, and Chinese cultures located in largely temperate continental climate zones have generally focused on the production potential of *Cannabis* for fiber and edible seed. Indian, Middle Eastern, African, and Southeast Asian cultures, living in a wide range of climates from alpine to tropical, have utilized *Cannabis* primarily as the source of psychoactive and medicinal drugs and only secondarily as a fiber or seed plant. Ancient humans discovered wild psychoactive varieties and selected potent individuals, which had evolved the metabolic machinery (coded by the B_T allele) to synthesize THC. Possibly all ancient *Cannabis* was somewhat psychoactive. However, since its prehistoric range was likely far to the north, where existing PA and NLHA populations are very low in THC and high in CBD, we do not believe that primordial *Cannabis*, although possessing the B_T allele, could have been very potent, at least prior to strong artificial selection for high-THC content.

Western Europe and East Asia both served as origins for dispersal of fiber varieties to North America. Initial dispersals of NLH were from Western Europe during the seventeenth century, and later BLH came from China and Japan during the late nineteenth and early twentieth century. Average chemotype ratios are lower in North America than Western Europe, and Eastern Europe has chemotype ratios slightly higher than Western Europe and lower than Central Asia. The decrease in chemotype ratios from Western Europe reflects continued positive selection for favorable hemp traits and negative selection for drug potency, leading to the near loss of THC biosynthesis (see also Chapters 4, 7, and 9). Intensive breeding of fiber varieties has continued from the late 1800s until the present (see Chapters 5 and 10).

NLD varieties originated in southern Asia and likely spread into parts of Southeast Asia and South and East Africa before 2,000 years ago. Average chemotype values have remained high even though changes in latitude have been extreme. However, these derivative varieties exhibit even higher chemotype values than those from India and have nearly CBD-less populations that are very high in THC. This may be explained by founder effects and continued, rigorous selection for potency after dispersal into these regions.

Western Europe also served as the origin for early dispersals of NLH varieties into Central and South America during the sixteenth and seventeenth centuries. Indian NLD varieties were introduced into Central America and the Caribbean during the middle 1800s and from there introduced into South America during the middle 1900s. Central American and Caribbean varieties are potent with high chemotype values, although not as high as Indian varieties. South American drug varieties are derived from those in Central America and the Caribbean and have even lower chemotype values. South America, in particular Colombia, has a recent history of large-scale, commercial marijuana cultivation. In a rapidly expanding market with new fields being turned to *Cannabis* each year, seeds are generally all sown without selection, since all the seeds are needed to expand production. Lack of selection leads to gradual lowering of THC levels and decreased cannabinoid ratio. Moderate chemotype values expressed by present-day Latin American and Caribbean varieties may originally have resulted from hybridization between European-derived NLH fiber varieties and Indian-derived NLD varieties. However, genetic drift in concert with expansion of range for economic reasons has been a more important determinant of present-day cannabinoid chemotypes.

instances, cannabinoid compounds may also be isolated from noncarbonized remains. Cannabinoid remains originate undisputedly within the *Cannabis* plant and provide some of our most positive evidence for *Cannabis*'s antiquity; yet controversial cannabinoid remains have been reported from apparently pre-Columbian pipes in Ethiopia (van der Merwe 1975) and in Egyptian mummies (Balabanova 1992). *Cannabis* in the form of charcoal has been reported from at least one Neolithic site at Senuwar in Bihar, India, dated to ca. 1800 to 600 BCE (Saraswat 2004; also see Fuller and Madella 2001; and the section on South Asia in this chapter).

PHYTOLITHS

Phytoliths ("plant stones" or "plant opals," e.g., silicon phytoliths) are rigid microscopic bodies that vary in size and characteristic shape and are found in many plants. Analysis of phytoliths has become an important source of botanical and ethnobotanical information used to identify plants and in some cases aspects of their ethnobotany. Soluble silica in groundwater is absorbed by roots and eventually deposited in certain aerial parts depending on the species, with different species of plants forming phytoliths in different tissues. Phytoliths are durable and tend to be released into the soil or other medium as the plant decays; thus phytoliths can be used to identify ancient agricultural areas or associated uses of the plant from which the phytoliths are derived, even after all organic traces of the plant have disappeared.

It should also be noted here that ancient DNA recovery and analysis can provide, now and perhaps more so in the future, another method for positive identification of *Cannabis* remains as well as insights regarding cultural and chemotaxonomic biodiversity within the genus over time, origins and dispersal of the genus, and the status of domestication; questions regarding the origins, dispersal, and biodiversity of *Cannabis* "cannot be answered by morphological archaeobotanical studies alone" (see Schlumbaum et al. 2008 for a review of the use of ancient plant DNA in archaeobotanical research; more specifically, see Russo et al. 2008 and Mukherjee et al. 2008 for studies that involve genetic analysis of ancient *Cannabis* DNA from Central Asia discussed later and in Chapters 7 and 8).

From a historical point of view, the size and quality of archaeobotanical collections varies among different regions of the world and are often determined by the interest and scope of archeological research in a particular area. Indeed, at present, the archaeobotanical record for ancient *Cannabis* remains is overwhelmingly more detailed in Europe than it is across the rest of Eurasia. However, the written record helps fill in and flesh out thin sections of the discontinuous archeological and archaeobotanical records.

Written Records of *Cannabis*'s Presence and Use

Extensive searches through the literature, including diverse sources from many different cultural traditions, were employed to reconstruct the history of *Cannabis* use by humans. Although some secondary sources are used, we rely heavily on primary written sources. It is important to recognize that differential levels of verifiable information have been established regionally for historical and especially archeological evidence. During modern times, scientific and to a lesser extent scholarly study in general has varied from area to area. Until recently, scientists in countries strongly influenced by Western civilization have undertaken more empirical research; cultural differences and biases have also affected interpretations of history.

For example, in South Asia, until the latter half of the twentieth century there was a relative dearth of archeological science as well as a different set of circumstances affecting historical research. The reconstruction of early history of South Asia has been hindered by lack of an adequate translation of ancient Harappan written records, and there has also been a comparative deficiency of ancient artifacts recovered and analyzed by the largely Aryan dominated society that followed. In addition, until recently, remnants of the *Vedas*, an extraordinary body of Aryan religious literature and mythology, have been more or less the sole source of information for the early Aryan period and the South Asian region as a whole. Just as the paucity of authenticated Aryan remains has baffled archeologists, the Vedic literature, with its symbolic obscurities and disregard for sequence of events in time, has frustrated historians. Therefore, any dating and sequential arrangement of written data remains tentative. For the prehistoric and Vedic periods, students of the cultural diffusion and history of *Cannabis* can rely on only vague reports and scanty information passed down over centuries. Thus historically minded scholars encounter chronological chaos when delving into the antiquity of early Indian civilization. This situation is a manifestation of Indian metaphysics: the cyclical concept of nature (central to Indian thought) indirectly de-emphasizes the importance of an individual historic event and its chronological position. However, ancient Indians were not completely lacking in their appreciation of historical perspective as manifested by carefully preserved lists of teachers in various Vedic, Buddhist, Jain, and other religious texts. Aside from these historically meager religious chronicles, there was a lack of enthusiasm or ingenuity to collect and organize scattered yet pertinent evidence into a critical historical text popular enough to ensure its preservation. Ancient Indian civilization did not produce scholars whose historical methodology paralleled the notable early Western historians such as Herodotus or the ancient Chinese chroniclers.

It must be emphasized that the ancient Indian cultural milieu was, and to a large extent still is, thoroughly interwoven with philosophical or religious ideals and oriented toward their oral transmission. The ancient Vedic tradition of the Aryans was a hymnal in its earliest form. The *Rg Veda*, India's earliest literary record, was considered as *sruti*, the revelation of truth by Brahma—the universal and omnipotent entity out of which all things emanate and eventually return in cyclic order—and therefore not to be "desecrated" by the written word. In this tradition, the *sruti* was to be orally transmitted from guru to acolyte in orderly succession according to meticulous discipline. As the oral tradition gradually broke down, the *sruti* became secularized through their popularization among the masses, and the sacred Vedic legends and mysteries were eventually written down. This written record was essentially the antithesis of the pristine Vedic tradition, and thus we find the only concrete result of historical study of the most ancient period can be found in the long lists of kings and associated legends preserved in the sacred *Vedas*, the popular *Puranas*, and the epics of the *Mahabharata* and *Ramayana*.

In addition to a negligible and often faulty chronology and lack of substantial, documented historical and cultural information, research concerned with ancient Indian geography and history is further impeded by the use of similar or identical names for various physical and cultural phenomena and the use of different names for the same place or material. What is more, propagation over generations of fanciful geographical legends and the frequent disregard for distinction between real and imagined geography also raises formidable obstacles for inquisitive and rigorous students of ancient India. We find that the most accurate information for the historic period comes not from indigenous sources but from writings and observations of travelers from Greece, Persia, and China. In sum, the written records from South Asia have many limitations that can hinder scientific analysis of the cultural diffusion of Cannabis throughout the region. Indeed the fundamental challenge when dealing with questions of historical focus, and especially where Indian antiquity is concerned, is to accept the fact that some speculation can hardly be avoided. These caveats must be kept in mind during scholarly interpretation of ancient traveler's records and oral traditions.

In this chapter and those that follow, ancient remains and historical references for Cannabis and its use are discussed in terms of accuracy and reliability in the three areas of archeological interest outlined earlier: artifact interpretation, archaeobotanical analysis, and dating techniques.

Nonhuman Agencies Affecting the Geographical Range of Cannabis

Ultimately, the geographical range of Cannabis, like other organisms, involves both dispersal and migration, which are closely related but different concepts. According to Polunin (1960), "Dispersal merely involves dissemination from the parent and distribution (in the dynamic sense) to a new spot, whereas migration implies also successful growth and establishment. Thus dispersal is a necessary forerunner of migration, which is actually accomplished only on establishment in a new place." However, migration also implies intent and therefore the "migration" of plant species and even the earliest humans may more accurately be described as "dispersal" and/or "diffusion."

Like many other widely distributed crop plants, Cannabis makes poor use of both physical and biological dispersal agents, other than humans, for the dissemination of its seed. Indeed, the seeds of Cannabis are smooth, relatively heavy, and not very buoyant. In addition, they lack attractive flesh and are devoid of wings, burrs, or other dispersal mechanisms. Streams and rivers are physical agents that may pick up a fallen seed and carry it downstream. If this occurs, it might be deposited near the surface of finely sorted alluvium, which as it drains, provides an excellent medium for germination and growth. However, low buoyancy of seeds seems important in limiting their dispersal by stream flow very far from the parent plant.

Wind serves to dislodge seeds from mature flowers so they drop to the ground. This is especially true for wild and self-sown feral Cannabis plants with readily shattering inflorescences. Since Cannabis seeds are relatively small but not as tiny as other widely dispersed species such as red poppy (Papaver rhoeas L.), wind dispersal to any great distance would require gale force winds. Wind, therefore, is a very unlikely or only occasionally effective physical agent to move Cannabis seed more than a very short distance. Water and wind mitigated long-range dispersal are certainly possible, but like most nonbiotic dispersals would have to occur over short distances and repeat through many generations.

The mobility of animals, and their activity among and dependence upon plants, makes them a significant agent for dispersal across longer distances. Animals disperse seeds by two basic methods, transported either internally (after swallowing) or externally, adhering to their bodies. In Cannabis, resin glands covering the protective bract surrounding the seed repel some animals, preventing them from removing seed from inflorescences. Most rodents find seeds fallen on the ground and crack and destroy them while feeding, which may have led to natural selection for a sturdy pericarp (seed shell).

A common natural means of seed dispersal involves transportation in the digestive tracts of birds. Birds kept as pets feed on commercial hemp seed (Steyermark 1963; Booth 2003), and this is a strong indication that some wild species forage for it (e.g., the hemp linnet, Carduelis cannabina L., a widely distributed migrating bird). The common Chinese name for the field sparrow is má qiāo (麻雀) or "hemp bird," alluding to its preference for eating hemp seeds. Most Cannabis seeds are cracked by feeding birds and are thus destroyed, although some species may occasionally swallow them whole. The efficiency of internal seed transport to a new location is dependent upon three basic factors: the range of the bird, the resistance of the seed to digestion, and the retention time of the seed in the bird's body. In any case, all ingested seeds certainly do not remain viable. Some may be regurgitated intact and fertile; some may be fully or partially digested and rendered infertile and others may pass through the alimentary tract unharmed, to be deposited in a small pile of nutritive feces some distance from their source. We should note that movement through the alimentary tract of an animal might even benefit the seed by promoting faster germination and providing nutrients, thereby producing stronger plants (Polunin 1960). The relative rates of seed degeneration seem to vary according to factors that are not completely understood. Sir Charles Darwin (1881), a great student of natural history and the father of modern evolutionary theory, studied this specific problem and uncovered some interesting evidence concerning the Cannabis plant:

> After a bird has found and devoured a large supply of food, it is positively asserted that all the grains do not pass into the gizzard for twelve to even eighteen hours. A bird in this interval might easily be blown to the distance of 500 miles, and hawks are known to look out for tired birds, and the contents of their torn crops might thus readily get scattered. Some hawks and owls bolt their prey whole, and, after an interval of from twelve to twenty hours, disgorge pellets, which, as we know from experiments made in the Zoological Gardens, include seeds capable of germination. Some seeds of the oat, wheat, millet, canary [grass], hemp, clover, and beet germinated after having been from twelve to twenty-one hours in the stomachs of different birds of prey.

Ridley (1930) discovered Cannabis seeds in the stomach contents of the European magpie, and Polunin (1960) pointed out that after gorging, birds sometimes regurgitate seeds without passing them through the alimentary canal. In addition, fallen Cannabis seeds stuck in mud could have become attached to the webbed feet of water birds and moved from one area near a water body to another.

These references regarding the dispersal of seeds by birds raise the possibility that long-range avian transport carried Cannabis out of Central Asia into Europe, China, and South Asia prior to human dispersal. If and when this process influenced the diffusion and distribution of Cannabis are difficult concepts to verify. Documentation and techniques for collecting pertinent data are extremely rare, and therefore, theories must be tentative and highly speculative. Darlington (1969) proposed hypothetical routes of Pleistocene bird migrations leading across Central Asia into Arabia, Persia, and as far as Southeast Asia. Although the Himalaya Mountains appear to block migrations of many birds into the Indian subcontinent, modern studies have established the natural Indo-Asian flyway, one of eight major north/south migratory bird routes in the world, which traverses the eastern Hindu Kush Mountains connecting Central and South Asia.

Several species of ducks (*Anatidae*), wagtails (*Motacillidae*), starlings (*Sturnidae*), and sparrows (*Ploceidae*) are known to use the Indo-Asian flyway during their annual migrations. Since they fly south in autumn, after *Cannabis* seeds are mature, the timing would be correct for them to carry ripe seeds south from Central Asia across the Himalaya and Hindu Kush Mountains into India. East-west migrations are less common, but northern summer ranges cover most of northern Eurasia (McClure 1974; Boere et al. 2006).

Direct adhesion is another means of seed dispersal that may have affected the diffusion of *Cannabis*. The seeds of some plant species are better adapted to adhesive transportation than others, but many can, by one physical means or another, attach themselves to the skin, claws, toes, hooves, hide, feathers, or furry coats of animals. *Cannabis* seeds, however, are nearly ovoid, smooth, and dense, and must therefore be trapped in the hooves, fur, or feathers of an animal, rather than, in most cases, actually adhering; for example, the hooves of wild horses roaming across the vast Eurasian steppe during the late Pleistocene and early Holocene could have provided just such a vehicle for *Cannabis* dispersal.

The chances that a dispersed seed may come to rest in a favorable environment are enhanced by the natural foraging of animals. Polunin (1960) states that "animals, like plants, tend to keep, as birds tend to alight, within a single habitat range, so increasing the chances a dispersed seed would have of coming to rest in a place suitable for germination and successful establishment." A hoofed mammal, such as a horse, goat, or pig, for instance, browsing around *Cannabis* plants, may step on a fallen seed, and the seed may become trapped in the hoof and be dispersed to a similar habitat in another location as foraging continues. If such mammals move out of their range under environmental pressure, they could disseminate seeds into new habitats. Wild horse ancestors may have spread seeds either internally or externally throughout the Eurasian steppes (see later in this chapter).

Human Impact on the Dispersal and Expanding Geographical Range of *Cannabis*

As man moves about the earth, consciously and unconsciously, he takes his own landscape with him. (ANDERSON 1967)

People are clearly the most ubiquitous agents of environmental disturbance, and our vast interference with the world's vegetation has wrought profound ecological changes. From prehistoric times, humans have disturbed mature vegetation formations previously under the exclusive control of nonhuman biological, climatic, and soil factors. By cutting, burning, and grazing, we have directly and indirectly transformed the face of the earth with ever-increasing speed. Traveling about the world, humans have been, and still are, dominant dispersal agents.

Herding of grazing animals and disturbance of natural herbivore communities vastly influenced the diffusion and establishment of many plants. Some of these plants, the so-called weeds, thrive in the wake of human disturbance. This is a key aspect in terms of where *Cannabis* fits into our model of early human contact and especially its subsequent spread with people during their trade and colonizing activities. Indeed, the means and routes of *Cannabis*'s dispersal are closely related to our cultural histories. Du Toit (1980) aptly noted that the great majority of the dispersal pattern of *Cannabis* is not natural: "In essence, hemp is a social plant; it is basically associated with human settlements." The human-*Cannabis* relationship is so ancient and strong that the dispersal of *Cannabis* was in many, if not most cases, directly associated with the spread of humans and the creation of their settlements. First we review the contemporary consensus regarding the early spread of humans out of Africa.

Recent studies indicate that anatomically modern humans (*Homo sapiens sapiens* or AMHs) dispersed from northeastern Africa about 60,000 BP and then began their spread across Eurasia and eventually the entire world. AMHs first reached Southeast Asia by about 45,000 years ago, Central Asia around 40,000 years ago, and Europe by 30,000 to 35,000 years ago (Carmichael 2007; also see Wells 2002; Stix 2008; Finlayson 2009; and Armitage et al. 2011). According to human genetic evidence initial dispersal into Eurasia spread north into Central Asia before moving eastward into East Asia and westward into Europe. This vast human dispersal entered Central Asia via the southwestern Asian corridor, "a wide geographical area that extends from Anatolia and the trans-Caucasus area through the Iranian plateau to the Indo-Gangetic plains of Pakistan and northwestern India." Today this region consists of "a patchwork of different physical anthropology types with complex boundaries and gradients and by the coexistence of several language families (e.g., Indo-European, Turkic, and Sino-Tibetan) as well as relict linguistic outliers" (Quintana-Murci et al. 2004). The southwestern Asian corridor, positioned at the intersection of key early human population expansions, was the first noncoastal area in Eurasia reached by modern humans venturing from northeastern Africa, and it was from this "corridor" that modern humans spread across Eurasia and to the rest of the inhabitable world (Quintana-Murci et al. 1999, 2004; also see Tishkoff et al. 1996; Watson et al. 1997; Tishkoff and Kidd 2004). Sometime after humans entered Central Asia they encountered *Cannabis* and a very long ethnobotanical relationship began. Human's ancient and diverse association with *Cannabis*, and vice versa, is the focus of the following chapters of this book. Here we address probable first human contacts with *Cannabis* and then focus on the evolving relationships of humans with the plants in this genus in Central Asia and beyond. This includes the associated spread of people and *Cannabis* throughout Eurasia and eventually the more or less worldwide human dispersals of *Cannabis*.

The genetic basis of the grand human dispersal hypothesis referred to earlier rests primarily upon study of nonrecombining portions of the human male Y chromosome, which has shown to be a very useful tool in understanding population history (e.g., Renfrew et al. 2000; Jobling and Tyler-Smith 2003). More specifically, it is recognized that the preservation of "extended haplotypes [genetic constitution of one of a pair of genes] characteristic of particular geographic regions, despite extensive admixture," lets us comprehend multifaceted demographic events. In fact, a recent study focusing on Central Asia used this approach in combination with a contextual interpretation of "Eurasian linguistic patterns" and revealed this region "to be an important reservoir of genetic diversity, and the source of at least three major waves of dispersal leading into Europe, the Americas, and India" (Wells et al. 2001; also see Chaix et al. 2008; Mellars 2006; Karafet et al. 1999; Santos et al. 1999). However, much needs to be sorted out regarding the genetic and cultural impact of later human migrations and their effects on *Cannabis*'s dispersal, especially since these subsequent movements of people in Eurasia were "overlaid upon previous population ranges" (Underhill et al. 2001; Quintana-Murci et al. 2004).

According to Wells and his colleagues (2001), the assumed age of a key Y chromosome haplotype (M45, dated to ca. 40,000 BP) matches logically with "the first appearance of anatomically modern humans and their toolkits in southern Siberia, during a period that saw the desertification of southern Central Asia and the disappearance of human remains from the southern Central Asian lowlands." Based on previous studies, these researchers suggested that a broad-based AMH dispersal from southern to northern Central Asia may have occurred during this later Pleistocene period, people following migrating herds of hoofed animals into the Eurasian steppes, and this early spread northward could have stimulated subsequent movements to the east and west, as partially supported by evidence of another Y chromosome haplotype: "The M173 haplotype is thought to delineate the earliest expansion into Europe, during the Upper Paleolithic ~30,000 years ago. It is likely that M173 arose initially in Central Asia, and that M173-carrying subpopulations migrated [actually dispersed] westward into Europe soon thereafter. The extremely high frequency of this haplotype in Western Europe is probably the result of [genetic] drift, consistent with an inferred population bottleneck during the Last Glacial Maximum" (Wells et al. 2001).

In any case, it appeared clear "that an ancient M45-containing population living in Central Asia was the source of much modern European and Native American Y-chromosome diversity" and that "the pattern of Y-chromosome diversity indicates strongly that Central Asia has played a critical role in human history" (Wells et al. 2001). More recent molecular biological research (Quintana-Murci et al. 2004) focusing on early human dispersal in Eurasia involved mitochondrial DNA (mtDNA), genetic material passed on only through the maternal line: "The phylogeographical cross-comparison of mtDNA and Y-chromosomal data is very useful for tracing differential male and female histories" and indicates a more complex human history with multiple dispersals going various directions out of, into, and perhaps through Central Asia. In any case, it is clear that Central Asia was a major corridor for human movement and the associated expansions of certain plants and animals. Assuming that *Cannabis* has long been a native of Central Asia, at least well back into the Pleistocene, humans probably began their long-term association with it somewhere within this huge region, utilizing one of more of its natural products. Human-related dispersals and subsequent range expansion of *Cannabis* is described in more detail later.

Even though evidence indicates human occupation in Central Asia was scattered sparsely over an enormous area, within a number of areas in this huge region, "settlement was continuous from the Middle Pleistocene to the Holocene"

(Madeyska 1990; also see Bolikhovskaya et al. 2006; Agadjanian and Serdyuk 2005; Agadjanian 2006). If humans first discovered *Cannabis* during early hunting and gathering forays into Central Asia, where the assumed natural origins of this genus are to be found, they could have carried the putative ancestors of modern *Cannabis* taxa to glacial refugia in both Europe and southern East Asia. Thus it is possible that even 30,000 years ago or more *Cannabis*'s biogeographical range and botanical evolution was already affected by human intervention, and this may have led to subsequent speciation (see Chapters 2 and 12 for discussions of the evolution of *Cannabis* species).

Here we assume that early AMHs first moved into Eurasia via a route northward and eastward through the southwestern Asian corridor referred to earlier. Some certainly entered through lowlands to the east of the Caspian Sea (in what is today part of Iran) and subsequently migrated north into Central Asia (e.g., present-day Turkmenistan, Uzbekistan, and Kazakhstan). On the other hand, other early peoples who migrated into Central Asia could have done so moving eastward from Europe via the East European Plains. For example, based on archeological and other evidence, Dolukhanov (2004) concluded that "the initial colonization of Northern Eurasia by anatomically modern humans occurred during the Last Ice Age and took the form of three consecutive waves stemming from Africa and Western Asia and consequently spreading from the west to the east." In any case, identification of the general location where early humans might have initially encountered *Cannabis* is not easy to determine, especially given our limited data base for the geographical range of humans and *Cannabis* during the climatically variable Pleistocene. Perhaps with increased use of molecular genetic evidence, better recognition of DNA differences between *Cannabis* and *Humulus*, and more archeological and archaeobotanical verification, we will be able to more closely identify this location and thus confirm or reject our contention that the ethnobotanical origins of human-*Cannabis* relationships began in Central Asia.

Earlier in this chapter we discussed some of the problems associated with identification of Cannabaceae pollen. Using ancient pollen remains as proxy evidence for the prehuman presence of hemp and/or hop in Central Asia can be seen as inconsistent and even questionable, depending on the effectiveness of the identification process. Nevertheless, there is intriguing pollen evidence indicating that Cannabaceae was present in the central part of Eurasia a very long time before humans arrived. For example, Granoszewski et al. (2005) reported "*Humulus/Cannabis*" pollen among the nonarboreal species recovered from Lake Baikal during the Last Interglacial period, ca. 120,000 to 130,000 years ago (see Tarasov et al. 2007; also see the following description for *Cannabis* pollen found in Lake Baikal dating to the Holocene). It is worth noting here that Soviet agricultural agents successfully cultivated hemp in the Trans-Baikal area in 1941, contrary to a common belief at that time that *Cannabis* "could not be raised under local conditions of climate and soil" (Mandel 1944), even though climatic conditions were likely quite different tens of thousands of years earlier.

Another recent study reported "*Cannabis*" pollen dated to well over 100,000 years ago (Molodkov and Bolikhovskaya 2006); in this case, pollen assumed to be hemp was recovered from soil deposits of a "loess–palaeosol series" found in the center of the East European Plain in what is today Russia. These deposits are associated with the third ("Late Moscow") interstadial warming period dating back to about 155,000 BP in northern Eurasia. Molodkov and Bolikhovskaya (2006) characterize the Pleistocene flora and vegetation associated with their excavated pollen assemblage as "periglacial birch woodlands with *Betula fruticosa* [dwarf cherry] in the shrub layer and the ground layer of *Arctous alpina* [alpine bearberry], *Cannabis*, *Artemisia* [mugworts], *Seriphidium* [daisies], *Thalictrum cf. alpinum* [meadow rue] and others." During the Pleistocene, and to a lesser extent in the Holocene, the climate changed dramatically over time, and thus the ecological distribution of plants, and biotic communities as a whole, expanded and contracted as well (e.g., see Bolikhovskaya and Molodkov 2006; Tarasov et al. 2005).

AMHs have lived in the Altai Mountains of southern Siberia since well back into the Paleolithic, and recent fossil and archeological research indicates that dynamic Pleistocene environmental changes probably had significant impact on hunting activities (e.g., Derevianko 2001; Agadjanian 2006; Bolikhovskaya et al. 2006; Agadjanian and Serdyuk. 2005; Scott et al. 2004; also see Kuzmin and Keates 2002 for critical comments about when, where, and which hominids were early occupants of Eurasia). Remarkably, evidence also indicates that Neanderthals (*Homo neanderthalensis*) lived in the Altai region many thousands of years ago. This conclusion is based on mitochondrial DNA extracted from skeletal remains of an adolescent Neanderthal found in Okladnikov Cave located in the Altai Mountains and dated to between 38,000 and 30,000 years old (Krause et al. 2007). If confirmed, this evidence indicates that the Neanderthals traveled much farther afield, perhaps more than 2,000 kilometers (1,250 miles) more to the east than previously thought. Whether these ancient Neanderthals or AMH groups moved from west to east or vice versa, the DNA from Okladnikov Cave matches closely with that of Neanderthals found in Belgium. Some scientists, however, remain skeptical that the skeletal remains from Okladnikov Cave really belong to Neanderthal (e.g., see Khamsi 2007). Obviously, because of the undoubtedly complex processes of human colonizations of the Old and New Worlds, a critical evaluation of this and other studies must be undertaken. Nevertheless, we assume that hominids (either Neanderthal or AMH groups or both) have been in several regions of Eurasia since up to 35,000 or more years ago and could have encountered and come to utilize *Cannabis* in one or more locations for a single or, more likely, multiple uses. In the intriguing case of Neanderthals, there is some evidence for their existence from about 400,000 years ago in Europe (from Swanscombe, England; see Stringer and Hublin 1999) and as much 150,000 years ago in Western Asia (Tabun, Israel; see Grun and Stringer 2000), although any conjecture regarding their use of *Cannabis* would be highly speculative, even provocative, at present.

Some additional paleoenvironmental studies based on *Cannabis* seed and/or pollen evidence from Eurasia are worthy of brief description here. The discovery of quite ancient *Cannabis* pollen has been reported in two studies of vegetation patterns in those Eurasian areas covered by the former Soviet Union and Mongolia. One study (Tarasov et al. 2000) focused on the LGM and the other (Tarasov et al. 1998) on the more recent, mid-Holocene period. In both studies, *Cannabis* is listed as being found within the "Steppe forb" (herbaceous open grasslands) group of flowering plants. Furthermore, in their mid-Holocene reconstruction, the authors argue that *Cannabis* can be seen realistically as a natural component in two of the broad-based biomes (regions of climatic similarity)

identified as the "Steppe" and "Warm Mixed Forest." A sediment core study from Lake Kutuzhekovo in the Minusinsk Depression, north of Tuva in the Central Asian, Russian Republic of Khakassia, carried out by Dirksen and van Geel (2004a/b), reported finding "*Cannabis ruderalis*" pollen dating from well before 4000 BP until the present, but not continuous over this whole time period. The land surrounding the lake was originally steppe, with forest-steppe (formerly more extensive) located in "up-slope" mountainous areas in this region. According to Dirksen and van Geel, the vegetation represented by the microfossils recovered from zone KTH-I (dated at 4310 ±195 BP and at the upper half of the zone) is typified by the dominance of pollen of arid-adapted shrub flora. These researchers referred to the extraordinary "synchronous fluctuations of *Artemisia*, Chenopodiaceae and *Cannabis ruderalis* pollen curves" and thus argued that these "xerophytic [arid adapted] components taken together prevail over the moist-demanding Cyperaceae and Poaceae indicating dry conditions during this time interval, when steppe and desert-steppe persisted at the depression" (Dirksen and van Geel 2004a/b; also see Dirksen 2000). These researchers classified *C. ruderalis* in their ancient vegetation reconstruction as belonging to the "upland herbs" group. In response to research by van Geel et al. (2004a/b) discussing the effects of climate change that occurred roughly 2,850 years and how this affected Scythian cultural history and migration, Riehl and Pustovoytov (2006) took issue with the classification of *Cannabis* as a xerophyte (a terrestrial plant adapted to dry conditions). Riehl and Pustovoytov argued that "*Cannabis* is a typical mesophyte" (a terrestrial plant adapted to neither particularly dry nor particularly wet conditions) and referred to it in Khakassia as a common ruderal plant (a plant species that invades disturbed areas). The authors of the original paper (van Geel et al. 2006) answered; they agreed that *Cannabis* is a characteristic mesophyte but explained why they chose to classify it as a xerophyte in their study. They argued that *Cannabis* in Khakassia is indeed a common ruderal but "climatically nonspecific." They reemphasized that it was the similarity of the pollen curves of *Cannabis* and the "xerophytic *Artemisia* and Chenopodiaceae" found in the lower levels of the excavated site in Lake Kutuzhekovo that led them to "consider *Cannabis* to be a xerophyte in this situation." However, van Geel and his colleagues (2006) also argued that the precise environmental adaptation of *Cannabis* in the past, at least in the Kutuzhekovo area, remains uncertain:

> Despite its modern preferences, *Cannabis* cannot be definitely regarded as a ruderal taxon for the lower part of the Kutuzhekovo Lake record, for the same reason as with *Artemisia*. There is no reference in the literature concerning natural plant communities with *Cannabis*. However, according to recent observations, *Cannabis ruderalis* grows in mature forest (with *Betula pendula*, *Populus nigra*, *Padus racemosa*, *Ribes hispidulum*) along the Abakan River without traces of anthropogenic disturbance (V.G. Dirksen, unpublished data). *Cannabis* is also present in natural communities in adjacent Tuva (E.A. Volkova, pers. com.). In the case of *Cannabis* pollen in the lower part of the Kutuzhekovo record, the question of its climatic significance indeed remains open.

Another more recent and relevant study was undertaken in the Buguldeika Saddle of Lake Baikal in southern Siberia, a region considerably east of Lake Kutuzhekovo (Tarasov et al. 2007; see their table 2); here, Holocene pollen extracted from sediments further supports the inclusion of *Cannabis* in the Eurasian steppe biome. Historically *Cannabis* was an important fiber, seed, and oil crop in large areas of Russia (Encyclopaedia Britannica 1911). In Russia today, hemp is still a significant crop in two large agricultural regions: the South Taiga, located roughly between about 50 and 60 degrees north latitude, and the Forest Steppe, situated in the center of Eurasia, south of the taiga zone (boreal forest), between roughly 43 and 50 degrees north latitude (Blagoveshchensky et al. 2002). During recent paleobotanical wetland research in the western Siberia region of Russia, Beilman (2008) observed *Cannabis* growing along rivers and ditches in several locations even at relatively high latitudes.

As noted previously, except for human and possibly avian or mammalian intervention, *Cannabis* has relatively poor seed dispersal, and this tends to reinforce the isolation of geographical varieties. Although bird and hoofed animal dispersals most likely played a significant role early on, people have been the most effective dispersal agent for *Cannabis* owing to their attraction to one or more of its diverse and useful products. Thus we assume that geographical diffusion diffusion of *Cannabis* from its probable center of species formation in Central Asia was to a large extent aided by human agency, both knowingly and unknowingly, over time. This could have included wandering nomads, friendly or militant migrants, traders on land and inland waterways, fleeing victims of natural and cultural calamities, and common or elite travelers.

Early Relationships among Humans and *Cannabis* in Central Asia

It is known to archeologists that Central Asia was an important center for the transmission of new discoveries and religious ideas from prehistoric times onwards. The hemp plant, being of major technological importance as a fibre [*sic*] and being one of the most influential psychoactive plants in human culture was most likely a key trade item from a very early date. (RUDGLEY 1998)

The evolutionary origin of *Cannabis* was somewhere in Eurasia, probably in Central Asia in the southern margins of the taiga or forest steppe regions (see Chapter 3). This location effectively put native *Cannabis* populations in position for early relationships with humans, especially in areas located near fresh water (see Chapter 1). These aquatic environments included streams, rivers, lakes, and other wetlands and possibly open areas associated with environments known as *tugia* in Central Asia, which are ecological communities associated with wetlands "consisting of mature woodland, thicket and meadow in river valleys" (Dolukhanov 1986).

We know that *Cannabis* is a sun-loving plant that thrives in open, nitrogen-rich environments, including rubbish piles created by humans (Anderson 1967; Merlin 1972). Close associations between humans and *Cannabis* stimulated its early cultivation, and over time this eventually led to its domestication. Although Blumler (1996) pointed out that the rubbish pile or dump-heap hypothesis for incipient cultivation has "limited validity" in explaining domestication among the majority of plants in most places (see also Abbo et al. 2005), he did concur with Anderson's early suggestion that *Cannabis* "is a natural for dump-heap domestication." In fact, Blumler tells us that hemp would have been among

those plants that "are most likely to attract the attention of hunter gatherers," which includes those that "produce edible vegetables, potential containers (e.g., gourds), spices or drugs—supplements and luxuries rather than staples." In addition, we suggest that plants that provided useful fibers to make cordage for ropes, nets, and even fishing lines, as well as cloth, were noticed, utilized, and even cared for with special attention early on, especially those found in areas disturbed naturally or by humans in and around camps (also see Gilligan 2007). Once humans had *Cannabis* relatively close at hand, they began a long-standing ecological relationship, purposefully or unwittingly contributing to its growth near their settlements. As an early, multipurpose plant resource, *Cannabis* could have provided food and oil from its seeds, fibers for cordage from its stalks, and psychoactive and physiologically active substances from the resin of its flowers (see Chapters 5 through 9 for detailed discussions of the ethnobotanical histories of these uses of *Cannabis*).

Although human-*Cannabis* relationships may not have begun until the end of the LGM or during the beginning of the Holocene, we think it more probable that this series of plant-people associations began in some environmentally favorable areas of Central Asia before the LGM. Adovasio et al. (2007) referred to the profusion of archeological sites in Europe and in the steppes of Russia that have led to many studies regarding the subject of "modern humanity," perhaps with an over emphasis on European perspectives. Consequently, in a recent trend, numerous scholars are focusing on regions east of Europe: "In fact, Russian archaeologists have recently turned up evidence in the Altai Mountain regions of Siberia as well as Central Asia of the early arrival of modern technologies as well—some time before 45,000 years ago" (Adovasio et al. 2007). However, this early date precedes the proposed dispersal of AMHs into Central Asia around 40,000 years ago determined by human genetic analyses.

It is in Central Asia, perhaps in the Altai Mountains that humans first encountered and eventually started using wild *Cannabis* in or near their early temporary settlements. *Cannabis* can be referred to as a "habitation weed" favored by high nutritive conditions in the soil that developed around these settlements, which to a large degree were unconsciously augmented by human waste and rubbish as well as manure from domesticated animals; as Hawkes (1969, citing Englebrecht 1916) has noted, "Such plants sought man out as much as he sought them out, because of their specific manorial requirements" (see Chapter 13 for a theoretical discussion of the reciprocal evolution of *Cannabis* and humans).

Hawkes (1969) refers to Sinskaja (1925), who studied *Cannabis* firsthand and observed "that weed hemp, just as cultivated hemp, required a very richly fertilized soil and that it was always to be found around the camps of the nomads in the Altai where the soil had been enriched by cattle during the winter, as well as in kitchen gardens and in rubbish heaps." Sinskaja focused on the broad diversity of spontaneous hemp, "which followed man's wanderings through the Old World," and hypothesized that people selected among this diversity during famine times for "forms with less shattering fruits and higher oil content." Sinskaja (1925, as quoted by Vavilov 1926) referred specifically to her observations in the Altai region, where "one could see all the details of hemp cultivation," and suggested four stages of a developing human-*Cannabis* relationship: "(1) The plant occurred only in the wild. (2) It spread from its original wild [centers] to populated places. (3) Hemp then began to be [utilized] by the population. (4) It was finally cultivated."

In the discussion that follows, we look at some possible associations between *Cannabis* and humans during the Upper Paleolithic. This is followed by an examination of late Pleistocene and early Holocene human-*Cannabis* relationships during more advanced, early stages of cultural development in Central Asia. Subsequently, we investigate how early ethnobotanical relationships helped spread *Cannabis* through much of Eurasia, from Central Asia to East Asia and Eastern Europe and later into South and Southeast Asia, Southwest Asia, the Mediterranean, and other areas of Western Europe. Later, we summarize the phases of *Cannabis*'s cultural dispersal within and beyond Eurasia and present a model for early use and diffusion of *Cannabis* by humans. "To some extent, the history of Eurasia as a whole from its beginnings to the present day can be viewed as the successive movements of Central Eurasians and Central Eurasian cultures into the periphery and of peripheral peoples and their cultures into Central Eurasia" (Beckwith 2009). Although *Cannabis* was most likely distributed naturally and/or as a cultivated crop across parts of Europe, Central Asia, and East Asia for millennia prior to the beginning of the Metal Age (ca. 4,000 BP), intensive cultivation and use of *Cannabis* by settled cultures likely began around this time.

Cord-marked pottery, some of which was most likely impressed with hemp fibers, has been discovered in pre–Bronze Age sites from the Beaker cultures of northern Europe dated to around 6,000 BP, across much of Eurasia, and as far to the east as Japan, involving the early Jōmon culture dated to around 7000 BP and perhaps earlier (e.g., see Kudo et al. 2009; Okazaki et al. 2011; Crawford 2011; also see Chapter 5 for a detailed discussion of the history and spread of fiber use). These early dates for fiber impressions are one of the indications that *Cannabis* may have been widespread across Eurasia by the beginning of the Holocene or earlier, long before nomad-pastoralists began to expand across the steppe with woolly varieties of sheep after 6000 BP (see Sherratt 1983; Barber 1991; Harris 1996; Chessa et al. 2009). La Barre (1980) suggested that *Cannabis* was already important in the Mesolithic period of Central Asia (ca. 10,000 BP) as part of a religious-shamanic complex "in parallel with an equally old shamanic use of soma, the hallucinogenic mushroom" celebrated in the ancient *Rg Veda* of India (see Chapter 7 for a discussion of the history of *Cannabis* use for ritualistic psychoactive purposes).

Fishing and Hemp

Many social anthropologists have made the point that agriculture cannot originate until man takes up a sedentary life, presumably engaged in fishing or restricted by the terrain of mountains from roaming for great distances. Since few cultivated plants and certainly none of the primary crops are derived from maritime plants it can therefore be assumed that the fishing was carried out in fresh-water rivers or lakes. (HAWKES 1969)

First we will focus on relationships between *Cannabis* and ancient fishing. Although the roles of various fishes in the diets of early Central Asian peoples are quite diverse, in some environmental situations, we can imagine, if not actually document, how early fishing activities in this huge continental realm may have involved *Cannabis*. We argue that human use of natural resources in streams, rivers, lakes, and other wetlands relied on fibers and even possibly seed and resin from *Cannabis* to attract, trap, catch, and collect aquatic organisms. We believe this began prior to the Mesolithic period when fishing became quite important in some parts of Eurasia.

What some have referred to as the "oldest net in the world" was found buried in the mud of an ancient lake in Karelia, a region with a lengthy history as part of Finland but now a region of Russia (e.g., Honkanen and Mäkelä 2005). Although this ancient net, which is 30 meters (about 100 feet) long and ca. 10,000 years old, is reportedly woven with willow fibers, we assume that true hemp fibers were also employed early on in Eurasia for twining cordage to make ropes, nets, traps, and fishing lines and even was used as a caulking material for simple canoes or rafts. Hemp nets do have a very lengthy history in northeastern Europe (Honkanen and Mäkelä 2005, also see Chapter 5), and *Cannabis* seeds may also have been useful in fishing. They have certainly been utilized as fish food and may have been used early on to attract fish (for contemporary use of hemp seed for fish bait, see, e.g., Wheildon 2000; Bailey 2001; Karus 2004).

The dietary contribution of fish for Upper Paleolithic hunters and gatherers, especially from anadromous fish such as salmon and eels that are born in fresh water, spend most of their life in the ocean, and return to fresh water to spawn, has had much archeological attention. A number of researchers agree that fish made up a much larger proportion of calories in ancient, Stone Age diets than previously thought. The low estimates stem from flawed archeological recovery methods and lower preservation rates for fish remains compared to those of mammals (e.g., see Casteel 1976; Jochim 1979; Marshack 1979; White 1980; also see Soffer 1985 for a critical discussion of this proposition as it applies to the Central Russian Plain). Fish were definitely a significant resource for these early people, especially seasonally where streams drained into lakes, oceans, and seas. Referring to natural resources of the Russian Far East (northward and eastward of Lake Baikal), Mandel (1944) describes the general scene: "Timber stood in an almost unbroken forest from the [southern Russian] border to the beginning of the arctic tundra, hundreds of miles to the north. Fish were so thick in the rivers and the coastal waters that gulls were, and are, known to stand on the backs of shoals of salmon while they pecked for their dinner. These resources did not have to be sought out. They were evident to the naked eye."

In the Altai Mountains of southern Siberia, early humans using stone tools were living in both open and forested environments at least by the Late Pleistocene. Unlike more northern latitudes of Eurasia where humans had to deal with extreme climatic fluctuations and related subsistence problems with availability of food, the Altai was covered with forest and forest-tundra. The region remained a key refuge for tree and other plant species, making the environmental situation for humans in the Altai Mountains during the early to middle Late Pleistocene reasonably good. The abundant mammalian and fish resources, as well as the less severe weather of the lower Altai mountains, allowed prehistoric people to settle permanently in these areas a very long time ago (Derevianko 2001; Derevianko et al. 2005). Early Altai hunters and gatherers exploited a wide variety of herbivorous and some carnivorous mammals as well as several types of birds. Larger mammals may have been more easily hunted in the colder months, while during the summer "people probably collected plants, small mammals, and perhaps fish"

(Derevianko et al. 2005). Here is another situation (summer time) in which *Cannabis* would be ripening and ready for fiber extraction and seed harvesting; both hemp fiber and seed could have been used to attract and/or catch fish.

Even though fish bones in some Altai sites may have been introduced by bears, substantial fish ribs, vertebrae, and scales, indicating fishing, were recovered from remarkably old Pleistocene cultural horizons in the Altai, such as the Mousterian layers (ca. 35,000 BP) of Ust-Kanskaya Cave. According to Derevianko et al. (2005), "The fish remains are of particular interest, also in view of studies which found that aquatic resources did not constitute an important part of the human diet in Europe until the mid-Upper Palaeolithic" (also see Richards et al. 2001). Overall, the extensive network of rivers in the Altai Mountains during the Late Pleistocene seems to have stimulated long-distance human dispersal, and in turn encouraged various types of human activity throughout the Middle and Upper Paleolithic. Consequently, land use changed more or less in the direction of "intensification and greater complexity in the Upper Palaeolithic," with people hunting hoofed mammals in winter and most likely foraging for plants and catching small mammals and fish in summer (Derevianko et al. 2005). Although *Cannabis* "grows wild throughout Russia today, the Altai region remains . . . a primary area of cultivation, albeit illicitly" (INCSR 2008).

At the end of the Pleistocene, ecological changes stimulated the human penetration into the southern Ural Mountains from the Caspian region to the west of the Altai Mountains. These included the unique Yangelskaya and Romanovsk-Ilmusrin cultures who fished in rivers and lakes, which, according to Matyushin (1986), "probably played a role in the development of more permanent settlement." Matyushin also pointed out that fishing among these people involved nets, some up to 45 meters (about 150 feet) long. Interestingly, he also suggested that Early Neolithic peoples in the southern Urals, living in a "mosaic of forest and steppe ecozones," exploited a variety of fauna including the horse, and this is where early horse breeding might have begun at least by the end of the seventh millennium BCE. Although this date is much earlier than most zooarcheologists accept, we propose that hemp cordage could have been useful in the development of ancient human-horse relationships, which are discussed in detail later. Indeed, we argue that hemp cordage has a remarkable ancient history of use in hunting horses and several other animals, including fish. Matyushin (1986) also referred to Mesolithic hunters and gatherers in ancient Central Asia who fished, such as those who utilized the site of Ust-Belaya on the Angara River west of Lake Baikal; here many fishing tools were found, including large sinkers, "suggesting the introduction of net-fishing." Again we propose that at least some of these nets could have been made of hemp. In Primorye (Maritime) Territory of the Russian Far East, fishing began at least by the time of the early transition to the Neolithic in the seventh through sixth millennia BCE, as revealed by the archeological recovery of many pebble fishnet sinkers. Here it appears that fish caught during their spawning were jerked, dried, smoked, and stored during winter (e.g., see Kuzmin 1995). The Primorye region also is a primary area of illicit cultivation of *Cannabis* in Russia today (INCSR 2008).

Even though fish was an important dietary component among some Paleolithic peoples, the significance of fishing increased markedly with the onset of the Mesolithic, when it became a key activity of many peoples in Eurasia, including those living in parts of Central Asia. This rise in fishing activity involved technological innovations such as dugout canoes and paddles, fishhooks, fish traps, nets, and lines. O'Connell and colleagues (2003) pointed out that fibers used to make fishing lines and nets must have been important among foragers living in the Eurasian steppe regions, especially in wetland environments. In fact, stable isotope analysis of human bones recovered from four steppe sites indicates that fish provided a significant proportion of food consumed by these incipient farmers. Peoples taking up fishing as part of their hunting and gathering economy were apt to find it easier to remain in places longer, often settling down near water to form semipermanent communities. This led to experimentation with plants growing nearby, perhaps in dump heaps (Sauer 1952, 1967, 1969; Anderson 1956, 1967), and in at least some areas of Central Asia wild hemp would have been available as a fiber resource (see Merlin 1972). Early fishing and farming peoples probably cultivated or at least selectively protected and harvested wild *Cannabis* plants, extracting the fibers to make some of their fishing gear as well as using the seeds as bait. The growth and spread of broad spectrum foraging, hunting, and fishing economies in the postglacial Mesolithic period laid foundations for domestication of plants and animals, and the rise of farming communities was stimulated by this integrated livelihood. In sum, early ethnobotanical relationships of humans and hemp in fishing activities most likely began in or near the Eurasian steppe and became important in the development of at least semipermanent settlements.

Dolukhanov (1986, citing Vinogradov 1981) specifically refers to the importance of fishing during the Holocene, as well as other forms of animal exploitation in the Neolithic economy (sixth to late third millennia BCE); for example, substantial fish remains were recovered from two sites on the Akcha-Darya Delta of the Aral Sea in western Central Asia, with a great majority belonging to pike (86 percent) and carp (9 percent). In addition, Dolukhanov (again citing Vinogradov) refers to evidence that hunting of waterfowl also played a major role in the economy of some ancient peoples in western Central Asia, and, in this case, nets and cordage were undoubtedly used. Evidence indicates that early farmers living in permanent or semipermanent settlements in delta regions such as those of the Amu-Darya and Akcha-Darya Rivers continued to rely heavily on hunting, fishing, and gathering. This would have favored human contact with and use of *Cannabis* for its fiber, edible seeds, and even its psychoactive resin.

Additional strong evidence for the importance of fish in the diet of third-millennium BCE Bronze Age peoples was recently reported from excavations of the *kurgan* living sites of three roughly contemporaneous cultural sequences that included the ancient Yamnaya, Early Catacomb, and East Manych Catacomb peoples of the northwestern Caspian steppe region (Shishlina et al. 2007; also see Gavriljuk 2005). Remarkably, these same people also utilized *Cannabis* for funeral rituals, and abundant *Cannabis* phytoliths have been found in bowls left in burials (discussed in more detail later). Evidence for fishing and relatively heavy consumption of fish comes from stable isotopes in human fossils, as well as fish remains in the same burial. For example, a pot recovered from one of these Bronze Age burials (kurgan 1, grave 5, of the Zunda-Tolga-2 burial ground) contained a mass of river or lake fish scales, which clearly indicated "that the vessel contained a fish soup, prepared using two or three species," perhaps pike, sturgeon, and carp, along with stream mollusks. Ancient communities were made up of small pastoral family groups that lived in varying environmental situations. As mobile herders who also relied on foraging, they utilized watersheds of small steppe rivers or plateaus during warm seasons, and then shifted during the cold seasons to a variety of regions including the deltas of Don and Volga Rivers, the Caspian to Black Sea steppe areas north and south of the Caucasus Mountains, the desert areas of the Black Lands (east part of the Caspian steppe), or the mountainous northern Caucasus region (Shishlina et al. 2007; also see Gavriljuk 2005). These migrating peoples moved themselves and their herds across large areas of the vast northwestern Caspian steppe within which they encountered a diversity of aquatic (or associated) ecosystems, most likely harvesting *Cannabis* along or near freshwater sources in the summer and autumn. This region is characterized by a variety of ecological niches including "numerous small steppe rivers, small and large steppe lakes, large valleys of the Lower Don and Volga Rivers, and large tributaries such as the Sal, and the East and West Manych Rivers where fishing, hunting and gathering and watering their domesticated animals could take place" (Shishlina et al. 2007). Throughout central and eastern Eurasia, fishing was an important food-procuring strategy for early peoples, and with the aid of both hemp fiber and seed, it formed an early aspect of human-*Cannabis* relationships.

Hemp, Humans, and Horses in Eurasia

We argue that eastern Eurasian subsistence economics are best understood not as a single "type" of production but as a productive process based on multiresource capacities (agro-pastoral, hunting, gathering, fishing) and the flexibility to readily adjust resource emphasis, degree of mobility, and specialization relative to a changeable environment. (HONEYCHURCH AND AMARTUVSHIN 2007)

The Central Asian steppes have long been characterized as marginal to ancient centers of civilization in Europe, the Near East, and China. In fact, the steppes were much more than an economically backward region with a long history often characterized as dominated by nomadic hordes of mounted or chariot-driving "barbarian" warriors wreaking havoc on their more "civilized" neighbors. Most importantly, Central Asia served as a setting for crucial communication routes and trade networks that eventually connected the civilization centers of Europe, the Near East, and the Far East. By the time of the Bronze Age, the vast and relatively homogeneous steppe region stretching across much of Eurasia with its numerous nomadic herding societies was facilitating crucial contact between more culturally advanced and settled cultures to the east and west. Far from being only a "one-way corridor leading from west to east," the Eurasian steppes served as a "bridge" across Central Asia, and following the opening of this long distance connection and the domestication of horses and their use for transportation, "the dynamics of historical development changed permanently, not just for the societies east of the Tian Shan [Mountains], but for all the peoples of Eurasia" (Anthony 1998). In the discussion that follows, we trace the crucial development of relationships between humans and horses in the central regions of Eurasia and introduce our hypotheses concerning the

FIGURE 14. Horses played a key role in the dispersal of both humans and *Cannabis* throughout Eurasia. The horse above is feeding on a large feral hemp plant in the Altai Mountains of eastern Central Asia (photo ©Shelly Benoit).

ethnobotanical associations that linked hemp with people and equids (horses) in this part of the world.

Up until about the fifth millennium BCE, almost no peoples living in the steppes had domesticated animals. They were foragers, hunters, and fishers, with their camps located almost entirely in riverine forest environments where terrestrial and aquatic game were plentiful, and useful plant resources were abundant, including, we assume, *Cannabis*. It is within these ecological conditions that humans were best able not only to fish but also to trap other animals such as horses, and here again we suggest that hemp ropes and nets played a significant role in these activities.

Drews (2004) explained why locations along moving or standing fresh water bodies were the best areas to hunt horses within the Eurasian steppe region, especially in the drier, more or less treeless zones where the horse is very well adapted:

Thanks to its large eyes, set at a diagonal on the front corners of the skull, the horse has a visual field of more than 300 degrees, and in the flat and open steppe was able to see a predator approaching from almost any direction. The dryness of steppe, much of which receives less than ten inches of rain annually, was in fact an advantage for the horses. Because of their ability to go for relatively long periods without water (even in summer, a horse on the steppe could manage with one drink a day), and to cover long distances quickly, horses were less dependent on nearby water sources than were cattle, pigs, sheep or other relatively slow animals. Horses could come to river to drink, and then quickly return into the steppe, where predators were few.

In large areas of the arid Eurasian steppes, such as Kazakhstan, Bactria, and Mongolia, only camels were better adapted to survive than horses. In addition, before domestication of the horse, the abundance of wild equids on the steppe was facilitated by the widespread absence of humans: "People found the steppe, except for its river valleys, difficult terrain in which to hunt, and too dry and too intractable for agriculture. As a result, until men learned to ride domesticated horses well enough to hunt on horseback, the wild horses of the steppe had little to fear from human hunters" (Drews 2004).

Anthony (2007) pointed out how numerous wild horses were in the Eurasian steppes and how commonly humans hunted these equids, which supplied the bulk of land-based meat in the diet of people living in this vast arid region: "The most efficient hunting method would have been to ambush horse bands in a ravine, and the easiest opportunity would have been when they came into the river valleys to drink or to find shelter." Early relationships between humans, hemp, and horses in the Eurasian steppes may have furthered the use and spread of *Cannabis* within the region and beyond. If so, where and when was the horse domesticated, and did hemp fiber really play a role?

After the end of the last Pleistocene ice age, during the beginning of the Holocene, climatic warming in Eurasia allowed forests to replace grasslands; consequently, horses lost much of their preferred habitat. In Europe, for example, horses disappeared from many areas including the British Isles, France, and Spain, perhaps due to some degree to overhunting as well as environmental change. Eventually, horses became more or less restricted to the steppes of present-day Ukraine and neighboring Central Asia. Some researchers believe that horses would have disappeared entirely if it were not for domestication. In fact, some presume that all domestic equids today are descendent from those domesticated north of the Black and Caspian Seas about 6,000 years ago (e.g., Anthony 2007); it is from this western steppe region that *Cannabis* spread further west into Europe and perhaps south into the Near East.

During the latter part of the Pleistocene, people hunted Eurasian wild horses (*Equus ferus*) in many areas of Eurasia for their meat, skin, and bones, "certainly by Neanderthal (Middle Paleolithic) times, but probably even during *Homo erectus* times" (Olsen, personal communication 2012; also see Olsen 1996; Clottes 1996). A number of magnificent, very ancient depictions of equids in cave paintings have been found in Eurasia, principally in France and Spain. These bear witness to ancient interactions between AMH people and wild horses that took place for thousands of years over widespread areas of the northern hemisphere (e.g., at Les Eyzies de Tayac in France; Merlin, personal observation, 2007).

As human population increased during the latter part of the Pleistocene, meat and hides of wild horses remained important, but in some areas these resources were progressively dwindling. The fossil record tells us that wild, native horses became extinct in North and South America sometime near the end of the last ice age ca. 10,000 BP (MacFadden 1994; Meltzer 2009); in Europe their range was undergoing a slow but steady decline by about 7000 BP. However, horses continued to persist in northern Germany, Denmark, and across the Eurasian steppe (Olsen 1996). Indeed, horses may never have become extinct in parts of Europe; archeologists just may not yet have recognized early remains in many areas. It is possible that horses were more widespread than we might assume, but perhaps they did not occur in very large numbers or were hunted infrequently, and, if so, in some areas truly wild horses may have remained widespread during the Holocene. This probably included the Eurasian steppe region as well as throughout some very large areas of northern and eastern Europe but perhaps not "in Turkmenistan and other areas generally considered [to be part of] Central Asia" (Olsen, personal communication 2012).

Approximately 7,000 to 6,000 years ago, peoples began to focus their attention on transitional zones between the forested and steppe areas in central Eurasia. With the growth of human population in this region, more food was needed. This may have stimulated adoption of plant cultivation and herding of cattle and sheep, which were definitely domesticated prior to horses. Nevertheless, with the decline in wild populations of pigs, deer, and aurochs (the larger ancestors of domesticated cattle), hunting of horses increased. This probably stimulated early horse herding and progress toward horse domestication, and as suggested previously, in the process of hunting as well as herding, people probably used hemp ropes and nets (e.g., see Adovasio et al. 2007 for a discussion of the importance of nets in the cultural evolution of humans before and in the early stages of farming).

Captive horses are not the most efficient animals for meat production because they have lengthy gestation periods (about 11 months) and infrequently generate more than one foal per pregnancy. Breeding and rearing horses for meat therefore would have been troublesome. On the other hand, horses can graze herbaceous vegetation at times when cattle and sheep have difficulties obtaining enough food. For example, horses can use their hard hooves to break through snow crusted with ice to feed on the grass below, whereas cattle and sheep, using their noses, can only push fresh snow out of the way. Therefore, horses could have been a valuable source of food during cold winter months in a wider range of climates, and this probably stimulated development of horse herding and eventually initial domestication of horses (see Anthony 2007).

As the Holocene Epoch began in western Asia, humans began to develop profound new methods of obtaining food through cultivation and eventual domestication of edible plants, such as certain cereals and legumes, and herding and breeding of selected animals, such as goats, sheep, pigs, and cattle. This series of significant changes in the way humans manipulated the environment and its components began about 10,000 years ago or more; however, horses were not domesticated until much more recently. Some have argued that this domestication occurred during the sixth millennium BP and that the oldest excavated remains indicate significant effects of artificial selection and consequent genetic manipulation in the western steppe region of Eurasia, especially the Ukraine (e.g., Anthony 1986, 1991, 1998; also see Anthony and Brown 1991), while others have dismissed these early dates. Levine (1999a/b) pointed out that we really do not know where the horse was first domesticated. Thus the notion that it occurred in the western Eurasian steppe remains as only one theory. Citing many sources, Levine (2002) argued that the earliest undeniable dates for "textual and artistic evidence for horse domestication probably only dates back to the end of the third millennium BCE." She further indicated that direct evidence is even more recent since "horses in graves, accompanied by artifacts unambiguously associated with riding or traction" only date back to the early second millennium BCE. On the other hand, more recent indirect geochemical research results from Bronze Age sites in Kazakhstan indicate corralling if not domestication of horses about 5,600 years ago (Outram et al. 2009; also see Olsen 2006a/b/c and Lovett 2006). Utilizing mitochondrial DNA evidence, Vilà et al. (2001) and Jansen et al. (2002) provided insight into the geographical range of horse domestication. Their research supports the hypotheses that domesticated horse lineages have widespread origins. Rather than springing from one local population, domesticated horses appear to have arisen from wild stock of several distinct populations distributed over a relatively large area such as the Eurasian steppes. In any case, the domestication of horses probably occurred in or at the edge of the Eurasian steppe because horses feed on grasses or other low-lying herbaceous species typically found in this vast region; however, we do not know for sure how far east or west in the steppe region this took place. According to Anthony (2007), the lengthy history of human hunting of wild horses in the Eurasian steppes "created a familiarity with their habits that would later make the domestication of the horse possible." He argues that horses were probably first ridden in the Pontic-Caspian steppes at least by 3700 BCE but perhaps earlier in the fifth millennium BCE. The Pontic steppes stretch from north of the Black Sea as far east as the Caspian Sea, extending today from eastern Romania across southern Moldova, Ukraine, Russia, and northwestern Kazakhstan to the Ural Mountains. Anthony tells us that horse riding "spread outside the Pontic-Caspian steppes between 3700 and 3000 BCE, as shown by [an] increase in horse bones in southeastern Europe, central Europe, the Caucasus, and northern Kazakhstan." In his recent book, Anthony (2007) provides an in-depth, comprehensive discussion based on historical linguistics and archeology of his views about the early relationships of horses and humans, as well as how horses, wheels, and language utilized by Bronze-Age riders from the Eurasian steppes played an important role in shaping the modern world. He asserts that the "mother tongue" (i.e., the Proto-Indo-European language) evolved over time into the extensive and far-flung Indo-European language family that spread rapidly several millennia ago across Eurasia from its origin in the Pontic steppes. Although a reliable winter food source may have stimulated early herding and eventually domestication of horses, once humans were able to breed relatively docile horses and use them for fast-mounted transport across the vast steppes, interregional cultural relationships increased significantly and likely initiated or at least intensified the rapid spread of Cannabis and its use. Slow walking dispersal of early modern humans into new environments suddenly changed with the discovery of the horse. Movement quickened and isolated cultures that had diverged ages before were reintroduced across the steppe lands.

The great Eurasian steppe is a huge continental region with an irregular character. The European steppe is an almost continuous, monotonous belt of varying breadth, while the Asian steppe territory contains considerable environmental diversity including steppe islands, forest-steppes, and semi-deserts along the foothills of Kazakhstan, the southern Ural Mountains of western Siberia, and the Altai Mountains, as well as the Tuva (Uyuk) and Minusinsk hollows in southern Siberia (Zaitseva and van Geel 2004 and Zaitseva et al. 2004). The notion that Cannabis had a connection with early horse exploitation and eventual domestication of horses is admittedly speculative. However, this idea is supported by the fact that hemp has a propensity to grow around Central Asian nomad camps, even today, especially where there is nutrient-rich soil (Olsen, personal communication 2012); thus early Central Asian people could have encountered Cannabis just where the early horse-tamers would have found it when they needed fiber to make ropes and nets.

In a recent Science article by Outram et al. (2009), one of the three independent lines of evidence indicating domestication occurred among the "Eneolithic Botai Culture of Kazakhstan, dating to about 3500 B.C.E.," and suggested "that some Botai horses were bridled, perhaps ridden." If bridles were used in these early times, what materials were used to make them? Did Cannabis really have a role in the domestication

of the horse? Trapping stallions with nets and ropes, perhaps made of hemp, would have allowed humans to come closer and make more intimate contact with the stallions' mares. This eventually may have encouraged herding and selective breeding. Since horses cannot be controlled in any numbers without a mounted herdsman, it has been suggested that horses were ridden from the beginnings of domestication (Clutton-Brock 1992, citing Anthony 1991). More recently, Anthony (2007) asserted that "horseback riding began in the steppes long before chariots were invented, in spite of the fact that chariotry preceded cavalry in the warfare of the organized states and kingdoms of the ancient world." Whenever and wherever riding began in the Eurasian steppes, early riders may not have depended upon harnesses to control their horses as most modern riders do. However, if they did, what materials were used to make these early harnesses? Leather and horse tail hair were important resources for early equine tack or gear (for reference to evidence of leather thong use, see Olsen 2006a/b/c and Lovett 2006); however, hemp fiber also could have been utilized, if not in the earliest sites indicating horse herding and domestication, then subsequently as fiber arts developed in association with Cannabis cordage. For example, intricate, Eastern European, nineteenth-century hemp harnesses are in the possession of the ethnobotanical collections of the Center for Economic Botany at the Royal Botanical Gardens in London (Merlin, personal observation, 1985); although these are relatively recent in age, they represent an ancient use and technology. There are also many early bridles from Central Asian nations that are known to have been made from plaited horse mane and tail hair, as well as leather, which would have been readily available from slaughtered horses. Oddly enough, a unique and perhaps relevant recent discovery has been made in eastern Central Asia involving ancient plant fibers used to represent a horse tail on a 36 cm (15 inch)-tall ceramic figurine. The model depicts a well-dressed rider upon a horse with Cannabis identified as the fiber source for the tail. The artifact, dated to 658 CE, was found in an Astana cemetery grave in the eastern region of Central Asia, now part of Xinjiang province in western China (Hongen Jiang, personal communication 2013).

In addition to the early use of hemp fibers to make ropes, nets, and other cordage products, at some point, people began to use these fibers to weave cloth for a variety of purposes. Barber (1991) suggested that linguistic data and some archeological evidence indicate strongly that Cannabis fibers (and perhaps seeds for food) were known and used by people from Europe to northern Asia, during the Neolithic. As we previously suggested, one or more products derived from Cannabis may have been utilized well before the development of agriculture in the Upper Paleolithic period. However, successful spinning of hemp and other bast fibers relies on long, smooth, and parallel fiber bundles, and due to their branched form wild Cannabis plants produce only short sections of coarse bark. It was only later in settled communities that the advent of hemp cultivation and artificial selection to promote long, straight, and uniform fibers allowed the widespread weaving of hemp cloth.

According to Olsen (2012, personal communication; also see Olsen and Harding 2005) evidence from the Botai culture (ca. 3500 BCE) in ancient northern Kazakhstan indicates that hemp may have been used to make simple cloth and two-ply, S-twist cordage impressions in Copper-Age pottery. Olsen and Harding (2005) reported that female figurines carved from horse phalanx bones associated with the cloth and cordage impressions are incised with a diversity of designs representing ancient feminine clothing of the time. Some features on the figurines signify the garment configuration, while others reveal decorative elements. As part of their investigative research, Olsen and Harding fabricated a hemp garment to duplicate "a Botai woman's dress and computer images were made to reconstruct several of the dress designs based on the incised phalanges." They chose hemp "because it grows all over the place now and is the best fiber for making cordage in that region" (Olsen, 2012 personal communication).

During the Pleistocene and Holocene Epochs, cultural diffusion of *Cannabis* within and out of Central Asia was definitely associated with movements of people. Even early on from 8000 to 5000 BP social contact and acculturation generated by successive, outward-radiating dispersals of nomadic and largely pastoral groups likely intensified the spread and use of hemp. Sherratt (1981, 1983) pointed out that a "secondary products revolution" fostered a change in strategies for using plant and animal resources, which occurred when humans began to understand they could obtain more than one hide and one feast from a single animal (also see Barber 1991). If they kept their domesticated animals alive rather than slaughtering them, they could glean a steady supply of food (milk, yoghurt, and cheese), clothing (wool), and even labor energy (pulling plows, wagons, and sleds). Barber suggests that this "total revolution" in uses of domesticated livestock (deducible from the slaughter and lifestyle patterns seen in animal bones) occurred from about 4000 BCE in the Near East and quickly spread in all directions, including northward into the steppes. This included the introduction of woolly sheep, which provided people with "grass-eating machines of a specific mode that could provide them with enough food and clothing to make a go of living in the deep steppe." Using horses to carry materials and control the herds of "sheep (and whatever else) efficiently was what finally made it *possible* to make a living in the steppe as nomadic or transhumant herders." If the early dates for *Cannabis* use in Europe and China (perhaps a millennium before 4000 BCE or earlier) are valid, then people were probably spreading *Cannabis* before the emergence of the nomad-pastoralists in Eurasia (see later for a discussion of the origin and spread of *Cannabis* in China and other regions of East Asia). Nevertheless, the rise of horse-riding, nomad-pastoralists seems to have played a significant role in further spreading the use of *Cannabis*.

Horse domestication allowed for greater systematic utilization of steppe resources by larger groups of people. Horse riding and driving facilitated a significant reduction in transportation time between more diverse, concentrated resources such as those found in river valley watersheds separated by large expanses of relatively uniform grasslands; horse riding also increased the territorial ranges of resource allocation. Furthermore, the use of horses also accelerated trade and aggressive tribal behavior. Not incidentally, this helped disperse important culture traits such as feast and funeral rituals that included *Cannabis*—not only for its fiber and seed but also for its psychoactive resin, which we discuss later (also see Sherratt 1991, 1995b).

By the Bronze Age, horses were spreading rapidly across Eurasia for purposes of food, transportation, farm work, and warfare. Perhaps as early as 3500 to 3000 BCE, and definitely during the period between 2500 and 2000 BCE, human reliance upon domesticated horses extended across much of Eurasia, and *Cannabis* was spread consciously and perhaps unwittingly with the movements and activities of ancient Central

Asian herders as well as farmers. In some cases, *Cannabis* and its uses spread into areas such as China where it was already well known and into other areas such as Western Europe and perhaps the Near East where it previously was not known.

Sometime after cereal farming culture entered the steppes north of the Black Sea, early farmers added at least one new crop from Central or East Asia to their introduced complex of cultivated plants—*Cannabis*, grown for seed, fiber, and/or resin. Harlan (1992) viewed the steppes as marginal to the innovations of farming and as an unlikely place for cultivation of plants to have begun; however, he believed that a small number of domesticated plants could have had their origins in the Eurasian steppes (also see Harlan 1975). Ancient evidence of hemp seed use in the steppes, dating back more than 6,000 years, was found in the Dniester-Prut region north of the Black Sea in what is now part of Moldova located between Romania and the Ukraine. This ancient *Cannabis* seed evidence includes six potshard fragments bearing a total of nine "*Cannabis sativa*" seed imprints recovered from the fifth-millennium BCE site of Dantcheny I and are associated with the early Neolithic Linearbandkeramik (LBK) farming culture (Yanushevich 1989). There is also recent indication that ancient seed evidence of "*Cannabis* sp." has been recovered from the "cultivated plant remains . . . found in Moldova sites of the Sabatinovka culture" (Pashkevych 2012), west of Black Sea coastal steppe region and dated to the Bronze Age in the middle of third millennium BCE (Kuzminova and Petrenko 1989). Although hemp seed may have been used as food or as an oil source, other plant parts may have been utilized for ritual purposes by the Neolithic period. The steppes are among the regions of central Eurasia where *Cannabis* was probably growing wild and/or cultivated, at least by the Neolithic period, if not long before that time.

A recent paleogeographic study in the northeastern area of the Black Sea coast uncovered pollen of both *Humulus* and *Cannabis* (Bolikhovskaya et al. 2004). The pollen was extracted from a sediment core excavated on the Taman Peninsula within the delta region of the Kuban River and dated from ca. 6000 to 1000 BP. The fact that this evidence, albeit pollen of both *Humulus* and *Cannabis*, was recovered in the lowlands adjacent to river tributaries is noteworthy since it fits with the idea that early temperate zone farming was a kind of floodplain horticulture, as opposed to the once-prevailing model of shifting cultivation. Zernitskaya and Mikhailov (2009) reported the recovery of "*Cannabis*" pollen and that of other "specific weeds" along with forest clearance indicators such as a high frequency of charcoal particles associated with cereal farming dated to more than 4000 BP from sediments in Lake Neropla in Belarus (located in eastern Belarus, on the interfluve between the Drut and Dnieper Rivers). Much more recent evidence for the presence of *Cannabis* in northern Ukraine comes from Ovruch Ridge where ancient pollen was tentatively attributed to "*C. sativa*" and "*C. ruderalis*" were recovered and dated to the "Early Middle Ages" (Bezusko et al. 2009); although this much more recent evidence may not contribute greatly to the very ancient presence of *Cannabis* in this region, it does suggest that its use has been occurring here over a number of millennia.

Evidence for existence of *Cannabis* in the western Eurasian steppes region thousands of years ago is relevant because it has been suggested by a number of authors that hemp and its use spread west from this area into eastern Europe and perhaps east and south through other areas of Eurasia during the later Neolithic or soon after the development of metallurgy.

For example, de Candolle (1967) and Schultes (1970) suggested that *Cannabis* was spread west out of what is now southern Russia by nomadic herders (e.g., Scythians) and Forbes (1964) proposed that *Cannabis* arrived in "prehistoric Europe from southern Russia, as is also evident from the etymology of the terms for hemp in Indo-Germanic languages" (see Anthony 2007 for a detailed discussion of the introduction of Indo-European languages into Europe by Bronze Age nomads; also see Barber 1991; Beckwith 2009). *Cannabis* was already present in parts of Europe by Scythian times (ca. 2500 BP) and by then was widely used in East Asia. However, it is possible that Scythians introduced the psychoactive use of *Cannabis* to Eastern Europe and possibly even brought in more potent varieties from Central or East Asia (see the following text).

In any case, *Cannabis* use was well established in Eurasia by the time of the Iron Age (eighth century BCE in Europe), ranging from Western Europe through the steppes all the way to China and beyond to Japan. There is well-known historical evidence for Scythian (in the steppes) and the related Pazyryk (in the Altai Mountains) cultural use of *Cannabis* dating to the first millennium BCE (see Herodotus, Book VI: 74 for literary evidence regarding the Scythians and Rudenko 1970 for archeological evidence concerning the Pazyryk culture). However, Sherratt (1987, 1991, 1997) suggested that peoples occupying the Eurasian steppes possessed and used *Cannabis* long before the Scythian nomadic tribes rose to power and hypothesized that *Cannabis* was available and utilized in the steppes prior to the third millennium BCE, when it was introduced into Europe; Sherratt (1997) also suggested that the use of *Cannabis*, at least for ritualistic if not utilitarian purposes, more or less disappeared in Europe during the Bronze Age "and only became common again under Scythian influence" in the Iron Age.

According to Sherratt's interpretation, early inhabitants of the Eurasian steppes, such as people belonging to the Sredni Stog culture, which flourished from about 4300 to 3500 BCE, used *Cannabis* to make a "socially approved intoxicant," celebrating its significance "by imprinting it on their pottery." Sherratt argued that the ingestion of *Cannabis* was a fundamental aspect of Eastern European mysticism (see Chapter 7 for a detailed discussion of ancient psychoactive *Cannabis* use in ritual and other contexts and Chapter 5 for a discussion of *Cannabis* and cord-marked pottery remains). We believe that *Cannabis*, along with cord-marked pottery and domesticated horses, were dispersed together as parts of a general cultural complex that developed in the Eurasian steppes during the Copper Age and Early Bronze Age (ca. 6000 to 5000 BP).

In the northern Caucasus Mountains, southeast of the western Eurasian steppes, sometime during the fourth millennium BCE, high-ranking chiefs rapidly ascended to power from what had been small-scale farming cultures (Anthony 2007). This rise of ostentatious leaders was associated with lucrative long distance trading that took place from about 3700 to 3500 BCE between members of the Maikop culture located in the northern Caucasus foothill region and people in advanced cities of Mesopotamia via middlemen in Anatolia. It has been argued that the Maikop culture was the filter through which innovations from southern areas in the Near East entered areas to the northeast in the Eurasian steppe, and these possibly included wagons and certainly incorporated innovative metal alloys that stimulated a more advanced metallurgy. Uncertainty exists regarding what items were traded from north in the steppes to areas south in the Caucasus and possibly further on into more southerly regions of Southwest Asia or the Near East.

Based only on circumstantial evidence, Anthony (2007) suggests that some of these exchange items could have included wool, horses, antelope hides, and even *Cannabis*. Indeed, Sherratt (1997b, 2003) proposed that psychoactive *Cannabis* specifically was among the more significant exports coming out of the steppes during this period.

In the late Neolithic or Early Bronze Age (ca. 3500 to 2300 BCE), people associated with what archeologists refer to as the "Yamnaya Horizon," essentially a pastoralist *kurgan* culture, developed out of eastern origins in the steppes of the Don and Volga River regions. The herding people or societies of this culture were most likely speakers of "classic Proto-Indo-European" and were the first in the Eurasian steppes to generate a pastoral economy requiring regular seasonal migrations to fresh grazing land. They used wagons pulled by cattle to carry their tents and supplies far into the steppes of Central Asia when necessary to graze their animals. They also used horses to survey huge amounts of territory and to drive their large herds of domesticated animals (Anthony 2007). Beginning about 3100 BCE, people associated with the Yamnaya herding culture spread swiftly across the steppes carrying *Cannabis* and its use with them, eventually broadening their range to include areas to the west in the Danube Valley and then into other areas of Eastern Europe including Serbia and Hungary, where they encountered settled farmers.

Even if the idea that the Yamnaya peoples were using hemp for fiber is speculative at this point, there is evidence that they and other contemporary peoples in the region used *Cannabis* for ceremonial purposes. Indeed, a major characteristic of the Yamnaya cultural horizon was their funeral ritual and how this is manifested in their *kurgans*. For example, in Eastern Europe there are two sites that have yielded hemp seeds more than 4,000 years old. One is a grave at Gurbanesti, east of Bucharest in the Danube Valley region of Romania where a clay vessel (brazier or "pipe-cup") with carbonized hemp seeds was discovered, perhaps the earliest evidence for the burning of *Cannabis* (Ecsedy 1979). The second site where Early Bronze Age seeds of *Cannabis* have been found is located in the northern Caucasus region where a similar "smoking vessel" with charred hemp seeds was discovered in a burial (Sherratt 1991, citing personal communication with István Ecsedy; also see Ecsedy 1979). Sherratt pointed out that this kind of vessel used for burning plant material, which is frequently found in tombs, is not "tomb pottery." In other words, such vessels were not produced specifically for burials, but "occur as *offerings*, usually in the forecourt, and should thus be seen as ritual food containers in the context of a feast that included the dead" (Sherratt 1997, author's italics). Hemp seeds are not psychoactive. However, they are the most heat-resistant part of the plant, and the two discoveries of braziers with charred hemp seeds referred to previously suggests that the inflorescences and leaves with their resin had burnt away. According to Sherratt (1991), these charred seeds are the earliest evidence for intentional burning of *Cannabis* and suggest ritualistic, perhaps psychoactive use. Sherratt (1995b) proposed that smoking or inhaling *Cannabis* fumes was introduced into the Danube Valley by immigrants of the Yamnaya culture dating back approximately 5,000 years (also see Anthony 2007). Indeed the relationship between *Cannabis* and the Yamnaya culture is important in understanding major aspects of the origin and spread of *Cannabis* into and throughout much of Europe.

In addition to these findings, many other relatively small, decorated pottery objects called polypod bowls (small dishes on three or four feet), which originate in the southwestern steppes during the period from about 4000 to 3600 BCE (Sherratt 2003), are also believed to be braziers for smoke inhalation. These kinds of vessels have been found in several parts of eastern Europe dating at least from the early third millennium BCE and are associated with the rise of the Pit-Grave culture on the steppes where the earliest of these bowls have been found; here they are also coupled with the westward penetration of the Yamnaya culture into Romania, Bulgaria, and Hungary. Examples of these braziers have thus far been found "in the Carpathian Basin and then in Czechoslovakia and southern Germany somewhat later, indicating that this type of pottery spread from east to west" (Rudgley 1995). These artifacts prompted Sherratt to suggest that Bronze Age temperate Eurasia, resembling that of native North America (see von Gernet 1995), was fundamentally a broad region of smoking cultures rather than drinking cultures and that "this agrees with the botanical data for a steppe distribution of *Cannabis* . . . and its spread from this region as a cultivated plant into Europe and China" (Sherratt 1997).

Another interesting if controversial discovery of ancient traces of *Cannabis* in ritual vessels was made in the ancient Bronze Age fortresses of the Bactria Margiana Archeological Complex (BMAC) in southern Central Asia. These walled settlements are found in the Merv Oasis, which is located within the delta of the Murghab River in what today is southeastern Turkmenistan; this oasis is thus located within the broad region east and south of the Caspian Sea, which Emboden (1972) referred to as the origin area of "*Cannabis sativa*." It was in this region that the Russian archeologist Sarianidi (1994, 1998, 2003) studied the temples of Bronze Age Margiana and the remains located within them. Sarianidi (2003) alledgedly found seeds of *Cannabis* (mixed with those of *Ephedra* and the opium poppy, *Papaver somniferum*) in "fire-temples" within walled structures at Togolok 21, Togolok 1, and Gonur dating to about 4000 BP; he suggested that the use of *Cannabis* and other drug plants could be compared with the roles of sacred, psychoactive *soma* of the *Rg Veda* or *haoma* of the *Avesta* (see also Meyer-Melikyan and Avetov 1998). However, later examinations by a series of European paleobotanists of the seed and stem impressions recovered from the "white rooms" at Gonur and Togolok 21 resulted in a tentative revision of the identification of the seeds in question. These scientists claimed that the vessels contained seeds and stems of broomcorn millet (*Panicum miliaceum*), not *Cannabis* or *Ephedra* (Bakels 2003; Merlin 2003; Anthony 2007; also see Hiebert 1994 and Parpola 1998 for other dissenting views). However, not everyone has dismissed the identification of these BMAC seeds as *Cannabis*. Russo (2007) referred to the drug plant remains from the BMAC culture as the "actual physical remnants of *Cannabis* [*sic*] flowers and seeds, along with opium poppies and ephedra," and pointed out as did Sarianidi (2003) that the examination by Bakels and other archaeobotanists occurred "after several years of exposure of the material to the elements." Although Bakels determined that the residual seed impressions in gypsum were too small to be *Cannabis*, Russo compared the average seed length of the feral *Cannabis* seeds from Kashmir, which "average 2.2 mm [0.09 inch] in length," with those of broomcorn millet, which "average 2.8 mm [0.11 inch] in diameter." The photo in the *Electronic Journal of Vedic Studies* is not definitively *Cannabis* and can just as easily be millet. Thus these putative 4,000-year-old remains of *Cannabis* from southern Central Asia are still contentious.

Sherratt (2003) forged a hypothetical connection between the contemporary rise of the polypod braziers in the southwestern steppes and the development of a cord-impressed style of pottery ornamentation new to the region at that time. He suggested that while the clay of the braziers was still moist during their production by potters, cordage was wrapped around them in order to make a patterned impression. Sherratt proposed that markings were a celebration of the contents, with hemp cord decoration indicating the botanical source of the material to be burnt and inhaled. Sherratt (1995b) also pointed out that the use of cord-impressions to decorate pottery is characteristic both of China and of the Eurasian steppe region from the fourth millennium BCE onward, and indeed cord-impressed pottery has an apparently even older history in the ancient Jōmon ("cord-impression") culture of Japan, which was essentially a hunting and gathering culture but one with evidence of small-scale cultivation (or management) of a few plants, including *Cannabis* (see Crawford 2011 and the section on Japan in this chapter).

In our earlier discussion regarding prehistoric relationships between fishing and the use of *Cannabis*, we referred to Bronze Age *kurgans* (pit-grave mounds) located in the northwestern Caspian steppes where remarkable evidence was uncovered for *Cannabis* and its connection with funeral ritual among a people who relied heavily on fish in their diet. The evidence for *Cannabis* from these same Bronze Age *kurgans* was found in the form of phytoliths recovered from many "vessels" in the burials of the Catacomb culture populations. According to Shishlina et al. (2007, also see Gavriljuk 2005), the people of these cultures consumed "common drug plants like wormwood (*Artemisia*) and strong drug plants such as hemp (*Cannabis*)." Karlene Jones-Bley (2007) hypothesized that vessels of a particular type "unique to the Middle Bronze Age Catacomb culture of the Eurasian Steppe area" often contained "evidence of burned *Cannabis*" and were "undoubtedly ritual vessels due to their distinctive shape, which is quite uniform and usually includes a small internal section."

During the Bronze Age, pastoral migrations also occurred seasonally across the northern Caucasus foothills through which flow large rivers such as the Kuban, Kuma, Kalaus, and Egorly. In addition to the use of aquatic resources found in this large area, a new, extensive trade and exchange network developed during the third millennium BCE. This Bronze Age trade led to the appearance of "exotic imported items" in the steppe region. These imported goods included metal objects such as tools, weapons, and ornaments (some made of silver and gold), along with textiles; in return, the peoples of the steppe traded their stock of domesticated animals and craft products, perhaps explaining "why they had to exploit all the food resources of their territory, including aquatic resources" (Shishlina et al. 2007; also see Gavriljuk 2005). As noted earlier, the psychoactive products (inflorescences or resin), as well as the fiber of *Cannabis*, may also have been among the significant exports moving south out of the steppes during this Bronze Age period, perhaps reaching beyond the northern Caucasus and even down into the Near East (see Sherratt 1997, 2003; Anthony 2007).

Historians and other scholars believe that the domestication of the horse; the development of carts, wagons, and mobile war machines such as the chariot; and the evolution of highly effective cavalry techniques all, in sequence, were key processes that influenced the success and continuation of migrations of nomadic peoples across Eurasia. However, we are not yet sure if during the early periods people were actually moving beyond their territories or whether only materials and technologies were being transferred. In any case, peoples in Central Asia were on the move, at least toward the southwest, migrating into what is now Turkey and other northern Mediterranean areas as new, high-ranking overlords spoke languages novel to these areas, and during the Middle Bronze Age (ca. 2000 to 1600 BCE), these migrating people created a ripple effect to the south of them as the people they displaced southward in turn displaced others to the south of them, and so on, all the way into Egypt. Once the use of horses for pulling loads of humans and their goods in wagons and carts began, this new transport system diffused quickly, some would say "explosively," through Western Asia and Eastern Europe, as well as eastward into China. "By the middle of the 2nd millennium BCE horses were being used to pull chariots from as far afield as Greece, Egypt, Mesopotamia, Anatolia, the Eurasian steppe, and in China by the 14th century BCE" (Levine 2002; also see McGovern 1939). This intensification of human migration, with its associated entourage of biota and trade goods, helped continue the spread of *Cannabis* and its uses farther and wider, even into many areas where the species could already be found, but in different forms put to different uses. *Cannabis* seeds were being spread with migrating peoples, perhaps both purposefully and unintentionally.

The dispersal of *Cannabis* as a cultivated crop and a camp-following weed, along with knowledge of its various uses, undoubtedly involved several Late Bronze Age peoples in Central Asia, especially those that utilized riverine environments to combine hunting, gathering, and fishing with agriculture. Examples include the Andronova culture, which covered a vast area of Central Asia during this period, and the Oxus Civilization, which flourished in the area drained by the Syr Darya and Amur Darya River systems. However, the most well-known association of *Cannabis* and ancient peoples of the Eurasian steppes involve the peoples belonging to the Scythian cultures during the first millennium BCE.

Scythians and *Cannabis*

Ancient nomadic tribes traditionally connected with "Scythian" cultures occupied a wide area of Eurasian steppe and forest-steppe zones during the ninth to third centuries BCE. The majority of archeological sites associated with these cultures are located in lands near the northern Black and Aral Seas, along the Pamir, Altai, northern Caucasus, southern Ural, and Tian Shan Mountains, as well as in Semireche ("Land of the Seven Rivers," south of the Tian Shan), central and northern Kazakhstan, southern Siberia, Tuva (extreme southern Siberia), Mongolia, Ordos (the desert and steppe region located on a plateau south of Inner Mongolia), and the lower Volga River (Alekseev et al. 2001, 2002; van Geel et al. 2004a/b).

The broad distribution and early migration of Scythians is now thought to be associated with climate change around 800 BCE, which increased the humidity in various regions of Eurasia and may have changed the range of pasture lands. For example, the oldest Scythian monument in all of Eurasia is located in the Central Asian valley of Uyuk (Republic of Tuva), an area principally unoccupied until the arrival of the Scythians in the Early Iron Age (Zaitseva and van Geel 2004

and Zaitseva et al. 2004, 2005; van Geel et al. 2004a/b). Climatic and related ecological changes during the first millennium BCE most likely also affected the natural and cultivated distribution of *Cannabis*.

The Scythians (as they were known by the Greeks but by many other names elsewhere) were essentially equestrian pastoralists with warlike tendencies, and as noted earlier, they were quite familiar with *Cannabis* (e.g., Herodotus, Book IV; Rudenko 1970). Scythian use of *Cannabis* is further supported by a more recent discovery of ancient seeds found in ceramic pots at the Pastyrske site, located in the Tiasmyn River valley (a tributary of the Dnieper River) in the Cherkasy Oblast region of the Ukraine (Pashkevich 1998a; also see Pashkevich 1997, 1998b, 1999, 2003).

Van Geel et al. (2004a/b) argued that a climate shift to more humid conditions, caused by a temporary decline of solar activity about "850 calendar years BC[E]," brought on a rapid increase in carrying capacity (e.g., a higher biomass production) in various parts of central Eurasia; they hypothesized that because of climate change the area now known as the Republic of Tuva was part of a vast, east-west situated belt (the southern part of the temperate climate zone) that rather quickly "became available as an attractive living area" and as a result had an indirect influence on "the cultural blooming and expansion of the Scythian culture." Zaitseva et al. (2004a/b, 2005) tell us that the archetypical Scythian culture began in the Central Asian Uyuk Valley and southern Siberian Minusinsk Valley "after increased humidity and occupation capacity of the steppe zone during the 9th century BC[E]." It is worth noting here that Diakonova (1994), in her discussion of shamans in traditional Tuvan society during the twentieth century, refers to a crucial ritual involving the cutting of the sacred larch tree for wood to make the shaman's drum: "Before cutting the tree a sacrificial feeding rite had to be performed around it, which included the smoking of sacred herbs for the spirit master of the trees of the forest." Whether such "sacred herbs" included *Cannabis*, which is a common wild plant in Tuva today, remains to be verified. It should be pointed out here that petroglyphs located near Lake Baikal (in southeast Siberia) and dated to about 2000 BCE depict a shaman-like figure "flying" and holding a drum like those still used for inducing ecstatic trance over a wide area of Eastern Eurasia and the New World (Devlet 2001; also see Chapters 7 and 9).

Southwest of the Republic of Tuva in southeastern Kazakhstan, remarkable, ancient phytoliths resembling *Cannabis* trichomes were found in an ancient archeological site located on the Talgar alluvial fan (Chang et al. 2003); the site is believed to be a hearth associated with the Saka people dated to the Iron Age (775 BCE to 100 CE). The Saka can be described as eastern variants of Scythian populations who were supported by a dual economy based on farming and herding livestock from one grazing ground to another in a seasonal cycle between lowlands in winter and highlands in summer. Arlene Rosen (2000) studied the phytoliths recovered from this Kazakhstan site, describing the hearth samples from Tsegonka (Tse-00-4 #1, Unit D-13, Floor #4) as an assemblage containing a high density of phytoliths including those from a sedge (Cyperaceae), a millet species (*Setaria*), and wild grasses, along with layers of wood ash that are likely the "remains of fuel for the hearth fire." Rosen suggested that phytoliths from the husks of the millet and wild grasses could be an indication of "dung fuels." However, she was impressed with the very large numbers of dicot leaf phytoliths, an intriguing characteristic of this sample, especially because of the species they seem to represent:

Dicotyledons (woody herbaceous shrubs and trees) do not generally produce abundant phytoliths. Leaves of dicots produce the most diagnostic phytoliths, but usually they are less distinctive than phytoliths produced in the husks of grasses. Sample Tse-00-4 contained very large numbers of polyhedral forms from leaves, and also large numbers of quite distinctive complex hair cells. These hairs are identical to the silicified hairs formed on leaves of modern samples of *Cannabis sativa* (the marijuana plant). *Cannabis* is a common shrub in the region today and was probably also abundant in the area during the Saka period. It is very likely that it was used for spiritual, ritual, or medicinal purposes. This might point to a special function of the structure in which this hearth was found.

The sedge, millet, and wild grasses as well as *Cannabis* "would have been found close to the site along the water course of the Tseganka River" (Chang et al. 2003; also see Rosen et al. 2000). Drawing on the ancient account of Herodotus (Book IV) and other sources, Emboden (1972) argued that the use of *Cannabis* by the Scythians in their funeral customs allegedly originated with their defeat and brief domination by Thracian Getae (northeast of Greece in the western steppes region) approximately 2600 BP, about 200 years before Herodotus's famous historical report. According to Emboden, *Cannabis* was used by the Getae shamans (*Kapnobatai*) "to induce visions and oracular trances." The ritual use of hemp, however, did not start with the Thracians (see also Chapters 7 and 9).

Recently, remarkable direct evidence for *Cannabis* use was discovered along the eastern margins of arid Central Asia near Turpan in western China and has been dated to the Iron Age, confirming our assertion that hemp was known and used over a very wide area of Eurasia, likely well before but certainly by the first millennium BCE. Jiang et al. (2006; also see Russo et al. 2008; Mukherjee et al. 2008) reported the recovery of a surprisingly large stash of *Cannabis* flowers, stems, and seeds from the burial of a probable shaman in the Yanghai Tombs near Turpan in the Xinjiang Uighur Autonomous Region. The intact plant parts, identified conclusively as *Cannabis*, were found in two containers, one placed near the head of the deceased man approximately 2,700 years ago providing the "oldest known documentation of *Cannabis* as a pharmacologically active agent" and also contributes to the archeological and ethnobotanical record of the "Gūshi" culture (Russo et al. 2008). This discovery of very unique and remarkably preserved ancient flowers, stems, and seeds of *Cannabis* in association with a putative shaman is discussed in more detail in Chapter 7. It should be noted here that ancient "Gūshi" nomads raised horses and other grazing animals and were cultivators of some crops (e.g., see Mallory and Mair 2000).

Although *Cannabis* appears to have been present in Europe and China well before major migrations of people assisted by horses, we maintain that people moved or exchanged goods such as *Cannabis* in and out of the Central Asiatic steppes before, during, and after the reign of the horse-riding Scythian cultures and that they probably brought *Cannabis* and ideas regarding its uses with them as they invaded, traded, and eventually settled into new homelands. Horse domestication further stimulated migrations of nomadic groups out of and across Central Asia on a significantly larger scale, even though mass movements of people probably depended upon later introductions of woolly sheep and the herding culture that developed along with the secondary products that the sheep provided.

Archeological and Historical Evidence for the Spread of *Cannabis*

In the previous section we described how *Cannabis* spread into diverse regions of Eurasia at varying times following the migrations of and contacts between early cultures, with the earliest evidence of *Cannabis* appearing in archeological sites associated with early human habitation and burial. In the following discussion, we continue our survey of early archeological finds of *Cannabis* seed, fiber, and pollen and evaluate their bearing on historical references and our understanding of *Cannabis*'s dispersals.

Diffusion throughout East Asia

The initial phase of *Cannabis*'s dispersal out of Central Asia was an active period of expansion across most of Eurasia including China, South and Southwest Asia, and Europe (see later this chapter for a summary model of the phases of dispersal within and beyond Eurasia). Here we begin with the historical and archaeobotanical evidence for the occurrence and spread of *Cannabis* into China, Korea, and Japan. As we noted earlier, *Cannabis* probably spread naturally during the Pleistocene into various areas now included in modern China, especially in western regions, but may not have spread further east and south into more humid regions of East Asia until it was forced to migrate by advancing ice sheets and aided in doing so by humans. Human genetic research indicates that early Denisovian populations spread as far as Southeast Asia and Oceania prior to the southerly spread of East Asians (Reich et al. 2011), and these migrations could also have carried *Cannabis* from its Central Asian region of origin prior to the Holocene. Recent phylogenetic research concerning the spread of AMHs into and throughout China via western regions in Central Asia relates to the early spread of *Cannabis* in many areas of East Asia. Based on genetic Y-chromosome evidence extracted from 76 Chinese men covering 33 ethnic minorities as well as the Han majority surveyed throughout China, Deng et al. (2004) reconstructed a Chinese nonrecombinant Y-chromosome phylogeny and tentatively suggested that "haplogroup O" originated in western or northern China and evolved mainly within China from where it dispersed further all over eastern Eurasia, with "a multilayered, multidirectional, and continuous history of ethnic admixture that has shaped the contemporary Chinese population." Accompanying this early movement of people into northern China from the north and west was *Cannabis*, but how long ago and if migration of hemp via nonhuman dispersal vectors had already occurred, we still do not know.

Although the remarkable ancient Chinese historical record has important references to *Cannabis* and its use in East Asia, and some macrofossil evidence has been documented for the ancient presence of the genus in the form of seed and fiber remains, the archaeobotanical record for microfossils in Asia suffers generally from a lack of reliable pollen samples (e.g., see Chen et al. 2009). Ancient pollen

TABLE 8

Earliest archaeobotanical evidence of *Cannabis* organized by biotype and country
or region where found with conservative age estimates BP (rounded low).

Evidence types are ordered left to right by decreasing certainty with the most reliable (seed evidence) on the left.

Region	Type of archaeobotanical evidence					
	Seed	THC	Phytoliths	Pollen	Fiber	Charcoal
Asian *C. indica* subspecies (BLH, NLDA, or NLD)						
Japan	10,000 BP				7000 BP	
China	4800 BP					
Israel	pre-2500 BP					
Korea	2500 BP				5000 BP	
South Asia	1900 BP		3300 BP			3200 BP
Kazakhstan			775 BCE to 100 CE			
European *C. sativa* subspecies (NLHA or NLH)						
Germany	7000 BP				2500 BP	
Baltic Region	5000 BP			5000 BP		
Romania and Bulgaria	4000 BP			7000 BP		
Poland				5500 BP		
Hungary	2000 BP			2000 BP		
Sweden	2000 BP			1500 BP		
Italy	1900 BP			3400 BP		
Switzerland	1800 BP			4700 BP		
Netherlands	1750 BP					
Middle East		1600 BP				
British Isles	1600 BP			1600 BP	2800 BP	
Norway	1150 BP			1500 BP	1150 BP	
Denmark	1100 BP			4000 BP		
Northern France	1100 BP			2900 BP		
Austria, Czech Republic and Slovakia	900 BP			9000 BP		
Finland	900–500 BP			2000 BP		
Iberian Peninsula				1350 BP	4200 BP	
Belgium	300 BP					

(See TABLE 1 at the beginning of this book for explanations of *Cannabis* gene pool acronyms.)

data for *Cannabis* is presently limited due to two main factors: (1) a relative lack of palynological studies in the whole region until recently, and (2) the difficulty of differentiating *Cannabis* and *Humulus* pollen, as noted earlier. For example, Chou (1963) reported a pollen curve for what he interpreted as *Humulus* from Banpo (Pan-p'o, near Xi'an in Shaanxi province) dated to ca. 4500 BP. However, the ancient pollen grains in this case, like so many discovered in various European sites, could, just as reliably, be ones produced

by *Cannabis* (e.g., see Li 1974b). Currently this is the only ancient pollen record of *Cannabis* for China because researchers have paid more attention to fruit and fiber remains than pollen (Jiang et al. 2006). Recent research by Chen and his associates (2009) in the Lake Chaohu area of Anhui province in eastern China focused on Holocene vegetation history with an emphasis on indications of human impact over time. Although they claimed that pollen data revealed anthropogenic vegetation change, they admitted that additional clarification of the full interaction between environment and human activities is needed, including better pollen identifications (in particular of the "cereal pollen") along with a "high-resolution pollen study of the lower part of the deposits." In this context, the relatively substantial amounts of "*Humulus* type" pollen remains from the lowest levels (early Holocene, ca.10,000 BP) up to at least about 3,000 BP tells us that wild hop (*Humulus*) or wild, partially managed, or cultivated hemp (*Cannabis*) occurred in that region at least since early Holocene times. Since it is not clear if the pollen is *Cannabis* or *Humulus*, we must look further in the Chinese archeological record for additional direct evidence of *Cannabis* (e.g., actual fiber and seed remains or impressions of fibers and seeds).

Historical evidence indicates that *Cannabis* has been cultivated in almost every province and climatic zone in China and in a number of cases from ancient times right up to the present. Even today, in several regions of China, hemp fibers are still used to make cordage, cloth, and paper, while its seeds are pressed for their oil or are eaten raw or roasted as snacks between meals (see Chapters 5 and 6); in 2004, China reportedly produced 38,000 metric tons of hemp fiber (Lewin 2006). The uses of *Cannabis* for medicinal and even ritualistic mind-altering purposes are now rare; however, they were important during China's ancient past (see Chapters 7 through 9).

We have suggested that *Cannabis* was indigenous to some areas of Western China well before the arrival of Central Asian nomads and therefore prior to any long distance trading contact between Central and East Asia. It was certainly an early cultigen in northern China. Chang (1986) referred to fiber hemp as one of the principal cultivated plants of ancient China, while Harris (1996) pointed out that in spite of the dearth of exact identification and dating of plant and animal remains, "there is incontrovertible archeological evidence for the establishment in northern China, at least by the mid-seventh millennium BC[E], of substantial villages, with pottery and storage pits, supported by a mixed economy of hunting, fishing, gathering, the cultivation of millets and a few other crops, and the raising of domestic pigs, dogs and probably chickens." By about 7,000 to 8,000 years ago, the Huang He (Yellow River) Valley was dotted with developed agricultural communities marked by different cultural features (Zhimin 1999), and it is in this context that we envision early *Cannabis* cultivation in northern China.

More than a half century ago, Vavilov (1949–51) suggested northern China as one of the original homes of hemp. More recently Harlan (1975) agreed that the Eurasian steppes have been more or less marginal in terms of plant domestication and are thus an unlikely place for agriculture to have begun, but, as noted earlier, hemp cultivation may have begun here: "The few domestic plants that might be of steppic origin are *Panicum miliaceum* [broomcorn millet], *Setaria italica* [foxtail millet], and *Cannabis sativa* [probably *Cannabis indica*]" (see Chapter 3; also see Harlan 1995).

There is some archeological evidence for fiber hemp in China a millennium or two before nomadic pastoralism became possible. For example, Barber (1991) refers to circumstantial evidence for fiber use in China in the late seventh and early sixth millennium BP, concluding that hempen cloth impression evidence is "strong for northern China," and also refers to "actual pieces of cloth more directly identifiable as hemp" dated to the latter part of the third millennium BP. In addition, there is more recent archaeobotanical evidence that *Cannabis* was grown in China during the well-developed Neolithic period and perhaps much earlier. Underhill (1997, citing Crawford, 1992) argued that a methodical recovery of archaeobotanical macrofossil remains could establish the function of cultivated plants such as *Cannabis*: "This plant probably was first domesticated during the earlier Neolithic Period. People in more than one region of northern China could have used the fruit for food or oil and the coarse fiber from the stems for clothing or mats."

Underhill (1997) referred to evidence from the ancient Neolithic Majiayao culture site of Linjia in Dongxiang county, Gansu province, in the upper reaches of the Huang He Basin, where ancient carbonized hemp seeds, dated to ca. 3400 to 2800 BCE, were recovered along with those of broomcorn millet (see Teacher's College of Northwest China 1984). Underhill also referred to the base of a pottery bowl recovered from the early Neolithic Yangshao culture site of Banpo (Pan-p'o) that exhibits trace impressions of a woven fabric, possibly hemp (see Banpo Museum 1982). According to Barber (1991), "hemp had already spread far and wide by early Neolithic times to Europe, Tibet and China" and may have somehow even moved "with the Paleolithic net-makers" (see Chapter 5 for more detail on ethnobotanical and historical evidence to suggest early cultural dispersal of *Cannabis* use for fiber; also see Adovasio et al. 2007; Kühn 1987).

The oldest historical records of *Cannabis* use come from ancient Chinese texts. For example, although *Cannabis* was not referred to in Shang (1700 to 1027 BCE) oracle bones or Zhou bronze inscriptions (1027 to 221 BCE), it is mentioned seven times in the *Shih-Ching*, or "Book of Odes." *Shih-Ching* is among the earliest collections of Chinese literature and is an assemblage of fragments dating from the eleventh to midsixth century BCE (Ho 1969). According to Chinese mythology, the legendary emperor and patron divinity of Chinese herbalists, Shen Nung (ca. 4800 BP), invented agriculture. The *Lu Shu* or "Book of Pitches" of the Song dynasty (960 to 1279 CE) indicated that Shen Nung first instructed the people to cultivate *má* (麻), which is the Chinese word for *Cannabis* hemp (Dewey 1914).

In spite of the rush to industrialize modern China today, the great majority of Chinese people are still farmers and gardeners, and it is almost inconceivable to think of the Chinese way of life without agriculture. Even when we consider more recent archeological evidence that pushes the birth of agriculture in China back to shortly after the end of the last ice age (Crawford 2009), we must remember that over the long time span members of genus *Homo* have existed in China, maybe one million years or more, only during the last 10,000 years or less (less than 1 percent) have modern Chinese peoples practiced agriculture, while the Paleolithic ancestors of Chinese farmers engaged in hunting, fishing, and gathering for a much longer period of time in the region. Based on archeological evidence supplemented with ethnographic records of modern hunters and gatherers, Paleolithic ancestors of the present-day Chinese had a vast knowledge of their environment and exhibited "cunning and resourcefulness that enabled them to compete with fellow creatures in that environment, and cosmology and religion that rounded out their life in the wild" (Chang 1986).

Recent paleoecological evidence suggests that from Approximately 8,000 to 4,000 years ago the climate of northern China was significantly warmer, thus allowing forests to thrive and consequently providing humans a more diverse environment to exploit than the present semiarid conditions (also see Chapter 3 for a theoretical discussion of how early hunters and fishers learned to use *Cannabis* and laid the foundations for its domestication in China and elsewhere in Eurasia): "Recognition of a warm, humid and forested China—insofar as its low-lying and waterside areas are concerned—enables us to envision an environment—or a vast area with numerous microenvironments—with diverse and plentiful animal and plant resources, which were at the turn of the Holocene being effectively exploited by the terminal Paleolithic inhabitants" (Chang 1986).

Students of agricultural origins, who have hypothesized about the transformation of hunter-fisher-gatherers into early herders and cultivators, tend to agree that diverse and plentiful environmental resources were a prerequisite, for only among people living under these favorable conditions could the necessary experiments toward plant and animal domestication take place. In other words, necessity is not necessarily the mother of invention, and abundant "free" time may have been necessary for the experiments needed to develop successful herding of animals and cultivation of plants (Sauer 1952). In order to understand and demonstrate transformation into agricultural life in China, archeologists, and biologists investigate in detail the faunal and floral changes that took place during several millennia before and after the beginning of the Holocene. N. I. Vavilov (1949/51) recognized that "in the wealth of its endemic species and in the extent of the genus and species potential of its cultivated plants, China is conspicuous among other centers of origin of plant forms." Li (1966) lists the distinctive cultivated plants of both northern and southern China such as *Cannabis* that must have been derived from native wild ancestors. Wild plants abounded in the early Holocene environment, and Neolithic inhabitants used them for food, material, fiber, medicine, and other purposes and also experimented with them as early cultivars.

China is now recognized as one of the primary regions where agriculture originated, and Chinese farming has been described as more advanced than anywhere else in the world before the rise of modern science and technology. Since 1920, when the early Neolithic village site of Yangshao near Loyang city in Honan province was first excavated, hundreds of additional contemporary sites have been discovered and studied in various parts of the middle Huang He Basin from central Honan to Gansu and Qinghai provinces. The Neolithic culture these sites represent is now the best characterized in China. According to Chang (1986), the majority of Yangshao culture sites are remnants of ancient villages located in relatively dense distribution on the lower loess terraces alongside the banks of the three largest rivers of the region, the Huang He, Weishui, and Fenho, or more commonly next to the banks of their tributaries: "The time period of their occupation was within the climatic optimum of the northern Chinese postglacial, and the Pan-p'o [Banpo] stage [the earliest] occurred during the warmest and wettest recent interval in the area."

Abundant vegetation and rich supplies of wild fish and terrestrial animals from which to acquire food and other useful materials were available to incipient and early swidden (slash-and-burn) farmers. Although wild plant foods were collected, the archeological record clearly indicates that farming was practiced, especially the cultivation of millet but also of other crops such as wheat, broomcorn millet, soybeans, and hemp (Chang 1986; Chen 1984). Te-K'un (1964) asserted that early Neolithic people in China must have found suitable locations within woody uplands that had "hilly flanks, swampy basins and a climatic optimum," which provided adequate natural resources in order for them to "experiment with agriculture."

Early farming in central China began in two major areas, south in the Yangzi River Basin, and north in the Huang He Basin, the core areas of Neolithic cultures in East Asia. The incipient period of farming in East Asia is referred to as the Mesolithic, a general age during which we believe *Cannabis* was brought into cultivation as a camp following weed via its spontaneous growth in human dump-heaps, principally near streams and lakes where drinking water, terrestrial game, and fish could also be found. As we will discuss later, there is evidence that hemp was available and in use in ancient northeastern China, probably Korea, and certainly in Japan many thousands of years ago; furthermore, it appears to have been cultivated, or at least its growth encouraged, very early on (e.g., in Japan by the people of the ancient Jōmon culture). This evidence suggests that *Cannabis* may have been native across a much broader geographical Eurasian range than heretofore believed, or at least it was naturally dispersed beyond its putative origins somewhere in Central Asia by the beginning of the Holocene, possibly already restricted to refugia in southeastern Europe and southeastern Asia.

Although *Cannabis* may have long been present and used in the ancient Far East, by the time of the Bronze Age in Central Asia (ca. 4000 to 3000 BP), trading took place between sedentary farmers on the western frontier of China and nomadic herders in more remote parts of inner Asia. This exchange occurred as people spread east from what now is Russia, across Siberia, into Mongolia and northern China. Although early trading contacts via horse riding nomads may not have involved the movement of *Cannabis* and/or its products, certainly by the time of the Iron Age, people were moving back and forth across Eurasia and engaging in lots of trading. Barfield (2001) refers to cloth made of "hemp, ramie [*Boehmeria*] or kudzu [*Pueraria*]" among the goods received by nomads from Chinese farmers during the time of imperial state formation along the Chinese-nomad frontier more than 2,000 years ago. It becomes apparent then that the "introductions" of *Cannabis* by Central Asian nomads into eastern Asia discussed earlier must have been subsequent introductions of *Cannabis* rather than initial ones.

Even though *Cannabis*'s presence and use in China most likely reaches back in time to the Mesolithic or even Paleolithic period, modern evidence indicates that subsistence patterns in ancient China went through a period of diversification during the later Neolithic (Underhill 1997) and that these changes affected the distribution and relative use of *Cannabis*. For example, it has been suggested that early Chinese farmers purposefully imported new species of domesticated plants or animals to diminish risks such as poor yields due to unfavorable weather conditions or other uncertain factors (e.g., see Morrison 1994). *Cannabis* or special cultivars of the genus could have been among these crop plants, brought in from areas to the west for its nutritious seeds. During the later part of the Chinese Neolithic at least some *secondary products* seem to have become more significant, such as wool harvested from herded sheep. In addition, some supplemental products derived from plants and animals such as "hemp, silkworms, and the horse" were most likely "raised for uses other than food" (Underhill 1997; also see Lee et al. 2007 and Sherratt 1997 for discussion of the secondary product revolution). The spread of *Cannabis* into or out of East Asia remains to be fully understood, but we

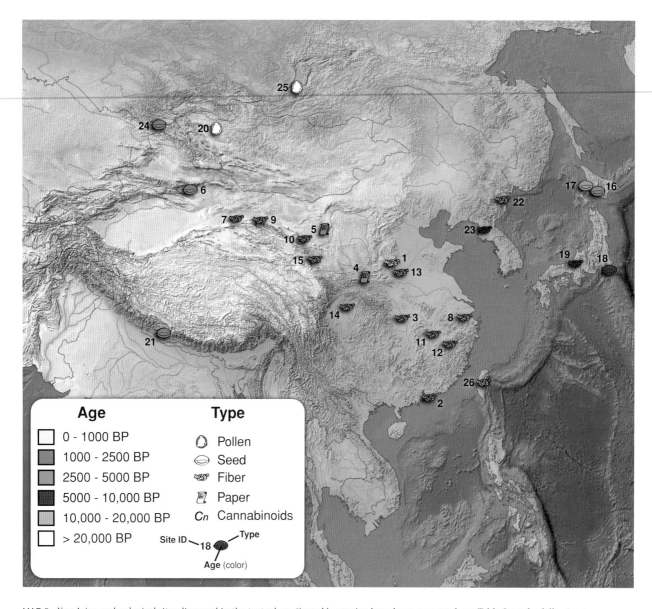

MAP 5. Key Asian archeological sites discussed in the text where *Cannabis* remains have been recovered; see Table 9 on the following two pages for the numbered locations and additional information (cartography by Matt Barbee).

can find a good deal of evidence indicating its significant use in ancient Chinese culture (see Chapters 5 through 9), much of which predates evidence of the migrations of pastoral nomads, especially into the Chinese realm. Of all the earliest large-scale civilizations, including those in Mesopotamia, Egypt, and the Indus Valley, the Chinese cultural tradition was least influenced by external stimuli. This is understandable if we consider the several thousand kilometers that lay between China and the other important civilizations to the west and southwest. Besides these geographical restraints, after about 3700 BP, political conditions restricted travel across the vast steppes separating China from other influential cultures. However, there are scholars of Chinese antiquities who downplay the importance of trade routes across this natural highway (e.g., see Jixu 2003; Chang 1988), parts of which became known as the "Silk Roads." These scholars envision a more indigenous growth of cultural ideas and activities in China. However, by 2500 BP, the Eurasian steppe zone, which is vast but

relatively easy to traverse, undoubtedly served as a route of cultural movement and diffusion of knowledge between high cultural centers in the East and West, and during this time, or perhaps much earlier, the Eurasian steppes may have provided a diffusion route for *Cannabis* seed and technologies.

Despite its isolation, Chinese culture did not develop solely under its own inner momentum. In fact, there was limited but significant external contact that undoubtedly affected the diffusion of hemp, or at least some cultivated varieties, and specific uses and farming techniques for *Cannabis*. As we described earlier, there are a number of Neolithic sites with evidence of *Cannabis* cultivation located along or near the natural corridor of the Central Asiatic steppes connecting western China and areas much further west in Eurasia; these sites provide us with grounds for speculating about the cultural diffusion of hemp. In the archeological record, we see two radiating centers, one in the Iraq-Iran area of the Near East and the other in the Huang He Basin of China, which

TABLE 9
Key asian archeological sites discussed in the text where *Cannabis* remains have been recovered.

Site #	Country	Location	Age	Type	Source	Reference
1	China	Shaanxi province, near Jinzheng city	1040 to 720 BP	Cordage statue cores	Song dynasty statuary	Kao 1978
2	China	Hong Kong, Tuen Mun, So Kwun Wat	2200 to 1800 BP	Woven bamboo mat	Han dynasty, rescue excavation	Hong Kong Year Book 2000
3	China	Hubei province, Jiangling Xian	2173 to 2167 BP	Cloth document wrapper	Tomb	Giele 1998
4	China	Shaanxi province, Xi'an city, Baqiao Tomb site	2138 to 2085 BP	Paper	Western Han	Bloom 2001
5	China	Inner Mongolia, Alxa League, Tsakhortei	1900 BP	Oldest surviving piece of paper with writing	Watchtower	Temple 1986
6	China	Xinjiang province, Turpan prefecture, Yanghai Tombs site	2700 to 2500 BP	Seeds, shoots, and leaves	Shamanic tomb	Jiang et al. 2006; Russo et al. 2008
7	China	Gansu province, Dunhuang, Mogao Cave 17	1650 to 710 BP	Fabric paintings	Buddhist religious site	Whitfield and Farrer 1990
8	China	Zhejiang province, lower Yangtze River	5500 to 4200 BP	Textile, rope, and yarn remains	Liang-Chu culture, coastal plain	Cheng 1966.
9	China	Gansu province, Juyan region	2200 to 1800 BP	Hemp strings	Han dynasty	Lao 1943–44; Tsien 2004
10	China	Gansu province, Wuwei Xian, Hantanpo Cemetery site	2000 to 1935 BP	Document sack	Tomb	Giele 1998
11	China	Jiangxi province, near Nanchang city	2100 to 1900 BP	Textiles	Han dynasty wooden pit burial	Kuo 1978
12	China	Fujian province, Wuyi Mountains	3600 to 3300 BP	Fabric funeral shroud	Boat coffin	Li 1984
13	China	Henan province, Anyang city	3764 to 3120 BP	Bronze weapons, impressions of cloth wrappers	Shang dynasty	So and Bunker 1995
14	China	Sichuan province, Taipingqíang village	4000 to 3200 BP	Cord-impressed pottery	Late Neolithic	Cheng 1982
15	China	Gansu province, Yongqing, Tahozhuang village	4150 to 3780 BP	Fabric	Chhi-chia culture	Kühn 1987
16	Japan	Hokkaido Island	1400 to 800 BP	Seeds	Satsumon period	Imamura 1996a; Crawford and Takamiya 1990
17	Japan	Hokkaido Island, Sapporo city, N-30 site	3500 BP	Seeds	Late Jōmon	Clarke personal observation 2006
18	Japan	Honshu Island, Boso Peninsula, Okinoshima site	10,000 BP	Seeds in sediment attached to pot shards	Early Jōmon	Okazaki et al. 2011

(continued)

TABLE 9 (continued)

Site #	Country	Location	Age	Type	Source	Reference
19	Japan	Honshu Island, confluence of the Hasu and Takase Rivers, near Lake Mikata, Torihama, Shell Midden site	7000 BP	String	Early Jōmon	Nunome 1992
20	Mongolia	Southern Altai Mountains	130,000 to 120,000 BP	Pollen, Cannabaceae	Prehistoric soil cores	Tarasov et al. 2007
21	Nepal	Jhong Khola Valley	2400 to 1960 BP	Seeds	Artificial cave site	Knörzer 2000
22	North Korea	North Hamgyong province, Musan county, Pomuigusokin site	3000 to 2300 BCE	Rope	Archeological site	Anonymous 1990
23	North Korea	South Pyeongan province, Oncheon county, Unhari, Goongsan village	5000 BP	Threads along with bone needles	Archeological site	Sim 2002; Kim 1979; Nelson 1993; Korea News Service 2002
24	Russia	Siberia, Ukok Plateau, Ak-Alakha Valley	2400 to 2500 BP	Seeds in dishes	"Ice Maiden" tombs	Davis-Kimball 1997
25	Russia	Lake Baikal	150,000 to 100,000 BP	Pollen, Cannabaceae	Prehistoric lake sediment cores	Molodkov and Bolikhovskaya 2006
26	Taiwan	South China coast and Taiwan Island	12,000 BP	Cord-impressed pottery	Early postglacial fishing sites	Chang 1963

spread their influences across the intervening steppes from opposite directions and made scattered contacts with indigenous peoples. However, there is little evidence that the steppe zone during the sub-Neolithic and Neolithic stages served as a route of significant cultural transmission from one region to the other as these cultural centers were yet to flourish. Indeed, it appears that it was during the Bronze Age that significant contacts began between the steppe zone and either the Near or Far East.

By 3300 BP, the use of the horse and chariot, together with bronze armor, the composite bow, and rectangular fortifications, were established in northern China. The diffusion of these traits is attributed to the nomadic warriors and traders of the Eurasian steppe who came to the Far East from either the margins of the Iranian Plateau or the Altai Mountains. Furthermore, similarities among certain pottery styles may reveal even earlier contact between China and western Asia. However important these limited contacts, indigenous cultural influences in China were always strong, and if the Chinese did adopt some Western ideas and techniques, they were soon incorporated into the distinct Chinese culture. In the discussion that follows, we will first deal with key linguistic connections and ancient written records in China, which go back remarkably far and give us insight into the early use of *Cannabis* in Chinese society. Then we will review some representative archaeobotanical evidence for the early presence and spread of *Cannabis* in China. Linguistic connections between peoples in Central and East Asia may provide clues to understanding when and why *Cannabis* and its use spread from one of these large areas into the other. Until relatively recently most students of the history of Chinese language have not seriously considered

probable connections with, and perhaps even origins from, the languages of peoples living in Central Asia far west of what today is China. In fact, the associations of Old Chinese with Proto-Indo-European are strong and, when linked with important historical and cultural traditions, indicate that the Chinese language of today owes much of its vocabulary and other characteristics, if not its characters, to neighboring Indo-European languages. Articles published in the *Sino-Platonic Papers* have provided challenging and creative studies covering a wide variety of philological subjects dealing with the study of languages in Eurasia, in particular those manifested in ancient texts, and especially related to the development of civilization. In 1988, Tsung-tung Chang published an article reviewing the history of comparative linguistics; he pointed out that during the past two centuries numerous propositions were offered to explain relationships of Indo-European languages to several other languages such as Semitic, Altaic, Austronesian, Korean, and so on, noting that students of Indo-European languages generally rejected such attempts because of unconvincing evidence. Even though scholars of historical linguistics (e.g., Ulenbrook 1968 and Ulving 1968, cited by Chang 1988) have argued that a rather large number of word equivalents connect the older Chinese language with the older English and German languages, skepticism among "Indo-Europeanists" remained strong. Disbelief or uncertainty is based mostly on the rigid view that word form alone is important, and vocabulary is insignificant; Chang (1988) described the situation of linguistic scholarship: "Since the typology of Chinese seems to preclude a cognate relation to Indo-European, they are inclined to discard any lexical correspondences as merely accidental or onomatopoeic. Besides, prehistorical contacts and mixtures between these languages have been doubted because the Indo-Europeans are supposed to have originated in northern Europe or at best in the Central Asian steppe, thousands of miles away from East Asia. Hence, any research into a relationship between Old Chinese and Indo-European languages would be but futile from the outset."

However, not all earlier Indo-Europeanists held this view, and some even suggested that the Chinese language corresponds closest to the hypothetical prototype of Indo-European (e.g., see Karstien 1936). An additional major objection to acceptance of significant historical relationships between Indo-European and Chinese languages has been the existence of tonal accents in Chinese, which led many linguists to consider this language as "highly exotic." However, the tonal, rather than expiratory, accents of most contemporary Chinese languages can be largely dismissed because "the use of tonal accents as a means of lexical differentiation is a result of comparatively recent development in the long history of Chinese language, the earliest monuments of which date back to 1300 BC[E]" (Chang 1988; also see Chang 1970).

More recently, Jixu (2006) shed light on the historical linguistic problem by asking why this seemingly close relationship was not discovered earlier, especially if there are in fact a large number of word equivalents between Old Chinese and Proto-Indo-European. Jixu argued that the answer to this question lies with the unique Chinese character writing system that has been in constant use for "at least 3300 years." Jixu refers to this system as acting "like a heavy curtain that covers all the differentiations and evolution of Chinese language, because this kind of writing system has almost never been revised since the second century BC[E]." Earlier, Chang (1988) pointed out that tonal accents in Chinese were only first noted in the fifth century CE by the poet Shen Yiieh (lived 441 to 513 CE); Chang argued that "Old Chinese emerged as a mixed language, though spoken with Proto-Chinese native tongue, using mainly the Proto-Indo-European idiom which seems to have stretched from Mongolia to Europe during the 3rd millennium BC[E] in the northern part of the temperate zone." This extremely wide-ranging language association was most likely also paralleled by the exchange of ideas and materials, including useful plants such as *Cannabis*.

Chang (1988) connects the appearance of Old Chinese language with the establishment of the Chinese Empire by Huang-ti, the "Yellow Emperor" who is still considered to be the founder of China and thus the initiator of its highly sophisticated civilization. Furthermore, the first chapter of the *Shih-chi* or "Records of the Grand Historian" tells us that toward the end of Shen Nung clan rule, the "Divine Farmer," Huang-ti, was victorious in the momentous defeat of his enemies in northern China, and thus he became the first Chinese emperor. The decisive battle, not insignificantly, occurred along the main road between present-day Beijing and Inner Mongolia, and following his victory Huang-ti ordered new roads to be built and was continually "on the move with treks of carriages." This and other nomadic characteristics indicate the origin of Huang-ti from a tribe of "stock-breeders" in Inner Mongolia (Chang 1988). Furthermore, Chang tells us that the introduction of wagons pulled by horses or cattle revolutionized transport and traffic in Northern China and that this major technological innovation strongly assisted in the founding of a state controlled by central government. Here we should remember the associations of hemp, horses, and humans raised earlier in our discussion of the origins and spread of *Cannabis* within and to the margins of the central regions of Eurasia.

Huang-ti is mentioned in the *Li Ki* or "Book of Rites" as the founder of Chinese language; in chapter 23 ("Rule of Sacrifices"), which provides the rationale for worshiping ancient rulers and heroes, it is said that "Huang-ti gave hundreds of things their right names, in order to illuminate the people about the common goods," thus stimulating a standardization of the Chinese language. Through this process, Chang (1988) argues that the native peoples in northern China learned new foreign words from the emperor and suggests that this explains how "the Proto-Indo-European vocabulary became dominant in Old Chinese." Referring to traditional Chinese dating, Chang states that "we may assume that the founding of the first Chinese empire took place at the latest about 2400 BC[E]." This places the supremacy of Huang-ti with the archeologically established start of the classical Lung-shan culture ca. 2400 to 2000 BCE, which was based in the eastern valleys of northern China. Lung-shan culture was typified by a significant increase in stock-breeding, which not only included pigs, dogs, and poultry as in the preceding Neolithic cultures but also involved sheep, cattle, and horses. Chang argues that cattle and horses were especially "important for their usage in transport service and warfare, and for improved protein supply for the warriors" and that the resultant combination of crop cultivation and animal herding "laid a sound economic basis on which a great empire could function and be maintained," helping raise Chinese civilization to regional greatness. We assume that the use of *Cannabis* fiber for making cordage and its seed for feeding poultry played a significant role in this stage of Chinese cultural development and precipitated a further geographic expansion and local intensification of hemp cultivation. Although the earliest archeologically established use of *Cannabis* from China dates to about 4700 BP (Li 1974a), we believe its earliest use, at least in some areas of China, began much earlier.

Based on archeological and other evidence, Chang (1988) assumes that "the culture in the northern steppe was once superior to that of northern China." Chang thinks that more favorable climatic conditions in Inner Mongolia, due to warmer and more humid conditions associated with longer hours of summer sunshine at the start of the third millennium BCE, produced a "richer economy and higher civilization" than in northern China. Following Chang's interpretation, starting in the mid-third millennium BCE, climatic changes took place in Northern Eurasia that were most likely unfavorable for pastoralists. These changes stimulated a series of southward emigrations of "stock-farmers" into present-day China, and during the second half of the second millennium BCE, "the dominance of Indo-European vocabulary in Chinese was already consolidated" (Chang 1988). Corresponding to the emergence of the Chinese Empire and language in East Asia, warriors speaking Indo-European languages, and utilizing *Cannabis* in one or more ways, invaded and occupied areas in China where settled agrarian life including *Cannabis* cultivation had likely existed for several millennia.

Jixu (2006) also expressed an iconoclastic view of the prehistory of the Huang He area. His research indicates that nomadic herders entered the great Huang He watershed and eventually settled down to become the legendary first farmers of ancient Chinese civilization; however, these nomads-turned-farmers were in reality newcomers that learned from earlier cultivators in the region and subsequently were credited with the development of an agrarian society based primarily on millet but also associated with early cultivation of "hemp and beans." According to Jixu, the "Huang Di" (Huang-ti) people migrated into the Huang He region ca. 4,500 years ago, over a thousand years before Aryan peoples entered the Indian subcontinent. Therefore, the connection between the Proto-Indo-European and Old Chinese languages is harder to discern because more changes have occurred over this long time period. Jixu (2006) also argued that the classic Chinese agricultural society and the civilized way of life that characterized the Xia, Shang, and Zhou dynasties developed within a little more than a thousand years. This involved the replacement of nomadic subsistence with settled agriculture and resulted in the archaic Chinese people achieving an affluent large-scale society based on farming before the time of the Western Zhou dynasty (1021 to 771 BCE). This ancient Chinese tradition honoring the beginning of cultivation by the descendants of the "Yellow Emperor clan" is elaborately described in the *Shi Jing* or "Book of Odes." Within this ancient text are poems Jixu (2006) tells us present "a reliable account of the historical fact." One of these, the great Chinese hymn *Sheng Min* or "She Bore the Folk" in the *Da Ya* section of the *Shi Jing,* is an epic ode handed down by the Zhou people for more than a thousand years; it praises the achievements of Hou Ji (the "Millet King") in creating and developing farming. The following verses describe the circumstances that stimulated Hou Ji to plant "cereals and other plants such as hemp," as well as import select varieties of seeds, develop cultivation techniques, and gather the harvest:

When he [Hou Ji] was able to crawl,
He looked majestic and intelligent.
When he was able to feed himself,
He fell to planting large beans.
The beans grew luxuriantly;
The rows of his paddy shot up beautifully;
His *hemp* and wheat grew strong and close;
His gourds yielded abundantly.
(JIXU 2006, italics added by the authors)

The birth of Hou Ji and his invention of cereal cultivation are also described in the *Shi Ji* or "Records of the Grand Scribe" (ca. second century BCE), but in a somewhat different way: "When he played, he loved to plant *hemp* and beans. The *hemp* and beans he planted were luxuriant. By the time he became an adult, he loved to farm. He would observe what was suitable for the land. Where it was suitable, he would plant and harvest grain. The people all modeled themselves on him. When Emperor Yao heard of this, he brought Qi into service as the 'Master of Agriculture'" (Jixu 2006, citing *The Basic Annals of the Zhou* from Sima Qian 1993, italics added by the authors).

Sima Qian, writing sometime after 221 BCE, was considered the "Grand Historiographer" and most famous of ancient Chinese historians. The key point for us here is that the contact and cultural assimilation of the nomads to the north, south, and west with the earlier settled farming cultures of the Huang He valley probably produced associated changes in *Cannabis* use in ancient China. Additional new archeological evidence supports the suggestion of early trans-Eurasian connections during the second and first millennia BCE. According the Barber (2007),

Long before the Chinese established the famed Silk Road from the east (around 110 BC[E]), Caucasoid people were moving into Central Asia from the west, bringing such western

domesticates as wheat and wooly sheep and eventually horses. The naturally mummified and spectacularly clothed bodies of some of these Bronze Age people (dating roughly to 2000 to 500 BC[E]) have provided much new evidence as to their origins, and have spurred further efforts to analyze more thoroughly the linguistic fossils they left behind. This evidence proves that Iranian-speakers had ridden all the way to northern China during the Shang Dynasty (1500 to 1100 BC[E]), spreading not only the use of the spoke-wheeled chariot but also a number of rituals. Some of the horse-riders' rituals also ricocheted westward, leaving fascinating traces in the cultures of Great Britain and hence the United States-including the magician's pointed cap and the child's hobby-horse.

As we have noted earlier, in ancient China, *Cannabis* was used for a number of purposes including fiber, seed, and resin early on, and the history of these uses is discussed in the chapters that follow. Cheng (1982) tells us that hemp was very important among the late Neolithic people of China: "The Proto-Chinese dressed in tailored clothes of hemp cloth and silk fabrics. Spindle whorls are common articles in their settlements, and loom weaving may well have been practiced. The discovery of textile imprints on Yangshao pottery and in burials, and an artificially cut cocoon of the *Bombyx mori* in Shansi, indicates that the fibre [*sic*] was hemp and that mulberry might have been planted for the silk industry."

The *Sheng Min* hymn also makes reference to a hemp crop growing "strong and close." This indicates that the hemp crop was cultivated primarily for fiber rather than seed or drug. Early yarn spinning was accomplished by joining strands of fiber by hand and feeding them onto a twirling, weighted stick or spindle suspended in the air and kept spinning with hand motion. The spindle weights, referred to as "spindle whorls," are often made of clay, ceramic, or stone, and "they are just about the only evidence that survives of ancient thread making" (Anthony 2007); although these spindle whorls have been more frequently associated with producing flax in Europe or woolen thread across Eurasia, some of the early ones, at least in China if not elsewhere, were used to add twist to hemp cordage, a technique still used today. Is it possible that the idea of spinning yarn, and even early back strap loom weaving technology in China and other areas in eastern Eurasia, became a stimulus increasing the recognition of *Cannabis* for additional uses? Or was the ubiquitous occurrence and frequent use of *Cannabis* as food and drug a seminal influence in the initial development of spinning and weaving? Sheep may have been slaughtered for food and hides and their hair spun and woven long before hemp fiber was utilized. It seems most likely that spinning yarn with a weighted spindle was initially a wool-processing technique that spread to East Asia with the nomadic sheep raising cultures, long after hemp was already utilized for spinning and weaving in China, Korea, and Japan, where techniques specific to bast fibers were developed (Clarke 2008, 2010a/b). As described earlier, wild growing hemp plants are usually multibranched, and thus they are poor fiber sources for spinning yarn. Although short sections of rough yet useable fiber can be extracted from wild plants, only hemp cultivated with close field spacing, and consequently without limbs, is suitable for fine spinning and weaving. Therefore it is possible that spinning and weaving were stimulated by the availability of animal fibers such as sheep's wool rather than vegetable fibers, and that vegetable fibers became more popular after settled agriculture led to large populations and animal herds diminished. However, the ancient Chinese may have developed methods for utilizing bast fibers prior to the introduction of wool-bearing animals from Central Asia (see Clarke 2010a/b for a comparison between European and East Asian hemp processing, spinning, and weaving strategies).

In sum, both ancient Chinese texts and archeological evidence suggest that *Cannabis* use in China for medicine, fiber, and food, as well as ritual and ecstatic purposes, reaches back into the early Neolithic period and perhaps to the late Paleolithic. Although *Cannabis* likely originated farther to the west than East Asia, in pre-Pleistocene times, it may have taken refuge in southern East Asia during the later Pleistocene glaciations (see Chapters 2 and 12). China was certainly one of the early regions of elaboration of its many uses, if not the birthplace of *Cannabis* agriculture. Over time, the emphasis on medicinal, ritual, and ecstatic use of *Cannabis* declined in China, while use for fiber and food remained widespread well into the twentieth century and survive today at a more limited level. Some cultural uses (e.g., medicinal, psychoactive, and ritual) may first have been practiced in Central and/or East Asia and were later disseminated into Europe via Eurasian steppe nomads.

By the end of the first phase of dispersal (out of its putative natural range of origin along with early ethnobotanical relationships in Central Asia), *Cannabis* had spread throughout much of its present-day range in China and evolved from a hypothetical proto-*indica* putative drug ancestor (PDA) into *C. indica* ssp. *chinensis* or broad-leaf hemp (BLH; see Hillig 2005a/b). BLH varieties express both the B_D and B_T alleles (de Meijer et al. 2003) for cannabinoid biosynthesis but show no evidence of recent human selection of either allele (a form of a gene) and produce only moderate amounts of both CBD and THC. However, although they cannot be considered drug varieties, BLH produces considerably more THC than European *C. sativa* narrow-leaf hemp (NLH) varieties. We are unsure whether the broad-leaf trait has arisen independently in China as well as in Afghanistan (*C. indica* ssp. *afghanica* broad-leaf drug, or BLD, varieties) or possibly the two taxa are interrelated through *C. indica* ssp. *kafiristanica*, the putative narrow-leaf drug ancestor (NLDA, also see Chapters 1, 2 and 12 for discussions of the evolution of *Cannabis* biotypes).

Since Genghis Khan founded Karakorum in the early thirteenth century, the city, with its semiarid, continental climate characterized by short, hot summers and long, very cold, dry winters, has been the capital of the Mongolian empire (Barkmann 2002). Recent archaeobotanical research at Karakorum by Rösch et al. (2005) exposed the ancient cultivation and utilization of *Cannabis* for fiber and seed oil at least from the thirteenth to the fifteenth centuries, along with a number of grains including millets (*Panicum miliaceum* and *Setaria italica*), barley (*Hordeum vulgare*), and some imported rice (*Oryza sativa*) in irrigated fields. This suggests that the only oil seed or fiber plant "more or less common" in this region was hemp: "This species also occurs today rather frequently and in large amounts in Mongolia, especially on moderately moist soils in valleys where the grazing pressure is not too high. Past or recent usage, for example as a drug, is not known from written or other sources. But the number of identified seeds indicates use. If this medieval use was based on cultivated or wild gathered material and in which way the plant was used is unknown." In addition to *Cannabis* seed, Rösch and his colleagues (2005) also extracted "*Humulus/Cannabis*" pollen in "small amounts over the full range of the core" in their excavations in the lake sediments of Ugii Nuur, 50 kilometers (31 miles) to the north of Karakorum.

FIGURE 15. Three female figurines from the lid of a Dian Kingdom (fourth to second centuries BCE Red River culture) bronze cowry jar depicts weaving on body-tension looms similar to those used until recently in southwestern China to weave hemp (adapted from Luo and Zhong 2009).

Diffusion from Northeastern China into Korea and Japan

Tarasov et al. (2006) used pollen records and archeological data to reconstruct environmental and human dynamics in northeastern China around 6000 to 5000 BP. They reported finding pollen of "*Cannabis/Humulus*" at the Taishizhuang site located in the transitional forest–steppe zone of northern Hebei province and associated it with the vegetation of the steppe biome. Hemp cultivation could have begun relatively early in northern and northeastern China and southeastern Siberia, and it is the only fiber species of any substantial significance in this temperate East Asian region today. In the early days of cultivation in northeastern Asia, millets and other annual crops such as *Cannabis* could effectively be grown in openings along rivers or streams in areas of periodic flooding, which, due to sediment deposition, had relatively fertile and humid soils; this is where *Cannabis* is presently found growing spontaneously.

Based on reliable seed evidence, we know that several domesticated plants, including *Cannabis*, were being cultivated in northeastern China at least by 3000 BP in a number of sites within a region formerly known as "Manchuria" and referred to simply as Dongbei (northeast) in Chinese today. For example, during archeological field work in the northern Changbaishan Mountains (Jilin province), Jia (2005) used flotation methods (one of the rare applications of this methodology in China) to recover ancient seeds associated with people who lived in the region approximately 2,000 years ago and employed a mixed economic strategy of hunting, gathering, fishing, and farming. Among five seeds recovered from soil samples excavated from the Qiaohexi site of the Tuanjie culture were two of wheat, one of broomcorn millet, one of hemp and one unknown. Based on comparative artifacts, Jia assumed that these seeds date "to around 1900 to 2400 BP." Jia also referred to carbonized "domestic hemp" seeds found in a house at the Guntuling site in the northern Changbaishan Mountains, which were dated to ca. 2000 BP, and suggested that some species such as soybean and hemp may have been domesticated first, or at least cultivated relatively early in northern and northeastern China. In addition, Jia pointed out that "remains of hemp fabrication were found in the Xingxingshao and Houshishan sites of the Xituanshan

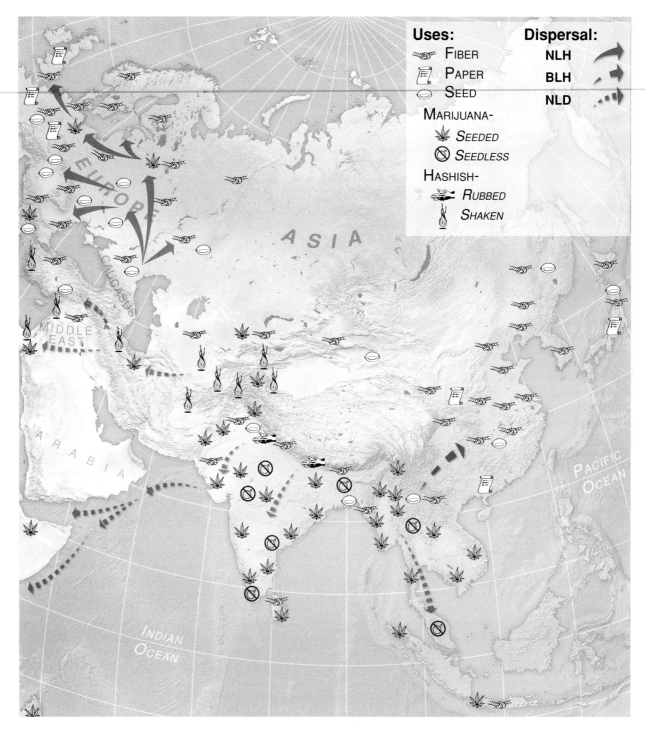

MAP 6. Dispersal of *Cannabis* for differing uses across Eurasia (cartography by Matt Barbee). (See TABLE 1 at the beginning of this book for explanations of *Cannabis* gene pool acronyms.).

culture" (ca. 3000 to 2500 BP) located in central Jilin. It is probably from northeastern China that *Cannabis*, or at least its use, was spread into the Korean Peninsula.

In Korea and Japan, as in China, *Cannabis* has been used ritually and to make cordage and cloth for thousands of years. Today *Cannabis* fiber is still produced on limited acreage in Korea (12,800 metric tons, or 14,100 tons in 2004 according to Lewin 2006; see also Clarke 2006a) and to a much smaller extent in Japan, and at a few locations on the Korean Peninsula it grows spontaneously. The use and

cultivation of *Cannabis* spread over much of northern East Asia, including what today is Korea and likely from there to present-day Japan. *Cannabis* probably arrived about 10,000 or more years ago (e.g., Kudo et al. 2009; Okazaki et al. 2011). We first review the evidence for the early presence and use of *Cannabis* in ancient Korea and then Japan.

Nelson (1993) suggested that around 5500 BP *Cannabis* may have been an early cultivated fiber crop among Korean "Chulmun" (or Jeulmun) coastal farmer and/or fisher groups. Based on earlier spindle whorls found in late Neolithic "Mumun" sites, she raised the question whether a "whole complex of plant domestication and its consequent industries" probably based on hemp, ramie, or native fiber plants was spreading across the Korean Peninsula during this period (Nelson 1993). The earliest archeological evidence for hemp textiles in Korean culture dates to ca. 5000 BP and includes cloth as well as clothing production equipment including sewing needles and spinning tools that were found in shell mounds at Gimhae, South Gyeongsang province in South Korea (Hyo-Soon 1995). Ancient bone needles and hemp threads were also found at an archeological site in Goongsan, Unha-ri, in Oncheon county of South Pyongan province, North Korea; this site is contemporaneous with the Gimhae site (Sim 2002; also see Kim 1979; Nelson 1993). Hemp fibers may also have been used in ancient Korea to weave bags used to collect and detoxify (in moving water) wild or semidomesticated nut crops (e.g., chestnuts and acorns), a cultural trait still practiced in some remote parts of Japan (see the section on Japan). Most of the plant remains from the Chulmun culture horizons in Korea, even before 7000 BP, consist of nut remains, especially acorns. Millet crops appear to have been introduced into Korea from China during the Middle Chulmun more than 5,000 years ago (Crawford and Lee 2003; Ahn 2004), about the same time Cannabis was also introduced into Korea or used more intensively as part of an early incipient farming complex. It is also possible that hemp or additional varieties of this crop were introduced into Korea during the Later Chulmun period along with rice, wheat, and barley. Seed evidence from the Daecheon-ri site at Okcheon-gun, South Korea, includes rice, wheat, barley, foxtail millet, and Cannabis dated from 5000 to 4800 BP and no later than 4500 BP, based on the recorded chronology of bulk charcoal and ceramics (Central Museum of Hannam University 2003). Although no direct dates for the seeds have been determined, and the possibility exists that some intrusion (mixing) may have occurred at the site because it is very near to the surface, there are published photographs that "indicate a good morphological basis for identification" (Fuller et al. 2007).

Possible cordage and fabric evidence of early Cannabis use in Korea includes an ancient rope, most likely made of Cannabis, which was recovered from the site of Pomuigusokin in Musan county, North Korea, and dated to the Bronze Age (ca. 3000 to 2300 BCE, Chösen iseki ibutsu zukan 1990); in addition, remains of a mat reportedly made of silk, hemp, and rushes was found with bodies buried in a wooden painted basket tomb in the Eastern Han Chinese colony of Nangnang in North Korea, near the present-day capital of Pyongyang and is dated to about 1900 BP (Koizumi 1934). There is, in fact, plenty of evidence that Cannabis was used as a fiber source for cordage and clothing, as well as a ritually significant plant, in Korea spanning the past 2,000 years (e.g., Choi 1971; Min 1985; Sim 2002; also see Chapters 5 and 9).

In Japan, as on the Korean Peninsula, Cannabis has been used as a fiber source and a ritually important plant for thousands of years. Olson (2002a; also see Olson 1997) tells us that a painting in a "Neolithic cave" from western coastal Kyushu Island closest to the Korean Peninsula shows "tall stalks with hemp-shaped leaves" along with oddly dressed people, horses, and waves that may depict "Korean traders bringing hemp to Japan"; although this interpretation is interesting, it is also highly speculative. In any case, we can be sure Cannabis was introduced to Japan long before extensive, large-scale rice farming began in that archipelago and that this early introduction occurred via the ancestors or descendents of the people who established the early and long-lasting Jōmon culture (ca. 12,500 to 2300 BP). Hemp was one of the earliest cultivated or at least managed plants in this part of East Asia, with evidence dating to at least 6,000 to 7,000 years ago (Imamura 1996a; Matsui and Kanehara 2006), and probably much earlier as supported by Cannabis seeds dated to ca. 10,000 BP that were found in sediment attached to potsherds in the Okinoshima archeological site on the Boso Peninsula in eastern Honshu (Kudo et al. 2009; Okazaki et al. 2011).

In contrast to other prehistoric peoples who made and used pottery, the early Jōmon hunted terrestrial animals, fished extensively (both coastal and inland; see Yoneda et al. 2004), and gathered wild plant foods; they also appear to have developed settlements at an early time and farmed to some extent (Imamura 1996b; Matsui and Kanehara 2006). However, none of the cultivated plants found in early Jōmon sites apparently served as their staple food resource, which they obtained mainly from wild native plants (Habu 2004). Warming climatic conditions following the most recent ice age resulted in significant increases in the number and range of edible nut-bearing, broadleaf deciduous and evergreen trees in Japan, including Japanese horse chestnuts (Aesculus turbinata and Castanea crenata), acorns (Quercus), and walnuts (Juglans sieboldiana). Access to and storage of these nuts, especially in parts of Honshu Island (Kanto and Tohoku districts) and southern Hokkaido Island, along with processing techniques to remove the toxic tannins in the nuts, allowed intensification of a "storage economy" that sustained sizeable human settlements. This in turn stimulated the growth of a complex ritualistic, spiritual culture with a social hierarchy (Matsui 1996). In fact, the later Jōmon included some of the "most affluent hunter-gatherers known to archeology" with a sophisticated material culture that was "spectacular both in quantity and quality" and included, among other elaborate items, "decorated pottery, personal ornaments and ritual objects" (Pearson 2006).

The course of Jōmon cultural development differed greatly from that of China where farming appeared very early on and formed the subsistence basis for development of Chinese civilization. While it is agreed that the Jōmon carried out a limited amount of farming for perhaps thousands of years, fully agricultural communities did not appear in Japan until the first millennium BCE with the introduction of the Yayoi culture based on relatively large-scale rice cultivation. In addition to their gathering of wild nuts and fishing, the Jōmon people cultivated a relatively small number of plants such as Cannabis at some early sites. Although Pearson (2006) argued that Jōmon "villages of collector hunter-gatherers show no signs of reliance upon cultivation," Crawford (2011) has further articulated Jōmon relationships with key plants, including Cannabis: "Jōmon populations engaged in niche construction/anthropogenesis that ranged from annual plant encouragement and probably management, lacquer tree (Toxicodendron verniciflua) and nut tree (Castanea crenata and Aesculus turbinata) management, and probable domestication of barnyard millet and soybean as well as cultivation of bottle gourd and hemp and possible cultivation of Perilla and adzuki."

The ritual, spiritual, and/or social status importance of cultivated plant products were most probably quite substantial, otherwise the Jōmon people would not have continued to grow them. For example, at least two of their known cultivated plants, hemp, and paper mulberry (Broussonetia papyrifera) are notable fiber plants, and clothing was probably an important symbol of status as well as providing personal protection from the elements. Indeed, hemp textiles have long played a highly significant role in Japanese culture with direct relations to the highest-ranking individuals in Japanese society right up to the present time (see Chapters 5 and 9), and cultivated *Cannabis* would have provided resources that may have been associated with status, ceremonies, and important nonstaple resources such as medicines and other luxury items (see Hayden 1990; Thomas 1996; Gilligan 2007). Relationships between humans and *Cannabis* in Japan stretch from pre-Neolithic to modern times, although very restricted from the middle twentieth century on, providing useful resources other than its edible, oily seeds, although these seeds may have been consumed as they were in early China (see also Chapters 5 through 9). In sum, as a reliable fiber and food source and perhaps an early ritually important resource for clothing, and possibly even for mind-altering and medicinal purposes, *Cannabis* has a long multipurpose ethnobotanical history in Japan.

The Japanese word "*asa*" is used collectively to refer to a number of bast fiber crops including jute, ramie, wisteria, elm, mulberry, sisal, and flax as well as hemp. *Cannabis* hemp in Japan is more explicitly referred to as "*taima*" (tall or large hemp), which is derived from the original Chinese *dà má* (大麻) and shares the same character "*má*" (麻), another indication that *Cannabis* spread into Japan from continental East Asia. In Chinese, there are many fiber plants known as "*ma*" and in English many known as "hemp." As in Chinese and English, when interpreting written Japanese records, collective nouns, in this case "*asa*" meaning *Cannabis* as well as other bast fibers, can lead to considerable confusion.

During the late Paleolithic period, up to approximately 18,000 years ago, a land connection between the Asian continent and what was to become the Japanese Archipelago formed a peninsula. This terrestrial bridge allowed people from Eurasia to migrate into areas known today as Japan. People who hunted marine and terrestrial animals and gathered wild plants as their main foods probably moved into these unoccupied areas following herds of game or due to pressures from other humans. As temperature increased, glaciers retreated, and sea level rose during the terminal phase of the last ice age, the Japanese islands formed, more or less isolating local hunters and gatherers, and over time they became progressively more sedentary, exploiting a variety of food and other resources throughout the year (Pearson 2006). As noted earlier, it appears the Jōmon people practiced limited cultivation beginning at a relatively early date, but the extent of their farming activities has been a point of contention:

> The most debated question about [Jōmon] subsistence concerns the possible contribution of agriculture. Many [Jōmon] sites contain remains of edible plants that are native to Japan as wild species but also grown as crops today, including the *adzuki* bean and green gram bean. The remains from [Jōmon] times do not clearly show features distinguishing the crops from their wild ancestors, so we do not know whether these plants were gathered in the wild or grown intentionally. Sites also have debris of edible or useful plant species not native to Japan, such as hemp, which must have been introduced from the Asian mainland. (DIAMOND 1998; see also MATSUI AND KANEHARA 2006)

Crawford and Takamiya (1990) referred to hemp as one of the "indigenous cultigens" in ancient Japan; in fact, hemp is one of a suite of cultigens including perilla (*Perilla frutescens*

var. *japonica*), buckwheat, cock's foot (*Echinochloa* sp.), and burdock (*Arctium lappa*) identified in Jōmon period sites from Japan and Korea as well as China and eastern Russia (Yamada 1993; also see Crawford 2011). Although it is possible that *Cannabis* was native to the Japanese Archipelago, we argue that people more likely introduced it from the eastern Asian mainland during the late Pleistocene or early Holocene.

Sometime during the latter part of the first millennium BCE, rice farmers migrated to Japan from the Korean Peninsula. These "Yayoi" immigrant cultivators mixed with the Final Jōmon period people who were primarily hunter-gatherer inhabitants of the Japanese Islands and thus formed "the roots of the Japanese people and language of today" (Bellwood 2005). According to Rathbun (1993), hemp was brought to Japan before or along with paddy rice from the Asian mainland around 2300 BP. However, as noted earlier, there is evidence that *Cannabis*, and perhaps its cultivation, were introduced to Japan much earlier, maybe more than 10,000 years ago.

Imamura (1996a), for example, refers to macrofossil remains of plants that may have been cultivated, including hemp, which were deposited in the coastal Early Jōmon Torihama shell midden, which lies at the confluence of the Hasu and Takase Rivers, near Lake Mikata in Fukui prefecture on Honshu Island in southwestern Japan (Okamoto 1979; D'Andrea 1999). Archeological excavation at Torihama began in 1962, and archaeobotanists were among those who eventually devoted much attention to this special site—unusual in that it is a wetland site where the Jōmon people regularly dumped considerable amounts of organic rubbish over the side of a hill upon which this early settlement was located. The rubbish fell into the lake below, helping explain the excellent preservation of the organic waste over time. Indeed, early scientific investigations of macro and microfossil remains from the Torihama site "encouraged the establishment of archaeobotany in Japan" (Matsui and Kanehara 2006). From this wetland site, fragments of ancient hemp cordage, about 2.0 millimeters (0.08 inch) in diameter (ca. 7000 BP), and knitted fabric (ca. 7000 to 5500 BP) were excavated (Nunome 1992; also see Okamoto 1979; Kasahara 1981, 1984; Unemoto and Moriwaki 1983). More recently, Matsui and Kanehara (2006) refer to *Cannabis* as an example of a plant cultivated by the ancient Jōmon culture, with archaeobotanical evidence for its presence found at both the ancient Matsugasaki and Torihama sites "dating from 6000 to 5220 BP."

The evidence described earlier linking hemp use to Early Jōmon cultivation is not unique. For example, pollen evidence from Ubuka Bog in southwestern Honshu Island indicates that hemp and a number of other cultigens "were present in Japan or undergoing domestication during the Early Jōmon" (Crawford and Takamiya 1990; also see Morikawa and Hashimoto 1994; Matsui and Kanehara 2006; Crawford 2011 for more recent reviews of these early cultivated plants in the Jōmon culture). The Jōmon period has a very lengthy time span, broadly dated from 12,500 to 2300 BP. Among the people of the transitional Final Jōmon-Tohoku Yayoi period, crops besides rice were cultivated; for example, in the "pit house contexts at Kazahari," large amounts of *Cannabis* seeds, as well as those of broomcorn and foxtail millets, were recovered, indicating that once farming was well established by the Yayoi culture, hemp was commonly grown (D'Andrea 1999; also see D'Andrea 1992; D'Andrea et al. 1995).

The word "Jōmon," used in reference to the long-lasting, ancient, first peoples of Japan, can be translated literally as a "pattern of ropes." According to Imamura (1996a), this word means "cord-mark on pottery" in Japanese, as distinguished from other kinds of pottery marking produced by a paddle and anvil technique, which is widely distributed in China and Southeast Asia. "Jōmon" refers specifically to "an impressed pattern created by rolling a cord or cord-wrapped dowel on the still-soft surface of a pot." Olson (1997) describes the Jōmon people as "hunting and collecting people who lived a civilized, comfortable existence and used hemp for weaving [actually knitting or knotting] clothing and basket making." Jōmon pottery was a characteristic cultural implement of hunters and gatherers as well as farmers in early Japan, and at least some of the pottery cord-impressions were probably made with hemp cordage. Archeologists have uncovered ancient cloth artifacts that were pressed into ceramic ware for designs (e.g., see Imamura 1996a, figures 2.1 and 2.2); these early fabrics, along with ancient spindle whorls and loom parts, were found in many farming sites of the later Jōmon culture (ca. 3000–2500 BP to about 1700 BP), suggesting that clothing made of hemp and ramie (karamushi) fiber was widely available at least by the third century CE (Farris 1998b; also see Aikens and Higuchi 1982). In addition, these artifacts suggest that loom weaving may also have been introduced from the Asian mainland during this time. For example, hemp cloth was recovered from sediment levels dated to the Yayoi period at Nabatake in Saga prefecture, northeastern Kyushu (Kasahara 1982, 1984; also see Crawford and Takamiya 1990; Hudson and Barnes 1991).

However, as we noted earlier, evidence indicates that Cannabis was introduced to Japan several thousand years before the Yayoi period, perhaps first entering the central part of the archipelago from the Korean Peninsula and spreading from there. On the other hand, hemp could have been first or also brought to Japan along a northern route from the Sakhalin Peninsula of eastern coastal Siberia more than 3,500 years ago, as a continuous record of seeds have been recovered from Jōmon sites on Hokkaido. Imamura (1996a, citing Yamada 1993) refers to ancient remains from the Satsumon period in Hokkaido (ca. seventh to thirteenth century CE), indicating cultivation of several food species, along with "hemp." Recent archeological interpretation suggests that the Satsumon people were ancestors of the Ainu and were, like the Jōmon people to the south, not only hunters and gathers but also farmers, including hemp cultivation. Utilizing floatation methods, Crawford and Takamiya (1990) recovered Cannabis seeds from the Satsumon period, dated to the ninth century CE.

We think the use of hemp fiber has been important in Japan since early Jōmon times not only in clothing production but also for a variety of cordage uses, and as we noted earlier, "Jōmon" refers to cord-marked pottery, a key artifact of the Jōmon people. However, it must be remembered that Japan hosts a wide range of wild bast fiber sources such as kudzu, elm, and wisteria, which are known from historical and ethnobotanical records to have been used for making cordage and rough cloth.

Another important and longstanding use of hemp cordage was to make sacks, some of which had special shapes and were used to carry and detoxify certain wild foods such as horse chestnuts, and such sacks are still being used to carry chestnuts in some rural areas of Japan (see Chapter 5). Hemp sacks have also been used for other purposes in Japan such as extracting seed oil in a traditional wedge press. One of the plants used as an oil source at least until the nineteenth

century in Japan was Perilla, an early cultigen among the Jōmon (Huang 2000; Matsui and Kanehara 2006). Cannabis seeds may also have been pressed for oil by the Jōmon people. This traditional use of Cannabis fiber for sack production is perhaps another reason it was cultivated early on in Japan.

Furthermore, it should be noted that Jōmon ceramic ware was probably used for, among other things, storage of wild or managed plant foods such as water chestnuts and hemp seed. Although archeologists and anthropologists have devoted much attention to food storage among food-procuring societies, particularly as it related to human movement patterns and social class development, they have given much less thought to the practical features of storing food. Obvious features that differentiate various plant foods are the degree of processing needed to store them, as well as how long they will remain edible and desirable. Cannabis achenes with their thin but tough outer husk or shell can be stored for relatively long periods of time; this may explain the presence of early Holocene Cannabis seeds found attached to potsherds found at the Okinoshima archeological site on the Boso Peninsula in central Japan and dated to about 10,000 BP (Kudo et al. 2009; Okazaki et al. 2011).

The Sanguo Zhi or "Records of Three Kingdoms," written by the Chinese historian Chen Shou between 280 and 297 CE, includes the Wei-chih tung-I-chuan or "Accounts of the Eastern Barbarians," also known as the "Annals of the Wei." In the second part of this ancient literary work, the Wo Zhuan or "Account of Wo", Chen Shou tells us that Japanese people cultivated grains such as rice, along with "hemp plants and mulberry trees" (Farris 1998b). Support for this written record comes from the famous fortified village site at Yoshinogari on northern Kyushu Island, which was excavated in detail in the 1980s. Evidence indicates that this relatively large settlement site was occupied throughout the Yayoi period, with about 1,000 to 1,500 residents at the height of its urban importance during the first three centuries CE. In the Yayoi cemetery at the Yoshinogari site, archeologists recovered both silk and hemp cloth (Hudson and Barnes 1991).

As noted earlier, when rice farming came to Japan more than 2,000 years ago from China and Korea, Cannabis was probably already being cultivated, and likely farming of hemp crops increased along with the introduction of rice and its culture (Rathburn 1993). It may be that hemp was utilized as a wild or semicultivated food and fiber source before the Yayoi period; subsequently, after loom weaving and more advanced cultivation techniques were introduced from the Asian mainland, increased need for fiber sources led to an expansion of hemp growing. However, the first use of Cannabis in Japan and elsewhere in East Asia may have been for food. Hemp seed possesses a high content of both protein and vegetable oil containing essential fatty acids (EFAs) and therefore is a perfect dietary compliment to rice (see Chapter 6). In any event, very early on, Cannabis became a basic fiber source used to produce ropes, clothes, and, somewhat later, Japanese paper (washi; see Chapter 5).

Among the most significant traditional uses of Cannabis in ancient Japan was for its special ceremonial symbolism and protective spiritual power. The ritual use of hemp in Japan is, to a certain extent, still very important and is discussed in detail in Chapter 9. However, it is important to note here that Cannabis hemp fibers have long been used ceremonially by Shintō priests because of their association with purity. For example, in the Kogo shūi (1:550), a document written in the Heian period by Hironari in 807 CE, we are told that in the early period of the "first legendary emperor Jimmu," Prince Ame no Tomi, "with the assistance of the various branches of the Imbe [Inbe] clan," was assigned to produce "sacred treasures" for Shintō religious use; these included "mirrors, jewels, spears, shields, cotton and *hemp cloths*" (Coward et al. 2007; also see Brinkley and Kikuchi 1915; italics added for emphasis)

The use of *Cannabis* for shamanic or other cult use may have been a culture trait of significant importance in ancient Japan, especially before the Yayoi culture began spreading through the islands. During the long-lasting hunting and gathering period of the Jōmon culture, shamanism was undoubtedly important, since it lies at the roots of Shintō and is still practiced in some popular forms in Korea. Cult use of *Cannabis* for mind-altering purposes to communicate with ancestors and in divination was practiced at least in remote western China as recently as the third millennium BCE (see Jiang et al. 2006; Russo et al. 2008; Chapter 7). We suggest that after the rise of Confucianism, which spread from China through East Asia to Japan, the ingestion of *Cannabis* resin for psychoactive, ritualistic purification rites was eventually suppressed in Japan, as it was in China. However, the vestiges of the religious and ritualistic importance of *Cannabis* remain in varying degrees in China, Korea, and Japan, primarily for its symbolic fiber use (see Chapter 9). Although, traditional use of *Cannabis* for a number of purposes was an important part of Japanese culture during the Jōmon, Yayoi, and more recent periods, the present use of *Cannabis* has been reduced to a mere fraction of its former status.

Diffusion into South Asia

Having explored the antiquity of *Cannabis* use in China, Korea, and Japan, we now turn our attention to early evidence for *Cannabis* use in South Asia. In modern times, *Cannabis* has been reported growing spontaneously in many locations across extensive areas of South Asia, including northern India, northwestern Pakistan, and along the slopes of the Himalayas from Kashmir in the west through northern Burma (Myanmar) in the east up to 3,000 meters above sea level (Chopra and Chopra 1939). *Cannabis* also grows spontaneously across the Yungui Plateau of southwestern China (Clarke personal observations, 1995 to 2003). Although feral and cultivated *Cannabis* grows spontaneously over a vast area of South Asia, according to the comprehensive *Report of the Indian Hemp Drugs Commission 1893–1894*, it is not indigenous to India (also see Watt 1908; Kaplan 1969). Assuming that *Cannabis* did not have its origin in South Asia, from where and when did it enter ancient India? Here we briefly review the historical, chemotaxonomic, and archaeobotanical record that indicates the presence and use of *Cannabis* in the South Asian region.

A variety of evidence suggests that the Indo-Aryan tribes of western Central Asia introduced *Cannabis*, or at least its use as a drug plant, into South Asia from somewhere to the north or west around 4000 BP. Historical interpretations have generally suggested that after about 3700 BP, militant nomads commanding effective, mobile armies moved out of the vast Central Asian steppes, mountains, and deserts. These "Aryan" nomads traveled southwest onto the eastern Iranian Plateau of Persia and then southeast into the Indian subcontinent. Some of them may have crossed the Hindu Kush and Himalaya Mountains in relatively large numbers by pushing through a few remote passes and eventually down into the Punjab, or "the Land of Five Rivers," in northern Pakistan, although there has been little evidence to support this specific route. Some Aryan migrations and the associated introduction of *Cannabis* into South Asia may also have come directly from the north. At least one psychoactive preparation of *Cannabis*, hashish, probably diffused into South Asia from this northern direction and ultimately, we believe, from southern Central Asia. Although it seems that the Aryans produced most of their clothing from wool (Burkill 1962) and depended upon sesame (likely native to India) as their chief oil source (Basham 1959), there is evidence that these people came into India using hempen ropes, with *bhanga* as their name for the *Cannabis* plant, fiber, and drug and that they may have used hemp seed oil before they contacted and began using sesame oil (Majumdar 1952). However, within South Asia, evidence suggests that hemp was not a significant fiber source, probably because of the abundance of native fiber sources and the long-standing use of cotton and later flax (Fuller 2008).

More recently, genetic, archeological and archaeobotanical evidence has supplemented traditional historical interpretations of the spread of Indo-European-speaking peoples into South Asia and their probable initial introduction of *Cannabis* in this region. Modern research tracing genetic markers on human Y chromosomes supports the traditional belief that Indo-European-speaking peoples migrated into South Asia from north and east of the Iranian deserts and that the introduction of agriculture to this region preceded the introduction of the Indo-European language group (Wells 2002). Beginning about 5,000 years ago, the domestication of animals, especially the horse, provided some Central Asian steppe peoples a chance to enlarge their geographical range in diverse directions (Zvelebil 1980; also see the relevant discussion on relationships between horses, hemp, and humans earlier in this chapter). These Central Asian peoples, probably derived from the Andronovo and Srubnaya cultures, moved through what is now Iran and Afghanistan to reach South Asia. The appearance of these immigrants in what is Pakistan, India, and Nepal today more or less coincided with the decline of the extensive agricultural civilizations of South Asia at that time, including those that controlled the Indus Valley (e.g., the ancient "Harappans"). In contrast to the sedentary agricultural economy of the ancient Harrappan civilization (ca. 2600 to 1900 BCE) the Aryans were originally nomadic herdsmen. Recent studies of mtDNA samples from the southwestern and Central Asian corridors reflect the large amount of genetic variation occurring in the peoples of this large area and also indicate it was in the Indus Valley and Central Asia that genetic lineages of western Eurasia encountered those of both southern and eastern Eurasia: "The amalgamation of different genetic components in this area may have resulted from the successive and continuous waves of migration from diverse geographical sources at different time periods, from the early human settlements in the region after the 'out of Africa' dispersal to migrations associated with the diffusion of new technologies, such as farming and/or pastoral nomadism, and accompanied by new languages, like the incursions of Indo-Iranian speakers from the northwest" (Quintana-Murci et al. 2004).

We suggest that these migrating Aryan tribes brought to South Asia a number of new materials and innovations, among which were *Cannabis* and some of its various uses. As both a camp-following weed and a useful plant consciously transported by these newly arriving peoples, *Cannabis* eventually escaped human cultivation and colonized open habitats, especially near rivers and streams in the mountainous uplands of the Himalayan foothills. Thus at least in the early phases of recorded South Asian history, *Cannabis* was probably a relatively obscure plant for the mass of population centered on the lowland Gangetic Plain and riverine valleys of southern India.

On the other hand, chemotaxonomic evidence (Hillig and Mahlberg 2004; Hillig 2005a) leads us to another possible scenario for the introduction of *Cannabis* into South Asia. Genetically distinct plant species (or taxa within a species) are largely the products of reproductive isolation, either natural or human induced. The spread of *Cannabis* along with migrating peoples and its later evolution as a crop plant were influenced by human choices of which *Cannabis* products were important and therefore were the key determinants of the variation we observe in genus *Cannabis* today. Plant dispersals can result in both genetic isolation and recurring selection, which reinforce the varietal characteristics of populations; however, humans also transport *Cannabis* to new environments with the added possibility of the newly introduced *Cannabis* hybridizing with extant *Cannabis* populations in the area, and different uses may also be introduced. *C. indica* narrow-leaf drug or NLD varieties and their feral relatives (NLDA) found in South Asia today are more closely related to other members of *C. indica*, such as BLH and BLD taxa, than they are to present-day European *C. sativa* NLH and narrow-leaf hemp ancestor (NLHA) populations.

As noted earlier, the Aryans entered South Asia basically from the northwest from Central Asia; however, if they carried *C. sativa* seeds with them, their genes were apparently swamped out by the dominant *C. indica* gene pool extant today. Alternatively, *C. indica* might have traveled all the way from ancient China across the steppe to the Indo-Aryan territories and then dispersed into South India, all of this occurring before about 3500 BP when Central Asian nomads could have encountered psychoactive *C. indica* during their eastward travels. Another possibility is that *C. indica* arrived in South Asia directly from a Pleistocene refuge in southern East Asia. All these hypothetical scenarios would be in agreement with taxonomic divisions proposed by Hillig (2005a; also see Chapter 11 for a detailed discussion of *Cannabis* taxonomy).

A hypothesis suggesting the original diffusion of *Cannabis* into ancient India with Aryan tribes around 3500 BP, either via the Iranian Plateau or directly from the Eurasian steppe via Central Asia across Himalayan passes, presupposes the nonexistence of this plant in the Indus Valley or elsewhere in South Asia before the Aryan migrations. Until recently there has been a dearth of archaeobotanical or archeological evidence that can substantiate the presence of *Cannabis* in ancient South Asia prior to about 3500 BP. This may be because archeologists, up until a few decades ago, have not looked long and hard enough for plant remains, and it is likely that additional and more substantial *Cannabis* evidence will be discovered in the future. Satisfactory interpretation of the enigmatic Indus script and comprehensive, reliable pollen analysis of ancient flora may also present evidence for the presence of *Cannabis* in the Indus Valley and elsewhere in South Asia before 4000 BP. But until such time, we suggest that hemp was not growing in northwestern India, or anywhere else in South Asia, until the influx of migrating Indo-European speaking tribes from the north approximately 3,500 years ago.

Archaeobotanical Evidence from South Asia

Recent archaeobotanical research at the ancient Lahuradewa mound, the site of an early lakeside settlement in the Middle Ganges Plain region (Sant Kabir Nagar district) of Uttar Pradesh, has yielded evidence of *Cannabis* dating to "Period II," ca. 4000 to 3200 BP (Tewari et al. 2006). This evidence includes fragments of "wood charcoals" recovered by archeological flotation methods. Among the 34 taxa of species identified from charcoal fragments, the only "nonnative" reported is *Cannabis*: "Except *bhang* (*Cannabis*), which is a Central Asian element and [has] been spread in the Indian region in [the] wake of human preferential treatment, all the [other] 33 taxa were a fraction of indigenous vegetation in the surrounding forested zones of Lahuradewa settlement" (Tewari et al. 2006). In addition, ancient *Cannabis* pollen was reportedly recovered from Lahuradewa Lake, where it was extracted from "the upper part of the profile" and said to be consistently associated with pollen of cereals and other economic plants (Saxena et al. 2006, citing Chauhan et al. 2005). Although the ancient date and identification of this pollen as *Cannabis* may warrant caution at this point, according to Tewari et al. (2006), pollen identified as "*Cannabis*" was recovered from the Period II level dating to around 5000 BP.

As we noted earlier, the problematic differentiation of *Cannabis* and *Humulus* pollen makes this evidence somewhat questionable, although *Humulus* may not have become established in South Asia until much more recently. The Neolithic site at Lahuradewa is one of the earliest so far discovered in India with indications that agriculture including rice cultivation was established here as early as the seventh millennia BCE. Indeed, new evidence has led some to suggest that the Middle Ganges Plain might have been an area where agriculture developed independently. Nevertheless, dwarf wheat (*Triticum sphaerococcum*), bread wheat (*Triticum aestivum*), lentil (*Lens culinaris*), and other plants such as *Cannabis* indicate the diffusion of crops into the region "between [the] north-western Harappan zone and the Early Farming zone of the Middle Ganga Plain" (Tewari et al.2006), with perhaps some of these and other plant species being introduced from differing directions.

Additional ancient remains of *Cannabis* charcoal from the South Asian subcontinent have been found in the Neolithic site of Senuwar in the northern Indian state of Bihar (Saraswat and Chanchala 1995; also see Saraswat 2004 and Fuller 2002). Fuller (2008, citing Saraswat 2004) refers to indications for hemp in the form of ancient seeds and wood charcoal recovered from Chalcolithic Senuwar and dated to ca. 3300 to 2600 BP. In Pakistan, *Cannabis* remains have also been recovered from the famous Harappa site, the remarkable ancient urban settlement associated with the early Indus civilization (ca. 5000 to 3500 BP). Harappa itself is located on the banks of the Ravi River in the Punjab region of Pakistan; it is here that some of the earliest archaeobotanical remains of *Cannabis* in South Asia have been found in rare opal phytoliths (Fujiwara et al. 1992). As noted earlier, opal phytoliths are tiny, three-dimensional inclusions of plant cells formed as a product of absorbing water with dissolved silica. They are very durable and can serve as markers of past occurrence of specific plant species. Analyses of phytolith assemblages found at ancient Harappa provide evidence for a number of crops and other useful plants, including a few derived from the hairs on the leaf epidermis of *Cannabis*. These silicate plant deposits were found in a soil sample taken from the floor in the "citadel" of Harappa and have been dated to the second millennium BCE (Madella 2003). This places them in the earliest part of the Late Harappa Phase ca. 3800 to 3300 BP: "In sample H36B [from an excavated trench] has been observed the typical epidermic structures related to long hairs of dicotyledonous plants. Some of these silica skeletons still retain the hair while others only preserve the ring of hairs surrounding the hair base. These epidermal hair structures have the same structure as the ones produced in the leaves of hemp (*Cannabis sativa* [probably *C. indica*])" (Madella 2003).

The extraordinary recovery of *Cannabis* phytoliths in Harappa or charcoal remains at Lahuradewa and Senuwar do not indicate use of the plant for a particular purpose. *Cannabis* may have been used during the latter period of the Indus Civilization in one or more of these sites for fiber or seed rather than for ritual or medicinal uses; however, this is doubtful (e.g., see Royle 1855; Fuller 2008; and Chapter 5, which focuses on fiber use). Zohary and Hopf (1988) suggested that the psychoactive properties of *Cannabis* appear to have been recognized and utilized in India at least by 3000 BP (see Chapter 7 for more detail about the mind-altering use of *Cannabis* in South Asia).

Fuller (2006) refers to the technological diffusion into the Harappan culture of such things as "Chinese like stone harvesting knives" found in Kashmir, which seemed to have occurred "during the later Harappan horizon, after 2000 BC[E]," and it was early in the second millennium BCE that *Cannabis* most likely spread into South Asia directly from China or via Central Asia. Fuller (2006) surmised that the northern Pakistan and Kashmir region had "developed contact with cultural groups to the north [and] east in the Chinese cultural sphere, indicating either long-distance trade or immigration into adjacent Himalayan zones of Sino-Tibetan speaking groups." In this context, Fuller also refers to the presence in ancient South Asia of foxtail millet, which is not native to the subcontinent but rather to some part of Asia to the north. This grain crop was apparently domesticated in Neolithic northern China, possibly along with hemp, and perhaps also in other areas in Central Asia or even the Caucasus region (Le Thierry d'Ennequin et al. 2000). If the identification of foxtail millet is also correct, then it can be assumed that this species was introduced through northwestern South Asia and eventually reached Gujarat by the beginning of the Mature Harappan Phase ca. 4600 to 3700 BP. *Cannabis* could have entered South Asia by cultural diffusion as well, only perhaps during the later period of the Indus Valley Civilization after about 3800 BCE. Indeed, Fuller (2006, 2011) thinks it possible that foxtail and broomcorn millets were brought with a number of additional "new technologies and species from Central Asia or the West at the end of the Harappan period." This would have included horses, donkeys, and camels as well as "Central Asian fruits like apricots, peaches, almonds and walnuts, and possibly a field crop variety of *Cannabis*" (Fuller and Madella 2001). It is during this same time frame that there is cultural diffusion from central China, as suggested by ancient Chinese-style harvest knives found in northern India.

In western Nepal, *Cannabis* evidence was discovered during an investigation of ancient highland agriculture in the Jhong Khola River Valley, which is located north of the main chain of the Himalayas, between 3,000 and 4,000 meters above sea level. The Jhong Khola is a tributary of the Kali Gandaki River whose watershed forms the deepest gorge in the world and contains extraordinary biodiversity from subtropical forest to a high, dry landscape similar to the Tibetan Plateau. Ancient river terraces in the upper zone of the watershed are still cultivated. Archeological excavations in a series of artificial caves hewn into the rock face on the northern bank of the Jhong Khola River, as well as excavations in settlement ruins at the base of these rock faces, provided an extraordinary chance to examine the history of human settlement and agricultural impact on the landscape (Knörzer 2000). Excavations revealed a series of archeological periods, the first dating from ca. 1000 to 400 BCE, with fossil seed evidence for only three cultivated plants including barley, buckwheat and the fiber and oil crop flax. The second period of use at these sites dates from 400 BCE to 100 CE, with fossil seed evidence from Mebrak Cave in the Jhong Khola Valley for many more cultivated species. Among the macrofossils from this second period were 41 seeds of flax from five samples and 32 seeds of "*Cannabis*" recovered from seven samples. Knörzer (2000) refers to flax as one of the traditionally cultivated plants in the area but identifies hemp as one of the "imported cultivated plants" along with several others including rice (*Oryza sativa*). He suggests that these plants were probably brought in from the north since "the Kali Gandaki Valley is one of the main corridors which connect the Tibetan [Plateau] with the Indian subcontinent and it was always an important trade route, evidently already in these early times." No fossil seed evidence for either hemp or flax was found in Jhong Khola Valley samples dated after 100 CE. The seeds of hemp may have been used for oil or food or could have also been cultivated for fiber as well as ritual and medicinal purposes. However, based on our review of its historic use for fiber in South Asia, hemp would seem to have been of relatively minor importance in the ancient cultures of this region. The overwhelming use of *Cannabis* in South Asia has been for its ritualistic, spiritual, and medicinal attributes.

By the end of the first phase of its dispersal, *Cannabis* had moved into northern South Asia with the aid of humans, probably long before its initial introduction onto the Indian subcontinent; this introduced, hypothetical putative ancestor (PA, possibly *C. ruderalis*) evolved into the putative drug ancestor (PDA) capable of producing THC. (PA, possibly *C. ruderalis*) had evolved into the putative drug ancestor (PDA) capable of producing THC. The mutation of the cannabinoid synthase gene giving rise to the B_T allele allowing the biosynthesis of THC was the primary evolutionary step leading to drug *Cannabis*. All modern marijuana varieties are descendants of traditional South Asian *C. indica* narrow-leaf drug (NLD) and broad-leaf drug (BLD) landrace varieties that evolved from a THC-producing common ancestor.

Diffusion into Southwest Asia and Egypt

Definitive evidence for the appearance of *Cannabis* in the early stages of Mesopotamian civilization is lacking. However, we propose that *Cannabis* use moved into the general area of Southwest Asia and eventually the Mediterranean region relatively soon after domestication of the horse several thousand years ago. As noted earlier, about 3700 to 3500 BCE, chiefs of the ancient Maikop culture of the northern Caucasus Mountains began trading, via "Anatolian middlemen," with recently established urban areas of Middle Uruk in Mesopotamia. Items traded north from the Maikop culture to

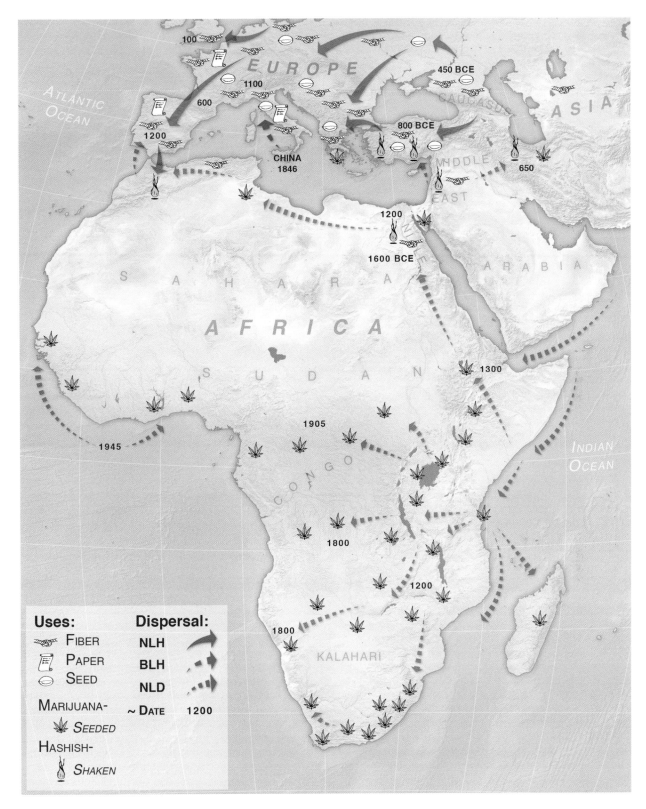

Uses:

〰 FIBER
📄 PAPER
⬭ SEED

MARIJUANA-
🍁 SEEDED

HASHISH-
🔥 SHAKEN

Dispersal:

NLH ➤
BLH ⇢
NLD ⇢

~ DATE 1200

MAP 7. Dispersal of *Cannabis* for different uses across Europe and Africa based largely on historical evidence (cartography by Matt Barbee). (See TABLE 1 at the beginning of this book for explanations of *Cannabis* gene pool acronyms.)

the Pontic steppes included new metal alloys that allowed for more sophisticated metallurgy and perhaps the first wagons (Anthony 2007). In return, circumstantial evidence indicates that wool, horses, antelope hides, and *Cannabis* may have been moved south as exchange items. Regarding the possible movement of *Cannabis* through trade from the Eurasia steppe to Mesopotamia, Anthony (2007, citing Sherratt 2003, 1997) pointed out the following: "Greek *kánnabis* and Proto-Germanic *hanipiz* seem related to Sumerian *kunibu*. Sumerian was dead as a widely spoken language by about 1700 BCE,

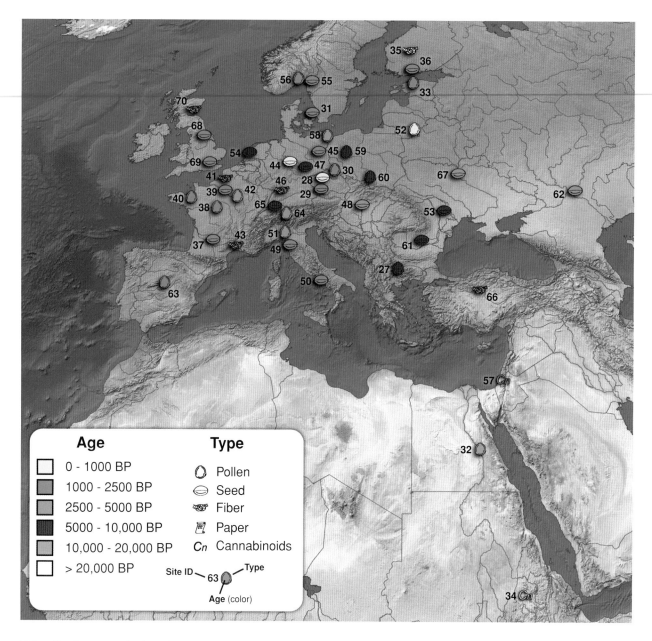

MAP 8. Key European archeological sites discussed in the text where *Cannabis* remains have been recovered (cartography by Matt Barbee).

so the connection [among these ancient names from *Cannabis*] must have been a very ancient one, and the international trade of the Late Uruk period provides a suitable context."

The impact of chariot warfare on the history of civilization and indirectly on the cultural diffusion of *Cannabis* was quite significant in that it allowed successive waves of militant, chariot-driving Eurasian steppe nomads to spread their often dominant influence over vast areas of the Eurasian continent, including Southwest Asia. Where barbarian war bands settled, they altered the lives of native peoples, sometimes drastically and in other cases only superficially. Although *Cannabis* was not commonly known nor widely used in the early stages of Mesopotamian civilization or elsewhere in Southwest Asia, we assume that its use, perhaps for fiber and medicinal and ritual purposes was spread into this large region and subsequently into the eastern Mediterranean regions relatively soon after regular contact with the nomads of the Eurasian steppes.

There is evidence that *Cannabis*, at least as a fiber source, was known in Palestine well before 2000 BP. According to the Polish historian and ethnographer Sarah Benetowa (1936, also known as Sula Benet 1975), the assumption that hemp was not known or used in ancient Palestine (e.g., see Dewey 1913; Moldenke and Moldenke 1952; Bennett 2011) is incorrect; Benetowa suggested that *Cannabis* is a term "derived from Semitic languages and that both its name and forms of its use were borrowed by the Scythians from the peoples of the Near East." Benetowa also proposed that *Cannabis* was known, traded, and used in the ancient Near East and Europe at least a millennium prior to the first reference to this useful plant by Herodotus, the famous Greek historian (lived ca. 484 to 429 BCE). In addition to its fiber applications, Benetowa referred to Old Testament references to its use in ancient Palestine for ritual incense and psychoactive purposes; she also refers to Meissner (1925), who indicated that *Cannabis* incense was employed in Assyrian and Babylonian temples "because its aroma was pleasing to the Gods."

The ancient presence and use of *Cannabis* in what is now Israel is supported by unique evidence for medicinal and/ or ritual use discovered in the 1990s. *Cannabis* was found in a tomb from the late Roman period (315 to 392 CE) located in Beit Shemesh near Jerusalem. In this burial site, Israeli researchers discovered carbonized material associated with a 14-year-old girl dated to 1600 BP. Initial microscopic investigation indicated that the material resulted from burning a mixture of *Cannabis* and other plants, and subsequent chemical analysis revealed the presence of Δ^8-THC. Researchers concluded that *Cannabis* had been burned to facilitate the birth process (Zias et al. 1993), an early medicinal use of plants in this genus (see Chapter 8). Recent polymerase chain reaction (PCR) analyses of DNA extracted from two cordage samples recovered from "Christmas Cave" in the Qidron Valley near the Dead Sea and Qumran dated to Roman times but possibly much older indicates that they contain hemp fiber (Murphy et al. 2011).

In Egypt, *Cannabis* pollen was reportedly recovered from the mummy of "Ramsés the Great" (Ramsés II, lived ca. 1303 to 1213 BCE), who ruled for 67 years during the Nineteenth dynasty. Leroi-Gourhan (1985), the paleobotanist who identified seven pollen grains of *Cannabis* in association with the Pharaoh's mummy, also found pollen of cotton (*Gossypium*) in the same ancient context and suggested that these crops were not native but introduced species. Leroi-Gourhan also referred to *Cannabis* as having such a variety of potential uses in ancient Egypt that it could have been cultivated in different types of environments from large fields to smaller gardens. Citing an oral communication from Aline Emery-Barbier of the Laboratoire d'Ethnologie Préhistorique in Nanterre, France, Leroi-Gourhan also referred to ancient *Cannabis* pollen found at "Nagada Khattara near Luxor in Egypt which was dated to ca. 2600 BC[E]."

In ancient Egypt, we know that many drugs of animal, vegetable, and mineral origin were used. However, only a few of the several hundred medicinal sources mentioned in existing papyrus texts have been identified, and *Cannabis* is one of them. Moreover, the early Egyptians, whom Homer regarded as a nation of druggists (Birdwood 1865), likely were familiar with opium, which was presumably the active and powerfully psychoactive ingredient of the famous drug *Nepenthe* (Merlin 1984). It is possible that the ancient Egyptian knowledge and use of opium precluded the need for *Cannabis* as a euphoric drug even if the Egyptians had access to it. We find this situation today among traditional Southeast Asian ethnic groups with a long history of *Cannabis* use for fiber and seed but use the opium poppy as their mind-altering drug source. It should also be noted that even though some have referred to the ancient use of hemp in Egypt (e.g., Booth 2003; Rudgley 1998), no confirmed traces of hemp fiber have yet been found there. The early Nilotic civilization most likely did not know of *Cannabis* until relatively late, after it had diffused through Arabia, Asia Minor, Italy, and Greece (Laufer 1919), and even then perhaps as a drug plant rather than a fiber source (see Chapter 5 for a discussion of hemp in ancient Egypt and see Chapter 8 for ancient Egyptian medical use).

Mystic Muslim devotees from Syria appear to have carried hashish into northern Africa, first bringing the drug use of *Cannabis* to Egypt in the twelfth century CE (Khalifa 1975). In the case of eastern and southern Africa, Arab traders likely brought *Cannabis* directly from India (La Barre 1980).

According to Du Toit (1980), some of the dispersal of *Cannabis* into north and east Africa was "due to its association with Muslim migrant communities." *Cannabis* was grown in Egypt for hashish production well into the twentieth century (Clarke 1998).

Diffusion into Europe and the Mediterranean

Long ago, the plant geographer Alphonse de Candolle (lived 1806 to 1893), in his seminal book *Origin of Cultivated Plants* (first published in 1882), suggested an approximate date for diffusion of hemp out of Western Asia to Europe. Although de Candolle's time estimates may be inconsistent with earlier human introductions of *Cannabis* into Europe in the broadest sense, the direction of dispersal is corroborated by linguistic evidence: "It seems probable that the Scythians [more correctly proto-Scythians] transported this plant from Central Asia and from Russia when they migrated westward about 1500 BC[E], a little before the Trojan War. It may also have been introduced by the earlier incursions of the Aryans into Thrace and Western Europe; yet in that case it would have been earlier known in Italy" (de Candolle 1967).

Schultes (1970), following de Candolle, supported the hypothesis that the Scythians, or their ancestors, brought *Cannabis* into Eastern Europe from Central Asia and western Russia approximately 3,500 years ago, long before it became established in the Mediterranean as a fiber crop (also see Chapter 5). Elaborating on this chronological scenario for the spread of *Cannabis* westward from Central Asia, Damania (1998) identified its origin in the area north of the Caspian Sea. He also suggested that as a crop species it reached western Asia around 2000 BCE and later was taken to Europe about 1500 BCE. Zohary and Hopf (1988) proposed that this species reached Anatolia and Europe much later, with its archeological remains appearing from the "8th century BC[E] onwards."

We and others such as Sherratt (1997) believe that *Cannabis* (in our view, *C. indica*) and its use, along with a series of shared cultural traits, were dispersed across the steppes of Eurasia and then into Eastern Europe during the opening centuries of the Bronze Age (between 4000 and 3700 BP). Similar nomadic pastoral economies, pottery styles, weapon types, house and settlement forms, and mortuary rituals that included *Cannabis* (at least in some cases) were among the traditions spread over the Eurasian steppes during the Early Bronze Age. By this time, generally comparable and broadly interacting cultures occupied the steppes from the borders of China to the middle of Europe. As we describe later and in the following chapters, there is archeological and other evidence indicating that *Cannabis* (in this case, *C. sativa*) was cultivated and used in Western Asia and much of Europe certainly by the time of the Iron Age, and in some areas of Europe it was present by the Bronze Age. Furthermore, its use may go back to times prior to the Neolithic and even Mesolithic periods in certain regions of Europe, for example, northwestern Russia and the Baltic.

In the following section, we provide a comprehensive survey and critical evaluation of evidence for *Cannabis* and its use in Europe from ancient times through to the modern era with reference to archeological, archaeobotanical, and written records.

Among the more comprehensive recent surveys of micro and macrofossil evidence for the distribution and diffusion

TABLE 10

Key european archeological sites discussed in the text where *Cannabis* remains have been recovered.

Site #	Country	Location	Age	Type	Source	Reference
27	Bulgaria	Northern Pirin Mountains, Lake Dalgoto	10,200 to 8500 BP	Pollen *Humulus-Cannabis* type	Lake sediment cores	Stefanova and Ammann 2003
28	Czech Republic	Bohemian region	530 BP	Seeds	Medieval well	Opravil 1979
29	Czech Republic	Prachatice region	1600 to1400 BP	Seeds	Late medieval water pipe works	Beneŝ 1996
30	Czech Republic	Ķrkonoše Mountains, Cerná Hora Bog	2100 BP	Pollen *Humulus-Cannabis* type	Lake sediment cores	Speranza et al. 2000
31	Denmark	Zealand, Vallensbaek Nordmark	1120 BP	Seed	Bog sediments	Robinson et al. 2001
32	Egypt	Luxor, Nagada Khattara	4600 BP	Pollen *Cannabis*	Embalmment	Leroi-Gourhan 1985
33	Estonia	Pirita River, Lake Maardu	2000 BP	Pollen *Cannabis* type	Lake sediment cores	Veski 1996
34	Ethiopia	Lake Tana, Lalibela Cave	640 to 500 BP	Cannabinoids—Δ^6-THC	Ceramic water pipe bowls	van der Merwe 1975
35	Finland	Keuruu, Suojoki site	800 to 700 BP	Tarred hemp caulking	Boggy meadow bordering a lake	Laitinen 1996
36	Finland	Kastelholma	1200 to 950 BP	Seeds	Archeological site	Kroll 1998
37	France	Midi-Pyrenees region, Fontanes, Lot, Al Poux	2100 to 2000 BP	Seeds	Waterlogged assemblage	Bouby 2002
38	France	Erdre River, Nantes, Carquefou	2450 to 2300 BP	Pollen *Cannabis-Humulus* type	Peat bogcores	Cyprien and Visset 2001
39	France	Northern, Serris-Les Ruelles	1400 to 1200 BP	Seeds	Early Medieval manor house	de Hingh and Bakels 1996
40	France	Northwestern, Malingue (Mayenne)	2000 BP	Pollen Cannabaceae	Sediment cores	Corillion and Planchais 1963
41	France	Northwestern, Fontvielle à Vareilles (Creuse)	1900 to 1700 BP	Woven textile material	Tomb graves	Lorquin and Moulherat 2002
42	France	Brittany, Finistère	2900 BP	Pollen *Cannabis-Humulus* type	Sediment cores	van Zeist 1964
43	France	Provence, Adaouste	4000 BP	Hemp and linen fiber	Snagged in a bone tool	Barber 1991
44	Germany	Lower Saxony, Göttingen	800 to 700 BP	Seeds	Two cesspits and a pond	Hellwig 1997
45	Germany	Wilmersdorf	2500 to 2400 BP	Seeds	Vase in a tomb	Reininger 1967
46	Germany	Hochdorf	2550 to 2500 BP	Cloth	Celtic tomb	Delaney 1986; Biel 1981; Körber-Grohne1985
47	Germany	Thuringia, Eisenberg, Bandkeramik site	7400 and 6900 BP	Seeds	Neolithic early farming community	Willerding 1970
48	Hungary	Budapest	4500 to 4000 BP	Seeds	Celtic settlement	Dálnoki and Jacomet 2002

(continued)

TABLE 10 *(continued)*

Site #	Country	Location	Age	Type	Source	Reference
49	Italy	Mediterranean coast between San Remo and Montecarlo, Ventimiglia	1500 to 1300 BP	Seeds	Late Antique period	Arobba 2001
50	Italy	Naples, Pompeii	1932 BP	Seeds	Storage vat in the cellar of a farm house	Ciaraldi 2000
51	Italy	Po Plain	3500 to 3400 BP	Pollen *Cannabis*	Middle Bronze Age village site	Fleming and Clarke 1998
52	Lithuania	Lower Vilnius Castle	700 to 600 BP	*Cannabis* pollen	Cleaned wall of archeological ditch	Mažeika et al. 2009
53	Moldova	Dniester-Prut region, Dantcheny I site	7000 to 6000 BP	Seed impressions	Neolithic Linearband-keramik culture	Yanushevich 1989
54	Netherlands	Holland, Gouda-Oostpolder	9000 to 8000 BP	Seeds	Country farm	Bakels et al. 2000
55	Norway	Vestfold, Slagen	1150 BP	Fabric and seeds	Viking ship	Holmboe 1927; Vindheim 2002
56	Norway	Oslo Fjord area	2300 to 1500 BP	Pollen Cannabaceae	Sediment cores	Hafsten 1956
57	Palestine	Near Jerusalem	1700 to 1600 BP	Cannabinoids in ashes	Tomb of a 14-year-old girl	Zias et al. 1993
58	Poland	Baltic Sea, Wolin Island	3000 BP to present	Pollen "hemp"	Sediment cores	Latalowa 1992
59	Poland	Bory Tucholskie Lake	5500 BP	Pollen Cannabaceae	Lake sediment cores	Miotik-Szpiganowicz 1992
60	Poland	Lake Gosciaz	7000 to 3900 BP	Pollen Cannabaceae	Lake sediment cores	Ralska-Jasiewiczowa and van Geel 1992
61	Romania	Frumușica, Botoșani county	8600 to 5500 BP	Seeds	Archeological site	Willerding 1970
62	Russia	North of Caucasus Mountains	5000 to 4000 BP	Seeds	Early Bronze Age site	Ecsedy 1979
63	Spain	Lake Estanya	1400 to 1350 BP	Pollen Cannabaceae	Lake sediment cores	Riera et al. 2004
64	Switzerland	Fuorn Valley in Swiss National Park	3000 BP	Pollen *Humulus-Cannabis* type	Sediment cores	Stähli et al. 2006
65	Switzerland	Thayngen-Weier site	8600 to 5500 BP	Seeds	Archeological site	Willerding 1970
66	Turkey	Ankara, Gordion site	2700 BP	Fabric	Phyrgian Kingdom grave mound	Godwin 1967b
67	Ukraine	Tiasmyn River Valley, Pastyrske site	3000 to 2500 BP	Seeds	Possibly Scythian	Pashkevich 1998a
68	United Kingdom	England, York, Skeldergate	1800 to 1600 BP	Seeds	Roman era well	Hall et al. 1980
69	United Kingdom	England, London	1900 to 1600 BP	Seeds	Roman period	Gray 2002; Davis 2003
70	United Kingdom	Scotland, St. Andrews	2800 BP	String and fabric	Bronze Age site	Ryder 1999

of *Cannabis* in Europe are those offered by Godwin (1967b), Merlin (1972, 2003), Körber-Grohne (1985), and Dörfler (1990). Although we have referred to the problematic differentiation of *Cannabis* and *Humulus* pollen earlier, the palynological record indicates a two-pronged spread of *Cannabis* throughout Europe in ancient times. This interpretation, supported by archeological and historical records, suggests that first *Cannabis* was dispersed by people from Western Asia into the Balkan states and was carried from there both north and west through Europe. After hemp spread from Eastern Europe into Central Europe and the Baltic region, its cultivation and in many places its spontaneous growth dispersed farther north, west, and south through Europe, eventually reaching the British Isles, Scandinavia, and the Mediterranean.

Where *Cannabis* pollen (assuming it is *Cannabis* pollen and not that of *Humulus*) has been found in ancient Europe, its record in association with other plant species suggests that cultivation of field crops was occurring and was strongly associated with local deforestation for farming purposes; this included, for example, the clearing of deciduous trees and at times pines. Increased *Cannabis* pollen levels are often linked with pollen records of other plants including weeds associated with cultivation such as plantains (*Plantago*), docks and sorrels (*Rumex*), ragweeds (*Artemisia*), and several Chenopodiaceae, as well as cultivated crops such as flax and rye; the combination of these plants is a strong indication that farming was practiced. Both *Cannabis* and *Humulus* thrive in open environments created by cutting down and/or burning vegetation in order to plant crops. This relationship is well documented in pollen studies throughout Europe, Scandinavia, and the British Isles. Literature reporting ancient *Cannabis* pollen in various regions of Europe, especially from western and northern areas, is extensive. On the other hand, confirmed reports of ancient hemp seed discoveries are much less widespread yet provide more accurate evidence. Pre–Iron Age (i.e., before about 2800 BP) archaeobotanical records of *Cannabis* seed remains are rare. However, those seeds that have been recovered provide insight into the antiquity of *Cannabis* in Europe, but on a relatively small scale compared to the cultivation, use, and trade of hemp that expanded significantly across Roman Europe during and after the Iron Age.

Now we review the geographical distribution of archaeobotanical, artifact, and literary evidence for *Cannabis's* spread across modern-day European nations. In rough chronological sequence, we follow the distribution of the evidence suggesting hypothetical routes of dispersal into and across Europe, first from southeastern areas and eventually radiating across much of the subcontinent.

ROMANIA

Romania is located primarily on the lower Danube River in southeastern Europe with a coastal area bordering the Black Sea. Neolithic archaeobotanical evidence for *Cannabis* was reportedly discovered at the Frumuşica site in northeastern Romania near Moldova (Willerding 1970; also see Renfrew 1973). Earlier we referred to ancient evidence for *Cannabis* use in the form of a "pipe-cup" holding charred *Cannabis* seeds discovered in a pit-grave (*kurgan*) at Gurbanesti near Bucharest and dated to the later third millennium BCE (Sherratt 1991). Although this cultural context indicates a ritualistic, perhaps psychoactive use, the discovery documents the the presence of *Cannabis* in southeastern Europe over 4,000

years ago where it was likely cultivated for fiber as well as other uses; Kroll (1999) referred Comşa (1996) for evidence of ancient *Cannabis* in Romania from the seventh to fourth millennia BCE.

Early evidence for *Cannabis* reported from both Frumuşica and Gurbanesti is probably associated with ancient Yamnaya cultures, which are related to what archeologists refer to as the Corded Ware horizon (e.g., see Anthony 2007). The people represented by this culture spread rapidly across greater northern Europe beginning about 4900 to 4700 BP, all the way from Ukraine to Belgium. In addition, peoples who deposited their debris in the Corded Ware horizon developed a pastoral, mobile economy that precipitated an almost complete loss of settlement sites (as had happened with the spread of similar Yamnaya cultures in the steppes of Western Eurasia). Furthermore, the great majority of the widely dispersed people of the Corded Ware horizon adopted funeral rituals related to single graves under mounds (see Chapter 7 for a more detailed description of such pit-grave burials or *kurgans* in Central Asia).

Peoples of the Yamnaya (also called Kurgan) and derivative cultures also spread a special trait of drinking linked to specific types of cord-decorated ceramic cups and beakers. Regarding this beverage tradition, Sherratt (1997) argued that people belonging to the Sredni Stog culture, which began in the Pontic steppes and preceded the Yamnaya cultures, included psychoactive parts of *Cannabis* (possibly inflorescences and leaves extracted in heated butter or milk) in a mind-altering drink. In appreciation of this culturally important brew, Sherratt suggested the peoples decorated their drinking vessels ubiquitously with cordage impressions, much as polypod bowls were decorated, as discussed earlier. Sherratt tells us that the people of the Sredni Stog culture spread *Cannabis* and their knowledge about this *Cannabis*-based brew to their neighbors. This cord-impressed ornamentation was initiated in the steppe region and subsequently was carried into Europe, first reaching the "eastern wing of the Globular Amphora culture and eventually becoming much-imitated (as false cord and pit-and-comb decoration) by cultures of the North European Plain, from the Pontic steppes to the British Isles" (Hornsey 2004).

Pottery decorated with cord-impressions was made extensively in ancient eastern Asia as well, beginning relatively early in the Holocene period. Take, for example, the Shengwen culture (ca. 12,000 to 10,000 BP) from the lower Yangzi River in China and the Incipient Jōmon culture (ca. 12,500 to 10,000 BP) of Japan. The terms "Shengwen" and "Jōmon" both refer to "corded ware" or cordage impressed ceramics. People associated with the Neolithic Shengwen horizon likely cultivated *Cannabis* for fiber and seed and perhaps for mind-altering purposes as well (e.g., see Sherratt 1997; Hornsey 2004). As noted earlier, Sherratt believed that at least some of the steppe peoples in the third millennium BCE honored their highly esteemed *Cannabis*-based drink by impressing ceramics with hemp cord imprints. In this way traditional use of the plant, including its associated psychoactive properties, were graphically represented (for comparison, see the section earlier in this chapter on the ancient Jōmon peoples of Japan). Sherratt (1997) also believed that cord-impressions acted as a kind of marketing technique for the contents of the container; he compared this presumed advertising method to the design concept inherent in the Late Bronze Age Cypriot opium juglets traded throughout much of the Eastern Mediterranean region, which mimicked the relative size, shape,

elevated percentage and concentration values of *Humulus/Cannabis* with raised pores ('*Cannabis*-type')" between 1560 and 1260 BP indicated that this taxon could have been introduced and cultivated as a new crop during this time. Rasmussen tells us that the "oldest reported macro-remains of *Cannabis* in Denmark were found in an archeological site at Vallensbæk Nordmark, Zealand and dated to ca. 1110 CE" (see Robinson et al. 2001; also see Robinson and Karg 2002). Rasmussen and Anderson (2005) also found considerable evidence for cultivation of hemp and flax in Denmark among ancient samples cored from a shallow lake, Gundsømagle Sø, in Zealand, northeast of Copenhagen. This small lake is located in a catchment area that had long been used for farming with dated macrofossil remains reaching back about 7,000 years. Interestingly, the archaeobotanical record indicates that the remains of aquatic plant remains were limited before about 3300 BP, after which "submerged and emergent 'macrophyte' plants increased greatly," and that this development was "paralleled by an increase in sediment minerogenic [soil-forming] matter and non-arboreal pollen." For more than 3,000 years, until the mid-twentieth century, water plants appear to have been plentiful and also macrofossils of flax (*Linum usitatissimum*) and "high pollen percentages of '*Cannabis* type' (hemp) were recorded in periods between ca. 1150 BCE and 1800 CE." Rasmussen and Anderson concluded that clearing, probably for farming purposes, increased in the watershed surrounding the lake starting about 5,000 years ago, and the lake experienced direct human impact from retting flax and hemp. In sum, this study suggests that after about 1300 BCE vegetation changed radically as the lake became progressively shallower due to a combination of natural and anthropogenic causes—a change in climate induced lower water levels that were enhanced by human-induced erosion that filled the lake basin. If Rasmussen and Anderson's conclusions regarding retting activities associated with hemp are correct, then *Cannabis* may have been cultivated in northeastern Europe for more than 3,000 years (see Chapter 5 for more discussion of the cultivation of hemp for fiber and possibly seed in Denmark at least from early medieval period).

BRITISH ISLES

In the mid-1960s, the English botanist Sir Harry Godwin (1967a/b) interpreted data available to him regarding *Cannabis* cultivation in Great Britain; he concluded that hemp farming first began at the end of Roman times when the Anglo-Saxons invaded the British Isles (ca. 1500 BP). The traditional use of *Cannabis* in these islands has been mainly for fiber to produce cordage, sails, fishing nets, and clothing, and to a lesser extent for pressed hemp seed oil. Godwin provided early archaeobotanical evidence for what he identified as *Cannabis* pollen, indicating that broad cultivation of hemp began, for example, at Old Buckenham Mere in Norfolk county, England, about 400 CE. According to his pollen curves, the rise for Cannabaceae is mirrored by increases in other important crops such as rye, flax, and wheat. Mounting emphasis on arable cultivation during Anglo-Saxon and Norman times was manifested in continuous pollen curves for arable crops. Based on his palynology, Godwin suggested that hemp farming was relatively important in Great Britain at least from ca. 1200 to 800 BP but suffered a considerable setback around the fourteenth century and then appeared

to have a high point in the early part of the sixteenth century following the order by King Henry VIII to raise hemp needed to supply the growing English navy. We expand on this synopsis of hemp cultivation in England in the following, updated discussion of the early presence and use of *Cannabis* throughout the British Isles.

The occurrence and utilization of *Cannabis* may go back as far as the Bronze Age in the British Isles (Ryder 1993, 1999). This is supported by the discovery of hemp string and fabric specimens along with a hoard of metalwork in a Bronze Age site (ca. 2800 BP) at St. Andrews in Scotland. Identification of the natural cordage source of the ancient string was based on measurements of fiber diameter that were compared with samples of flax (*Linum usitatissimum*). The resulting measurement showed that the fibers comprising the Bronze Age strings are more similar to those of *Cannabis*, which led to the conjecture that hemp was utilized to make cordage and textiles in Britain much earlier than had been thought. Recently, a carbonized seed fragment possibly belonging to *Cannabis* was recovered from a Bronze Age "round house" at Grange Rath, Colp West, in County Meath, Ireland (Jaques and Hall 2003). More seed and woven fiber evidence of *Cannabis* is needed to confirm it was grown and utilized in the British Isles more than 2,000 years ago.

Evidence that *Cannabis* was present in Britain during the Roman Age is more substantial than for earlier times. Confirmation is found in the *Edict of Diocletian*, which was published in 301 CE. This document served as "a prospectus of trade goods and services" available across the Roman Empire. Among the items produced in Roman Britain was hemp fiber used to make rope and sailcloth (Wild 2002). Other evidence for *Cannabis* during this time includes various pieces of rope, more than 2,000 years old (ca. 2140 to 2180 BP) and tentatively identified as made of hemp; these cordage fragments were found in a Roman fort well in Dunbartonshire in Scotland (Godwin 1967b). *Cannabis* seeds, dated to 1800 to 1600 BP, were recovered from a Roman era well at Skeldergate near York (Hall et al. 1980), and a review of the archeological record for food and oil plants in the British Isles by Tomlinson and Hall (1996) indicated that four ancient *Cannabis* seeds dated to the Roman Age came from waterlogged sites in or near York. Another example from Roman Age York includes seeds found at Tanner Row, Rougier Street, and Bedern (Hall and Kenward 1990). As of 1996, 34 *Cannabis* seeds, from Roman times up through the Middle Ages, had been recovered from sites in the British Isles, and no confirmed hemp seeds have yet been uncovered that date to the period before written records in Britain (e.g., Bradshaw et al. 1981; Tomlinson and Hall 1996; Hall 2007). Tomlinson and Hall (1996) pointed out that evidence of *Cannabis* from the post-Roman period is almost always found in ancient urban sites and that "hemp is quite likely to have become a ruderal in the vicinity of habitation sites, as it is today."

The data base for Roman Age hemp seed discoveries in Great Britain expanded during the early twenty-first century. Hemp seeds have also been found in northwest Southwark, London, that are dated from the early second to late fourth century CE (Davis 2003; also see Gray 2002). In addition, discoveries of *Cannabis* seeds dated to the Roman (or Romano-British) period in England have been made, for instance, in the Swinegate sites in York and at Kingswood II in Hull (Hall 2007). The presence of these Roman-age seeds, albeit not in great numbers, indicates that *Cannabis* cultivation, and also perhaps the early processing of hemp fiber, took place in

England during Roman times as well as later, in the Middle Ages (e.g., Gearey et al. 2005).

We now focus on *Cannabis* evidence in the British Isles from the end of the Roman occupation through Anglo-Saxon times in the premedieval period. At Stafford in central England, ancient "*Cannabis*" pollen was recovered from King's Pool, an elongated depression with up to 21 meters (69 feet) of organic sediments (Bartley and Morgan 1990); in these deposits *Cannabis* pollen was identified in association with "arable agriculture including cultivation of cereals," which seems to have increased at this location near the end of Roman times with the initial peak somewhat after 650 CE. This evidence supports suggestions by Beales (1980) and Barber and Twigger (1987) "that *Cannabis* began to be cultivated in this region in Anglo-Saxon times." Tomlinson and Hall (1996) addressed the question as to why there is not much macrofossil plant evidence in England before about 800 CE:

> It is not until the Anglo-Saxon/Anglo-Scandinavian period, around the mid 9th century that the way people were living and depositing their rubbish around them in the rapidly developing towns began to produce ideal conditions for the preservation of plant remains, namely, the widespread practice of depositing faecal material in cess pits. Coupled with this is the increase in the excavation of these well-preserved, waterlogged, deposits and the extensive sampling programmes [*sic*] which have been carried out on these towns. This work has produced a wealth of evidence, showing the range and variety of plant material used by ordinary people (traders and craft workers) living in towns.

Discoveries of waterlogged *Cannabis* seeds in England dated to premedieval times, including the Saxon, Late Saxon, Anglo-Saxon, Anglo-Scandinavian, and Saxon-Norman periods (all before ca. 1150 CE), have been reported from several locations. These include ancient seeds found in York (Hall et al. 1983; Hall and Williams 1983; Kenward et al. 1986; Tomlinson 1989; Kenward and Hall 1995), Gloucester (Green 1979a/b), Thetford (Murphy 1984), and Norwich (Murphy 1983, 1988). Seeds were also found at Micklegate, York, that have been dated to tenth century Viking (Anglo-Scandinavian period) levels (Hall 2007).

Waterlogged *Cannabis* seeds dated to medieval or postmedieval periods (ca. 1150 to 1700 CE) have been reported in the literature from many sites in England including those in Southampton (Green 1986), London (Jones et al. 1990), Norwich (Murphy 1988), Chester (Greig 1988), Hull (Williams 1977), Kingston-upon-Hull (McKenna 1987), Beverley (Allison et al. 1996; McKenna 1992), Dudley Castle, West Midlands (Moffett 1992), Doncaster (Hall et al. 2003), York (Hall and Kenward 1990; Williams 1977), Askham Bog south of York (Bradshaw et al. 1981), Hartlepool (Huntley 1987), and Newcastle upon Tyne (Nicholson and Hall 1988; Huntley 1987; Grinter and Huntley 2002).

Hall et al. (2004) found fragments of *Cannabis* seeds in excavations undertaken in Aberdeen, Scotland, dating to the late twelfth to thirteenth centuries (also see Dickson and Dickson 2000 for a review of ancient *Cannabis* use in Scotland). *Cannabis* seeds dated to the early medieval period have also been recovered in Ireland from Dublin and Drogheda, which is situated only 45 kilometers (28 miles) from Dublin (Mitchell et al. 1987; Mitchell and Dickson 1985; Dickson and Mitchell 1984). There are also many reports for sites in England where *Cannabis* seeds dated to the medieval period have been found, including, for example, those in Canterbury,

Chester, Liverpool, Doncaster, Scunthorpe, Grimsby, Howden, Selby, Sherburn-in-Elmet, Beverley, York, and Hull, as well as for another site in Aberdeen, Scotland, and some in Dublin, Ireland (Hall 2007).

As noted earlier, many archaeobotanical and archeological studies in the British Isles have reported evidence of ancient *Cannabis*, several in the form of seeds or fibers, others in the form of Cannabaceae pollen with various authors referring to such pollen as *Cannabis*, *Humulus*, *Cannabis/Humulus*-type, and so on. To a large degree, this archaeobotanical record is similar to what has been reported across much of northern and central Europe, as described earlier in this chapter; the strongest evidence for hemp presence and use in this region is dated to the Iron Age and subsequent periods through the Middle Ages, and in some cases up into modern times. These reports are also referred to in Chapter 5 as they apply to evidence indicating hemp fiber cultivation. We now focus on the ancient *Cannabis* evidence from the Mediterranean Region.

MEDITERRANEAN REGION

The geographical area where *Cannabis* initially appeared in the Mediterranean region is uncertain. This is largely due to the incomplete archaeobotanical record, which is problematic, especially as it applies to Cannabaceae pollen and remains of woven materials. In the case of woven remains, for example, an ancient piece of "hemp" fabric dated to approximately 2700 BP was recovered from a Phyrgian Kingdom grave mound site at Gordion, near Ankara in Turkey (Bellinger 1962; Godwin 1967b); however, the fiber source of this ancient artifact remains unverified. In Greece, ancient cloth fibers dated to ca. 2500 BP were also found at Trakhones, Attiki province, but positive identification of these ancient fibrous materials also has not been confirmed (Barber 1991). It should be remembered that the major fiber plant of the ancient Near East, Mediterranean, and various European regions was flax, which also produced valuable oil seed. Indeed, fragments of possibly the oldest human-associated fibers ever found were recently recovered from the Caucasus region at Dzudzuana Cave in the Republic of Georgia; these fragments have been dated to about 30,000 years BP and identified as "wild flax" (Kvavadze et al. 2009). Although these fibers have not been universally recognized as derived from wild flax, it is probably valid to assume that *Cannabis* became available and utilized in various areas of Europe somewhat later than flax.

Cannabis spread into the Mediterranean region from one or more of the following areas: Southwest Asia, Asia Minor, the Eurasian steppes north of the Black Sea, or somewhere in Northern Europe. Godwin (1967b) argued that "historical evidence as a whole clearly points to cultivation of hemp in the Middle East," with a rapid spread of *Cannabis* cropping in the Mediterranean region during "classical times." Furthermore, Godwin suggested that there was little evidence that hemp cultivation expanded northward within and during the time of the Roman Empire. On the other hand, *Cannabis* may not have been cultivated in Southwest Asia, at least on a relatively large scale, until more recent historical times; however, over the past 40 years, historical records and archaeobotanical research indicate that hemp cultivation appears to have taken place in the Mediterranean region earlier than previously thought. The use and trade of hemp in Europe, within and beyond the Mediterranean region, was well established at least by the Classical period (ca. seventh century BCE to fifth century CE). The first strong evidence we have that ancient Greeks knew about *Cannabis* comes from the now well-known remarks by the famous historian Herodotus, in the fifth century BCE, who reported that Scythians used *Cannabis* for fiber and inhaled its smoke but claimed it was scarcely known in Greece (see the section on Central Asia in this chapter, as well as Chapters 5 and 7 for more detail on the Scythian uses of *Cannabis*). Written accounts also tell us that other classical Greeks and Romans knew about *Cannabis* and used hemp products more than two thousand years ago. These included rope and sailcloth made of hemp fiber, some confections that appear to have included its psychoactive resin, and parts of individual male and female *Cannabis* plants that were used for medicine. These products are mentioned later and are described in more detail in the chapters that follow.

We now focus on both archaeobotanical and historical records for early local use of *Cannabis* within the Mediterranean region beginning in Greece and moving westward from there. This geographical orientation assumes that *Cannabis* was spread by people across the Mediterranean region starting from some area east of Greece. Willis (1992) studied the pollen stratigraphy of late Quaternary deposits in northwest Greece but did not observe any occurrences of *Cannabis* or *Humulus* and only very small quantities of other cultivated or weedy species associated with agriculture. Willis hypothesized that the region was not appropriate for cultivation and that local tree clearances indicated in the pollen record were made to support herding. This explanation seems to concur with historical evidence for this area of ancient Greece, where we have no evidence yet for early hemp farming.

Recent archaeobotanical research in Greece at Nisi Fen and in the Boras Mountains has uncovered evidence of Cannabaceae pollen (Lawson et al. 2005). The wetland of Nisi Fen occupies a 12-square-kilometer (four and one-half square miles) basin, 10 kilometers (6 miles) west of Edessa in north central Greece. The upper two zones of the Late Holocene record at this site (dated to ca. 5000 to 500 BP) revealed a decline of forest in the region that could have been brought on by increased grazing. On the other hand, farming alone, or in tandem with grazing, also could explain the forest decline. According to the interpretation offered by Lawson et al. (2005), archaeobotanical evidence from Nisi Fen shows an expansion of herbaceous taxa about 5,000 to 4,000 years ago associated with agricultural activity; the recovered pollen indicates that the following weedy and cultivated species were present: "*Artemisia, Cannabis* type, cereals, *Plantago,* Umbelliferaceae, *Urtica,* and grasses." Modification of vegetation, a reduction of forest diversity, expansion of maquis species (associated with dense scrub growth characteristically found in Mediterranean vegetation), general decline in airborne pollen, and the spread of wayside herbs and crop plants are all unmistakably evident in palynological diagrams from other low- to midelevation archaeobotanical sites in the Nisi Fen area. Human impact is suggested by decline in woodland and appearance of crops in the pollen record and is supported by evidence found in environments at all elevations in this region in the later Holocene (Lawson et al. 2005). However, the "*Cannabis* type" pollen recovered at Nisi Fen *could* belong to either *Humulus* or *Cannabis*. As of now, we have no fully confirmed evidence in Greece for *Cannabis* before the Classical Greek period. The situation is different in Italy.

The record for ancient hemp evidence in Italy is relatively abundant, especially in the middle and northern areas. Although Godwin (1967b) referred to the early occurrence of *Cannabis* in the northern provinces of Italy as rare, more recent evidence suggests that *Cannabis* was present and cultivated in northern Italy during the Neolithic, Bronze, Iron, and Middle Ages (e.g., Castelletti et al. 2001). For example, *Cannabis* pollen was recently recovered from the Po Plain in northern Italy, a region famous for hemp cultivation during the historical period. Ravazzi (cited in Fleming and Clarke 1998) reported *Cannabis* pollen, positively identified based on pore structure and grain size, from a Middle Bronze Age village site in this region (ca. 3500 to 3400 BP). Here *Cannabis* pollen percentages were 30 percent next to the river and 8 percent in sampled locations distant from the river (Mercuri et al. 2006, citing Ravazzi and his colleagues; also see the pollen record diagram, figure 5, in Ravazzi et al. 2012 for evidence of "*Cannabis/Humulus*" from the lower Mincio River valley in the central Po Plain).

According to Mercuri et al. (2002; also see Accorsi, Bandini Mazzanti, and Mercuri 1998; Mercuri et al. 2011), pollen extracted from cores of Lake Albano and Lake Nemi, both near Rome in central Italy, provides "unambiguous" evidence that both *Cannabis* and *Humulus* were present in the region even as early as the Late Pleistocene, and both became abundant cultivated crops by the later Holocene. A palynological synthesis of core data from the Italian lakes indicates that humans were increasingly disturbing vegetation in this region during the Holocene with forest clearing for farming and other purposes. The pollen record here shows a noticeable expansion of certain tree types at the beginning of the Holocene, followed by "progressive deforestation, accompanied by the increasing importance of so-called 'anthropogenic indicator plants' (e.g., cereal grasses and weeds of cultivated areas)" in the mid-Holocene and subsequently a rise in the pollen percentages of olive, chestnut, and grape from the Roman period onward (Lowe et al. 1996).

This interpretation is supported by more recent studies of ancient fire and plant life evidence provided by accelerator mass spectrometry (AMS) radiocarbon-dated charcoal and sedimentary pollen remains in the Colli Euganei (Euganean Hills) and northeastern Po Plain from ca. 16,500 BP (Kaltenrieder et al. 2010; also see Kaltenrieder et al. 2009 in which the authors conclude that the Euganean Hills were one of the northernmost refugial areas of temperate taxa in Europe during the late Pleistocene). Based on pollen and charcoal records, vegetation in this area was mixed coniferous-deciduous forest at least since about 14,500 BP, with significant anthropogenic change (i.e., fire and cultivation disturbance) starting about 6400 BCE that involved increases of European alder (*Alnus glutinosa*), chestnut (*Castanea sativa*), and walnut (*Juglans regia*), along with hop, hemp, and flax. The close link between crops, weeds, and fire activity suggests human impact as the main source of Neolithic vegetation changes. The authors suggest that "these are the oldest palaeobotanical data suggesting the cultivation of *Castanea* and *Juglans* in Europe and elsewhere." Both chestnut and walnut produce edible food resources, and hemp and flax offer useful fiber and seed products. The multiproxy evidence recovered in this study also documents the subsequent cultivation of chestnut, walnut, olive (*Olea*), and grains (along with increased amounts of *Cannabis* pollen) during the Bronze Age 4150 to 2750 BP; then, during the Iron Age, Roman Era, and medieval period, intensification of land use continued. Change in vegetation and presence of *Cannabis* is indicated by a noticeable reduction of total tree pollen suggesting a significant decrease of forest areas: "Forest openings were related to intensification of land use, as inferred from the increases of pollen of *Cannabis*, Cerealia and *Plantago lanceolata*" (Kaltenrieder et al. 2010). The authors also pointed out that the site of their study was naturally a more open area due to "its location near river banks and floodlands," which would be a type of environment in which *Cannabis* as well as some of the other taxa recorded are well adapted.

Research by Mercuri et al. (2006) at Terramara di Montale in the central Po Plain has also been used to support the assumption that *Cannabis* was cultivated in this region well before the Roman Era. Terramara di Montale is a significant archeological site within the widespread "Terramara cultural system," which was the dominant culture complex of the Po Valley Plain in the Middle-Late Bronze Age (1650 to 1200 BCE). Archaeobotanical evidence from this site indicates rapid change during the Late Bronze Age, from a largely natural forest that included mixed oak woodlands and conifers to a more open and anthropogenic landscape typified by cultivated grain fields, grazing land, and open meadows. Pollen assumed by Mercuri et al. (2006, 2011) as belonging to hemp has been found throughout the Middle-Late Bronze Age period at the Terramara di Montale site. Terramara culture was characterized by large, fortified villages with surrounding embankments and ditches, supporting a large socioeconomic system with an estimated population approaching 150,000 around the fourteenth to thirteenth centuries BCE. Subsequently, it collapsed and disappeared at the end of the Late Bronze Age about 1350 to 1200 BCE, "possibly through a combination of climatic, ecological and socio/economic causes" (Mercuri et al. 2006; also see Bernabó Brea et al. 1997). The archaeobotanical record for *Cannabis* use among the Terramara culture is supported by pollen deposits found in the Middle Bronze Age site at Santa Rosa di Poviglio in the province of Reggio Emilia (Mercuri et al. 2006, citing Ravazzi and his colleagues), as well as scarce pollen records from other Bronze Age sites in northern Italy such as Canar-Rovigo (Accorsi et al. 1998b).

According to Mercuri et al. (2002, 2006), evidence from the Bronze Age sites in northern Italy referred to earlier suggests "the use and cultivation of hemp in the area based on the whole archaeobotanical-archeological context." Furthermore they hypothesize that the pollen record, in general, indicates that *Cannabis* was growing wild in Italy long before its usefulness was recognized and that cultivation of hemp started much later. According to Wick (cited in Mercuri et al. 2002), hemp was brought into the region in the Neolithic period along with grain crops. However, *Cannabis* seeds have not yet been recovered from any of the Bronze Age sites presently identified in northern Italy. This is due to a lack of uncharred remains and resulting poor preservation of seeds; therefore, *Cannabis* and some other cultivated plants are "most probably underrepresented," and for now "we can only infer that *Cannabis* was cultivated at [Terramara di] Montale through comparison with the pollen records from Santa Rosa di Poviglio" (Mercuri et al. 2002).

Aside from problems associated with *Cannabis* pollen and the lack of macroremains found in Italy, hemp appears to have been an important component of the ancient Central Mediterranean agricultural complex as a fiber source first and possibly for other purposes later. Evidence for cultivation and use of *Cannabis* during the Roman period is more substantial and includes historical written references and ancient seed discoveries (described later), as well as relatively strong pollen records described in the following summary of palynological information gleaned from research at Lake Albano and Lake Nemi near Rome (Mercuri et al. 2002):

Hop pollen values rise during the mid Holocene, while hemp pollen becomes more abundant from *ca.* 3000 BP onwards. The highest earliest hemp peak (21 percent) is dated to the 1st century AD [*sic*] [2000–1900 BP]. This "*Cannabis* phase", with the abrupt rise of hemp pollen soon after the rise of cultivated trees (*Castanea*, *Juglans* and *Olea* [chestnuts, walnuts, and olive]) is associated with the increase in cereals and ruderal plants. This unambiguous proof of cultivation by Romans around 2000 BP occurs as well as a long lasting pre-Roman presence of hemp in the area, which is natural and possibly also anthropogenic. Subsequent clear episodes of cultivation in the medieval period were [also] found.

Pollen evidence of hemp cultivation and use during the Roman Era was also reported by Valsecchi et al. (2006), who utilized lake-sediment records to reconstruct human impact on the landscape around Lago Lucone east of Milan in northern Italy. In this study, the sediment zone dated from ca. 350 BCE to the twentieth century shows chestnut and walnut pollen appearing together with what these authors specifically refer to as "*Cannabis sativa* and *Secale* [rye]," while key native forest trees such as hornbeam (*Carpinus*), oak, and alder decline. According to these authors, this pollen distribution is evidence of "strong human impact," such as farming and agroforest activities typical of the Roman Era. This indicates that grazing and weedy communities were established as farming of cereals, and "*Cannabis sativa*" increased from the later Roman Era through the Middle Ages and modern times. During this same period, pollen of grass species and weedy indicators of agricultural disturbance such as "*Plantago lanceolata* [plantain] reached their highest values," and this included hop, which was recorded over thousands of years at the Lake Lucone site as a probable native species, which, like *Cannabis*, cereal crops, and certain weedy species, increased significantly with an associated increase in human disturbance after about 350 BCE. It should also be noted that Caramiello et al. (1992) reportedly found pollen of both *Cannabis* and *Humulus* at sites in southern Italy dating to ca. 2500 to 2200 BP.

Ancient references to *Cannabis* by two Greek historical figures are of special interest here. One of these comes from Pedanius Dioscorides, a Greek from Cilicia near the Mediterranean Sea in southern Asia Minor. Dioscorides lived in the first century of the current era (ca. 50 to 90 CE) and referred to both the medicinal and fiber utility of *Cannabis* (Gunther 1959). During much of his adult life Dioscorides lived in Rome and is said to have been the personal physician of Nero. Dioscorides gathered information and wrote a book on medicinal herbs, one of the earliest ethnobotanical manuscripts in the Western World, an ancient pharmacopeia that has stayed in print over the centuries. Dioscorides's descriptions of the therapeutic uses of *Cannabis* are discussed in more detail in Chapter 8.

Another ancient Greek, also originally from Asia Minor, who deserves particular attention is Claudius Galen (lived 130 to 193 CE). Galen became the most famous physician in ancient Roman times and was the personal doctor for Emperor Marcus Aurelius. Galen's written work had widespread authority in medical practice until the sixteenth century. He was born in the ancient Greek city of Pergamos near the Aegean Sea in northwestern Anatolia. Galen emulated the great Greek physician Hippocrates and was a renowned scholar and author; unfortunately most of his literary works were destroyed by fire. We do know, however, that Galen referred to *Cannabis* as part of a commonly consumed food substance during the Roman Era (see Chapter 7 for detail concerning this apparently psychoactive *Cannabis* confection).

In addition to Dioscorides and Galen, two Greeks that became well-known Roman physicians, other writers and notables living during the Roman Age referred to *Cannabis*. Among these was Lucilius, the satirist who lived from about 180 to 103 BCE. Another is Hiero II (ca. 306 to 215 BCE), who lived during the period when the rise of the Roman Empire was under way. Hiero II the King of Syracuse in Sicily from 270 through 215 BCE reportedly purchased hemp from the Rhone River area in Gaul, which he had made into ropes and sailcloth used in his powerful fleet of ships. Pliny, the great Roman natural philosopher of the first century CE, also commented on the use of hemp in his book 19, which deals with the cultivation of flax, the most common Mediterranean fiber plant, and other plants used for cordage and fabrics (see Chapter 5 for more details about Hiero II and Pliny the Elder and their relationships with ancient hemp).

The presence of hemp in Italy during Roman times is supported by the discovery of nine *Cannabis* seeds found along with many seeds of other useful plants at the bottom of a storage vat in the cellar of a farm house near Pompeii in southern Italy. These seeds were dated to 79 CE by volcanic fragments (lapilli) associated with the eruption of Mount Vesuvius that year (Ciaraldi 2000). Ancient *Cannabis* seeds have also been recovered from sites in Italy that date to the sixth to seventh centuries CE; these include those found at Ventimiglia on the Mediterranean coast between San Remo and Montecarlo (Arobba 2001) and those discovered in the "Cogneto Hiding Well" at Modena in northern Italy (Accorsi et al. 1998a). In addition, ancient *Cannabis* seeds were discovered in the town of Ferrara in northern Italy and dated to the tenth to twelfth centuries of the Middle Ages (Mercuri et al. 1999) and in the "the Mirror Pit," also in Ferrara, dated to the fourteenth to fifteenth centuries (late Medieval and Renaissance periods), where hemp was cultivated for its fiber from the time of the Roman Empire to after World War II (Bandini Mazzanti et al. 2005). Arobba et al. (2003) presented a review of *Cannabis* history in Liguria near the western Mediterranean coast of northern Italy ranging from the Roman period to the Middle Ages (see Chapter 5 for more detail on the long-term history of hemp cultivation for fiber in Italy).

Further west in the Mediterranean region of southern France there is possible artifact evidence for ancient *Cannabis* use. Barber (1991) reported that both hemp and linen fibers were found snagged in a bone tool recovered from a late Neolithic (ca. 4000 BP) site at Adaouste in southern France in the region of Provence; however, confirmed identification of some of these fibers as belonging to *Cannabis*, like those referred to earlier from Turkey and Greece, has not been fully substantiated. Boyer and Encart (1996) referred to the discovery of fragments of hemp rope dated to the fourth century BCE found among carbonized remains of a burned house in the area of Lattes, near Montpellier in southern France. This evidence for hemp cordage also requires further confirmation.

Bouby (2002) described his discovery of six *Cannabis* seeds in southwestern France at the site of Al Poux near Fontanes,

Lot, located in the lower part of the Boulou Valley in the southern region of Quercy. According to Bouby, these seeds have been dated to the Roman Era, sometime from the end of the second century to the beginning of the first century BCE; therefore, these may be the oldest *Cannabis* seeds yet found in Western Europe. Bouby acknowledged that the ancient history of *Cannabis* is "poorly documented in France and western Europe" and reminded us that the precise identification of both *Cannabis* pollen and fibers is difficult, and therefore seed remains are the best archeological and archaeobotanical evidence for documenting the origin and the early spread of hemp (see the methodology section at the beginning of this chapter). Based on the fact that these six, "essentially noncarbonized" hemp seeds represent the only cultivated plant in the archaeobotanical record from the Al Poux site, Bouby proposed that hemp was being farmed in southwestern France at least during the latter part of the Iron Age. Furthermore, Bouby (2002) offered two hypotheses that may explain the presence of ancient hemp seeds in valley bottom sites such as Al Poux. On one hand, he suggested that hemp could have been brought to the river bank and cultivated there because it requires fertile soil and plenty of water. Alternatively, Bouby proposed that *Cannabis* plants might have been brought to the river from inland fields for purposes of retting the fibers free from the stalks. He also noted that the only other waterlogged, valley bottom site that has been studied in southwestern France, Douville, Pont-Saint-Mamet, produced hemp seeds in a medieval layer dated to the sixteenth to seventeenth centuries CE, and this second record for *Cannabis* "is in agreement with a peculiar relation between bottom valleys and hemp cultivation." Bouby also pointed out that only a few waterlogged assemblages of ancient prehistoric and protohistoric plant remains have been examined in southern and western France. Consequently, our understanding of ancient hemp farming could be enhanced with additional systematic studies of the kind involved at Al Poux, especially if confirmed evidence of *Cannabis* pollen and fiber records were recovered to support his hypothesis that hemp was grown in southern France at the end of or even before the Iron Age (also see Carozza et al. 2002).

Cannabis has not been identified very often in seed assemblages from sites in southern France prior to ca. 2000 BP and there are only a few references to palynological evidence indicating its presence in Late Iron Age sites. Argant et al. (2006) have, however, provided some evidence for early *Cannabis* from the high elevation occupation site at Lake Lauzon, almost 2,000 meters (6,500 feet) above sea level in the Alps Mountains of southern France. The site is now in the midst of grassland apparently created by people through repeated burning of the original forest. Among pollen of species spread by humans, Argant et al. (2006) reported the presence of cereal crops throughout the time period covered in their sediment cores with a small rise in Cannabaceae pollen recorded ca. 2020 BP. Similar to Bouby, they recognized the "problem of distinguishing *Humulus* and *Cannabis*" but were convinced that *Cannabis* was present at least two thousand years ago, based on their observations of the pollen grain size, the protruding pores, and the steeply sloped annulus (exine wall thickening), supported by the associated indicators of arable cultivation (cf. Whittington and Gordon 1987; Fleming and Clarke 1998).

Argant et al. admitted that evidence for cultivation is limited to cereal pollen that was found continuously at the site and extracted from sediment layers much older than that of "hemp found during the Gallo–Roman period." In addition, they raised the possibility that pollen assemblages from high elevation lakes such as Lauzon may be deposits of "pollen and spores that are transported from the valley by strong upward flowing air currents" and therefore may not "reflect real vegetational changes in the vicinity of the lake." However, they also pointed out that "cereal pollen is scattered with difficulty outside cultivated fields," and thus they argue that "the probability of finding it in samples collected in Lake Lauzon should be very low, especially because the forested environment at the level of the lake would have functioned as a barrier for the arrival of . . . pollen."

On the other hand, Argant et al. presented another reason cereals and hemp could have occurred together with ruderal plants such as plantain, docks and sorrels, and nettles. They referred to Moe and van der Knaap (1990), who pointed out that sheep that are moved upland for summer grazing can carry seeds from the lowlands in their hair and that seeds dispersed by these animals can germinate in the uplands near their tracks: "In this way the pollen of ruderals and cultivated plants restricted to the lowlands would only indicate the presence of mountain trackways and would not be directly associated with intensively used areas." This explanation that the supposed *Cannabis* pollen came from feral plants is supported by the lack of archeological sites around the lake. Nevertheless archaeobotanical evidence (nonpollen palynomorphs or small organic granules) recovered from Lake Lauzon "clearly indicate fire, erosion and the enrichment of the lake water with nutrients." Therefore, it is difficult to explain these other anthrogenic indicators based solely on natural phenomena or long distance dispersal. Because the evidence typically corresponds with "anthropogenic pollen assemblages," Argant et al. concluded that humans "occupied the surroundings of the lake and may have cultivated crop plants," possibly including *Cannabis*.

Cannabis populations expanded across Eurasia early on, but which taxa do they represent? Following the taxonomic research of Hillig (2004, 2005a/b) and Hillig and Mahlberg (2004), archaeobotanical evidence from eastern Asia (e.g., China, Korea, and Japan) and South Asia (e.g., India, Nepal, and Pakistan) represents biotypes or subspecies of *C. indica* (BLH, BLD, NLDA, and NLD), while European evidence belongs to *C. sativa* biotypes or subspecies (NLHA and NLH). We have no reliable archaeobotanical evidence of sufficient age for *C. ruderalis*, the putative ancestor (PA) of other *Cannabis* species, and although some regions of Eastern Europe include the range of the proposed taxon, no seeds with wild-type characteristics (other than small size) have been recovered from archeological contexts. In light of their different taxonomic classifications and native ranges, we will discuss the dispersals of Asian *C. indica* and European *C. sativa* separately.

Seed evidence is the most reliable for all regions and both species. Examples of early dates for seeds recovered in East Asia are quite old (e.g., 4800, 3500, and 2500 BP) yet with the one recently uncovered exception from Japan (10,000 BP) not as old as some from northern Europe (7000 and 5000 BP). This may simply reflect that European archaeologists and archaeobotanists have worked longer and more intensively in a smaller geographical region and have recovered more data than their Asian counterparts. However, this is changing as manifested in the discovery of the oldest *Cannabis* seeds, indeed the earliest direct evidence for the genus yet found and dated; these are reportedly about 10,000 years old and

were recovered in a very ancient site in central Japan discussed earlier (see Kudo et al. 2009; Okazaki et al. 2011). The oldest pollen evidence for *Cannabis*, albeit tentative, is from the Czech Republic (9000 BP) in Europe where Cannabaceae pollen finds are regularly reported; on the other hand, early pollen evidence is almost entirely lacking in Asia where palynology has only recently been employed in archaeobotanical research. In each region pollen and/or fiber evidence is of comparable age to or older than seed evidence, but because of the confusing differentiation of ancient microfossils of *Cannabis* and *Humulus* pollen, as well as unverified confirmation of the taxon to which fiber remains belong, we defer to seed evidence wherever possible. The oldest South Asian evidence is in the form of rarely reported phytoliths (3300 BP) and charcoal (3200 BP). The ancient Middle Eastern evidence is presently limited to two reports, the chemical remains of THC from Jerusalem dated to 1600 BP (Zias et al. 1993) and the pre-2500 BP rope and fabric made of *Cannabis* found in Christmas Cave in Qumran Israel (Murphy et al. 2011); both are considered quite reliable. Up until the recent discovery of 10,000-year-old seeds found in Japan, the earliest Asian *C. indica* seed evidence was recovered from central China (4800 BP) but is unsupported by other ancient archaeobotanical evidence. The earliest seed evidence from Korea is at or before 2500 BP, but much earlier fiber evidence from both Korea and Japan (5000 and 7000 BP respectively) has been reported.

South Asia's oldest seed evidence (1900 BP) is supported by older, unique phytolith (3300 BP) and charcoal (3200 BP) evidence, so *Cannabis*'s arrival in South Asia may have come considerably earlier than the seed data indicates. We would expect the first human-related appearance of *C. indica* throughout Korea and Japan as well as in northern South Asia to have occurred relatively quickly after its early dispersal from China.

The antiquity of European *C. sativa* is well represented by its seeds, although the overwhelming majority of our data come from less reliable pollen studies. The earliest reported seed evidence comes from Germany (7000 BP), the Baltic (5000 BP), and Romania (4000 BP), and all predate or roughly coincide with the earliest date (4800 BP) from China but are much later than Japan (ca. 10,000 BP); however, it should be noted that the very early date for *Cannabis* seed remains from Germany (Willerding 1970) is not fully confirmed and must remain provisional. The next oldest seed data in Europe is dated to about 2,000 years later from Hungary, France, and Sweden (ca. 2000 BP); Italy (1900 BP); Switzerland (1800 BP); the Netherlands (1900 to 1750 BP); the British Isles (1600 BP); Norway and Denmark (1150 to 1100 BP); and likely followed by the remainder of Europe within 200 years. Although European and Asian *Cannabis* represent different taxa, due to climatic factors and perhaps later because of continuing human dispersals, we would expect that *Cannabis* reached both sides of Eurasia more or less simultaneously. Initially spreading out of Central Asia in both southwesterly and southeasterly directions, *Cannabis* eventually found refuge from the advancing Pleistocene ice sheets in southeastern Europe (*C. sativa*) and southeastern Asia (*C. indica*) and then began their rapid colonization of Eurasia as the Holocene began. See Chapters 2 and 12 for discussions of the natural origins and early evolution of *Cannabis*.

In this chapter we have presented an extensive review of the evidence for ancient *Cannabis* in Eurasia, much of which is problematic pollen and the more reliable seed fossils.

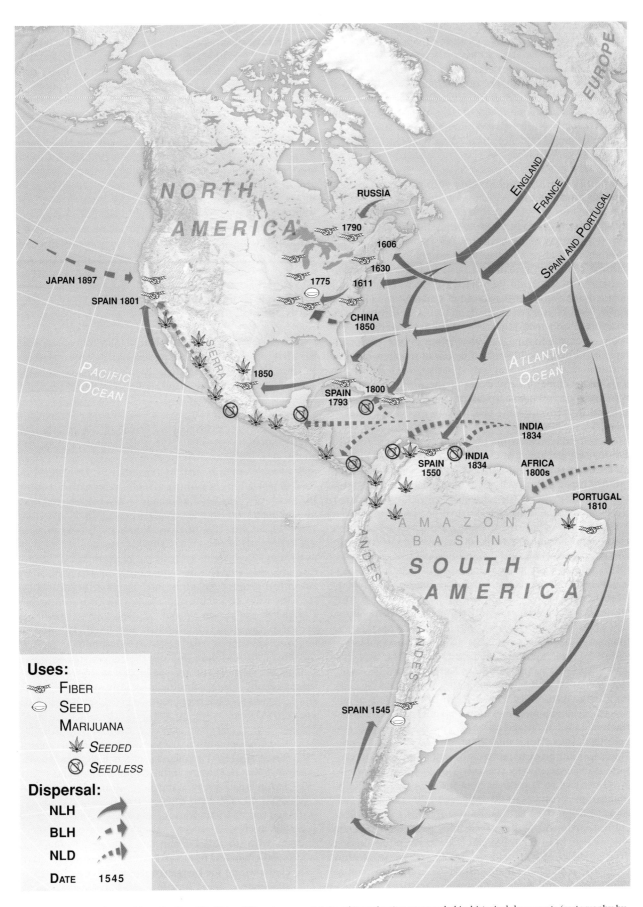

MAP 9. Dispersal of *Cannabis* to the New World for different uses and dates of introduction as recorded in historical documents (cartography by Matt Barbee). (See TABLE 1 at the beginning of this book for explanations of *Cannabis* gene pool acronyms.)

Cannabis was undoubtedly an ancient group of plants utilized by humans throughout much of Eurasia from the extremes of its far eastern distribution in Japan to the western reaches of the British Isles. However, more exact determination of its natural range and earliest human use is needed; this should be provided through additional research into the botany, evolution, and ecology of the genus, along with further archeological investigation and advanced study of the archaeobotanical evidence already discovered.

Summary of Dispersal Phases within and Beyond Eurasia

By combining various archeological, historical, and linguistic evidence, we have constructed a generalized history of human use of hemp, suggesting regional outlines for its cultural diffusion throughout Eurasia and eventually much of the rest of the world (also see Chapters 5 through 9 for detailed discussions of the spread of Cannabis use for fiber, paper, seed, recreation, religion, medicine, and ritual). The widespread and relatively recent dispersal history of Cannabis, especially beyond Eurasia, was almost entirely facilitated by humans. Following our review of the archeological and historical evidence discussed earlier, we now present a six-phase model for the stages of the worldwide human dissemination of Cannabis taking into account different cultural influences on the evolution of Cannabis as a crop plant.

In the second chapter we proposed a hypothetical model for the early evolution of Cannabaceae and genus Cannabis up to the early Holocene about 10,000 years ago when early humans began to spread it across limited parts of Eurasia. In this section we divide the subsequent spread of Cannabis as the climate warmed into six principal phases. During Phase 1 leading up to the current era, Cannabis was spread from glacial refugia throughout Eurasia and evolved under human and natural selection into the taxa encountered today—as a fiber and seed plant throughout much of East Asia (broadleaf hemp, BLH) and parts of Europe (narrow-leaf hemp, NLH) and as a drug plant in South Asia (narrow-leaf drug, NLD) and Afghanistan (broad-leaf drug, BLD). It is important to realize that very early, prior to the current era, these four basic taxonomic groupings—NLH in Europe, BLH in China, NLD in South Asia, and likely also BLD in Afghanistan—had already evolved through dispersal into new habitats in concert with human efforts to domesticate Cannabis for differing uses. During Phase 2, from about 2,000 to 500 years ago, human-mediated dispersal continued throughout the remainder of Europe and Asia, and NLD Cannabis also expanded into Africa and Southeast Asia. Subsequently, Phase 4, from the early nineteenth until mid-twentieth century, is characterized by the arrival in the New World and West Africa of NLD varieties from India, as well as the dissemination of Chinese BLH varieties to Europe and North America. Until the end of World War II, NLH and BLH fiber cultivars were widely distributed throughout the temperate climate regions of the world and trade in the mind-altering drug products of NLD varieties (ganja and charas) within Central Asia and India flourished. Phase 5, from the end of World War II until about 1990, was characterized by a huge expansion of illegal cultivation and illicit trade in marijuana worldwide, while a decrease in use of hemp products caused a steep decline in fiber cultivation. During the period between

1970 and 1990, the worldwide hemp fiber industry collapsed to near extinction and only began its present renaissance during the early 1990s. Concurrently, many NLD and BLD traditional farmer's landrace seeds were brought to North America and Europe, where they were used to establish the now widespread cultivation and breeding of domestic North American and European marijuana cultivars. Cannabis is currently in Phase 6 of its cultural diffusion, resulting on the one hand from a marked increase of clandestine drug Cannabis production in artificial environments (indoor growing of sinsemilla marijuana under electric lights using vegetatively reproduced cuttings) largely as a reaction to Cannabis prohibition and on the other hand from the resurgence of legal industrial hemp cultivation and the spread of European and East Asian hemp cultivars (see the map accompanying Table 1 at the beginning of the book for present-day Phase 6 distribution of differing biotypes).

Phase 1: Primary Dispersal across Eurasia– ca. 10,000 to 2000 BP

Central Asia's original contribution to cultural history has thus been far-reaching. Still more important has been the role that this region has played in the transmission of culture traits from one part of the Old World to the other. (MCGOVERN 1939)

Phase 1 Cannabis dispersals by humans were *not* the initial introductions into China, Europe, Southwest Asia, and South Asia, the genus having become established in these regions very early on with its range largely determined by the last glaciations beginning 100,000 years earlier. During the Pleistocene glaciations Cannabis was forced into warmer microclimates serving as refugia where it could have survived the extreme cold (for a discussion of theories pertaining to the early evolution of Cannabis gene pools, see Chapters 2 and 12). From its refugia, the prehistoric dispersal of Cannabis was aided either consciously or accidentally by Paleolithic peoples expanding as the ice sheets retreated. By the early Holocene, these nomads reoccupied warming Eurasia and in the process spread Cannabis and/or knowledge of its uses, radiating from differing glacial refugia into various areas of Central Asia, East Asia, southwestern Europe, the Hindu Kush Mountains, and the Indian subcontinent. This set the stage for the four major Cannabis taxa to evolve independently. Proposed refugial origins, dispersal routes, and early uses for each taxa proposed by Hillig (2005a/b) are summarized here:

1. C. sativa ssp. sativa or narrow-leaf hemp (NLH) originated in refugial populations of the putative hemp ancestor (PHA) living in the Caucasus Mountain foothills and spread as the narrow-leaf hemp ancestor (NLHA, C. sativa ssp. spontanea) early on into Europe and the Mediterranean via open steppe grasslands and along river valleys.

2. C. indica ssp. indica or narrow-leaf drug (NLD) Cannabis originated from a putative drug ancestor (PDA) living within the Hengduan Mountains of present-day southwestern China from where it spread across the Himalayan Foothills and eventually onto the Indian subcontinent, where it was used primarily as a drug plant.

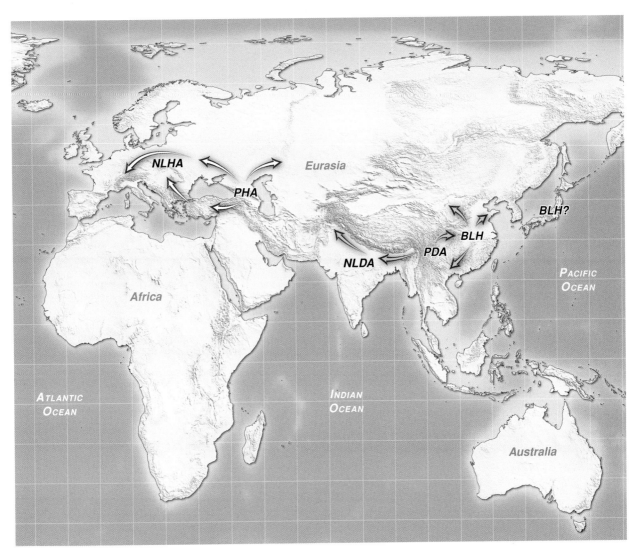

MAP 10. Primary diffusion of the putative hemp ancestor (PHA) from the Caucasus region into Europe and Central Asia and both broad-leaf hemp (BLH) into eastern Asia and narrow-leaf drug (NLD) populations into southern Asia, from the Yungui Plateau-Hengduan Mountain region during Phase 1 dispersal from 10,000 to 2000 BP (cartography by Matt Barbee). (See TABLE 1 at the beginning of this book for explanations of *Cannabis* gene pool acronyms.)

3. *C. indica* ssp. *afghanica* or broad-leaf drug (BLD) *Cannabis* was derived from PDA populations that spread along the Himalayan foothills from southwestern China and evolved in isolation within the Hindu Kush, Pamir, and western Himalaya Mountains of present-day Afghanistan, where it was eventually used as a drug plant for making hashish.

4. *C. indica* ssp. *chinensis* or broad-leaf hemp (BLH) *Cannabis* also may have originated from refugial populations of a putative drug ancestor (PDA) living in either the Hengduan Mountains of present-day southwestern China or refugia along coastal regions of East Asia and was spread early on into the nuclear areas of Chinese civilization in the Huang He and Yangzi River basins as well as the Korean Peninsula and Japanese Archipelago for use primarily as a fiber and seed source.

Additional discussion regarding the more recent spread, cultivation, and use of *Cannabis* for fiber, paper, seed, medicinal, and ritual applications is presented in Chapters 5 through 9. In the remaining part of this chapter we summarize the spread of *Cannabis* beyond Eurasia. Throughout the following five phases of *Cannabis*'s diffusion by humans, our model relies increasingly on historical rather than archaeobotanical evidence (presented in detail in the earlier sections of this chapter), supported by modern chemotaxonomic research.

Phase 2: Spread into Africa and Southeast Asia— ca. 2000 to 500 BP

The second phase in the dispersal of *Cannabis* by humans covers the period from the beginning of the current era until the fifteenth century and is characterized by expansion of the Arab Empire into Africa, the Indian Empire into Southeast Asia, and the dissemination of *Cannabis* primarily by Arab and Hindu adventurers and traders. Usage by people belonging to Arab and Hindu cultures was predominately for drugs; they spread both psychoactive varieties and knowledge of their use. Historical records hardly mention dissemination

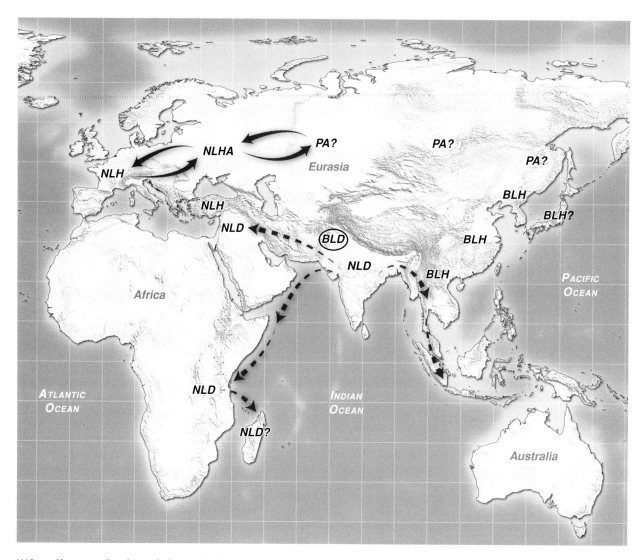

MAP 11. Human-mediated spread of narrow-leaf drug (NLD) *Cannabis* into the Middle East, eastern Africa, and southeastern Asia during Phase 2 dispersal from 2000 to 500 BP. Narrow-leaf hemp ancestor (NLHA) and putative ancestor (PA?) populations form an interbreeding complex (NLHA/PA?) stretching across eastern Europe and western Central Asia (cartography by Matt Barbee). (See TABLE 1 at the beginning of this book for explanations of *Cannabis* gene pool acronyms.)

for fiber or seed use during this period. Modern taxonomic research supports *Cannabis*'s introduction into Africa and Southeast Asia from West and South Asia, as taxa extant in these regions today are all NLD varieties. We first review the early introduction of *Cannabis* into Africa and then cover its initial entry into Southeast Asia.

Drug *Cannabis* was introduced into Africa at an early date, although accounts differ greatly and no certain time period has been agreed upon. Schultes (1970) estimated that the first introduction of *Cannabis* into Africa occurred between approximately 4000 and 3000 BP; however, the first physical evidence from the African continent does not appear until sometime between ca. 2000 to 500 BP (e.g., see Merlin 2003; also see Burney et al. 1994; van der Merwe 1975, 2005). Because of a relative lack of archeological, archaeobotanical, and historical evidence the path(s) through which *Cannabis* was brought into Africa remains ambiguous, but Emboden (1972) presumed that it was introduced "from India or Saudi Arabia" and appears to have been present "in the valley of the Zambezi in pre-Portuguese times, that is, before AD [*sic*] 1500."

Although archaeobotanical evidence for the early presence of *Cannabis* in sub-Saharan Africa is limited, a review of available historical, linguistic, and archeological references indicates that dispersal of *Cannabis* into Africa, at least into most of the eastern and southern regions, originated from South Asia before European contact. However, *Cannabis* may have been present in northeastern Africa much earlier, and these populations could have been derived from PHAs living in Caucasus Mountain refugia via the Middle East. Some authors report that hemp cordage was used in ancient Egypt more than 2,000 years ago or even earlier. For example, Rudgley (1998) indicated that ropes made of *Cannabis* fiber were used extensively in ancient Egypt as early as the Eighteenth dynasty (1550 to 1291 BCE), citing evidence from the crypt of the Pharaoh Akhenaten (Amenhotep IV) located at el-Amarna, and Booth (2003) referred to hemp cordage impressions discovered in ancient Egyptian tombs. Furthermore, according to Nelson (1996, also see Russo 2007), the ancient Egyptian word for hemp (*shemshemet*) can be found in "Pyramid Texts" in connection with rope making as early as the third millennium BCE. Yet, since Rudgely, Booth, and

Nelson provide secondary sources of information regarding *Cannabis* history, the confirmation of supposed remains of *Cannabis* fiber in Egypt will depend on the support of verifiable primary written sources presently unavailable and more significantly on reliable fiber analyses. In addition, Leroi-Gourhan (1985) reported her identification of *Cannabis* and *Gossypium* (cotton) pollen found in the mummy of Ramsés II (lived 1279 to 1212 BCE) and referred to both species as introduced crops in Egypt. She also referred to *Cannabis* pollen found at "Nagada Khattara near Luxor in Egypt which was dated to ca. 2600 BC[E]" (see Chapter 5 for further details about these intriguing but problematic discoveries). It should also be noted here that controversial evidence in the form of psychoactive *Cannabis* resin residue was reportedly found in the body tissues of several ancient Egyptian mummies that may support medical or ritual use by early Egyptians (Balabanova et al. 1992).

Northern and southern areas of Africa are separated by the extensive Sahara Desert and therefore the cultural dispersal of *Cannabis* basically followed two paths, one above and one below this great natural obstacle. Although there is some circumstantial evidence that hemp fiber may have been used in ancient Egypt, in areas both north and south of the Sahara, *Cannabis* use was focused primarily, if not exclusively, on production and consumption of its mind-altering resin rather than its fiber and seed products. North of the Sahara, *Cannabis* spread through Egypt, where it has been cultivated since at least the twelfth century and eventually westward with Muslim traders across sub-Mediterranean North Africa all the way to present-day Morocco. Muslim traders also carried *Cannabis* down the eastern coast of Africa from where it spread inland throughout many tropical and semitropical sub-Saharan regions. Its cultural diffusion across Africa was likely slow because sedentary farming cultures were sparsely distributed within this huge region, only reaching West Africa following World War II (Phase 4). Emboden (1972) pointed out that complex methods of preparing and using *Cannabis* in the Mediterranean or Near East regions (e.g., sieving hashish) were not transferred along with NLD plants into Africa, and techniques of consumption in the central region of the continent in the thirteenth century were very basic.

Direct evidence of ancient *Cannabis* use in Africa comes from "two ceramic smoking-pipe bowls, excavated in Lalibela Cave located in the Begemeder province of Ethiopia" and dated to 1320 ± 80 CE. Specialized thin-layer chromatography (TLC) of the pipe residues revealed "*Cannabis*-derived compounds" (trace amounts of Δ^8-THC), which suggests that "some variety of *Cannabis* . . . was smoked around Lake Tana in the 13th to 14th century [CE], in much the same way as it is today" (van der Merwe 1975; cf. Dombrowski 1971). Related dates for charcoal at ca. 630 BP and residue in one of the pipes at ca. 510 BP support this interpretation (van der Merwe 2003; also see van der Merwe 2005). The anthropologist Weston La Barre (1980) pointed out that "the hypothesis of African pipe smoking in the fourteenth century must, however, run the gauntlet of Americanist opinion that the smoking of plant narcotics would be post-Columbian, after the pattern of Amerindian tobacco smoking." On the other hand, smoking may have taken place in parts of Africa and Eurasia well before European exploration. Du Toit (1975) proposed that *Leonotus leonurus* (wild "*dagga*") and various species of *Salsola* ("*ganna*") and most commonly *Cannabis* were smoked as far back as the Iron Age in South Africa, chiefly among Bantu speakers.

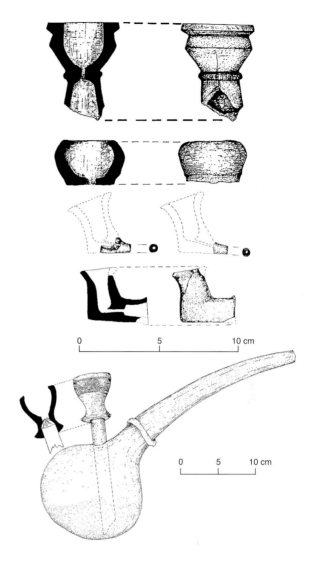

FIGURE 16. Pipe fragments recovered from Lalibela Cave, Ethiopia (top), and containing *Cannabis* residue resemble the bowl of a modern Zambian gourd *Cannabis* pipe (bottom). The Lalibela Cave remains are controversially dated to 1320 ± 80 CE, and the European-style pipes from Sabanzi, Zambia (center), are dated to the Bronze Age, both prior to the early sixteenth century introduction of tobacco smoking from the New World (illustrations courtesy of N. J. Van der Mewre).

Excavations by Fagan and Phillipson (1965) at the Iron Age site of Sebanzi Hill located on the southern edge of the Central Kafue Basin in southern Zambia exposed four smoking pipes of non-Arab design made of baked clay. Based on radiocarbon dating in association with related pottery sequences, the oldest two of these pipes appear to have been in use during the eleventh and twelfth centuries, and the others seem to have been smoked during the thirteenth and fourteenth centuries (van der Merwe 2003). Chemical tests of the residues have not yet been undertaken. Although it is possible that other substances were smoked in the pipes, the researchers believe they were used for smoking *Cannabis* since tobacco could not have been known in Zambia at the time (Phillipson 1965; also see Chapter 7 for more detail concerning the psychoactive use of *Cannabis* in Africa).

The *Lüshi Chunqiu* or "Annals of Lü" were compiled in 239 BCE by several scholars under the patronage of Lü Buwei, prime minister of Qin state (221 to 207 BCE). These annals include records of six crops, "*he, su, dao, shu, ma* [hemp], *and mai,*" that were commonly planted by the ancient Chinese (Lu and Clarke 1995). In book 26, chapter 6, part 5 of the *Lüshi Chunqiu,* (Knoblock and Riegel 2000), a translation of the section called "When You Plant Hemp at the Proper Time" reads as follows:

> Its seed heads are certain to be . . . long,
> It will have widely-spaced nodes and be brightly colored,
> Roots are small and stems hard,
> The male plant stems are thick and even.
> Ripening late, it produces in profusion,
> Maturing a second time by the autumnal equinox [seed plants]
> And so is not affected by locusts.

The *Er Ya*, regarded as the first Chinese dictionary with cultural, agricultural, and social contents, was written about 2,200 years ago by Fu Kong (lived ca. 264 to 208 BCE). This ancient text includes the following translated sentence: "Female hemp grows tall and straight. Its fiber is very thick and strong, and its seed can be eaten. The fiber of male hemp is thin and soft, and can be used to spin cloth" (Lu and Clarke 1995). Thus we know that the dioecious trait of *Cannabis* was initially recorded at an early time in China and that the different uses of male and female fibers were certainly well known by Han times, probably much earlier.

The *Ch'i Min Yao Shu* or "Essential Ways for Living of the Common People," an agricultural encyclopedia of the sixth century CE written by Jia Sixie and translated by Shih (1974) tells us how uniquely important the male plants of *Cannabis* have long been in East Asia. According to Shih,

> The only fibrous plant recorded in the *Ch'i Min Yao Shu* is hemp. Although allusion has been made, in a quotation from the *Book of Odes* [*Shi Jing*], to China grass [ramie] or *Boehmeria nivea*, the *Ch'i Min Yao Shu* gives no description whatever of that plant. Sesame was cultivated for its oily seeds, the use of its fiber is not mentioned. Utilization of wild plants such as *Pueraria* [kudzu], *Zizania*, *Andropogon* [both grasses], etc. are also cited *en passant*, but none of them has ever been cultivated as a fibrous crop. Even *tsu*, the female hemp plants, are primarily cultivated for seeds, both for reproduction and for oil; the stalks are good only for torches, not for spinning or textile purposes. The male hemp plants, on the other hand, are cropped primarily for fiber, yet their function in pollination was clearly understood long ago, as manifested in the *Fan Shêng Chi Shu* [an agricultural treatise of the first century BCE]. Chia Ssu-hsieh [Jia Sixie], however, went further in emphasizing the importance of timely reaping, "When (male hemp) is puffing [releasing pollen] like dust, reap quickly. When not yet puffing, the bast [fiber] is not well-formed; long after puffing, the bast darkens."

In crop rotations referred to in the *Ch'i Min Yao Shu*, adzuki beans (*Phaseolus angularis*) were recommended as a crop to precede hemp, which in turn was recommended as a crop to be cultivated before foxtail millet. Much more recently, in seventeenth century Guangdong province of southeastern coastal China, hemp was grown in significant quantities on flat hills and ridges alternating with sugar cane, cotton, beans, and other crops (Bray 1984).

The *Nung Shu* or "Book of Agriculture" written in 1313 CE by the "scholar-official" and agronomist Wang Chen is regarded as an outstanding, ancient treatise on agriculture in China. According to the *Nung Shu*, hemp and ramie, the two most important sources of textile fibers in early China, were divided geographically as indicated: "The appliances used for hemp and ramie are not the same in the south and in the north of China. How can the customs of people be changed? As the people of the south do not understand the reaping of hemp, so the people of the north do not know about the treating of ramie" (translation from the *Nung Shu* text by Kühn 1987).

Fiber use of hemp and ramie probably developed independently in the north and south of East Asia, and, as previously noted, hemp was most likely one of very few, if not the only, fiber plants cultivated in Neolithic northern China during the Yangshao period about 5,000 to 7,000 years ago (Andersson 1923; Chang 1977; Li 1974b; also see Huang 2000). However, there is evidence that the uses of hemp and ramie began to intermix geographically much earlier than has usually been suggested. For example, fragments of hemp textile, rope, and yarn remains have been recovered from an eastern Chinese Neolithic archeological site in Zhejiang province along the lower Yangzi River in the eastern coastal region of China (Cheng 1966), and by the Song dynasty (960 to 1279 CE) hemp was being raised "in the deep south" of China, while, at the same time, ramie was being grown in the north (Kühn 1987). For example, three pieces of ramie cloth were recovered from the Neolithic Chien-shan-yang site (dated 4850 to 4650 BP) in Zhejiang (Wang and Mou 1980; Anonymous 1960, cited in Kühn 1987). Although it is possible that ramie may have occurred as naturalized populations "in a few districts of the north" of China in these early times, certainly by the Western Qin period (221 to 207 BCE), the use of ramie fiber was spreading from the south through the Yangzi River region, into the Central Plains. According to the scholar Sung Ying-Hsing (lived 1587 to 1661) ramie was grown during his time in "every region of China" (Kühn 1987, citing the ancient *Thien Kung Khai Wu* or "Exploitation of the Works of Nature"; also see Ying-Hsing 1966).

According to interpretations of ancient Chinese agricultural literature by Kühn (1987), the use of ramie fiber for producing fabrics became more significant than hemp during medieval and premodern times. If, over time, ramie eventually did receive more attention as a useful source of textile fibers than hemp, why did this occur? Kühn points out that ramie is not difficult to grow, and with adequate nutrition will produce abundantly; while hemp is an annual species, ramie can be grown as a perennial under favorable conditions, yielding about 20 to 30 kilograms (44 to 66 pounds) of fiber per 1,000 square meters (about 0.25 acre) per harvest, which is two or three times the quantity of fiber that hemp can produce per year. In addition, once a plantation of ramie is established it can be cultivated without significant loss of productivity for 15 to 20 years (Kühn, citing Wagner 1926). Ramie is also well adapted to a fairly wide range of climatic conditions, from semitropical to warm temperate, and overlaps the more northerly cool temperate to warm temperate range of hemp. Furthermore, ramie fibers have excellent properties, including superior tensile strength and a brilliant luster that is nearly equal to silk: "The quality of the ramie thread produced varied from coarse to extremely fine, and all qualities of fineness were within the range of the [ancient Chinese] spinners' abilities" (Kühn 1987).

On the other hand, if ramie is superior to hemp in many ways, why did *Cannabis* remain such an important fiber plant for so long? Overall, it was probably due to *Cannabis*'s many uses, providing a wide range of other natural materials beside strong and pliant fibers for textile and paper pulp production. Hemp seeds have provided food for humans and livestock for millennia, and when crushed they provide valuable vegetal oil that can be eaten or burned to provide light. References in ancient Chinese texts date back thousands of years, probably many centuries or even millennia before these texts were written (e.g., as noted, there is the tradition of the mythical emperor, Shen Nung, and his associations with hemp use about 4,700 years ago). Also, the notoriety of *Cannabis* as a modern-day source of mind-altering drugs should not mask the fact that in the ancient past of China and several other areas of the world it provided important psychoactive drugs used for ritual as well as medicinal purposes (see Chapter 7 for a history of psychoactive *Cannabis* use including remarkable, recently discovered, macrofossil evidence from northwestern China; also see Chapter 8 for a historical discussion of medicinal use in China). In addition, at least since the time of Confucius, hemp fiber played a special role in mourning garments, a use that continues into modern times (see later in this chapter, and in Chapter 9 about the nonpsychoactive ritual use of *Cannabis*). Over time, emphasis on medicinal, food, ritual, and ecstatic use declined in China, while fiber use remained important well into the twentieth century. From an agroeconomic perspective, it is also important to point out that *Cannabis* can be cultivated in colder, higher latitude and higher elevation regions than ramie. Ramie would compete with food crops for arable land in the populous river valleys of northern China, while hemp could be grown throughout higher elevation areas much farther from urban centers.

Because of the physical structure of hemp fiber bundles, its yarn is normally coarser than that of other fibers including ramie. However, hemp yarn of extreme fineness, rivaling that of ramie, is presently produced in Korea (Clarke 2006a) and Japan. Kühn (1987) assumes that in ancient China hemp fibers were used primarily to "weave rough textiles and to manufacture ropes." He points out that there is no mention of hemp in the fabric section of *Thien Kung Khai Wu* and only a minor reference in "The Growing of Grains" chapter, which indicates that the rough cloth "woven from its bark fibres [sic] is of very little worth." However, the economic value of hemp in ancient China is evident in many ways, including its use as payment of tax by commoners of the Warring States period (475 to 221 BCE) and generally as a product of trade during Eastern Zhou times (770 to 221 BCE) when people in less developed states lying beyond the "Central Kingdom" were said to have worked with "wooden carts" and wore "tattered hempen clothes to bring the hill and forests under cultivation" (Cheng 1963). Later in Han times, after about 200 BCE, hemp remained a notable trade item in China (Hsü 1977). In Mawangdui Han grave 1 in Changsha, Hunan province, bamboo slips used to record wills of dead people were found attached to "hemp sacks" (Fuminori 1998; also see Hunan Sheng po-wu-kuan 1973).

All indications suggest that in the past, as well as present times cultivation of hemp, ramie, and other bast fiber crops, as well as the raising of silkworms all occurred away from urban areas due primarily to the lack of inexpensive land and labor in or around Chinese cities (Kühn 1987). Traditionally, spinning of bast fibers was "a seasonal cottage industry" and exclusively the domestic work of women, especially those in the peasant class. This rural activity normally occurred in homes of the poor themselves or in places owned by wealthy local people (Kühn 1987). Thus we are told in the *Mo Tzu*, or "Book of Master Mo," written by the philosopher Mozi (lived ca. 490 to 403 BCE), that women had to rise early in the morning and retire late in the evening because of their obligatory spinning and weaving activities with hemp, silk, and kudzu (Schmidt-Glintzer 1975). After taxes were paid in cloth and household needs were met, any surplus fabric was sold in neighborhood markets, and because of a number of natural and cultural factors "the socio-economic history of the yarn producing peasant households is the history of the peasants of pre-modern China" (Kühn 1987).

Although everyday life of peasant women in early China was preoccupied with textile production, this cultural pattern was also typical of many upper class females, including concubines, young wives, and girls (Kühn 1987, citing *Chou Li Chu Shu*). The *Li Chi* also tells us that

> "a girl at the age of ten ceased to go out (from the women's apartments). Her governess taught her the arts of pleasing speech and manners, to be docile and obedient, to handle the *hempen fibres* [sic], to deal with the cocoons, to weave silks and form *fillets* [narrow bark strips], to learn (all) woman's work, how to furnish garments, to watch the sacrifices, to supply the liquors and sauces, to fill the various stands and dishes with pickles and brine, and to assist in setting forth the appurtenances [subordinate adjunct parts] for the ceremonies." (LEGGE 1885a, italics added)

The important role of *Cannabis* fiber production for ancient Chinese textile industries is emphasized in the key traditional concept of *favorable balance*; in relation to major fiber crops, this is noted in the *Yen Theih Lun* or "Discourses on Salt and Iron" dated to the early Han dynasty (ca. 80 BCE), where it tells us that there will be a great harvest if cultivation is carried out according the seasonal changes of the climate: "If silkworms and *hemp* are raised according to the seasons, cloth and silk will be more than what is required to wear" (Gale 1931, italics added). The dire consequences of insufficient agricultural production were also well known; for example, according to a proverb in *Nung Shu*, "A man who does not plough, may suffer from hunger, a woman who does not weave may suffer from cold" (see Kühn 1987).

During the reunification of China during the Sui and Tang periods, beginning in 581 CE, the "equal field system was extended to everyone, with 80 *mu* of land (one *mu* = 666 square meters, and 80 *mu* = 5.3 hectares or about 13 acres) given to each ordinary male householder to work during his active lifetime, and 20 *mu* or 25 percent of this allotted land for fiber growing (mulberry and hemp land) a strong sign of the importance of fiber in the Chinese economy and the significance of these two crucial plants" (Anderson 1988; also see Wu and Raven 1994). At least by the time of the Southern Song dynasty (1127 to 1279 CE), when farming techniques were already quite advanced, hemp plants were cultivated along with mulberry bushes because their respective root systems had "different lengths and exploited the soil in different strata" (Theobald 2000).

Archaeology provides many fragmentary insights into ancient uses of hemp with some indications that it was used in China as a fiber source before, or at least soon after, farming began, perhaps many thousands of years ago by some Mesolithic peoples.. Although we have no direct evidence of this, the recent discovery of ca. 10,000-year-old seeds in association with pottery shards on Honshu Island in Japan is intriguing (Okazaki et al. 2011; also see Chapter 4). Cord-impressed pottery has been discovered in early postglacial fishing sites along the southern China coast and dated to approximately

12,000 BP; hemp cordage may have been used to create the markings (Chang 1968). Additional evidence of cord impressions were recovered from early Neolithic sites on Taiwan (e.g., those of the Tapenkeng culture ca. 7000 to 5500 BP; see Tarling 1999); however, the identity of fibers in the cordage used to make these markings has not been clearly established. Cheng (1966) refers to the lower cultural stratum of the Yuan-shan culture (also ca. 7000 to 5500 BP) at one of the richest and most important archeological sites in eastern Taiwan: "The characteristic remains here as elsewhere on the island, are the cord-marked pottery, a coarse sandy ware, handmade, thick-walled, with a plain surface covered with cord marks and, occasionally, with lineal impressions . . . There is also a stone beater, a rod-shaped implement with incisions, which might have been used for the linear impressions in pottery decoration, or for pounding hemp fiber, a common raw material for rope and textile."

It should be noted that in Taiwan today, a residual ramie weaving tradition exists but no hemp cultivation for fiber. Cord-marked pottery also appears in ancient New World sites where impressions must have been made from a fiber other than hemp. Marking pottery with cord impressions is a simple, and therefore a widespread technique, making it difficult to imply any special meaning.

More cord impressed ceramics dating from the seventh millennium BP to about 1200 BP have been found in several regions of mainland China (Li 1974a/b). For example, ancient coarse paste pottery, with marks impressed by using rope, twisted cord, or cord-wrapped sticks and paddles, has been excavated from early Holocene agricultural sites; these include Cishan dated to ca. 7400 to 7100 BP in the Huang He valley (IAASS 1984), fishing sites on the south China coast, in northern parts of Southeast Asia, and at T'ai-pingch'ang, a Late Neolithic site in Sichuan province of western central China dated ca. 4000 to 3200 BP (Cheng 1982). Most of these sites were close to bodies of water from which archaic bands collected shellfish and trapped fish, and the fibers used in the cords most likely came from plants nearby, where, we argue, wild *Cannabis* would have grown.

The majority of early fiber impressions originate in northern China. Li (1974b) refers to cordage impressions attributed to *Cannabis* recovered from various later Yangshao culture sites dated to ca. 4410 to 3100 BP, as well as from an older Yangshao site dated to ca. 5000 to 4000 BP in Henan. Imprints of textiles found on numerous pottery shards were recovered from many early Neolithic Yangshao sites such as Pan-p'o, dated to ca. 6225 to 5430 BP, and pottery fragments bearing rope imprints were found at a Lung-shan culture site at Hsichou in Hunan province of south central China dated to between ca. 3170 to 2230 BP. Another Yangshao site in Shaanxi province yielded pottery spinning whorls, fine bone needles, and textile impressions in the soil of one grave that were interpreted as remains associated with hemp (Li 1974b). The pioneering archeologist, paleontologist, and geologist Johan Gunnar Andersson (1923) was the first to surmise that woven textile impressions found on Neolithic (ca. 6000 to 4000 BP) pottery sherds in Yangshao archeological contexts actually represent impressions of hemp cloth. Bronze weapons of the Shang times recovered from excavations at Anyang in Henan display impressions left by cloth wrappers believed to be made of hemp, and a dagger with fiber impressions recovered from a ca. 2900 to 2800 BP burial at Ning Xian in Gansu was also wrapped in hemp or a similar coarse fabric (So and Bunker 1995). Some if not many textile and cordage imprints might

have been made from fibers other than hemp; however, *Cannabis* remains the most likely choice based on biogeographical parameters and historical records (Lu and Clarke 1995).

Although imprints of cordage and cloth do not allow for fiber identification, actual remains of fiber in cloth or other textile forms come from a variety of sites over many millennia, and several examples from ancient China are provided in the discussion that follows. In 1960, part of a rough hemp fabric, dated to ca. 4150 to 3780 BP, was recovered from an archeological site at Tahojuang in Yungjing county of Gansu belonging to the Chi-chia (Chijia) culture (Kühn 1987). Hemp textile, rope, and yarn remains dated between ca. 5500 to 4200 BP were reported from a Neolithic Liangju culture site located in the coastal plain around the lower Yangzi River in Zhejiang (Cheng 1966). *Cannabis* fiber remains have also been discovered at a Shang period site dated to ca. 3520 to 3030 BP at Anyang in Henan, where thousands of funerary objects were found, along with inventory lists that included references to hemp textiles (Cheng 1963). In addition, large quantities of preserved "silk and hemp fabrics" were excavated from the Zhou period Changuo tombs in and around Changsha in Hunan dated to ca. 3122 to 2021 BP (Cheng 1963). For example, in a broad rectangular Changuo pit tomb, the inner wooden coffin was painted with brown lacquer and "reinforced on the outside with bands of hemp cloth each fourteen centimeters wide, two horizontally, three along the long sides and two more on the short sides," and these were "fastened to the coffin with lacquer" (Cheng 1963; also see Cheng 1966). Another Zhou tomb in Shanxi province of northeastern central China contained lightweight hemp cloth, indicating that hemp weaving had reached a high standard more than two thousand years ago (Li 1974b). Fragments of hemp and silk cloth, dated to ca. 2655 to 2615 BP, were discovered in a tomb excavated in Anhui province in eastern central China, where they were found adhering to the outer surface of bronze sculptures, indicating that the relics had been wrapped in cloth (Yin 1978). Another relevant ancient burial—a boat coffin carved from a single tree trunk found in the Wuyi Mountains of Fujian province in coastal southeastern China and dated to ca. 3600 to 3300 BP—contained the body of an elderly man wrapped in a cloth funeral shroud said to be made of jute, hemp, silk, and cotton (Li 1984).

In 1930, a Sino-Swedish expedition in China discovered a large cache of original Han period inscribed artifacts in the Juyan region of Gansu. This included a sizeable amount of bamboo tablets used primarily as official documents excavated from a series of sites along the east bank of the Edsen-gol River. Perhaps the most remarkable artifact among this cache was an ancient bundle of 77 linked tablets—each approximately 23 centimeters (9 inches) long and 1.3 centimeters (one-half inch) wide—tied together by two strands of "hemp thread" at each end and rolled up to facilitate convenient transport and storage. The text on these tablets was inscribed between 93 and 95 CE and presents an account of armaments stored at the military settlement where it was found (Tsien 2004). A wooden paint brush holder, discovered at Mudurbelijin near the ancient town of Juyan in 1927 by Folke Bergman and dated to the Han dynasties (206 BCE to 220 CE), was made of "four vertical pieces of wood that are fastened into a rod with two hanks of hemp string" (Tsien 2004).

Traces of ancient hemp fibers were also found on the outer surface of bamboo tablets discovered at the "desert oasis" of Dunhuang in Gansu (Tsien 2004). Dunhuang is located at the edge of the Gobi Desert in the very western end of the Hexi Passage near the juncture of three provinces, Gansu, Qinghai, and the Xinjiang Uighur Autonomous Region. This armed camp was established by the Han Emperor Wudi in 111 BCE as one of four military posts to guard the northwestern frontier enabling the Chinese to control trade routes to the "western regions" and command the "Silk Roads" (Whitfield 1995). During the centuries following its establishment, Dunhuang developed as an important center of communication between China and Central Asia, and for a thousand years was also an important Buddhist center.

At Qianfodong, a site some 25 kilometers (16 miles) to the southeast of Dunhuang, are the Mogao Caves also known as the "Caves of the Thousand Buddhas," excavated out of a tall gravel conglomerate cliff cut by stream flow. The cliff provided space over a kilometer (two-thirds mile) long in which more than seven hundred grottoes, many set in three tiers, were hollowed out and painted with frescoes during the fourth through fourteenth centuries. The Mogao Caves complex is considered to be one of the great ancient sites of Buddhist art, presenting a first-rate glimpse at a millennium of Chinese painting with Indian, Greco-Roman, and Iranian influences. A translation (Whitfeld et al. 2000) from a tablet at the Mogao Caves, dated to 698 CE, gives an account of the first identified activity at these caves: "Even in remote antiquity the sand-swept height overlooking the desert river valley was known as the 'Wonderful Cliff' or the 'Precipice of the Immortals.' It had probably been considered a locus of spiritual power for centuries when in 366 [CE] a wandering monk named Yezun, resolute, calm, and of pure conduct . . . traveling the wilds with his pilgrim's staff, arrived at this mountain and had a vision of a golden radiance in the form of a thousand Buddhas. Thereupon he erected scaffolding and chiseled out the cliff to make a cave." This grotto and others at Mogao, with their frescoes and special artifact caches, remained sealed and essentially unknown to the outside world until the arrival of Sir Marc Aurel Stein (1862–1943), a Hungarian-born British archeological explorer who made three successful expeditions to Central Asia over a 14-year period beginning in 1900. His collections remain important to the study of Central Asian history and the art and literature of Buddhism. In 1900, a Daoist abbot, Wang Yuanlu, the self-styled custodian of Mogao discovered a sealed chamber, Cave 17, the now famous "Library Cave," inside the larger Cave 16. In 1907, Stein, as the first foreigner to gain access to Mogao, was able to convince Wang Yuanlu that he should enter this sealed chamber, which had been the memorial chapel for an important ninth-century monk. Stein found it packed with an estimated 50,000 ancient manuscripts and paintings (Whitfield 1999a; Whitfield et al. 2000), including the Diamond Sutra, reputed to be the first (868 CE) book printed on fine paper. This manuscript and many others appear to have been made of hemp pulp paper, apparently the first paper to be used widely by the Chinese (e.g., Pan 1979; Shimura 1980).

Thousands of artifacts, including large numbers of ancient Chinese texts, were removed from Mogao and taken to various foreign countries by a series of British, French, Japanese, and Russian explorers. Many thousands of Buddhist manuscript scrolls remained in fine condition because of the arid desert climate and lack of use. This popular cave site is now designated a world heritage site by UNESCO, which includes an ongoing environmental monitoring and conservation program (Whitfield et al. 2000). Numerous artifacts found in the Dunhuang region were made of hemp fiber. For example, a compilation listing 11,839 items from the Sir Aural Stein collection originally gathered from the Dunhuang area and now held at the India National Museum in New Delhi includes reference to 244 banners from Dunhuang made of "silk and hemp" (IDPNI 1995).

A number of paintings found in Mogao Cave 17 were made by stitching together pieces of silk, "with the edges protected by a sewn-on border of silk or hemp cloth" (Whitfield and Farrer 1990). Others, such as a double-sided painting, 84 centimeters tall and 76.5 centimeters wide (about 3 feet tall and 2.5 feet wide, no. 68 in the Oriental Antiquities Collection of the British Museum), were created with ink and colors on paper (possibly made of hemp fiber) that were glued onto both sides of a coarse piece of hemp cloth. This painting was intended to be inserted in a window lattice showing flowers on the side that could be seen from outside, and "Bhaisjyaguru, the Buddha of Healing," flanked by two attendants, was to be viewed on the side seen from inside the cave (Whitfield and Farrer 1990).

Ancient embroideries in silk thread on hemp cloth have also been found in this cave complex, "some of which were so long that they could only have been intended to hang from the Mogao cliffs" (Whitfield et al. 2000; also see Whitfield and Farrer 1990; Whitfield 1999b). One example discovered in the Library Cave shows Shakyamuni preaching the Lotus Sutra at Râjagrha on Vulture Peak. Shakyamuni in Sanskrit means "Sage of the Shakyas," which was the title of the original Buddha, Siddhartha Gautama, a prince of the northern Indian Shakya kingdom. Dated to the Tang dynasty (618 to 907 CE), this beautifully crafted religious artifact is among the largest known examples of Chinese embroidery. Now held in the British Museum's Department of Oriental Antiquities study collection (CH. 00260), this famous embroidery measures 241 centimeters tall and 159.5 centimeters wide (about 8 feet tall and more than 5 feet wide) and was fabricated from three widths of hemp cloth covered entirely with thin, closely woven silk fabric. Although there is some obliteration of the disciples and bodhisattvas shown attending to Shakyamuni, this artwork is in extraordinary condition. Inevitably, parts of the silk under the embroidery have worn away, exposing the backing of hemp fabric (Whitfield and Farrer 1990; also see Stein 1921; Zwalf 1985).

Another interesting artifact found in Cave 17 at Mogao is a hemp cloth banner measuring 94 centimeters tall and 27 centimeters wide (about 3 feet tall and less than a foot wide) that has been dated to the early tenth century; it was donated to the British Museum by Sir Marc Aurel Stein. Using basic ink lines painted directly onto hemp cloth without any additional coloring it depicts Avalokieshvara—the popular bodhisattva of compassion—in a frontal position, standing under a parasol, holding a lotus bud with his left hand and making the gesture of teaching with his raised right hand. Since hemp was less expensive than silk, these lower-quality banners were probably custom built and bought by less affluent donors (British Museum Compass Collections Online 2000; see Whitfield 1982–85; also see Whitfield 1999a).

Excavations of Han period tombs in Gansu dated ca. 100 BCE to 100 CE produced complete specimens of hemp cloth used to cover corpses. The shrouds were wrapped around silk dresses and tied with hemp ropes, and hemp fibers were also used for reinforcing the plaster covering of the brick walls of these crypts (Kansu Museum 1972). In addition, Giele (1998) referred to a "hempen bag" containing 92 documents found "above the head of the corpse" in an "unspecified tomb" (VI.34) at Hantanpo cemetery in Wuwei Xian, also in Gansu. This bag was "dated through stylistic comparison and *wushu* coins" to the later Han period between 12 and 75 CE. Giele also reported that "hempen cloth" wrapped around several hundred documents was recovered from a tomb at Jiangling Xian (Zhangjia Shan) in Hubei province of central China dated to ca. 2173 to 2167 BP.

Remains of Song period (960 to 1279 CE) statuary excavated from a site near Chincheng in Shaanxi show that hemp cords were used to wrap the core of the statues before the clay exterior was applied (Kao 1978), and fragments of a hemp and bamboo mat were recently recovered in Hong Kong from a pit found in the cultural layer of a large-scale archeological rescue excavation at So Kwun Wat in Tuen Mun district (Hong Kong Year Book 2000).

An ancient styled wig made of hemp twine at least 2,000 years old was unearthed in southern Sichuan (People's Daily Online 2005; China Heritage Newsletter 2005). This ancient hairpiece was found on the lower part of the skull of a skeleton excavated from a tomb in Zhaojue county in the Daliangshan Mountains and dates to the Qin period. The wig was uncovered in the general region of the Yi ethnic group who migrated into this region from central Yunnan province in ancient times. Although the identity of the people who built and used the tomb is still unverified, some historians assume they were the ancestors of Yi peoples who still reside in the area and continue to grow hemp. A heritage repair technician with the local museum in Liangshan prefecture consulted with a number of experienced, local hemp weavers prior to concluding that the hairpiece is indeed composed of hemp cordage.

In addition to wigs, hemp was also used in the construction of shoes and sandals. Stein's collections include articles such as ancient footwear combining features characteristic of both a shoe and a sandal. Constructed with "tapestry-woven uppers" and "peculiar open-work nets" (see Stein 1921) they were stitched with horse hair and hemp, according to Andrews (1935), who tells us that "the entire shoe, other than the [hemp] string sandal, was composed of at least three thicknesses of [fabric] material, of which one was hemp."

Li (1974b) refers to a grave discovered at Turpan (Turfan) in Xinjiang province of far western China dated to 721 CE where hemp cloth and shoes woven with hemp fibers were found. Another odd and somewhat ambiguous ancient reference to even older Chinese hemp shoes comes from one of many bamboo tablets excavated from a Qin period tomb at Shuihudi in Yunmeng, Hubei, that dates to 217 BCE. This tablet of the *Rishu* or "Day Book" reads in translation: "If without any reason one's hair stands erect like worms, whiskers or eyebrows, he will have met with evil air. [changes includes new paragraph] To counteract, one can boil a hemp shoe with [or into?] paper [*zhi*, a kind of fine paper], and the evil will be expelled" (Tsien 2004). Although hemp fiber waste or rags were used as raw material to make paper in ancient times, the association of paper with hemp shoes is unclear.

Many other ancient fabric samples reportedly woven with hemp fiber were recovered from additional sites across China, but these are undated and have been omitted from this survey (e.g., more evidence for hemp use in ancient China is provided by Stein 1921; Andrews 1935; Cheng 1959, 1963, 1966; Li 1974a/b; and Yin 1978).

Scholars generally accept that many early Chinese manuscripts, fabric banners, paintings, embroideries, wigs, and shoes are made of *Cannabis* hemp fiber. However, few if any of these ancient textiles have been analyzed to determine their actual fiber content. To the naked eye, and certainly under the magnification of a hand lens, one can readily discern between hemp and kudzu fibers (see Clarke 2006b), but differentiating between hemp and ramie remains problematic. We can only hope that modern analysis techniques will be applied to make positive identifications of past and future textile finds.

Hemp Fiber Use in Clothing, Lacquerware, Weapons, and Ships in Ancient China

As our previous discussion shows, *Cannabis* was an important crop in ancient China, serving as a major fiber source for several millennia. By the time of the Han dynasties about two thousand years ago, the Chinese population had increased significantly and intensive farming was practiced within a controlling feudal system. Under these conditions, hemp fiber became a valuable commodity in the newly forming rural market economy and tribute system. Commercialization of agricultural products tied local economies to the regional and national economic system, and hemp cloth from some ancient Han states began to be traded across China. This occurred, for example, in places such as Lu (in present-day Shandong province of eastern China where hemp is grown today). Prices for hemp varied according to the distance to market from production areas; in urban areas hemp cloth was relatively costly, yet rural peasants across China wore hemp clothing throughout the Han era (Hsü 1978). Traditionally, silk fabrics were worn by the wealthy people, while hemp cloth was the "textile of the masses," at least in the northern regions (Li 1974a/b).

During mourning rites for deceased parents, which lasted for three years, the *Li Chi* tells us that the "king did not sacrifice (in person), excepting to Heaven, Earth, and the Spirits of the land and grain; however when he went to transact any business, the ropes (for his chariot) were made of hemp (and not of silk)" (Legge 1885a). In addition to hemp rope, we also know from the *Li Chi* that ancient Chinese bandages, corpse shrouds, and simple sheets were usually made of hemp cloth, which was also used to fabricate the special clothing worn during mourning rites for the deceased, and "a system existed for rating bean-creeper [kudzu] cloth and hemp cloth as mourning robes" (Kühn 1987). Basic mourning garb usually consisted of a coarse hempen sackcloth robe and headband. The *Li Chi* refers to five degrees of hempen mourning dress, some of which were made solely of fibers from female plants of *Cannabis* and others only using fibers from male plants. Differing hemp grieving garments were worn by specific people, and for varying time periods, according to their familial relationship to the deceased, with their particular relationship determining the coarseness of cloth used to make the mourner's costume (Kühn 1987; also see Legge 1885b). The custom of wearing hemp clothing as a form of respect for the dead is still followed by millions of East Asian people (also see Chapter 9 for more discussion of the relationships of *Cannabis* to rites and religion in China). As the archeological evidence indicates, hemp fiber and especially hemp fabric have a long association with death and mourning and have commonly been found in Chinese tombs, often along with lacquerware, an iconic Chinese craft.

Hemp fiber was commonly employed in lacquerware, dating back at least to Shang times more than 3,000 years ago. Chinese lacquerware is well known for its hard finish, excellent construction, and beautiful forms and has remained popular throughout history. In fact, lacquerware is not made entirely from lacquer itself; it is only coated with a special lacquer varnish. The core material varies, ranging from bamboo, rattan, or wood to metal or even plastic in the modern era. In most cases, traditional lacquerware utilized bamboo to provide the core shape, and strips of hemp cloth were integrated into the object as it was varnished. According to Cheng (1963), bamboo and hemp cloth were "the chief material[s] used with plain lacquer as the foundation before the colored fluid was applied."

Very early examples of lacquer work were found in tombs of pre-Shang cultures living before the sixteenth century BCE. Later, during the Warring States period and the Han dynasties, great amounts of lacquerware were produced. Excavation of the early, wooden-chambered pit burial tomb of Laofu Shan, near Nanchang city in Jiangxi province of southeastern China (dated to the Western Han period, 206 BCE to 8 CE), uncovered more than 200 grave goods, and among these were several lacquerware winged cups built on hemp cloth cores (Rudolph 1978). In addition, large numbers of "realistic clay sculptures were found in a Tang monastery in Shaanxi, made on wooden cores reinforced with iron wire and nails, hempen cords, and minced straw" (Rudolph 1978).

Lacquer's durability was crucial for the afterlife needs of important people in ancient China whose burials were expected to last indefinitely. Long-lasting lacquer continued to be a popular adhesive and protective material at least into the Western Zhou period when artisans crafted household and decorative items praised for their lightness and smooth colorful finish. During the Song dynasty artisans invented lacquer carving with its attractive forms and patterns that precipitated a kind of artistic renaissance in lacquerware construction that continues today. The use of hemp cloth in lacquerware manufacture has also long been popular in Korea and also Japan, where layers of hemp cloth soaked with lacquer are applied over a core in the traditional *kanshitsu* or "dry lacquer" technique (Nihon Kogeika 2005).

Since ancient times traditional lacquer for such work was obtained from the sap of *Rhus verniciflua*, a native East Asian tree that grows up to 20 meters (65 feet) tall and belongs to Anacardiaceae, the cashew or mango family. Called "lacquer sumach," "varnish sumach," or "lacquer tree," like many other species in Anacardiaceae, it produces a poisonous secretion; this toxic sap can be pigmented to coat objects in different colors. After initial varnishing, one or more layers of lacquer are applied, and the application of up to 30 or more layers is common in fine lacquerware. The crucial component of the sap is called urushiol, based on *urushi*, the Japanese term for lacquer. When exposed to oxygen, urushiol polymerizes, and when fully dried, it is extremely resistant to water, acid, and, to a certain extent, heat. Consequently, lacquer is an ideal adhesive and binding agent and provides a durable, protective varnish on many materials, particularly wood, bamboo, textiles, and leather. During drying and curing, the poisonous sap loses its toxicity and lacquerware can then be used for various purposes.

Unprocessed lacquer sap is a peculiar substance, grayish-white with the consistency of molasses. Cheng (1963) explains how durable lacquer products can be and part of their connection with hemp cloth:

FIGURE 19. Imperial Chinese Qing dynasty oceangoing ships relied on locally produced hemp sails and rigging (adapted from Needham et al. 1971).

On exposure to the air [the white sap] turns yellowish-brown and finally black. After being strained through a sheet of hempen cloth to remove the impurities, it is stirred gently to give the required uniform fluidity. It is then heated over a very low fire or in the sun and stirred again to evaporate the excess moisture. The liquid is finally stored in airtight vessels for use. In its fluid form it is weak, easily acted on by a small quantity of salt, vinegar, oil or the weakest acid, but when hardened, it is insoluble in all solvents and becomes most resistant to salt and strong acid and can be preserved underground and in water for thousands of years.

As the previous quote indicates, hemp cloth, therefore, played a role in the preparatory filtering of raw lacquer, as well as in forming the cores on which liquid lacquer was applied to give it form and function. Thus the strong and pliant fibers of *Cannabis* are found at the "very core and epitome of Chinese culture, in their lacquerware" (Robinson and Nelson 1995).

We now shift our focus from the artistry of fine Chinese craft to learn how hemp was involved in waging war. Artifacts recovered from archeological sites in China indicate that hemp fiber was used in early weapons construction. An excellent source of information concerning warfare in ancient China is the *Wujing Zongyao*, or "Essentials of the Military Arts," completed in 1043 CE during the Northern Song dynasty by Zeng Gongliang, Ding Du, and others at the order of Emperor Renzong (Gongliang and Du 1988; also see Tseng 1935; Kierman and Fairbank 1974; Yates 1982; Needham and Yates 1994). In the *Wujing Zongyao* and other ancient Chinese texts we learn that the military was significantly enhanced when they discovered that longbow and crossbow strings made of hemp fibers were much stronger than those made of bamboo. Consequently, hemp cordage became an important component of potent Chinese bows because of the power it helped provide. Crossbows were probably invented in China and were in use by the fifth century BCE, subsequently becoming significant weapons in armed conflicts of the Warring States period. In addition to its use in ancient bowstrings, hemp twine was used as the bridle to attach the bow portion to the crossbow stock. First century CE documents uncovered by Stein (1921) in the Dunhuang network of military walls and watchtowers indicate that "silk and hemp were both used for bow strings" (Andrews 1935). Abel (1980, citing Eberland 1968) tells us that hemp fiber was such an important crop for bowstring production that "Chinese monarchs of old set aside large portions of land exclusively for hemp, the first agricultural war crop." Hemp crossbow strings are still favored by traditional hunters in southwestern China and northern Southeast Asia.

Catapults with a simple lever-operated mechanism were the "siege artillery par excellence" in medieval China, and hemp fiber was also an integral part of these large assault weapons since they all had a number of tightly twisted elastic ropes used to arm them: "The ropes must be made from hemp and leather twisted together, because if the weather is fine, leather shrinks and hemp distends, whereas in rainy weather leather tends to become soft and hemp shrinks. A mixture of the two materials therefore will guarantee uniform operation throughout the year" (Kierman and Fairbank 1974, translation from the *Shou-ch'eng lu* or "Records of the Defense of Towns" ca. 1130 CE).

The most significant use of catapults in medieval Song times was for flinging fiery materials at the enemy's towns, battlements, and wooden war instruments such as siege towers, ladders, platforms, catapults, and multiple crossbows. Some of the incendiary bombs hurled outward by the catapults contained hemp or cotton fiber soaked in oil and ignited, and one recipe for an ancient Chinese gas bomb "prescribed paper, hemp-bark, resin, yellow wax, yellow cinnabar, and charcoal powder," which was "a simple device for producing a suffocating smoke when ignited" (Kierman and Fairbank 1974). Cruel as it seems, animals with unpredictable behavior such as birds were also used to deliver burning torches to enemy camps, and oxen "with burning hemp lashed onto their tails" were stampeded toward enemy troops (Gongliang and Du 1988).

Once again hemp fiber played an apparently minor yet ultimately important role in directing human history; in this case adding efficiency and killing power to tools of warfare, empowering ancient Chinese armies, and possibly determining the outcome of key battles. The Chinese also used hemp fiber to build and rig ships that explored much of the world and further spread Chinese influence.

Although there is no known outstanding historical manuscript dealing with Chinese shipbuilding, some insightful, ancient texts describing traditional nautical technology in China have been identified. One of these is found in the *Tian Gong Kai Wu* or "Exploitation of the Works of Nature," written in 1637 by Song Yinsheng, who compiled extensive information concerning various aspects of Chinese technology. Song described a typical vessel that carried grain on the Grand Canal in the latter part of the Ming dynasty (1368 to 1644 CE), and the halyards used to hoist the sails of such a ship were made of hemp fibers twisted together until they had a diameter of about three centimeters (more than an inch; Ronan 1978).

Marco Polo, who is said to have traveled in China from 1275 to 1292, described with admiration the construction of ships of many kinds; for example, he tells us how the cracks between nailed siding planks are caulked (sealed or waterproofed):

> They are not pitched with pitch, because they have not of it in those regions, but they oil them in such a way as I shall tell you, because they have another thing which seems

FIGURE 20. Raw Chinese hemp fiber (top left) is combed, carded, and bleached to produce fiber (center) for spinning yarn (top right), which is used to weave sturdy fabric (bottom right).

> to them to be better than pitch. For I tell you that they take lime, and hemp chopped small, and they pound it all together, mixed with an oil from a tree. And after they have pounded them well, these three things together, I tell you that it becomes sticky and holds like birdlime. And with this thing they smear their ships, and this is worth quite as much as pitch. (RONAN 1978)

Hemp was grown for fiber until recently in various regions of China. The finest *Cannabis* fiber produced in China during the last century is said to have come from Shandong and the greatest yields were obtained in the eastern and southwestern regions; hemp was normally cultivated "in fertile soils, on the banks of rivers and in areas where there are alluvial soils with a water-table near the surface" (Kirby 1963). Our discussion concerning the utilization of hemp fiber in ancient China underscores the multipurpose nature of this important *Cannabis* product. Strong and pliant hemp fiber was also used extensively in other areas of East Asia.

Traditional Korea

Cannabis fiber has been used to make textiles for thousands of years in Korea where it is known as *dae ma* (from Chinese *dà má*) or *sambae*. According to Ree (1966) its early cultivation began "in order to have a long vegetable fiber suitable for spinning and cordage." Today *Cannabis* is still grown on limited acreage and also is self-sown in a few locations throughout the Korean Peninsula. Fiber hemp is cultivated primarily in paddy fields prior to the rice crop but is also grown in upland areas, mostly on terraced hillsides.

Cannabis use and cultivation in Korea was probably introduced from China, at least by the Neolithic farming period that is characterized at many archeological sites by "combware"—incised pottery decorated on the exterior with geometric patterns (Nelson 1993). The earliest evidence of clothing in Korean culture dates to ca. 5000 BP and includes artifacts of cloth and clothing production such as sewing needles and spinning tools, as well as personal accessories such as earrings, bracelets, shell necklaces, and rings excavated from ancient Neolithic shell mounds at Gimhae town in South Gyeongsang province of far southern South Korea (Hyo-Soon 1995). Ancient hemp threads along with bone needles were recovered from a contemporaneous site in South Pyongan province, in western central North Korea (Sim 2002; also see Kim 1979; Nelson 1993; Clarke 2006a). Early use of hemp in Korea is also supported by evidence from a Bronze Age ca. 1000 to 300 BCE horizon at the site of Pomuigusokin in North Hamgyong province of North Korea along the border with China, where a rope, most likely made of *Cannabis*, was recovered (Anonymous 1990).

Nelson (1993) raised questions regarding the role of *Cannabis* among early Korean coastal farmers or fisher-gatherers around 3500 BCE. For example, were early Neolithic west coast "Chulmun" (combware incised pottery) groups growing and weaving hemp during this period? Was a "whole complex of plant domestication and its consequent industries" spreading across the Korean Peninsula? Nelson suggested that spindle whorls found in late Neolithic Mumun culture (ca. 1500 to 300 BCE) sites may represent a "textile industry, probably based on hemp, ramie or similar native plants."

Although hemp and ramie are not assumed to be native to Korea, *Cannabis* was likely brought into the peninsula long before the Bronze Age (ca. tenth to third centuries BCE). Beginning in the Han dynasty two thousand years ago, Chinese documents provide a variety of information about ancient Korea, and Nelson (1993, citing Parker 1890) tells us that the Ye-Maek, a tribal group who resided in the north of the Chinhan region in southeastern Korea, were "agricultural, raising hemp and silkworms." The ancient remains of a "yellow silk, hemp and rush mat" were discovered in the wooden "Painted Basket Tomb," which contained corpses of one male and one female adult along with a child, interred around 100 CE at the Eastern Han Chinese colony of Nangnang in North Korea, near the present-day capital of Pyongyang (Koizumi 1934).

According to the Korean Central News Agency of the Democratic People's Republic of Korea (KCNA 2007), various types of paper have been made for daily use since early times. During the ancient, indigenous Three Kingdoms (Goguryeo, Paekje, and Silla) period, paper was made from hemp (*maji*) and paper-mulberry (*joji*), and hemp paper made during the Goguryeo period (277 BCE to 688 CE) was famous for its thoroughly bleached color and smooth finish. The technology of paper making initiated during the Three Kingdoms period developed further in the Goryeo period, from the early tenth to the late fourteenth centuries, and Korean *paekchuji* hemp paper—also revered in China during this time for its excellent "snow-white color and high quality"—was used in the publishing of important Korean books (KCNA 2007).

During the Silla Kingdom (57 BCE to 668 CE) farmers produced hemp fiber to be woven into fabric for a variety of uses (Choi 1971), and Suh (2008) discusses the ancient traditional textile use of hemp fibers during the Paekje Kingdom (18 BCE to 660 CE) in southern Korea. Since ancient times in Korea, hemp cloth has been classified based on its thread count, color, production area, and use. Fine, high yarn count hemp fabrics approximately 50 centimeters (20 inches) in width were an essential item of trade with China during the Unified Silla dynasty (668 to 935 CE). During that time, higher count fabric could only be used by members of noble families and superior hemp fabric was reserved for the emperor's crown. Adjusting to pronounced seasonal differences in climate, upper-class Koreans traditionally wore warm wool and silk in the cold winter, and hemp and ramie in the hot humid summer. Commoners, as in China, wore coarse, lower yarn count fabrics throughout the year (Min 1985; also see Sim 2002).

Cotton was introduced from China during the late Goryeo dynasty (918 to 1392) and quickly cotton farming became important, especially in the southern part of the peninsula. With mass production of cotton, costume styles of Korean people changed, but hemp remained a significant fiber, as socioeconomic differences continued to determine clothing choice. Following the spread of cotton, common Korean people traditionally wore clothes made of lighter weight hemp or cotton in the summer time. In the spring and fall they wore warmer cotton, and in the cold winter they put on padded cotton clothing. Affluent people, on the other hand, were able to acquire and wear clothes made of very fine, high yarn count hemp and ramie fabric during the hot humid summer and padded silk garments stuffed with cotton or fur in the winter. High ranking men dressed in sandals made of hemp cords, while commoners wore sandals made of rice straw (Choi n.d.). Sim (2002) shows a pair of fourteenth-century, late Goryeo leather boots recovered from the Taesa shrine in Andong city of North Kyongsang province of South Korea. These are lined with cotton and have hemp fabric "interlined between the leather and cotton" similar to the construction of Chinese slippers.

Seo-Geung, a Chinese envoy sent to Korea during the Song dynasty, informs us in his *Goryeo Dogyeong* or "Illustrated Book of Goryeo" that Korean hemp and ramie (*moshi*) "products are as clean and white as jade" (Sim 2002). Traditionally, Koreans have appreciated bleached hemp and ramie, as white was the color of choice for clothing (Min 1985). The Goryeo dynasty was followed by that of the Joseon (1392 to 1910 CE), which was first established in the basins of the Liao and Daedong Rivers of present-day South Korea; these areas combined with other walled-town states to form a large confederation in which hemp remained an important source of fiber.

As in traditional China, the weaving of fabric in Korea, including that made of hemp, has long been the work of women. During the Joseon dynasty, girls learned to weave before the age of 10 and continued to work at looms throughout their lives. Weaving contests became events during which women momentarily overlooked the fatigue of their work and focused on friendship; this included group singing of songs associated with weaving. During Joseon times each region had its own distinctive hemp cloth. The finest hemp fabric came from the Yukjin area of Hamgyong and was called *pukp'o*, and that from Kyongsang was called *yongp'o*; the coarse cloth called *kangp'o* from Kangwon was used mostly for farmers' and fishers' clothing, and high-quality cloth known as *andongp'o* was woven in Andong (Hyo-Soon 1995).

Historically, hemp fiber was also used to weave the canvases for Korean paintings. Although Buddhist images were painted on silk during the Goryeo period, they were more

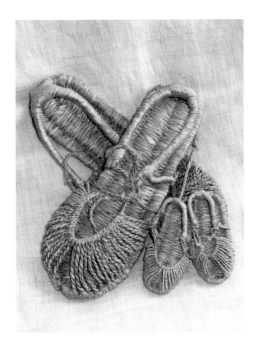

FIGURE 21. Hemp bark strips were commonly used to fashion sandals across eastern Asia. These two pairs for mother and child are from Korea.

commonly painted on hemp cloth or heavy paper (also made of hemp fiber) during Joseon times: "The predominant use of hemp (or sometimes thick paper) in 16th and 17th century monumental Buddhist paintings is undoubtedly due to economic circumstances: silk and gold had become too expensive to use for Buddhist images, especially under the conditions of austerity imposed by the early [Joseon] rulers and as a result of the significant change in patronage, from the court to wealthy provincial donors and local monasteries. Finally, painting on hemp required less demanding craftsmanship" (Hammer 2001).

Increasing use of hemp was encouraged as meditative Sôn Buddhism (Chan in China, Zen in Japan) rose in popularity and sparked a boost in production of portraits of venerated Buddhist priests (Hammer 2001). For example, a huge Joseon hanging scroll (214.6 × 224.8 centimeters or roughly 7 feet square), dated to the late sixteenth century and now in possession of the Metropolitan Museum of Art in New York City, is painted with "ink and color on hemp" and shows the Indian god Brahma who was included in the Buddhist pantheon of the times.

Two additional examples of Joseon Buddhist paintings on hemp cloth are currently held in the Oriental Antiquities Collection of the British Museum in London, England. One, titled "The Great Renunciation" (150 × 122 centimeters or five by four feet, and part of a set of eight paintings), depicts the life of the historical Buddha, Prince Siddhartha Gautama, and was produced during the early eighteenth century (see Portal 2000). The other painting, dated to the late eighteenth or early nineteenth century, was also produced on hemp cloth and depicts "Dhratarastra, Guardian King of the East." This latter painting probably comes from Daegu in South Korea and is quite large (301 × 207 centimeters or about 10 × 7 feet), with dynamic, decorative lines and the malachite green coloring characteristic of eighteenth century Korean Buddhist paintings (see Zwalf 1985; Portal 2000). Also, a late nineteenth-century Buddhist colored

sinjungdo or "Buddhist deity painting" (119 × 95.2 centimeters or about 4 ×3 feet) composed on hemp fabric was found at Mitasa temple in Seoul (Korean National Heritage Private Museum 1996).

It should also be noted here that lacquerware has been produced in Korea since ancient times and, as in some Chinese and Japanese lacquer work, depended primarily on hemp cloth to serve as a supporting core for coats of lacquer varnish. *Chaehwa* lacquerware is a traditional, sophisticated Korean craft that involves forming various colors and patterns by applying powdered lacquer and pigments onto layers of raw lacquer. According to Korean craftsman Kim Hwangyeong, part of this process involves the use of a mixture of raw lacquer (30 percent) and glutinous rice powder to paste hemp cloth onto a wooden article with a spatula, which is then used to fill the hemp cloth two or three times with bone ash before completing the rest of the production process (Seoul Metropolitan Government 2006).

Contemporary South Korea and North Korea

Contemporary Korean hemp agriculture provides us with a representative example of the techniques followed by farmers both past and present in other regions. Hemp crops are grown in Korea today by much the same means as they have been for thousands of years across much of temperate East Asia; traditionally, the majority of hemp fiber was used in the Korean farmer's household to weave cloth for sacking, summer clothing, and funerary costumes. During the 1950s, the industrial uses of hemp included twine, sewing thread, packing materials, canvas, and sailcloth (Ree 1966). Hemp fiber and textile output reached a high during the late 1920s, but by the early 1970s—as petrochemical-based alternatives to industrial hemp, as well as affordable imported cotton and synthetic clothing came to dominate the Korean marketplace—production plummeted to less than one tenth of 1 percent of levels 40 years earlier. Hemp made a slight resurgence in the 1990s and early twenty-first century, but production levels fluctuate annually, and future trends are impossible to predict (Clarke 2006a).

Korean varieties are reproduced from local landraces traditionally maintained by farmers rather than created by hybridization and rigorous artificial selection for certain traits (e.g., registered European cultivars selected for extremely low THC content). All Korean hemp varieties, as well as those grown in China and Japan, are dioecious (Ree 1966; Clarke 2006a). Generally Korean hemp varieties are moderately branched and 2.5 to 4.0 meters (8 to 13 feet) in height at maturity. The foliage is medium to dark green, and the leaves have seven or nine medium to broad leaflets. The inflorescences are relatively sparse, and seed yield is low compared to improved European varieties. Although some individuals produce elaborate resin glands, they apparently produce little, if any, of the primary psychoactive cannabinoid THC, and the local landraces are not considered to be drug varieties (Clarke 2006a). Formerly, East Asian hemp varieties from China, Korea, and Japan were considered by the vast majority of taxonomists to be members of *Cannabis sativa*. More recently, taxonomic research by Karl Hillig (2005a/b) at Indiana University provided compelling data indicating that East Asian broad-leaf hemp (BLH) is more accurately circumscribed by *C. indica,* and he proposed the designation *chinensis* as the biotype or subspecies name (also see Chapter 11).

All hemp growers in South Korea are licensed by the government through their local agricultural extension department where they must register their names and addresses as well as the sizes and locations of their fields. There is one license required for fiber cultivation and a second for seed cultivation. A farmer who harvests hemp fiber in June and decides to leave the plants along the margins of the field for later seed harvest in October is required to have both licenses. There is no license fee and farmers are not required to have their crops monitored for cannabinoid content (Clarke 2006a). Practical agricultural control of hemp crops is generally casual in most parts of Asia where *Cannabis* has a long tradition as a valuable fiber and seed plant.

The first European to report on travels in the southern Han River region of Korea was Isabella Bird (1898), an intrepid explorer, who observed and recorded a unique method of hemp processing still used in its modern form today:

> At the bottom of a stone-paved pit large stones are placed which are heated from a rough oven at the side. The hemp is pressed down in bundles upon these, and stakes are driven in among them. Piles of coarse Korean grass are placed over the hemp, and earth over all, well beaten down. The sticks are then pulled up and water is poured into the holes left by them. This, falling on the heated stones, produces a dense steam, and in twenty-four hours the hemp fibre [sic] is so completely disintegrated as to be easily separated.

The steaming of hemp stalks is unique to the Korean Peninsula and one of the defining features of Korean hemp processing. According to Nelson (1973), this method in itself argues for local discovery of the use of hemp, for it is quite different from methods used elsewhere, such as water and dew retting (see Chapter 4 for a discussion of hemp retting and its relationship to archaeobotanical evidence for *Cannabis* processing). However, we suggest that the steaming technique described here was developed long ago within Korea following hemp's early introduction and was further modified following its arrival in Japan. Some processing techniques (e.g., steaming the stalks) are apparently indigenous to Korea while other techniques are shared only with Japan (e.g., scraping the bark strips) or with the Hmong ethnic group of Southwest China and northern Southeast Asia (e.g., loom construction and other spinning and weaving tools), the ramifications of these relationships are yet to be investigated (see Clarke 2010a/b).

Traditional division of labor in Korean agrarian society stipulated that men worked outside the home and women worked within, and this traditional system is echoed in present-day division of hemp labor. Men are generally responsible for growing crops, processing stalks, and transporting them back to the home, where women take over the duties of stripping bark from the plants, spinning yarn, and weaving cloth. To our knowledge, there are no male hemp weavers in Korea, and men are only rarely called upon to assist in other aspects of hemp work within the home, such as constructing and maintaining wooden spinning and weaving equipment. Presently, women also assist in weeding fields, harvesting stalks, removing leaves, and anxiously monitoring their precious hemp crop during postharvest processing. Therefore, hemp labor is predominately the work and responsibility of women, and men play only a seasonal role assisting with fieldwork (Clarke 2006a). Similar hemp-related gender roles are played out across much of Eurasia.

Processing hemp fiber, from harvest through weaving, involves many stages that vary only slightly from region to region; they are summarized by Hyo-Soon (1995):

> Hemp is planted in the third lunar month and harvested in the sixth lunar month. After harvesting, the hemp is steamed and the bark peeled off. . . . The hemp bark is torn into three strands from the top. . . . The fibers are then bundled together and placed in water to soak. Next the fibers are spun into strands and, to keep them from becoming brittle, placed in a warm room for five to seven days, covered with a straw mat to keep the heat in. The strands are then boiled in water, rinsed and dried in the sun. The weaver then prepares threads of different density, depending on the intended use. One *sae* consisting of 80 strands, five *sae* were usually used for work clothes, seven *sae* for regular clothes and three *sae* for mourning clothes. After the strands are prepared, the hemp is tied together and dried again by a bonfire. Then the threads are wound on a reel. After this process is over, the thread is placed on the loom and woven into fabric.

Traditionally, hand weaving was a fundamental task in every Korean home, especially in those communities that were more or less self-sufficient. Today weaving is undertaken in only a small number of locations in Korea, and there are only a few remaining master weavers of hemp fabric. In response, the South Korean government has designated their craft, and some of the weavers themselves, as cultural assets. Hemp weaving is considered to be Intangible Cultural Asset No. 1 of the Andong city district. Kumso village in Imha township near Andong is the home of about 70 weavers, and several master weavers have been designated as Intangible Cultural Assets of North Gyeongsang province. The oldest as of 2004 was 100-year-old Bae Bull-young, by which time she was no longer an active weaver. The most well know as of 2006 was Kim Jeom-ho, who was 79 that year, and was nominated for national status as a master weaver; a book had been written about her life and craft. Hemp weaving is also recognized as Intangible Cultural Asset No. 32 of South Jeolla province, where Kim Jeom-sun lives in Gokseong village and is another recognized Cultural Asset. She has been featured in a documentary about traditional Korean hemp weaving and was 90 years old in 2006. Next to her home stands a small studio with a room for weaving and a museum collection of traditional tools along with photos taken at the time the documentary was made. Her daughter-in-law, Yang Nam-suk, and two other women in the village are also recognized as master weavers. The vast majority of hemp weavers are women of at least 60 years of age, and presently few, if any, younger women are learning the many complicated and rigorous steps involved in hemp processing, spinning, and weaving. Modern young women are encouraged to go to college and to seek employment in urban areas rather than learn traditional crafts in their rural hometowns, a situation common across modernizing Asia. However, the aging weavers are stoically optimistic and feel that as long as there are people to farm the land, hemp will still be grown and woven (Clarke 2006a).

Today, ramie clothing is especially favored for summer wear because it is very white and somewhat less costly than hemp. However, hemp was the traditional favorite during the warm season and is still worn, especially by the elderly. Traditional handwoven fine hemp fabric continues to be woven today, with the specialized knowledge handed down through generations. For example, *map'o* from the Gokseong region and *andongp'o* from the Andong region can still be purchased today (Sim 2002).

Andongp'o, a fine yet stiff, yellowish hemp cloth, provides good airflow, wicks away perspiration, and is particularly durable; therefore, along with highly rated ramie cloth, Andong hemp fabric is used today in Korean summer clothing. The hemp *top'o*, a traditionally well-respected long outer coat, continues to be a symbol of the gentleman-scholar and remains a matter of pride for people living in the Andong region where these coats are still sewn with the finest hemp cloth. The climate and soil in the Andong area are said to be ideal for cultivating hemp and the local techniques for weaving textiles have long been known to be superior; *andongp'o* hempen materials were traditionally sent to the royal palace and presented to the king.

All handwoven hemp cloth in Korea today is approximately 35 centimeters (14 inches) wide from selvage to selvage (the uncut margins of the fabric along the right- and left-hand edges as it comes out of the loom), although in ancient times, hemp cloth was more commonly 50 centimeters (20 inches) wide (Koh 2004, personal communication). The grade of hemp cloth is designated by a number, usually ranging from No. 4 to No. 9, corresponding to the number of *sae* (sets of 80 warp yarns) in the warp. For example, grade No. 5 is woven from five *sae* (400 warp threads) and No. 8 quality from eight *sae* (640 warp threads). Master weavers in the Andong region produce very fine cloth of up to No. 15 quality containing 1,200 warp threads, which equals approximately 34 threads per centimeter or 85 threads per inch (Clarke 2006a). Numerous weaving cooperatives scattered throughout Korea assist farmers by providing advice, procuring seed, and marketing yarn, cloth, and cloth products. In the Andong region, the Jeojeon Hemp Cooperative consists of about 30 weavers and the Imha Hemp Cooperative represents more than 100 weavers.

The *hanbok* has been Korea's iconic traditional costume for more than 2,000 years, originally made of silk, hemp, and more recently white cotton. Today Koreans wear *hanbok* in many colors and types of fabrics, especially during traditional holidays and social affairs. There are separate *hanbok* costumes for men and women, both with simple lines and no pockets, largely modeled after those worn during the Joseon dynasty. Modern designers are still turning to *hanbok* and other ancestral costume designs for inspiration to develop contemporary clothing, integrating the lines and cuts of customary clothing and utilizing traditional fabrics such as hemp and ramie. Today there are numerous shops in South Korea specializing "in a new generation of *hanbok* for everyday wear" (Korea Overseas Information Service 2003).

Today the majority of Korean hemp cloth is used to sew *sui*, funeral clothing for dressing the corpse prior to burial. A set of *sui* usually consists of up to 30 garments completely covering the corpse from head to toe and may require up to 100 linear meters (about 330 feet) or 35 square meters (350 square feet) of fabric (see Chapter 9 for detailed discussion of ritualistic funeral use of hemp fiber). In addition, the dressed corpse is wrapped in a coarse hemp cloth shroud before placement in the coffin. In recent years, the cost of a set of hempen *sui* purchased in the region where it was produced varied from 2,500,000 Won (about US$2,250) when sewn from coarser grade Boseong cloth and up to 5,000,000 Won (about US$4,500) when made from average quality Andong cloth. *Sui* made from progressively higher grades of cloth would cost proportionately more, and in urban areas, *su-ui*

are even more costly, constituting the largest single expenditure at a funeral, even more than the coffin (Clarke 2006a).

Several commercial Korean textile factories weave pure hemp and hemp blend fabrics as a portion of their production, and much of the yarn they weave is purchased from China to lower production costs. Contemporary woven hemp products include bedding items (e.g., covers for bamboo pillows, pillow cases, sleeping mats, and blankets), traditional *hanbok*, men's underpants, shirts and trousers, women's blouses, skirts and trousers, other apparel items (e.g., slippers, sport shoes, satchels, and handbags), and household items (e.g., wall coverings, cushion covers, chair pads, curtains, table cloths, place mats, tea mats, coasters, and food covers). All these items are relatively expensive, costing from 2 to 10 times more than similar products made from silk, ramie, and cotton (Clarke 2006a). Additionally, hemp fabric imported from China, of nearly comparable quality, sells for less than 25 percent of the price of Korean hemp and has flooded the market. "Only rarely do people with a discerning eye appreciate handmade hemp, and the number of people with an interest in buying it has dwindled over the years" (Lee 1998). Today only wealthy Koreans with a strong appreciation for traditional crafts can afford to buy Korean hemp, and the majority of Koreans cannot distinguish local hemp cloth from imported. Although ramie cloth is also relatively expensive to produce in Korea, as noted previously, it is much more popular for summer clothing; weaving of ramie cloth in Korea is mechanized on a factory scale, the production volume is much higher, competition is keen, and the price is about half that of hemp cloth. Even so, traditional handwoven Korean ramie cloth production is also threatened by Chinese imports. (For additional detailed information on Korean hemp production, see Clarke 2006a.) Now we turn our focus to the North Korean approach to *Cannabis* agriculture in the twenty-first century.

The *Rodong Sinmun* or "Newspaper of the Workers," official publication of the Central Committee of the Workers' Party of Korea, is widely read in North Korea. First published in 1945, the contents of its articles are often quoted by international media since it is regarded as a source of official opinions from the North Korean government on many topics. On August 25, 2004, *Rodong Sinmun* published a front-page editorial calling for a patriotic "mass movement" to greatly increase hemp cultivation on a nationwide scale. This appeal was preceded by a series of articles about national research efforts to identify new breeds of hemp, develop innovative cultivation technologies, and secure seed supplies, all under orders from the country's former leader Kim Jong-il. The editorial also praised the usefulness of hemp, pointing out that hemp seed oil could be used for cooking or to make soap; hemp waste could feed livestock; and fibers would be available to weave clothes, awnings, mosquito nets, rope, and sandbags and as a component in paper. The *Rodong Sinmun* also reported the "hemp cultivation plan" as elevated to equal status with the "revolutionary potato cultivation plan" and "revolutionary two-crop farming plan" that together form the mainstay of current agricultural policy in North Korea. Apparently this campaign to encourage and greatly expand hemp cultivation is a high priority for the centrally controlled government, which went so far as to instruct the ruling party organization to accept responsibility for the "hemp cultivation plan and bring it to fruition" as a way of producing "raw material for a number of light industries like weaving, paper and foodstuffs, and hence could solve shortages of cloth and paper required by North Korean citizens." This campaign also recommended

FIGURE 22. Korean farmers steam fresh hemp stalks (A) to loosen the bark before drying the stalks. The bark is peeled off by hand and then rewetted and scraped with a knife blade (B) to remove the outside layer of cortex from the phloem fibers. A weaver from the Andong region of South Korea demonstrates the traditional East Asian technique of making hemp cloth. Dried bark strips are split into narrow strips (C), joined together end-to-end (D), and spun (E) to secure the joins and form yarn. The tensioned warp yarns are carefully prepared (F), and the hemp fabric is woven on a body-tension loom (G).

that people plant hemp on idle land, not only on collective farms but also on land associated with factories and schools (Choson Iibo 2004).

In the early part of the twenty-first century, hemp has begun to play a role in the Korean reconciliation movement. In October 2005, the South Korean Andong Hemp Textiles Company signed the first ever joint-venture agreement with a North Korean company, their mutual intent to grow vast acreage of hemp. Andong Hemp Textiles Company financed the construction of a modern hemp fiber processing, spinning, and weaving factory near North Korea's capital city of Pyongyang and planned to contract with North Korean farmers for production of hemp fiber on 10,000 hectares for the domestic and export markets. If successful, it would have made this pioneering joint venture the world's largest hemp producer.

On October 30, 2008, three years after development of this cooperative economic concept called "Money for New Rope," Pyongyang Andong Hemp Textiles was established in North Korea as the initial equal partnership joint venture between the South Korean Andong Hemp Textiles Company and the North Korean Saebyol General Trading Company. Each side invested US$15 million, and the new venture employed as many as 1,000 North Koreans on a 47,000 square meter site near Pyongyang (Foster-Carter 2009). Unfortunately, this economic activity was banned in May 2010 following the sinking of a South Korean ship in March 2010, which, according to a Seoul government declaration, was torpedoed by a North Korean submarine (Yonhap News Agency 2010). As of 2009, China was the leading producer of hemp with about half of the world's production, followed by Europe, Chile, and the Democratic People's Republic of Korea.

Ancient Japan and Hemp

Cannabis was most likely introduced into present-day Japan from the Asian mainland very early, perhaps by the late Paleolithic period, about 18,000 years ago. At this time, sea level was much lower than it is today and a land bridge connected the Asian continent and Japanese Archipelago, allowing Eurasian peoples to migrate overland. These early immigrants were undoubtedly hunters and gatherers who probably also engaged in fishing, and their descendants practiced similar means of subsistence for many generations, even after they began to settle down in areas with abundant resources close to the ocean or freshwater sources. When the ancestors of modern Japanese people began to cultivate plants including *Cannabis* remains somewhat uncertain, although it seems to have started with only a few species and not necessarily involving those that provided staple foods but rather those used for fiber, seed oil, and perhaps medicine and ritual. Ancient *Cannabis* seeds more than 10,000 years old were discovered in association with early Jōmon pottery remains at the Okinoshima archeological site on the Boso Peninsula in central Japan (Okazaki et al. 201; see Chapter 4 for a more detailed discussion of ancient subsistence activities in Japan and the origins and early use of *Cannabis*). *Cannabis* was first introduced into Japan most likely from the eastern Asian mainland during the early Holocene, and it may have been employed and even cultivated for more than one use. In any case, its fiber use in Japan began at an early date, perhaps during the early Neolithic period (see Crawford 2011).

The Jōmon period in Japan lasted a very long time, from about 10,500 to 300 BCE. The word "Jōmon" refers to both the characteristic "cord-mark" pottery of this early culture and the people themselves. The Jōmon were "hunting and collecting people who lived a civilized, comfortable existence and used hemp for weaving [actually knitting or knotting] clothing and basket making" (Olson 1997, citing personal communication with H. Mayuzumi 1996). The Jōmon are best known for their impressed ceramic ware, which bears a pattern created "by rolling a cord or cord-wrapped dowel on the still-soft surface of a pot (Imamura 1996a)" and likely some of the patterns were made with hemp cordage.

The presence of *Cannabis* in Japan since early times is supported by a continuous record of seeds recovered from Jōmon sites on Hokkaido, the northernmost island (Yamada 1993), and by artifacts made of hemp fiber found in the Early Jōmon Torihama shell midden, which lies at the confluence of the Hasu and Takase Rivers, near Lake Mikata in Fukui prefecture on Honshu (Imamura 1996a; also see Okamoto 1979; Kasahara 1981, 1984). Here fragments of ancient hemp thread or string, about 2 millimeters (one-twelfth of an inch) in diameter (dated ca. 7000 BP), and knitted fabric (dated ca. 7000 to 5500 BP) were recovered (Nunome 1992). Pollen evidence from nearby Ubuka Bog indicates that hemp and a number of other cultigens "were present in Japan or undergoing domestication during the Early Jōmon" (Crawford and Takamiya 1990). In addition, archeologists have found ceramic ware with impressed cloth designs, along with ancient spindles and parts for looms, in a number of Yayoi farming sites dated from about 1000 to 500 BCE to approximately 300 CE (e.g., see Imamura 1996a). For example, hemp cloth has been recovered from sediment levels dated to the Yayoi period at Nabatake in northeastern Kyushu (Kasahara 1982, 1984; also see Crawford and Takamiya 1990) and from the famous fortified village site at Yoshinogari in northern Kyushu (Hudson and Barnes 1991). These artifacts suggest that hemp and ramie (*karamushi* or *cho ma*) clothing was widely available in Japan at least by the third century CE (Farris 1998), about the time when loom weaving was probably introduced from the Asian mainland.

Although a strong case can be made for the use of hemp fiber to make textiles and cordage in Japan since early Jōmon times, it should be pointed out that Japan hosts a broad range of wild, bast fiber sources, including kudzu, *ohyo* (elm), and *fuji* (*Wisteria*), which were also traditionally used for making cordage and rough cloth. Another significant and probably ancient use of hemp was for woven bags, some of which were purposely designed to carry and detoxify certain wild foods such horse chestnuts (*Aesculus turbinata*) or used to extract seed oil. Domestic and commercial uses of *Cannabis* present additional reasons why hemp was cultivated early on in Japan.

Ainu people have long inhabited Hokkaido, as well as the Kurile Islands and Sakhalin Peninsula of Russia. The Ainu differ in physiognomy, language, and culture from the modern Japanese, who only explored Hokkaido fully and first colonized it in the late nineteenth century. Although the Ainu are often recognized as the aborigines of Japan because of their ongoing traditional folkways, they may have arrived at approximately the same time as the Yayoi, perhaps replacing earlier Jōmon people, or may even be descendants of the original Jōmon. Presently, the origin of the Ainu is a source of much scholarly debate. For example, while the Jōmon made

FIGURE 23. In the Andong region, South Korea, weavers make a very fine, open weave, stiff hemp fabric traditionally worn during hot summer weather because it allows cooling air circulation.

clay pottery impressed with rope markings as noted earlier, the Ainu ancestors produced smooth pottery in different shapes than those of the Jōmon, and in contrast to the earliest Jōmon who are not known to have had agriculture, the Ainu practiced some farming along with hunting, fishing, and gathering and may have also cultivated hemp.

The Ainu made heavy cloth from fibers extracted from native elm trees (*Ulmus davidiana* var. *japonica*) and fabricated a finer cloth from various species in the nettle family (Urticaceae). Although these plants were apparently the primary sources of fiber used to make their distinctive coat-like garment, known as *attush*, hemp fiber may have been used for weaving some articles of clothing as well as for cordage, although *Cannabis* may have been more important as a food source. The Ainu are known for their shamanistic beliefs and rituals derived from Siberian traditions, and shamanism is one of the original sources of Shintō, the native Japanese religion. Although hemp is widely used in present-day Shintō rituals, it is not used in Ainu shamanic rituals. Nevertheless, it can be argued that among the most significant traditional Japanese uses of *Cannabis* were for its special ceremonial symbolism and protective spiritual power. The ritual use of hemp in Japan is still very important, and hemp fibers have long been used ceremonially by Shintō priests because of their association with purity; this includes Japan's emperor "who acts as a kind of chief priest of Shintōism" (Olson 1997).

Zen, the contemplative, Taoist-influenced branch of Buddhism, was also affected significantly by hemp (Olson 1997). For example, the military nobility of preindustrial Japan, known as *samurai*, as well as scholarly specialists who followed the elusive doctrines of Zen, manifested the inspiration of *Cannabis* fiber in arts such as "*haiku* (short poems), *aikido* (a martial art), *kyudo* (archery) and *chanoyu* (tea ceremony)." The strong linkage of hemp to these traditional Japanese arts is reflected in the customary use of strong, malleable *Cannabis* fiber to make bowstrings, as well as in the clothing used by elite warriors, martial artists, and formal meditators. Based on Japanese informants, Olson summarized a familiar adventure story for children in which *ninja* (warriors) use *Cannabis* plants to increase their jumping abilities, a tale that vouches for the vitality of hemp as much as the skill of the *ninja*: "The student *ninja* plants a batch of hemp when he begins training and endeavors to leap over it every day. At first this is no challenge, but the hemp grows quickly every day and so does the diligent ninja's jumping ability. By the end of the season, the warrior can clear the 3–4 meter high hemp."

Another, quite different version of this old tale about ninjas jumping over hemp plants was described by Kenji Nakagami, one of Japan's great postwar literary masters, and can be translated from the Japanese as thus: "A long time ago, it is said that hemp plants were cultivated in gardens and you would jump over them. I learned all of this from an old man, and was told that if you were not spirited when you jumped over the hemp plant you would die around that time, but this is the way to fly like a ninja" (Nakagami 1995).

From a more general cultural perspective it is interesting to point out that *noren*, ritual curtains traditionally made of hemp, are hung today over the entrances of countless Japanese businesses and homes; these special curtains still function in purification of entrants and protection from harmful forces, even though contemporary *noren* are usually made of ramie fabric (see Chapter 9 for more discussion of the ritual uses of hemp fiber).

The Japanese word "*asa*" is used to refer collectively to fiber crops including jute, ramie, sisal, and flax, as well as *Cannabis* hemp. *Cannabis* hemp in Japan is more explicitly referred to as "*taima*" (tall or large hemp), which is derived from *dà má*, the original Chinese name for hemp. As in Chinese and English, collective nouns representing *Cannabis* as well as other bast fibers can lead to considerable confusion when interpreting written records. However, Asian sources referring to Japan contain many references to what we consider to be *Cannabis* hemp. The Chinese *San Guo Zhi* or "Records of the Three Kingdoms," written ca. 297 CE, includes the *Wei Zhi* or "Records of Wei," which covers the history of the Cao Wei kingdom (220 to 265 CE); the section titled *Wei-chih tung-I-chuan* or "Encounters with Eastern Barbarians" tells us that the *Wa* or *Wo* (Japanese people) cultivated grains such as rice, along with "hemp plants and mulberry trees" (Farris 1998). Archeological support for this written record comes from the famous fortified village site at Yoshinogari, northern Kyushu, which was excavated in the 1980s. Evidence indicates that this relatively large settlement was occupied throughout the Yayoi period with about 1,000 to 1,500 residents at the height of its urban importance during the first three centuries CE, and archeologists have recovered both silk and hemp cloth from this site (Hudson and Barnes 1991).

As previously noted, when rice farming came to Japan more than 2,000 years ago from present-day China and Korea, *Cannabis* was present and was either already cultivated in Japan, or its cultivation came along with rice and its culture (Rathbun 1993). Hemp appears to have been utilized in Japan as a wild or semicultivated food and fiber source before the Yayoi period. Subsequently, after loom weaving and possibly more advanced cultivation techniques were introduced from the Asian mainland and the Yayoi population grew, increased need for weaving fiber led to expansion of hemp cultivation. In any case, *Cannabis* became an important fiber source early on in Japan used to produce clothes, ropes, nets, and, somewhat later, Japanese paper (*washi*).

The ancient Nara city tax codes sent to the Yamato court indicate that hemp, ramie, and silk fabrics were part of the local products taxed prior to 645 CE. Taxation, an essential part of the Yamato tribute system, followed the Chinese system using cloth levies as tariffs. Throughout the eighth and into the ninth century CE, large "Chinese-style" capital settlements were constructed using pools of workers. These corvée laborers were paid with hemp cloth and other items that were collected from adult males among the commoners (as "corvée exemption revenue") by the imperial court at Nara and then redistributed to the hired laborers of the surrounding Kinai Plain (Farris 1998).

Hemp fabric was also used as a surface upon which to draw ancient Japanese maps. For example, microscopic and X-ray studies of samples taken from an extraordinary ninth-century CE pictorial map, *Nukatadera garan narabini jorizu*, have revealed that this "national treasure," showing the Nukatadera temple buildings and surrounding area in Nara with its checkerboard grid reflecting the ancient system of land division, was drawn on hemp cloth woven using earlier, eighth-century CE techniques (Japan Foundation 1993).

Matsui (1992) used flotation methods to excavate late seventh century CE seeds from an ancient "toilet" at the Fujiwara Palace site in Nara. Among seeds recovered from this open sewer were those of melons, eggplant, wild grapes, and hemp. "The discovery of the open sewage toilet at Heijo gives new reading to the ancient poem *The Fragrant Capital of Nara*" (Barnes and Okita 1999). This discovery is relevant not only to the presence of *Cannabis* in ancient Japan but also because it indicates that at that time hemp seeds were eaten as well as grown for fiber.

In spite of disorder and distress during the Japanese era of warrior dominance (dating back to 794 CE), there was significant growth in many economic and cultural aspects during this time. In order to build up minor states for long-term stability through this warlord period, the more adroit *daimyo* (powerful landlords) promoted a series of initiatives to strengthen agriculture. These included the cultivation of certain crops, such as hemp, in order to produce more paper, which like other preferred products such as silk, dyes, and vegetable oils were very profitable commodities. Therefore hemp provided additional sources of family revenue as well as income for the *daimyos* through taxation. This is how the feudal lords of the Edo period (1603 to 1867 CE) encouraged the common people to cultivate hemp fiber (Stearns et al. 1992) as they were often in debt to wealthy city merchants to whom they sold their hemp cloth; they were also in debt to *shoguns* (rulers appointed by the emperor) to whom they owed either rice or cloth as tax payments. Although feudal merchants were of low social standing, they soon became the wealthiest class and learned from foreigners to deal in coin rather than barter, which like Chinese coins shared a hole in the center and were carried on hempen strings (Olson 1997).

In ancient Japan, rich merchants and powerful *samurai* (military retainers of the *daimyo*) often wore silk clothing until the introduction of cotton in the seventeenth century CE. On the other hand, hemp cloth was used to make formal uniforms and leisure kimonos of *samurai* warriors, as well as training clothes for meditative and martial arts practice. Hemp clothing was also used by the military in combat. For example, a fine example of an ostentatious surcoat (*jimboari*), worn over armor for extra weather protection during the Azuchi-Momoyama period (1573 to 1603 CE), is now held in the study collection of the British Museum's Department of Japanese Antiquities (Acc. No. 1897.3–18.6). The body of this 72-centimeter-long (29 inch) soldier's coat is covered with feathers from two species of Japanese pheasants and a duck, which are glued onto hemp fabric (see Smith et al. 1990).

Among common people during ancient times in Japan, *Cannabis* was generally the main source of fiber used to weave cloth (Hughes 1978). Japanese hemp fiber processing follows much the same procedure as in Korea with one major exception; rather than steaming hemp stalks, the Japanese briefly plunge bundles of long stalks into boiling water, spread the stalks to dry in the sun, and at a later time, employ a controlled fermentation retting to loosen the bark (Clarke 2010a/b). During the early Edo period Japan was isolated from foreign trade and underwent significant social change, including widespread rural poverty, which sparked a period of adaptation and local inventiveness. Resources formerly plentiful became scarce, and those previously rare became highly valued. Throughout the Edo period various utilitarian clothes evolved into well-crafted works of art. While larger agricultural plots in the southern regions were used to grow expensive cotton reserved for the upper classes, hemp was grown throughout the colder northern areas on smaller, more irregular out-of-the-way sites not utilized for rice or cotton. Alternative cultivated fibers like mulberry and ramie along with wild wisteria, elm, kudzu, and nettle were also used locally to weave functional and attractive fabrics for utilitarian use. The elite classes normally wore *kimono* garments of silk or polished cotton, and peasants were forbidden from wearing these two kinds of cloth. On the other hand, the common people could and did wear hemp clothing. *Cannabis* was relatively easy to cultivate in small home gardens, and its fiber was renowned for its strength, durability, and lower price than even cotton rags. Consequently, hemp was by far the most widespread fiber plant among the populous working classes such as fishermen, farmers, and trades people during the Edo period (Yoshida and Williams 1994).

Eventually the use of cotton clothing among the emerging urban working class became widespread, and cotton began to replace hemp as the major fiber crop. In the countryside, however, cotton cloth was only available in the form of rags collected in urban areas, and these were combined with local hemp to make peasant clothing most commonly as sewn patchwork or by weaving wefts of recycled cloth strips into hemp warps (see Yoshida and Williams 1994). Following the Meiji period (1868 to 1912) and the related rush to modernize Japan, increased fertilizer use and implementation of mechanized processing methods resulted in high cotton yields. Although the new class of city workers switched to cotton garments, hemp clothing was still worn daily by rural peasants and by certain people for ritual uses (see Chapter 9). In fact, hemp was used to make an array of specific products, "including the straps of *geta* (wooden sandals), long-lines for eel fishing, and packaging ropes," as well as kite string and twine used to sew *tatami* floor mats (Olson 1997, citing personal communication with H. Mayuzumi 1996). Hemp cordage is still used in these ways but on a much reduced scale, but hemp cloth is rarely woven in Japan today.

Even with the rise of the urban workforce, farmers in the Japanese countryside were still responsible for the greatest amount of labor. Daily they toiled for hours under difficult conditions in order to supply necessary commodities for an increasingly urban population. Hemp clothing continued to provide rugged and functional garments during yearly climate extremes from humid summers to freezing winters. In the later nineteenth and early twentieth centuries, during the Meiji and Taishō Eras (1912 to 1926), seaweed and broom-straw were mixed with hemp fiber to produce a solid, circular helmet with a conical peak that facilitated the relatively easy shedding of heavy summer downpours and winter snowfall. Farmers also used comparable natural materials including hemp to make supporting pads for tumplines to carry large loads (Rathburn 1993).

During this same period, following the Meiji restoration, specialized weavers utilized fine hemp yarn to which only slight twist had been applied to make exceptional, highly refined cloth in a sheer weave with both flexibility and a stiff hand. The talent of these Japanese textile makers is seen in traditional hemp summer *kimono* frequently dyed in large containers of fermented indigo. In many cases, the yarn is so fine that it resembles fine flax or even raw silk. Some have argued that considerable experience with silk thread inspired Japanese weavers to form extraordinarily fine diameter hemp strands into a fabric more functional and less costly than silk (e.g., Olson 1997, citing personal communication with Kolander 1996; also see Kolander 1995). However, it must be remembered that ramie fibers are finer, longer, shinier, and more silklike than hemp, and therefore even more lustrous fabric can be woven from ramie than hemp, and it is with ramie that Japanese weavers have taken their craft to the greatest heights.

Special clothing made of fine hemp fabric did become fashionable but was restricted in use to the elite of Japanese society. Because of the relatively high labor costs to produce these unique hemp textiles, the common people attempted to imitate them by using cotton when it became widely available in Japan about 200 to 300 years ago (eighteenth century). For example, "the modern summer *yukata* (cotton *kimono*) was the common person's adaptation of the *yukatabira* (absorbent hemp bathrobes) that the wealthy wore after soaking in the hot springs" (Olson 1997, citing personal communication with H. Mayuzumi 1996). The availability of machine-harvested and processed cotton increased rapidly, causing labor-intensive hemp cloth to become less common, and eventually it came to be used solely for producing ritual outfits for the wealthy upper classes. Over the course of the twentieth century, people throughout Japan increasingly wore cotton clothing, and only people in the upper classes continued to wear traditional hemp garments. This was a complete role reversal for hemp, formerly the fabric of commoners, as well as for cotton, formerly the fabric of the elite.

Although once common, hemp weaving is now rare in Japan. Traditional Japanese hemp weaving is documented in the film *Grandma Haru and Hemp Cloth: A Record of Life in Harihata, Kutsuki Village*. This film shows us that people living in Shiga prefecture traditionally carried horse chestnuts down from the mountains to town in cylindrically shaped bags made from heavy hemp cloth. All households in the village had their own bag dyed with Chinese indigo, a plant also known as the dyer's knotweed (*Polygonum tinctorum*) and persimmon (*Diospyros kaki*) juice. Hemp cloth, known locally as *noh-noh*, was also commonly used for making clothing. With each approaching spring every household would plant four to five furrows with hemp seeds, and all facets of hemp cultivation, fiber extraction, and weaving were the work of women. The main focus of the documentary mentioned earlier is one old woman, Haru, who had spent 93 years of her life in Harihata village. Using her diary, personal interviews, and film documentation, the producers showed how she worked throughout the year and most of her life making hemp cloth and sacks (see Marutani et al. 2001 for several photographs, the full text discussion in Japanese, and a summary in English). Traditionally, sacks made of hemp were used for many purposes in Japan.

Beyond the urban areas of Japan, until about a century ago, almost all households possessed at least one working loom. Grandmothers and mothers taught girls the arts of weaving as handwoven textiles were a part of everyday life in old Japan, and this aspect of common culture and craft work continued (at least in part) until the middle of the last century. As noted previously, there are four principle fibers utilized in traditional Japanese textiles—hemp, ramie, cotton, and silk. Hemp and ramie were common fibers for the average person until cotton cultivation and use rose dramatically by the eighteenth century. Even so, hemp and ramie both remained important particularly in rural areas where local traditions continued well into the twentieth century.

Currently, hemp cultivation is highly restricted in Japan, but the general ban on its growth did not begin until after World War II. Following that conflict, the occupying allied forces revised the Japanese constitution and in the process enacted a prohibition of *Cannabis* cultivation through the *Taima Torishimari Ho* or "Hemp Control Act" because hemp had been a strategic war crop for the Japanese during World War II, as it was for the United States and Europe (Constantine 1992; Yamada 1995). In addition, during the postwar reconstruction period in Japan, overseas foreign businesses capitalized on market opportunities by introducing many novel synthetic products that replaced traditional Japanese items, and consequently hemp cultivation was nearly eliminated in all but more remote regions. *Cannabis* continued to grow spontaneously in urban areas, particularly in disturbed, open environments along railways, at least into the middle 1950s, and still persists as a "weed" in the vast rolling hills of the cold northern island of Hokkaido.

Today hemp is cultivated in Japan on a very small scale, following a strict licensing system similar to Korea. The lustrous golden hemp bark strips commonly seen in Shintō ritual settings are highly revered in Japan (see Chapter 9), and hemp grown on Japanese soil, rather than imported from China, commands extremely high prices. The majority of Japanese hemp farmers grow the 'Tochigi Shiro' or "White Tochigi," variety developed after WWII and distributed by the government to farmers, who multiply the foundation seed in small gardens so they will have enough seed to sow a hemp fiber crop the following year. Comparative studies by the United States Department of Agriculture in the late nineteenth and early twentieth centuries indicated that the large, hearty Japanese strains were significantly more productive than those being grown in Europe and China at that time and could be highly productive if grown in North American locations such as California (e.g., Dodge 1896; Dewey 1920). By far the largest commercial crops in Japan today are grown by a handful of families in Tochigi prefecture north of Tokyo but rarely exceed 10 hectares (approximately 25 acres) in total. The few remaining licenses, totaling less than a hundred, are almost all held by Shintō shrines where small gardens are grown and processed for ritual use. A small amount of hemp cloth is woven by a craft cooperative near Kyoto using much the same techniques as Korean weavers and is of equal quality and price to the finest Korean hemp.

Our review of the history of traditional *Cannabis* use in Japan presented earlier clearly indicates that hemp fiber and seed formed an important part of Japanese culture during the Jōmon, Yayoi, and even more recent periods. However, within the past 60 years or so, following thousands of years of cultivation for food, cordage, and cloth, the major role played by hemp in the history of Japan has been largely forgotten. Next we focus on South Asia, where *Cannabis* hemp has always met stiff competition from a vast array of differing wild and cultivated plant fiber sources.

Ancient Evidence from South Asia, Southwest Asia, and Egypt

Although scant archeological evidence of ancient *Cannabis* use has been discovered in all of South Asia, some fiber remains have been found in western Nepal near where *Cannabis* is still cultivated today (see Clarke 2007). During a recent archeological investigation of ancient highland agriculture along the Jhong River, in the arid Muktinath Valley high in the Himalaya Mountains, many naturally mummified human bodies and well-preserved woven fabrics were discovered in a community burial cave and dated to between 400 BCE and 50 CE (Alt et al. 2003). While the majority of the delicate textiles were made of cotton or wool, a few were woven with plant fibers that could possibly have been hemp or wild Himalayan nettle. The wide assortment of weaving materials and methods provides evidence of sophisticated techniques of textile production more than 2,000 years ago. The Muktinath Valley, as part of the larger Kali Gandaki River system, lies at the crossroads of traditional north-south and east-west trade routes and thus serves as an important connection between lowland India and the Qinghai-Tibet Plateau, as well as a link on the larger geographical scale between South and Central Asia. Furthermore, the size and shape of the skeletons and the styles of the artifacts recovered from the cave indicate that these people or their ancestors probably came to the Himalayan region from western Central Asia, principally Xinjiang (Alt et al. 2003). Even if these ancient but as yet not fully identified plant fibers are *Cannabis*, based on our review of its historic use in South Asia discussed later, hemp fiber seems of relatively minor importance to the ancient cultures of this region (see Chapter 4 for ancient flax and hemp seed evidence from Nepal).

During the nineteenth century, British rulers of the Indian subcontinent established the Hemp Drugs Commission mandating the collection of information on the habitat, cultivation, and products of *Cannabis* and published in 1893–1894. George Watt described various traditional uses of *Cannabis* as they existed at the end of the nineteenth century, first in his extraordinarily comprehensive two-volume *Dictionary of the Economic Products of India* (1889) and then in his *Commercial Products of India* (1908) based on additional information gleaned from the commission's reports and a variety of other data. *Cannabis* was cultivated in the Indian subcontinent for two main products, most importantly for the mind-altering drugs secreted from its glandular hairs and less significantly for bast fibers extracted from its stems. For example, in the Sindh (now the southern province of Pakistan in the lower Indus River Basin), "wild hemp" was referred to as *kohi bhang* and reportedly grew in hilly areas (Watt 1908). These feral populations of *Cannabis* were probably alien and weedy in origin and were mainly used as a drug source; however, sometimes plants were gathered and utilized for fiber and seed oil as well. In addition, Watt (1908) noted that in some mountainous areas of upper India, in particular in the northwestern region of the Himalayas, *Cannabis* has been grown for its fiber for a long time and that the native people wove it into clothing or twisted it into cordage and roasted its seeds for food (see Chapter 6 for traditional seed uses). Among northern Indian peoples who traditionally used hemp are the Kumaoni who live in the Kumaon region of Uttaranchal state, they refer to the plant as *bhangau*. Shah (1997) described the extraction and use of *Cannabis* fibers and stalks among these people:

> The leaves are removed from the stems by hand and sickle. The stems are piled against the terrace walls of the field to dry. After several days of exposure, they become dry and brown and are tied into bundles. These bundles are retted for a couple of weeks by soaking in water in a pool or running stream. The bundles are taken out and beaten with wooden batons and poles and again dried in the sun. The fibres [*sic*] are then peeled off. The fibres [*sic*] so obtained are cleaned and washed [and] are ready for spinning and weaving. They are used for making ropes, known as *jeor*, coarse canvas of great strength and durability known as *bhanga pakhuli* (*Cannabis* canvas) and stout sack-cloth known as *kothla*. The pith and wood left after separating the fibres [*sic*] are used as an excellent torch wood and as a fuel igniter.

Although hemp was a significant fiber and seed crop among the Kumoani people up to at least 60 years ago, Shah (2002) noted that it is now only cultivated and used for its seeds, while modern fabrics have superseded the use of its fiber. Nylon sacking and synthetic cordage are now readily available even in the more remote areas of the region, and consequently the cultivation of hemp "has been greatly reduced."

Watt (1889) also referred to ancient Indian texts for evidence of hemp fiber products, and, for example, pointed out that hemp cordage is referred to in the *Kausitaki Brahmana* of the *Rg Veda*, the first of the Vedic Hindu scriptures (ca. 1700 to 1100 BCE), where reference is made to the "male and female forms of *Cannabis*: *bhangajala*, meaning a hempen net, and *bhangasayana* meaning a bed-stead woven with hempen cords." Furthermore, according to Watt (1908), in the *Sabha Parva*, book 2 of the *Mahabharata* Sanskrit epic, it is said that the Sakas (Scythians of Turkestan who migrated into ancient India) brought gifts of thread spun by "worms" (silk) and "*patta* (*Cannabis*) fiber" (also see Shah 1997 for his theory concerning the origins of *Cannabis* among the Kumaoni people).

Cannabis is still occasionally utilized for its fiber by minority groups inhabiting western Nepal; however, except for Watt's reference to the longevity of its use for fiber, we have found no other indication of its ancient general or widespread use on the Indian subcontinent. Chopra (1958) claimed that the "hemp plant [as a fiber source] has never been cultivated in India to any great extent." Apparently usage of *Cannabis* for its psychoactive and/or ritual properties in India as a whole have been much more important than fiber and seed usages, which traditionally were secondary or minor ethnobotanical applications (see Chapter 7 for a detailed discussion of the widespread and important traditional use of *Cannabis* for psychoactive purposes). The lack of hemp fiber use in most of South Asia can also be explained by the natural availability and subsequent development of other traditional fiber plants such as cotton and flax (Fuller et al. 2004), as we will see again in the next section concerning the Middle East. The Indian subcontinent is also rich in other, seemingly preferable, bast fiber crops such as ramie, jute (*Corchorus* spp.), and sunn hemp (*Crotalaria juncea*), as well as a wide variety of wild plant fiber sources.

Hemp fiber was not commonly known nor widely used in the early stages of Mesopotamian civilization or elsewhere in Southwest Asia or the Near East. However, the use of *Cannabis* for fiber or other purposes spread into Southwest Asia and the Mediterranean regions at least as soon as frequent contact with the Eurasian steppes began, sometime after the sixth century BCE. This probably occurred after farming spread into Central Asia and was greatly facilitated by domestication of the horse and other beasts of burden. Hirth (1966) tells us about a relevant Chinese reference to hemp fiber cultivation and its use in the "Roman Orient" in the twelfth chapter of the *Weishu* or "Book of Northern Wei," compiled by Wei Shou in 554 CE; this text describes the country known as Ki-kan (Syria), which then ranged from Babylonia west to the eastern Mediterranean, where "all kinds of grain, the mulberry tree and hemp" were produced, with smaller dependent states within its control that fashioned "fine hemp fiber for cloth."

Sula Benet (1975) cites historical evidence suggesting that hemp was known in the Near East well before the Christian era. For example, a religious obligation existed among ancient Hebrews that required that dead persons be dressed in hemp shirts as part of their burial attire and not until hundreds of years later did flax linen shirts replace those made of hemp (see Chapter 9). According to Benet (1975) the assumption that hemp was not known or used in ancient Palestine or Egypt (e.g., see Dewey 1913; Moldenke and Moldenke 1952) is incorrect; Benet claimed that "*Cannabis*" is a term "derived from Semitic languages and that both its name and forms of its use were borrowed by the Scythians from the peoples of the Near East." Benet also proposed that *Cannabis* was known and used in the ancient Near East and Europe at least a millennium prior to the first written reference by Herodotus, the famous Greek historian (lived ca. 484 to 425 BCE; Herodotus's description of the use of hemp by people living northeast of Greece during the fifth century BCE is discussed later). Benet (1975) pointed out that trade routes between Egypt, the Near East, and Asia crossed through ancient Palestine, and she refers to evidence for *Cannabis* fiber use in this area recorded in Ezekiel 27:19 in the original Hebrew version of the Bible. Benet directed attention to the biblical description of Tyre,

the Phoenician royal city whose markets were frequented by Jewish merchants and customers and that was famous in ancient times for its widespread trade. The passage in Ezekiel referred to earlier tells us that "Vedon and Yavan traded with yarn for thy wares; massive iron, cassia, and *kaneh* [*Cannabis* hemp] were among thy merchandise" (this biblical quotation comes from *The Holy Scriptures* published by the Jewish Publication Society of America, e.g., see Margolis 1908). According to Benet (1975, citing Salzberger 1912), "King Solomon, a contemporary and friend of King Hiram of Tyre (lived 971 to 939 BCE), ordered hemp cords among other materials for building his temples and throne."

Hehn (1885) pointed out long ago that few if any traces of hemp fiber were found in the wrappings of ancient Egyptian mummies during nineteenth century excavations. On the other hand, Rudgley (1998) asserts that *Cannabis* was used extensively as a fiber to make ropes in Ancient Egypt, with hemp remains having been found "in the 19th Dynasty [ca. 1550 to 1292 BCE] tomb of [Pharaoh] Akhenaten (Amenophis [Amenhotep] IV) at el-Amarna." Booth (2003) referred to imprints on artifacts, allegedly from hemp fibers, discovered in Egyptian tombs. According to Nelson (1996), *Cannabis* was known in Egypt by the third millennium BCE, where the "ancient Egyptian word for hemp (*shemshemet*) occurs in the Pyramid Texts in connection with rope making." However, since Rudgley, Booth, and Nelson rely on secondary sources of information, the primary sources for these claims of hemp fiber remains and hemp's connection to rope making in ancient Egypt must be investigated further.

There is some pollen evidence that may indicate the presence of *Cannabis* in ancient Egypt supposedly recovered from the mummy of "Ramsés the Great" (i.e., Ramsés II, lived 1279 to 1212 BCE), who ruled Egypt for 67 years during the Nineteenth dynasty. Leroi-Gourhan (1985), the paleobotanist who identified the Egyptian pollen as *Cannabis*, also found pollen of *Gossypium* (cotton) in the same ancient context and referred to these species as introduced crops; she also referred to *Cannabis* pollen found at Nagada Khattara near Luxor in Egypt and dated to ca. 2600 BCE (see Chapter 4 for further details about these intriguing but problematic discoveries).

In any case, the relative lack of hemp in ancient Egypt might be explained, as the ancient historian Pliny the Elder indicated, by the fact that fine linen cloth from the flax plant was available and plentiful (see Pliny 1950). As noted earlier, flax had importance in the ancient West comparable to that of hemp in ancient China, at least until cotton was introduced across Eurasia; furthermore, linen was considered the appropriate textile for the ceremonial dress of the Pharaohs and their entourages. Cultivated flax, *Linum usitatissimum*, is derived from the wild progenitor *L. bienne* (= *L. augustifolium*), which, according to Zohary (1989), has the widest distribution of the wild progenitors of Neolithic crop plants with a geographical range that included the "Atlantic coast of Europe, the Mediterranean basin, the Near East, northern Iran and Caucasia" and was spread as far as the western Himalayas as early as 400 BCE. A recent report that 30,000-year-old "wild flax fibers" were found in a late Pleistocene site in the northern Caucasus Mountain region of the Republic of Georgia (Kvavadze et al. 2009) is intriguing; however, we emphasize again that the species identification of these remarkably old fibers needs further investigation and verification. In fact, more archaeobotanical research is needed to build a more complete understanding of ancient fiber use of both hemp and flax in Southwest Asia and Egypt.

Ancient Mediterranean Region

European cultivation and use of hemp for fiber and other purposes seems to have begun, at least on a recognizably large geographic scale, during the Roman Era. Later, during the medieval period, hemp fiber production and utilization intensified significantly. The early appearance, use, and farming of *Cannabis* in this region seem to have begun over a wide time range in a variety of areas. These broad temporal and spatial differences reflect, at least in part, a growing yet incomplete and problematic archaeobotanical record, especially in terms of pollen finds assumed to be *Cannabis*. As noted in the previous chapter, researchers face a distinct and challenging problem differentiating *Humulus* and *Cannabis* pollen, forcing us to rely primarily on *Cannabis* macro remains such as fibers or, more reliably, seeds. Archeological hemp fiber remains and historical records for hemp use in Mediterranean Europe are described in the following discussion.

It is generally believed that *Cannabis* originated and evolved naturally somewhere in Central Asia. Damania (1998), following de Candolle (1967), suggested that *Cannabis*, as a crop, reached Western Asia about 4,000 years ago and Europe about 500 years later. Others, such as Zohary and Hopf (1988), proposed that *Cannabis* reached Anatolia and Europe later, sometime after the ninth century BCE. Godwin (1967b) suggested that *Cannabis* was grown in the Middle East at an early time and that farming of this crop spread rapidly in the Mediterranean region during the Roman classical period; Godwin also pointed out that evidence of expansion of hemp cultivation northwards in Europe during the time of the Roman Empire was limited. However, archaeobotanical research over the last few decades clearly reveals that *Cannabis* cultivation, especially for fiber production, was taking place north of the Mediterranean region during the classical period, or even earlier in the Iron Age, from about the seventh century BCE to the fifth century CE. It is also possible that the early spread of farming and use of hemp came into at least parts of the Mediterranean region from northern areas of Europe (see Chapter 4 for archaeobotanical evidence for the spread of *Cannabis*).

An ancient piece of "hemp" fabric dated to approximately 700 BCE was found in a Phyrgian Kingdom grave mound site at Gordion, near Ankara in Turkey (Godwin 1967b); however, whether or not this ancient artifact is in fact made of *Cannabis* remains unverified. In Greece, very old cloth fibers dated to ca. 500 BCE were found at Trakhones, in Attiki province. However, similar to the ancient cloth from Turkey, identification of these ancient Greek fibrous materials is not confirmed (Barber 1991). Recently, the Greek Culture Ministry revealed that archeologists recovered a yellowed, brittle, 2,700-year-old piece of fabric placed inside a copper urn in a burial in the southern town of Argos. The remains and their context are speculatively portrayed as a possible imitation of the elaborate cremation of soldiers described in the *Iliad of Homer* (Paphitis 2007). The ancient cylindrical, corroding copper urn played a role in the rare preservation of the organic materials it contained, including dried pomegranates, ashes, and charred remains of human bones and the piece of cloth, all dated to the early seventh century BCE. This crematory artifact and its contents need further scientific investigation, which may allow positive identification of the ancient textile remains.

Based on archaeobotanical evidence available to him in the mid-1960s, Godwin (1967b) pointed out that in the northern provinces of ancient Italy *Cannabis* was rare. However, over the past 40 years or more, paleobotanical evidence indicates that hemp was present and cultivated in northern Italy on an increasing scale from the Neolithic through the Bronze, Iron, and Middle Ages (Castelletti et al. 2001). Some of the most recent archaeobotanical research in central Italy, such as the freshwater sites at Lake Albano and Lake Nemi, provides "unambiguous" evidence that both *Cannabis* and its close relative *Humulus* were present (and possibly grown) in the region as early as the Late Pleistocene and were abundant as cultivated crops by the early Holocene. The proposed first use of *Cannabis* in this region was for fiber and subsequently for other purposes (Mercuri et al. 2002).

Ample evidence of ancient hemp seeds indicates long term farming of *Cannabis* for fiber in Northern Italy (e.g., Accorsi et al. 1998a; Mercuri et al. 1999; Arobba 2001). Hemp was regularly cultivated for fiber in the lowlands of northern Italy, principally in Bologna and Ferrara provinces, from the time of the Roman Empire until after World War II (Marchesini 1997; Bandini Mazzanti et al. 1999; Bosi 2000). The area around Ferrara was especially appropriate "because of [an] abundance of wet environments suitable for hemp retting" (Bandini Mazzanti et al. 2005).

Cannabis pollen was recovered in sediment from Lake Lavarone in the Dolomites of northeastern Italy dating to the onset of the early medieval period; it is associated with a large decrease in tree pollen, a huge increase in charcoal concentration, and anthropogenic plant indicators including those related to hemp fiber production (Filippi et al. 2005; also see Arobba et al. 2003 for a review of *Cannabis* in Liguria near the western Mediterranean coast of northern Italy from the medieval period to the Middle Ages and Bosi et al. 2011 for related discoveries in northern Italy). In fact, the cultivation and use of hemp fiber has a long and significant history in Italy. For example, rope production and the trade of hemp products in and out of Venice during the fifteenth and sixteenth centuries are among the more important manifestations of its historical role in the Mediterranean region, but its cultivation and use has a much more ancient history in this area of southern Europe.

Literary evidence for the antiquity of hemp in areas bordering the Mediterranean region can be found in works by the Greek historian Herodotus of Halicarnassus, who described the utilization of *Cannabis* by the Scythians in areas northwest of Greece. Herodotus noted that hemp fiber was used in Scythian textile and cordage production as well as in their funeral rituals (Herodotus, Book 4, sections 74 and 75; also see Lehmann and Slocum 2005).

Herodotus also referred to the Scythian use of hemp for ecstatic purposes and also stated that it was cultivated by Thracians living in the region of the eastern Balkan peninsula and in Dacia (in present-day Romania) and that they used it to make fine cloth as well as ropes: "There is in that country *kannabis* growing, both wild and cultivated. Fuller and taller than flax, the Thracians use it to make garments very like linen. Unless one were a 'Master of Hemp,' one could not tell which it was. Those who have never seen hemp would think it was flax." The Thracians were close relatives of the Scythians who lived in the Balkans southwest of the Eurasian steppe around 3000 to 2000 BP, and they were excellent hemp weavers.

Another regional, historic literary reference to hemp can be found in *Aesop's Fables*. Aesop's life, like that of the great Greek poet Homer, is shrouded in obscurity and even his birthplace is contentious. He is generally believed to have been born a slave in one of the ancient Greek republics of the eastern Mediterranean about 620 BCE. After gaining freedom in compensation for his knowledge and wit, he traveled through many countries learning and teaching. One of Aesop's many long-admired fables refers to sowing hemp seed for fiber production and how hemp cordage was important in making nets used to trap wildlife; as a metaphor the fable tells us to "nip a danger in the bud," or in this case, be on the lookout for hemp seed if you are a flying, feathered creature:

THE SWALLOW AND THE OTHER BIRDS

It happened that a Countryman was sowing some hemp seeds in a field where a Swallow and some other birds were hopping about picking up their food. "Beware of that man," quoth the Swallow. "Why, what is he doing?" said the others. "That is hemp seed he is sowing; be careful to pick up every one of the seeds, or else you will regret it." The birds paid no heed to the Swallow's words, and by and by the hemp grew up and was made into cord, and of the cords nets were made, and many a bird that had despised the Swallow's advice was caught in nets made out of that very hemp. "What did I tell you?" said the Swallow. "Destroy the Seed of Evil, or it will grow up to be your ruin." (JACOBS 1894)

This fable underscores the idea that *Cannabis* was a key source of fiber for cordage used to make nets for hunting small game—a utilization of hemp's durable fiber that probably goes back many millennia.

Historical evidence indicating the use of hemp fiber to make cordage and fabric for sailing vessels in the Mediterranean region dates back to at least the fourth century BCE. For example, Athenaeus, a Greek rhetorician and grammarian who lived in Naucratis, Egypt, during the end of the second and into the third century CE, refers to hemp use centuries before his time. In volume 1, book 5 of his *Deipnosophists* or "The Banquet of the Learned" also known as "Philosophers at Dinner" (Athenaeus of Naucratis 1854 translation), he tells us that hemp fiber was produced in the western Mediterranean and imported into the island of Sicily for use in ship construction. According to Athenaeus, Hiero II (lived ca. 306 to 215 BCE), King of Syracuse, an independent Greek city-state at the time of his reign (in what is now Sicily), imported hemp fiber from Spain and the Rhone River area in Gaul (present-day France) to make sailcloth and ropes for his ships. In addition, hemp from the Rhone Valley was used along with pine pitch to caulk the ships of the King's fleet, such as the great *Syracusia*, which was 55 meters (180 feet) long and one of the largest transport ships in ancient see Hehn 1885): "For material [to build the huge merchant vessel *Syracusia*, King Hiero] caused timber to be brought from Aetna, enough in quantity for the building of sixty quadriremes [ships powered by both oars and a sail]. In keeping with this, he caused to be prepared dowels, belly-timbers, stanchions, and all the material for general uses, partly from Italy, partly from Sicily; for cables of hemp from Iberia, hemp and pitch from the river Rhone, and all other things needed from many places" (Athenaeus 1927–41).

The satirist, Gaius Lucilius (lived ca. 180 to 103 BCE), is said to be the first known Roman writer to refer to *Cannabis* plants (Hehn 1885). Another Roman satirist, Persius (Aulus Persius Flaccus), definitely refers to the use of hemp cordage and pitch

on an ancient Roman ship in his *Satire V*; according to Evans (1871), this is Persius's longest and finest satirical poem, upon which he may be said to "rest his claims to be considered a philosopher and poet." It is in this satire that Persius defends the "stoical paradox," as he states that only a philosopher can be of "*sound mind*," and in passing he reminds us of the notable role of hemp fiber on ships during his time:

Nothing prevents your sweeping over the Aegean in your big ship, unless cunning luxury should first draw you aside, and hint, "Whither, madman, are you rushing? Wither? What do you want? The manly bile has fermented in your hot breast, which not even a pitcher of hemlock could quench. Would *you* bound over the sea? Would *you* have your dinner on a thwart, seated on a coil of hemp ['*torta Cannabis*']? While the broad-bottomed jug exhales the red Veientane [booty] spoiled by the damaged pitch." (EVANS 1871; also see COVINGTON 1874)

During the Roman Empire, large amounts of hemp were utilized for a variety of purposes. Much of it was imported from Sura, a city in ancient Babylonia, as well as from ancient cities in Asia Minor such as Alabanda, Cyzicus, and Ephesus in present-day Turkey and Colchis in present-day Georgia that also served as "early major centers of hemp industries" (Frank 1959; see also Nelson 1996). A recent report on human environmental change in ancient western Turkey at Bafa Gölü suggests that during the Roman period (first century BCE to fourth century CE) deforestation and extensive farming took place, with sediment cores revealing pollen grains of the "*Humulus/Cannabis* type that could be determined as *Cannabis*, therefore, cultivation of hemp is likely" (Knipping et al. 2008).

Marcus Terentius Varro (lived 116 to 27 BCE), among the most knowledgeable Romans of his time, wrote a large number of manuscripts, yet only two have survived. In the third volume of his *Rerum Rusticarum* (*De Re Rustica*; Cato and Varro 1935), which has endured in its entirety, Varro described an aviary (*ornithone*) he designed for his estate near Casinum (modern-day Cassino in Italy). His aviary, which was located on an island and held many kinds of birds, had large columns that "supported netting made from hemp" (Johnson 1999).

The extraordinary Roman scholar and natural historian Pliny the Elder (Caius Plinius, lived 23 to 79 CE) referred to several key aspects of hemp cultivation and use in his *Naturalis Historia* (books XIX and LVI):

Hemp . . . is exceedingly useful for ropes. Hemp is grown when the spring west wind sets in; the closer it grows the thinner its stalks are. Its seed when ripe is stripped off after the autumn equinox and dried in the sun or wind or by the smoke of a fire. The hemp plant itself is plucked after the vintage, and peeling and cleaning is a task done by candlelight. The best is that of Arab-Hissar [probably the ancient city of Alabanda in present-day Turkey], which is specially used for making hunting-nets. Three classes of hemp are produced at that place: that nearest to the bark or the pith is considered of inferior value, while that from the middle, the Greek name for which is "middles," is most highly esteemed. The second best hemp comes from Mylasa [another ancient city in present-day Turkey]. As regarding height, the hemp of Rosea in the Sabine territory [located near Praeneste in modern Italy] grows as tall as a fruit-tree. (PLINY 1950)

Pliny also refers to the use of flax for nets and ship's rigging, and points out that esparto grass (*Stipa tenacissima*) came into use following the first Carthaginian invasion of Spain in 237 BCE, after which it became the preferred fiber source for making maritime cordage. In book 19, Pliny also refers to retting of flax and esparto grass but only mentions peeling of hemp (Pliny 1950). Lucius Junius Moderatus Columela, a Hispano-Roman agronomist born in Cádiz, Spain, in the year 3 BCE, also wrote about hemp in his agricultural treatise, which like that of Varro's, is named *De Re Rustica* or "On Rustic Affairs," which, in the case of Columela's work, is a famous multivolume book about farming in Andalusia (Columella 1941–55). Recent "multiproxy" evidence (archaeobotanical remains of pollen, sediments, flora, and fauna) from northern Spain indicates that water retting of hemp took place by at least 600 to 650 CE (Riera et al. 2006; see the section on Iberia in this chapter).

Many centuries later, Monsieur Marcandier, an eighteenth-century magistrate of Bourges in the central-eastern region of France, produced a small book titled *Traité du Chanvre* or "A Treatise on Hemp," first published in French in 1758 and later translated into German (1763) and English (1764). Marcandier used the earliest written records of Roman historians to describe the ancient history of hemp in the Mediterranean and some southern European regions. He pointed out that people of the Roman Empire utilized so much hemp for their needs on land and at sea that they kept arsenals of it in some of the main urban areas of the eastern empire, with huge amounts also accumulated at Ravenna in Italy and Vienna in Gaul, both following the orders of the emperor. The officer who supervised the gathering of hemp on the further side of the Alps resided in Vienna and was referred to as the "procurator of hemp manufactures in Gaul." Great quantities of hemp fiber were amassed in these areas during the Roman Era for a wide variety of reasons, including the fact that

> their husbandmen used it in fixing their oxen to the yoke and other purposes of agriculture; that their laws and their annals were written on hempen cloth; that the use of it was very common in adorning their theatres, covering their streets and public places, their amphitheatres and their arenas for the gladiators, to shade those who assisted at their public shows; that the Romans had their table linen of hemp, and that each guest brought his napkin with him; whence we may infer that it was known to the ancients as a material of cloth for the common service of their families, as well as for the purposes of agriculture, shipping, etc.
> (PAINNE 1766)

Italian hemp was known for its superior strength and durability and its multipurpose importance continued until modern times.

Undoubtedly one of the most important historic uses of *Cannabis* fiber, especially for those European countries with coastlines and navigable rivers, was for cordage and cloth used on ships. Ship fittings included anchor ropes, cargo and fishing nets, flags, shrouds, and essential oakum. Oakum consists of long strands of loosely twisted hemp fiber impregnated with tar used to caulk seams and pack joints between wooden planks on the hulls of sailing ships. Oakum has long been a crucial part of building watertight, safer boats. The coarse, strong, durable cloth covering and sailcloth referred to as "canvas" received its name from *Cannabis*, its original and main plant source. The etymological derivation of canvas being from the Anglo-French word *canevaz*, which comes from the Old North French *canevas*, which in turn is assumed to have been derived from the Vulgar Latin *cannapaceus* ("made of hemp"); the Latin *Cannabis* is believed to have its origin in the Greek *kannabis* ("hemp"), a Scythian

or Thracian word, or as Benet (1975) has suggested, the name "*Cannabis*" may be originally derived from Semitic languages.

Even recently in Rochefort, an old French river port about 375 kilometers (230 miles) southwest of Paris, a group of seafaring Frenchmen utilized hemp in their recreation of the *Hermione*, a 45-meter (almost 150-foot), 32-gun, 3-masted frigate that carried the famous Marquis de Lafayette, a young French nobleman, on a 38-day voyage to Boston in 1780. The original lean, speedy warship subsequently played a significant role in a number of sea battles in the American Revolution but then met an ignoble end, sinking off the coast of Brittany after striking a sandbar in 1783. While under construction, visitors to the new *Hermione* could watch modern boat builders at work employing the ancient method of using chisels and hammers to pack long strands of hemp fiber oakum "as filler between the wooden planks" (Sciolino 2007; also see Hermione-La Fayette Association 2011). The reconstructed Frigate *L'Hermione* was launched at Rochefort, France, on July 6, 2012.

From the end of the twelfth century until its fall at the end of the eighteenth century, the Republic of Venice, strategically located on a lagoon in the northern region of the Adriatic Sea, was the focal point of a classic colonial empire that ruled over much of northern Italy and established numerous overseas possessions. The Venetian Empire was a maritime commercial force with citizens "above all money-people," leading Pope Pius II to declare in the fifteenth century that all Venetians were beholden to "the sordid occupations of trade" (Morris 1980). Fundamentally, based on imports and exports, and dependent upon their maritime power, the safety and security of the Venetian shipping fleets was of paramount importance, and among the necessary aspects of fitting and maintaining their boats was manufacturing and providing large amounts of sailcloth and cordage. Indeed, supplying their ships with adequate amounts of strong and flexible, water-resistant, high-quality rope was of utmost consequence, and it was here that hemp fiber was so very important. Because of a monopoly on hemp cultivation and manufacturing that developed in northern Italy in the Middle Ages, specifically in Bologna, the Venetian government built and maintained a huge rope factory in Venice, known as the Tana from the early fourteenth to the late eighteenth century. According to the Venetian book *Cuore Veneto Legale* or "Heart of Veneto Laws" written during this period, the Venetian senate declared that the production of hemp cordage in their Tana factory "is the security of our galleys and ships and similarly of our sailors and capital" (Lane 1932). The monumental Tana structure in Venice, more than 300 meters (nearly 1,000 feet) in length, became the place where hemp rope, primarily premium cordage, was manufactured under state ownership and control for almost five centuries. This cordage was mainly reserved for fine grades of ship rigging, especially for superior mooring and anchor cables, but also for the manufacture of excellent bowstrings. Much high quality *Cannabis* fiber came not only from Bologna but also from Montagnana, southwest of Venice, and to a lesser degree from Treviso, north of Venice (see Lane 1932 for a lengthy, historical discussion of the role of the Tana and hemp fiber in general in the Venetian Empire). According to Abel (1980), by the nineteenth century Italy was among the major areas where hemp was produced, and from there thread and fabric for clothing were exported to a number of Western European countries: "In skilled Italian hands, hemp fiber was turned into a thread that almost equaled silk in its delicacy. It was much finer than cotton and certainly much stronger."

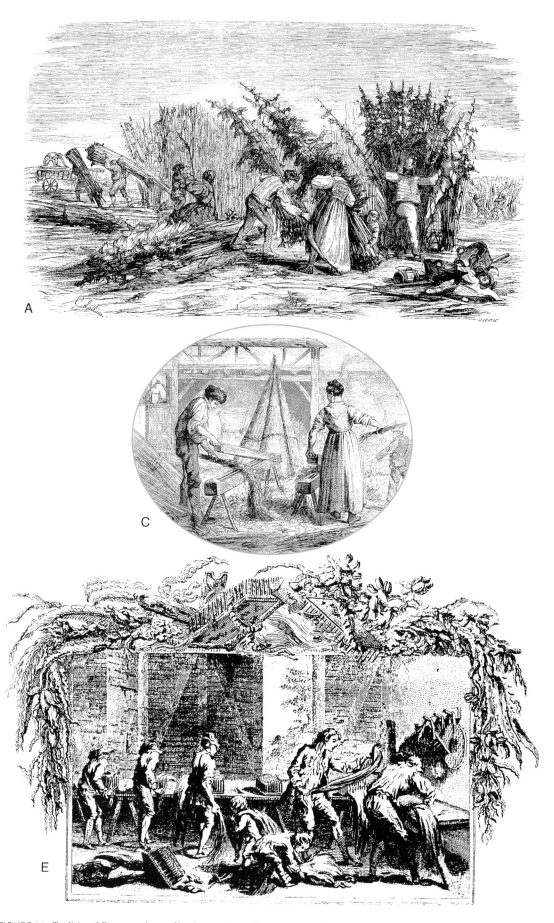

FIGURE 24. Traditional European hemp fiber harvesting and processing are illustrated in these nineteenth-century British and French engravings and an early twentieth-century Hungarian ethnographic report. Fiber hemp crops are harvested (A) in late summer when they begin to flower, and then the stalks are bundled together and submerged in water to ret (B). After soaking for a few days, the stalks are removed from the water and dried in the sun.

FIGURE 24 (*continued*). The dry stalks are broken (C) to free the fibrous bark, the bark strips are scutched (D) to remove bits of woody pith and hackled (E) to split the strips and remove short fibers. The individual fiber bundles were combed and spun from a distaff much like wool (F). Images A–C are from *L'illustration Journal Universel* (1860), D and E are from Wissett's *A Treatise on Hemp* (1808), and F is from Domokos (1930).

FIGURE 25. The world's great navies and merchant sailing fleets relied on strong hemp rope rigging for hoisting sails and lashing down cargo. Engraving from *The Illustrated London News*, November 23, 1861.

Hemp fiber also played a culinary role in Italian history. A special kind of pasta known as *garganelli* was developed some hundreds of years ago in the administrative region of Emilia-Romagna, the heartland of northern Italian food, which has its capital in Bologna. This pasta owed its special ribbed form to hemp weavers. *Garganelli* ("little weeds"), also known as *maccheroni al pettine* ("macaroni of the comb") in Modena to the north, is usually served with braised sausages as its traditional accompaniment. The pieces of this pasta are small sinewy cords with a thin hole running through them, made via a process that is more than 200 years old. This "pasta of the loom" originated specifically from the plain of Romagna, north of Cesena, where the majority of homes had enough land with rich soils to cultivate a patch of hemp.

> *Garganelli* owe their creation to hemp. They were first made near Castel Bolognese, about 25 miles [40 kilometers] south of Bologna, and are rarely eaten much farther south. Hemp flourished on that particular part of the Po Plain, and every house had its hemp loom. Originally, the ribs were pressed into *garganelli* by rolling the pasta [wrapped around a thin dowel] over the loom's comb. (KASPER 1992; also see ALEXANDER 2000)

> I think about those women's hands, coarsened and thickened over years of working with wiry hemp, that then shaped such delicate pastas. Now the looms are all but gone. But the combs remain, and today regional women gather to gossip, laugh, and make great piles of *garganelli* for special occasions. (KASPER 1992)

Occasionally, a small square frame with rough hemp cordage stretched across it was used as a substitute for the hemp loom comb, and now small plastic washboards are bought to accomplish the same result.

To the west of Italy, on the Iberian Peninsula, production of hemp cordage and other textiles also has a long history. According Guerra Doce (2006), the early presence of *Cannabis* in the Iberian Peninsula is supported by textile remains from at least two locations. One comes from Abrigo de los Carboneros or "Shelter of the Coal Miners" at Murcia in southern Spain, dated to the third millennium BCE (also see López García 1988, 1991); the other, dated to the Iberian Iron Age (ca. 800 BCE to 50 CE), was found in the site of the necropolis of Coll del Moro, six kilometers (about three and a half miles) west of Gandesa in the province of Tarragona (also see Guerra Doce and López Sáez 2006; Alonso and Juan-Tresserras 1994; and Rafel and Blasco 1995). In the case of the Chalcolithic (Copper Age, ca. 3000 to 2200 BCE) site at Abrigo de los Carboneros, a female corpse was found laid upon a mat of esparto grass that covered supporting planks of walnut wood, and wrapped around her head was an elaborate bandage of woven "*cáñamo*" (hemp) fiber. At the second Coll del Moro site, an Iron Age hemp textile fragment was recovered from an ancient weaving workshop. In both cases, further confirmation that the textiles were in fact made of *Cannabis* is required.

On the other hand, there is some substantial, albeit indirect, evidence that hemp fiber was produced in northern Spain at least by the early medieval period. This evidence was uncovered during a recent multidisciplinary study of aquatic sediments undertaken at Lake Estanya and two other associated karstic lakes in the External Ranges of the Mediterranean Pre-Pyrenees, at the northern boundary of the Ebro River on the Iberian Peninsula where substantial pollen remains associated with *Cannabis* have been found (Riera et al. 2006). The sedimentary context in which this pollen was discovered provides strong evidence of the retting of hemp fibers, along with the environmental consequences of this activity, indicating that relatively large-scale cultivation of hemp occurred in this area during the Middle Ages. Here it is noted that the retting process alters the aquatic chemistry of the water body used in the retting process "by increasing acidity, eutrophication and toxicity" (Riera et al. 2006; also see Gearey et al. 2005). In addition, the written record from Spain and elsewhere tells us that hemp retting creates pollution problems (Sanz 1995; Sharma and van Sumere 1992a/b). Retting can indeed induce eutrophication, the process whereby a body of water becomes overenriched with nutrients resulting in an overgrowth of some organisms such as algae. This in turn depletes oxygen levels in the aquatic environment, which eventually leads to death of certain fresh water animals. Recent paleoecological research involving diatomaceous deposits indicates that eutrophication occurs when hemp is retted now as in the past (e.g., see Bradshaw et al. 2000; Lotter 2001; Enters et al. 2006). Artifactual changes associated with hemp retting include alterations of the aquatic species assemblage, increases in Cannabaceae pollen percentages, variation in the sediment components, and artificial development of water channels, all of which can be detected in the archeological record (Riera et al.2006).

The study of Lake Estanya and the two other lakes shows that hemp retting left its characteristic "multiproxy" evidence of environmental impact in these lakes (Riera et al. 2004): "The perturbation caused by the introduction of hemp retting practices in Lake Estanya led to changes in all the proxies such as sedimentology, non-pollen palynomorphs, pollen and faunal communities" (Riera et al. 2006). Furthermore, man-made channels connecting the Estanya Lakes probably indicate water management related to hemp retting and irrigation (Riera et al. 2006). "Although fed by underground springs, water levels in the three lakes are to some extent controlled by these artificial channels which were constructed using a dry-stone method frequently employed in both medieval and modern times." The initial peaks of Cannabaceae pollen (5 percent of the total recovered pollen) occur at a sediment level dated to about 600 to 650 CE, which suggests that hemp cultivation in the region started during this time period, and corresponds "with the spread of hemp between the 5th and 9th centuries in other European regions" (Riera et al. 2006). The pollen record recovered from Lake Estanya also indicates that "Cannabaceae values increased after 1360 CE and reached a maximum of 25 percent in 1760 CE," demonstrating that hemp farming and processing were important activities from medieval times up until the nineteenth century (Riera et al. 2006; also see, e.g., Godwin 1967b; Gaillard and Berglund 1988; Laitinen 1996; Fleming and Clarke 1998; Cox et al. 2000).

Hemp cultivation in medieval Spain is also mentioned in several historical references. For example, in Ibn al-'Awwam's *Kitab al-Filaha* or "Book of Agriculture," written in the second half of the twelfth century, the Hispano-Muslim horticulturist (probably from Seville) discussed the methods of cultivating cotton, flax, and hemp and notes that hemp was used in his time for making coarse fabrics, ropes, and paper. In this text Ibn al-'Awwam also points out that it was grown under conditions similar to that for flax but did not need as much water (Bolens 1992, 1981). Other twelfth-century documents, belonging to the medieval Catalan noble family of Montcada, give us insight into the productive activities that occupied peasant workers in "Catalonia." These included the cultivation of grape vines, olive trees, and cereals for food as well as flax and hemp, which supplied fibers for cordage and cloth (Shideler 1983). Additional written evidence indicates that hemp was used by the later medieval Catalan weaving industry (Rui 1983). By the end of the twelfth century, a workshop in Lérida (inland from Barcelona) was producing cloth that was most likely a combination of hemp and wool, and weaving soon became a regular activity in other areas of Catalonia (Altisent 1970, 1967–68). The government "exchange-brokers' tariff" of Barcelona, drafted in 1271, declared that a specific tax was to be paid by both the buyer and vendor per quintal (46 kilograms or almost 100 pounds) of hemp or wool yarn (Corominas 1959). Blended hemp and wool cloth, marketed among the mountain communities of northeastern Spain and southeastern France, may well have been introduced by the Gallic entrepreneur Nicolau de Sarlat from Périgord (Dordogne, France), "where the production of such cloth was already known at this period" (Rui 1983).

Hemp fiber was also an important component in production of shoes and probably crossbow strings at least by the latter part of fifteenth century; in the case of shoemaking there was a thriving trade between the Christian cord makers of Valencia and the "Muslims of the Vall de Uxó, who specialized in the fabrication of hemp sandals" (Meyerson 1991, citing the Archivo de la Corona de Aragón, Cancillería Real 3665: 18v-19v, 23 November 1486, and Archivo del Reino de Valencia, Bailía General 1431: 529r-v, 20 May 1495). The cord makers of Valencia supplied the hemp cordage and sent it a relatively short distance north to Vall de Uxó, where it was used in the production of sandals (*espardenyas*) that were then carried back to Valencia, the largest market for this footwear. In 2004, Spain produced approximately 15,000 metric tons of hemp (Lewin 2006), largely for use in the production of paper.

Ancient Europe North of the Mediterranean

The historical record clearly shows that *Cannabis* fiber was in use in the Mediterranean region at least by the later period of classical Greek and Roman times, and as described in the previous chapter, recent archaeobotanical evidence from macroremains found in waterlogged soil sediments indicate the *Cannabis* was already being used in several other areas of Europe around 800 to 400 BCE, primarily for fiber and seed. This suggests that ancient use of *Cannabis* north of the Mediterranean, for fiber or otherwise, predates the Roman period. In fact, much older evidence of possible hemp fiber use has been reported from European archeological and paleobotanical sites. Of special interest here are impressions of plant fiber cordage on clay fragments found on the floors of dwellings at Gravettian sites in the Czech Republic dated to ca. 26,980 to 24,870 BP that were said to represent the "earliest forms of fiber arts known, some 10,000 years earlier than anything found before" (Adovasio et al. 2007; also see Kvavadze et al. 2009 for even older putative evidence of flax fiber production found in the northern Caucasus region of the Republic of Georgia dating back about 30,000 years ago.) The Gravettian people were spread over a huge area stretching from Spain to southern Russia "some 29,000 to 22,000 years ago." These people were very distinctive in that they "used nets, rather than speed and might, to capture vast numbers of hares, foxes, and other mammals" (Pringle 1997). If this interpretation is valid, it suggests that the Gravettian people were among the first known net hunters, and this could explain why the Gravettian culture is characterized by bigger more settled human populations.

James Adovasio, an expert on prehistoric fiber technology, identified cordage imprints on four Gravettian clay fragments, suggesting that "the impressions were created from fabrics woven of fibers from wild plants, such as nettle or wild hemp, that were preserved by accident" (Pringle 1997, citing Adovasio et al. 1996; also see Adovasio et al. 2007). David Hyland, another researcher who studied these remains, reported that the cordage ranged in diameter from 0.31 to 1.15 millimeters (0.01 to 0.05 inches), certainly within the range of yarn sizes used in other regions to weave traditional hemp cloth. The ability to catch animals would have facilitated the acquisition of comparatively "huge windfalls of food regularly at very low risk of injury to human participants" and may represent "the key to the Gravettian cultural development" (Pringle 1997). The ancient netting suggested by the cordage impressions could also have been used to carry gathered materials. The small mesh (estimated four millimeters or 0.16 inches) of these ancient nets indicates that although they were possibly used for trapping birds and small game, they are more likely fragments of net satchels like those made by many tribal peoples well into the twentieth century (Clarke 2008). Even very small game would be much easier to trap in a lighter, less visible, more open mesh net. If positive identification could be made that these fiber impressions were made by *Cannabis* cordage, they would be by far the oldest archeological evidence for the use of hemp, but until there is further substantiation, the researchers' interpretation is only speculation.

In a recent, collaborative publication (Adovasio et al. 2007), the simple technology of knotting cordage to make nets in Paleolithic Eurasia is emphasized as a key aspect of human cultural progress and early resource utilization in the Central European region (also see Hardy 2008). These authors present a scenario depicting the underappreciated role of women in fabricating small diameter cordage for many utilitarian purposes. They suggest it was nettle (*Urtica diocia*) fibers, which were used by their hypothetical group of (Gravettian?) people living in a temperate valley of what is now the Czech Republic about 26,000 years ago. These people could have extracted fibers from *Cannabis* as well as, or rather than, from nettle or other bast fiber plants such as willow trees (*Salix* ssp.) or lime bass trees (*Tilia cordata*, also known as small-leaved lime or small-leaved linden, which today is the national tree of the Czech Republic and Republic of Slovenia). In any case, the imaginary scenario presented by Adovasio and his colleagues is worth our consideration. This description of the extraction and manipulation of wild plant fibers focused on an essential aspect of early human adaptation to the environment, especially among hunters and gathers and eventually among early farmers. For the sake of argument, it should be reiterated that although Adovasio and his colleagues suggest that this imaginary group carried long hunting nets made out of nettle fibers, these nets, and other forms of cordage, could have been made from hemp fibers if *Cannabis* was available:

> By late afternoon, each family's net has been unrolled and carefully inspected, and tied together to form two long nets, each some 80 feet across. Now the children, some of the women and men, and a few elders set out with the nets. The children carry sticks, which they brandish bravely as they run along behind the adults. Several of the adults carry clubs fashioned from fallen branches. Led by the oldest in the party, they pause after a half hour's walk on the slopes that are covered by underbrush. Carefully the oldsters unfurl the nets, unwinding them from the carrying poles, which are then used to anchor the nets to the ground. Several of the younger women, the men, and all the children silently circle around upslope until they reach nearly to the top of the hill. There they form into a wide arc and on a signal begin the charge down the hill, shrieking wildly, whacking trees as they go by, setting up a terrifying din. Rabbits, foxes, and other small mammals emerge from the underbrush and dart back and forth, trending downhill to escape the mayhem coming their way. Within minutes several dozen of these creatures have leaped into the nets to be quickly dispatched by people swinging their clubs. As the sun drops down to the western horizon, the people head back with more meat and fur than they will be able to use for days. (ADOVASIO ET AL. 2007)

The circumstances described earlier for the activity of this band of early hunters and gathers may seem overly speculative to some. However, the underlying evidence of fiber impressions on ancient earthenware artifacts from this Upper Paleolithic site is also of interest because the relic itself is "the earliest ceramic object ever found," which, the authors argue, supports the "crucial role of the fiber arts" in human cultural evolution (Adovasio et al. 2007). The early people depicted in this scenario were well advanced into what has been referred to as a revolution in Eurasian fiber and cordage production. Elizabeth Barber (1991, 1994) articulated the importance of the crucial fiber and cordage innovations that led to the remarkable invention of various forms of cordage. She has effectively argued that this produced major effects on human

history, perhaps more significant than advances made in the early production of many stone tools such as scrapers, knives, or spear points. Fiber use was a crucial step in human utilization of natural resources, allowing people to spread out and occupy a great diversity of environments. The production of cordage allowed a whole set of novel work for people, principally women, and consequently enhanced human survival by allowing people to better fasten, catch, hold, and carry things. This led to early development of "snares and fish lines, tethers and leashes, carrying nets, handles, and packages, not to mention a way of binding objects together to form more complex tools" (Barber 1994; also see Chapter 4 for possible use of hemp nets in ancient fishing). From this point of view, ancient fiber remains found in ancient sites in the Czech Republic are a good deal more significant than one might think. Cordage, in its various forms, is generally quite perishable, breaking down relatively quickly over time, and thus fiber artifacts are rarely found in archeological deposits, especially in very old sites, and if they are recovered, their identification can be problematic depending on the degree of preservation and the context of their deposition. Normally evidence is in the form of fiber impressions, with the fibers themselves having rotted or burned away. It has been pointed out that when fiber artifacts are discovered in places such as dry caves, water bodies, or permafrost areas—where because of the arid, anerobic, or frozen conditions they do not break down and vanish—these relics are present in much greater abundance than are stone artifacts (Adovasio et al. 2007).

In sum, the ancient people in this European area, and perhaps other regions, were most likely making a variety of cordage products—bags, baskets, nets, mats, and possibly clothes—well back into the Late Pleistocene, and in at least certain parts of Eurasia, *Cannabis* fiber could have been used to make manufactured goods starting many thousands of years ago. However, it should be pointed out that our hypothesis suggesting that the Gravettians used wild *Cannabis* to spin such fine yarns has some fundamental problems. Spinning fine yarn from *Cannabis* is challenging, even from properly grown fiber hemp crops, and is impossible when using wild, well-branched hemp plants with many nodes. If these ancient people had access to wild hemp and used it as a fiber source to spin such small diameter yarn, then they would likely have had to find rare stands of immature *Cannabis* plants (with strong, flexible fibers) growing very close together (rough hemp bark from mature branched plants being unsuitable for spinning fine cordage). Young plants could have been found near trash heaps or threshing areas and more or less intentionally managed, which raises interesting questions regarding the origins of agriculture.

In an article on Eastern European use of hemp and nettle for fiber, food, and medicinal purposes, Zajaczkowa (2002) referred to hemp as a crop that has been cultivated for centuries, primarily for its fiber, seeds, or both, but also pointed out that hemp stalks must be collected in August during flowering before the seed is set. Generally, as seeds form, stalks yellow at their bases, leaves begin to fall, and fibers become coarser. Fiber hemp crops are traditionally harvested before they go to seed, and consequently crops cultivated specifically for fiber or seed are usually grown in different plots. When both fiber and seed are desired from a common plot of hemp, male plants are normally collected first by hand pulling them, and female plants are left in the field so the seeds can ripen. In this way fine fiber can be extracted and spun to weave sheer cloth. Körber-Grohne (1987) stated that the earliest documented discoveries of hemp cordage and cloth in Europe were not yet made from processed hemp fibers but rather from the bast stripped from the plant's stem, as is the tradition in the Far East (also see Clarke 2010a/b; Hai and Rippchen 1994).

Although historical evidence tells us that *Cannabis* use for fiber was spread to, or at least intensified in, various areas in Europe during the era of the Roman Empire, true hemp was apparently known at least to some early northern Europeans long before then. In the final centuries before the "Christian" or current era (CE), significant cultural changes occurred in large parts of central and northern Europe affecting social and economic activities. This included the spread of Roman influence as far as the Danube and Rhine Rivers and the construction of many "Roman limes" or defensive forts to secure their frontiers. These fortifications split Europe into two major cultural regions—the "civilized world" and what lay beyond, which was referred to by the Romans as Barbaricum. As noted earlier, Herodotus referred to hemp being cultivated by "barbarian" Thracians in the region of the eastern Balkan Peninsula and in Dacia in present-day Romania, where *Cannabis*, despite its notorious reputation for coarseness, was used to make fine cloth as well as ropes: "They have hemp growing in their country, very much like flax except for thickness and height. In this respect the hemp surpasses flax by far. This grows by itself and sown, and out of it the Thracians even make clothing very much like linen" (Herodotus, Book 4, section 74; see Lehmann and Slocum 2005).

Ardelean (1893) described traditional Romanian woman's clothing based on a blouse or shirt of woven linen, wool, or hemp and belted at the waist and male clothing as "made of strong hemp-linen" and also indicated that it was women who grew and wove hemp.

In Hungary, Török (1954) reported finding hemp textile fragments dated to ca. 1050 CE in a Magyar graveyard at Halimba-Cseres and suggested that hemp fibers from this site were differentiated from those of linen by "microscopical observation." As noted in the previous chapter, *Cannabis* seeds were found in Late Iron Age pits of the ancient Celtic settlement of Corvin tér in Budapest, indicating that hemp farming took place in the surrounding region, probably for fiber or seed (Dálnoki and Jacomet 2002; Jacomet 2007). Hemp was likely cultivated at Corvin tér by the middle of the first century BCE before the Roman conquest. High levels of *Cannabis*-type pollen with dates covering the past 1,000 years have also been reported from sites in Hungary. Willis (1997) attributed this to a lengthy history of hemp rope production; Hungary also certainly has a long tradition of peasant hemp cloth weaving.

In northeastern Hungary large amounts of "hemp (*Cannabis*)" pollen were recovered in sediment cores from a small *Sphagnum* moss peat bog called Kis-Mohos Tó located in the undulating hills and valleys of the Kelemér region, about 60 kilometers (37 miles) north of the Great Hungarian Plain (Willis et al. 1998). *Cannabis* pollen begins to show up chronologically in this bog sometime toward the latter part of the first millennium BCE, and quantity levels rise through the past two millennia. Willis and his colleagues attribute the large increase in *Cannabis* pollen in the bog, which was then a lake, to "barbarian groups" who were using Kis-Mohos Tó to ret hemp plants, which "unintentionally introduced large quantities of *Cannabis* pollen into the lake." In addition, charcoal deposits in the bog cores indicate that burning of the surrounding wild vegetation was carried out intentionally, "probably for clearance of land for cultivation of *Cannabis*." During the period from about 900 to 1600, "the Magyars continued to use the lake for the retting of *Cannabis* and evidence suggests that if anything, production became more intense." These authors concluded that after about 300 CE, woodland clearance and associated "land degradation" became irreversible and continues up to modern times; they argue that this forest removal is associated with the manufacture of hemp rope, a significant economic practice, which persisted into the medieval period. They also emphasize that it was intensification of agricultural activity for production of introduced industrial hemp—rather than new technologies or a massive population increase—that caused changes in the landscape around Kis-Mohos Tó. Increased demand for rope could be explained by its association with increased herding of horses and cattle in the region.

At Karcynz, a second- to third-century CE Barbaricum settlement situated beyond the Roman realm of control in the Kuiavia region of central Poland, archeologists made an extraordinary discovery related to ancient fiber production. While undertaking salvage work at this site, in preparation for the construction of the remarkably long natural gas pipeline from Siberia to Europe, they found large wooden retting vats built within a shallow body of water. The vats were dug into the ground below the water table, and at the bottom of one they found exceptionally well-preserved bundles of both flax and hemp stems approximately 1,700 years old: "This was the first time that archeologists were able to record this initial stage of plant fibre [sic] processing; the sum of previous knowledge on this subject having stemmed from indirect evidence in the form of much later ethnographic analogies and written sources" (Chłodnicki and Krzyżaniak 1998).

Another example of ancient hemp cultivation in Poland involves a study of human impact on the vegetation history of Wolin Island in the Baltic Sea by Latalowa (1992), who recorded a large amount of *Cannabis* pollen occurring continually over the past 3,000 years, which, along with high values of winter rye and barley pollen, may indicate crop rotation, a method practiced in Asiatic hemp cultivation (also see Latalowa 1999).

In Germany, according to Körber-Grohne (1987), the cultivation and use of *Cannabis* for fiber has a relatively long history dating back at least to the pre-Roman period, ca. 800 to 400 BCE. Probably the earliest material remains of hemp fiber in Europe come from southern Germany in the Early Iron Age grave mound of Hochdorf. In this impressive Celtic tomb dated from the late Hallstatt period (ca. 2550 to 2500 BP), an ornate bronze couch was put in place with a "chieftain" laid on it, and the thick cushioning of the couch was covered in "horsehair, hemp, wool, and the fur of badgers" (Delaney 1986; also see Biel 1981; Körber-Grohne 1985, 1987; Enters et al. 2006). Further confirmation of this reported hemp fiber is needed.

In Switzerland, scanning electron microscopy has been used to identify textiles made of hemp, wool, and linen from Roman Age sites; however, the structure of the woven materials suggests that they may not have been manufactured locally (Rogers et al. 2001). More recent evidence for use of "*Cannabis sativa*" for fiber in Switzerland was discovered at Winterthur in the northern part of the country and dated from the thirteenth to fifteenth centuries (Windler and Rast-Eicher 1999/2000). Brombacher (1998) reported finding *Cannabis* seeds near the medieval settlements of Develier and

Courtételle in the canton of Jura in northwestern Switzerland that are dated to the Middle Ages, and in this case, since many fragments of hemp stalk were also recovered, we can assume that hemp processing took place here (see Irniger and Kühn 1997 for a discussion of the economic importance of hemp and flax in the late Middle Ages and early modern times in Switzerland). More evidence for the long-term importance of *Cannabis* in Switzerland is supported to varying degrees by pollen, seed, and fiber remains referred to later.

Welten (1952), for example, was able to date *"Cannabis"* pollen recovered from a site at Spitzierbucht, Switzerland, to ca. 2200 BP; based on the presence of indicator species associated with probable cultivation Welten concluded that hemp was commonly grown nearby during the twelfth through the seventeenth centuries (also see Chapter 4). Ancient Cannabaceae pollen was also recorded in the western and southern parts of Switzerland; this includes evidence from 15 lakes, 7 mires, 1 soil profile lake, and additional mire deposits, distributed in several areas, including the Jura Mountains (3 sites), Swiss Plateau (2 sites), northern Pre-Alps and Alps (6 sites), central Alps (5 sites), southern Alps (3 sites), and southern Pre-Alps (4 sites). Pollen from these sites indicates widespread hemp cultivation in historical times and shows clearly that hemp became an important and widespread fiber crop in various parts of Switzerland from the medieval period until at least the nineteenth century; in some areas hemp cultivation probably dates back more than 1,000 years (van der Knaap et al. 2000; also see Lotter 1999). For example, evidence from three peat profiles in pasture woodland within the small valley of Combe des Amburnex indicate the early presence of *Cannabis* dating to about 850 CE (Sjögren 2006), and a paleobotanical study at Praz-Rodet bog in the southwestern area of the Jura Mountains shows that *Cannabis* pollen first occurs in small amounts dated to the eighth century CE, a period when relatively limited clearance and early cultivation in the forest began (Mitchell et al. 2001). The amount of *Cannabis* pollen in the Praz-Rodet study increases up until modern times. Species interpretations and usage patterns are based on chronostratigraphic pollen markers, regional patterns, and local histories (van der Knaap et al. 2000; Sjögren 2006; Mitchell et al. 2001).

More reliable medieval period archaeobotanical evidence from the Jura region includes hemp seeds found at Develier and Courtételle referred to earlier (Klee and Brombacher 1996; Brombacher 1998; Brombacher and Rachound-Schneider 1999), as well at La Neuveville along the Bieler See in northwestern Switzerland, dated to the fourteenth century (Brombacher 1999), and at Laufen in northwestern Switzerland, near the French and German borders, dated to the thirteenth to fifteenth centuries (Karg 1996). Brombacher and Klee (2007) recently reported more about their long-term excavation in the Delemont Basin, indicating that early medieval plant remains from the site at Develier-Courtételle provide evidence not only for nine different cereal crops but also for flax and hemp: "A lot of finds of *Linum usitatissimum* and also *Cannabis sativa* point out the great importance of textile manufacture." As noted earlier, hemp and flax are traditional raw materials that reached significant economic importance in the late Middle Ages and early modern times in Switzerland (Irniger and Kühn 1997; also see Chapter 4).

Both hemp and linen fibers were reportedly found snagged in a bone tool recovered from a late Neolithic site at Adaouste in the Provence region of southern France dated to about 4000 BP (Barber 1991). If the fibers are in fact *Cannabis* and

are as old as reported, they would represent very old macrofossil evidence for hemp fiber use. Fragments of *Cannabis* hemp rope dated to the fourth century BCE were also reportedly found among carbonized remains in a house that burned in the area of Lattes, near Montpellier, also in southern France (Boyer and Encart 1996). In addition, a piece of string also allegedly made of hemp was discovered in a first century BCE structure on the site of Béthisy, north of Paris (Malrain et al. 2002), and discovery of hemp textile material in Gallo-Roman (second to third centuries CE) graves from a Late Iron Age tomb at the Fontvielle à Vareilles site in Creuse, west of Amiens in northwestern France, has also been reported (Lorquin and Moulherat 2002).

More evidence of ancient hemp use comes from the limestone tomb of the Merovingian queen Arnegunde, who was laid to rest in Paris about 570 CE underneath the cathedral in St. Denis. Adorned in her burial with many precious items, including silver, gold, and a silk dress, the queen's body was ultimately covered with a shroud of hemp cloth (Werner 1964; also see Godwin 1967b; France-Lanord 1979). Abel (1980) suggested that this hempen mantle was made of "a material apparently deserving of this place of honor among the rich and elegant burial wardrobe of the French nobility during the early Middle Ages" (see Chapter 9 for a discussion of hemp funerary traditions). Analysis of these and other ancient, mineralized fibers preserved in burial contexts in France are said to provide evidence of the use of hemp well before the Roman period in this region of Europe, but identification of these "hemp" fibers, like many others, has not been fully confirmed.

In northern France, Ruas (1988, 1998) provided strong evidence for the early medieval use of *Cannabis* in the form of carbonized seed remains recovered from a site at Baillet-en-France (cf. Bakels 2005). Ruas (2000) also found ancient *Cannabis* seeds in central France at Auvergne, Saint-Germain-des-Fossés (Allier), dated to the early Middle Ages (tenth century). Early medieval period *Cannabis* seeds have been excavated from sites in northern France dated from the seventh to ninth centuries (e.g., from Serris-Les Ruelles, an early manor house; de Hingh and Bakels 1996; also see Bakels 2005).

Cyprien and Visset (2001) found "*Cannabis-Humulus* type" pollen during their paleoenvironmental study of core samples from a peat bog near the town of Carquefou (northeast of Nantes) on the Erdre River in western France. This pollen type begins to appear in sediment dating to at ca. 478 to 281 BCE, during the Iron Age, when, they argue, hemp was being cultivated near the settlement and continued to be grown until about 543 to 681 CE. Cyprien and Visset suggest that the curtailing of hemp production in this area may have occurred in association with the decline of Roman economic and demographic influences; however, they also reported that during the Middle Ages (887 to 1149) the forest in Carquefou disappeared almost completely, probably due to clearance or a change in water level associated with the intensified cultivation of crops such as hemp and especially cereals. In addition, Cyprien and Visset pointed out that crop farming, including hemp cultivation, lasted only until the later Middle Ages, when cattle herding became more important.

In the introduction to her book *Fine Linen*, Françoise de Bonneville (1994) provides us with insight into the relative importance of hemp and flax fibers in France during the past millennium. She tells us that a hemp weavers' guild was established in France in the Middle Ages, well before the founding of a linen weavers' guild, indicating that hemp was more common until the late fourteenth century. Although the best quality bed sheets in France were being made of flax linen by about 1322, most were still being made of hemp cloth, and less desirable but commonly used bed sheets were woven from scrap hemp or combings of flax until the end of the seventeenth century. Posthumous inventories dating from the thirteenth century onward offer some insight into the quality and quantity of highly valued flax linen clothes kept in the closets of medieval royalty and the upper classes. On the other hand, clothes belonging to the impoverished were generally soiled, worn out, and discarded, or the rags were used to make paper. Although flax was more frequently used from medieval times until the nineteenth century, hemp was still important. For example, underclothes, household cloth, and rope in France were still generally made of hemp, and religious groups and the underprivileged frequently favored hemp cloth because it was considerably less expensive. De Bonneville (1994) tells us that Maria de Medici, queen consort and second wife of the Bourbon King Henry IV, "kept two hemp blouses among her magnificently rich trousseau-perhaps for doing penance." Until relatively modern times "hemp was used more widely in the countryside than in towns, since every farm had its field of hemp (and perhaps another of flax) designed to meet the daily household fabric requirements and to supply part of the daughters' bridal trousseaux." France has a long history of hemp production and use that continues today, but hemp weaving is very rare, and the vast majority of hemp fiber is presently destined for specialty paper production, while the hurds (inner pith) are used as building insulation and animal bedding (e.g., see Nigro 2012).

In Belgium, archaeobotanical evidence also indicates that *Cannabis* has been used as a fiber source for centuries. For example, Meersschaert et al. (2007) recently completed archaeobotanical research related to reconstruction of the seventeenth century ramparts of Damme, a municipality located in West Flanders, northeast of Brugge; here they found remains of a wide spectrum of plants used by people hundreds of years ago including seeds of *Cannabis* (also see Cooremans 1995/96a/b for evidence from Oudennaard dating to about 1350 to 1450 and Oostende dating to the fifteenth century and Gelorini et al. 2003 for evidence from Beveren from the late Middle Ages to early modern times).

The lengthy Belgian history of hemp fiber cultivation and use is amply supported by written records, old photographs, and even antique machines used to process hemp that date from the latter part of the nineteenth century. Many of these items can be found in the "impressive collection" of the Musée Gaumais located at Virton in the Belgian Lorraine region, about 200 kilometers (125 miles) from Brussels, near the borders of both France and Luxembourg (Jensen 1996). A large part of this museum draws attention to the significant, historic role of hemp farming among members of a whole community, including the men who cultivated and harvested the crop and the women who spun and wove its fibers. A whole room in the museum is devoted to working of the hemp crop (*travail du chanvre*) and includes a reconstructed nineteenth-century workshop exhibiting various traditional procedures used to extract hemp fibers from their stalks, spin them into thread, and weave them into cloth. These processes include breaking, scutching, combing, and spinning the fibers on machines, which are on display in the museum, along with a loom and hemp cloth. "For families in the Belgian and French Lorraine region at the turn of this century, hemp was an essential crop in meeting their needs for strong, durable cloth and linen—an important criteria for a population that only did laundry once a year! The families in Gaume reserved the best parcels of land, generally less than one hectare, for their hemp crops, which in turn provided for all of their clothing and feedstock needs, and furnished oil for their lanterns" (Jensen 1996).

In 1986, Myriam Pezzin, director of educational services at Musée Gaumais, published a well-researched educational brochure titled *Daily Life in Gaume in the 19th Century: Working with Hemp*. Her booklet features the household activities of rural women's lives, in particular two chores that took up a large part of their time and energy—the yearly laundry and the annual transformation of hemp from crop to woven fiber. The hemp plant and its fiber products were so admired in the everyday lives of the Gaumais people that most females longed for "a hand break or scutching board with a hand-carved heart for their anniversary present" (see Jensen 1996 for a brief discussion of the revitalization of hemp farming in the area near Musée Gaumais in the late twentieth century).

In neighboring Netherlands, *Cannabis* was present and perhaps even cultivated during Roman times; indeed a few waterlogged seeds that date to the latter part of the Roman period have been reported. For example, a hemp seed fragment found at Woerden-hoek Molenstraat at Kazernestraat has been dated to between 50 and 250 CE (Brinkkemper and de Man 1999). Evidence for *Cannabis* and its use for fiber and seed in the Netherlands during the medieval period is much more substantial. In fact, *Cannabis* seeds were found in a large number of medieval period sites, as well as in more recent archeological contexts with dates ranging from about 700 CE to the early twentieth century. These ancient to more modern seeds, including a few Carolingian finds from the ninth century and many more from the twelfth century onward, are often found in cesspits, and frequently they have been interpreted as indicating the use of hemp fiber for cordage or hemp seed for its oil (see Chapter 6 for a detailed discussion of hemp seed use). A few representative examples of the many ancient *Cannabis* macroremains listed in the Dutch RADAR database (e.g., see van Haaster and Brinkkemper 1995), along with some additional, more recent discoveries, are described briefly in the following section.

Bakels et al. (2000) carried out archaeobotanical excavations at a twelfth-century country farm near Gouda-Oostpolder in the Netherlands peat lands where medieval farmers practiced mixed farming based on the raising of cattle, pigs, and crops, including emmer wheat, barley, oats, flax, and hemp; nine ancient waterlogged *Cannabis* seeds were found at this site. In a more recent study, Bakels (2005) listed discoveries of carbonized and waterlogged *Cannabis* seeds in the southern region of the Netherlands dated to the early medieval period from the sixth to twelfth centuries (also see de Hingh and Bakels 1996). Other medieval Dutch sites where hemp seeds were recovered include those in Lieshout-Nieuwenhof, with more than 200 seeds dated from the thirteenth and fourteenth centuries (de Man 1996); Gorinchem-Krijtstraat, with about 2,000 waterlogged seeds dated to the fourteenth century and 20 seeds dated from 1575 to 1650 (van Haaster 2003); Oldenzaal-St. Agnes cloister, with 57 seeds dated to the fifteenth century (Brinkkemper and de Man 1999); and Haarlem (Haarlem-Kokstraat, Haarlem-Korte Begijnestr and Haarlem-Spaarne), with about a hundred seeds dated from ca. 1400 to 1675 (Pals 1983; van Haaster and Hänninen 1998).

The past importance of hemp rope in the Netherlands is apparent in the recent, highly acclaimed reconstruction (1985 to 1995) of an early seventeenth-century CE sailing ship, the well-known Dutch East India Company merchantman *Batavia*. As a nation historically dependent on seafaring and trade, the Netherlands needed a substantial and seaworthy fleet, and among the essential components for these vessels was cordage, including many tons of strong, sturdy, and flexible hemp rope. The original *Batavia* was built at Peperwerf (Pepper Wharf, a name celebrating the historic spice trade) in Amsterdam and completed in 1628, during the Dutch "golden age," when their fleets established and maintained long distance trade with various parts of the world. This ship ran aground and sank during its infamous maiden voyage followed by well-documented mutiny and human massacre (e.g., Dash 2002; Drake-Brockman 1995; Leys 2005).

Authentic materials were used in reconstruction of the *Batavia*, including the rigging, which was "mainly made of long-fibred hemp" prepared in a Dutch ropeyard. The modern replica carried approximately 21 kilometers (13 miles) of hemp ropes on its maiden voyage (van der Zee 2002). However, their longevity has become a problem for long-term use: "Hemp is a natural product and must be protected against extreme weather conditions through a series of processes. The rigging of the *Batavia* has been repaired and renewed several times during the past few years. Despite years of experimenting, no solution has yet been found to the problem of effectively protecting and properly conserving hemp; therefore, important parts of the rigging have been replaced by more modern and more durable materials. The reason for this change is mainly a safety consideration, as hemp can deteriorate slowly and then suddenly break" (van der Zee 2002).

The new, longer-lasting material used to replace ropes on the *Batavia*, as on many other modern ships, is synthetic "Hempex," a strong, weatherproof artificial fiber made to look like natural hemp. However, it "is too elastic for real standing rigging, which needs a certain stiffness" (van der Zee and Klein 2005).

We now shift the geographical focus of our hemp fiber history to Russia, then move on to the Baltic Region and Scandinavia, and complete our discussion of ancient *Cannabis* fiber use in Europe with a review of evidence from the British Isles. As we suggested in Chapter 4, the narrow-leaf hemp ancestor (NLHA) was spread from the Caucasus region both east and west through western Eurasia. Moving west it likely entered Europe from one or more areas north and/or south of the Black Sea. According to Rubin (1975b), *Cannabis* was cultivated in southern Russia as early as the seventh century BCE. Given the evidence presented in the previous chapter, hemp farming was probably undertaken much earlier in other regions than that north of the Black Sea—that is, in the Pontic Steppes and northward into the temperate forest, including much of the area from west of the Ural Mountains to parts of Eastern Europe. Early hemp farming was probably for fiber and seed, but in most areas, especially in the northern portion of this region, it most likely favored fiber production. Forbes (1964) claimed that hemp entered Europe from southern Russia suggesting that use in Russia predates that of any part of Europe. He also argued that the Goths introduced hemp into eastern Europe from western Russia in the second and third centuries CE, and only after that cultural spread did use of *Cannabis* fiber begin in central Europe. Forbes went on to suggest that ninth century CE Slavic migrations produced an increased interest in farming hemp in western Eurasia and

that it then began to displace flax as a fiber crop in some European regions; in this case he may be correct. Goodrich (1856) reported that flax and hemp were cultivated for both domestic use and export mainly in the Russian provinces along the Black Sea. In the following section we review evidence for the early presence, use, and spread of *Cannabis* within and out of Russia, which in all likelihood took place well before the dates suggested by Forbes (see Chapter 4).

Cannabis fiber has a very lengthy history in Russia, which eventually became the world's leading producer of hemp rope (e.g., Rubin 1975; also see Crosby 1965). In addition to a limited written history compared to many areas in Europe, we also have scant archaeobotanical evidence to support the antiquity of *Cannabis* use and cultivation in Russia. One fragment of evidence comes from northwestern Russia, where Vuorela et al. (2001) reported the discovery of one ancient *Cannabis* seed from the post–Iron Age (after ca. 1150 to 1300 CE) layer of lake sediment from the eastern part of Igumeeninlampi, a small twin lake in the northwest area of Valamo Island, which is located within Lake Ladoga in northwestern Russia. Also found in this same site were flax seeds discovered in a lower, older Late Iron Age layer (ca. 800 to 1150–1300); in addition, seeds of hop, considered a possible native in this area, were found in an even lower Early Iron Age layer (ca. 300 to 800) of this site along with pollen identified as flax, hop, and hemp.

Medieval archaeobotanical and paleoenvironmental evidence for *Cannabis*, dated to the ninth and tenth centuries, comes from an ancient Viking Age settlement, Staraya Ladoga. Founded in 753 on the southern shore of Lake Ladoga, it is considered by some to be the first capital of Russia, and in Old Norse sagas, it was known as *Aldeigjuborg*. It served as an ancient, wealthy trading outpost in the eighth and ninth centuries for people from many ethnic groups, largely Scandinavians, and remained one of the more significant ports in Eastern Europe until about 950, with merchant vessels using it to ship goods, such as hemp, to diverse areas (Aalto and Heinäjoki-Majander 1997; see also Heinäjoki and Aalto 1997; Alenius 2007; and Kroll 1999). Delusina (1991) investigated Lake Ladoga's Holocene pollen stratigraphy in an attempt to correlate local pollen zones with the neighboring Russian region of northern Karelskaya (Karelia) and also southern Finland. She found no Cannabaceae pollen, but noted great quantities of pondweed (*Potamogeton*), an indicator of hemp retting at many sites in England (Bradshaw et al. 1981; also see selected sites in Spain discussed earlier and in the British Isles discussed later). On the other hand, Miettinen et al. (2002) did find "*Cannabis*-type" pollen dating to the first millennium CE in Lake Pieni-Kuuppalanlampi west of Lake Ladoga, which they suggest appears "possibly as a result of retting of fiber hemp in the lake at that time" (see also Kaukonen 1946; Grönlund et al. 1986).

Hemp was a widely cultivated seed oil and fiber crop in other areas of Russia at least since the eleventh century; in fact, during the medieval period the two most important cultivated fiber plants in Russia were hemp and flax. According to Zajaczkowa (2002), flax was grown in the northern part of the country and hemp in the southern regions, although hemp is reported to have been cultivated in Russia near its "northern border of agriculture," approximately 66°N latitude (Grigoryev 2007). By medieval times hemp had become a relatively important plant in the Middle Pre-Ural region of Russia, stretching roughly from the East European Plain to west of the middle part of the Ural Mountain range. Analysis of archaeobotanical remains recovered from buried soils at the ancient settlement site of Idnakar and at neighboring monuments at Gur'yakar and Ves'yakar has been used to reconstruct the agroecosystems and cultivated plants utilized in this area from the ninth through the thirteenth centuries (Tuganaev and Tuganaev 2002). Idnakar is located near the present-day city of Glazov, in the Kama region east of Kirov in the Cheptsa Basin of the Udmurt Republic. The natural vegetation of this temperate to boreal area of continental Russia is coniferous forest, and people who inhabited Cheptsa Basin in the Middle Ages combined farming with hunting, gathering, and fishing (Ivanova 1998). Recent archaeobotanical data, dating from the ninth to the thirteenth centuries, indicates that wheat, rye, barley, oats, peas, lentils, flax, and *Cannabis sativa*" were among 16 species of cultivated plants grown. Samples included mixtures of seeds or other remains of different plants: "Only two of the samples of ancient seeds found in the Glazov site contained seeds of only one species, *Cannabis* in one case and barley in the other" (Tuganaev and Tuganaev 2002).

Working at two archeological sites in the far north of the West Siberian forest tundra, Korona (2006) discovered ancient seeds from a variety of cultivated and wild species, including hemp. At the Russian settlement site of Mangazeya dated to the seventeenth century, Korona recovered evidence of cultivated oats and barley and many species of "ruderal plants," mostly typical of nearby natural communities, including ball mustard (*Neslia paniculata*) and "*Cannabis sativa.*" Korona proposed that these weedy species were probably introduced along with cultivated grasses. Today *Cannabis* can be found growing wild along river banks and canals in western Siberia (Bielman 2008). It is possible that Russian immigrants first arriving in the eleventh and twelfth centuries brought *Cannabis* to this cold northern area.

The Muscovy Company, also known as the Russia Company, was English owned and operated over a large area of Russia from 1555 to 1649. This business took advantage of the discovery of a near Arctic water route via the White Sea to increase access to Russian resources, including hemp cordage and sailcloth needed in Britain for outfitting and refurbishing naval and commercial ships (Willan 1953, 1956). In 1710, Russian forces took control of the Baltic region, replacing Swedish rulers. The newly installed Russian government of Czar Peter the Great (Peter I) made a series of changes in the politics, economics, and social aspects of the Baltic region but kept up political contact, and especially trade, with Britain (e.g., see Weber 1968; Kirby 1998). Among these "improvements" were efforts to move large numbers of people to Archangelsk, located on the banks of the Dvina River near where it empties into the White Sea in the far north of European Russia. However, when the Czar ordered the trade center of Vologda, located east of St. Petersburg, to move to Archangelsk near the Arctic Circle, businessmen tried to dissuade him because of a number of problems, including the fact that three German merchants employed thousands of workers in Vologda to sort and prepare hemp fiber for foreign trade (Weber 1968). The exportation of hemp from Russia, principally from Baltic seaports, continued into the nineteenth century CE, by which time the Unites States had also become a major importer of Russian hemp.

By the early 1800s, American traders were shipping a number of products including tobacco and furs to Russia. In return, relatively large amounts of well-known Russian products such as iron, flax, and hemp were exported to North America. The lengthy process of cultivating, processing, and moving bulky hemp to ports was supported in Russia by a large amount of cheap labor provided by serfs of the Russian Empire (Crosby 1965).

Hemp ropes resisted stretching, and up until the Civil War, Russian hemp was an essential raw material for American shipbuilding, where it was used for laying the "bolt-rope and standing rigging" required on large sailing vessels; thus it was Russian hemp that "upheld the lofty spars of our [American] clipper ships and indeed all of our vessels until wire rigging was introduced in the [1860s]" (Morison 1921). An interesting example of a nineteenth-century American trader who imported large amounts hemp from Russia was William Ropes of Salem, Massachusetts. Ropes established a foreign office for his trading company at St. Petersburg in 1832 and, until his death in 1859, traded American products such as mineral oil, for Russian goods such as hemp rope. Because of his connection with ropes and oil, he was known as "*Ropeski kerosin*" (Kerosine Ropeman) throughout the Russian Empire, where he traveled thousands of miles into its vast interior to buy hemp rope and other products, accompanied by his head clerk and a large dog named Tiger, who helped protect him from outlaws (Morison 1921). Hemp cultivation in Russia remained important in some areas until modern times. For example, at least from the latter part of the nineteenth century through the first decade of the twentieth century in the central Russian agricultural area of Orel (Orlov), hemp was one of the principal crops with both hemp fiber and seed oil being "extensively exported . . . to Riga, Libau and St. Petersburg" (Kropotkin 1911).

Local cultivation, use, and export of *Cannabis*, mainly for fiber and to some degree for its seed, have a lengthy history in the Baltic region, probably beginning well before the medieval period. Both archeological and archaeobotanical research provide evidence for the early presence of *Cannabis* in the Baltic. According to Rimantiene and her colleagues (1999), the "earliest known undeniable evidence of *Cannabis* cultivation [in the Baltic] are remains from the Šventoji area in coastal Lithuania, dated to the Middle Neolithic, ca. 4000 BCE." This ancient evidence, in the form of a "hemp cord," was found in the Šventoji-23 site, which is associated with settlements of the Early Neolithic Narva culture (Rimantiene 1992b). Although the inhabitants of this ancient Stone Age site were mainly hunters, gatherers, and fishers, evidence indicates that they also farmed and raised domesticated animals. Similar to many other areas of the world, there is very little direct evidence for textile use in the Baltic, especially in prehistoric times; the oldest identified textile finds only date to the Roman Iron Age (50 and 450 CE) in Estonia, with much older Late Mesolithic evidence of common lime or bass tree bark fibers discovered in a number of northern European areas, including the bog at Narva Siiverts in Estonia (Indreko 1931), at Antrea Korpilahti in Finland (Pälsi 1920), and in the bogs of Nidlöse and Ordrup in Denmark (Becker 1941; Hald 1980). We suggest hemp could also have been an ancient Mesolithic fiber source along with wild nettle and bass, followed much later by its more widespread use when agriculture was introduced into the Baltic region.

Sillasoo (1995, 1997, 2002) reported archaeobotanical evidence for *Cannabis* from the fourteenth and fifteenth centuries. Salvage archeology during the early 1990s, carried out in a suburban area of medieval Tartu in southern Estonia, produced two pollen cores and a number of soil samples with micro and macrofossil evidence (see Hiie and Kihno 2008). Archaeobotanical data indicated that *Cannabis* was among the "few remains of cultivated plants" recovered from medieval layers of this Tartu site that also included buckwheat, barley, flax, and rye (Hiie et al. 2007).

More recently, Niinimets and Saarse (2007) reported their discovery of *"Cannabis"* pollen in their "high resolution" stratigraphical and lithological study of a sediment core almost 10 meters deep from Lake Lasva in southern Estonia. The first attempt at small-scale farming in this region occurred around 5100 BP near Lake Lasva, as detected in the initial recovery of wheat pollen. However, it is not until much later that evidence of *Cannabis,* along with barley, wheat, and rye, starts to show up in the palynological record and date these pollen grains to about 1600 to 1400 BP during the Middle Iron Age when these crops became the main means of subsistence in this area. By the end of the Roman Iron Age (about 1550 years BP), barley, wheat, and rye pollen, along with several other archaeobotanical and cultural clues, indicate that limited, swidden cultivation was taking place and that *"Cannabis* and *Linum* were cultivated, obviously to obtain fibre [sic] for ropes, clothes and nets" (Niinimets and Saarse 2007; also see Jaanits 1992).

Veski et al. (2005) recorded *"Cannabis*-type" pollen from Rõuge Tõugjärv, a small Estonian lake with "annually laminated lake sediments situated in a dense prehistoric setting." This pollen type begins to show up in association with barley, wheat, and rye in strata dated to about 1000 to 1200 CE and then becomes much more common, along with rye and abundant ruderal species, between 1350 and 1650 CE as part of a gradual clearing of the forest (as shown by a rise in recovered charcoal). After about 1350, *"Cannabis* pollen may, however, reflect local retting in the lake, as retting, which may be the source of hemp pollen, gives a bias to the pollen record, but confirms the cultivation of hemp in any case" (Veski et al. 2005). Hemp cultivation has dropped over the past 200 years due to the introduction of artificial fibers (Veski et al. 2005, citing Joosten 1985).

Expansion of farming began in the fertile soils of the Estonian Lasva Lake area about 800 years ago following the invasion of German crusaders and construction of the Kirumpaa stronghold. This agrarian expansion continued—with some gaps in time due to plague and/or climatic calamities—right up into historical times with thriving crops such as rye, wheat, barley, buckwheat, flax, and hemp, and the cultivation of hemp and rye in the area seem to have been replaced by wheat after about 1650 (Niinimets and Saarse 2007; also see Tarvel 1992).

In eastern Lithuania (not far from present-day Belarus), ancient seeds including millet, rye, flax, buckwheat, and *Cannabis* were collected from a wall of an archeological ditch in the Vilnius Royal Castle complex; these seeds date from about the fifth or sixth to twelfth centuries CE, but mostly from the seventh to ninth centuries (Kisielienė et al. 2007). Later, at the start of the fourteenth century, a period of intensive human activity in the area involved further clearing of vegetation cover accompanied by the increasing importance of agriculture revealed by large amounts of millet, rye, and *Cannabis* pollen (Mažeika et al. 2009). Intensive farming and settlement continued during the first half of the fifteenth century. The pollen complex, including *Cannabis,* recovered from the Vilnius Royal Castle area is more or less similar to that found in many other areas of temperate and boreal Eurasia, indicating a close association with human environmental alteration. As in other cases in medieval Europe, this largely anthropogenic palynological mix, along with similar

weed and ruderal pollen flora, is the result of intense farming activity, which regularly involved *Cannabis.*

As noted earlier, hemp was produced in the Baltic region into recent times and became very important as an international commodity. Goodrich (1856) referred to hemp as one of the "chief products" grown on fertile lands along with wheat, rye, oats, barley, flax, hops, tobacco, and fruits in a considerable portion of the Baltic region that was part of Prussia through the nineteenth century. During the great age of sailing, spanning the centuries leading up to the modern era, the eastern Baltic region was a major source of naval supplies, including Latvian pitch, pine, and hemp. On October 18, 2006, Queen Elizabeth II of the United Kingdom attended a state banquet during her visit to Riga, the capital of Latvia. At this feast honoring the role played by Latvia in the crucial Battle of Trafalgar (which occurred when "Latvia" did not yet exist), Queen Elizabeth acknowledged the crucial role that Latvian hemp played in this famous sea battle, remembering that "British ships at that time were waterproofed with pitch from Riga, rigged with ropes of hemp from Riga, and their masts were of pine from Riga." According to the German Press Agency (2006), "During the Napoleonic wars, the French attempted to bar Britain's Royal Navy from the Baltic. British admirals organised [sic] convoys of up to a thousand ships at a time to ensure the vital hemp got through. And at the battle of Trafalgar British warships rigged with Baltic hemp broke the power of the Franco-Spanish fleet. The victory is often viewed as the most decisive naval battle in history."

With the rise of long distance shipping and trade, from the seventeenth until the twentieth century, *Cannabis* played important multipurpose roles in many regions involving many products including its fibers to make cordage and canvas. It can be argued that hemp, at least in some areas, was a strategic commodity comparable in its heyday to petroleum in modern times. Crosby (1965) provides an in-depth discussion of the eighteenth and nineteenth century hemp trade from the Baltic region and Russia to America and Britain before and after the American Revolution (cf. Saul 1969).

Nearby Finland and Scandinavia also have deep historical records of hemp production lasting into the twentieth century (e.g., see Lempiäinen 1999a/b). Many elderly Finns still remember hemp fiber. However, hemp farming declined in Finland in the 1950s (Ahokas 2007), and hemp has not been cultivated since the 1960s (Callaway and Hemmilä 1996), dropping from more than 1,500 hectares in 1910 to fewer than 100 hectares in 1950 (Laitinen 1996).

Ahokas (2002) traced the cultivation of hemp by ancient Finnic peoples using linguistic, historical, and ethnological data, pointing out that Uralic languages, such as Finnish, indicate an old and important use of hemp fiber in clothing; for example, the etymological origins of *kangas,* a Finnic word for cloth is derived from Uralic words meaning hemp (for additional old Finnish names referring to hemp or hempen materials, see Ahokas 2003). The term *sini-sinn* is one of the native words for hemp used by the Saami, the indigenous "reindeer people" of northern Fennoscandia and the Kola Peninsula of northwestern Siberia. The origins of Saami people are still uncertain, but they may be some of the oldest postglacial people living in the northern regions of Fennoscandia and parts of northwestern Siberia. Originally hunters, fishers, and gatherers, many Saami became herders by the Middle Ages, and their pastoral relationship with domestic reindeer remains their most distinctive form of livelihood. Did they have an early ethnobotanical relationship with *Cannabis*? Ahokas (2007) suggested that the most ancient significance of *Cannabis* for the peoples of the "Finno-Ugric nations" may have been for bowstrings, since "the word for tendon, hemp and bowstring is identical or nearly so in some Finno-Ugric languages."

According to Ahokas (2007), during its earlier agricultural history, hemp was cultivated on a "small scale by almost all Finnish peasants, at least at intervals in the southern and middle part of the country, mainly for fiber and also grains and oil," and hemp fiber and cordage became especially important during the eighteenth century for use in the ship building industry with large boats being extensively caulked with tarred hemp oakum. Even though hemp was an important agricultural plant, Ahokas (2007) noted that hemp was considered by Finns to be the most repulsive and "bad-smelling of all the crops," and Finnish workers "tried to avoid even breathing the dust from hemp while handling it."

Laitinen (1996) pointed out that only recently has ancient hemp been identified in Finland and that the oldest known site of hemp cultivation is located on Ahvenanmaa (a large island between Finland and Sweden) where ancient hemp seeds were found in a habitation and burial complex near the present castle of Kastelholma and dated to the Viking Age (800 to 1050 CE). Laitinen also referred to *Cannabis* seeds found in medieval sites in southwestern Finland, dating from 1100 to 1500 CE, which suggests that hemp was cultivated locally, probably as a profitable fiber crop, rather than imported. According to Laitinen hemp farming, at least on a large scale, began around the fourteenth century, entering from Karelia, which lies to the east, adjacent to modern Finland. Even though Laitinen argued that hemp farming was introduced into eastern Finland from Russia, he points out that some ancient seed evidence suggests it could have also entered "southwest Finland through Central Europe, the Baltic or Scandinavian regions and not necessarily from Russia," and in any case, hemp farming in Finland "is older than that of flax."

An instructive example of probable hemp cultivation comes from Likolampi, a small groundwater kettle-hole lake near Ilomantsi in eastern Finland. Here pollen analysis revealed that the first deposition of sediment corresponds with the start of "intensive fiber plant retting with large quantities of *Cannabis/Humulus*-type pollen and even regular occurrence of the insect-pollinated *Linum usitatissimum*," which were dated to ca. 1590 to 1900 CE by the varves (annual layers of silt and clay; Grönlund et al. 1986). The retting association, as we note elsewhere, indicates probable cultivation of fiber hemp and not hop.

In the case of evidence from Lake Pönttölampi, pollen data, along with sedimentary microscopic charcoal particle records and dendrochronological (tree growth ring) data from the same annually laminated sediment of this small lake in a presently unpopulated part of eastern Finland, indicate that intensive slash-and-burn clearing and swidden cultivation of rye and hemp began there in the seventeenth century (Pitkänen and Huttunen 1999; Pitkänen et al. 1999). *Cannabis* pollen appears in the latter part of the nineteenth century. According to historical and paleoecological studies, swidden cultivation increased from the sixteenth century onward, and in the latter part of the eighteenth and early nineteenth centuries shifting cultivation was at its zenith, resulting in major alterations of forest structure and landscape near greater human populations.

Archaeobotanical evidence for *Cannabis* in Finland also comes in the form of oakum. For example, during excavations in central Finland early in 1990, a large quantity of boat remains and associated materials were discovered in a boggy meadow bordering a lake near Suojoki in Keuruu county and dated to the thirteenth century CE. These ship relics included planks, oars, and other artifacts, and cracks between the ancient planks were "caulked with tarred hemp," which "serves this purpose well, since it does not rot as quickly as flax" (Laitinen 1996). The people who built and used these ancient ships are believed to be among the early Finnish "fishers and hunters" inhabiting the interior wilderness. However, recent archaeobotanical research indicates that they farmed and gathered along with hunting and fishing (Laitinen 1996).

In spite of a relative dearth of early written records concerning hemp farming in Finland, and the problematic situation associated with Cannabaceae pollen, *Cannabis* may have been cultivated even earlier than previously suggested. For instance, Cannabaceae pollen curves from southern Finland published by Tolonen (1978) show *"Humulus"* pollen in the palynological record from ca. 3530 to 2639 BP and *"Cannabis"* pollen from ca. 2000 to 500 BP. According to Callaway and Hemmilä (1996), cultivation of hemp fiber and seed "has a very long tradition in Finnish agriculture" dating back thousands of years, perhaps to the Finnish Stone Age. Although *Cannabis* cultivation for fiber or other purposes remained important for centuries, because of economic and political conditions noted earlier, hemp has not been widely grown in Finland since the 1960s, following a major decline during the first half of the twentieth century (Laitinen 1996). As is the case in many other areas of Eurasia, more extensive and thorough paleoecological research may reshape the complex settlement and agriculture history of Finland and should help resolve the early history of hemp.

Fiber hemp also has a long history in Scandinavia supported by archaeobotanical and historical evidence. For example, hemp seed remains were found in some ancient sites in Sweden dating from as early as ca. 2000 BP (Påhlsson 1982), indicating early cultivation occurred locally for fiber and seed. Rasmussen (2005) argued that very large amounts of Cannabaceae pollen discovered at the lakes of Bjäresjösjön and Bussjösjön in southern Sweden present a clear sign that hemp retting took place in these lakes "from around CE 700," suggesting that Vikings or their ancestors may have introduced hemp into Scandinavia (also see Gaillard and Berglund 1988; Regnéll 1989; Björkman 1999).

There is a limited amount of palynological evidence for *Cannabis* in Norway in prehistoric times. High Cannabaceae pollen percentages reported by Hafsten (1956) from excavations in the inner Oslo Fjord once again could be seen as an indication that ancient hemp retting took place (Bradshaw et al. 1981). It is possible that the Vikings could have been importing foreign-grown hemp stalks and then retting them in their home territories; however, it is much more likely that the hemp was grown nearby.

According to Falk (1919), Old Norse sagas tell us that the traditional tunic (*kyrtill*) was always worn over a short garment (*skyrta*) with long sleeves made of flax (*hor*) or hemp (*hampi*), and the use of hemp fibers in ancient Norway is documented by hempen textiles that have been recovered from Viking archeological contexts along with fabrics made of wool, flax, and nettle fibers. For example, in the burial site at Birka, a garment including a trimming strip of beaver fur lined with hemp was found on the body of a woman laid to rest in an early ninth century CE grave, and a wool caftan lined with hemp cloth was found with a woman buried in the mid-tenth century (Geijer 1938. also see Hald 1980). In her discussion of cloth and clothing in medieval Europe, and more specifically her work with the ancient Birka cloth remains, Geijer pointed out that "linen" has been used as a "collective designation for plain fabrics in tabby weave where the yarn consists of either flax or hemp, closely related bast fibres [sic] which are, to some extent, customarily interchangeable the world over." Geijer also notes that similar to all types of plant fibers, flax and hemp decompose quickly when exposed to moist conditions. However, traces of flax or hemp fabrics in archeological contexts can occasionally be found in close proximity to "metal objects." After more or less complete disintegration of the fabric's fibers, some parts may form hardened remnants or be recognized in impressions on the metal's surface. Geijer used "the cryptic designation FH (Flachs Hanf, that is, flax hemp)" in her original classification of fabric finds in approximately 45 Birka graves. Thus we cannot be absolutely sure if any of the cloth remains recovered from the Birka graves were actually made of or included *Cannabis* fiber. On the other hand, Vindheim (2002) tells us that during medieval times in Norway, hemp fiber was used to make clothing normally worn "closest to the body, as underwear," as well as occasionally for finer textiles and sometimes "even for decorative weavings" that can be seen in the Nordenfjeldske Kunstindustrimuseum in Trondheim.

One of the oldest industries in Norway was the production of hemp maritime cordage at small factories built at a number of locations along the Norwegian coast. These rope-making factories became a typical feature of many towns such as Bergen where a "new rope walk" was installed in 1607 (Vindhiem 2002). In several areas of Norway, streets and public areas have names that refer to the former existence of their rope walks. Well into the twentieth century, the principle use of hemp was for rope needed by the Norwegian merchant navy and fishing fleet.

The first definite evidence for *Cannabis* fiber use in Norway is in the form of a hemp textile and seeds found in a grave dated to the Viking age over a thousand years ago. This discovery occurred in 1903, when a farmer uncovered a Viking ship while digging in a large burial mound on his farm, Lille Oseberg in the town of Slagen located along the Oslo Fjord in Vestfold, Norway. The ancient boat contained the bodies of two women buried around the year 850 CE, along with other items including a small piece of hemp cloth, perhaps for ritualistic or other uses that have not yet been determined (Vindheim 2002). Four *Cannabis* seeds were also found (Holmboe 1927).

Vindheim suggested that the hemp fabric found in the early Viking boat at Oseberg was most likely a sail fragment. He also argued that before the superior flexibility of the long *Cannabis* fiber was recognized and hemp became the principal material to make cordage, "ropes and lines were mainly produced from the bast of lime [basswood] trees." According to Vindheim, hemp cordage, with its superior qualities, eventually became a necessity for lengthy Viking sea journeys, and consequently hemp fiber became a significant trade item and source of bounty gathered in "their regular armed raids," with hemp principally used as "raw material for ropes and rough textiles." Vindheim assumed that hemp was cultivated in a number of areas in Norway by 1000 CE but also noted

that importation has always exceeded homegrown production (also see Prins 1975).

Vindheim also pointed out that textiles made of hemp were common in Norway by the Middle Ages and indicates that names for hemp in Norwegian such as "*hamp*" are commonly integrated into place names in many areas and occur in "Norwegian forms of speech like '*inn i hampen*' ('in the hemp field', meaning 'beyond credibility')." Hemp farmers were taxed by the kings of Norway from the thirteenth until at least the sixteenth century when government income records show that hemp was cultivated in Vestfold, where the Oseberg boat mound is located. Vindheim (2002) described traditional Norwegian hemp production and gives us a glimpse of the workload and gender responsibilities of this ancient European occupation:

> Traditionally the male plants have been harvested first, the females a little later. The plants were pulled up with the roots and dried in bundles. The seeds could then be shaken off before the plants were soaked in water for retting. When the soft parts of the plant have been dissolved, the stalks are dried once more and pulled bit by bit through the *hampebraak* [hemp brake]. This has a "mouth" that breaks the stalks, and makes it possible to remove the fibres [sic] from the wooden parts. This is heavy work that was usually done by the menfolk. The next step is pulling the fibres [sic] through the hackle, a kind of comb that separates the fibres [sic]. This was considered lighter work and was done by the women. The hackled fibres [sic] could be spun and then woven, or used for making cordage.

Although many Norwegian farms in recent centuries had a hemp field, greater amounts of hemp could be cultivated in inland valleys than along the coast because of more favorable climatic conditions, and trading must have been important because the largest demand for hemp was in coastal settlements where it was used to caulk hulls and make all kinds of lines, ropes, nets, and additional cordage materials needed on ships. As noted earlier, in more recent times, up into the early part of the twentieth century, the principle use of hemp "was for cordage to supply the Norwegian merchant navy as well as the fishing fleet" and the quantity required was massive; great lengths of various kinds of rope were needed for a single sailing ship, and a large proportion of this was made of hemp imported from Russia: "The pinnacle of Norwegian naval rope making was the anchor cable for the frigate *Kong Sverre*, made from four tons of Russian hemp. It was carried through the streets of Tønsberg, in the year 1864, by 120 sailors led by a marching band" (Vindheim 2002).

In addition to *Cannabis* seeds and a textile recovered from Oseberg, old sewing threads or possibly fishing lines made of hemp fiber have been discovered in other sites in Scandinavia. For example, hemp twine for sewing boats was recovered in Sweden from two northern sites at Norra Volmsjö in the parish of Fredrika and at Giltjaur (Jiltjer) in the parish of Sorsele, both in Lappland, and at Degerfors in the parish of Ekträsk in Västerbotten (Westerdahl 1985a/b). The "hemp thread" found in a "stone heap" in the forest at Norra Volmsjö is especially intriguing; this cordage was found with root fibers from an unidentified plant that were used to sew three strakes (horizontal strips of wooden planking) together. Most likely they are fragments of a small boat, like the one Linnaeus (Carl von Linné) sketched being carried by his Saami guide during his trip to Lappland in 1732. Interestingly, a small six-oared "Skolt Saami cod fishing boat" constructed

FIGURE 26. A huge, newly commissioned hempen anchor cable is carried through the streets of Tønsberg, Norway, in 1864, led by a marching band (adapted from Vindheim 2002).

well over a hundred years ago in the Varanger region near the Finnish border with Russia is built with "three strakes a side and running stitches of hemp twine, wedged with treenails, with hemp strings all over, even in the stem/stern and the joints at the keel" (Westerdahl 1895b).

This suggests that Viking and other early sailors in the region now known as Scandinavia were familiar with the strength, suppleness, and water-resistant qualities of hemp cordage and used it to stitch and caulk hulls and lash their boats together, as well as for rigging, anchor cables, and fishing lines and nets and used hemp cloth for at least some sails. Although wool was the fiber of choice for making sails in much of Scandinavia before historical times, unmistakable remains "of linen or hemp sails are known from the *Wasa* (*Vasa*), the 64-gun royal Swedish warship that sank right after leaving the shipyard at Stockholm in 1628" (Svensson 1965). According to Bengtsson (1975), the sails of the Vasa were "made from hemp . . . for both warp and weft." More recently, archeological marine research revealed that the sails of this ship were mainly fabricated from hemp and only in part from flax, and all the rigging was made completely of hemp imported from Latvia via Riga (Cederlund 2006; Möller-Wiering 2005). *Cannabis* fiber has also been identified in a number of similar old sail fragments in the Marine Museum at Karlskrona and the Statens Sjohistoriska Museum in Stockholm that date to the eighteenth and nineteenth centuries (Westheden Olausson 1988).

The long tradition of using hemp in sails and rigging of maritime vessels in Sweden was rekindled during recent construction of a replica of the *Götheborg*, a large Swedish commercial sailing vessel that ran aground in 1745 and sunk with all its valuable cargo at the entrance to its home port of *Göteborg*. More than two centuries later, archeological marine excavations to recover what remained of the ship were carried out for several years (1986 to 1992), and this provided the incentive for building a replica of the *Götheborg* that began in 1995. When it came to the production of ropes and

sails for the modern construction of this traditional eighteenth century ship, hemp once again played a major role, as it had for centuries throughout much of coastal Europe. Both hemp and flax are believed to have been used to make the sails for old Swedish East India Company vessels such as the *Götheborg*.

Large amounts of appropriate hemp cloth were difficult to obtain during the reconstruction, and consequently, sails of the replica ship were fashioned from semibleached flax linen. Nevertheless, enough hemp was obtained for the sailmakers to utilize hemp cordage for certain applications. Following eighteenth-century tradition, the sails were stitched by hand with waxed hemp cord and the reinforcements and ferrules were wrapped with tarred hemp cord; in addition, all the rope used in the standing and running rigging was laid from hemp twine imported from Hungary. In Sweden, under the expert supervision of Ole Magnus, a master Danish rope maker noted for his traditional knowledge, the twine was tarred before it was twisted into rope, which helps prevent decay and increase longevity. The recreated *Götheborg* was launched in June of 2003, with Swedish royalty and a large audience in attendance (Arensberg 2006).

In a rather remarkable study, Cook et al. (2002) began a program of reconstructive maritime archeology in Denmark to test the seaworthiness of traditional woolen sails and to compare them with those made with hemp and flax. Their study was carried out under the leadership of the Viking Ship Museum of Roskilde and was based on the recovery of parts of five ships that were scuttled in the Danish Roskilde Fjord during the second half of the eleventh century when threatened by seaborne attack. The sinking of the ships was designed to narrow and control the passage to Roskilde, and it apparently worked; however, once the peril had past, the sunken vessels vanished as they settled into the fjord bottom, only to be discovered in 1924 when fishermen found a keel of one of the boats. Recovery and archeological work started in 1957 and was completed by 1962. Outstanding preservation kept the ancient boats essentially intact, which inspired an innovative and instructive rebuilding effort: "The entire process of shipbuilding down to the placing of the woolen luting [the substance used to seal a porous surface] and the insertion of each and every treenail set in train the long process which was to lead to the reconstruction of four of the five ships and the immensely rewarding process of learning how to equip, rig, trim and sail these ancient ships" (Crumlin-Pedersen and Olsen 2002).

Among the more revealing aspects of experimental archeology and the reconstruction process involving the Roskilde boats was quantitative testing and analysis of the yarns and fabrics of the replica sail of one of the Roskilde boats, the *Skuldelev 1*, which yielded insights concerning the utility and seaworthiness of different types of sail fabric, including hemp: "The experience of the Vikings showed that woolen sails needed to be much thicker and heavier than hemp sails," and in sum, "the critical design parameters for hemp sails are adequate tensile strength to resist tearing together with adequate cover. . . . Unfortunately such wool fabrics in their loom state have a major design fault: they are too permeable to airflow to function as efficient sails" (Cook et al. 2002). However, the study of verified archeological remains indicated that most ancient Viking square sails were made of wool, which could, if treated properly, beat "into the wind and most likely outperform linen and hemp sails" (Cook et al. 2002).

Hemp and flax are regarded as the oldest industrial crops in Denmark (Rexen and Blicher-Mathiesen 1998), with flax cultivated by the Iron Age period (ca. 400 BCE to 800 CE) and hemp at least by the Middle Ages (ca. 800 to 1400 CE). A traditional belief of Danish sailors refers to hemp cultivation being initially brought to Denmark from the Baltic area during medieval times (Brøndegaard 1979). In fact, Godwin (1967b) referred to *Cannabis* seeds found in Viking settlements in Denmark dating back at least to the medieval period, thus providing substantial evidence for the ancient importance of hemp cordage and cloth.

Archaeobotanical research involving aquatic plant development in Dallund Sø, a shallow Danish lake on Funen Island, tells us that *Cannabis* was already grown for fiber in Denmark much earlier, with evidence that "suggests intensive retting of hemp plants" (e.g., Rasmussen and Anderson 2005; also see Chapter 4).

Noting that one pollen grain alone documents cultivation of flax at the end of the Late Iron Age in the area of his study at Dallund Sø, Rasmussen (2005) also suggested that elevated percentages of *Cannabis* pollen indicate use of the lake for hemp retting. *Cannabis* seeds are generally the strongest evidence for retting (e.g., French and Moore 1986); however, the sediments from Dallund Sø did not yield any hemp seeds. Nevertheless, Rasmussen pointed to the scattered finds of flax seeds as evidence of the use of the lake for retting of fiber plants during the period around 700 CE and suggested that the absence of hemp seeds could result from "utilization of male (pollen) plants rather than female (achene [seed]) plants." Rasmussen also noted that during historic times "male *Cannabis* plants were often preferred for fibre [*sic*] production as they are considered to produce better quality fibres [*sic*] than the female ones" (e.g., see Hegi 1957; Brøndegaard 1987; Edwards and Whittington 1992). Almost all European and East Asian hemp fiber crops are harvested before the seeds are ripe, because as the seeds ripen the fibers become more brittle, and in rare situations where mature seed plants are used for fiber, the valuable seeds are first removed. Therefore, few if any seeds are released into retting ponds.

According to Rasmussen (2005), during the early part of the medieval period at Dallund Sø, a huge growth in arable farming occurred in association with major deforestation: "Extensive *Linum*-retting in the Medieval Period is documented by the macrofossil record and high values of *Cannabis*-type pollen in the sediments also suggest intensive retting of hemp plants." Evidence of hemp retting is also demonstrated by high diatom concentrations between ca. 380 and 530 CE, which is indicative of a major eutrophication of the lake. Even without clear confirmation from "other aquatic proxy data to support this conclusion, a likely cause of increased nutrient loading is the retting of hemp in the lake, indicated by finds of *Cannabis*-type pollen in the sediment in this period," and from about 1100 CE onward, high levels of *Cannabis*-type pollen strongly suggest that the lake at Dallund Sø continued to be used intensively to ret hemp (Bradshaw et al. 2005).

Yeloff et al. (2007) studied late Holocene vegetation and land use history in Denmark. Based on ancient pollen recovered from Lille Vildmose in northeast Jutland, they argued that removal of native forest in the medieval period increased the occurrence of typical weedy species. This suggested that the remaining "open areas had a far greater cover of cereal crops" indicated by a large percentage of rye as well

as "*Humulus/Cannabis* pollen (probably from hemp) which also become evident about 1140 CE." Forest clearance accelerated again after about 1540 CE, associated with another increase in arable agriculture represented by pollen of *Humulus/Cannabis*, Brassicaceae (mustard or cabbage family), and Caryophyllaceae (carnation family). Although rye was cultivated in Denmark from about the fourth century (Henriksen 2003), "the diversification of farming practices during [the medieval] period is shown by the cultivation of other field crops in the Lille Vildmose area, including hemp (*Cannabis*) after ca. 1140 CE." However, the cultivation of cereals, as well as hemp for fiber and perhaps seed, probably dropped off some time during the later Middle Ages about 1360 to 1540 CE due to the arrival of the "Black Death" plague, which "reached a peak in Denmark during the summer and autumn of 1350, and may have been responsible for the loss of 50 percent of the population" (Vahtola 2003). This precipitated the abandonment of farmland in the Lille Vildmose area, which in turn resulted in the regeneration of woodland and a significant drop in agricultural production (also see Yeloff and van Geel 2007). During the postmedieval period, after about 1540, deforestation occurred again in the vicinity of Lille Vildmose, linked to intensification of farming due to in part to technological innovations, and the reopened areas were mainly used for farming rye and hemp (Yeloff et al. 2007).

Since rope production was a necessity in all places where ships and boats were constructed or repaired, by the seventeenth century, Danish-Norwegian kings ordered the organization and taxation of the rope making industry. Within the Danish region of the old Danish-Norwegian single unified state or realm (1380 to 1814), the conditions for growing hemp were much better than those in what is today Norway. As a consequence, in 1629, King Christian IV ordered Danish farmers to supply his naval forces with cultivated hemp, and sheriffs distributed the necessary hemp seed. In the Danish Law of 1683, King Christian V formalized the obligation of growing hemp: "Every farmer who holds a full farm, and does not sow a bushel of hemp seed, and he, who holds half a farm, half a bushel, should by his lord be charged and punished as an obstinate and reluctant servant, unless he proves that he has no suitable soil therefore" (Vindheim 2002).

Although the equivalent Norwegian law at the time omitted the passage on hemp growing, the king's officials probably encouraged *Cannabis* fiber production in those parts of Norway where it was feasible. Indeed there was cause for concern because huge amounts of hemp fiber were being imported during these times, primarily from the Baltic region as alluded to in a familiar poem written in 1702 by Norwegian priest and poet Petter Dass (lived 1647 to 1707):

To Revel and Riga your Voyage did go
Your Hemp and your Flax to acquire.
(Quoted in VINDHEIM 2002)

In Denmark, most of the hemp fibers were traditionally used by the Danish navy or for clothing production, and by the eighteenth century hemp was being cultivated on "approximately 3500 hectares [8650 acres]" (Rexen and Blicher-Mathiesen 1998).

As in many countries, there was a renewal of interest in hemp cultivation in Denmark during World Wars I and II; in 1942, fiber hemp was grown on about 2,000 hectares (about 5,000 acres) in this country. However, following World War II, the Danish government banned hemp production because psychoactive resin is produced by some forms of *Cannabis*. Subsequently, in April 1988, the government announced that hemp cultivation for industrial purposes could resume but only using hemp fiber and seed cultivars that produce low levels of psychoactive resin as certified by the European Union; however, because of their northern latitude, most European Union–approved hemp fiber cultivars ripen quite late as the weather turns cold, thus limiting seed reproduction. On the other hand, early maturing seed cultivars such as 'Finola' produce high seed yields at northern latitudes, while fiber yields appear to be in the same range as other northern European countries. In 1998, the Danish Institute of Agricultural Sciences located at Flakkebjerg, Rønhave, and Grønt Center in Holeby planted about 60 hectares (150 acres) of EU-approved French, German, and Hungarian varieties in field trials. Industrial hemp is well thought of as "an environmentally friendly crop" with little need of artificial fertilizers and pesticides and therefore has potential as a raw material for production of several nonfood commodities with a "green label" certification (Rexen and Blicher-Mathiesen 1998).

Following the spread of *Cannabis* into Northern Europe including Scandinavia, the plant and its useful products were transported to remote Iceland during the Middle Ages. According to Godwin (1967b), there is no solid historical or etymological evidence that hemp was known to Iceland until quite late in medieval times, first noted in the *Kornungs Skuggsja* written in ca. 1240 CE. However, Abel (1980, citing Godwin 1967b) tells us that ropes made with true hemp fiber were discovered in Iceland among artifacts that date back to the early Middle Ages: "These ropes were carried there by the intrepid Vikings, for whom strong rope often meant the difference between survival or disaster in the vast uncharted Atlantic."

Cannabis also has a lengthy history as an important crop in the British Isles, where it had become significant by at least 800 CE. Its fiber had long been used to make cordage, clothes, sails and fishing nets, and its seeds provided vegetal oil (Gearey et al. 2005), and evidence of hemp fiber and seed use in Britain reportedly dates back further to the Roman period. Godwin (1967b), for example, pointed out that various pieces of "hemp" rope were found in a Roman fort well in Dunbartonshire, Scotland, and dated to ca. 2140 to 2180 BP.

Although evidence for the presence of *Cannabis* in the British Isles before the Roman period is not yet confirmed, the earliest use and possibly even cultivation of *Cannabis* in the archipelago may have begun even earlier in the Bronze Age. This is supported by the "probable" fibers from hemp "*Cannabis sativa*" string and cloth fragments found along with a hoard of metalwork in a Bronze Age site in St. Andrews, Scotland, that date to about 2800 BP (Ryder 1999). Identification of the fiber content of these ancient strings was based on measurements of fiber diameter. In this case, comparisons of the fiber diameters of the fabric and twine fragments were made with modern samples of flax, which differ significantly, while comparisons with those of modern hemp fiber diameters were much more similar. This led to the proposition that the first use of hemp as a textile fiber in Britain occurred a good deal earlier than previously believed (Ryder 1999; also see Ryder 1993).

The problematic identification of *Cannabis* vis-à-vis *Humulus* pollen, as the reader knows, has been raised often in our survey of the origin and spread of *Cannabis* through the ages. In the British Isles this issue has been investigated in detail

in a number of studies, reminding us again that most reliable evidence for the early presence of *Cannabis* in any area is still its seeds (e.g., see Wodehouse 1935; Erdtman 1943; Walker 1955; Gearey et al. 2005). Nevertheless, as noted in a number of cases referred to previously in this chapter, as well as in Chapter 4, we can argue effectively that *Cannabis* was present when there is strong evidence for ancient fiber retting, as manifested by one or more proxy forms (e.g., see Dumayne-Peaty 1999; Cox et al. 2000; Schofield and Waller 2005). Some British archaeobotanists argue that evidence from fossil insect remains (e.g., indicating water pollution), as well as other ancient data, can enhance the precise identification of ancient *Cannabis* retting (Gearey et al. 2005). The use of evidence demonstrating the retting of hemp is discussed in detail in the methodology section of Chapter 4. It will suffice to remember here that the process of water retting produces odorous, fetid, decomposed products that are unpleasant and can contaminate local water supplies. Consequently, retting is usually removed from human settlement areas.

An example of evidence of hemp retting can be found in a study by Dumayne-Peaty (1999) that provided a pollen diagram from Dogden Moss in Berwickshire. This research presented the initial, high-resolution record of anthropogenic impact on plant life in southeastern Scotland from the Bronze Age to modern times. Based on corroborative radiocarbon dating and historical records, the first extensive clearance of woodland in this region appears to have occurred in the Iron Age. After this deforestation, involving almost total removal of the forest, renewed clearance associated with monastic settlement and agriculture occurred in early historical times. Dumayne-Peaty (1999) referred to "Anglican advances in the 7th century CE," when British authority in control of southeastern Scotland was reduced dramatically. Such conditions in this region during the premedieval period (700 to 1100 CE) led to greater political and social stability, which in turn stimulated settlement and increased agriculture including the appearance of Cannabaceae in the pollen diagram; this is strong evidence of ancient hemp cultivation and hemp retting, which may in part explain why woodland clearance took place during the fifth through tenth centuries CE.

Historical records show that the center of the traditional hemp industry in England was located in the southeastern region (Fleming and Clarke 1998). For instance, Klinglehöfer (1991) described the significant use of *Cannabis* fiber during the premedieval period at Micheldever Hundred, Hampshire, England, for production of canvas sails and ropes used in land and water transport. He pointed out that during this period, exchange of produce and craft goods increased from local to regional importance, and in some areas of England, hemp fiber production expanded because it was more lucrative to raise fiber than food crops. In a study of hemp as "a forgotten Norfolk crop," Pursehouse (1961) pointed out that until relatively modern times about 15 percent of all cultivated areas in the Waveney Valley located on the border between Norfolk and Suffolk counties were customarily devoted to hemp.

Peglar (1993a/b) studied the archaeobotany of two nearby lakes, Quidenham Mere and Diss Mere, in Norfolk County from which she was able to make a comparison between human impacts over time on surrounding landscapes. Peglar concluded that pollen records, showing very high values of Cannabaceae occurring in the catchments of these two lakes from premedieval to modern times, "are indicative of retting of hemp during the past 1500 years." Cox et al. (2000) also described evidence for early medieval hemp-retting pools at Glasson Moss in Cumbria in extreme northwestern England (also see Wimble et al. 2000).

Recently, a team of British scientists used evidence from environmental archeology to reveal the broad geographical range of hemp fiber processing during the mid- to later medieval period (ca. fourteenth to seventeenth centuries) at three different sites in eastern England; these places include a monastic environment at Ellerton Priory in East Riding, Yorkshire; an urban site at Morton Lane in Beverley, also located in old East Riding; and Askham Bog, a rural peat land area south of York (Gearey et al. 2005). In all three of these medieval sites, there is evidence of hemp retting that took place comparatively close to the hemp fields. This use of water bodies close to agricultural plots occurred in part because the fiber removal process produces a foul smell and can pollute the aquatic environment but also because this decreases the quantity of material transported to other places for further processing by eliminating unwanted, nonfibrous materials in the stems. After retting, the lengthy bast fibers of hemp can then be bundled and transported without the waste parts to village areas for the next steps in processing.

Gearey et al. (2005) used a variety of methods to date the three sites referred to earlier, including ceramic artifacts found at Morton Lane; these dating techniques indicate that hemp retting began to take place in the fourteenth to fifteenth centuries and that "Cannabis-type pollen" at Askham Bog shows "hemp was certainly being grown locally from the earlier Medieval Period, with the dates on the hemp achenes [seeds] themselves suggesting retting in pools on the mire surface continued into the 17th century." The dating of the Priory at Ellerton is less certain, but most likely hemp was retted prior to surrender of the monastery during the sixteenth century. In sum, environmental archeology can play an important role in the study of medieval economy, even as it relates to hemp fiber production, especially in tandem with additional forms of evidence. However, Gearey et al. (2005) concluded their report with a cautionary note:

> The presence of a few hemp seeds in apparently waterlaid deposits should not be used alone as evidence that retting was carried out—complementary evidence, usually in the form of [fiber] or large concentrations of pollen are needed for confirmation. Hemp seeds are very frequently recorded in small numbers from archeological occupation deposits of Roman to post-Medieval date doubtless having been used for food (perhaps mainly for animals) and perhaps representing part of the background scatter of biological remains which had been blown, trampled, washed in, or dumped with redeposited sediment. Secondly, every large shallow feature cannot be suspected of being a retting pit. A pond-like cut on the fringes of post-Medieval Doncaster, South Yorkshire, for example, was—in view of its form and location in an industrial area—a prime candidate for identification as a retting pit, but bioarcheological analysis produced no more evidence for hemp than is generally found in urban occupation deposits with good preservation by anoxic waterlogging.

Schofield and Waller (2005) have interpreted very high amounts of "Cannabis sativa" pollen dated to the Middle Ages (ca. 1000 to 1400 CE) found in the natural cavity of Muddymore Pit, near the seaside in the Dungeness Foreland of southeast England, as evidence of ancient hemp retting at this coastal site. These archaeobotanists point out that this old retting site is located near the former settlement and harbor of Lydd, Kent, which was a flourishing port town during the medieval period, and suggest that it played an important role in filling the demand from the port for hempen fiber products "such as rope . . . and cloth."

In another recent study, Dark (2005) used pollen, charcoal, and sediment analyses of a radiocarbon-dated, medieval to postmedieval sequence from Crag Lough near Hadrian's Wall in northern England to reconstruct changes in vegetation and land use over time. Dark suggested that Cannabis was cultivated locally and probably retted in the waters of Crag Lough throughout the medieval and early postmedieval periods.

French and Moore (1986) studied Llyn Mire in mid-Wales, a small (15 hectares, or 37 acres) mire complex located in a natural hollow on the eastern side of the Wye Valley close to the village of Newbridge-on-Wye. French and Moore used the pollen density of "Cannabis/Humulus" as a marker horizon in their research but realized the need to separate these pollen taxa adequately. The large quantity of pollen grains of the Cannabis/Humulus type that they recovered from the mire sediment in the floating peat bog (Schwingmoor) formed at Llyn lake allowed them to employ "numerical methods based on pore protrusion" (see the discussion of problematic differentiation of Cannabis and Humulus pollen in the methodology section in Chapter 4). French and Moore came to the conclusion that before the Schwingmoor was formed at Llyn Mire, "Cannabis sativa was cultivated around Llyn Lake" during Tudor times when hemp farming was relatively widespread.

In their paleoecological-based study of human impact at Abbeyknockmoy Bog in County Galway, Ireland, Lomas-Clarke and Barber (2004) excavated peat deposits and utilized pollen fossils to reconstruct land-use change surrounding the bog over the last two millennia. Palynological evidence indicates that during the Middle Ages continued woodland clearance was accompanied by an increase in pastoral indicators, especially grasses, plantain, and to a lesser extent sorrel and dock species, along with an increase in ragweed, followed by a slightly later rise in arable indicators with cereal-type pollen being accompanied by rye and occurrences of Cannabis-type pollen. Cereals appear to be the dominant cultivars, although hemp and flax (Linum bienne type) pollen also appear; Cannabis sativa pollen starts to show up in sediment deposits dated to the twelfth century and continues to be present until modern times.

Although Cannabis use in the British Isles may have begun well before 2000 BP, hemp was certainly utilized for fiber and/or seed during the Roman period. And later, during the Middle Ages, cultivation and use of hemp fiber became more important in this region, with a peak occurring in the Tudor period under King Henry VIII (lived 1491 to 1547). Several types of written evidence are also available to help us identify the locations and processing activities that were associated with hemp farming in the British Isles during the medieval period. These include "parish records and government reports, place-name evidence (e.g. Hempholme and some instances of Hempstead), and features on old maps, such as Hempisfeld (hemp field)" (Gearey et al. 2005, citing Gelling 1984; Ekwall 1960; Payne 1838).

Hemp was also used in medieval English weapons. Longbow strings were made from twisted hemp and flax fibers (Zygulski 1999; see also Mann 1957 and Reid 1976 and compare with the use of hemp cordage on ancient Chinese

FIGURE 27. Before the "Age of Steam," both commercial and naval vessels relied entirely on hemp rigging. Sea battles were often fought over control of trade in strategic shipments carrying raw hemp fiber and cordage. The Battle of Trafalgar fought in 1805 between the British Royal Navy and the combined fleets of the French and Spanish Navies proved to be the most decisive British naval victory of the Napoleonic Wars. Engraving from *The Illustrated London News*, October 24, 1874.

bows). Longbows designed and refined in the workshops of English town guilds were made from a variety of woods and reached lengths of about 1.8 meters (6 feet); these bows were used to launch a series of arrow barrages at eye level. Needless to say, the force generated by bending the bow determined the range and power of projectiles, and thus the strength, lack of stretch, and durability of hemp fiber would have made these bows more effective. Such weapons played a crucial role in the success of English foot soldiers fighting against French cavalry in the Hundred Years' War, a crucial series of battles that took place over a period of 116 years from 1337 to 1453 and is frequently regarded as one the more important series of events in medieval history. From a military standpoint, this period of armed conflict witnessed the introduction of innovative weapons and strategy that undermined the traditional system of feudal armies dominated by heavily armored, mounted troops. Powerful new longbows, strung with hemp or flax fibers, allowed effective deployment of standing armies in Western Europe for the first time since the era of the Western Roman Empire. A number of these old bows are preserved in British museums, and identification of fibers used to make the ancient strings may further corroborate the medieval military use of hemp in Western Europe.

Historical records indicate that cultivation of *Cannabis* in Britain reached its peak during the early sixteenth century in response to the seafaring needs of the English military. England's maritime supremacy during Elizabethan times greatly increased the demand for hemp fiber, and in Tudor times (1509 to 1603) English countryside farmers were required to plant a certain percentage of their arable land with *Cannabis*, as was required in parts of Scandinavia. For example, King Henry VIII of England strongly encouraged hemp cultivation in the early sixteenth century, particularly for its use by the expanding English navy (e.g., see Godwin 1967b). Well before their momentous naval clash with the

Spanish armada in the English Channel in 1588, the military leaders of England were acutely aware of the importance of *Cannabis* fiber, clearly recognizing the need to increase and then maintain a supply of hemp. In the beginning they tried to force citizens to grow hemp locally, as manifested in the first royal decree issued by King Henry VIII in 1533, which ordered each farmer to set aside a quarter acre of land for every sixty acres he controlled to cultivate hemp. The fine for failing to comply was set at "three shillings and four pence" (Abel 1980); 20 shillings equaled an English pound in Tudor times when a laborer earned only about 5 to 10 pounds a year.

William Bulleyn (lived 1500 to 1576) was a farsighted sixteenth-century physician and relative of Anne Boleyn (the second wife of King Henry VIII) who praised the multipurpose importance of hemp fibber in Elizabethan England: "No Shippe can sayle without hempe . . . no Plowe, or Carte, can be without ropes . . . the fisher and fouler muste have hempe, to make their nettes. And no archer can wante his bowe string: and the Malt man for his sackes, with it the belle is rong, to service in the Church" (quoted in Rudgley 1998).

The "Bard of Avon" himself, William Shakespeare (lived 1564 to 1616), referred to the common presence of hemp rope on English ships in the lines of the chorus in act III of his play, *The Life of King Henry the Fifth*, first published in 1599:

Thus with imagin'd wing our swift scene flies
In motion of no less celerity
Than that of thought. Suppose that you have seen
The well-appointed king of Hampton pier
Embark his royalty; and his brave fleet
With silken streamers the young Phoebus fanning;
Play with your fancies, and in them behold
Upon the *hempen tackle* ship-boys climbing
Hear the shrill whistle which doth order give

To sounds confus'd; behold the threaden sails,
Borne with the invisible and creeping wind,
Draw the huge bottoms through the furrow'd sea,
Breasting the lofty surge.
   (Italics added for emphasis)

Hartley (1979) pointed out that a single sailing ship in the time of Henry VIII needed more than a mile (1.6 kilometers) of rigging cordage, and although twisted straw rope could be used on land to secure stacks; pack carts; and make chairs, baskets, and beds, "strong hemp rope was required for binding bales for export goods, and often merchant adventurers and traders were better supplied than the Navy."

In 1563, 30 years after King Henry VIII imposed the stiff fine for those who did not cultivate hemp, his daughter, Queen Elizabeth I, used the Letters Patent to license agents who enforced the planting of hemp and later reissued the rule, raising the penalty to five shillings (Lipson 1931). However, she might not have had only the best interests of the realm in mind, since most of the hemp seed in England was sold by a Lawrence Cockson, a man who enjoyed the queen's favor, and if English farmers complied with the royal proclamation, Cockson stood to make a lot of money (Youngs 1976). Despite the law, few Englishmen respected the royal decrees, resulting from the simple fact that by this time any landowner or small farmer could make more money by raising almost any crop other than hemp. Not only were the prices they received for hemp fiber too low for them to make a good profit, but some farmers also complained that hemp exhausted the soil and made it unsuitable for growing other crops. Furthermore, in response to the foul odor associated with hemp and flax retting, legislation was introduced in England in 1541 to control the nuisance caused by this method of extracting *Cannabis* fibers, which further limited the regions where hemp could be grown: "It shall not be lawful . . . to water any manner of hemp or flax in any river, running water, stream, brook, or other common pond, where beasts be used to be watered, but only on the grounds or pits for the same ordained . . . or else in other their own several ponds" (Cox et al. 2000). Later in the sixteenth century, Thomas Tusser noted in his agricultural treatise *Five Hundred Points of Good Husbandry* (1984, originally published in 1573) that hemp retting, "commonly done in standing Plashes, or small pools . . . leaves a loathsome Smell in the Water." Tusser also described in verse what to avoid when gathering and retting winter hemp:

Now pluck up thy hemp, and go beat out the seed,
And afterward water it, as ye see need;
But not in the river, where cattle should drink,
For poisoning them, and the people with stinke.

In William Mavor's 1812 edition of Tusser's sixteenth century classic, the reader is reminded that hemp retting will produce an offensive odor if not done carefully, and rotting hemp should be removed from the water "as soon as it begins to swim," since the "smell left by hemp is extremely noisome." In addition, Tusser pointed out that hemp (and flax) may be retted either in water or by dew. However, the latter method of open air retting would leave little if any trace in the fossil record. Instructions for the cultivation, harvesting, and processing of hemp in England during the early seventeenth century can be found in *The English Housewife, Containing the*

Inward and Outward Virtues which Ought to Be in a Complete Woman, written by Gervase Markham and originally published in 1615. Markham also tells us that hemp can be retted by two methods, either by immersing the stalks in ponds or streams (water retting) or by simply laying the stalks in the fields after the harvest (dew retting). Markham indicates that retting is finished when "the fiber bundles appear white, separate from the woody core, and divide easily into individual, finer fibers for their full length." Subsequent to this the stalks are dried and then broken apart to free the fibers using a "breaker." The fibers are then separated through a "scotching" (scutching) or beating process to remove bits of wood and eventually "hanckled" (hackled or combed) into individual fiber elements that can be spun.

It should be pointed out that hemp clothing was not the most desirable garb in England, as in some other parts of Europe, at least as a fashion statement of social status. In fact, William Shakespeare poked fun at low-class pretenders to higher rank when he has Puck mock the arrogance of rustic, strutting intruders wearing homemade hemp clothes in his comedy *A Midsummer Night's Dream* (Act III. Scene 1, line 65): "What hempen homespuns have we swaggering here, so near the cradle of the fairy queen?" Interestingly, in 1843, Felix Mendelssohn was inspired by this Shakespearean reference to compose "What Hempen Homespuns," a sprightly, humorous instrumental musical composition, as part of his *A Midsummer Night's Dream* (Incidental Music, Op. 61).

The importance of the British hemp crop declined by the early nineteenth century in response to importation of hemp from Russia and the Baltic Region (e.g., see Crosby 1965). Mavor (1812) described the waning importance of hemp cultivation in England during this time, as well as underscoring the danger of relying too much on importation of strategic materials such as fiber hemp: "It is evident that hemp was formerly cultivated here to a great extent. The neglect of this valuable plant is one of the misfortunes arising from a dependence on foreign trade, which war and other casualties may interrupt. Our soil in many places is excellently adapted for the culture of hemp; and in fact, we possess the means within ourselves of raising or manufacturing almost everything necessary for our domestic wants, or public defense."

In spite of the decline in local hemp cultivation in England during the nineteenth century, hemp fiber was still vital for ships ropes and sails as well as for many other purposes. European sailing and steam fleets, like many others throughout history, continued to rely heavily on strong, durable, water-resistant hemp well into the twentieth century.

In Europe, hemp is still cultivated in relatively limited amounts as a source of raw fiber for textile and paper production ranging from Spain and France through much of Eastern Europe and the Baltic region. Contemporary fashion preferences for ecologically sustainable products have stimulated a limited revival of hemp and ramie use in clothing manufacture in a few regions of the world including parts of Europe, North America, China, and elsewhere. However, for many centuries before the modern period, flax and hemp were widely used to make household textiles, at least until cotton became popular, originally unadulterated and more recently mixed together with synthetic fibers. Given the contemporary pace of increasing archaeobotanical research, we can be reasonably sure that future discoveries of additional ancient evidence will allow us to reassess the earliest uses and ancient geographic distributions of hemp cultivation for fiber and other purposes in Europe and elsewhere in Eurasia.

Spread of Hemp Fiber Use to the New World

Well before either England or France envisioned massive hemp fiber imports from their New World colonies, the Spanish government was encouraging hemp cultivation in the Western Hemisphere. According to Mosk (1939), the *Recopilación de las leyes de las Indias* or "Compilation of the Laws of the Indies" (León Pinelo 1756) indicates that by 1545, Spanish officials in their New World colonies were given orders to encourage the cultivation of hemp and flax. Hemp fiber was especially important to the Spanish Crown, so much so "that in many parts of the Spanish New World, orders were imposed requiring that farms produce their quota of hemp to be delivered to the Viceroy" (Forster 1996). It was also in 1545 that hemp cultivation began in Chile, north of Santiago in the Quillota Valley, where it was continuously grown for the following 400 years. Most locally produced Chilean hemp fiber was used domestically for army and naval needs or shipped to the colonial capital in Lima, Peru, and even though the Spanish government also encouraged hemp cultivation in Peru and Colombia, only the Chilean trials proved successful in South America (Mosk 1939; also see Husbands 1909; Vasquez De Espinosa 1960).

In areas in Chile with favorable climatic, soil, and water resources, hemp cultivation has remained viable and active up until the present time. Hemp is a "dual-purpose crop" produced for both fiber and seed, and the sale of seed brings the farmer the funds necessary to raise the fiber crop, which in turn ultimately provides profit. In earlier times, hemp was cultivated almost completely in Aconcagua province, north and west of Santiago, but after World War II, the geographical area of hemp farming shifted southward to small farms in central Chile, where it was always grown with the use of irrigation (Kirby 1963). Although farms with hemp could be found to some degree throughout much of Chile, eventually the industry associated with fiber production became "concentrated in a few central areas with ideal climatic factors and nearby processing factories" (Forster 1996). According to Kirby, there was only one commercially cultivated variety of hemp, simply known as 'Chilean Hemp'. Developed from an Italian cultivar it produces roughly the same weight of seed as fiber but has a shortcoming in that dead male plants with overmatured fibers were harvested along with females. Nevertheless, hemp fiber has been a traditionally important resource in Chile: "Hemp was not only for military and strategic uses, but was an extremely important item in everyday farm life, being used to make sacks, shoes, canvas, clothes, rope, cable, twine, bow strings, lassos, saddles, animal feed and seed oil. To this day, many a Chilean farmer still gets a gleam in his eye when he sees a good quality hempen sack" (Forster 1996).

In Chile, a hemp sack or rope is valued more than plastic because it is more durable. On the other hand, because hemp is more costly, synthetic alternatives for traditional hemp products are more common today, and this has had a strong effect on hemp fiber production. In addition, as international markets for hemp shrank and exports declined, hemp farmers focused sale of their crops to markets within Chile but were furthered affected by the passing of stricter environmental laws controlling where water retting of hemp stalks can take place. Forster describes the traditional *Cannabis* retting process in Chile with its environmental problems and legal constraints:

> The stalks are stacked in retting ponds and covered with stones. Then the ponds are filled with water to immerse the stalks. Retting ponds are usually about one to two meters deep, and as long and wide as needed for the quantity of stalks processed at any one time. Stalks remain in this retting pond for a period of 8 to 10 days in the summer, and 12 to 18 days during the winter, during which they complete the retting process. As the stalks begin to ferment, the water becomes putrefied. If it was to be directly released into rivers and lakes, it could cause injury to the animals and people who drink water from these sources. Chilean Law # 3133 prohibits the retting of hemp in stagnant waters for this very reason. The Law of Fishing also prohibits the release of this effluent into rivers and lakes. Thus careful attention must be paid to the disposal of the retting effluent. (Also see KIRBY 1963)

As noted earlier, the cultivation of hemp in the Spanish New World was restricted to certain areas. This was most likely due to environmental limitations. For example, some areas of Chile have Mediterranean and temperate climates similar to those of Spain, largely due to their location in similar latitudes but in a different hemisphere with opposite seasons; this, to a large degree, explains why hemp farming became an important activity in Chile and to a more limited extent in eighteenth- and early nineteenth-century California, when it was funded completely through government subsidies.

Bowman (1943) produced a comprehensive report on hemp cultivation at the California missions, indicating that cultivation and use of hemp was introduced at Mission San Jose in 1795 with the support of Diego de Borica a Spanish explorer and the seventh governor of "Las Californias" (1794 to 1800). In 1801, the Spanish government sent farming specialists from Mexico to their New Spain colonies in California to promote hemp cultivation with some promising results. Within six years the new hemp plantations were producing many thousands of pounds of fiber, mostly in areas around Santa Barbara, San José, Los Angeles, and San Francisco: "The 1809 crop of 123,000 pounds [55,792 kilograms] was four times that of the preceding year, while in 1810 almost 220,000 pounds [100,000 kilograms] were raised." Substantial crop yields were mostly attributed to production by the Spanish Missions, which accounted for about two-thirds of the total during these years (Mosk 1939). However, 1810 was also the year that Mexico seceded from Spain, and consequently California was cut off from subsidies that had stimulated hemp growing. As a result, commercial hemp production ceased in California in the early nineteenth century and was never restarted, although the Spanish missions continued to produce small amounts for their own use (Mosk 1939, citing the *Archivo de la misión de Santa Barbara*; also see Bancroft 1886). On the other hand, hemp was also cultivated in the northern coastal area of California during at least part of the first half of the nineteenth century on a local, subsistence scale by the Russians at Fort Ross, which was established in 1812 and abandoned in 1841 (Thompson 1951).

Mexico and the United States signed a treaty that ended the Mexican War in February of 1848; this gave the United States a huge portion of southwest North America, including what today is California. On September 9, 1850, California was admitted as the thirty-first state in the Unites States of

America. Shortly thereafter, John Bigler, an early governor of California (1852–56), proposed cultivation experiments for hemp as did Governor Leland Stanford in 1863, but these proposals failed (Hittell 1897). Nonetheless, hemp production in California continued to have supporters into the twentieth century. For example, the *San Francisco Call* newspaper, on April 1, 1907, argued that California should become a large grower of hemp, and by 1916, the Statistical Report of the California State Board of Agriculture reported that 300 acres of hemp were under cultivation in Butte county, and the following year in the Imperial Valley experimentation began with new hemp decortication equipment developed by George W. Schlichten (Wirtshafter 1994).

In Cuba, hemp production was common as early as 1793; however, the demand for sugar outgrew the need for hemp and sugar cane became the main agricultural crop (Schafer 1958). The Spanish introduced *Cannabis* into Guatemala about the same time as they did in Cuba, but large-scale hemp fiber production was not successful there either and was also soon replaced by sugar cane (Abel 1980).

Around the end of the nineteenth century, a Russian hemp expert was sent to Jamaica by the British government to see if hemp could be cultivated in their colony, but the effort was unsuccessful. Apparently, *Cannabis*'s use for mind-altering purposes developed in Jamaica after African slaves were emancipated in the British Caribbean in the mid-nineteenth century. Indentured laborers imported from India to work in Jamaica either found *Cannabis* (which they knew from their homeland) already growing and/or brought NLD *ganja* varieties with them from southern Asia (Rubin and Comitas 1975).

The Portuguese introduced *Cannabis* into Brazil in the early 1800s, but for what purpose is unclear. Portuguese colonists and/or their government probably tried growing hemp originally as a fiber resource, and the slaves the Portuguese brought in from western and southern Africa knew how to use *Cannabis* for psychoactive purposes (Hutchinson 1975), but it is unclear from where these psychoactive varieties came. In 1830 the Municipal Council of Rio de Janeiro banned importation of *Cannabis* plant material into the city, and any person caught "selling the drug was liable to a large fine and any slave found using it could be sentenced to three days' imprisonment" (Abel 1980; see Freyre 1946). According to David Livingstone (1858), the famous English explorer, the colonial Portuguese in Angola felt so strongly about the "deleterious effects" of psychoactive *Cannabis* "that the use of it by a slave [was] considered a crime."

In the European colonies of North America, hemp became a very significant fiber crop. As Western European colonization and commercialization of the New World progressed, the hemp fiber crop became a major focus of English government. In 1585, Thomas Heriot, a friend and tutor of Sir Walter Raleigh, noticed a wild plant that superficially resembled *Cannabis* growing in what today is Virginia. The weedy plant he observed was *Acnida cannabinum*, a member of the amaranth family (Amaranthaceae), referred to in English as "water hemp." Although this plant produces fiber that can be woven, it is not nearly as strong as that extracted from *Cannabis*. Regardless of differences between these two species, Raleigh and others became excited about the possibility of cultivating true hemp in the American colonies, and by 1611, King James I issued hemp cultivation orders for the colony at Jamestown, Virginia, which had only been founded in 1607. Although the colonists' initial enthusiasm for raising hemp was relatively low, within five years, one

of them bragged that the people of Jamestown were already producing hemp better than any grown in England or Holland (Gray 1958). The English government realized early on that it could probably free itself from heavy commercial debts "if only the energies of the colonists could be directed toward raising hemp" (Abel 1980).

Jamestown colony grew hemp under contract with the Virginia Company. By 1699, British parliament had passed the infamous "Wool Act," which deprived American colonists of the right to export wool to any nation outside of Britain. This attempt to prevent independent clothing manufacture in North America created a strong impetus for the colonists to grow their own raw materials and increase production and domestic use of hemp and flax (Booth 2003). But there were other cash crops that would become very important in the British colonies.

The significant rise in demand for American tobacco, for example, was much larger than anybody could have expected. Even though Jamestown colonists were cultivating hemp of high quality, raising tobacco became more important for financial aspirations as much as for personal consumption (Bishop 1966). Consequently, throughout the seventeenth century, colonial governments in Virginia and later Maryland repeatedly offered pounds of tobacco for a lesser weight of hemp, and eventually Virginia (1682), Maryland (1683), and Pennsylvania (1706) colonies established hemp fiber as "legal tender for as much as one-fourth of a farmer's debts" (Abel 1980). These legal maneuvers and financial incentives helped raise hemp output, but not much fiber was shipped to England, as Yankee merchants purchased almost all the hemp produced locally. The scarcity of hemp fiber in the northern colonies was so intense during the period preceding the American Revolution "that supply could not keep up with demand and New England merchants were prepared to buy all the available hemp they could get their hands on" (Abel 1980). In his dissertation for the University of Virginia, Herndon (1959) presents a description of the importance of *Cannabis* fiber in colonial Virginia:

Prior to the Revolution, hemp had become a major staple in the [Shenandoah] Valley and one of a number of crops supplementary to tobacco in other areas. It was also a source of supply for a certain amount of home manufacturing carried on to some degree by all families, and at least a limited supply was needed for various and sundry purposes on the farms and plantations. Pre-Revolutionary inventories and accounts indicate that most families owned equipment for spinning and weaving cloth and also for rope making. The warp was always of hemp or flax, though the filling was frequently of wool or cotton. It is also said that the finest laces of the olden days were always made of hemp in preference to any other fiber. Because of the time of uneasiness on account of the Stamp Act, a strong movement got underway for families to make hemp into *osnaburgs*, a kind of coarse linen originally made in Osnabruck, a town in North Germany noted for its manufacture of linen. This linen was used in North America in the making of shirts, jackets, and trousers. In 1767, 1355 1/2 yards of linsey-woolsey (cloth with a hemp or flax warp and wool filling) and linen (hemp or flax) were made at Mount Vernon (George Washington's estate).

In Massachusetts colony, *Cannabis* was one of the initial crop plants to be introduced, and it was hoped that hemp production would become an important commercial enterprise, as well as providing fiber for local weaving. Hemp cloth became common in rural colonial households, and hempen clothing was worn by all members of farming families. Towels, napkins, handkerchiefs, and tablecloths were among the many other items woven from hemp, and around the farm hemp was used for sacking and cordage. When shipbuilding began in Salem village in 1629, hemp fiber was insufficiently produced in New England to meet local demands. As a crucial resource for this industry, especially for rope production, hemp fiber or its products had to be brought in from overseas (Clark 1929). Consequently, in 1639, by court order, all householders in Salem were required to cultivate hemp, and during the next year, the local Connecticut government tried to convince its colonists to grow hemp, at least enough to provide adequate fiber to make clothing for protection against freezing New England winters (Abel 1980). Most, if not all, households in the northern colonies during this period contained at least something made from hemp.

In addition to critical clothing needs that hemp fiber could help satisfy, the North American colonists also had other very important uses for hemp, including most significantly the manufacture of rope. As noted earlier in reference to Scandinavia, the strong preference for hempen ship-rigging cordage and oakum has been one of the driving factors of the ethnobotanical history of this genus for thousands of years. Factories that produced rope from hemp in colonial America were known as ropewalks. The first one was established in Salem in 1635, followed relatively soon afterward in Boston in 1642 when an English rope maker, John Harrison, was invited to immigrate and set up a ropewalk. Before long, ropewalks proliferated rapidly along much of the eastern seaboard of midlatitude North America to meet the constant needs of early shipbuilding and fishing industries, and by the time of the American Revolution Boston had 14: "The early ropewalks were relatively primitive industries. All that was needed was a large open field, a number of posts to wrap the rope around, and of course, a good supply of hemp fiber" (Abel 1980). Strips of hemp bark or groups of fibers were added into the rope and twisted tightly in the same direction to secure them. As each cord was twisted, the paired strands nearest the point where the cord was forming were relaxed and backed off slowly by the rope maker in the opposite direction, each strand wrapping around the others: "The cords were then twisted with another set, and on and on, until thick strong rope was created" (Abel 1980).

Although rope making developed into a major industry in early America, with the open field ropewalks becoming long, enclosed, covered structures eventually powered by steam engines, the problem of shortages of hemp fiber in both England and the Americas was chronic. The ongoing shortages were due in large part to the lack of labor required to harvest and process the hemp crop, largely because it was a relatively difficult and labor-intensive task, and, along with peculiarities of cultivation, extraction of the fiber itself was especially problematic. Both Thomas Jefferson and George Washington struggled unsuccessfully to make money as hemp farmers (Betts 1953).

After planting, harvesting, and drying, stalks were retted by the controlled rotting process we referred to earlier, which is designed to allow biological decomposition of pectin compounds bonding the bark to the stalk without weakening the fibers. Retting was carried out in three different ways in North America, including water retting (believed to be the best method but also the one that required the most labor); winter retting (simply laying out the stalks in the field to rot over the winter months), which produced inferior fiber; and dew retting (turning the stalks on the field and allowing dew moisture to promote decomposition),

which produced the weakest fiber but became the most prevalent practice in Kentucky. Dew retting was favored by cotton farmers who could buy rough dew-retted hemp burlap at low cost and use it to wrap their cotton bales for delivery to market (Abel 1980).

Once the hemp stalks were fully retted and dried again, the outer fibers had to be broken free from the stalks. This was very tedious and wearisome work, and for millennia, this process was accomplished by slowly peeling the bark from each stalk by hand. By the time *Cannabis* was being grown and processed for fiber in the American colonies hand brakes, traditionally used in Europe, had been introduced into the hemp industry; although more efficient, they were brutal to operate. As Thomas Jefferson noted in his farm record book, "The shirting for our laborers has been an object of some difficulty. Flax is so imperious to our lands, and of so scanty produce, that I have never attempted it. Hemp, on the other hand, is abundantly productive and will grow forever on the same spot. But the breaking and beating it, which has always been done by hand, is so slow, so laborious, and so much complained of by our laborers, that I have given it up" (Betts 1953).

In some areas prior to the American Revolution, slaves and prisoners often did the arduous work of breaking hemp stalks, and the English government even considered sending prisoners to Virginia to supplement the hemp work force. Ironically, the sturdy fibers extracted by prisoners, when twisted into cordage, were used on the gallows to break their necks (see Chapter 9 for a discussion of hempen hanging rituals). However, much raw hemp fiber was broken free by pioneer American colonial families:

> After her husband brought her the broken hemp fiber, the farmer's wife placed it across the top of a "swingling" block, a strong wooden board three to four feet high mounted on a sturdy wooden frame. She and her older daughters now began to pound the fibers as hard as they could with wooden paddles until it was beaten free of woody particles. The long fibers that survived this beating were then drawn through a hatchel, a wooden comb that removed remaining short fibers. Hatcheling [hackling] was done several times, each time with a comb with teeth set more closely together than the previous one. After the final combing, the fine soft pliable threads were spun into cloth. Short fibers removed during the preliminary hatchelings [hacklings] were called tow, and were made into heavy thread for burlap and cord. (ABEL 1980)

In reaction to restrictions on imports and high costs of imported goods, and given the relatively low income of colonials, a great percentage of necessary household and other goods were produced locally in North America rather than being brought from England or elsewhere in Europe. This became not simply an economic campaign but a political statement, especially in reaction to legal maneuvers such as the Wool Act that forced colonists to export their wool. The export of wool encouraged domestic use of plant fibers such as hemp and flax (O'Callaghan et al. 1860), resulting in a great increase in home use of spinning wheels to produce hemp yarn. In New England, "spinning bees" were formed, such as those of the "Daughters of Liberty," a women's group that was instrumental in the campaign to boycott British goods, and eventually the spinning bees spread to Pennsylvania, Virginia, and as far south as South Carolina. As a result of the American Revolutionary War (1775 to 1783), textile imports from England were completely cut off, but by this time the colonists had become self-sufficient (Abel 1980).

However, the American Revolution put great demand on supplies of many resources—among the most important of these was hemp fiber for rope, sail, and paper products—and consequently the price of hemp fiber rose dramatically. Ropewalks and sail making factories became so crucial to the Revolutionary War that men who were employed in the manufacture of rope or sail for a minimum of six months became exempt from military service for the rest of the war (Herndon 1966). In addition, hemp became more valuable than cash and developed into the "standard commodity" for more than a quarter of a century following the American Revolution because paper money had little or no value in the colonies; while hemp, due to its relative uniformity, durability, and widespread constant demand, was greater in value than any other raw produce (Moore 1905).

Hemp production started in Kentucky in 1775 and by 1810 had developed into the "grand staple" there. By 1850, only cotton and tobacco plantation production was greater than hemp in the United States, and most of the hemp farms were in Kentucky with the rest dispersed over areas in Tennessee, Missouri, and Mississippi (Abel 1980; also see Eaton 1966). Apparently, by the mid-1830s, "naturalized *Cannabis sativa*" (NLH) was already quite common in the eastern regions of the United States (Haney and Kutscheid 1973, citing Schweinitz 1836). The demand for hemp also increased the demand for slaves, especially in Kentucky and Missouri: "Without hemp, slavery might not have flourished in Kentucky, since other agricultural products of the state were not conducive to the extensive use of bondsmen. On the hemp farm and in the hemp factories the need for laborers was filled to a large extent by the use of Negro slaves, and it is a significant fact that the heaviest concentration of slavery was in the hemp producing area" (Hopkins 1951).

In 1841, the US federal government approved "payment of not more than $280 per ton for American water-retted hemp, provided it was suitable for naval cordage," and as a result numerous plantation farmers retted their hemp harvests in large pools of water; however, the strenuous labor required took its horrible toll on slaves and this forced activity was soon eliminated: "Many Negroes died of pneumonia contracted from working in the hemp pools in the winter, and the mortality became so great among hemp hands that the increase in value of the hemp did not equal the loss in Negroes" (Hopkins 1951).

The American Civil War (1861 to 1865) brought on the collapse of the American hemp industry, when trade between north and south ceased, and southern suppliers of hemp fiber products lost their major market in northern states. Following the hostilities, hemp production did not fully recover, in large part due to rising use of iron wire cables and bands in shipbuilding, as well as the import of cheaper agricultural sacking made of jute fiber. Thus large numbers of farmers simply gave up growing hemp, refocusing their agricultural attention on other staples such as wheat (see Gates 1965). However, "as late as 1890, thirty-three million dollars" worth of cordage was manufactured in the United States, and during World Wars I and II the hemp industry experienced temporary revivals. We are told in the *Children's Encyclopedia* of 1909 that the "great hemp region of this country is the Blue Grass region in Kentucky, where a rich, moist, well-drained loam overlies limestone," and after hemp is cultivated "the

land is left in better condition than before the hemp was planted" (see West 2005). Following the development and rise in importance of synthetic fibers, along with the US Marijuana Tax of 1937, the huge hemp plantations in Kentucky and some other states would be "gone forever" (Abel 1980).

England was not the only Western European country that envisioned the New World colonies as environments where hemp could be extensively cultivated. The French also believed that their own New World colonies held great opportunity for supplying their navy with crucial hemp resources. However, unlike the English, the French also wanted hemp to export to other countries. France had been selling hemp fiber abroad since the fifteenth century, and by the sixteenth century they had become a major exporter. Nevertheless, like the English, the French had trouble convincing their colonists to grow hemp. In the seventeenth century there was a labor shortage in the French New World colonies in Canada, and as a result many colonists had problems producing sufficient amounts of food for their families and showed little interest in growing fiber crops. An example of how the French colonial administration dealt with this situation involved the finance minister of Québec colony, Jean Talon, who took possession of all the thread in the colony and told the colonists that they could only get it back in exchange for hemp fiber that they were thus encouraged to produce. Along with his confiscation of thread declaration, Talon provided hemp seed without cost to farmers as an additional incentive. Along with the essential need for clothing, this crafty administrative manipulation of the colonists' access to hemp thread and seed created a demand for hemp and an industry to supply that demand (Costain 1954). However, following the English takeover of New France in 1763, they failed to obtain significant amounts of hemp fiber from the Canadian colonists. Even when it appeared that the French under Napoleon could possibly get control of Russia and thus usurp their main supply, the English still could not persuade many Canadians to ship their locally grown hemp to England.

Although *Cannabis* had been a significant fiber crop for centuries in so many areas of the world, by the 1930s hemp was becoming much less important in regions such as the United States even though it still had many valuable applications. Dewey (1931) lists a remarkable number of uses for hemp:

Wrapping twines for heavy packages; mattress twine for sewing mattresses; spring twines for tying springs in overstuffed furniture and in box springs; sacking twine for sewing sacks containing sugar, wool peanuts, stock feed, or fertilizer; baling twine, similar to sacking twine, for sewing burlap covering on bales and packages; broom twine for sewing brooms; sewing twine for sewing cheesecloth for shade grown tobacco; hop twine for holding up hop vines in hop yards; ham strings for hanging up hams; tag twines for shipping twines; meter cord for tying diaphragms in gas meters; blocking cord used in blocking men's hats; webbing yarns which are woven into strong webbing; belting yarns to be woven into belts; marlines for binding the ends of ropes, cables and hawsers to keep them from fraying; hemp packing or coarse yarn used in packing valve pumps; plumber's oakum, usually tarred, for packing the joints of pipes; marine oakum, also tarred for calking the seams of ships and other water craft.

The Marijuana Tax Act, passed by the United States Congress in 1937, placed all production of *Cannabis* under the control of the

United States Treasury Department. The act made it obligatory that all hemp farmers register with and obtain licenses from the federal government and, along with further legislation that created penalties for production, sale, and possession of marijuana and hashish, was an attempt to restrict growth and use of mind-altering *Cannabis* in the United States (Dempsey 1975). Several additional nations also established prohibitions around this time. However, *Cannabis* cultivation, including that undertaken specifically for producing hemp fiber, was not limited in much of Asia, South America, Eastern Europe, and several countries in Western Europe (Ehrensing 1998).

During World War II, shipment of tropical jute (*Corchorus* spp.) and abaca (*Musa textilis*) fibers to the United States was disrupted, and this stimulated an emergency program to produce hemp domestically. This plan was developed and put into action rapidly by a special division of the Department of Agriculture (USDA) Commodity Credit Corporation, War Hemp Industries Inc., which contracted the production of hemp seed and fiber. This federally initiated corporation built hemp fiber processing mills in the Midwest, with American hemp production peaking during 1943 and 1944. However, after World War II the sanctioned cultivation of hemp declined quickly under the pressure of legal restrictions, increasing use of synthetic fibers and the returning supply of less expensive imported tropical fibers. Although a small hemp fiber industry persisted in Wisconsin until 1958, the production of hemp fiber in the United States has been negligible since the middle of the last century (Ehrensing 1998; also see Ash 1948 and Dempsey 1975).

Some Aspects of the Modern History of Hemp

The use of *Cannabis* hemp fibers to make rope, cloth, and other textiles has been very important for thousands of years over a huge geographical range. Although cotton and artificial fibers have taken over much of the world fiber market, it was only in the latter part of the twentieth century that true hemp production dropped significantly or was abandoned, especially in the United States and some other industrialized countries such as Canada, the United Kingdom, and Germany. Nevertheless, from about 1950 to 1980, the world's largest *Cannabis* fiber producer was the Soviet Union with approximately 3,000 square kilometers (300,000 hectares, or about 741,000 acres) under cultivation in 1970, and the main production areas were in Ukraine, the Russian regions of Kursk and Orel, as well as near the border with Poland (Ehrensing 1998).

China, Korea, Serbia, Montenegro, Romania, Hungary, Poland, France, Italy, and Chile also have had sizeable areas of *Cannabis* cultivation for fiber and seed in more modern times, and Canada legalized hemp production in 1999. France and Spain produce hemp mainly for the paper industry. Commercial production of "industrial" hemp for various nonwoven fiber applications and seed products resumed in Germany, the United Kingdom, and Canada in the last decade of the twentieth century, and relatively small-scale *Cannabis* fiber production continues in several other Asian countries where it has been grown traditionally (e.g., Korea, Japan, and Nepal). Commercial fiber hemp cultivation has yet to reach most of Central and South America, South and Southeast Asia, and Africa. In fact, the cultivation of *Cannabis* for fiber in the United States has been nonexistent since the 1950s, and industrial hemp cultivation remains illegal.

Until the "rediscovery" of *Cannabis* as a potentially important commercial fiber crop in the late 1980s, when it began to be promoted as an environment friendly "crop for the future," the use of hemp had been in decline (Vantreese 1998), and the following economic situation prevails in all Western-style nations with increasing labor costs: "Hemp appears slightly more profitable than traditional row crops, but less profitable than other specialty crops. An important constraint to a viable commercial hemp industry is the current state of harvesting and processing technologies, which are quite labor intensive, and result in relatively high per unit costs" (Fortenbery and Bennett 2004).

In some regions of the world, hemp will continue to retain its significant status among natural fibers largely because it is strong, durable, and unaffected by water but just as importantly because of hemp's associations with traditional cultures (e.g., Korean, Japanese, eastern European, and minority ethnic cultures of southwestern China, South Asia, and Southeast Asia). Up until relatively recently, Chinese consumers, however, had not yet warmed up economically or culturally to their ancient ally and generally hold hemp in low esteem, as it reminded them of harder times before the inception of "Modern China" in 1949. On the other hand, in recent times, the cultivation, production, and distribution of hemp fiber products have remained widespread in China. In the late 1990s, China produced over 38 percent of the total world supply of hemp textiles using European processing machinery (Vantreesse 1998). Much of this production involved blends made with cotton or ramie, and overall, the quantity of this fabrication has been much smaller than in various historical periods. One of the more encouraging present-day uses for hemp fiber is in pressed, composite, molding panels for the automobile industry. The woody bits of stalk (hurds) remaining after the fiber is extracted are sieved, cleaned, and packaged, and these woody bits are finding a ready western European market for housing insulation (mixed with mineral lime and water) and as animal bedding.

Between 2000 and 2006, world production of hemp fiber nearly doubled with about half of it produced in China. A review of an online commercial listing of hemp spinning, weaving, and product companies in China in 2013 displayed well over 1,500 businesses in operation (yellowpageschina .com). Today, the world's leading producer of hemp is still China, with smaller production in Europe, Chile, and the Democratic People's Republic of Korea. In the European Union the major produces are France, Germany, and the United Kingdom. China is also presently the largest exporter of hemp textiles, most of which are shipped to Europe and North America, where the market for hemp clothing has been growing rapidly.

According to an FAO report in 2009, the vision of the director general of China's Hemp Research Center in Beijing is to see the development of large plantations of hemp "growing across 1.3 million hectares [3.2 million acres]" of the country's farmland, which would be ample enough to "produce up to 10 million tonnes [sic] of hemp plants a year and, with it, around two million tonnes of hemp fibre [sic]." This vision is said to offer huge benefits for China: "First, it would provide a major new source of fibre [sic] for the textile industry, reduce dependency on cotton and, in the process, free large areas of cotton-growing land for food production. In addition, hemp cultivation would generate extra income for millions of small-scale farmers in some of the country's poorest rural areas."

FIGURE 28. In the 1930s, hemp was mowed with tractors (A), but the stalks were still broken by hand in the field (B). Kentucky hemp mill workers place skeins of hemp bark into a scutching and hackling line (C) (Images Courtesy of University of Kentucky Archives).

China currently grows industrial hemp on around 20,000 hectares (49,420 acres), just a fraction of the 5.6 million hectares (13.8 million acres) dedicated to cotton (China is the world's biggest cotton grower, with a harvest of some 6.6 million tons in 2006). Among natural fibers processed for use in Chinese textiles, hemp output ranks far behind that of wool and silk and of other bast sources such as flax, jute, kenaf, and ramie.

Today, a small quantity of pure hemp fashion fabric is produced in China for high-value niche markets. One key to hemp's future in fabrics is "cottonization": removing gummy lignin that binds hemp fibers and gives stalks their rigidity but prevents them from being spun and finished on modified cotton or wool processing equipment. Using specially developed machines and an array of degumming technologies, Chinese scientists have successfully reduced the lignin content in hemp fibers from 8 to 10 percent to as little as 0.2 percent so that it can be spun on cotton equipment and blended with other natural as well as synthetic fibers.

China's Hemp Research Centre recognizes that most parts of the hemp plant can be used in a variety of applications. The seed is an excellent source of edible oil also suitable for cosmetics and lotions, while the leaves and flowers can be used in medicine. The Hemp Research Centre has also made viscose from hemp hurds, which, because of its short fiber length and low density, is usually treated as waste. Hemp hurds were used in the wood/plastic composite outdoor flooring of the Beijing Olympic Park (FAO 2009).

Cannabis and Paper

Humans have been around for five million years. For 99.9 percent of that time, or until about five thousand years ago, they did not write at all. The invention of writing around 3000 BCE transformed human society by enabling people to transmit greater quantities of knowledge more accurately across vast distances of space and time. (BLOOM 2001)

The word paper is derived from the name of the Egyptian paper reed papyrus (*Cyperus papyrus*), yet the methods of production of paper and *papyrus* are dissimilar. *Papyrus* is made from sliced sections of the inner pithy bodies of papyrus flower stems that are laid in two layers at right angles, pressed together, and dried. Paper is defined by the *American Heritage Dictionary* (2000) as "a material made of cellulose pulp, derived mainly from wood, rags, and certain grasses, processed into flexible sheets or rolls by deposit from an aqueous suspension, and used chiefly for writing, printing, drawing, wrapping, and covering walls."

Although paper is used for a wide variety of important tasks, by far its most significant role in human society has been to facilitate the full development of writing, and *Cannabis* played a seminal role in the origin and development of the written record. Hemp was a very early source of cellulosic fiber used to fabricate paper and remained as a major paper resource for centuries. In many countries, *Cannabis* stalks have been processed and made into excellent paper; today, however, true hemp is no longer readily obtainable at a competitive price. This is regrettable since hemp provides an excellent alternative to softwood pulp paper as Dewey and Merrill (1916) pointed out a century ago. In fact, without hempen paper we would not have had the medium to record parts of the ancient history of *Cannabis* itself!

Although pulp for paper in ancient East Asia was variously prepared from the fibers of hemp, rattan, mulberry, bamboo, rice straw, and even seaweed, archeological research in China indicates that hemp was the first fiber used to make paper (e.g., Barrett 1983). From China, where papermaking first began about 2,000 years ago, the process spread to Korea at least by the sixth century CE, where *Cannabis* had already been growing for millennia. In Japan, where the earliest direct evidence of *Cannabis* presence and use goes back approximately 10,000 years (e.g., see Kudo et al. 2009; Okazaki et al. 2011), paper production using hemp fibers started at least by the sixth century CE. At first, the Japanese used paper only for keeping official records, but with the rise of Buddhism demand for religious manuscripts printed on paper grew rapidly. Chinese papermakers also spread their expertise into Tibet, Central Asia, and Persia, from where it was introduced into India. Subsequently, papermaking spread from Asia to Europe and then on to other parts of the world. Either weedy or cultivated *Cannabis* would certainly have facilitated papermaking in new regions as it would have provided a ready source of pulp, and the dispersal of *Cannabis* itself may have accompanied the spread of papermaking, at least into some areas.

ADVENT AND EARLY HISTORY OF PAPERMAKING IN CHINA

The great development of making pulp paper, in combination with the inventions of the compass, gunpowder, and printing, are outstanding technological advances that occurred in ancient China, indicative of the industriousness and cleverness of the early Chinese people. However, people in what is China today were using other materials to make written records long before true paper was invented. In fact, writing, in the form of inscriptions, has a quite ancient and widespread history in China. During the Shang and Zhou dynasties (1384 BCE to 256 BCE), inscriptions on tortoise shells, mammal bones, and bronze-ware artifacts were used to make written records over a broad area of East Asia, and from the Shang through to the Wei dynasty (386 to 557 CE), a period of more than two millennia, bamboo and wood were used extensively to produce written documents or inscriptions. This extended period of Chinese history can be referred to as the "Era of the Bamboo Slip." However, these slips were heavy and bulky, and they decayed relatively quickly. Eventually such materials were abandoned by Chinese scribes who began to write on woven materials after the development of the hair brush used for both drawing and writing (Hunter 1943, citing Carter 1925).

Silk is perhaps the most desirable fabric upon which to write. It is light and compact, but unfortunately it is also expensive. The significant disadvantages of bamboo's rapid decay and the relatively high cost of silk were serious, especially coupled with rapid economic and cultural expansion in China. This greatly increased demand for durable writing materials that were inexpensive and more readily available and stimulated the invention of papermaking technology. Years of experimentation eventually resulted in development of paper made of vegetable fibers, and one of the raw materials used was recycled hemp, salvaged from "rope ends, rags and worn out fishing nets" (Pan 1983). According the Chen et al. (2003), early paper made in China was primarily composed of hemp rag because these old cloth materials woven of *Cannabis* were easier to pulp than raw plant fibers, and Hughes (1978) referred to Chinese hemp paper as the "oldest kind of true paper in the world."

More than a thousand years ago Su I-Chien (lived 957 to 995 CE) wrote the initial or at least a very early treatise dealing with *Cannabis*, in which he referred to the use of "hemp" in Sichuan province (Needham and Tsien 1985). However, the use of *Cannabis* fiber for making paper in China is a much more ancient tradition. Bloom (2001) referred to the "utilitarian role" of early paper as "the first use of facial tissue," citing a very old Chinese story dating to 93 BCE in which "an imperial guard advised a prince to cover his nose" with piece of paper, which we presume contained hemp fibers. A generally better-known reference to early Chinese paper production can be found among the biographies of famous eunuchs in the *Hou Hanshu* or "Book of Later Han." In this official history of the Han dynasty (206 BC to 220 CE), written by Fan Yeh in the fifth century CE, there is a relevant account of the court eunuch Cai Lun and his involvement in the history of paper production. In 105 CE, Cai Lun reportedly told the emperor about a process of papermaking using plant fibers, including those extracted from hemp stalks, and he is said to have received much praise for his accomplishments and abilities in promoting the development of papermaking. Fan Yeh pointed out that paper use was ubiquitous by the fifth century and that it was known as the "Paper of Marquis Cai."

This well-worn account of Cai Lun's invention of papermaking occurred well after its actual invention and may indeed be mythical (Pan 1983). Recent archeological discoveries indicate that paper was already in use during the Early Han dynasty about 200 years before the time of Cai Lun. Hemp fiber, old silk floss, and the bark of trees were used to make this early paper (Barrett 1983). Nevertheless, making pulp paper from plant fibers did solve problems associated with writing on quickly decaying bamboo slips and cumbersome wooden tablets, as well as the costs associated with using brushes to write on silk fabrics. The main method to produce paper involved the chopping and crushing of hemp rags and cordage scraps, as well as hemp stalks and the inner bark of paper mulberry, and then macerating the mix into pulp in a tank of water. The paper mulberry tree has its own ancient, widespread history as a fiber source to make bark cloth, such as *tapa* in traditional Polynesia, where it was introduced by ancient seafaring peoples. The addition of old hemp fiber fishnets to paper pulp "since time immemorial" in China is also noted (Li 1974a, see also Bloom 2001). During pulp processing, the slurry of hemp and paper mulberry fibers would eventually rise to the top of the tank as a tangled mat. Portions of the fibers in liquid combination were then spread thinly on a flat, stretched cloth frame to drain and the fibers formed into sheets, which after further drying could be used to make written records.

Hemp rag paper can be torn when wet but regains its full strength when dry and usually has a low acid content. It can be preserved for centuries as long as it is not subjected to extended periods of high humidity or strong light. Chinese scientists carried out simulations in attempts to reconstruct the early manufacture of hemp "rag" paper. Pan (1983) presents a rough description of the ancient but relatively sophisticated process:

> Raw materials such as rope ends and rags are soaked in water and after expansion cut to pieces with an axe and then cleansed in water. The pieces are then cooked in a grass-stalk ash solution, which may be cited as the earliest alkaline treatment by chemical method. When such impurities in the raw materials as lignin, fructose, [coloring] matter and fat have been further removed, the pieces are washed in clear water and pounded in a mortar. The fine fibres [*sic*] after pounding are mixed with water to form a liquid pulp suspension. Paper moulds [*sic*] are used to scoop up portions of the liquid substance, which become paper sheets when dewatered and dried. If the surface of the paper is rough and creased, it requires calendering [smoothing] before it is fit for writing on.

The popularity of recycled fiber use in early paper mills resulted not only because it was readily available and low in price but also because used hemp cloth and cordage, made of pure hemp fiber and lacking some of the gummy compounds removed during retting and yarn processing, are much easier to process than raw hemp bark. Although we have found few specific mentions of "rag" papermaking traditions for China, this may be because there was always a surplus of raw hemp, mulberry, cereal straw, and bamboo to use for paper pulp. Needham and Tsien (1985) tell us that *Cannabis* fiber was used less to make paper after the thirteenth century; however, they did suggest that it was the first fiber source for papermaking and was especially important during the Western Han dynasty (206 BCE to 9 CE). Following this period, other fibers became as much or more significant than hemp in ancient China, with paper mulberry being heavily utilized during the Eastern Han dynasty (25 to 220 CE), rattan during the Chin dynasties (265 to 420 CE), bamboo during the middle of the Tang dynasty (618 to 907 CE), and straw probably from before the Song dynasties (960 to 1279 CE).

China is thus the original home of papermaking, and paper scraps containing hemp fibers have been discovered at several ancient Chinese archeological sites. For example, Bloom (2001) referred to a very old fragment of "hemp paper found at Ejin Banner in Inner Mongolia" dated to the Western Han period, but it "is too coarse to be suitable for writing." In 1957, a scrap of paper was discovered in an early Western Han grave at the Baqiao Tomb site, located near Xi'an in Shaanxi province and dated to ca. 2138 to 2085 BP (also see Li 1974a; Barrett 1983). This very old surviving piece of paper is described by Temple (1986), who also explains how it was fabricated:

> It is about 10 centimeters square and can be dated precisely between the years 140 and 87 BC [actually between 188 and 95 BCE]. This paper and similar bits of paper surviving from the next century are thick, coarse and uneven in their texture. They are all made of pounded and disintegrated hemp fibers. From the drying marks on them, it is evident that they were dried primitively on mats woven as pieces of fabric [also likely hemp], not on what we know as paper molds. In these early days, the water just drained slowly through the underlying mat of fabric, leaving the paper layer on top. This was then peeled off and dried thoroughly. But so thick and coarse was the result, that it could not have been very satisfactory for writing.

According to Bloom (2001), who studied this same paper scrap, it and other specimens from Baqiao were produced using fibers "from hemp and dated no later than the reign of the Western Han ruler Wudi (141 to 87 BCE)." Bloom also tells us that one of these ancient paper specimens contained "some microscopic loops of fiber, and another contained a small remnant of thin, two-ply hemp cord, suggesting that they had been made from reused fibers."

Temple (1986) refers to the most ancient surviving piece of inscribed paper that was found beneath the remains of a very old "watchtower" in "Tsakhortei near Chü-yen" (Juyan, formerly Karakhoto) in Inner Mongolia in 1942. This paper scrap has approximately 25 readable characters written on it and is dated to between 109 and 110 CE based on the fact that the watchtower was deserted by its Chinese defenders during this time period; in 1974, two more larger fragments of hemp paper "dated to 55 BC were discovered in a different watchtower in Juyan" (Bloom 2001; also see Cotterell 2004).

Maps excavated in 1986 from a tomb at the Fangmatan site in Tianshui (Gansu province) are significant because they are also among the oldest paper items ever found in China, again dating to the early years of the Western Han dynasty (e.g., see Buisseret 1998). The paper is remarkable for its smooth texture, demonstrating that fine papermaking technology was already well developed by that time (see Bloom 2001, figure 15 for a photographic glimpse of a remarkable fragment of "hemp paper bearing traces of a map").

Bloom (2001) also referred to numerous inscribed specimens of hemp paper dated to before 220 CE that were recovered from a Han tomb in Hanatanpo, near Wuwei, in the western province of Gansu: these fragments of hemp paper "are said to be more technically advanced than other examples, being white and much thinner, and perfectly suited to writing with brush and ink."

Although the archeological discoveries referred to earlier demonstrate that papermaking technology had already been mastered in China before or soon after the beginning of the current era, the contributions of Cai Lun to the development of papermaking in ancient China should not be underappreciated. Although he almost certainly was not the original inventor of the paper production process, Cai Lun ought to be viewed as one who improved the production of coarse hemp paper during the Western Han dynasty. Through his sponsorship by the royal court and the organization of their abundant human and natural resources, he was able to produce high quality paper by 105 CE. Subsequently, the technique of papermaking was spread throughout China and eventually the world.

As noted in the previous chapter, ancient manuscripts, documents, and paintings were found in 1900 in the secret Mogao Caves near the frontier town of Dunhuang in Gansu province in western China. In addition to paintings on silk and hemp found in one of these caves, printed materials were recovered, including the earliest dated printed book in the world. Most of the early papers from Mogao were produced using a mixture of raw fibers including those of hemp, paper mulberry, laurel or ramie, and rags of "flax, hemp and ramie" (Tsien 2004). This conclusion contradicts some literary records indicating that each of the raw materials was used separately to produce a different kind of paper. For example, the third-century CE Chinese author Dong Ba was specific when he pointed out in his treatise on agriculture and food titled *Yufu zhi* that paper made of hemp was referred to as "hemp paper," while paper made of tree bark was called "paper mulberry paper" and paper composed of fibers from old nets was known as "net paper." Dong Ba's comments were quoted in the *Taiping yulan* or "Imperial Readings of the Taiping Era," an imperial encyclopedia compiled during the Taiping reign of the Song dynasty, 960 to 1279 CE (Fang 1935). The scholar Su Yi-jian (lived 957 to 995 CE) distinguished between sources of fiber for ancient papermaking in China according to geography; he pointed out that hemp fiber was used in "Sichuan," paper mulberry bark was used "in the north," rattan was used "in Yanxi," lichen was used "in the south," and "husks of grain" were used in "Zhejiang" (Yi-jian 1936). Ultimately, paper may have been produced without practical problems from a combination of fibrous materials or from a single fiber source (Tsien 2004).

Archeological research at the ancient Astana-Karakhoja Tombs, now an underground museum located at Turpan in Western China's Xinjiang Uygur Autonomous Region, has uncovered approximately 2,700 paper documents. These date from the time of the Western Chin to the Tang dynasty (265 to 618 CE). Turfan is located at the northern edge of the arid Taklimakan Desert about 900 feet below sea level, one of the lowest depressions in the world (Anderson 1988). Li (1974b) describes the discovery of an uncommon, incomplete script of the *Lun Yü* or "Analects of Confucius" (originally written in 716 CE), which was inscribed on white hemp paper and found in a grave dated to 1100 CE. Li also referred to the discovery of shoes in this grave made of pasted layers of white hemp paper sewn together with hemp threads (also see Lu and Clarke 1995).

The earliest examples of printing have also been found in China and date to the eighth century CE. Although printing in China is thought to date from the seventh century, a long tradition of graphic arts including stone seals, rubbings of stone engravings, and bronze casting led up to the actual printing of script on paper and the development of moveable type elements in the eleventh century (Temple 1986; also see Needham 1954, volume 5). It is clear that hemp played a crucial role in the development of papermaking technology. It is one of those relatively rare plants that can provide abundant fibers that are rich in "long cellulose but lower in binding substance, which must be eliminated in the process of maceration" (Needham and Tsien 1985). The invention of hemp paper in China more than 2,000 years ago revolutionized the record-keeping process and may therefore be regarded as an indispensable contribution to the rapid development of civilizations throughout the world. The ancient papermaking techniques were first developed in China and subsequently spread to various regions of Eurasia, and these early manufacturing processes have been replaced almost completely by modern production techniques. Old methods of papermaking can still be found in China in three counties, Huishui, Changshun, and Danzhai, all situated near Guiyang, the capital of Guizhou province in the southwest. Changpei (no date) recently described ongoing contemporary paper production in Huishui county:

> The local records say that the papermaking industry there began 300 years ago, in the late Ming and early Qing dynasties. Later it flourished during Qing emperor Qian Long's reign (1736 to 1796), and the number of papermaking households increased from three to a dozen. Every household had eight kilns, with a monthly rough-straw paper (or toilet paper) output of 150 kg [330 pounds]. At that time, hand production included husking bamboo and hemp with pestles and mortars. Today, such primitive means of production can still be seen in Xieyao village, Changshun, 30 kilometers [18 miles] from Huishui. (Also see ANONYMOUS 2010)

The Chinese kept the papermaking process secret for many centuries; eventually, however, people in Korea learned to produce paper, sometime before the sixth century CE, and the Japanese most likely started making paper by this time as well (see Pan 1983).

FIGURE 29. Traditional Chinese papermaking began by cutting hemp stalks or other plant materials (e.g., bamboo or mulberry) into short pieces (A) and trampling them in water (B), followed by thorough washing (C). Next the fibers were boiled (D) and pounded (E) to macerate the fibers into paper pulp before blending (F) with other fibers. Hemp cloth stretched on wooden frames was used to lift, drain, and spread (G) a thin layer of liquid pulp, the frames were dried in the sun (H), and the finished paper (I) was peeled from the framed cloth (adapted from Needham and Tsien 1985). Original adapted from Phan (1979).

The production of hemp paper in Korea also has a very long history. The oldest paper manufactured on this East Asian peninsula is said to have been made from *Cannabis* fiber and called *maji*, which is produced, more or less, by the following method, probably derived from Chinese techniques: "Scraps of hemp or ramie cloth are soaked in water for some time and then shredded into tiny pieces. These pieces are ground in a grindstone to produce a slimy pulp, which then is steamed, cleansed with water, ground and placed in a tank. This raw material is pressed onto a frame and sun-dried while being bleached. This method of papermaking was most popular during the Three Kingdoms period [57 to 668 CE], probably towards the latter part of this period" (Korea Overseas Information Service 2003).

However, some scholars have conjectured that paper was already being produced in Korea by the third century CE. Yum (2003, citing Pan 2002) refers to a piece of paper discovered in 1931 by the Choson Archeological Site Research Group in the ancient tomb of Chechubchong, belonging to the Naknang period (108 BCE to 313 CE), as evidence supporting the hypothesis that paper was being used in Korea prior to the fourth century CE, although it may have been imported from China. Bloom (2001) referred to the "earliest Korean papers to survive," including "a glossy white paper made of hemp fibers discovered at a North Korean site dating to the Goryeo era."

Yum (2003, 2005) tells us that handmade paper was a traditional and essential product of daily life throughout Korean history and that it was utilized to produce many items: "For calligraphy, books, and envelopes; for doors, walls, and windows; for furniture, such as wardrobes, cabinets, and chests; for craft objects, such as writing brush holders, umbrellas, lanterns, boxes, baskets, fans, and kites; and for clothing and shoes." It is not clear exactly when the Korean people began to produce this plethora of items from paper, many of which remain significant in their traditional culture today. In some of the more important historical books and other written materials that remain from the late Three Kingdoms period, hemp paper is said to have provided the primary medium for documenting events that took place by the fifth century CE (Korea Overseas Information Service 2003).

The *Nihon Shoki* or "Chronicles of Japan," published in 31 volumes and brought together in 720 CE, tell us that Damjing (Donchō in Japanese), a Korean Buddhist monk, physician, and painter, introduced the process of papermaking to the Imperial Japanese palace in Nara, Japan, in approximately 610 CE, 60 years after Buddhism was introduced (Hughes 1978). The earliest known Buddhist text found in Korea: the *Mugujeonggwang* or "Great Dharani Sutra" was discovered inside a stupa at the Bulguksa Temple at Gyeongju in southeastern South Korea. This stupa was sealed in 751 along with a record telling us exactly when this printing project began, as well as who the papermaker was and when the conservation of the text took place; in addition, "it was confirmed that the paper had been made of paper mulberry, which Koreans term *dac*" (Yum 2003). By the time of the Goryeo Kingdom, people had begun to make relatively large amounts of paper from mulberry. This facilitated the increased publication of books of many kinds. Paper money was also being printed, and by the eleventh century, Korea had begun to export paper to China. This was the golden age of papermaking

in Korea, and with this rising market for paper, in 1145 CE the Goryeo government began pressuring farmers throughout the country to grow paper mulberry and encouraged it citizens to produce paper (Yum 2003). Consequently, by this time hemp fiber use in papermaking had become much less important for general use, but it was still used to some extent. Moon (1996) tells us that during the Joseon dynasty (1392 to 1910), Korean paper made from paper mulberry was produced in greater quantities than that made of hemp because the cultivation as well as processing of pulp from paper mulberry was easier. Moon also referred to papermaking experts from Korea who were sent by the dynasty leaders to China to "study techniques for producing hemp paper," although paper mulberry continued to be the major raw material for paper production in Korea. On the other hand, Yum (2003) used a polarizing microscope to examine old Korean paper samples (dated to the fourteenth and eighteenth centuries) and was able to observe "well-defined and smooth cell walls with partial longitudinal splitting" of the fibers in at least one of her samples. This is a distinct "characteristic of hemp" and thus indicates that *Cannabis* was at least occasionally still being used in papermaking. Presently, hemp pulp is occasionally included in Korean handmade craft papers. Paper was first introduced into Japan from the northwest Asian mainland as a precious import item about the middle of the third century CE. Subsequently, as expert papermakers from the Korean Peninsula, such as the Buddhist monk Donchō, started to settle in Japan in the early sixth century, this new and important technology became established in the island kingdom, and the production of handmade paper (*washi*) began in earnest. Crown prince Shōtoku (regent from 593 to 622) began the promotion of paper mulberry and hemp cultivation in Japan, and eventually "using the inner white bark of certain bast fiber plants, the Japanese took the basic methods brought overseas from China by Koreans, perfected the techniques, and improved upon them to render their paper distinctly Japanese" (Hughes 1978; also see Bloom 2001). The paper that Donchō revealed to the Japanese imperial court was most likely fabricated using a combination of hemp rags, along with fibers of two species of paper mulberry trees—introduced *kōzo* (*Broussonetia papyrifera*) and native *kazinoki* (*B. kazinoki*), as well as native *gampi* (*Wikstroemia gampi*).

After the Japanese government adopted the Chinese-style bureaucratic system of penal laws and ordinances, the use of paper increased dramatically as the number of government offices grew. In addition, the rise of Buddhism in East Asia including Japan encouraged an increasing demand for copied religious texts (*sutras*), especially for people of the upper classes in court, temple, and noble ranks. Consequently, the industry of papermaking in Japan developed quickly and spread widely.

Over time, papermaking experiments with fibers from several species were attempted, and by the early part of the eighth century hemp and the paper mulberry species, *kōzo* and *kazinoki*, had more or less become established as the foremost *washi* pulp materials, used alone or in combination. However, hemp was already beginning to lose importance, and following the shift of the capital from Nara to Heinakyō in 794, slowly but surely the use of paper mulberry and *gampi* replaced hemp (Hughes 1978; Bloom 2001). The Chinese method of papermaking brought by the Koreans produced paper that was not very sturdy; in addition, this type of paper was prone to insect damage and could not be preserved for lengthy periods of time. Furthermore, the Japanese people, who preferred purity and freshness in paper, were not attracted to recycling rags to make paper. Japanese papermakers were inspired by the availability of many wild plant species that provide sturdy fibers, and therefore they developed techniques for making handmade paper, utilizing only raw fibers extracted directly from wild or cultivated plants in their own environment.

Although papermaking technology spread first to Korea, the earliest large-scale printing in the world may have occurred in Japan with printing blocks made of wood, metal, stone, or porcelain. This early printing enterprise was associated with Empress Shōtoku, who reigned over Japan, with much Buddhist influence, from 749 until 769 CE. In 735 CE, a horrible epidemic of smallpox seriously reduced the population in Japan, and the suppression of a dangerous rebellion in 764 CE also resulted in a large loss of life. These two catastrophic reductions in the labor force are believed to have motivated the invention of text printing upon sheets of paper. To overcome these human calamities, Empress Shōtoku had a "million" *dhāranī* (chanted prayer) sheets printed on pure hemp paper, each enshrined in its own individual three-story wooden pagoda, which was approximately 20 centimeters (8 inches) in height and 10 centimeters (4 inches) in diameter at the base. Shintō, the indigenous religion of Japan, with a history more ancient than Buddhism, attributes purifying powers to hemp, and therefore hemp paper is still used in Shintō purification rituals. (See Chapter 9 for a more thorough discussion of hemp in Shintō ritual.) The production of these hempen prayer papers and their small pagoda receptacles was completed and documented on May 24, 770 CE. Although the actual number of prayer papers and pagodas produced may not have been one million, there was a definite intention, if only symbolic, to reassure the Japanese people that a huge number of prayers had been offered to purify their realm and help ward off dangers. This coordinated work of papermaking and printing was a very early example of mass production (Hunter 1943), and it was based on paper made of *Cannabis* fiber.

Sukey Hughes (1978), a Japanese paper expert, provides a description of part of an ancient *dhāranī* belonging to Hiroshi Haneishi, a traditional Japanese painter and member of the Japan Art Academy: "Haneishi was kind enough to show me one of his most prized possessions, a piece of the *dhāranī*, one of the oldest extant examples of printing on paper, dating from 770. The hemp paper, backed for protection, was delicate, slightly brittle, and a tender brown color."

Hughes also pointed out that the ancient Japanese *dhāranī* ordered into production by Empress Shōtoku were long believed to have been the world's first printed texts until 1966 when "a Korean *dhāranī* sealed into a stupa in 751 was found at Bulguksa Temple in Gyeongju, and this is now the oldest known example of printing."

Japanese papermakers refined their craft and elevated it to an artistic tradition. Papermaking became an essential part of Japanese culture, and paper has long been used in Japan for writing material, fans, garments, and dolls and as an important component of house partitions (Olson 1997; Hunter 1943). Although *Cannabis* fiber was a key ingredient in much of this papermaking tradition, today it has only very minor importance. Hemp fibers can be processed relatively easily and make flexible, smooth, radiant, and sturdy paper. However, hemp fiber is no longer readily available in Japan, and presently only a very small amount of hemp *washi* is made. Traditionally produced hemp *washi* is considered a "high-class paper," and since ancient times has been "symbolic of

the sacred" (Barrett 1983); it is still seen today in traditional Shintō ceremonies (see Chapter 9).

DISPERSAL TO NORTH AFRICA AND EUROPE

Rags make paper,
Paper makes money,
Money makes banks,
Banks make loans,
Loans make beggars,
Beggars make Rags.
(AUTHOR UNKNOWN, ca. eighteenth century, from HUNTER 1943)

The technology of making paper using hemp fiber pulp also spread westward out of East Asia into Central Asia at least by the middle of the eighth century and possibly as early as the third century CE. Carter (1925) referred to paper produced in Turkestan dated from the third to eighth centuries that was made from the bast fiber of paper mulberry and hemp fiber extracted from rags (also see Bloom 2001). Papermaking spread to the Indian subcontinent between about 800 and 900, and eventually the technology reached Europe in the late eleventh or early twelfth century. We assume that *Cannabis* fiber, in its raw form or in the rags of hemp clothing, was associated with most, if not all, of these early papermaking efforts.

During the period from the Han dynasty to the end of the Yuan dynasty (from ca. 220 BCE to 1368 CE), the crucial trade routes known collectively as "The Silk Road" helped foster contact between the Chinese empire and Western civilization. Silk was not the only thing that was moved and traded along this lengthy maze of caravan trails. These also served as a conduit for many other goods and ideas to and from China across the often politically and environmentally dangerous terrain of Eurasia, eventually reaching to the eastern Roman Empire on the shores of the Mediterranean, as well as southern Asia. Among the ideas that were spread far and wide was the use of hemp fiber and rags in the historically profound activity of papermaking.

According to an oft told story, when Arab troops captured Chinese papermakers during a battle near Samarkand in 751 CE, they effectively helped spread the art of papermaking into Central Asia. Bloom (2001) refers to this story as only "a story" but points out that we should not completely discredit the tale claiming that Chinese captives introduced papermaking expertise to Samarkand in 751 CE: "Much as Cai Lun's invention of paper began to be used for writing in early second-century China, the story of captured Chinese papermakers metaphorically describes how paper was introduced to the Islamic lands through Central Asia just at the time when, under Abbasid [descendant from the prophet Muhammad's uncle Abbas] rule, this region began to play an increasingly important role in Islamic civilization."

In a recent study, high resolution digital microscopy was used to analyze the morphology and raw material contents of paper from more than 50 ancient Chinese documents dating from the sixth through eighth centuries (mostly Tang dynasty artifacts in the Otani Collection, Japan) from Turpan (Turfan) and about 1,000 kilometers (600 miles) away in Dunhuang, in the remote west frontier of vast territory of the Tang Empire (Enami et al. 2010). This research revealed a clear difference between the documents of the military "Turfan brigade" in Western China and those of official, local government. Most of the analyzed old military documents were

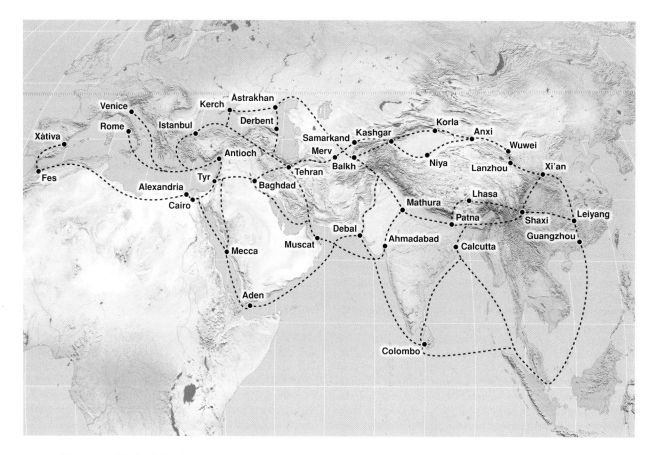

MAP 15. The ancient "Silk Roads" trade routes carried papermaking technology and the use of hemp fiber in papermaking from China to the Middle East and Europe (cartography by Matt Barbee).

written on paper of rag origin, most likely made out of cloth at least partially composed of hemp fiber, while the majority of documents utilized by the local government "were recorded on usual [more advanced] hemp or mulberry paper" during the Tang Dynasty period. Enami et al. (2010) concluded the rag paper was produced inside the brigade establishment, suggesting that the Turfan brigade created their own local supply of valuable paper from materials available to them:

> The fact that old fashioned rag paper of the primitive paper-making craft was found in the military documents of the Turfan brigade would give important and decisive information of the origin of the papermaking craft that was spread westwards. The story is widely accepted among historians of the transferring of paper craft from China to the West at the battle between Chinese army and Arab army in 751 at Talas [near Samarkand]. However the real reason why Arabs and then later Europeans had continued the primitive paper-making craft using cloth rag until the 19th century is yet unclear, in spite of the fact that sophisticated papermaking technology using tree bark, shrub or other plant fibre [sic] had been widely spread in China, Central Asian oases kingdoms and Asian countries already in the 8th century. The present results would give an answer to this question. The papermaking craft, obtained by Chinese craftsmen captured by the Arab army at the battle, was not the sophisticated technology using plant fibres [sic], but the primitive craft using cloth rag adopted by the Turfan brigade logistics.

In any case, within a half century of the battle near Samarkand, the fifth Arab Abbasid Caliph, Harûn-el-Rashid, had introduced Chinese papermaking technology to Baghdad, where an industrial unit staffed with skilled workers began fabricating paper before 793. By the tenth century paper was produced in Damascus, Syria, and had also spread to Tiberias on the Sea of Galilee and Syrian Tripoli. The papermaking industry helped make Damascus famous, and associated factories in ancient Syria profited greatly from advantageous local conditions for cultivating hemp (Zaimeche 2005). Damascus paper was exported to India, Red Sea ports, Egypt, and Europe regularly. The first true pulp paper appeared in Egypt sometime in the ninth century, but paper was not manufactured there until about a century later. The rise of true paper production developed over time, concomitant with the decline of papyrus use in Egypt. Bloom (2001) refers to production of paper at Fustat (Old Cairo), where it was reported by a traveler to Egypt in 1216 that papyrus production "was by then quite forgotten." The dry climate of Egypt, like that in Western China, has allowed preservation of a great deal of ancient paper from medieval Islamic times; however, in Egypt the great bulk of these true paper relics were made of linen rags. In 1980 at a site in Fustat dating to the eleventh century thousands of textile fragments were excavated, quite possibly left by ancient ragmen for recycling into paper, and although the majority were linen, a small percentage were made of hemp (Bloom 2001).

Eventually the technology of papermaking spread across North Africa to Morocco (the western region of the Maghreb), and from there it was introduced to Spain by the Moors in the middle of the tenth century (Glick 1979; Bloom 2001). Therefore, from the beginning of the ninth through the thirteenth centuries, Muslim Arabs transferred the art of papermaking all the way from western China or Central Asia to Western Europe via North Africa (Blum 1934; Bloom 1999; also see Burns 1996). During the early medieval period in the western portion of North Africa, paper, and consequently books, were particularly expensive since paper was rare in this region, not being produced locally (Bloom 2001; also see Burns 1985). However, by the eleventh century, during the Almoravid period, more than one hundred mills were in operation in Fez alone, producing paper from either flax or hemp (Benjelloun-Laroui 1990), and by the end of the twelfth century nearly 400 paper mills were operating in this area (Burns 1981). Later, during the fourteenth century, paper was exported to the Balearic Islands in the western Mediterranean Sea; however, by this time, quality paper was also being produced in Spain at Xátiva (Játiva or Shātiba) in Alicante province and was also traded into western Africa (e.g., Burns 1996),

Indeed the center of paper manufacture in medieval Muslim Spain was Shatiba (Xátiva or Játiva) in al-Andalus (Andalusia), and it was here that a unique paper was produced that is "still known as *shātibī* in Morocco" (Glick 2005). Furthermore, it was in Xátiva that the first Western European paper factory was established in 1056 CE, located "next to the old irrigation-channel" near the city; however, other paper factories went into operation soon after in Valencia and Toledo (Bloom 2001; also see Hunter 1978). Paper made in Xátiva contained fibers of flax, hemp, or a mixture of the two (Blum 1934). Although Burns (1985) suggested that only linen rags and flax fibers were originally used to make paper in medieval Xátiva, Bloom (2001) refers to hempen threads that were boiled in oil to ensure stiffness and used in paper molds at least by the eleventh century or earlier. During the decline of Islamic civilization in Iberia and beyond, following the Christian conquest of Toledo in 1085 CE, collaboration among Muslims, Jews, and Christians resulted in the translation of much of Arabic science into Latin, especially in Italy, Sicily, and Toledo, and it was in these same cities, about the same time, that Muslims taught Christians to make paper (Egerton 2002; see also Bloom 2001; Benoît and Micheau 1995; Haskins 1927).

During this period, documents were generally written on paper made from rags of flax and some of hemp, softened by soaking in limestone water and pulped in paper mills. By the twelfth century, paper factories were common in a number of Moorish cities in Spain, and travelers' accounts provide testimony regarding this flourishing industry (e.g., see al-Hassan and Hill 1986; Bloom 2001). Many twelfth-century Catalan documents refer to paper mills, and by the thirteenth century "paper was exported to Sicily from both Barcelona and Valencia" (Glick 1979, 2005). In reference to outstanding medieval examples of Islamic paper from the Aragon archives, Bloom (2001) described a "blood-red paper" that was "made equally of linen and hemp," upon which in 1418 a letter was written by Muhammad VIII of Granada to Aragon King Alfonso V.

The medieval Moorish Arabs in Spain valued their monopoly on papermaking so much that they banned the exporting of rags to other countries in an effort to control competition. However, such competition surfaced by the end of the thirteenth century, especially from paper production in Italy, which first appears in the Aragon archives in 1291 CE, and relatively soon after this, there was a precipitous rise in Italian papermaking and a concomitant decline in this industry in both Spain and North Africa. This situation was compounded by a reputed decline in quality and size of "Játiva paper" when compared to that produced in

FIGURE 30. German engraving of a papermaker lifting and draining a layer of wet pulp following the methods introduced from China. Notice the mechanical macerator and sheet press in the background. The printer's apprentice carries away finished sheets (adapted from Needham and Tsien 1985). Original by Jost Amman printed in Frankfurt, Germany, in 1568.

Italy, "but by this time it was too late to improve the situation, and the brief heyday of the Spanish paper industry was over" (Bloom 2001).

During the fourteenth century many paper mills were established in Italy, and it was from here that papermaking spread across Europe and eventually the New World. The Italians exported large amounts of paper during the late medieval period, dominating the European market for many years: "Papermakers usually set up shop near urban centers to be close to the supply of rags, or near ports, where they could collect discarded linen sails and hempen ropes and nets to make papers" (Bloom 2001).

In the sixteenth century, the art of making paper spread to Russia and Holland and then to Britain in the seventeenth century. In Genoa, Italy, a decree was passed stating that hemp and flax rags as well as cordage were essential ingredients of paper, and their trade was strictly controlled. Anyone who was not a dealer in rags used in papermaking was forbidden to buy or sell used linen and cordage, under penalty of a fine of four florins for each offense (Blum 1934).

According to Hunter (1943), the earliest known paper mill in England was built by John Tate in Hertford in 1495 CE, and the role of hemp in the early production of paper became significant. In 1620 CE, John Taylor, in his odd but insightful *The Praise of Hemp-Seed*, outlines and honors the variety of uses of hemp, including its use to make paper:

And I in forme of paper speake to you.
But paper now's the subject of my booke,
And from whence paper its beginning tooke;
How that from little Hempe and flaxen seeds,
Ropes, halters, drapery, and our napery breeds,
And from these things by art and true endeavour,
All paper is derived, whatsoever.

In 1716, England's famed Society of Gentlemen published a treatise on the use of a practical material other than linen or cotton for papermaking, advancing "the idea of using raw hemp without spinning or weaving as a material from which to fabricate paper" (Dewey and Merril 1916). The Society of Gentlemen also provided a detailed description regarding the preparation of hemp hurds, the woody inner portion of the hemp stalk that is broken into pieces and separated from the fiber in the processes of breaking and scutching (the removal of wood bits from fibers). Hemp hurds also have value as a pulp source; they are basically crop waste, and although the fibers are very short, they are almost pure cellulose and can be added back to hemp fiber to make high-quality paper.

Hemp paper production in Finland began in the mid-seventeenth century (Laitinen 1996) and is representative of traditional hemp paper production in much of Europe. Paper was originally made from rags gathered from hemp and flax clothing (Ahokas 2007). Rags were torn into small bits, soaked in water, and beaten into a pulpy mass, which was rolled out into thin sheets, dried, and pressed into paper. Papermaking started as a craft but gradually enlarged in scale, and the first Finnish paper factory opened in 1818 in Tervakoski. Although soft conifer wood eventually became an important raw material for papermaking in the second half of the nineteenth century, superior, thinner, and sturdier paper was still made from rags in some areas of Europe including Finland. For example, Laitinen (1996) pointed out that up until the end of World War II, and even under some circumstances today, rag paper has been utilized in fine printing for such items as bibles, bonds, stamps, money, and very thin cigarette papers because of its special properties. However, if the fibers of hemp and flax are so suitable for making paper, why have hemp and flax rags been used rather than the raw fibers to make paper? As Laitinen (1996) explains,

Probably because [the fibers] would have become too expensive, since processing hemp and flax into fiber was so labor intensive. Rags and discarded clothes were cheaper fiber sources and more easily processed into paper mass than fibers taken directly from plants. Flax fibers were not bought directly from farmers by Tervakoski factories before or during the Second World War and sources do not reveal to what extent rags were used or what part of those rags were hemp or flax. Probably during the 19th century, and in the beginning of the 20th century, hemp was more widely used. Most of the hemp rags were brought into Finland from Russia, which at that time was the largest producer of hemp in the world. As the 20th century progressed, flax became increasingly more common.

The primary use of rags in papermaking lasted up until the end of the nineteenth century in Finland, and until the start of the twentieth century paper was made entirely by hand. The amount of old clothing required to feed the growing paper industry must have been tremendous. After 1900, a variety of paper machines were developed and put into use.

Since the mid-nineteenth century, large-scale commercial paper mills have made paper from a great variety of cellulosic fibers, including hemp.

HEMP PAPER PRODUCTION IN NORTH AMERICA

In colonial America, hemp rag salvage was relatively abundant, since *Cannabis* was the only widely used fiber source, and the majority of people still wore hempen clothing. Before the widespread use of softwood to make paper, the American Colonies used "discarded sails and ropes sold by ship owners as scrap for recycling into paper," and all the remainder was made from "worn-out clothes, sheets, diapers, curtains and rags . . . made primarily from hemp and sometimes flax" (Dewey and Merrill 1916).

In the *Boston News Letter* (1769) there appeared an advertisement stating that "the bell cart will go through Boston about the end of each month to collect rags," and added,

Rags are as beauties, which concealed lie,
But when in paper, how it charms the eye!
Pray save your rags, new beauties to discover,
For of paper, truly, every one's a lover;
By the pen and press, such knowledge is displayed
As wouldn't exist, if paper were not made.
Wisdom of things, mysterious, divine,
Illustriously doth on paper shine. (HUNTER 1943)

There are a number of reasons to favor the cultivation of hemp for a paper source; for example, consider the following quote from Dewey and Merrill in their 1916 *United States Department of Agriculture Bulletin* article: "Without doubt, hemp will continue to be one of the staple agricultural crops of the United States. The wholesale destruction of the supply by fire, as frequently happens in the case of wood, is precluded by the very nature of the hemp-raising industry. Since only one year's growth can be harvested annually the supply is not endangered by the pernicious practice of over cropping, which has contributed so much to the present high and increasing cost of pulp wood. The permanency of the supply of hemp hurds thus seems assured."

Unfortunately, this optimistic prediction proved wrong. Since the Marijuana Tax Act of 1937 was passed, many of North America's forests were cut for softwood pulp, and only monocrops of coniferous trees have been cultivated for paper production, leading to significant environmental degradation. Hemp cultivation remains illegal in the United States, but revising antiquated legislation may prove to be only a small hurdle in the future of hemp cultivation to provide paper pulp. The growing of industrial crops in general has been brought into question by dire predictions of future water supplies based on the decline in recent decades of agricultural and drinking water resources.

As the earth warms and temperate latitudes become more arid, the major production zone for hemp as well as much of the world's food will become drought stressed, there will be an increasing number of mouths to feed, and all industrial crops will face stiff competition for remaining arable land. Economics will determine whether the remaining industrial crop lands will be allocated for growing relatively low value hemp fiber crops or higher profit food, fuel, and pharmaceutical crops. The term "industrial crops" implies not only that plant products are used to fuel industry as in the past but also that they may be replaced by industrial products such as

THE TREATY WITH FRANCE. FRENCH RAGS AND FRENCH PAPER.

RAG-SORTER IN PARIS.　　　　THE PARISIAN RAG-PICKER ON HIS ROUND.

FIGURE 31. Nineteenth-century French rag collectors in Paris; rags of hemp fabric were commonly gathered to be recycled into paper products. Engraving from *The Illustrated London News*, June 8, 1861.

synthetic fibers and films in the future, and we may never see hemp grown for fiber on a large scale again.

Summary and Conclusions

Cannabis has provided humans with durable fiber for centuries. Once the predominate northern temperate fiber crop, hemp and its textile products were in widespread general use among rural and urban dwellers. The principal traditional uses of hemp fiber have been for twine, rope, nets, webbing, sacking, rugs, tarpaulins, heavy industrial canvas, fabric for clothing, and pulp for paper; today these uses continue in only a few traditional cultures. The use of natural fibers such as hemp declined only relatively recently in the late twentieth century as more affordable petrochemical fiber products became widely available; in addition, fiber crops were forced to compete with food and medicine crops for arable farm land. Hemp is now commercially grown only on a limited scale for specialty markets.

The future of hemp as a crop plant relies on its applicability to the demands of increasing human population and environmental constraints. Although fiber is the most valuable component of hemp stalks, it only composes about 25 to 30 percent of dry weight with the remainder being made up of core, hurds, and dust released by processing; today, sustainable hemp production relies on finding more uses for these fractions as well as raising biomass and stalk yields through breeding (see Chapter 10) and improving cultivation strategies. The dominant use of highly adaptable synthetic fibers for woven textile production relegates hemp to lower value nonwoven and composite uses, further diminishing its economic value as a crop (Weightman and Kindred 2005).

Hemp is biodegradable and recyclable, which enhances its desirability and image as a "green" market alternative. Faced with ever more stringent mandates for reuse and recycling of vehicles, the automotive industry will increasingly utilize hemp and other natural fibers to replace glass fibers in compression molded interior panels. The construction industry utilizes hemp fiber insulation matting as a biodegradable alternative to rockwool and fiberglass insulation mats, and hurd chips for insulation fill and press board manufacture. Hemp fiber can also be used for nonwoven agricultural fleece, matting, and mulch for weed suppression and erosion control presenting a viable alternative to plastic sheeting. Dust generated during fiber and hurd extraction can be pressed into briquettes made into charcoal. In addition, faced with depleted marine resources, hemp seed oil is increasingly attractive as a source of *omega*-3 and *omega*-6 essential fatty acids as well as highly digestible protein. Aromatic essential oils may also find increasing popularity as flavoring and scent ingredients in food and household products.

Demand for natural fibers is likely to increase, and due to its environmental advantages of requiring fewer agricultural inputs, hemp is increasingly viewed as a favorable alternative. However, hemp requires fertile, well-drained soils and regular irrigation, which places it in direct competition with food crops, and boosting yield relies largely on increasing applications of nutrients (especially nitrogen), which may in turn tarnish the perception of hemp as an environmentally friendly crop and decrease its marketability. Wastewater from retting must be appropriately treated before it is returned to the environment, and therefore the use of unretted hemp fiber in automotive and construction materials is attractive from both processing cost and environmental perspectives. The paper industry can also utilize unretted hemp in the manufacture of specialty papers. Sustainability of hemp as a crop of the future will depend on several factors, including an easing of legislative barriers, increasing economies of scale, improving cultivars, developing environmentally beneficial cultivation and processing strategies, investing in joint processing and manufacturing facilities, identifying markets for all parts of the plant, and preserving and promoting its environmentally friendly image (Weightman and Kindred

2005). Hemp presents a renewable energy source or biofuel and an environmentally friendly biodegradable alternative to plastics derived from fossil fuels and will become increasingly popular when industrial mandates begin to favor environmental concerns.

The rapid spread of papermaking technology provided a durable, compact, and convenient record keeping device. It was a development that revolutionized world cultures and rivals in importance the digital age of data management we are experiencing today. As Bloom (2001) states, "In the end, it comes back to paper—a simple material that truly changed the course of history." *Cannabis* was one of the earliest paper pulp sources and hemp paper has been recovered from several ancient archeological sites across Eurasia. Hemp played a seminal role in the development of early papermaking technology, and *Cannabis*'s rapid dispersal across Eurasia was likely accelerated by its high value as a paper pulp source. Hemp fiber can be used to produce paper of very high quality, and it still finds popularity today in specialty papers used for printing money and rolling cigarettes. Hemp pulp processing uses a more environmentally friendly chemistry than acid-based pulping, and hemp is often cited (along with bamboo and mulberry) as a renewable pulp source and a more environmentally appropriate substitute for softwood pulp. Nonetheless, even though softwood pulp production has an intense, arguably devastating impact on our environment, in the face of decreasing water supplies, resurgence in hemp fiber production on more than a cottage industry scale is unlikely.

Food, Feed, and Oil Uses of Hemp

And say, O Hemp-seed, how art thou forgotten.

(TAYLOR 1620)

INTRODUCTION

HUMAN FOOD AND ANIMAL FEED USES OF HEMP SEEDS

EARLY HEMP SEED USE IN CHINA: NEOLITHIC PERIOD THROUGH THE HAN DYNASTY

HEMP SEED OIL IN ANCIENT CHINA

ANCIENT EVIDENCE FOR TRADITIONAL PRODUCTION AND USE BEYOND CHINA
Korea
Japan

South and Southwest Asia
Central and Eastern Europe
Mediterranean and Western European Regions

PRESENT-DAY HEMP SEED PRODUCTION AND USE

SUMMARY AND CONCLUSIONS

Introduction

Humans were gathering a broad diversity of edible plant material much further back in time than has been generally accepted by scholars of prehistory (e.g., see Flannery 1969; Weiss et al. 2004; Dolukhanov 2004). As early humans spread out of Africa and into Eurasia, they must have experimented with a great number of different plants to find new and readily available food sources. It is therefore likely that humans first ingested *Cannabis* seeds far back in Paleolithic times. For millennia, seeds have been one of *Cannabis*'s more important products, used for a variety of purposes including food (see Chapter 4 for a discussion of archeological seed evidence supporting our *Cannabis* dispersal hypothesis and Chapter 9 for a review of ritual uses of hemp seeds). We propose that during humanity's long quest for wild food, seeds may have been the first of *Cannabis*'s valuable attributes to be recognized, and thus discoveries of the psychoactive and fibrous properties of *Cannabis* were most likely a byproduct of seed collection for nourishment.

Cannabis seeds can be harvested from wild plants and eaten without parching, grinding, or cooking. However, wild plants produce very small seeds that dehisce easily and fall to the ground; consequently, the widespread and substantial use of hemp seed as a food source would have been made possible only by the cultivation of *Cannabis* beginning in the late Mesolithic (see Chapter 3 for a discussion

of the evolution of *Cannabis* under domestication). No matter how highly we may rate the importance of hemp seed as a food and feed, the primary reasons for any plant to produce seed are procreation and dispersal. It is in the roles of protector and disseminator of the life packaged conveniently within its seeds that humans have had their greatest impact on the biogeography and evolution of *Cannabis*. In this chapter we will discuss the history of *Cannabis* hemp seed use for human food, animal feed, and industrial applications.

Human Food and Animal Feed Uses of Hemp Seeds

Most commonly, hemp seeds are eaten whole, either raw or parched (roasted) or milled and boiled into porridge, often with rice or another grain. Botanically speaking, *Cannabis* seeds are actually hard-shelled, single-seeded fruits (called achenes), and the soft, white kernel inside that nourishes either its own embryo or the animal that eats it is actually the "seed." Throughout this chapter we will refer to *Cannabis* fruits as "hemp seed." Many authors mistakenly refer to hemp seed as a "cereal" or "grain" because it is grown as a field crop and provides sustenance similar to true grains (e.g., rice, millet, wheat, barley, rye, etc.), but hemp is actually an oil seed of relatively high protein and low carbohydrate contents. The

In Latvia and Lithuania, as well as in Poland and Ukraine, a soup made from hemp seeds, known as *semientiatka*, is eaten ritually on Christmas Eve. After potatoes were introduced into Lithuania in the eighteenth century, they quickly became a staple food, and in the ethnic Zemaitija region, boiled potatoes were served with hemp seeds (Ambrazevicius 1996). In Estonia, hemp seeds were traditionally used in preparations of butter, milk, and porridge (Kokassaar 2003), and in Finland, hemp seeds were consumed as a ground meal mixed with barley, buckwheat, and salt. This preparation served as a "dipping foodstuff for boiled turnip roots." When cereal meal was in short supply, Finns sometimes added "hempen meal" with flour to produce bread (Ahokas 2007).

Oil derived from pressed hemp seeds was an important part of traditional societies in Finland, Russia, Poland, and other Eastern European countries. For example, hemp seed oil has been used in Finland as a fuel for lamps, a raw material for soap, and in the production of varnish (Laitinen 1996). In addition, the cooking oils obtained from hemp and opium poppy (*Papaver somniferum*) seeds were essential in Eastern Europe, especially where religious restrictions prohibited the use of animal fats in food preparation. Russians traditionally used few fats besides butter, hemp seed oil, and imported olive oil (Mack and Surina 2005). Hemp seeds or oil were used in a variety of dishes, either integrated into the meal directly or the oil was used as "the medium in which the dishes were cooked, and hemp seeds were a common part of food grants or donations to the needy in the 16th century" (Smith and Christian 1984).

Hemp seed is an ancient Russian crop with a range reaching far to the north. The *Domostroi*, which literally means "Domestic Order" or "House Rules," is a collection of 43 Russian manuscripts assembled during the sixteenth through eighteenth centuries but more probably originally written during the 1550s; this collective document provides details about how a proper Russian Christian home should be managed (see Pouncy 1994; Khorikhin 2001; Kolesov 2001). In the sections dealing with food products, it indicates that provisions of hemp seed and hemp seed oil should be stored in the home. The *Domostroi* also lists the contents of a traditional Russian spice chest in detail and includes three references to hemp seed used principally as an oil source but also to make hemp seed cakes (Pouncy 1994). Hemp along with cereal crops were important trade items in Russia for centuries and were specifically referred to as the main Russian exports from 1758 to 1762 (Davidyan 1972). Exportation of hemp seed oil from Russia finally ceased around the turn of the twentieth century due to increased domestic use and was reported as the major source of edible fats consumed by peasants in Central Russia during the mid-twentieth century in place of beef and pork fats, which were rare. The largest extent of *Cannabis* cultivation (about 816,910 hectares, or more than 2 million acres) in this region occurred in 1925 (Grigoryev 2007).

MEDITERRANEAN AND WESTERN EUROPEAN REGIONS

According to Butrica (2006), the first literary evidence that ancient Greeks consumed hemp seed cakes appeared around the middle of the fourth century BCE. During this time, Ephippus, a comic poet, composed a record of snacks (*tragêmata*) eaten during gatherings known as *symposia*, where males gathered to discuss, conspire, show off, or celebrate, and this often involved eating food and drinking wine. Eaten at such events was "*kannabides*," which Butrica (2006) translates as "a confection of *Cannabis* seeds and honey." Several centuries later, the famous Roman physician Claudius Galen (lived 129 to ca. 216 CE) tells us that cakes containing hemp seed were still being eaten, reportedly to appreciate their psychoactive attributes, which left consumers feeling warm and elated.

Galen was born at Pergamos, about 23 kilometers (15 miles) inland from the Aegean Sea in western Asia Minor (modern Turkey). Galen emulated the great Greek physician Hippocrates and was a renowned scholar and author, as well as personal doctor to Marcus Aurelius, the Roman emperor from 161 to 180 CE. Galen's writings had widespread authority in European medical practice from his own time up until the sixteenth century, but unfortunately most of his literary works were destroyed by fire (see Chapter 8). We do know, however, that he referred to a dessert containing hemp seeds in his *De Facultatibus Alimentorum* or "On the Properties of Foodstuffs" (Galen 2003; also see Brunner 1973; Butrica 2006). This dish was served after the main meal in the form of "small cakes" at banquets and was popular among the Romans of his time. Reportedly the effects of consuming these cakes included a sense of hilarity (Watt 1908; also see Gerard 1633), as well as increased thirst and sluggishness if taken in excess (Lewin 1964). If in fact this confection produced relaxation and euphoria, as Galen suggested, then it probably contained inflorescences of *Cannabis*, or parts of them, including both seeds and psychoactive resin (see Chapter 8 for historical references to the mind-altering use of *Cannabis*).

Although huge amounts of imported hemp fiber were utilized by the Roman Empire for cordage and other uses, *Cannabis* seed probably was not a major crop. No historical records from ancient Rome referring to the dietary value of hemp seed have been found, nor are there any Roman accounts of hemp seed oil. Nevertheless, hemp seed may have been a relatively common food item in early Italy. Carbonized *Cannabis* seeds were discovered in the ruins of Pompei in southern Italy, which was buried by the eruption of Mt. Vesuvius in 79 CE (Frank 1959; also see Nelson 1996). Additional hemp seeds were recovered more recently, along with numerous seeds from other useful plant species, in a storage vat associated with a rustic farm house covered during the eruption (Ciaraldi 2000). Whether or not these seeds were to be used to produce fiber crops or food is unclear.

Early hemp seed remains have also been discovered further west in Europe. For example, six seeds were found at the Al Poux site in the bottom of the Boulou Valley of southwestern France and dated to the Late Iron Age ca. 2100 BP; presently, these are among the most ancient *Cannabis* seeds recorded from Western Europe (Bouby 2002). Again it is unknown whether these seeds were used for food, oil, or even medicine rather than sowing seed for fiber crops. During medieval times, water-powered mills were used to extract oil from poppy, mustard, and hemp seed, especially in France and Italy, where it is claimed that three-quarters of these mills operated, while there are no examples of hemp oil mills from medieval England (Lucas 2005).

Meersschaert et al. (2007) reported the discovery of *Cannabis* seeds in the seventeenth-century ramparts of Damme in Belgium. These and other ancient hemp seed finds in Belgium, Netherlands, Germany, Switzerland, France, England,

and several other European countries (see Chapters 4 and 5) indicate that *Cannabis* was cultivated for seed and oil as well as its useful fiber at least by the medieval period. For example, in the western Netherlands, Bakels (2005; also see Bakels et al. 2000) found ancient *Cannabis* seeds in many early medieval contexts, and therefore she referred to hemp as a widespread crop in the more moist regions during this period. *Cannabis* seeds have been recovered from many sites dating from about 700 CE to the early twentieth century. These are often found in cesspits and indicate that hemp was grown for cordage fiber or hemp seed oil. The most ancient hemp seeds recovered in the Netherlands, reported by Van Haaster in 2003 (also see Van Haaster and Hänninen 1998), included more than 2,000 waterlogged seeds found in medieval period contexts at the Gorinchem-Krijtstraat site and dated to 1300 to 1400 and 1575 to 1650. An informative archaeobotanical discovery documented in an internal report for a consulting firm in the Netherlands tells us that crushed seeds were found in a medieval site at Lochem and thus can be "interpreted as evidence for oil pressing" (de Man 1996).

In the late sixteenth century, John Gerard (lived 1545 to 1612) published his great *Herball or General Historie of Plantes*, containing more than 1,600 pages. In his section on the effects of consuming hemp seeds Gerard refers to comments of Pier Andrea Mattioli (lived 1501 to 1577), who indicated that such seed when fed to hens will cause them to lay more eggs (Gerard 1633, edited by Thomas Johnson). Another seventeenth-century English document referring to the use of hemp seed as food for domestic fowl reported that farmers were hesitant to substitute hemp seed for traditional grain crops since they claimed the seed "gave an ill flavor to the flesh of the bird that feeds on it" (Thistle and Cook 1972). People also consumed hemp seeds in historic England. For example, hemp seeds along with evidence of other edible plants were found in the latrine at Dudley Castle in the West Midlands of central England and dated to ca. 1642 to 1647; at this time the castle was being used by the Royalist garrison during the English Civil War (Moffett 1992).

In 1620, a lengthy, rambling poem written by John Taylor (lived 1580 to 1653), titled *The Praise of Hemp-Seed*, was published in London and promoted the multipurpose uses and importance of hemp to his countrymen. Taylor was a prolific English pamphleteer and popular rhymer commonly known as the "Water Poet" (Taylor 1620). Although he was not a refined writer, Taylor was an ardent eyewitness observer of human behavior and fashions in the seventeenth century, and thus his writings have been frequently studied by social historians. He opened his odd and eclectic ode to the seed of *Cannabis* with this comment: "And say, O Hemp-seed, how art thou forgotten by many Poets that are dead and rotten?" Taylor ended his praise of hemp seed with his vision of its place in his own written works: "This worke of *Hempseed* is my Master-peece." Essentially, Taylor beseeches his fellow imbibers with a bawdy line, "O all you Bachinalian drunkards honour Hemp-seed," and reminds us that in the England of his time, people would be insufficiently supplied with rope, cloth, rags, paper, and other essential material if it were not for the necessary hemp seed!

The cultivation and use of hemp seeds in England continued into the twentieth century as demonstrated in the following passage from the popular *Children's Encyclopædia* of 1909 (see Mee 1909): "Hemp seed is not ripe when the canes are right for fibre, so special plots are grown for seed, which is valuable as poultry food. Oil for paint is extracted from the seed. The plants are best grown in hills so that they have room to branch and produce the greatest amount of seed. The seed crop often nets the farmer almost as much as if he grew hemp fibre [*sic*]."

Present-Day Hemp Seed Production and Use

Hemp was cultivated for fiber and seed in North America from colonial times until it was banned in the twentieth century. Eastern European settlers in Canada, for example, carried hemp seeds with them when they immigrated into the prairie regions, and they grew *Cannabis* and utilized the seeds "for fresh oil, baking and traditional dishes," while Canadians of Chinese ancestry "have also long eaten hemp seeds for medicinal and dietary reasons" (CHTA/ACCC 2004). After several years of concerted governmental study, in 1998 Canada began once again to allow the cultivation of hemp for food and fiber, but the US government (as of early 2013) remains reluctant to repeal the outdated Marihuana Tax Act of 1937 and reinstate hemp as an industrial crop.

In the late 1930s, Ralph Loziers, as general consul of the National Institute of Oilseed Products, representing paint manufacturers and lubrication oil processors, testified before the Ways and Means committee of the U.S. House of Representatives about the Marijuana Tax Act. Loziers noted that *Cannabis* seeds were utilized for food in many Asian countries as well as in part of Russia and referred to its significance in this regard: "It is grown in their fields and used as oatmeal. Millions of people every day are using hemp in the Orient as food. They have been doing this for many generations, especially in periods of famine" (quoted in Herer 1992; also see Booth 2003). Loziers also claimed that hemp seed was the finest bird seed available (e.g., see McKenny 1939). Although hemp seeds were not banned by the Marihuana Tax Act (if they were sterilized before being sold), it effectively blocked the possibility of developing an important food industry based on hemp and greatly affected the paint and varnish industries, as hemp seed oil was the drying agent of choice before the invention of petrochemical paints. During the period just before the Tax Act went into effect, the American paint industry imported more than 40,000 metric tons of hemp seed annually. Following its passage in 1937 and subsequent trade restrictions, access to hemp seed became severely limited, and thus use for any purpose in the United States was essentially prohibited.

North American and European markets for food-grade hemp seed and hemp seed oil have increased in recent years (e.g., see Callaway 2004). Before 1990, hemp seed was only available through pet feed suppliers as it is a popular ingredient in cage bird foods said to promote avian health and encourage singing. According to Khan et al. (2010) hemp seeds are fed to male birds in Iran during the breeding season to increase their vigor, "and seed-eating migratory birds are especially attracted to hempseed fields at harvest time" (also see Hayatghaibi and Karimi 2007).

The vast majority of hemp seed used for human food and animal feed in the twentieth century was grown in China, with some supplied by Eastern Europe, Turkey, and France. Interestingly, recent agricultural research in Pakistan has shown that feeding powdered *Cannabis* seeds to domestic chickens has a "remarkable impact on growth of broiler chicks and can help in alleviating feed expenditure" (Khan et al. 2010). In addition, reference data research compiled at the

University of Manitoba (House et al. 2010), involving a multiyear study to generate data to support registration of hemp seed as a feed ingredient in poultry diets, "provides support of protein claims for hemp seed products and provides evidence that hemp proteins have a [protein digestibility-corrected amino acid score] equal to or greater than certain grains, nuts, and some pulses."

Sowing seed of many hemp fiber cultivars from France and Eastern Europe was commercially available throughout the late twentieth century, but it was far too expensive to be used as food. Canada is North America's only hemp seed producer, and according to Health Canada statistics (Alberta Government 2007), the cultivated area of seed hemp grew from 2,370 hectares (5,857 acres) in 1998 to 19,458 hectares (48,060 acres) in 2006; however, there has been substantial fluctuation in acreage. In fact, in 2007, the area under cultivation in Canada decreased once again, dropping "more than fourfolds" to 11,569 hectares (28,588 acres), primarily because of the lack of processing facilities for hemp fiber and stock. In addition, a negative image of hemp seed remains among many due to its international ban in 1961 under the United Nations Convention on Narcotic Drugs and other associations with psychoactive drugs. "Hemp does suffer from the 'snicker factor,' largely because of its hippy-dippy image and close association with marijuana, its conscious-altering cousin" (Alberta Government 2007; also see Chapter 7). Nevertheless, since 1998 *Cannabis* has been grown legally for seed across Canada with most of the cultivated area concentrated in the prairie regions (CHTA/ACCC 2004). Furthermore, during the same period, the production of hemp seed also increased in France, Germany, Netherlands, and Spain. The introduction of hemp seed kernels, nutritious hemp nuts without their crunchy shells, has significantly increased consumer interest in eating hemp seeds and hemp seed oil is featured in a wide array of body care products (e.g., balms, lotions, soaps, and shampoos).

Summary and Conclusions

Hemp seed has a long history of use as human food, animal feed, and dietary and lighting oil, as well as for its medicinal values. It is one of the plant kingdom's richest sources of high-quality proteins and EFAs and has likely played key dietary roles periodically throughout human history. In China, where archeological finds of seed remains support numerous historical accounts, we have developed the most complete scenario for the multiple uses of hemp seed. Hemp seed was an important "grain" in ancient China, one of the agricultural cornerstones of Chinese culture, and an excellent source of protein and fat. Although it fell out of favor as new oil crop species were introduced, people have always relied on hemp seed in times of famine and still enjoy them as snacks. Europe, the Middle East, and South Asia also provide archaeobotanical remains as well as interesting historical references that further expand our understanding of early hemp seed use.

As our familiarity with hemp seed grew it became more widely recognized for its medicinal values. The easily digestible proteins in hemp seeds benefit the elderly and those otherwise nutritionally compromised, and their EFAs have found nutritional as well as topical uses. Hemp seeds and kernels are commonly found in "natural food" products and hemp seed oil is becoming more widely available through alternative food stores, as well as in body care products. The ultimate modern use of hemp seeds may be in processed food applications where the EFA-rich oil can be extracted and the remaining seed meal used to make protein isolates. As long as starvation remains a persistent problem, availability of nutritionally stable dietary supplements high in EFAs and easily digested proteins could have worldwide consequences, improving the quality of life for many. Hemp seeds have served humanity well for millennia and should continue to do so well into the future.

CHAPTER SEVEN

Historical Aspects of Psychoactive *Cannabis* Use for Ritual and Recreation

Of all that Orient lands can vaunt
Of marvels with our own competing,
The strangest is the Haschish plant,
And what will follow on its eating.

(WHITTIER 1882)

Introduction

A global survey of *Cannabis* use by the United Nations Office on Drugs and Crime (2007) reported that in 2005 approximately 160 million people world-wide, or 1 in every 43 persons on earth, indulged in *Cannabis* for psychoactive purposes. The report also claimed that about 10 percent were "first-time users" and 50 percent of "daily users" developed a dependence on *Cannabis* (also see Hall and Degenhardt 2007; Clapper et al. 2009; Brook et al. 2011). A similar world survey by the United Nations Office on Drugs and Crime (2011) estimated that in 2009, between 125 and 203 million people between the ages of 15 and 64 had used *Cannabis* at least once in the previous year: "*Cannabis* is the most commonly used drug in the world" (Richardson 2010). Regardless of the accuracy of these estimates, it is clear that the psychoactive use of *Cannabis* is presently a very widespread and common culture trait. In this chapter we investigate and compare the use of psychoactive *Cannabis* across several regions over time based on research presented in the archeological, historical, and ethnographic literature.

Today most of us live in modern societies where a vision or mental image attributed to divine inspiration is often dismissed as hallucination, and the desire to experience direct communion with a god is frequently interpreted as a sign of mental illness. Nevertheless, some scholars and scientists assert that such visions and communications are fundamentally derived from an ancient and ongoing cultural tradition (Smith 2000; Schultes and Hofmann 1979; Schultes et al. 2001; Merlin 2003). People from many cultures, past and present, have practiced traditions involving the use of psychoactive *Cannabis*, often motivated by a desire to produce profound, essentially spiritual, experiences that supported their particular religions and spiritualistic beliefs. In earlier periods as well as modern times, societies have assessed the psychoactive social value of *Cannabis* from different points of view depending largely upon "whether a society was interested in using the plant or interested in preventing its use" (Farnsworth 1968).

A thorough survey of the ethnographic literature of diverse societies in the early 1970s by Bourguignon (1973) showed that 90 percent of these cultures "institutionalized culturally patterned forms of an altered state of consciousness." A large percentage of these altered states are produced through consumption of psychoactive drug plant substances, supporting the idea that "the ubiquity of mind-altering agents in traditional societies cannot be doubted—just as the moods of industrial societies are set by a balance of caffeine, nicotine and alcohol, among many others" (Sherratt 1995a).

Traditional ingestion of most mind-altering drug plants through the ages has been strongly associated with ceremonial and religious activity. Ritualistic consumption, in

FIGURE 36. In respect to its impact on dispersal worldwide, the most evolutionarily significant trait expressed by *Cannabis* is the adaptation of the female inflorescence to exude large amounts of readily apparent and easily collected psychoactive resin (photo ©GrowMedicine.org).

various forms, may be unequivocally religious, "as in the Christian Eucharist [Holy Communion] or the complex wine offerings to the ancestors in the elaborate bronze vessels of the Shang and Zhou Dynasties of China" (Sherratt 1995b). Customary, or preindustrial, stimuli for ingestion of mind-altering organic materials have been predominantly dictated by spiritual and/or medicinal requirements. Conversely, modern utilization in many societies, particularly in those areas impacted by Western Civilization, is often motivated by individual "recreational" interest in experiencing relaxation, happiness, or euphoria and frequently is prompted by peer group pressure. *Cannabis* can produce various altered states of consciousness depending on the quality and quantity ingested, as well as the psychological mind set of the user (i.e., the social-environmental setting in which the substance is consumed) and the mode of ingestion (e.g., oral consumption vs. breathing vapors or smoking; e.g., see Julien et al. 2011; Earleywine 2002).

A number of authors have suggested that relationships between humans and psychoactive plants began early on, most often within a ceremonial context, believing that the use of consciousness-altering plants provided the inspiration for initial human religious experiences (e.g., see Schultes et al. 2001; Merlin 2003). Others also draw attention to the relationships of mind-altering drug plant use and origins of spiritual concepts (e.g., Rudgley 1995; Furst 1972, 1976; La Barre 1970). Relationships between animals that possess complex brains and plants and fungi with mind-altering powers are ancient and may have originated before the evolution of our genus. Certainly, this relationship is one of the defining characteristics of *Homo sapiens*.

La Barre (1970) suggested that as bands of early humans spread into new regions, including new ecological conditions, they maintained a culturally inspired motivation to find and use species of plants or fungi that would allow them to transcend their "normal" consciousness and enable them to communicate with their ancestors or gods—their spirit world. This tradition extends back into the Paleolithic Era more than 15,000 years ago when people were all hunters and gatherers, long before the invention of agriculture (ca. 10,000 BP). The early, possibly first use of *Cannabis* may well have been for such profound purposes.

Archeological and archaeobotanical records, along with historical information and anthropological evidence from the ethnographic present, offer support for this general hypothesis. Although the evidence currently available is limited, with the aid of the powerful tools of modern science and human insight, our understanding of our natural, deep-rooted desire to experience ecstasy in its original sense—to break the mind free from the body and communicate with the "gods" or the ancestors—will become more clear with time. In the words of Richard Evans Schultes (1969b), the "Father of Ethnobotany," the "unprecedented strides achieved in the study of hallucinogens in the past 30 or 40 years owe their spectacular success to interdisciplinary studies and consequent integration of data gleaned from many seemingly unrelated fields of investigation: anthropology, botany, ethnobotany, chemistry, history, linguistics, medicine, pharmacognosy, pharmacology and psychology."

In the following review, we utilize Old World historical sources as well as archeological and archaeobotanical data from central, southern, and eastern Eurasia, with their varying time depths in different regions, to document the antiquity of a widespread tradition of human association with the psychoactive properties of *Cannabis*, not always for strictly religious purposes. As Schultes (1969b) reminds us, "The use of hallucinogenic substances goes far back into human pre-history. There have been suggestions that even the idea of the deity might have arisen as a result of their weird and unearthly effects on the human body and mind. Narcotic and other drugs have been reported by many writers in many cultures, since the very invention of writing. A truly interdisciplinary scientific interest in narcotics, however, has developed only during the past century."

In contemporary societies, mind-altering substances are utilized in various ways. These uses include religious, medical, and secular applications, which in ancient times may or may not have been viewed as separate. For example, Sherratt (1991) argued that psychoactive substances were consumed in ritualistic contexts in Neolithic Europe and that these "religious" uses undoubtedly included "medicinal" uses, "since it would be artificial to separate physical healing from ritual observance." On the other hand, Sherratt (1995b) also pointed out that "evidence for the employment of substances such as opium or *Cannabis* at various times in the past should not immediately be interpreted as an indication either of profound ritual significance or of widespread employment for largely hedonistic purposes—they may simply belong to the *materia medica*." We argue that the use of psychoactive substances by preindustrial societies, including those derived from *Cannabis*, was often if not always in ritualistic, religious contexts and that healing was at times the intention of the rituals. For organizational purposes, and to reflect the present-day legalistic contexts of recreational and medical uses for *Cannabis* drugs, we have separated the ritual and recreational uses of psychoactive *Cannabis* from the ritual uses of nonpsychoactive hemp fiber and seed (see Chapter 9), as well as from more clear-cut examples of medical *Cannabis* use founded on its pharmacological activity rather than only ritual practice (see Chapter 8).

Discovery of the Euphoriant Properties of *Cannabis* in Eurasia

We believe that Paleolithic peoples discovered the euphoriant powers of *Cannabis* while foraging for food very early on following their first arrival in its natural home range in Central Asia; this discovery probably occurred before the end of the last ice age, but certainly by the early Holocene as early humans radiated from glacial refugia and long before the advent of farming (see Chapters 1, 2, and 12). *Cannabis* seeds are hidden within the resin-covered female flowers, and picking through them in search of edible seeds would certainly result in sticky resins adhering to finger tips. The sticky accumulation would eventually interfere with seed collection and would be removed, its sticky consistency and fragrance attracting attention. Curious humans might then have tasted and swallowed some experimentally. Early gatherers may also have cleaned their hands and flicked bits of *Cannabis* resin into the campfire, noticed the pungent aroma, and began to breathe it intentionally, or they may have cooked the seeds along with their resinous bracts, making them much more potent (see later in this chapter).

The discovery of *Cannabis*'s psychoactive nature could also have occurred when processing dried seed plants: "The observation that *Cannabis* seeds shatter readily from mature dried flowers provided early humans with a second option for collecting *Cannabis* seeds. Once sedentary agriculture was established humans began to cultivate and harvest plants. It was then possible to dry the plants when the seeds were mature, but before they fell to the ground" (Clarke 1998a).

Threshing seeds from dried *Cannabis* flowers releases both seeds and resin glands, and the separation of resin from plant material is the basic principle of hashish manufacture. Since resin glands are tiny, round, and dense, they tend to drift to the bottom of threshed plant material along with the seeds, and both seeds and resin powder inevitably fall with the broken plant material onto the ground, floor, or carpets or mats. After removing broken flowers and leaves, separating the seeds from the curiously sticky resin powder would be the next obvious step. This would leave the farmer with a large pile of crude hashish containing little plant material. Removing small plant debris by shaking the resin powder through a stretched cloth is still an easy way to clean and purify the psychoactive resinous residue; this is most likely the origin of sieved hashish. The rubbing and sieving techniques of resin collection used today are representative of two different traditional methods. Hashish production in the Himalayan foothills represents the oldest hand-rubbed (*charas*) technique, and either Middle Eastern or Afghan/Pakistani hashish is the oldest extant example of the sieved hashish technique (Clarke 1998a).

Some may be surprised to learn that fresh and even most dried *Cannabis* has little if any psychoactive potency. For *Cannabis* to become psychoactive it must be heated to above 100 degrees Celsius (212 degrees Fahrenheit). Burning *Cannabis* will obviously raise its temperature sufficiently, but so will a longer period of lower-temperature cooking.

FIGURE 37. An Indian *sadhu* or holy man smokes *Cannabis* from a *chillum* pipe, a traditional South Asian ritual technique for communing with deities (photo ©Rohit Markande).

Tetrahydrocannabinol (Δ^9-THC or Δ^1-THC), or simply THC, is the primary psychoactive compound in *Cannabis*. However, within living plant cells THC is in the water-soluble form called THC carboxylic acid or THC-acid (THCA), which is not psychoactive. THCA is also a very stable compound and breaks down into neutral THC very slowly even in dry marijuana and hashish. *Cannabis* must be heated to decarboxylate the THC-acid (remove the carboxylic acid side chain) and turn it into its psychoactive form. In addition, THC in its neutral form is not water-soluble but is readily dissolved in ethanol (drinking alcohol) and fatty liquids (e.g., milk products, animal lard, and vegetal oils). When *Cannabis* is eaten, one of these solvents is employed as a carrier to facilitate dispersal of the THC and deliver it into the bloodstream. Eating unheated *Cannabis* will have only a minimal effect, as THCA is decarboxylated very inefficiently by digestion. So when weighing the accuracy of legends and historical accounts of purported psychoactive *Cannabis* use, it is important that we mention two key points: first, *Cannabis* was either burned or heated by cooking, and second, either alcohol or a fat was added.

In any event, if early humans consumed enough resin, altered consciousness would result. Sherratt (1991) pointed out that, in Neolithic times, even weak varieties of drug plants would have been appreciated and that stronger varieties and preparations would probably have evolved later, especially favored by human selection and breeding soon after the dawn of agriculture. When potent psychoactive substances such as alcohol and hashish came into use, earlier and less potent varieties of some mind-altering plants were probably displaced and their usage has since been forgotten.

According to Louis Lewin, in his seminal study of mind-altering drug use (first published in 1924), from our earliest knowledge of humanity, we find our ancestors consuming substances of no nutritive value but taken for the sole purpose of producing a feeling of contentment, ease, and comfort. Today, as in the past, the most conspicuous use of *Cannabis* is for euphoric or psychoactive purposes; it is one of the oldest known and most widely disseminated mind-altering species. As with many contemporary, popular psychoactive substances, the drug use of *Cannabis* has been and is often prompted by the desire to relieve the monotony of daily life or to increase the delight of living and induce a transcending spiritual experience (for recent clinical research demonstrating the mystical experience associated with hallucinogens, see, e.g., Griffiths et al. 2006; Johnson et al. 2008). Among some traditional African ethnic groups and associated diaspora groups, *Cannabis* still plays a significant role in religion and music. In India, it has been used for many centuries by yogis to generate vivid imagination, as well as feelings of transcendence and psychic exaltation, which these consumers believe are god-given qualities of the plant. Presently, *Cannabis* is smoked on every continent by a relatively small but stable percentage of the population, largely with recreational intent, and shows no signs of decreasing in popularity.

Central Asia

Shamanistic traditions of great antiquity in Asia and the Near East had as one of their most important elements the attempt to find God without a veil of tears; that *Cannabis* played a role in this, at least in some areas, is borne out in the philology surrounding the ritualistic use of the plant. Whereas Western religious traditions generally stress sin, repentance, and mortification of the flesh, certain older non-Western religious cults seem to have employed *Cannabis* as a euphoriant, which allowed the participant a joyous path to the Ultimate; hence such appellations as "heavenly guide." (EMBODEN 1972)

The use of *Cannabis* to the west of China has an ancient history, perhaps older and certainly somewhat different from that within China, at least in more recent Holocene times. Sherratt (1987, 1997) suggested that early inhabitants of the Eurasian steppes, members of the Sredni Stog culture (named after the Ukrainian village of Seredny Stih where it was first located), used *Cannabis* to make a "socially approved intoxicant," celebrating its significance "by imprinting it on their pottery." According to Anthony (1991; also see Anthony 2007), the crucial relationship between horses and human riders originated in the Sredni Stog culture, which flourished in Ukraine 6,000 years ago. The origin of horse riding was the first significant innovation in human land transport predating the invention of the wheel, and hemp fibers may have played an important role in horse control (see Chapters 4, 7, and 9 for in-depth discussions of the origin and spread of this relationship and the cultural derivation and geographical dispersal of *Cannabis* use for ecstatic and ritualistic purposes).

The Greek historian Herodotus (Book 4.73ff, written ca. 450 BCE) reported, in a well-known section of his *Histories*, that the horse-herding Scythians living northeast of Greece burned *Cannabis* seeds in censers filled with hot rocks and breathed the vapors enclosed within a small tent suspended over sticks. The following is a translation of Herodotus's description of this shamanistic funerary purification ritual: "They make a booth by fixing in the ground three sticks inclined towards one another, and stretching around them woolen felts which they arrange so as to fit as close as possible: inside the booth a dish is placed upon the ground into which they put a number of red hot stones and then add some hemp seed . . . immediately it smokes, and gives out such a vapor as no Grecian vapor-bath can exceed; and the Scythians, delighted, shout for joy" (Hyams 1971; also see Bremmer 2001; Sélincourt 1965).

The Scythians (variously referred to as Indo-Scythians, Indo-Iranian Sakas, or Sakas) belonged to a culturally unified group of extremely widespread nomadic tribes that inhabited a huge area of the Great Eurasian Steppes. Herodotus referred to Scythian tribe(s) located, more or less, in the farthest western region of Eurasia occupied by this somewhat heterogeneous group; these are the Scythians who first moved into the transitional area between Western Asia and Eastern Europe (eventually stretching from the Carpathian Mountains to the Don River) more than 3,000 years ago. Based on Herodotus's description, Meuli (1935), in his classic article on "Scythia," argued that Scythian "shamans" guided the souls of deceased persons to their place in the other world or afterlife and at the same time purified homes or burial places thus protecting people from returning ghosts of the dead (also see Dodds 1951; Burkert 1972; Margreth 1993; Ginzburg 2012). Meuli also suggested that the Scythians' screaming and howling was the usual loud calling of shamans in trance and that *Cannabis* seeds heated on glowing embers in censers likely caused their ecstasy (see Bremmer 2001 and Zhmud 1997 for an alternative view, suggesting ecstatic use in the "vapour-bath" ritual but perhaps not involving classic shamanism).

The Massagetae ("great Getae"), a nomadic Central Asian tribe (which many regarded as related to the Scythians), used a certain plant because of its psychoactive properties. According Sélincourt's translation of Herodotus's *Histories* (1.202) they discovered the mind-altering effect of the "fruit" (seeds or inflorescences?) of a "tree" (tall female *Cannabis* plant?), which had "a very odd property." When the Massagetae "have parties and sit round a fire, they throw some of it into the flames, and as it burns it smokes like incense, and the smell of it makes them drunk just as wine does us, and they get more and more intoxicated as more fruit is thrown on, until they jump up and start dancing and singing" (Sélincourt 1965). According to some, Herodotus's reference to this plant and its strange power (which comes from information he was told by someone else) is "probably a garbled reference to the use of *Cannabis*, which can reach great heights" (Bremmer 2001).

One of the Thracian tribes (Getae) from Dacia in the Carpathian Basin, north of the Danube River and west of the Dniester, were followers of a curious mystical "shaman" cult known as the *Kapnobatai*. The name of this cult can be interpreted as those who "walk in the smoke clouds," presumably referring to their use of *Cannabis* for ecstatic purposes (Emboden 1972; Strabo 7.3.3, quoting Posidonius; also see Eliade 1972; Eliade and Trask 1972).

These ancient accounts were corroborated by a well-publicized and controversial find of *Cannabis* seeds in one of the frozen tumulus-shaped, noble graves or kurgans (Barrow 2) located near Pazyryk in Russia, about 200 kilometers (124 miles) north of the Chinese border and 150 kilometers (93 miles) west of Mongolia in the region north of the Altai Mountains in Siberia. This remarkable grave site was originally discovered and excavated under the direction of Sergei Rudenko in 1947 and later dated to the Iron Age (ca. 2430 BP; Rudenko 1970). We include Rudenko's account of the occurrence of *Cannabis* seeds in the tomb, even though much of it is highly conjectural:

> Thus in Barrow 2, two smoking sets were found: vessels containing stones that had been in the fire and hemp seeds; above them were shelters supported on six rods, in one case covered with a leather hanging and in the other case probably with a felt hanging, large pieces of which were found in the southwest corner of the tomb. Finally, there was a [leather] flask containing hemp seeds fixed to one of the legs of a hexapod stand. Consequently, we have the full

set of articles for carrying out the purification ritual, about which Herodotus wrote in such detail in his description of the Black Sea Scyths. There had been sets for smoking hemp in all the *Pazyryk* barrows; the sticks for the stand survived in each barrow although the censers and cloth covers had all been stolen except in Barrow 2.

> In each vessel besides the stones . . . there was a small quantity of seeds of hemp (*Cannabis sativa* L. of the variety *C. ruderalis* Janisch.). Burning hot stones had been placed in the censer and part of the hemp seeds had been charred. Furthermore the handle of the cauldron censer had been bound round with birch bark, evidently because the heat of the stones was such that its handle had become too hot to hold in the bare hands.

Rudenko's speculations on the "smoking" of *Cannabis* by these Iron Age nobles, based largely on Herodotus's accounts of the of the Black Sea region, led to many claims that the Scythians laid *Cannabis* inflorescences on the red-hot stones and inhaled the smoke used both to purify them and to allow them to communicate with ancestral spirits. However, all that we know with certainty is that *Cannabis* seeds were found in a ritual context in the Pazyryk tombs.

Herodotus's accounts only describe ritual burning of *Cannabis* seeds, and he never mentions Scythians placing whole boughs or even clumps of flowers and stems upon the hot rocks. Later interpretations of his travelogues assumed that he confused *Cannabis* seeds with flowering tops containing seeds, arguing that *Cannabis* flowers must have been burned to cause the cries of joy. Coriander (*Coriandrum sativum*) seeds were also recovered from the Pazyryk burials. Coriander is not native to the Tian Shan or Altai regions, and its seeds must have been imported from Asia Minor (Rudenko 1970). Although we cannot dismiss the idea that psychoactive resinous parts of *Cannabis* along with its seeds were used in the censers, it seems just as likely that both hemp and coriander seeds were sprinkled atop the hot rocks in order to produce a thick and fragrant if not also inebriating smoke for ritual purification (it is interesting to note that linalool, the primary constituent oil of coriander seeds, is pharmacologically active, providing pain and inflammation relief, and is a possible synergist with THC and CBD, produced by *Cannabis*, e.g., see Russo 2011).

The conclusion that the Scythians "smoked" *Cannabis* has been referred to not only as the earliest known example of *Cannabis* smoking but also as one of the few examples of "smoking" in Eurasia prior to the discovery of the New World. Rudgley (1999) and others tell us that smoking might easily have been discovered early on:

> The ultimate origins of the use of such substances remain obscure, but Louis Liebenberg [1990] has suggested that the practice of smoking could perhaps be traced back to the domestication of fire by early man. Many tribal peoples traditionally transported fire in a small container such as a shell or a wooden tube. Perhaps the burning of a fragrant grass or other plant in such a container led to the development of the smoking pipe in some remote period of prehistory. It therefore seems likely that the deliberate inhalation of smoke is a very old habit indeed and may even be as old as the human control of fire.

Here we differentiate between the inhalation of smoke or vapors and true smoking. By our definition "smoking" is the inhalation of smoke through a device that holds the burning plant material and provides a pathway to deliver the smoke to the respiratory system. By this definition, the inhalation of smoke from the air is not smoking but simply inhalation, or what today is called "side-stream smoking." The censers, rods, and leather flask may be the earliest evidence of intentionally inhaling *Cannabis* smoke and certainly predate any contact with the New World by nearly 2,000 years. If this finding could be verified by the detection of psychoactive cannabinoid residues, it would stand as important evidence for the independent appearance of the ritual breathing of smoke in the Old World and smoking with special pipes in the New World.

Similar bowls referred to as "pipe-cup," "polypod," or "footed" vessels have been found containing hemp seeds in two ancient archeological sites, one in Romania from a kurgan pit-grave at Gurbanesti near Bucharest and the other in a northern Caucasus Early Bronze Age site, both dated to about 5000 to 4000 BP (Ecsedy 1979). In addition, many comparable bowls not containing ancient *Cannabis* seeds or other parts of the plant have been found in a number of sites of relatively similar age. Do these vessels indicate that there was a widespread cult use of hemp? Based on Herodotus's quote concerning Scythian vapor baths and seed discoveries in widely separated locations from the Altai Mountains of Central Asia to the Black Sea, Sherratt (1991) argued that "the practice of burning *Cannabis* as a narcotic [*sic*] is a tradition which goes back in this area some five or six thousand years and was the focus of the social and religious rituals of the pastoral peoples of central Eurasia in prehistoric and early historic times."

An artifact interpretation by Sherratt (1987) offered an explanation for the ancient cord pottery impressions on many of these bowls. He argued that these corded impressions were made with hemp fibers and that peoples of the Bell Beaker cultural complex (third millennium BCE), located over a vast area of steppe stretching from Eastern Europe to Western China, spread an alcohol-based hospitality or "social lubricant" culture. Sherratt also argued that hemp, long cultivated for its fiber, was also "enjoyed" by steppe peoples, well beyond the range of the drinking complex, and suggested that alcoholic beverages were infused with *Cannabis* and that corded ware and its forerunners may have been impressed with hemp fibers as a means of honoring the vessels' contents (also see Sherratt 1991, 1995a/b, 1997; Chapter 4). Corded ware may have been reserved for rituals rather than being in common use.

As noted earlier, ritual use of *Cannabis* by ancient nomadic tribes in western Central Asia was described in some detail by Herodotus, but there is less historical evidence of ancient magico-religious and medicinal use of *Cannabis* from eastern Central Asia, especially northwestern China where shamanistic practices were not shared by a majority of people or openly mentioned in any depth in ancient Chinese texts (Touw 1981).

What we do know is that the ancient Chinese historical work *Hou Hanshu*, or "Book of the Later Han" (compiled in the fifth century CE by Fan Ye), describes the "Western Regions" of China from 25 to 220 CE, and during this period, China's Western Regions were known for their fertility and advantageous cultivation of "rice, two kinds of millet, wheat and beans, mulberry trees, hemp, and grapes" (see Hill 2003). Furthermore, because of this agricultural richness, "the Han have constantly struggled with the Xiongnu [people] over Jushi [Gūshī culture, also known as Jüshi or Cheshi] . . . for the control of the Western Regions" (also see Hill 2003). We will see later that the Han had different relationships with *Cannabis* than the ancient nomadic-based people of the "Western Regions" they referred to collectively as the Xiongnu, who formed a confederation that rivaled the agriculture-based empire of the Han dynasty. The *Xiyu juan* or "Chapter on the Western Regions" from *Hou Hanshu* 88 (translated by Hill 2003; also see Hill 2009) provides insight into the attitude of the Han toward these remote, culturally different peoples:

> The Western Hu are far away. They live in an outer zone.
> Their countries' products are beautiful and precious,
> But their character is debauched and frivolous.
> They do not follow the rites of China.
> Han has the canonical books.
> They do not obey the Way of the Gods ["Way of the Spirits"].
> How pitiful!
> How obstinate!

Although the Chinese have a very long association with *Cannabis*, using it for millennia to extract fiber, edible seeds, and medicine, its utilization as a source of a psychoactive substance used in ritual or recreational activities was frowned upon at least since the time of Confucius. However, the recent discovery of remarkably well-preserved floral and other remains of *Cannabis* in the Yanghai Tombs provides physical evidence of its ancient ethnobotanical significance in this region of western China for shamanic, psychoactive, or medicinal purposes.

The Yanghai Tombs, attributed to the Gūshī culture, are located in a stony desert area of the Turpan Basin of Xinjiang province (formerly known as Chinese Turkestan). The initial written record that refers to the Gūshī in the *Hou Hanshu* defined them as "nomadic light-haired blue-eyed Caucasians speaking an Indo-European language," who were known to have raised horses and other grazing animals and cultivated crops and were proficient archers (Russo et al. 2008; also see Ma and Sun 1994; Mallory and Mair 2000; Karpowicz and Selby 2010). The Gūshī people probably migrated thousands of years ago from the Eurasian steppes of Russia to what is now China.

The Yanghai Tombs site is centrally located in the Eurasian landmass in one of the lowest spots on earth. In grave M90, approximately 2,700 years old, the skeletal remains of a high-ranking Caucasoid male were discovered; this man was about 45 years old when he died and was found with grave gifts more elaborate than others at the site. Based on the burial offerings, Jiang et al. (2006) and Russo et al. (2008) suggested that the man entombed was a shaman (also see Xinjiang Institute of Cultural Relics and Archeology 2004; Academia Turfanica 2006). Most of the utensils found with the skeleton were connected with horsemanship, but a bow, arrows, musical instruments, and some wooden cups were also discovered. In addition, a large, lidless leather basket and a wooden bowl filled with a total of 789 grams (nearly two pounds) of female *Cannabis* inflorescences and seeds were found near the head and foot of the corpse, seemingly within close reach of the deceased in the afterlife. The wooden bowl shows prolonged use as a mortar because its inner surface is smooth with one side perforated. Presumably *Cannabis* flowers were pulverized in the bowl before use for psychoactive or medicinal purposes. The remains from the Yanghai Tomb are very well preserved due to the depth of the burial (two meters, or approximately six feet or more), the extremely dry climate with little or no rainfall (averaging about 16 millimeters, or about two-thirds of an inch per year), and the alkaline soils

(pH 8.6 to 9.1). Ancient *Cannabis* remains from this tomb include leaves, stems, seeds, and bracts with nonglandular and glandular trichomes, all intact after more than two and a half millennia.

The *Cannabis* flowers in the ancient leather basket could have been a gift for the deceased, but why would he need them in the afterlife? "If they were meant as a cereal or for oil extraction, the leaves and shoots would have been removed; if fiber production was intended, only the stems needed to be saved. In this connection it is significant that no hemp textiles have been unearthed in the Yanghai Tombs" (Jiang et al. 2006).

In fact, the scientists who studied this remarkable discovery found no evidence that the ancient people of the Turpan Basin utilized *Cannabis* for food, seed oil, or fiber. The presence of female inflorescences, the most potent portion of the plant, strongly suggests that the deceased was well aware of their psychoactive properties. However, ancient medicinal, ritual, and psychoactive uses of *Cannabis* are often difficult to separate (Merlin 2003; Sherratt 1991). Among the nomadic tribes of Siberia and Central Asia, shamans often acted as physicians as well as ritual practitioners and may have utilized the psychoactive properties of *Cannabis* in shamanic practices associated with purification and healing ceremonies, similar to healing rituals using nonpsychoactive *Cannabis* recorded across much of Eurasia during the early twentieth century (Touw 1981; also see Chapter 9). In any case, Jiang et al. (2006) concluded that there was a spiritual connection between the "shaman" and the *Cannabis* contents in the ancient basket: "The deceased, presumably a shaman, may have been mainly concerned with the ritual of communication between the human and the spirit world. The gift of *Cannabis* may have been to enable him to continue his profession in the afterlife."

More recently, interdisciplinary research by an international team of scientists (Mukherjee et al. 2008) "demonstrated through botanical examination, phytochemical investigation, and genetic deoxyribonucleic acid [DNA] analysis by polymerase chain reaction [PCR] that this *Cannabis* material contained tetrahydrocannabinol [THC]." This ancient chemical material was found along with "its oxidative degradation product, cannabinol, other metabolites, and its synthetic enzyme, tetrahydrocannabinolic acid [THCA] synthase, as well as a novel genetic variant with two single nucleotide polymorphisms" Mukherjee and his colleagues concluded that the Yanghai *Cannabis* "was presumably employed by this culture as a medicinal or psychoactive agent, or an aid to divination." It represents the most ancient record of *Cannabis* use "as a pharmacologically active agent," and therefore it is a significant addition to the medicinal and archeological documentation of the pre–Silk Road Gūshī people (also see Russo et al. 2008). Although no smoking pipes were discovered in the M90 Yanghai tombs, *Cannabis* probably would have been eaten or placed in a burning fire to produce fumes, and genetic analysis suggests that it was cultivated rather than gathered from the wild (see Mukherjee et al. 2008). This hoard of *Cannabis* is not the oldest example of early use, but it certainly is among the best studied.

It is now more than 20 years since Uighur farmers uncovered the huge ancient Yanghai cemetery (54,000 square meters, or 580,000 square feet) containing at least 2,500 burials, including the one containing the "shaman" and a substantial amount of well-preserved *Cannabis* inflorescences. Rudgley (1998) tells us that the local Muslim peoples of Xinjiang province today, mainly Uighurs, "are still associated

FIGURE 38. Excavations at the Yanghai site near Turpan in Xinjiang province, China, revealed a Gūshī culture shaman's tomb (A, B), with many of his personal belongings such as a bow, musical instruments, and remains of many female *Cannabis* flowers (C) stored within a leather basket (D) and a wooden bowl (E). The ball-shaped mass of ancient *Cannabis* found in the tomb contained leaves, stems, seeds, and flowers (F). Laboratory analyses revealed that the *Cannabis* was psychoactive and therefore may have been used as medicine or for religious rituals (illustrations courtesy of Hongen Jiang).

with *Cannabis* by the Han Chinese." Indeed the Uighurs have long been linked to the psychoactive use of *Cannabis* in the form of hashish, with recent documented use by Uighurs living in large Chinese cities such as Beijing and Shanghai (e.g., see Labrousse and Laniel 2001). According to Dikötter et al. (2004), except for the Uighur people, there is a common opinion among the Chinese that the smell of smoked *Cannabis* is "foul" and that this "popular perception" hampered the spread of its use as a euphoric substance. This negative popular and official attitude toward the psychoactive use of *Cannabis* among the majority of Chinese has prevailed for many hundreds of years, but such a low regard for its mind-altering use may have been quite different before the rise of Confucianism.

China

As one of the oldest domestic plants in human history, *Cannabis* has probably been utilized for 10,000 years or more in Eurasia (Schultes et al. 1974; Fleming and Clarke 1998; Merlin 2003). It was eventually cultivated and selected for multiple purposes including the use of fiber from its stems, food and oil from its seed, and the medicinal and psychoactive substances in its resin glands (e.g., see Merlin 1972), and the use of these *Cannabis* resources by the ancient Chinese has been well documented (e.g., see Kêng 1974; Li 1974a/b, 1975, 1978).

Chinese historical accounts, some quite ancient, offer evidence supporting the ritual use of *Cannabis* for its mind-altering properties. Even the etymology of the Chinese character for hemp (麻, *má*) provides us with evidence that the stupefying effect of *Cannabis* was well known from the earliest days of Chinese culture. *Má* has two meanings: (1) "numerous or chaotic," derived from the nature of tangled hemp fibers, also providing the source of its use as a collective noun to identify bast fiber plants in general, and (2) "numbness or senselessness," resulting from the physiological properties of the seeded inflorescences, which were used in early medicinal infusions (Li 1975).

The Chinese were apparently well acquainted with psychoactive *Cannabis* euphoria since the Zhou dynasty (ca. 1050 to 221 BCE) when it was used for "the enjoyment of life" (Creel 1937). The *Shih-i-chi*, or "Gathering Remaining Accounts," written by Wang Chia (died 386 CE), speaks of the "juice of hemp, the eating of which causes one to see spirits" (Hirth 1966). According to the *Ko-chih-ching-yuan*, or "Mirror Origins of Science" (a literary repository of new inventions published in 1735 CE), *Cannabis* was introduced into China by Zhang Qian (lived 195 to 114 BCE), who explored much of Central Asia as an emissary for Han dynasty emperor Wu. This probably refers to the reintroduction of *Cannabis* use for mind-altering purposes, which were largely dismissed and/or forgotten after Confucian ethics became widespread during the time of Han dynasty (e.g., see Li 1974a, 1975).

In the first century BCE, radical herbalist experimenters of the Mao Shan Taoist tradition created esoteric scriptures; according to Needham (1974), these herbalists were "aided almost certainly by cannabis." The famous Mao Shan Taoist physician and herbalist T'ao Hung Ching (lived 451 to 536 CE) authored the *Ben Cao Jing Ji Zhu*, or "Collections of Commentaries on the Divine Husbandman's Classic of Materia Medica," which was written upon republication of the legendary Emperor Shen Nung's pharmacopoeia (attributed to the twenty-eighth century BCE). In this materia medica, T'ao

Hung Ching contrasted the nonpoisonous seeds (麻子 or *má zǐ*) of *Cannabis* with its "poisonous fruits" (more accurately the bracts covering the seeds, known as *má fěn* (麻粉), or "hemp powder," possibly referring to resin gland powder or hashish). Of the latter he tells us, "*Má fěn* is not much used in prescriptions (now-a-days). Necromancers [shamans?] communicate with the spirits of the dead to reveal the future and used *Cannabis* in combination with ginseng (*Panax ginseng*) to set forward time in order to reveal future events" (translated by Li 1974a).

In the *Wu Tsang Ching*, or "Manual of the Five Visceras," from the sixth century CE, which is attributed to Chang Ching-Ching, we find the statement, "If you wish to command demonic apparitions to present themselves you should constantly eat the inflorescences of the hemp plant" (Needham 1980). There is a similar passage describing the temporal distortion caused by *Cannabis* in the ancient materia medica *Chêng-lei pên-ts'ao* written by T'ang Shêng-wei, who lived in the tenth century CE. According to the translation of this section by Li (1974a), T'ang tells us that *má fěn* has a spicy taste and is toxic: "If taken in excess it produces hallucinations and a staggering gait. If taken over a long term, it causes one to communicate with spirits and lightens one's body."

These qualities of the *Cannabis* plant were probably known much earlier, at least as early as the days of Shen Nung, which takes the mind-altering use of *Cannabis* back nearly five millennia (see Chapter 8 for more about Shen Nung and the history of *Cannabis* medicines). Although Shen Nung and his later interpreters certainly realized that *Cannabis* was a powerful euphoriant as well as anesthetic, the plant was eventually condemned in ancient China. The early "official" condemnation of *Cannabis* probably did not fully restrict its economic or euphoric use and, to a large extent, can be explained by the long history of stern moral conduct in Chinese cultural tradition. Taylor (1963) suggested that among the ancient Chinese—as reflected in some more modern Chinese—to be a little happy was suspect, and to be very happy was quite certainly sinful. Hence they referred to the resinous female plant as the "Liberator of Sin." A number of Chinese compound words that include the character for hemp refer to or make an inherent association with the effects of *Cannabis* use as viewed within Chinese Confucian culture (see Li 1974a). For example, the word for "troublesome" is *má fan* (麻妨), "numb" is *má mù* (麻木), "narcotic" is *má zuì jì* (麻醉剂), and "anesthetic" is *má zuì* (麻醉); all these words include the radical *má* (麻), meaning hemp. The special mind-altering effects associated with *Cannabis* use were also recognized by ancient Chinese physicians.

Li (1974a, 1975) explained how the ancient tradition of shamanism and its important association with the psychoactive use of *Cannabis* "slowly declined in China beginning with the Age of Confucius" and only continued in "scattered small areas" (also see Unschuld 1985; Dikötter et al. 2004; Dannaway 2009). Li argues that the ancient Chinese were not averse to consuming drugs such as wine and opium in order to alter their consciousness but have long believed that *Cannabis* is a "hallucinogenic drug" that stimulates "mental exhilaration and nervous excitation" along with a distorted "sense of time and space." According to the Confucian Chinese mind-set, *Cannabis* could be overused and "may cause rapid movements and under certain situations stimulate uncontrollable violence and criminal inclinations." *Cannabis* use for psychoactive purposes was unsuitable to Chinese personality and customs, more specifically to the moral system or philosophy of life embodied in Confucianism. Li (1974a) succinctly described the ethical doctrines of Confucianism and how they permeate the lives of all Chinese people, with an emphasis on belief in a universal "Natural Order" based on moral behavior: "Man, the superior being, is a moral entity who can refrain from wrong doing through education and through the observance of the doctrines of uprightness and moderation (the doctrine of the Mean)." Although some key aspects of Confucian ethics such as "filial piety, reverence for ancestors, and the respect for elders" are preserved in the ongoing importance of hempen mourning (see Chapter 9), the psychoactive use of *Cannabis* has long been frowned on and hence forgotten in most of China, probably due to its association with ancient Eurasian shamanic traditions: "The traditional Chinese philosophy of life is centered on humanism. It thus emphasizes in particular interpersonal relations. Compounded by its universally adopted doctrine of the Mean and its strong social system based on the family, these cultural influences seem to provide sufficient background for the universal failure to adopt a drug which causes hallucination and fantasy" (Li 1974a).

Li uses a comparison of the effects of opium versus *Cannabis* use, pointing out that an individual Chinese person's conformity to Confucian societal codes of behavior "is regulated by a culturally instilled sense of shame" and that "the adoption of opium and non-adoption of *Cannabis* reflect a behavioral response to traditional Chinese society. The opium user was more likely to remain pacific and sedated, and thus not challenge social norms. *Cannabis*, with its stimulation of erratic effects, was likely to induce acts that might bring shame upon the user or his family" (also see Dikötter et al. 2004).

Opium, the dried latex of unripe opium poppies (*Papaver somniferum*), which are most probably native to the western Mediterranean region, did not arrive in China until foreign traders introduced the psychoactive sap and eventually the plants themselves, possibly as early as the third century CE or at least by the sixth century (Merlin 1984). By this time the use of *Cannabis* for mind-altering purposes in purification and funeral ritual contexts was certainly in decline, if not already long forgotten (see Li 1974a, 1975).

Taoism and Tales of Ma Gu

Where is the flower of the Great Hemp? (HAWKES 1959)

The *Liao chai chih yi*, or "Strange Stories from the Liao Studio," provides an example of Chinese literature that makes reference to time distortion associated with *Cannabis* use. One of the stories in this collection, written by P'u Sung-ling (lived 1640 to 1715 CE), recounts the legend of two friends who, while wandering among the mountains gathering food, encounter two lovely maidens guarding a fairy bridge. At the maidens' invitation, the two friends crossed the "azure bridge" and were "regaled with *huma*" (said to be a Chinese version of hashish), when they fell "deeply in love with their hostesses" and spent with them what appeared to be only "a few blissful days" in the Jasper City. Eventually they became homesick and returned to their village to find that "seven generations" had passed and that they had aged more than a century (Folkard 1884).

麻姑献寿

FIGURE 39. Ma Gu, the legendary Chinese "Hemp Damsel" is often depicted with gardening tools, a basket of fruits and her faithful deer. The Chinese phrase *Má Gū xiàn shòu* means "Aunty Hemp offers longevity." Ma Gu's image is commonly seen on restaurant plates in northern Vietnam (adapted from Kohn 1993). Original from the *Zengxiang liexian zhuan*, or "Illustrated Biographies of Ranked Immortals."

Early Taoism added Confucian ethics and eventually some Buddhist beliefs to a core derived from earlier shamanic traditions. According to Seaman (1994), rudiments of belief related to shamanism were expelled from authorized involvement in state religion during and just after the Han dynasty, and consequentially shamanism "began to merge with the philosophical opposition represented by what eventually came to be known as Taoism." (Also see Needham 1956; Ripinsky-Naxon 1993; Boileau 2002; Pratt 2007 for differing opinions about shamanism in ancient China.)

The alchemist factions of mystical Taoism must have been aware of the psychoactive effects of *Cannabis* and the use of *Cannabis*-laced incense and elixirs. There are several references to hemp in Taoist literature, and one of the eight major deities of this tradition is Ma Gu (麻姑 or *Má Gū*), literally "Auntie Hemp," also translated as the "Hemp Damsel," "Hemp Lady," "Hemp Maid," or "Miss Hemp." Ma Gu is a rare personification of *Cannabis* who was revered during the fourth and fifth centuries CE by the Taoist sect of the Highest Clarity School of Mao Shan Mountain (south of Nanjing in Jiangsu province). Ma Gu is also the Taoist goddess of the slopes of Tai Shan Mountain, the summit of which represents the head of Pan Gu, China's mythical creator. Tai Shan is located in Shandong province near the present-day hemp-producing cities of Dong Ping and Laiwu, and today its slopes support populations of feral hemp said to result from pilgrims tossing hemp seed from the summit while asking Ma Gu for health and longevity (Clarke 1998a). Here the plant was traditionally gathered on the seventh day of the seventh month, a day of séance banquets in Taoist communities (Needham 1974; also see Bretschneider 1895). In Song dynasty times (960 to 1279 CE) or thereafter, Ma Gu was incorporated into the Taoist series of Eight Saints (Eberhard 1968).

The several legends of Ma Gu may have their origins in actual women who became immortal deities. The first and most popular incarnation of Ma Gu was as a native of Shandong province during the Eastern Han dynasty (25 to 220 CE) who was the younger sister of the immortal Wang Fangping. This Ma Gu was a sorceress famed for reclaiming a large area of coastal land from the sea and transforming it into mulberry fields through her magic arts (Werner 1961), and legend has it that she "had hands that looked like the claws of a bird" (cf. Kohn 1993, translated from the *Zengxiang liexian zhuan*). Chinese legends of the "Hemp Bird" m ay have some connection to Ma Gu and her birdlike claws. Sparrows are archaic shamanic spirit helpers called *má niǎo* (麻鸟), or "hemp bird," because they feed on hemp seeds; sparrows are also considered the most sensuous of all birds. This is why the consumption of sparrow flesh is believed to strengthen sexual power (Eberhard 1938, cited in Rätsch 2001). According to the *Zengxiang liexian zhuan*, or "Illustrated Immortals' Biographies," edited by Liu Hsiang (lived 79 to 8 BCE), Wang Fangping descended to earth and visited the family of Cai Jing: "Jing and his family duly paid their formal respects to the visitor. After that, Wang sent off someone to invite his sister, the Hemp Lady. When she arrived, all found her a young girl of about eighteen years. She wore her hair tied into a topknot on the top of her head, but some strands were left untied and flowed down well to her waist. She wore a robe of brocade and a wide embroidered skirt, with colors so bright and radiant they dazzled the eye" (Kohn 1993).

A second Ma Gu lived during the reign of King Chao Wang (ruled 328 to 332 CE), founder of the Hou-chao Kingdom, and was said to have been a bodily reincarnation of the first Ma Gu. She was the daughter of the barbaric general Ma Qiu, a tyrannical commander with a violent temper. After being threatened by her cruel father for taking pity on his overworked laborers she fled from her home, became a hermit on Tai Shan, and then ascended to Heaven (Werner 1961).

Yan Jinfen (2002) discusses the importance of a recently discovered classical Chinese poem, "A Plaint of Lady Wang," which has drawn the attention of many students of Chinese culture. Yan Jinfen was acutely attracted to the feminine aspect of the poem in which minute and exquisite narrative is permeated with mysticism, romanticism, and syncretism of a "women's struggle between *hun* (mind-heart/soul/spirit) and *po* (the bodyperson)." In the poem, Lady Wang encounters immortal personalities including Ma Gu. "Lady Wang met the fairy Hemp Lady during her wandering journey in the east heaven. She was invited to play chases with her. She was so happy that she forgot all her mortal misery." Sinologists have differed on their view of the connection between Ma Gu and *Cannabis* (e.g., see Wilhelm [1944] for a positive interpretation and Eberhard [1968] for a negative one); consequently, the cultural and linguistic origins and deeper meaning of Ma Gu and her role as the "Hemp Maiden" remains an open but intriguing question. It is worth noting here that the image of Ma Gu has been frequently seen on small restaurant dishware throughout Vietnam accompanied by the Chinese phrase *Má Gū xiàn shòu*, or "Aunty Hemp offers longevity," which refers to the healthful benefits attributed to Ma Gu (Clarke 1996, personal observation).

Renowned historian and sinologist Joseph Needham reported early Taoist spiritual uses of *Cannabis*. For example, Needham (1974) indicates that according to the Taoist encyclopedia *Wushang Biyao*, or "Supreme Secret Essentials" (ca. 570 CE), *Cannabis* was added to ritual censers. In addition, Needham refers to Yang Xi (lived 330 to 386 CE) as assisted "almost certainly" by *Cannabis* during his writing of the Shangqing Taoist scriptures when he was visited nightly by Taoist immortals. Furthermore, Needham, in his translation of *Mingyi bielu*, or "Supplementary Records of Famous Physicians," written by Shangqing (editor) and Tao Hongjing (recorder; lived 456 to 536 CE), tells us that "hemp seeds are very little used in medicine, but the magician-technicians [shamans] say that if one consumes them with ginseng it will give one preternatural knowledge of events in the future." It is important to point out here that Shangqing Taoism is the most mystical of Taoism's lineages, including practices such as spirit-travel and communication with deities residing in the "external" universe. According to Anderson (1988), for almost four centuries following the Han dynasty, medieval China remained isolated from outside influences: "The age was marked by an obsessive concern with ale. Rarely in the history of the world has alcoholism been so idealized. No doubt much of this was poetic license, but we need not believe that the bards were drunk all the time (as they would have us believe). Alcohol was however, definitely considered the great social facilitator and a proper part of all social gatherings (and is still considered so today)."

On the other hand, use of alcohol for escapist purposes was also common. Many people used hallucinogenic drugs as well, and the properties of *Datura*, *Cannabis*, and many other plants were well recognized (Li 1977). Drugs were ostensibly taken for quasi Taoist consciousness expansion, but a deeper escapist motive seems to have underlain their popularity. T'ao Yüan-ming (lived 365 to 427 CE), the great poet who lived a simple "almost Neolithic" life on his farm, cultivated several crops including hemp and drank large amounts of ale brewed from millet (Anderson 1988).

Abel (1980) referred to alcohol as a greater problem to the Chinese than either *Cannabis* or opium. He suggested that the Chinese who experimented with *Cannabis* did so as "more of a flirtation than an orgy" and that the Chinese "who hailed it as the 'Giver of Delight' never amounted to more than a small segment of the population" (also see Dikötter et al. 2004). However, incense smoke had ancient ritual significance in Taoism. According to Needham and Wang (1954), the sixth century CE Taoist literary collection known as *Wu Shang Pi Yao*, or "Essentials of the Matchless Books," clearly states that *Cannabis* was added to the contents of incense burners, used to "make a stink" and "drive away the demons." As Needham (1974) inform us, "There is much reason for thinking that the ancient Taoist experimented systematically with hallucinogenic smokes, using techniques which arose directly out of liturgical [public worship] observance. . . . At all events the incense-burner remained the center of changes and transformations associated with worship, sacrifice, ascending perfume of sweet savor, fire, combustion, disintegration, transformation, vision, communication with spiritual beings and assurances of immortality."

In ancient China, as in most early cultures, medicine had its origin in magical rites, and shamanic healers were essentially practicing magicians. While shamanistic traditions remained strong in Central Asia and Eastern Europe and the use of psychoactive *Cannabis* in particular was increasing in South Asia and Arabia, hallucinogenic practices slowly declined in China beginning in about the sixth century BCE during the Age of Confucius, and shamanistic traditions only continued in scattered small areas during later ages (Li 1974a). It is important to remember that the stimulation, fantasy, time disruptions, and erratic behavior sometimes associated with *Cannabis* use were considered disruptive to the Confucian Chinese doctrine of moderation that frowned on extremes and excess. The use of *Cannabis* for euphoric purposes by Central Asian warlike "barbarians" who harassed the Han may also explain the prejudice against its use in ancient China, as any barbarian habits "were frowned upon emphatically," and connections between *Cannabis* use and the militant nomads of western China probably influenced the early condemnation of *Cannabis* drugs in ancient China (Li 1974a).

Even though *Cannabis* gained popular recognition in South Asia as a "delight giver" (Farnsworth 1968), its main applications in China, at least after the time of Confucius, were for fiber and seed. *Cannabis* repression in China began with the rise of Confucianism and the suppression of shamanic, ecstatic ritual and religious activities, and *Cannabis* use in China came to be viewed by some as a dangerous path "to demon possession and insanity" (Dikötter et al. 2004; also see Li 1974b). As major city states grew across Eurasia, social control was provided by the edicts of religiously motivated centralized governments. Shamanism and its ritual traditions, independent of religious structure and promoting social autonomy, were suppressed by all major Eastern and Western religions because independent thinking posed a threat to societal organization. Simultaneously, many shamanistic rites were preserved more or less clandestinely and cast in a new light by modern organized religions seeking followers. (See Chapter 9 for an in-depth look at the use of nonpsychoactive hemp fiber and seed in ritual contexts.) The general suppression of mind-altering *Cannabis* use in China is in strong contrast with the long-lasting tradition of its spiritual use in areas of South Asia.

India and Nepal

To the Hindu the hemp plant is holy. A guardian lives in the *bhang* leaf. (CAMPBELL 1894)

South Asia was greatly influenced by migrations of peoples and ideas from the north. The Aryans invaded the Indian subcontinent from their Central Asian homeland and eventually settled in the Punjab Valley along the Indus River and throughout the margins of the Ganges Valley. Hypotheses proposing the original diffusion of *Cannabis* into ancient India by Aryan tribes around 3500 BP, either via the Iranian Plateau or directly from Central Asia across the Himalayas, presuppose the nonexistence of this plant in the Indus Valley prior to Aryan migrations. Until recently, no archeological evidence had been uncovered to substantiate the presence of *Cannabis* in ancient India prior to this time. Madella (2003) reported finding *Cannabis* phytoliths (silica castings of plant cells) at the ancient Indus River Valley site of Harappa dated to 4000 to 3000 BP, while Saraswat and Chanchala (1995) recovered *Cannabis* charcoal remains from the Senuwar site in Bihar state of northern India dated to 3800 to 3200 BP (see Chapter 4 for more archaeobotanical evidence supporting the early diffusion of *Cannabis*). This ancient although

somewhat tentative evidence comes from around the time of the Aryan migrations or just before and may indicate that *Cannabis* was already in South Asia when the Aryans arrived. However, even if the Aryans did introduce *Cannabis* to South Asia, it is very likely that some of the peoples in this region already had some experience with it and may even have introduced novel uses for the plant. Based on taxonomic research (Hillig 2005a/b) combined with hypotheses of likely Pleistocene refugia (see Chapter 12), we propose that *C. indica* and its subspecies evolved ultimately from a putative drug ancestor (PDA), likely originating in a refugium within the eastern Himalaya or Hengduan Mountains. In this case *C. indica* would have spread westward relatively quickly along the Himalayan foothills and could have reached the Indus River valley at their western terminus early in the Holocene. In the future, an effective translation of the enigmatic early Indus script and comprehensive analysis of ancient Indian archaeobotanical remains may uncover evidence for the presence of *Cannabis* in the Indus Valley before 4000 BP.

In any event the Aryans were likely well aware of *Cannabis* and its psychoactive potential well before they spread into South Asia. Over many centuries, a synthesis of the immigrating Aryan and indigenous Dravidian cultures developed. Eventually, the original poetic theology of the Aryans embodied in the *Rg Veda* came under the conservative control of a dogmatic priesthood that dominated the socioreligious organization and activity of Indian communities. A strong class organization, the caste system, reinforced by strict, formalized ritual leaders (*Brahamanas*), emerged as the most influential element in the ancient Indian social order. In reaction to this trend, certain disaffected individuals left the more heavily populated areas and retreated to the Himalayan foothill regions north of the Punjab and Ganges River valleys. In their upland refuges, these *sadhus* (ascetic holy men) sought wisdom according to their own visions and composed treatises called the *Aranyakas*, or "forest texts," that contributed to the formation of traditional Hindu philosophy. It is in this region and cultural context that *Cannabis* use may have taken on early philosophical and spiritual importance; such significance is reflected in the multitude of names for *Cannabis* in Sanskrit and Hindi (e.g., see Russo 2005).

The ancient Hindu scriptures called the *Vedas* (beginning with the *Rg Veda*) are related to the *Aranyakas* and are based on oral traditions that may date from as early as 3500 BCE. They compose the core of the Indian literary tradition. The fourth volume, the *Atharvaveda*, or "Science of Charms," compiled between 1500 and 1200 BCE, seems to have been most representative of the peasant classes. In fact, because of this heritage, the popular blessings, formulas, theological concepts, and mysteries of the *Atharvaveda* apparently gained recognition as sacred literature only after a long struggle (e.g., Majumdar 1952). The oldest-known Indian literary reference to *Cannabis* occurs in book XI (11:6.15) of the *Atharvaveda*. Here the *Cannabis* plant is referred to as "sacred grass" belonging to one of the "five kingdoms of plants, with *Soma* as their chief, we address: *soma, darbha, bhangas* [*Cannabis*], *saha, yava* may free us from distress" (translated by Aldrich 1977; also see Sharma 1977a/b, 1980; Russo 2005).

In a Vedic account, the celestial nectar Amrita was produced when the gods and demons caused Mount Mandara to churn the primordial Sea of Milk (Dutt 1900). When something was needed to purify the nectar, Shiva created *bhang* (*Cannabis*) from his own body. According to another explanation, some nectar fell to earth, and from the ground the

FIGURE 40. Multipurpose *Cannabis* crops are sown in dense spacing in small gardens (A) throughout the Himalayan foothills of western Nepal. Whole plants consisting of only a single stalk topped with an inflorescence are harvested and carried back to the village (B), where the psychoactive *charas* resin is rubbed from the flowers (C), the seeds removed, and the fiber peeled from the stalks. Three products (D) are derived from each plant—fiber for weaving cloth, seeds for eating and sowing, and thin sticks of psychoactive *charas* resin to smoke.

Cannabis plant grew (cf. Campbell 1894 quoted in Schleiffer 1979, see also Mikuriya 1994). Traditionally it is believed that demons tried to take control of the celestial nectar, but the deities blocked them and gave *Cannabis* its name as *vijaya* (victory) to celebrate their success: "Ever since, this plant of the gods has been held in India to bestow supernatural powers on its users" (Schultes and Hofmann 1979).

Another traditional Indian spiritual explanation for the presence and utilization of mind-altering *Cannabis* indicates that the Gods felt compassion for the human race and therefore sent *bhang* to earth. Through its consumption, humans could "attain delight, lose all fear, and have their sexual desires excited" (Dutt 1900); furthermore, because worshippers offered and consumed this "child" of Lord Shiva they became one with their deity (Campbell 1894 in Schleiffer 1979; see also Mikuriya 1994).

> The term for the drug, variously prepared, is *bhang* in northern India, and *siddhi* in Bengal—a term that also means "occult power"—but the classical word used in the Tantric manuals and in scholastic reference is *vijaya*, "victory", or "victory giver". *Cannabis* . . . creates a strongly euphoric mood and the term *vijaya* might have been coined to signify it. . . . There cannot be the slightest doubt that the Hindus and probably the Buddhists of earlier days [regarded] the taking of psychedelic drugs as part of the wide range of *sadhanas* [prayers] which lead to ecstasy, albeit perhaps only in the preliminary stages. (BHARATI 1965 in SCHLEIFFER 1979; see also ALDRICH 1977)

Another mystical sutra reports that Siddhartha, he who became known as Buddha, "the enlightened one," lived on nothing but a single *Cannabis* seed per day for six years prior to his spiritual awakening.

It is of interest here that *Cannabis* receives only a fleeting but rather significant reference in Buddhist legends. Prince Siddhartha, who later became Gautama the Buddha, lived near Lumbini in the Himalayan foothills of present-day Nepal, a region rich in cultivated and spontaneous *C. indica* ssp. *kafiristanica* narrow-leaf drug ancestor (NLDA) and *C. indica* ssp. *indica* narrow-leaf drug (NLD) populations. Following an extended fast, eating only a single hemp seed each day (e.g., Rätsch 2001), Siddhartha was so weakened that he resumed eating, upon which he began his legendary meditation under the "Bodhi Tree" (*Ficus religiosa*). This ultimately led to his enlightenment concerning the denial of desire and release from its accompanying sorrows, as well as his subsequent pilgrimage to spread these truths. Despite the influential role hemp seeds played in legends surrounding Siddhartha's enlightenment and the origin of Buddhism, *Cannabis* seems conspicuously absent in Buddhist rituals of today, likely due to the suppression of psychoactive *Cannabis* use by centralized religions; however, for some insight into the role of *Cannabis* in Tantric Buddhism, see Aldrich (1977) and Parker and Lux (2008). On the other hand, Hinduism continued to invoke *Cannabis* in its legends and to incorporate it into ritual contexts.

According to Hindu fables, Lord Shiva told his wife Parvati that one who is planting hemp seed should repeat the spell "*Bhangi, Bhangi*" to summon the guardian and protect the crop. The same holy name must also be repeated during the daily watering for one year:

> When the flowers appear the flowers and leaves should be stripped from the plant and kept a day in warm water. Next day, with one hundred repetitions of the holy name "Bhangi," the leaves and flowers should be washed in a river and dried in an open shed. When they are dry some of the leaves and flowers should be burnt with due repeating of the holy names. Then, bearing in mind Vagdevata or the goddess of speech, and offering a prayer, the dried leaves should be placed in a pure and sanctifying place. *Bhang* so prepared, especially if prayers are said over it, will gratify the wishes and desires of its owner. (CAMPBELL 1894)

This myth probably originated with the *sadhus* (holy men), who still use the plant for spiritual purposes in South Asia. *Sadhus* are well-known consumers of *bhang* (i.e., the *Cannabis* plant, its leaves, and a psychoactive drink made from them) and *ganja* (i.e., marijuana flowers), and they are also worshippers of Shiva (Watt 1908). There is also reference to the use of *Cannabis* in connection with ancient Indian *Srauta* rituals. These ceremonies were performed by professional priests in accordance with Vedic scriptures as part of a ubiquitous and highly codified ceremony to summon the powerful god *Agni* on behalf of specific clients (householders) and were therefore an extremely significant activity in ancient India, at least by the first millennium BCE (e.g., Majumdar 1952). We found the following reference to *Cannabis* use in the *Srauta* ritual concerned with the construction of a fire altar for animal and human sacrifice: "Fourteen days after this consecration took place, and thenceforth fire was kept in the pot, which was filled with *munga* grass [botanical identity unknown] and hemp" (Barnett 1914).

The *Mahabharata*, or "Great Stories," is an extensive collection of Indian histories and legends consisting of more than 74,000 verses started about the eighth century BCE (Brockington 1998) and completed in its present form by no earlier than the first century CE (McNeill 1963). It has been said that the *Mahabharata* represents a whole literature rather than a single homogeneous work and that it comprises "a veritable treasure house of Indian lore, both secular and religious, and gives, as does no other single work, an insight into the innermost depths of the soul of the people of Hindustan [the Indian subcontinent]" (Prabhavanda 1963). We should be careful to note that the *Mahabharata*'s lengthy evolution, which transpired before assuming its present form, limits its reliability as a source of historical detail, especially since it was likely altered by religious scholars to fit the ethics and values of different periods. It is in this context that we should consider reference to a hemp taboo in the *Mahabharata*. In one section of this lengthy "epic" it is stated that one who wishes to attain glory (or prosperity) should avoid the fruits of *pippali* (*Piper longum*, or Indian long pepper), *vata* (*Ficus bengalensis*, or Indian banyan, the national tree of India), *udumbara* (*Ficus racemosa*, or cluster fig), and the leaves of *bhang* (Prakash 1961).

Mechoulam (1986) suggested that shamanic use of *Cannabis* probably originated in the Zoroastrian traditions of ancient Persia (practiced today by the Parsi faith) and spread through Central Asia, eventually influencing Islamic mysticism (see Chapter 4 for an in-depth discussion of early shamanic associations with *Cannabis* in Eurasia). The *Avesta*, a several hundred-volume ancient collection of sacred Persian religious texts, is said to have been first compiled in the second century CE but originally authored by Zarathustra (also called Zoroaster), the prophet who founded Zoroastrianism about 1200 BCE. In the volume called the *Vendidad* (XV, 14), or "The Law against Demons," there is reference to *bhang* as Zoroaster's "good narcotic." The close linguistic affinity between the Indian Vedic tradition and the Avestan religious literary record of ancient Persia reflects a common cultural heritage. Therefore, we should consider the following Avestan references for *banga* (Avestan for *Cannabis*), which is related linguistically with Indian *bhanga*. In the *Din Yast*, an Avestan devotional treatise dedicated to the goddess Kista, we find another reference to *Cannabis* use, in this case for inducing euphoric feelings and righteous action:

> To whom the holy Hvovi (Zarathustra's wife) did sacrifice with full knowledge, wishing that the holy Zarathustra

would give her his good narcotic [*sic*] (*Bangha*, the so-called *Bang* of Zoroaster). What must have been its virtue may be gathered from the legends of Gustasp and Ardu Viraf, who are said to have been transported in soul to the heavens, and to have had the higher mysteries revealed to them on drinking from a cup prepared by the prophet—Sardust Namah—or from a cup of Gustasp—*bang* that she might think according to the law, speak according to the law, and do according to the law. (DARMESTETER 1883)

Was *Soma Cannabis*?

Aryan tribes arriving in northwest India during the second millennium BCE brought with them a polytheistic religious tradition consisting of a collection of poetic hymns (the *Rg Veda*) praising and deifying certain natural phenomena. Some of the more prominent powers of nature personified in the *Rg Veda* include the sacred force of fire (*Agni*), the vigor of storms and thunder (*Indra*), and the psychoactive potency of *soma* juice. Indeed the praise and use of *soma* formed a crucial part of the original cult of the Aryans. The cult, although it eventually became embellished with complex elaboration and ritual (*Srauta*), was fundamentally simple. It focused on an outdoor altar fire and involved the sacrificial slaughter of animals and the offering of such substances as *ghee* (melted butter) and *soma*, and ceremonial libations of *soma* juice were conspicuous. According to the hymnal, gods and priests were equally fond of *soma* juice, often lauding its exhilarating and wondrous spiritual, psychological, and medical powers. The *Mandala*, or ninth book of the *Rg Veda*, is almost completely devoted to the praise and use of *soma*.

However, knowledge of the pristine Aryan hymns and consequently the psychoactive *soma* sacrament were originally restricted to the Aryans themselves. Eventually the indigenous priestly elite usurped much of the religious authority held by the authors of the Aryan hymnal and gained control of theological interpretation and ritual. In its formal interpretation, "Vedism" was not a religion directed at the masses (Drekmeier 1962), and apparently for some time, members of the priestly clique limited knowledge and use of *soma* to their own esoteric activities. Thus a small, influential segment of ancient Indian society controlled the religion and distribution of the *soma* plant: "The ordinary *soma* sacrifice was clearly a sacrifice of rich patrons" (Mac-Donell and Keith 1958).

Even though long restricted to the ancient Aryan elite, the use of *soma* was an extremely important religious practice. In fact, some scholars have suggested that the link between shamanism (e.g., early Aryan nature worship) and traditional Hindu religion may have been *soma*. Eventually *soma* became

the repository of all the nourishing and fertilizing principles of nature. At the same time it was the food of the gods and the intoxicating drink of man, symbol of the immortality of the one and of the fleeting life of the other. But Soma, the narcotic [*sic*] drink, brought exhilaration and at least the momentary sense of immortality. It united the imbiber with the gods: "We have become immortal, We have entered into the light, We have known the gods" (*Rg Veda* 48.3). Here are the first vague hints of a concern for salvation. In time *soma*, the instrument, became confused with the divine life itself and *soma* became king of the Brahmans (the universal god). (DREKMEIER 1962)

Heightened senses and visions stimulated by the use of *soma* may have produced the dominating Indian philosophic concept of *Maya* (i.e., what one perceives in the normal state of consciousness is not the ultimate nature of reality but illusion). This traditional Indian association of illusion with "normal" consciousness has parallels in other cultures that sanctify the use of certain psychoactive plants and has obviously exerted a tremendous historic influence on the Indian life-style and beliefs regarding consciousness. *Soma* helped stimulate the desire to transcend normal conscious experience. Indeed, as suggested, it also may have served in early Vedic times as a connective between pristine nature worship and the primary development of the Brahman cult that advocated the acceptance of one all-encompassing entity. This monastic metaphysical ideal gave rise to, and still pervades much of, Indian religious philosophy.

Before continuing our discussion of the use of psychoactive *Cannabis* throughout ancient India, we should consider the long-lasting, puzzling, and significant question regarding identification of the famous drug *soma* that played such an important part in the formation and evolution of the Vedic civilization. Basham (1959) tells us that "the effects of *soma*, with vivid hallucinations and the sense of expanding to enormous dimensions, are rather like those attributed to such drugs as hashish . . . *soma* may well have been hemp . . . from which modern Indians produce a narcotic [*sic*] drink called *bhang*." First of all, what were the qualities and significance of *soma*? Was *soma Cannabis*? Probably not, but there is considerable evidence to suggest such an identification. If it were possible to substantiate, beyond any doubt, that the plant source of *soma* juice was *Cannabis*, the diffusion of *Cannabis* use into ancient India would be deeply significant. The circumstantial evidence that supports this identification is presented later. Although the case for other species, such as the psychoactive *Amanita muscaria* mushroom or others, may be stronger, we address the possibility of *Cannabis* being *soma* because of *Cannabis's* ancient use in India for mind-altering purposes.

The identity of *soma* has been a deep mystery for more than 2,000 years, ever since the Aryans abandoned the original plant source and forgot its exact identity. Western scholars actually began systematic studies of the Vedic tradition only as late as the eighteenth century (McNeill 1963; also see Wasson 1968). But in this relatively short time compared to how long ago *soma* was in use, more than 100 species have been suggested as its source. However, no unchallenged acceptance of any species identification has as yet prevailed. Some of the more noteworthy proposals include the following: ephedra (*Ephedra* spp., e.g., MacDonell and Keith 1958; Brough 1971; Flattery and Schwartz 1989; Naranjo 1990); wild rhubarb (*Rheum* spp.); two milkweeds (*Periploca aphylla* and *Sarcostemma brevistigmata*); yellow foxtail (*Setaria glauca*; Lewin 1964); *Cannabis* (Ray 1939; Basham 1959); Syrian rue (*Peganum harmala*; Flattery and Schwartz 1989); and, perhaps still the favorite among many scholars, the hallucinogenic red cap, fly agaric mushroom (*Amanita muscaria*; Wasson 1968; Schultes et al. 2001). Others have suggested that *soma* was simply a symbolic myth and did not represent a single species, and certainly over time this became true. There are still some who believe that the true identity of *soma* is *Cannabis*, which grows wild in many parts of Central and South Asia and from which, as we have noted, ancient and modern Indians produce a psychoactive drink called *bhang* (Basham 1959). Given the focus of this book, we enumerate several similarities between literary descriptions of *soma* and *Cannabis*.

1. Both *soma* and *bhang* were species growing spontaneously in the mountains of northern India. According to literary references, *soma* grew on the mountains, especially on Mount Munjavant, possibly in the northwest Himalayas (Prakash 1961). "As the *soma* plant was usually found in mountains, the forest of *soma* might be in the sub-mountain tracts of the Himalayas, from the Punjab to Bihar" (Ray 1939). It is no coincidence that *Cannabis* used in *bhang* "grows wild throughout the Himalayas from Kashmir to the east of Assam at an altitude up to 3000 meters (10,000 feet) above sea-level" (Chopra and Chopra 1939).

2. Geographical references in the *Rg Veda* also indicate that *soma* eventually diffused to locations along the banks of the mythical Sarasvati and Arjikiya Rivers of the Indus Valley. The fertile alluvial soils adjacent to these rivers, and others that have their headwaters in the Himalayas, "are exactly the situations of the wild growth of *bhanga*" (Ray 1939).

3. The *Vedic Index* states that the Indian hemp preparation *bhanga* is associated with *soma* in the *Rg Veda* (IX.6l.13), where the word "*bhanga*" is used to characterize *soma*, presumably in the sense of "intoxicating," which then came to designate *Cannabis* (MacDonell and Keith 1958).

4. The plant sources for both *soma* and *bhanga* also seem to have certain botanical characteristics in common. MacDonell and Keith (1958) associate the term *naicasakha* with the *soma* plant. This quality indicates branches (or twigs or leaves) hanging down, also a characteristic of the *Cannabis* plant.

5. The Vedic descriptions for the color of *soma* include the word *hari*, which may be interpreted as meaning "green or greenish yellow" (Ray 1939). At Indore in Madra Pradesh, India, the female form of the hemp plant is called *hari* (Watt 1908).

6. In the *Rg Veda* (IX. 97.19; 107.2), the *soma* plant is said to have a strong and pleasant smell. *Cannabis* also has distinctive aromatic qualities.

7. It is also possible that both *soma* and *bhanga* arise from an annual plant species, coming up at the beginning of the rainy season (Ray 1939). This generally defines the natural life cycle of *Cannabis*.

8. "In the *Sukla Yajurveda* (IV.10), *mekhala*, the girdle, is described as tying the knot of *soma*" (Ray 1939). Is this an implication that *soma* had the same fibrous qualities as the hemp plant? Consider the word *amsu* (hair or ray), which is also associated with *soma*. Does this indicate a fibrous quality or the source of glandular resin?

9. Striking similarities can likewise be found for the preparation of both the *soma* and *bhanga* drinks. The following is a composite description of one of the many *soma* preparations recorded in the *Rg Veda*:

   > The shoots bearing leaves (IX.82.3) were first cleaned and next moistened with, or steeped in, water when the stalks would swell (IX.31.4). The mass was then crushed and ground between a pair of stones (IX.67.19) or in a mortar and pestle (I.28.1). The ground paste was next mixed with water in a jar and the mixture poured from one jar into another causing sound (IX.72.3). Then it was strained over sheep's wool (IX.69.9). Thus prepared it was a "pure" drink. Often it was mixed with milk or yoghurt (IX.71.8), sometimes with honey and barley meal (IX.68.4). (RAY 1939)

The usual contemporary mode of consuming *bhang* is in the form of a drink that was traditionally prepared in the following manner: The leaves were pounded and mixed with water to form a thick paste that was rolled into a ball and dried. Later it could be mixed with water or milk and strained through cloth (Chopra and Chopra 1939). *Bhang* always includes fat (i.e., butterfat) to solubilize the THC. Watt (1908) presented the following description of the preparation of *bhang* over a hundred years ago: "When prepared for consumption the fragments of the plant are ground to a paste, and of this an emulsion is made which, after being filtered through a cloth, may be consumed in that form, or flavored with sugar, spices, cardamoms, melon seeds or milk."

10. The effects of *soma* drink are similar to those of *bhang*. *Soma* used to be drunk between the eating of food (IX.51.3). It is nourishing when taken with milk and food (IX.52.1). It is exhilarating (VIII.48), exciting (Il.41.40), and intoxicating (IX.68.3, 69.3). It stimulates the voice and impels the flow of words (IX.95.2, 101.6). It awakens eager thought (VI.47.3) and excites poetic imagination (IX.67.13). It induces sleep (IX.69.3) and the desire for women (IV.67.10). It bestows fertility (IX.60.4, 74.5). It cures diseases (VIII.48.5) and was believed to prolong life (VIII.48.5). None but the strong can tolerate it (IX.53.3, 81.1). It is constipating (IX.18.1) and sometimes causes bowel complaints. It was drunk before military engagement (IX.61.13; 85.12) and after victory (IX.101.I), for which Indra's favor was prayed for (Ray 1939). It was said that taking *bhang* in the morning cleansed the user from sin, freed him of punishment from scores of sins, and entitled him to "reap the fruits of a thousand horse sacrifices." Such sanctified *bhang* taken at daybreak or noon could also destroy disease (CAMPBELL 1894).

Even disregarding the possibility that *soma* and *Cannabis* were identical, it seems quite clear that both were used to induce religious and euphoric experiences from a very early time in ancient India. The uses of *soma* for such purposes are well known (e.g., Wasson 1968; Schultes 1969a/b; Schultes and Hofmann 1992), but we note it is quite significant that similar uses in India have been and are attributed to *Cannabis*.

South Asian Psychoactive *Cannabis* Products

Such holiness and such evil-scaring powers must give *bhang* a high place among lucky objects. That a day may be fortunate the careful man should on waking look into liquid *bhang*. So any nightmares or evil spirits that may have entered into him during the ghost-haunted hours of night will flee from him at the sight of the *bhang* and free him from their blinding influences during the day. So too when a journey has to be begun or a fresh duty or business undertaken it is well to look at *bhang*. To meet someone carrying *bhang* is a sure omen of success. To see in a dream the leaves, plant, or water of *bhang* is lucky; it brings the goddess of wealth into the dreamer's power. (CAMPBELL 1894)

Bhang and *ganja* have long been used in South Asia, largely in ritual contexts. The larger leaves of the female plant, as well as the male plant, contain relatively less and sometimes only meager amounts of THC compared to the female flowers and therefore provide a much milder preparation, which is usually swallowed. In Persian and Hindi languages this grade of drug is called *bhang*, named after the *Cannabis* plant itself, but it is not commonly consumed in the West. *Bhang* is taken orally in the form of small balls or, as described earlier, mixed with water or milk in a beverage known also as *bhang*. This drink is regularly used by the lower classes in India and by a wider cross-section of the population on certain religious holidays. Jean Baptiste Tavernier (1676–77, quoted in translation in Schleiffer 1979), a seventeenth-century French traveler to the Near East, offered the following account of Persian *bhang* drinking: "The Persians have another kind of brew, very bitter and disagreeable to the taste. They call this *bengue* [*bhang*]. It is prepared from hemp leaves with some other drug added which makes it stronger than any other brew they use. Those using this get strangely mad. It is forbidden by law, whereas other beverages are allowed. It would be difficult to find many in Persia who were not given to some of these beverages without which apparently they would find no pleasure in life."

In more recent times, *bhang* is most often consumed either ritually or recreationally by blending it into a beverage, which again is called *bhang*. Today it is imbibed by both local people and occasional foreign travelers. In a simple preparation, *Cannabis* leaves are pounded; mixed with sugar and black pepper; blended with a little water; and added to milk, thin yoghurt (*lassi*), or milk tea. More complex formulations may include ground nuts, spices, and aromatic resins (Sharma 1977a; also see Russo 2007). *Cannabis* is also added to various foods as a way of achieving altered consciousness. *Halva* (a sweet confection, but not to be confused with Middle Eastern *halva* made from sesame seeds) is made by first boiling *bhang* in jaggery (unrefined sugar) syrup. This hot liquid is then filtered to remove most of the plant matter and mixed with flour and clarified butter. Powdered *Cannabis* leaves are also added to curries. *Bhang* dumplings (*pakoras*) are made by mixing fresh or dried *Cannabis* leaves with chickpea (*Cicer arietinum*) flour, to which water and the desired condiments (e.g., salt, black or red pepper, ginger, cumin seeds) are added. Small balls of the dough are fried in mustard oil and eaten as snacks with beverages. "On festive occasions, these *pakoras* are sought with great felicity" (Sharma 1977a).

The Irishman William Brooke O'Shaughnessy (lived 1808 to 1889) is credited with introducing *Cannabis* medicines to the West. During his time in India with the British Bengal Army, he was posted in many locations and befriended a number of Ayurvedic and Islamic physicians, some of whom interested him in the value of *Cannabis* as a therapeutic drug (see Chapter 8). The following descriptions of various *Cannabis* preparations and their effects are adapted from his firsthand reports. O'Shaughnessy (1839) described in detail a simple and highly effective protocol for extracting the active ingredients of *Cannabis* with clarified butter fat (*ghee*), the process having been "repeatedly performed" by "Ameer, the proprietor of a celebrated place of resort for hemp devotees in Calcutta," who was considered to be the premier "artist" in his profession at the time:

> Four ounces of *siddhi* [*bhang*] and an equal quantity of *ghee* [clarified butter] are placed in an earthen or well-tinned vessel, a pint of cold water added, and the whole warmed over a charcoal fire. The mixture is constantly stirred until the water all boils away, which is known by the crackling noise of the melted butter on the sides of the vessel. The mixture is then removed from the fire, squeezed through cloth while hot, by which an oleaginous solution of the active principals

and coloring matter of the hemp is obtained; and the leaves, fibers, etc., remaining on the cloth are thrown away. The green oily solution soon concretes into a buttery mass and is then well washed by the hand with soft water, so long as the water becomes colored. The coloring matter and an extractive substance are thus removed and a very pale green mass, of the consistency of simple ointment, remains. The washings are thrown away; Ameer says that these are intoxicating, and produce constriction of the throat, great pain and very disagreeable and dangerous symptoms.

According to O'Shaughnessy's account, the fat-soluble THC was extracted into a relatively small proportion of *ghee*, and then any water-soluble contaminants were removed, so the concentrated extract could have been extremely potent. Confections called *majoon* were made from this extract, and O'Shaughnessy (1839) reported how his host Ameer

> takes two pounds of sugar, and adding a little water, places it in a pipkin [cooking pot] over a fire. When the sugar dissolves and froths, two ounces of milk are added; a thick scum rises and is removed: more milk and a little water are added from time to time, and the boiling continued for about an hour, the solution being carefully stirred until it becomes an adhesive clear syrup, ready to solidify on a cold surface; four ounces of . . . fine [milk] powder are now stirred in, and lastly the prepared butter of hemp is introduced, brisk stirring being continued for a few minutes. A few drops of attar of roses are then quickly sprinkled in, and the mixture poured from the pipkin onto a flat cold dish or slab. The mass concretes immediately into a thick cake, which is divided into small lozenge-shaped pieces. . . . One *drachm* [almost four grams] by weight will intoxicate a beginner; three *drachms*, one experienced in its use. The taste is sweet and the odor very agreeable.

This sweet confectionary use can be compared with the after dinner food offerings eaten in Roman times (see Chapter 5). Ball (1910, cited by Abel 1980) provided a list of additional ingredients combined into *majoon* candies, including poppy, cucumber and caraway seeds, almonds, ginger, cloves, cardamom, cinnamon, nutmeg, and rosebuds. These pungent spices may have been preferred by the well-to-do because they masked the distinctive herbal taste of *Cannabis* extracts or because they were expensive. *Bhang* extracts were also used to make a green ice cream known as *hari kulfi* (La Barre 1977).

Dried, unfertilized (and therefore seedless) pistillate inflorescences and adjacent leaflets from the top of female *Cannabis* plants form a very potent preparation known widely as marijuana or by many other names including its Hindi name *ganja*, which is usually smoked in pipes or cigarettes but can also be added to cooked foods. In modern India, *ganja*, which is to be smoked, is usually diced up with a little dried tobacco, and sometimes a few drops of water are added. A little tobacco is placed in the *chillum* (conical smoking-pipe) first, then a layer of the prepared *ganja*, and then more tobacco and a coal on top. The *chillum* is passed around, and each person takes a single draw. Effects differ from those occasioned by eating *bhang*: "Heaviness, laziness, and agreeable reveries ensue, but the person can be readily roused and is able to discharge routine occupations" (O'Shaughnessy 1839).

A legend from the Gond people, who live in central India on the Deccan Plateau, indicates their awareness of the differing effects of smoking as compared to eating *Cannabis*. According to this mythical story, Maisur Dewar made his living by spinning rope. One day when he was in the deep forest he fell in love with a black cobra in the form of a woman and thereafter went daily to massage her hands and feet. One day he massaged with such great vigor that he grew tired and sighed deeply. The cobra thought, "My Dewar is tired, how can I put his weariness away from him?" So she scratched her head, and two seeds appeared, which she gave to him, saying, "Sow these, and when the tree grows pick the leaves and make an earthen pipe and smoke." But Dewar mistakenly "took the leaves with water and this made him so drunk that he could neither see nor hear" (Schleiffer 1979; also see Elwin 1949).

The New World equivalent to seedless *ganja* is *sinsemilla* (colloquial Spanish for seedless) marijuana. *Ganja* is most often cultivated by removing male plants to prevent fertilization of the females and prevent formation of seeds. In actuality, removal of male plants is rarely complete and some, but significantly fewer, seeds form. Many more seeds would form if male plants were allowed to freely fertilize the females. However, intersex plants with both male and female flowers are fairly common in Indian varieties, and special workers called *podhars* were employed to remove them (Prain 1904). Throughout the nineteenth century, areas surrounding Khandwa and Gwalior in central India and Bengal in eastern India were the largest producers and exporters of *ganja* to the remainder of the Indian subcontinent. Much of this legal *ganja* was exported to the Northwest Territories and the Punjab region along the Indus River (Clarke 1998a); these were the ancient lands of the *Vedas* where *Cannabis* use was prevalent. One of the few remaining seedless *ganja* cultures was located in Kerala along the western shore of the southern tip of India. In 1985, *ganja* production and sales were still licensed in Madhya Pradesh, Bihar, Orissa, and West Bengal states, which compose the traditional *ganja* cultivating regions of India.

The most potent preparation of *Cannabis* consists of concentrated resin glands (the product of secretory hairs called glandular trichomes), which are collected from flowering female plants and compressed into solid lumps. This preparation, referred to as *hashish* in Arabic and *charas* in Persian or Hindi, can be quite powerful and is often smoked and mixed with tobacco. The first, and certainly the most widely used, concentrated *Cannabis* drug was hand-rubbed *charas*, collected in much the same way as it is today. Nomadic peoples could have easily accumulated it during their seasonal migrations and used it throughout the year, while the production of powdered hashish would have required at least temporary settlements where *Cannabis* plants could be dried and sieved. Large-scale production of hashish requires agricultural production and a concentrated work force, which could only be supplied by settled agrarian communities.

Nepal has a long tradition of harvesting *Cannabis* resin by rubbing the flowers of ripe female plants and remains the most internationally well-known source of hand-rubbed hashish. Many popular legends report that *Cannabis* resin was collected on leather aprons or naked skin (Clarke 1998a). In Nepal, hashish is known by the Hindi and Persian name *charas* and usually refers only to traditional hand-rubbed resin, which is by far the most common form found in the Himalayas. However, most Nepalese usually smoke *ganja* or swallow *bhang* concoctions rather than smoking *charas*. *Ganja* is still the preparation favored by *sadhus* and the lower classes in Nepal, while *charas* is more expensive and can be afforded only by upper-class Nepalese and foreigners.

The British-administered *Report of the Indian Hemp Drugs Commission 1893–1894* (see Kaplan 1969; Mikuriya 1994) was an extensive survey of *Cannabis* cultivation, processing, and

FIGURE 41. In the nineteenth and early twentieth centuries Indian narrow-leaf drug (NLD) marijuana (*ganja*) was grown on large farms in Bengal, India, which was also the source of *Cannabis indica* medical preparations. Original photos from the *India Hemp Drugs Commission Report* of 1893–94.

local usage performed with the intention of taxing the *Cannabis* trade. It described the preparation of hand-rubbed *charas* in India:

The female plants, having been cut in November, are spread out to dry for 24 hours. The people then sit around in the heat of the day, and pluck off the flower heads, which are now full of seed, discarding the coarser leaves. Each handful is rubbed between the palms for about ten minutes and thrown aside. In the course of time a quantity of juice accumulates on the palms, which is scraped off and rolled into balls. These are *charas*. Sometimes the plants are trodden instead of handled and the feet scraped. A more uncommon method, by which a choice kind of *charas* is obtained, is to pass the hands up the ripe plants while they are still standing in the field.

Charas was also rubbed from plants in a few scattered regions across India until the early 1980s and is still produced for commerce near Manali in the northern Indian state of Himachal Pradesh. The paucity of evidence concerning the history of hashish has resulted in much speculation as to the antiquity of the preparation itself. Many assume that hashish is ancient, although proof of early manufacture or use is sorely lacking. Archeological evidence and historical records from Eurasia may help determine more specifically where hashish originated, but this is unlikely to happen in the near future. However, the general origins of hashish may be traced through historical records to Central Asia, but precisely where or in which period is yet undetermined. All of Central Asia has been contested territory for millennia. Invaders and merchants from both East and West brought their cultural influences, various traditions, beliefs, and legends together, while leaving few written records. Although presently we do not know when it started, there was significant trade in hashish (*charas*), at least in more modern times, from Central Asia to India, moving along innumerable trade routes and over high Himalayan passes to South Asia from the region formerly referred to as Chinese Turkestan (now known as Xinjiang, the "new frontier" province in Central Asia, e.g., see Watt 1889). In 1937 and 1938 this trade accounted for 42 percent of the total value of products being moved from Xinjiang to India (Lattimore 1950).

Hindu Acceptance of Ritual *Bhang* Use

To the Hindu the hemp plant is holy. A guardian lives in the *bhang* leaf. . . . so the properties of the *bhang* plant, its powers to suppress the appetites, its virtue as a febrifuge [medicine used to reduce fever], and its thought-bracing qualities show that the *bhang* leaf is the home of the great Yogi or brooding ascetic Mahadev [Shiva]. (CAMPBELL 1894)

While a high caste Hindu could be put to death for drinking liquor, no religious penalty was attached to the use of *bhang*, and a single day's fast was considered enough to cleanse one from the "coarser spirit of *ganja*." Campbell, in the *Report of the Indian Hemp Drugs Commission 1893–94*, described how even the most conservative Hindus did not object to using *bhang* for recreational pleasure if it was associated with religious ritual; indeed, among appreciators of *bhang*'s positive influences, such as "raising [a] man out of himself and above mean individual worries" and making him "one with the divine force of nature," it was recognized as "inevitable that temperaments should be found to whom the quickening spirit of *bhang* is the spirit of freedom and knowledge." Campbell goes on to explain succinctly why *Cannabis* has been closely associated with religiousness and devotion: "Much of the holiness of *bhang* is due to its virtue of clearing the head and stimulating the brain to thought" (see Shamir and Hacker 2001 for a lucid discussion of the Indian Hemp Drug Commission and its 1893–94 report).

Cannabis-based drugs have also been used in India from very early times in order to overcome fatigue and worry, for production of euphoria, and to give courage to warriors during times of stress (Chopra and Chopra 1939). According to Chopra (1958), the ritual use of *Cannabis* was accorded great reverence in early India, and ancient literature "is full of references to the virtues of this drug." By the tenth century CE, *bhang* was extolled as *Indracanna*, the "Food of Indra," divine ruler of the Indian pantheon, and in the later Indian texts the virtues of *bhang* were highly praised, with the real fervor for *bhang* consumption among the Indians beginning in the sixteenth century (e.g., see Grierson 1894; Mikuriya 1994). For hundreds of years, ingestion of psychoactive *Cannabis* preparations of *bhang* and *ganja* has been a significant cultural activity and source of inspiration and mental focus for the Hindu *sadhus*, or ascetic, wandering holy men. Progressively, some *sadhus* developed sophisticated, functional systems of meditation that more or less superseded (for able and dedicated students) the need for *Cannabis*'s effects as a vehicle to reach the heights of religious experience. Thus over

time some spiritual teachers and students first renounced the superstitious and cumbersome ritual of the Brahmins, and subsequently they discouraged the use of mind-altering preparations such as those derived from *Cannabis*. But the cultural spread and environmental adaptation of *Cannabis* varieties probably progressed quickly after retreating holy men established the drug plant's association with euphoric and religious experience and spread their endorsement. Therefore, a significant portion of Indian people, especially those of lower classes, continued to use *Cannabis* for such purposes. Ritual *Cannabis* users have been held in relatively high esteem for centuries in large parts of South Asia and were protected de facto as guardians of ancient cultural traditions, as long as their motives were viewed as spiritual. Campbell (1894) expressed the extreme reverence for *Cannabis* users as follows: "He who scandalizes the user of hemp shall suffer the torments of hell so long as the sun endures. He who drinks *bhang* foolishly or for pleasure without religious rites is as guilty as the sinner . . . He who drinks wisely and according to rule, be he ever so low, even so his body is smeared with human ordure [feces] and urine, is Shiva. No god or man is as good as the religious drinker of *bhang*." Although Hindus differ in their opinions regarding the obligation decreed by their *Shastra* scriptures that devotees should consume *Cannabis*, Tantric spiritual texts approve of its use, "and the custom may now be said from immemorial usage to be regarded by many people as part of their religious observances" (Kaplan 1969). Generally *bhang* was revered and used by Hindus of all castes from a majority of ethnic communities throughout South Asia and is still used for religious purposes in many localities today. Orthodox high-caste Brahmins take *bhang* regularly on Fridays but not as a part of their formalized ritual. The worshipper places the bowl of *bhang* on the triangular *mandala* (mystical diagram) drawn on the ground in front of him and purifies it by chanting his *mantra*. He touches his heart three times, "holding the *vijaya* bowl and offering its contents to his chosen deity," and then the devotee either drinks the liquid *bhang* or, if his offering is *majoon*, swallows it (Bharati 1965).

As Campbell articulated over a hundred years ago in the Indian hemp report, *bhang*'s holiness and prowess against evil has also endowed it with an elevated standing as an object of good fortune. When a trip was to start, or a new task or enterprise was begun, the devotee, upon rising from his or her sleep, should look into liquid *bhang* so that bad dreams and wicked feelings would take flight and fortunate paths materialize for the voyager. Encountering a person transporting *bhang* provided a certain sign of a successful journey, and to conjure visions of the *bhang* plant and its ritual uses produced fine providence: "To see in a dream the leaves, plant, or water of *bhang* is lucky; it brings the goddess of wealth into the dreamer's power. To see his parents worship the *bhang*-plant and pour *bhang* over Shiva's lingam [phallus-like symbol of Shiva] will cure the dreamer of fever. A longing for *bhang* foretells happiness; to see *bhang* drunk increases riches. No good thing can come to the man who treads under foot the holy *bhang* leaf" (Campbell 1894).

Shiva Worship and *Cannabis*

The Hindu spiritual world is ruled by three lords—Brahma the Creator, Vishnu the Preserver, and Shiva the Destroyer—the Trinath, or "Trinity of Three Lords." Shiva is frequently called Mahadev in Hindu texts, or Shankar, a favorite of *chillum*-smoking *sadhus*. Here we refer to him as Shiva, unless quoting another source. Shiva is the deity most often associated with *Cannabis* and is known as the Lord of the Herbs (*Ausadhisvara*). Most ritual *bhang* offerings and invocations involve him. In Nepal, Shiva is frequently depicted with a bowl of herbs under his arm "as one of the emblems of the mendicant" (Bharati 1965, quoted in Schleiffer 1979). *Cannabis* was the plant of choice for Shiva, and his devotees regard him as the *Bhangeri Baba* (the god who heartily prefers the enjoyment of *bhang*). *Bhang* was almost universally used by *sanyasis* (devotees of Shiva), religious mendicants such as *sadhus*, fakirs, yogis, and ascetics of all classes who heartily enjoyed drinking it as a means of transporting them to the "seventh bliss of heaven" (Majupuria and Majupuria 1978). Venerated Shivaite ascetics teach that the *Cannabis* plant is a special attribute of Lord Shiva, and this belief is largely shared by the general population. Shiva's name is commonly invoked in adoration before placing the *chillum* of *ganja* to the lips. Besides ruling over the demons of madness, *bhang* is victorious over the demons of hunger and thirst. Ascetics fasted for days with the help of *bhang* (Campbell 1894; also see Mikuriya 1994).

During rituals at Shiva temples, *Cannabis* libations devoted specifically to Lord Shiva were poured over the Shiva *lingam*, and *bhang* drinks were offered in his name (Aldrich 1977; Sharma 1977a/b). Shrine managers regularly distributed *ganja* to all *sadhus* who assembled during festivals. There is no passage in the Hindu scriptures allowing for provision of *ganja*, but its supply is essential (Campbell 1894). Countless households still grow a plant or two to be able to offer *ganja* to a passing *sadhu*. *Cannabis* is smoked by nearly everyone present during daily devotional services at shrines and ashrams of certain Hindu sects (Fisher 1975). In Brij Mandal, in Western India, the entire *Cannabis* plant is presented as a religious offering to Lord Shiva (Singh and Chauhan 2004).

Tantrism was another manifestation of ritual *Cannabis* use. This religious movement arose around the seventh century in an "explosive mingling of the doctrines and practices of Shivaite Hinduism and Tibetan Buddhism" and is practiced by Hindu, Buddhist, Bonpo (shamanic Tibetan), and Jain sects (Aldrich 1977). Tantric literature is concerned with powerful ritual acts of body, speech, and mind and reached the height of its popularity in medieval Bengal and the Himalayan kingdoms. Chinese and Tibetan traditions were incorporated into Tantric religious icons in which the Buddha may be depicted with sharply serrated *bhang* leaves in his begging bowl. The asceticism of Buddhism, as well as Shiva worship in Tantric belief, is therefore associated with *Cannabis* use. In Tantric ritual, *Cannabis* use fosters increased suggestibility and causes time and space distortions where "minutes seem like hours, small rooms yawn into deep caverns, and every activity is imbued with a sense of timeless grandeur." Under such circumstances, worshippers traditionally sit through lengthy rituals held in flower bedecked temples illuminated by flickering lamp light and reeking of incense with their "intent upon mystical experience," and thus the function of *Cannabis* in Tantric ceremony is "to enable the worshippers to feel the divinity within and without themselves" (Aldrich 1977).

A traditional example of ritual *Cannabis* smoking involves Trinath worship. It began in Bengal around 1867 and quickly gained popularity among certain devotees of Shiva who performed a special ritual in which *ganja* use was considered essential. Trinath rituals were observed throughout the year by both Hindus and Muslims, the latter calling it Tinlakh Pir. When a special desire was fulfilled (e.g., a person recovered from illness, a son was born, or a marriage ceremony performed), people worshipped Trinath, grateful for their blessings. Formerly, one *paisa* (one-fourth of an *anna*) worth of *ganja*, one *paisa* worth of oil, and one *paisa* worth of betel nut (*Areca catechu L.*) was offered to Trinath. Near the end of the nineteenth century *ganja* alone was offered, sometimes in large quantities, and during the performance of ritual incantations all of those present smoked (Kaplan 1969).

Many Hindu festivals prominently featured the ritual expression of *bhang* for its religious significance. On Shivratri (thirteenth night/fourteenth day of February), commemorating the day Lord Shiva bestowed salvation on his followers, *bhang* was freely poured over the *linga* in his temples. "The wish of him who with pure mind pours *bhang* with due reverence over the Lingam of Mahadev will be fulfilled" (Campbell 1894). The popular image of Shiva is as an ascetic fond of *bhang*, "and on this day [Shivratri] it is considered a religious duty to offer him his favorite drink." *Ganja* was offered to Shiva and consumed by his worshippers and formed an integral part of the ritual (Kaplan 1969). On Shivratri day, the Nepali Shiva temples are filled with worshippers, particularly hermits and sadhus, who "smoke ganja" (Manandhar 2002).

Bhang also played an important ritual role in times of war. For example, before the outbreak of hostilities and during its progress, the *linga* in Shiva temples and shrines were traditionally bathed with *bhang* in the name of Vijaya, the unbeaten and victorious: "So a drink of *bhang* drives from the fighting Hindu the haunting spirits of fear and weariness. So the beleaguered Rajput [soldier caste], when nothing is left but to die, after loosing his hair so that the *bhang* spirit may have free entrance, drinks the sacramental *bhang* and rushing on the enemy completes his *juhar*, or self-sacrifice. It is this quality of panic scaring that makes *bhang*, the Vijaya or Victorious, especially dear to Mahadev" (Campbell 1894). Among the Hindus throughout South Asia, on the last day of the Durga Puja or Dashain festival consecrated to the goddess Kali (Aldrich 1977), male family members floated idols of Durga (also known as Parvati or Devi) upon the waters, and when they returned the family exchanged greetings and embraced their friends and relatives. During this rejoicing, a cup of *bhang* was passed around, and all were expected to drink some, or at least to ritually place the cup to their lips as a token of respect. Sweetmeats containing *Cannabis* were also distributed (Pandey 1989).

Other Occasions on which *Bhang* Was Used

The hemp with which we used to hang
Our prison pets, you felon gang,
In Eastern climes produces Bang,
Esteemed a drug divine.
(LORD NEAVES [lived 1800 to 1876 CE], cited in WASSON 1968)

The Sanskrit word *siddhu* is a term indicating a yogi (one who practices yoga) or monk in Hindu mythology; it also refers to a religious person's "perfect knowledge of the way to one true god" (Patil 2000). In fact, "*banga*" was used traditionally by the *Siddhu* to concentrate upon this god; hence, it is called "*Siddhu*" or "*Siddhi*" (*siddhi* means perfection).

Both Hindu and Muslim worshippers traditionally consumed *bhang* at ritual occasions and offered *Cannabis* at the shrines of various favored deities throughout the year. Many of these traditions continue today: "In each region *Cannabis* is given to the locally most favored form of God" (Touw 1981; also see Morningstar 1985). Some brief examples from Kaplan (1969) are presented as follows. At the birthday festival for elephant-headed Lord Ganesh, son of Shiva and Parvati, *bhang* was widely used by attendants and worshippers who offered *majoon* candies to Ganesh before eating them with their friends and relatives; confections and beverages containing *bhang* were also offered to Vishnu and Hanuman the Monkey King. During the festival for Kama, the Indian deity of love, copious amounts of *bhang* are made, offered up, and imbibed. In the *Bhavishya Puran*, or "Book of the Future" (ca. 550 BCE), it is stated that "on the 13th moon of Chaitra (March and April) one who wishes to see the number of his sons and grandsons increased must worship Kama in the hemp plant." During the harvest season, offerings to local farm field divinities included a small quantity of *ganja*, and a custom among travelers was to offer a little *ganja* at collections of stones marked with colorful rags where travelers paused to smoke.

In addition, *Bhang* was a favorite drink of Hindus on the eve of Holi, or "Festival of Colors" (Pandey 1989), and it was commonly consumed at other important Hindu holy festivals such as Diwali, or "Festival of Lights;" Sankranti, or "Winter Solstice"; Sripanchami, honoring the goddess of learning; and Ramnavami, the birthday of King Rama; as well as on occasions of weddings and many other family festivities (Kaplan 1969). During weddings in Bombay, people belonging to the bride's family, who are almost all among the higher classes of Gujarat Hindus, as well as those among the Jain and Brahmani sects, sent supplies to the bridegroom's party that included a provision of *bhang*. A father who neglected to send *bhang* was held in contempt and despised by his caste as mean and miserly. After the wedding, when the bridegroom and his friends were entertained at the bride's home, richly spiced *bhang* was served (Campbell 1894; Mikuriya 1994). The predominantly Hindu tradition of serving *bhang* on ritual occasions continues today on a somewhat reduced scale.

Cannabis was also frequently used in ritual contexts by members of other religions. Muslims from better classes were more likely to use *bhang* (O'Shaughnessy 1839, cited by Dutt 1900), but Muslim fakirs and mendicants revered *bhang* as the "lengthener of life" and "freer from the bonds of self" and believed that "*bhang* brings union with the Divine Spirit" but not in the sense of worship, "which is due to Allah alone." To the follower of Islam "the holy spirit in *bhang* is not the Almighty." It was thus seen by some Muslims in India as "the spirit of the great prophet Khizr or Elijah." Khizr is the patron saint of water, but "still more *khizr* means green, the revered color of the cooling water of *bhang*." Muslim poets honor *bhang* with the title *Warak al Khiyall*, or "Fancy's Leaf," and the *Makhazan* Arabic pharmacopoeia recorded many other names for *bhang* (e.g., "Joy Giver," "Sky Flier," "Heavenly Guide," "Poor Man's Heaven," and "Soother of Grief"). On occasions of holidays or gala days, and at the Mohurram festival honoring the grandsons of the Prophet Muhammad, it was usual for Muslims as well as Hindus to take *bhang* (Campbell 1894; Mikuriya 1994).

Among the Sikhs, the use of *bhang* as a beverage was also an essential part of their religious rites having the authority of the *Adi Granth* Sikh scripture. *Bhang* was prepared during the Dasehra festival of victory of truth over sin and the Chandas festival of the god Sheoji Mahadeva, and every Sikh was obliged to drink it mixed with water. It was said that "*bhang* quickens fancy, deepens thought, and braces judgment," and

thus "oaths were taken on the *bhang* leaf," but to one who casually swore the oath and lied, the *bhang* oath was certain death (Campbell 1894). Sikhs are prohibited from smoking, so the use of *ganja* and *charas* in this way was not practiced by them. In old times, the Sikhs were permitted annually to collect a boatload of *bhang* without interference, which was afterward distributed throughout the year to *sadhus* and beggars who were supported by Dharamsala (the court of justice, tribunal, charitable asylum, or religious asylum of Sikh religion; e.g., see Kaplan 1969; Sikh Encyclopedia 2012).

The use of psychoactive *Cannabis* preparations for spiritual or other purposes also has a long history among tribal peoples of India (e.g., see Saini et al. 2011). For example, among the pre-Aryan Santhali (Santal) peoples of West Bengal, resinous parts of *Cannabis* are burned and inhaled through an earthen pipe (*kolke*) as blessing "for divine pleasure" (Mandal and Mukherjee 2003). Among the peoples of the Kumaon region of the Himalaya Mountains of Uttar Pradesh state in India, *charas* (known locally as *attar*) is still smoked by the *attarchis*, which today usually includes men older than 30 years of age. These Kumaoni men often smoke their *attar* with *sadhus*, inhaling their hashish smoke through chillums to facilitate meditation and to keep the body warm during winter months:

> Preparation of an *attar* pipe is considered as an art and the pipe is usually prepared by an experienced *attarchi*. While charging, a small piece of charcoal is fitted into the hole at the proximal or smaller end of the *attar* pipe. Over the charcoal is placed some cured tobacco and over this the *attar*. The quantity of *attar* depends upon the number of smokers (usually, two to three grams [less than one-eighth ounce] of *attar* is sufficient for eight to ten smokers). Then, a small piece of wet cloth, which acts as a filter, is inserted inside the stem of the pipe. Lastly, small pieces of burning charcoal are placed in the pipe. (SHAH 1997, 2003)

Cannabis had a presence in South Asian funerals as it did across Eurasia. When a *bhang* consumer died, *bhang* was offered to the gods and consumed by the living, and a *bhang* plant was kept near the corpse. *Ganja* was also used in additional ways at the funerals of particular ethnic groups. For example, in the ceremony of the Gonds (central Indian Gondi people), "*kalli* or flat *ganja* is placed over the chest of the corpse, and when the funeral party returns home, a little of the *ganja* is burned in the house of the dead person, the smoke of which is supposed to reach the spirit of the dead" (Kaplan 1969).

Worship of the *Bhang* Plant

The extensive drug use documented earlier occurred in ritual South Asian settings, devotedly venerating deities who often entertained additional relationships with *Cannabis*. However, the custom of worshipping *Cannabis* itself was also practiced across much of South Asia, but not nearly so prevalently as that of offering *bhang* to Shiva and other Hindu deities. Although ritual and recreational *Cannabis* smoking was more common in the Himalayan districts and northern regions of India and Nepal, *bhang* worship was common in many locations across India. This occurred as far south as the hilly districts of Madras state, where the *Cannabis* plant is looked upon with veneration among the rural population. The

Hindu *Shastra* scriptures refer to Lord Shiva telling the goddess Parvati to know the benefits derived from *bhang* since its worship raises one to his deified status, and the Gonds in their central Indian hill homes, as well as the Kouls of Kumaon in Himalayan India, revered the *bhang* plant as one of their objects of animistic worship (Kaplan 1969). In most parts of Kumaon, Shah (1997) tells us that every year the festival known as "Khatarwa" is celebrated on the seventeenth of September:

> It is the victory-anniversary of a local battle fought by the Kumaonis against people of the neighboring territory. . . . On this day, in every village, a bonfire is lit in the night and each child brings to the place of the bonfire, from his house, a big shoot of *Cannabis* on which are tied maize, small cucumbers, and flowers; this is known as "*lanka.*" After the bonfire is lit, it is offered to the fire and then [the] maize, cucumbers, etc. are eaten. When the bonfire nears its end, it is beaten by the village children with *Cannabis* shoots.

Although this ritual use of *Cannabis* is not focused on its psychoactivity, the Kumaon live in a *charas* producing region of the Himalayas. Consequently, their reverence for *Cannabis* incorporates both psychoactive and nonpsychoactive ritual use, providing us with a contemporary example of the difficulty of separating ritual *Cannabis* uses into discreet categories.

The *Report of the Indian Hemp Drugs Commission* summarized the relative popularity of different South Asian *Cannabis* preparations at the close of the nineteenth century. According to this report, the use of *bhang* was generally ubiquitous in association with the social and spiritual customs of the people, and "its use is considered essential in some religious observances by a large section of the community." In specific reference to *ganja*, the report tells us that "there are certain classes in all parts [of India] who use the drug in connection with their social and religious observances." However, the number of people in India at the time of the report "who consider it essential are comparatively very few," and the use of hashish (*charas*), which was then reported as a relatively new *Cannabis* preparation, was not identified as having any connection with religious observance. Nevertheless, the prevalent use of *Cannabis* for ritualistic or religious purposes was said to be significant and should be protected.

J. M. Campbell, whom we have cited frequently, served as Collector of Land Revenue and Customs and Opium in Bombay (Mumbai) and reported to the Indian Hemp Commission on the religious use of *Cannabis*. Through his studies, he became sensitive to the ritual traditions involving *Cannabis* and offered insightful, culturally sensitive advice to the British colonial government:

> To forbid or even seriously restrict the use of so holy and gracious an herb as hemp would cause widespread suffering and annoyance and to the large bands of worshipped ascetics deep-seated anger. It would rob the people of a solace in discomfort, of a cure in sickness, of a guardian whose gracious protection saves them from the attacks of evil influences, and whose mighty power makes the devotee of the Victorious, overcoming the demons of hunger and thirst, of panic, fear, of the glamour of Maya [illusion] or matter, and of madness, able to rest to brood on the Eternal, till the Eternal, possessing him body and soul, frees him from the haunting of self and receives him into the ocean of Being.
> (CAMPBELL 1894)

The basic conclusions of the *India Hemp Drugs Commission Report* argued that *Cannabis* use was widely popular, deeply rooted in ancient religious traditions, culturally vital to a large portion of citizens, and generally considered harmless and at times beneficial. These conclusions, however, were largely ignored by the British government and judgmental citizens. By the time the report was published, Britain was embroiled in moral controversy over the sanctioned production of opium in India and the East India Company's lucrative trade in opium to China, and the findings of the India Hemp Drugs Commission were either overlooked or interpreted by critics as further evidence that the English government was aiding and abetting the international drug trade (Mills 2003).

During the early twentieth century, the use of *Cannabis* drugs in connection with religious and social practices was still common at many locales across India, though to a much smaller extent than in the past (Chopra and Chopra 1939). In the middle of the last century, Sastri (1950) described the ongoing popularity of *Cannabis* for mind-altering purposes in this country: "Even in recent times, *bhang* remains the social indulgent and religious drink of the lower classes in India; and the smoking of potent *ganja* remains as the almost universal psychoactive practice of certain classes of *sadhus* and mendicants."

In summary, it is obvious that knowledge of the mind-altering qualities of *Cannabis* and its use for such purposes formed an ancient ritual tradition for many South Asian peoples. Among the Hindus, *Cannabis* is regarded as a holy plant, and the origin of this conception can be traced to the early Vedic period. According to old Hindu poems, Lord Shiva brought the *bhang* plant down from the Himalayas and gave it to mankind (Chopra and Chopra 1939). Even into the twenty-first century a *sadhu* smoking *ganja* has not only been tolerated but also looked upon with some veneration; *sadhus* are even now considered by some to possess supernatural powers to heal disease and infirmities. Sects of *sanyasi*, *mahanta*, and *mantradata gurus* or religious preceptors are held in great respect, although they indulge freely in *Cannabis* drugs. In fact, offering *Cannabis* to them always was and still is considered to be an act of piety (e.g., Chopra and Chopra 1939, 1957; Sastri 1950).

Indian government prohibition of *bhang*, the mildest *Cannabis* preparation, has been resisted by a large segment of the Indian community who feel that such action would restrict their religious liberty. The enormous influence that *Cannabis* and its associations have on the minds of certain classes in India and the tradition that has been built around it is readily apparent. Any interference with traditional use of *Cannabis* drugs, especially *bhang*, in connection with religious customs and observances may be regarded as an encroachment upon the religious rights and liberties of Indians (Chopra and Chopra 1939; Sastri 1950; Mikuriya 1994).

Mongols and *Cannabis*

The Mongol "barbarian" invasions, beginning in twelfth to thirteenth centuries, spread across Asia and had a major impact over a huge geographical range from Siberia to the Mediterranean, greatly influencing the diffusion of ideas and products. Aldrich (1977) described the Mongol impact on the cultural diffusion and use of *Cannabis*: "The Mongol hordes also spread the secret [of hashish]. Sufi schools, including a colony of Haidaris on the Malabar Coast, spread hashish use in India. Tamarlane sowed hemp around Samarkand, and his descendent Babur, first Mogul Emperor of India (early 1500s), learned of it in Afghanistan. The Moguls planted *Cannabis* and the opium poppy throughout their dominions: it is likely that the two great hemp cultivation centers of Afghanistan, Central Asia, and Kashmir on one hand and Bengal and Nepal on the other, grew up during this period."

The Mongol invasions were concurrent with the spread of hashish use in Arabia, and the Mongols and the Sufis were often maligned by early Arabic religious scholars for introducing the hashish habit (Aldrich 1977; also see Aldrich 1971; Rosenthal 1971). Unfortunately, the exact order of *Cannabis* introductions has not been determined, and it is not yet clear whether the Mongols brought hashish with them from Central Asia or hashish was brought to Arabia by Persians and other peoples displaced by advancing Mongol hordes. It is also possible that the Mongols first encountered hashish during their conquests in Persia and Arabia, where *Cannabis* use was likely introduced much earlier by the Scythians and related Central Asian or European steppe peoples.

Muslim Central Asia supplied hashish to Hindu and Buddhist South Asia for many centuries, earning a reputation for producing a high-quality resinous product, a tradition that may date back several thousand years. However, the large-scale cultivation of *Cannabis* and mass production of hashish by mechanical sieving were innovations demanded by the burgeoning hashish trade of the last few centuries.

Southwest Asia, the Mediterranean, Africa, and Europe

There was once, my lord and crown upon my head, a man in a certain city, who was a fisherman by trade and a hashish eater by occupation. When he had earned his daily wage, he would spend a little of it on food and the rest on a sufficiency of that hilarious herb. He took his hashish three times a day: once in the morning on an empty stomach, once at noon, and once at sundown. Thus he was never lacking in his extravagant gaiety.
(from "The Tale of the Two Hashish Eaters" in *One Thousand and One Nights*, MATHERS 1923)

Southwest Asia is considered one of the earliest centers of hashish use, but its antiquity is difficult to establish accurately. Most likely, hashish and its production by sieving were introduced to southwestern Asia from Central Asia via Persia or India (e.g., Rosenthal 1971; also see Merlin 1972; Shah 1997; Clarke 1998a). Assyrian texts from the eighth to the sixth centuries BCE refer to *Cannabis* as "the drug which takes away the mind" (Mechoulam 1986). This translation could be interpreted as ecstasy or the separation of body and spirit in shamanistic ritual and was not necessarily a negative appraisal.

Most regions of southwestern Asia have arid climates essential for the sieved collection of *Cannabis* resin powder. In recent years, hashish has certainly been manufactured in Lebanon, Syria, Turkey, and Greece. Early Arabic texts from the thirteenth to the fifteenth centuries often refer to hashish as "the little dust-colored one" (Rosenthal 1971), accurately describing both the color and texture of sieved hashish. It is likely that powdered hashish was also produced early on in Egypt, Iraq, and Iran. During the late twentieth century, Lebanon ranked as the major hashish producer in

FIGURE 42. Throughout the 1970s, Westerners searching for *Cannabis* highs and exotic cultural experiences followed the "Hippie Trail" to destinations such as Lebanon's Bekaa Valley, which is still famous for producing hashish (from Clarke 1998a).

the Middle East, followed by Syria, Turkey, Greece, and Egypt. In recent decades, little hashish was being produced in these countries. However, a report in 2007 indicated that hashish fields were in fact being cultivated in the Bekaa Valley of Lebanon (Blanford 2007), which the Internal Security Forces of the Office of Drug Control in cooperation with the Lebanese Army claimed to have eradicated in 2011, at least in the northern Bekaa Valley (Now Lebanon 2011); on the other hand, Morocco, where hashish was only first produced on a large scale in the early 1970s, has become the world's largest exporter (Clarke 1998a).

The close proximity of Southwest Asia to ancient Persia and Arabia resulted in the spread of hashish production into this region at a relatively early date. Hashish is not mentioned in the *Koran* (the central religious text of Islam), which originated with the enlightenment of the prophet Muhammad (lived 570 to 632 CE). However, Muslim clerics and lawmakers have debated the prohibition of hashish since the eleventh century when hashish eating spread widely throughout Arabia (Rosenthal 1971; also see La Barre 1980). It is not clear when hashish came to be used recreationally rather than for mystical ceremonial use in ritual settings. The oldest known monograph on hashish is the *Zahr al-'arish fi tahrim al-hashish,* written in Arabic by Az-Zarkashi during the thirteenth century, but it is lost. Early Arabic texts refer to *Cannabis* and/or hashish by many names such as "shrub of emotion," "shrub of understanding," "peace of mind," "branches of bliss," and "thought morsel" (Rosenthal 1971). These names indicate common recognition of the psychoactive nature of *Cannabis*.

Even though the ancient Avestan texts refer only to *bhang* and not *charas*, Persia (present-day Iran) is often credited as the original home of hashish, more specifically, as the first area where *charas* was produced by sieving dried *Cannabis* flowers rather than hand-rubbing it directly from plants. The legends of Persian Sheik Haidar as well as the "Hashishin Assassins" from the twelfth and thirteenth centuries are the earliest written tales of the discovery of the psychoactive potency of *Cannabis* and the use of hashish. Juliette Wood (1988) described the present-day view of the Hashishin Assassins use of hashish, and how its utilization by this sect provided its leader with a kind of awe-inspiring and mystifying element: "Historians of the Assassins are not in agreement about how prevalent the use of this drug was among [them], or even why it was used. The frequency of use has undoubtedly been exaggerated, since a large group of drug addicts would not make a very effective terrorist organization. However, its importance on the perception of the Assassins and their activities is undeniable." According to an early Arab legend, Sheik Haidar, the founder of an early Islamic Sufi religious order, came across a trembling *Cannabis* plant while wandering in the mountains of Persia and ate some of its leaves. Usually a quiet man, when he returned to his monastery, his disciples were amazed that he was talkative and full of spirit. After they cajoled Haidar into disclosing how he had made himself feel so happy, they went out into the mountains, tried *Cannabis* themselves, and learned its pleasures (Abel 1980). Maybe Haidar ate lots of potent flowers, but even then he would have felt little, since *Cannabis*, as previously described, has almost no effect until heated, and possibly the leaves were somehow prepared to release their potency.

Many early accounts of hashish use in Arabia are concerned with debates over its legal status and lengthy comparisons to alcohol, the prohibited drug (Rosenthal 1971). An account of early Arabian efforts to eradicate *Cannabis* cultivation and hashish use, presented by Lewin (1964), serves as a historical example of a failed drug prohibition campaign: "It is recorded that in the year 1378 the Emir Soudoun Sheikhouni tried to end the abuse of Indian hemp consumption among the poorer classes by having all plants of this description in Joneima destroyed and imprisoning all the hemp-eaters. He ordered, moreover, that all those who were convicted of eating the plant should have their teeth pulled out, and many were subjected to this punishment. But by 1393 the use of this substance in Arabian territory had increased."

The first mention of the euphoric properties of *Cannabis* in Hebrew texts is found in the *Talmud*, written in the sixth century CE during the early Middle Ages (Abel 1980; see Benet 1975; Russo 2007; Bennett 2011 for further discussion relating to early *Cannabis* in the Near East). The Jews may have learned of hashish in Egypt before their exodus ca. 1220 BCE (see the discussion of ancient psychoactive use of *Cannabis* in Egypt later in this chapter). The Scythians, however, conquered lands all the way to Palestine and Egypt from 630 to 610 BCE (Mechoulam 1986). As noted earlier, the Scythians certainly knew of *Cannabis*, if not hashish production, and they may have been the first to bring *Cannabis* to the Middle East. However, there is no mention of hashish in any ancient religious texts nor is there any reference to hashish eating until the tenth century (Rosenthal 1971). Indeed it is just as likely that *Cannabis* came to the eastern Mediterranean from Persia several centuries after the Scythians or possibly even much more recently.

Across the Mediterranean, in ancient Greece, *Cannabis* was known as a fiber source for cordage production and its seeds were used for a few medicinal applications (e.g., see Dioscorides 1968; Brunner 1973; Russo 1998, 2001, 2004); however, the euphoric properties of *Cannabis* were either not commonly known or generally ignored, and its use for ecstatic purposes was largely restricted to the Scythians and other peoples north of Greece (e.g., see Arata 2004). Nevertheless, there is some evidence for the psychoactive use of *Cannabis* among the ancient Greeks. For example, Bremmer (2001) refers to ancient *Cannabis* seed evidence found at the famous Greek oracle of Thesprotia, in Epirus. This oracle was a *nekyomanteion*, a place where the dead could be consulted

or souls could be guided, and it is here that Bremmer tells us that archeological excavations "have brought to light many bones of cows, sheep, and pigs, which were probably meant for the living, but also the carbonised [*sic*] remains of barley, wheat, broad beans, lathyrus [perennial pea] and, probably, hemp, which were clearly meant for the dead."

Bremmer also refers to possible literary reference to *Cannabis*, which is "never mentioned in the relevant literature" and has a probable association with Thracian "ecstatic use of cannabis." Bremmer pointed out that the *Antiatticista*, a Greek dictionary in Roman times that documented words suitable for use by people who wanted to write Greek accurately, notes the use of the word "*kannabis*" by Sophocles in his tragedy *Thamyras*, concerning a singing match between Thamyras and the Muses that includes ecstatic dancing; thus it seems reasonable to conclude "that Sophocles somehow connected the Thracian *Thamyras* with an ecstatic use of cannabis." This is supported by Posidonius's reference to "Thracian smoke-walkers (*kapnobatai*)" and the suggestion by Pomponius Mela that the "use of certain seeds [inflorescences?] by the Thracians results in a *similis ebrietati hilaritas*" (similar inebriated hilarity). This association with cheerfulness or laughter is found in later, Roman references.

Ancient Roman aristocracy was familiar with *Cannabis* seed oil as a medicament (Dioscorides 1968) and perhaps also knew about the mind-altering potency of the inflorescences. The great Roman natural historian Pliny the Elder may have been alluding to hemp when he referred to the *gelotophyllis* (laughing leaf), which he said came from Bactria, an ancient country lying to the northeast of modern-day Afghanistan in the traditional hashish producing region of Central Asia. On the other hand, this "laughing leaf" may have also been Syrian rue (*Peganum harmala*), which is sometimes referred to as *haoma* or *soma* (see earlier in this chapter and Chapters 4 and 8) and is also common in Central Asia, although no plant since the time of Herodotus has a greater traditional linkage to laughing than *Cannabis*.

Claudius Galen referred to *Cannabis* as a substance consumed on the ancient Italian Peninsula. He states that, "at dessert, small cakes were passed round which increased thirst, but if taken in excess produced torpor (or sluggishness)" (Lewin 1964; also see Kühn 1965; Mikuriya 1973). In addition, he tells us how it was customary to offer guests hemp seeds as a promoter of hilarity *cum aliis tragematis* ("with sweet fruits"); he also reports that the seeds were ingested "to create warmth," which suggests that *Cannabis* was well known as a mind-altering substance. Subsequent to these early Western references to the probable use of *Cannabis* for mind-altering purposes, Mikuriya (1973) pointed out that there is a "paucity of references to hemp's intoxicating properties in the lay and medical literature of Europe [including the Mediterranean Region] before the 1800s" (also see Walton 1938).

Cannabis was introduced into Africa at an early date, although accounts regarding the time of its arrival differ greatly and no certain date has been agreed upon. Schultes (1970) estimated that the period ca. 4000 to 3000 BP saw the first introduction of *Cannabis* into Africa, and Arab traders may have brought *Cannabis* directly from India to eastern and southern Africa (La Barre 1980; cf. Du Toit 1975a), but the first physical evidence for it psychoactive use in the African continent does not appear until about 640 to 500 BP (van der Merwe 1975; Merlin 2003; see Chapter 4 for theories of early human-related dispersal and later in this chapter for a discussion of ancient pipes associated with possible *Cannabis* residue found in Africa).

Historical references to *Cannabis* in Africa only mention use for mind-altering purposes, and hemp appears not to have been grown for fiber or seed, at least not in sub-Saharan Africa. The huge African continent is separated into northern and southern regions by the great Sahara Desert, and thus the spread of *Cannabis* into and through much of the continent followed two paths, one above the desert across North Africa from east to west, and another below the massive arid Saharan expanse. In both cases, its cultural dispersal was facilitated by Muslim travelers and sea traders.

In Egypt, archaeobotanical evidence indicates that *Cannabis* was present during dynastic times more than 3000 BP; however, the antiquity of its use for mind-altering purposes is not yet clear. On the basis of many literary references, Arata (2004) referred to ancient Egypt as "the most 'pharmaceutical' of the lands of the entire world." For example, Homer (the legendary Greek poet of the eighth to seventh centuries BCE) regarded the early Egyptians as a nation of druggists (Birdwood 1865). We know that many medicinal drugs of animal, vegetable, and mineral origin were used in Egypt during ancient times, perhaps as early as the predynastic Badarian culture (ca. 4400 to 3250 BCE); however, the ingredients of only a few of the several hundred drugs mentioned in existing papyrus texts have been positively identified. The Badarian culture is distinctive in predynastic Upper Egypt. It was the first to show evidence of agriculture in the region, as well as being the first Egyptian culture with a marked degree of social status, which is reflected in their burials and grave goods (e.g., see Hassan 1988; Bard and Schubert 1999). *Cannabis* may have been one of their drug sources, but this has not been verified. There is, however, some controversial archeological evidence suggesting quite ancient *Cannabis* use in Egypt, including body tissues of several Egyptian mummies chemically determined to contain cannabinoids. However, further confirmation of the presence of *Cannabis*-based substances is needed (Balabanova et al.1992; also see Chapter 4).

Ancient Egyptians were probably familiar with opium, which may have been the principal psychoactive ingredient of the famous drug Nepenthe (Merlin 1984; see Arata 2004 for an opposing view). The ancient knowledge and use of opium may have precluded the need for *Cannabis* as a euphoric drug even if the Egyptians had access to it. We find this situation today among traditional Chinese and Southeast Asian ethnic groups with a long history of *Cannabis* use for fiber and seed but opium as their drug of choice. The early Nile River civilizations most likely did not know of *Cannabis* until relatively late, after it had diffused through Arabia, Asia Minor, and eventually into Egypt as a drug plant rather than a fiber source (Laufer 1919).

According to Du Toit (1980), *Cannabis* has been grown in Egypt "for almost a thousand years, while the domestic variety was used for the production of rope, the wild [foreign] variety (though planted in gardens) was specifically recommended for use as a drug." *Cannabis* has definitely been cultivated in Egypt for hashish production for at least eight or nine centuries. During this time its use was focused primarily on consumption of psychoactive resin rather than fiber and seed products. The use of hashish was reportedly introduced to Egypt from Syria by mystic Islamic travelers in the twelfth century CE during the Ayyubid dynasty, a Muslim reign of Kurdish origins, which ruled Egypt and the surrounding region during the twelfth and thirteenth centuries CE (Khalifa 1975; La Barre 1980). Muslim migrants and traders probably also brought hashish and the technology to make it into Egypt. It is also likely that more potent varieties of *Cannabis* originating farther east may have been introduced around the same time. *Cannabis* was grown in Egypt for hashish production well into the twentieth century CE (Clarke 1998a).

Rosenthal (1971) provided descriptions of psychoactive *Cannabis* from the thirteenth century CE written by Ibn al-Baytar (also known as Ibn Al-Bitar, lived 1179 to 1248 CE), an Arab traveler and outstanding medical herbalist born in Spain of wealthy parents who recorded his observations of hashish consumption by Egyptian Sufis: "There is a third kind of *qinnab* (*Cannabis*), called Indian hemp, which I have seen only in Egypt where it grows in gardens and is also known to Egyptians as *hashishah*. It is very intoxicating if someone takes as little of it as a *dirham* [small monetary unit] or two. Taken in too large doses, it may lead to lightmindedness."

Ibn al-Baytar also described the making of a *majoon* confection much as it has been traditionally prepared in South Asia for millennia: "First, they baked the leaves until they were dry. Then, they rubbed them between their hands to form a paste, rolled it into a ball, and swallowed it like a pill. Others dried the leaves only slightly, toasted and husked them, mixed them with sesame and sugar, and chewed them like gum."

The baking or toasting of the leaves would turn any nonpsychoactive THC-acid into psychoactive THC, and chewing the confections like gum would deliver the THC directly to the brain via absorption into the veins under the tongue. In Tashkent, the capital city of Uzbekistan in Central Asia, a confection called *guc-kand* made of boiled *Cannabis* mixed with sugar and spices was popular among the women to induce a "happy mood" (Benet 1975).

Soueif (1972) surveyed the early history of hashish use in Egypt, as chronicled in twelfth- and thirteenth-century poems, and revealed several desirable behavioral changes associated with *Cannabis* consumption including euphoria, acquiescence, sociability, feeling carefree, feeling important, freedom from disturbance of mind or passion, becoming meditative, activation of intelligence, jocularity, and amiability. By way of comparison with alcohol, the following qualities were attributed to hashish: it was less expensive, it was not prohibited by Islamic religious authorities (at that time), only a relatively small amount was required to achieve the desired effects, it did not smell like alcohol, it was not as easily detected, and lastly, it was never pressed with the feet (an Islamic taboo) to be made ready for consumption. Adversaries, on the other hand, claimed there were undesirable effects of hashish use such as submissiveness, debility, insanity resulting from organic brain damage, and a tendency toward prostitution.

The Garden of Cafour was a legendary, but also real, early Egyptian haven near Cairo frequented by hashish-using travelers, scholars, and poets (Abel 1980); it was destroyed by religious zealots in 1251 CE. According to the Egyptian historian Maqrizi (lived 1364 to 1442 CE), Sultan Nigm Al Din Ayoub forbade the planting of hashish in the Regouri Gardens and also punished hashish eaters by pulling out their teeth (Khalifa 1975). Despite harsh punishment, by the end of the fifteenth century CE hashish had grown in popularity and was openly sold in the bazaars (Sumach 1976). Prior to the sixteenth century CE, the only manner in which hashish was consumed in Egypt was by eating it mixed into various confections. Subsequently, hashish was also smoked in its pure form or mixed with more recently introduced tobacco

(Khalifa 1975). Hashish production continued in rural valleys throughout Egypt well into the twentieth century and seedless marijuana (*bhango*) is still grown there today.

During Napoleon's Egyptian campaign of 1778 and 1779, he discovered that habitual use of hashish was spread widely throughout the Egyptian lower class (Kimmens 1977); subsequently, he decreed a total prohibition against it. However, his soldiers took the hashish habit back to France with them, somewhat like the many American veterans of the Vietnam War who brought their newly acquired *Cannabis*-smoking habit back to the United States during the late 1960s and early 1970s. Cairo, along with Istanbul in Turkey, were the largest hashish markets in recent centuries. In 1879, the Egyptian government installed the first in a series of laws with increasingly severe penalties prohibiting the cultivation of *Cannabis* as well as production, sale, or use of hashish (Khalifa 1975). Eventually *Cannabis* cultivation and use spread westward all the way across North Africa to Morocco, where it became widely consumed.

Historical evidence indicates that *Cannabis* also followed a second route to sub-Saharan eastern Africa carried by Muslim sea traders from the Indian subcontinent, via Arabia, possibly as early as the first century CE (Du Toit 1980), spreading down the eastern coast and then inland throughout the semitropical and tropical sub-Saharan region. Du Toit (1975a) presents a detailed history and ethnography of *Cannabis* in southern Africa with maps suggesting paths of cultural dispersal of its use into and throughout much of Africa. La Barre (1980) referred to three distinct terminological provinces for *Cannabis* in Africa: "South Africa (*dagga* terms), east-central (*bangi* and variants, from "Bengal"), and West Africa (*diamba* and variants)."

Pipes thought to have been used for smoking *Cannabis* have been recovered from Zambia and Ethiopia. Excavations at the Iron Age Sebanzi Hill site on the southern edge of the Central Kafue Basin in southern Zambia by Fagan and Phillipson (1965) uncovered four baked clay, non-Arab designed smoking pipes. Radiocarbon dating, along with related pottery sequences, on the oldest two specimens indicates that they were in use around the tenth to twelfth centuries CE. Although the pipes were not chemically analyzed, it has been argued that they were used for smoking *Cannabis* because tobacco could not have been known in Zambia at the time (Phillipson 1965).

North of Zambia in Ethiopia, remains of two ceramic water pipe bowls were recovered from Lalibela Cave and dated to ca. 640 to 500 BP. Both contained trace amounts of Δ^8-THC according to modified thin-layer chromatography. Δ^8-THC (also referred to as Δ^6-THC) is produced by *Cannabis* plants in trace amounts during the biosynthesis of the primary psychoactive cannabinoid Δ^9-THC (also called Δ^1-THC, see Chapter 4). But whereas Δ^9-THC decomposes into unidentifiable compounds relatively quickly, Δ^8-THC is very stable and can be detected in archeological contexts. Van der Merwe (1975; also see Merlin 2003) therefore interpreted the occurrence of Δ^8-THC as an indication that psychoactive *Cannabis* containing THC was smoked in these pipes. These reports are controversial because both these dates precede the exploration of the New World by Spain and the supposed first date of introduction (ca. 500 BP) of tobacco, pipes, and smoking from the New World into Europe. More substantive data are needed.

However, *Cannabis* was probably smoked in Africa well before European exploration, at least in some regions, certainly as far back as the Iron Age in southern Africa,

predominantly among Bantu speakers (Du Toit 1975a/b). According to Emboden (1972) *Cannabis* entered Africa by obscure routes, but we may assume that it was probably NLD *Cannabis indica* and came originally from India or Saudi Arabia:

> None of the more elaborate techniques of using *Cannabis* in the Mediterranean or the Near East accompanied the plant into Africa, and practices in the central part of the continent in the 13th century were very simple. The confections which were known to Galen, such as *Cannabis* wine, or the date, fig, raisin, nut and *Cannabis* confection (*majun*) of the North Africans, apparently had not reached central Africa at this early date. Initially, the simple but efficacious practice of throwing hemp plants on burning coals of a fire and staging what might today be called a "breath in" seems to have been popular. This was elaborated into a ritual in which members of a given tribal unit would prostrate themselves in a circle around the fire and each would extend a reed into the fumes in order to capture the volatized resins, without the accompanying irritation produced by standing over the vapors and inhaling. At a later date the fire was elevated to an altar, where humans could sit or stand while inhaling through a tube extending into the smoke. (EMBODEN 1972)

Cannabis seems to have appeared in the Zambezi Valley of Zimbabwe before 500 CE during pre-Portuguese times and perhaps much earlier. In 1609, in a very early book on Africa, the Portuguese Dominican missionary João dos Santos reported *Cannabis* cultivation in the Cape of Good Hope region, in far southern Africa, and referred to indigenous peoples consuming the leaves and "becoming drunk as if on a surfeit of wine" (Booth 2003). In 1662, Jan Anthoniszoon van Riebeeck (lived 1617 to 1677 CE), a Dutch colonial administrator and founder of Cape Town, South Africa, wrote in his journal about the ritual tribal use of *Cannabis* and observed that native Africans "live permanently on the same spots, where they plant and dry a certain plant which they call *dacha* (*dagga*), which they bruise and eat, and which makes them very silly" (Schleiffer 1979, citing van Riebeeck 1900).

Cannabis continued to be used in various parts of Africa, often spreading into new areas via Arab traders, and in many regions was considered a medicinal herb (see Chapter 8). The attitudes of European colonists and explorers about *Cannabis* use by native African peoples changed over time, with some of the Dutch settlers even cultivating the plant for their native employees as an enticement to work for them, but not all tribes or colonial Europeans were in favor of *Cannabis* use for mind-altering purposes and had mixed opinions about its widespread consumption (e.g., see Du Toit 1975b; Booth 2003). Henry Stanley, the American journalist and explorer famed for his introductory statement when meeting the well-known British explorer and missionary David Livingstone, "Mr. Livingstone, I presume," often articulated his utter distain for native Africans, describing them as "wild as a colt, chafing, restless, ferociously impulsive, superstitiously timid, liable to furious demonstrations, suspicious and unreasonable," and in his mind almost subhuman; for example, in his book *Through the Dark Continent*, he reveals his racist perspective, telling us how he believed that the psychoactive use of *Cannabis* caused deterioration in the native peoples to such a degree that they were worthless as cargo carriers, and he considered the "almost universal habit of vehemently inhaling the smoke of the *Cannabis sativa* or wild hemp" to be the drug most hazardous to the "physical powers" of these people: "In a light atmosphere, such as we have in hot days in the tropics, with the thermometer rising to 140 Fahrenheit in the sun, these people, with lungs and vitals injured by excessive indulgence in these destructive habits, discovered they have no physical stamina to sustain them. The rigor of a march in a loaded caravan soon tells upon their weakened powers, and one by one they drop from the ranks, betraying their impotence and infirmities" (Stanley 1879).

Livingstone (1858) had other things to say about to the use of *Cannabis* among native African peoples. In the first example, he further describes its effects on his own porters:

> We had ample opportunity for observing the effects of this *matokwame* [*Cannabis*] smoking on our men. It makes them feel very strong in body, but it produces exactly the opposite effect upon the mind. Two of our finest young men became inveterate smokers, and partially idiotic. The performances of a group of *matokwane* smokers are somewhat grotesque; they are provided with a calabash of pure water, a split bamboo, five feet long, and the great pipe, which has a large calabash or kudu's horn chamber to contain the water, through which the smoke is drawn *narghille* [water-pipe] fashion, on its way to the mouth. Each smoker takes a few whiffs, the last being an extra long one, and hands the pipe to his neighbour [*sic*]. He seems to swallow the fumes; for, striving against the convulsive action of the muscles of chest and throat, he takes a mouthful of water from the calabash waits a few seconds, and then pours water and smoke from his mouth down the groove of the bamboo. The smoke causes violent coughing in all, and in some a species of frenzy which passes away in a rapid stream of unmeaning words, or short sentences, as, "the green grass grows," "the fat cattle thrive," "the fish swim."

In another passage he describes his views about psychoactive *Cannabis* use and its effects on the Bakota tribe:

> The Batoka of these parts are very degraded in their appearance, and are not likely to improve, either physically or mentally, while so much addicted to smoking the *mutokwane*. . . . They like its narcotic effects, though the violent fit of coughing which follows a couple of puffs of smoke appears distressing, and causes a feeling of disgust in the spectator . . . This pernicious weed is extensively used in all the tribes of the interior. It causes a species of phrensy [*sic*], and Sebituane's soldiers, on coming in sight of their enemies, sat down and smoked it, in order that they might make an effective onslaught. I was unable to prevail on Sekeletu and the young Makololo to forego its use, although they can not point to an old man in the tribe who has been addicted to this indulgence. I believe it was the proximate cause of Sebituane's last illness, for it sometimes occasions pneumonia. Never having tried it, I can not describe the pleasurable effects it is said to produce, but the hashish in use among the Turks is simply an extract of the same plant, and that, like opium, produces different effects on different individuals. Some view every thing as if looking in through the wide end of a telescope, and others, in passing over a straw, lift up their feet as if about to cross the trunk of a tree. The Portuguese in Angola have such a belief in its deleterious effects that the use of it by a slave is considered a crime.

Other authors have referred to the relationship of *Cannabis* drugs and warfare in Africa. For example, the missionary A. T. Bryant (1949), who started his work in South Africa in 1833, commented on the use of *Cannabis* (*Dagga*) by Zulu warriors prior to battle. Booth (2003) argued that *Cannabis*,

normally producing a relatively calm effect, would have more likely been combined as a "hallucinogen" with an "excitant" (stimulant) such as "*olkiloriti*," prepared from the bark and roots of a tree, which Maasai warriors are known to ingest prior to battle to suppress exhaustion and fright and perhaps even produce a "state of frenzy." *Olkiloriti* is derived from *Acacia nilotica* (syn. *Mimosa nilotica*), a member of the bean family, Fabaceae, that may produce known psychoactive substances such as tryptamine and harmaline alkaloids (Bhakuni et al. 1969).

A rather different, nineteenth-century reference to the association of *Cannabis* drug use and warfare as practiced among the Bashilange people (also known as the Baluba or Barua, who lived on the western shores of Lake Tanganyika) was reported by the German explorer Herman von Wissmann (1891). This European explorer observed them and their use of hemp (*riamba*) for mind-altering purposes during his 1881 expedition through Equatorial Africa. According to von Wissmann, the Bashilange were originally a warlike society: "One tribe with another, one village with another, always lived at daggers drawn. . . . The number of scars which some ancient men display among their tattooings gives evidence of this. Then, about twenty-five years ago . . . a hemp-smoking worship began to be established, and the narcotic effect of smoking masses of hemp made itself felt. The Ben-Riamba, 'Sons of Hemp,' found more and more followers; they began to have intercourse with each other as they became less barbarous and made laws."

The shift from warfare to peace among the Bashilange was only one transformation inspired by the new cultural trait of smoking *Cannabis*. Another new phenomenon was the rise of a novel religion based on *riamba*, which became strongly associated with tranquility, friendship, the supernatural, and security: "Tribesmen were no longer permitted to carry weapons in their villages, they called each other friend, and they greeted one another with the word *moyo*, meaning 'life' and 'health.' Although formerly cannibals, they abjured their previous custom of eating the bodies of their captured opponents" (Abel 1980).

The use of *riamba* was common among Bashilange men but almost exclusively nonexistent among women. During their nightly religious ceremonies village men would sit in a circle smoking *riamba* with their bodies naked and their heads shaved, and *Cannabis* was also smoked on significant holidays and when alliances were concluded (von Wissmann 1891; also see Abel 1980 and Booth 2003 for more discussion of the unique history of *Cannabis* use among the Bashilange). Cult use of *Cannabis* also occurred in other areas such as among members of a somewhat obscure sect in the Sudan focused on Sirdar, a strange woman under whose direction groups were established to promote the smoking of *Cannabis* (*dagga*). Similar to the *riamba* cult among the Bashilange, this one ultimately faded away.

When Europeans began colonizing Africa, psychoactive use of *Cannabis* by native Africans was very widespread. Consumption for both recreational as well as ritual purposes was tolerated by white residents, more or less, at least until certain segments of society became alarmed that use in cities might contaminate their culture and people. Use of *ganja* by indentured laborers imported from India to work in South African agriculture was particularly troubling to some people of European descent. Whites referred to Indian immigrants as "coolies" and considered them distinctly different racially and culturally. Once Indian immigrants fulfilled their contractual labor responsibilities they often moved to urban areas and indulged in *Cannabis* for mind-altering purposes as they had done in their homeland (Du Toit 1977). This worried enough white people that a law was passed in South Africa in 1870 banning use of *Cannabis* among "coolies." This legal maneuver was generally disregarded, and new anti-*Cannabis* actions were taken in 1887. South Africa went so far as to (unsuccessfully) advocate for a worldwide ban under the League of Nations in 1923. Subsequently, additional anti-*Cannabis* laws were passed in South Africa with similar poor consequences.

Although *Cannabis* was brought to the New World from Europe by individual immigrants or families, it was introduced under the auspices of colonial governments to encourage cultivation for fiber and seed production (see Chapters 5 and 6). *Cannabis* was also introduced from Africa surreptitiously or under more open conditions by slaves who brought *Cannabis* seeds during their forced emigration. *Cannabis* and its mind-altering use were probably transported on purpose to the New World by slaves from their homeland to the New World. For example, Rosado (1958, cited in Hutchinson 1975) indicated that *Cannabis* was introduced to Brazil from Africa beginning about the middle of the sixteenth century CE, if not earlier, and that 1549 was a significant year because Don João III of Portugal certified that the recently established sugar cane planters could bring in up to 1,200 slaves per sugar mill: "Rosado (1958), quoting Pio Correa, indicates that *Cannabis* seed[s] were brought to Brazil in cloth dolls which were tied to the rag tag clothing worn by the slaves. He further states that *Cannabis* was planted and adapted itself well to the entire areas from the state of Bahia all the way up to the state of Amazonas. . . . Most authors disclaim *Cannabis* use in southern Brazil until this century, in spite of evidence to the contrary" (Hutchinson 1975).

In nineteenth-century Brazil, *Cannabis* use was not only limited to the slaves. Apparently, Queen Carlota Joaquina, wife of Emperor Don João IV, then king of both Portugal and Brazil, picked up the habit of using *Cannabis* for psychoactive purposes while the Portuguese royal court spent a number of years in Brazil to avoid possible invasion of their country by Napoleon of France.

> The court spent approximately six years in Brazil, returning to Portugal at the end of the Napoleonic Wars. Queen Carlota Joaquina was dying in 1817. Her favorite Negro slave, Felisbino, who had accompanied her to Portugal, usually provided her with cannabis. On her death bed, she asked Felisbino to "bring me an infusion of the fibers of *diamba do amazonas*, with which we sent so many enemies to hell." Felisbino made an infusion of cannabis and arsenic and gave it to her. It is recounted by Assis Cintra (1934) that upon taking the infusion, Dona Carlota felt no pain because of the analgesic action of *diamba*, "thereupon taking up here guitar and singing" and later dying, "Her slave Felisbino had the same end, drinking a *diamba* infusion with arsenic."
> (HUTCHINSON 1975)

Freyre (1946; also see Doria 1986 and Iglésias 1986) referred to importation of cult objects into Brazil by slaves from Africa, including "sacred herbs for aphrodisiac purposes or pure pleasure." Prominent among these revered herbs was *Cannabis*, which probably came with slaves from Angola and had several local names in Brazil by the nineteenth century, such as *pungo* in Rio de Janeiro, *macumba* in Bahia, *maconha* in Alagoas and Pernambuco, and *diamba* or *liamba* among

"*aficionados*" (persons with a liking and enthusiasm for *Cannabis*). Freyre also suggested that *Cannabis* and its use for psychoactive purposes could have reached Brazil with Portuguese sailors directly from India or indirectly via Portugal. *Macumba* use was outlawed by the Rio de Janeiro government in 1830, and anyone caught selling it was subject to a very large fine, "while the slave who used it was to be condemned to three days imprisonment" (Freyre 1946, citing Querino n.d.). Freyre described his own response upon inhaling burning *Cannabis*:

> I have smoked *macumba* or *diamba*, and I disagree with Querino as to some of the qualities that he attributes to it. It really does produce visions, however, and a pleasing weariness; the impression is that of one who returns from a ball with the music still ringing in one's ears. It would seem, on the other hand, that its effects vary considerably with individuals. Inasmuch as its use has of late become general in Pernambuco, the police are rigorously prosecuting its venders and consumers—the latter smoke it in cigarettes and pipes, some of them even taking it in their tea.

According to Hutchinson (1975, citing Freyre 1967 and Moreno 1986), during the colonial period in northeast Brazil, *Cannabis* was commonly used as a mind-altering substance, especially during the annual periods of inactivity between sugarcane harvests when "the white man filled his empty days with perfumed cigars while the black man smoked *maconha* for its dreams and torpor!" By the first part of the twentieth century *Cannabis* was cultivated and smoked in many regions of Brazil, with some attributing mystical properties to it, including its association with sexual magic (Freyre 1946, citing Querino n.d.; also see Querino 1938). Additional sources of information regarding the origins, colonial history, and more recent psychoactive use of *Cannabis* in the New World can be found in reports by Comitas for Jamaica, Williams-Garcia for Mexico, Partridge for Colombia, and de Pinho for Brazil in the comparative, cross-cultural volume edited by Rubin (1975; also see Freyre 1981).

Cannabis, where and when it was formerly consumed in Europe for ritualistic or religious purposes, was typically eaten rather than smoked to attain an altered state of consciousness. Lacey and Danziger (1999) presented an intriguing account of mind-altering bread baked by rural British peasants around 1000 CE: "This hallucinogenic lift was accentuated by the hedgerow herbs and grains with which the dwindling stocks of conventional flour were amplified as the summer wore on. Poppies, hemp, and darnel [*Lolium* spp.] were scavenged, dried, and ground up to produce a medieval hash brownie known as 'crazy bread'. So even as the poor endured hunger, it is possible that their diet provided them with some exotic and artificial paradises." A combination of psychoactive parts of the opium poppy and/or *Cannabis* could have produced a wide variety of psychological and physiological effects, and psychoactive parasitic fungi sometimes infect grasses such as darnel. According to Polish anthropologist Sula Benet (1975), Eastern Europeans cooked "happy porridge," including the following ingredients in the recipe: almond butter mixed with hashish, dried rose leaves, Mt. Atlas daisy (*Anacyclus pyrethrum*) root, carnation petals, crocus bulb, nutmeg, cardamom, honey, and sugar. This mixture was the most expensive of all hashish preparations and was eagerly sought by men who considered it "the strongest aphrodisiac."

Archeological excavations of a third-millennium BCE pit-grave tumulus (*kurgan*) tomb at Gurbanesti near Bucharest in Romania revealed a pipe-cup containing charred remains of hemp seeds accompanying one of the burials, which has been interpreted as evidence of the ritual burning of *Cannabis*, possibly for its psychoactive potency (Sherratt 1991; see Chapter 4). Across Eurasia, there is considerable evidence of inhalation of *Cannabis* smoke prior to the introduction of tobacco smoking following discovery of the New World. While simple inhalation of vapors is not considered to be true smoking since it does not employ specialized equipment such as pipes or constructions such as cigarettes and cigars, it does demonstrate an understanding of the euphoric properties of *Cannabis* vapors and their utilization. Herodotus (ca. 500 BCE) described the Scythian and other Central Asian tribes' recreational use of *Cannabis* by breathing the vapors given off during its combustion and makes no mention of any medicinal benefit from inhaling *Cannabis* smoke. According to Mechoulam (1986), the ancient Assyrians employed "*Cannabis* fumes" as a cure for the "poison of all limbs," which was presumed to be arthritis. Near Jerusalem, ashes of *Cannabis* burned in a bowl were found in the tomb of a 14-year-old girl dated to the fourth century CE, apparently administered to her as an inhalant during childbirth (Zias et al. 1993; see Chapter 8). In the Chitral Valley of northern Pakistan and the deserts of Afghanistan, the nomads consumed hashish by roasting a small piece on top of a flat rock placed on coals near the edge of a charcoal brazier and then breathing in the vapors and smoke through a hollow stick or reed. The smoker held water in his mouth as he leaned over the hot stone and inhaled, forcing the smoke to bubble through the water just like in a water pipe (Clarke 1998a).

In all these cases, *Cannabis* was burned in an open vessel, brazier, or censer, rather than being smoked in a pipe or rolled into a cigarette, as is done today (Clarke 1998a). The breathing of smoke and vapors from the open air is more like breathing incense than it is smoking, yet in these cases it was still the conscious inhalation of *Cannabis* smoke. The dual intent was to appreciate the fragrance of burning *Cannabis* or hashish and to obtain mind-altering effects, often interpreted as the mystical benefits derived from inhaling the fumes. When dry *Cannabis* is placed on coals or hot rocks, most of the resin vaporizes, and the burst of aroma from the volatilized terpenoids, carrying along odorless THC, is at its strongest just before it catches fire. To a careful observer, this would indicate that the aromatic and psychoactive properties of *Cannabis* reside together on the surface of the plant. Perhaps this observation led some early *Cannabis* users to seek a method for isolating both the active and the aromatic components of *Cannabis*, which led eventually to the development of hashish production, first by rubbing living plants and then by bulk sieving dried plants (Clarke 1998a).

A thin stick of hand-rubbed hashish looks like a piece of stick incense and burns in much the same way, but a solid lump of resin, as is produced in hand-rubbing, would be difficult to blend with other precious aromatic ingredients. On the other hand, sieved resin powder would be easy to blend with other powdered ingredients and then shape into pieces of commercial incense. Thus it seems that incense makers would have had an early interest in the physical as well as aromatic and psychoactive properties of *Cannabis* resin, and

it is quite likely that they were among early hashish traders. Initially, hashish was probably a relatively rare ingredient in incense, but as the use of incense became more widespread, the demand for hashish would have increased. Incense makers likely influenced early hashish production and may have made choices of favored aromatic and potent *Cannabis* varieties long before tobacco and hashish smoking became popular (Clarke 1998a).

As we noted, in temperate regions such as western and central Europe, *Cannabis* was, and still is, grown primarily for fiber used to make rope and strong fabric. However, as we have also shown, it was probably used to some extent for ecstatic purposes around the Mediterranean region in antiquity and may have had limited drug use in more northerly regions of Europe (see Sherratt 1991, 1995a/b, 1997). In any case, as Zuener (1954) concluded, "In Europe and the Mediterranean world, where beer and wine were available, the narcotic [*sic*] properties of hemp were perhaps less appreciated, especially since flax provided fiber and an oil of at least equal quality."

The Advent of *Cannabis* Smoking: Tobacco Meets Hashish

The eating of hashish began many centuries ago, but there is some question whether hashish smoking in Eurasia predated Columbus's discovery of the New World in 1492. As we discussed earlier, three finds of pipes possibly used for smoking *Cannabis* may predate the Columbian Era. However, most other evidence indicates that smoking hashish in pipes has its roots in the relatively recent tradition of smoking tobacco (*Nicotiana* spp.) brought to Eurasia and Africa from the New World during the early sixteenth century. Prior to this, the most common method of hashish consumption was by combining it with food or drink, and we suspect its use was more widespread than previously recognized. Tobacco was exotic, addictive, and traded widely. Soon after tobacco was introduced into Eurasia, hashish was mixed with it and smoked as it still is today in much of Europe, the Middle East, and South Asia. The popularity of tobacco smoking was initially responsible for the rapid spread of hashish smoking where it was already eaten and soon after that into regions where hashish was hitherto unknown. Hashish tagged along on the international success of tobacco, and thus the hashish market blossomed anew within a few decades (Clarke 1998a).

Smoking potent *Cannabis* elicits a quick and normally pleasurable response, the peak effects usually occurring within the first 15 minutes, whereas absorption by the gastrointestinal tract is much slower. In addition, the psychoactive intensity of smoked *Cannabis* decreases markedly over a few hours and leaves the user with little if any residual physical effects. Any behavior that is pleasurable, nontoxic, and (apart from its illegality) does not disturb social structure—such as smoking *Cannabis*—is often repeated and can develop into a pattern of habitual use. Immediate and enjoyable rewards encourage positive reinforcement of habits (Bower and Hilgard 1997). In contrast, the oral consumption of *Cannabis* is most often associated with relatively long-lasting and more profound effects. According to the rather early but sophisticated interpretations of O'Shaughnessy (1839), the effects observed upon ingesting carefully extracted *Cannabis* drink preparations made by his Indian friend Ameer (noted earlier) were quite pleasurable.

From either of these beverages intoxication will ensue in half an hour. Almost invariably the inebriation is of the most cheerful kind, causing the person to sing and dance, to eat food with great relish, and to seek aphrodisiac enjoyments. In persons of a quarrelsome disposition it occasions, as might be expected, an exasperation of their natural tendency. The intoxication lasts about three hours, when sleep supervenes. No nausea or sickness of stomach succeeds, nor are the bowels at all affected; next day there is slight giddiness and vascularity [redness] of the eyes, but no other symptom worth recording.

This passage also could describe a typical pleasant experience following the smoking of *Cannabis*. On the other hand, the effects of orally imbibed *Cannabis* are sometimes much more potent and unpredictable. Smoking *Cannabis* vaporizes the THC, which is absorbed immediately by the lungs, passed without delay into the bloodstream, and straight to the brain within a minute. When a person eats *Cannabis* it passes first to the stomach, where the vast majority of the THC is absorbed and transferred directly to the liver; although rate of uptake is affected by the type and amount of previous stomach contents. The liver then metabolizes the THC, changing it into 11-*hydroxy*-THC, which enters the bloodstream an hour or more later and flows to the brain. A more significant factor than the delay in onset is that 11-*hydroxy*-THC is approximately three times as potent as plant-derived THC, and its profound effects can last for up to 24 hours (Clarke 1998a).

Effects experienced from smoking even very potent *Cannabis* are relatively mild and manageable compared to the more dramatic and long-lasting effects often experienced from eating *Cannabis*. Eating *Cannabis* can induce extreme lethargy, distort time and space, induce deep introspection leading to personality disintegration, and produce visions, and users often continue to get higher for several hours, the effects sometimes persisting until the next day. During the nineteenth and early twentieth centuries, several books were written by authors who experimented with orally consumed hashish and described just such vivid and at times frightening effects; examples include the Frenchman Theophile Gautier's *Club des Haschischins* (1846), the American Fitz Hugh Ludlow's *The Hasheesh Eater* (1857), and the German Walter Benjamin's *Über Haschisch* (1972, English translation 2006). The British folklorist Richard Folkard (1884) provided a nineteenth-century account of the potent effects commonly attributed to eating *Cannabis*: "The Arabians concoct a preparation of hemp, which produces the most varied hallucinations, so that those who are intoxicated by it imagine that they are flying, or that they are changed into a statue, that their head is cut off, that their limbs stretch out to immense lengths, or that they can see, even through stone walls, 'the colour [*sic*] of the thoughts of others' and words of their neighbors." Only the idle rich or the hopelessly poor and destitute could afford time eating *Cannabis* and lying about all day in a dream state. Everyone else worked for a living, and eating hashish is rarely conducive to working. The effects of smoking *Cannabis* are shorter acting and easier to control, both mentally and physically, and therefore fit more easily into the lives of the working class. *Cannabis* was consumed in much larger quantities by an increasing number of people once the idea of smoking caught on. Today, far more people smoke *Cannabis* worldwide than consume it orally (Clarke 1998a).

The primary difference in *Cannabis* consumption patterns between North America and Europe was and still

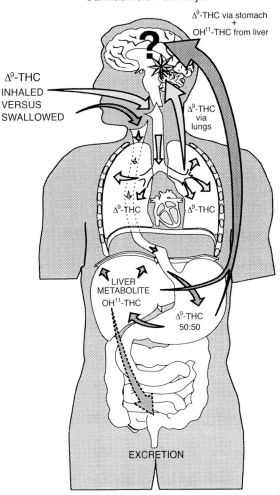

Cannabinoid Pathways

Δ^9-THC via stomach + OH^{11}-THC from liver

Δ^9-THC INHALED VERSUS SWALLOWED

Δ^9-THC via lungs

Δ^9-THC

Δ^9-THC

LIVER METABOLITE OH^{11}-THC

Δ^9-THC 50:50

EXCRETION

FIGURE 43. Vaporized Δ^9-THC enters the lungs, crosses into the bloodstream, and is carried directly to the brain, while the majority of orally ingested THC passes from the stomach to the blood stream, makes a first pass through the liver, and is converted to 11-*hydroxy*-THC (OH^{11}-THC). Human metabolite 11-*hydroxy*-THC is more potent and longer lasting than plant-based Δ^9-THC (adapted from Clarke 1998a).

is the form of *Cannabis* preferred by smokers. In North America, the majority of *Cannabis* smoked today is in the crude form of marijuana (normally the inflorescences of the female plants), while in Europe the preference has always been for a blend of tobacco and hashish. Only relatively recently has the spreading popularity of homegrown marijuana begun to change European smoking habits. The mass marketing of marijuana was primarily an American phenomenon beginning in the 1960s. Over the next 30 years marijuana imports into North America increased many fold. By shortly after the turn of the millennium, several hundred tons of *sinsemilla* were produced domestically each year across North America and Europe. (See Chapter 13 for present-day international *Cannabis* production statistics.) Traditional hemp-growing regions, formerly home to only *C. sativa* narrow-leaf hemp (NLH) cultivars and feral populations, now produce marijuana from *C. indica* BLD × NLD hybrid cultivars in commercial quantities, far exceeding the value of hemp crops.

FIGURE 44. As marijuana consciousness expanded throughout North America, early marijuana growers were beginning to save and sow imported seeds. This bumper sticker hails from the seminal yet failed 1972 California Marijuana Initiative.

Summary and Conclusions

Cannabis's early history as a mind-altering commodity was a period of great expansion into new climates and cultures, each with its own sets of selective pressures, setting the stage for its geographically localized and isolated evolution at the hands of humans. New *Cannabis* varieties arose through artificial selections directed by choice of preferred end product and type of effects, accompanied by natural ecological constraints imposed by respective geographical areas. Throughout the cultural dispersal of drug *Cannabis*, countless opportunities arose in terms of when and where it benefited humans, and subsequently its proliferation was encouraged. *Cannabis* could not have achieved nearly worldwide distribution, as demonstrated earlier, had it not been favored by humans time and again (e.g., see Benet 1975; Abel 1980; Grinspoon and Bakalar 1997; Clarke 1998a; Pollan 2001; Booth 2003).

Most likely the earliest mind-altering use of *Cannabis* occurred near the origin of highly psychoactive variants in Central or Southern Asia (see Chapter 11 for a discussion of *Cannabis* taxonomy). Well before contact with modern humans during the early Holocene, *Cannabis* was already present in several regions of Eurasia. Perhaps previous to and certainly during the movement of Aryan, Scythian, or other nomadic peoples from Central Asia, the knowledge of *Cannabis* use as a powerful tool in shamanic ritual was spread east, west, and south. Knowledge of the uses of psychoactive *Cannabis*, as well as seeds of selected variants with increased potency, were carried beyond Eurasia by later traders. In some areas, the inebriating properties of *Cannabis* were exploited later than its fiber and food uses, especially in some parts of Europe, East Asia, and more recently in North America. The first hashish was made by hand-rubbing living plants, while the method of drying and sieving plants to make hashish appeared much later, likely in response to commercial incentives.

The introduction of tobacco smoking to Europe and Asia during the early sixteenth century changed the usage patterns of both marijuana and hashish. Apart from ritualistic inhalation of vapors, until tobacco was introduced from the New World, marijuana and hashish were largely swallowed along with food and drink. Once people began to smoke rather than eat resinous *Cannabis* or concentrated hashish, many more began to consume it for its mind-altering effects. During the seventeenth and eighteenth centuries CE hashish smoking quickly gained popularity across Eurasia, from Turkey to Nepal, peaking during more modern times in the late nineteenth and early twentieth centuries. Throughout the eighteenth century, the technique of harvesting, drying, and sieving plants to make hashish became increasingly widespread as mass production became necessary to satisfy the rapidly increasing Eurasian hashish trade. During the late nineteenth and early twentieth centuries the patterns of hashish supply changed radically. Large-scale production of hashish in Central Asia shifted from Russian (Western) Turkestan into Chinese (Eastern) Turkestan, eventually into Kashmir, and finally into Afghanistan. The knowledge of hashish manufacture likely accompanied its importation across the Himalayas into India in ancient times (Watt 1889). Trade in hashish along innumerable trade routes over high mountain passes through the Himalayas from Chinese Turkestan (in present-day Xinjiang province of China) into India has a long history and remained economically important well into the twentieth century. By 1937 and 1938, trade in mind-altering *charas* represented an estimated 42 percent of the total value of Xinjiang's exports to India (Lattimore 1950). Greece, Syria, Nepal, Lebanon, and Turkey also became major hashish exporters. In the late 1960s, Morocco began to produce sieved hashish and by the early 1990s had surpassed all other regions in exports of this *Cannabis* product (Clarke 1998a). The international trade in *Cannabis*, as well as personal consumption for recreational and medical reasons, is far larger today than ever: "Cannabis remains the most widely produced and consumed illicit substance globally" (UNODC 2011).

Of the myriad effects *Cannabis* has exerted on the evolution of human culture, its ability to alter our consciousness is arguably the most profound. From a traditional, historical point of view, consumption of entheogens such as *Cannabis* has allowed some humans over the ages to recognize the god within and facilitate their communication with spirits. *Cannabis* was certainly used by shamans, and its increasing ritual use may have influenced aspects of worship adopted by the major religions of settled peoples. Nearly every cultural tradition of Eurasia was exposed to *Cannabis* consumption during its formation and evolution, and the psychoactive use of *Cannabis* is more widespread today than at any time in history; it said to be the most frequently used drug in the world today (e.g., Richardson 2010). If we are to accept the importance of *Cannabis* in molding our past and certainly influencing our present societies, then we should expect that it may well influence human cultural evolution in the future.

CHAPTER EIGHT

Ethnobotanical History and Contemporary Context of Medicinal *Cannabis*

Hemp, both because of its psychoactive properties and its mystical
significance, became a popular and widely-utilized plant in the
folk medicine of Europe and Asia. Since ancient times its soothing,
tranquilizing action has been known.

(BENET 1975)

Introduction

Earlier chapters in this book make it quite clear that *Cannabis*
has a very long history as a multipurpose resource with particular uses differing from region to region (see Chapters 4
through 7). How *Cannabis* was first used remains unresolved,
whether it was for fiber to produce cordage, seed for food, psychoactive resin for medicinal, ritual or spiritual applications,
or a combination of these. In fact, it is often difficult to separate early medicinal, psychoactive, ritual, and even food uses
of *Cannabis*, especially those involving mind-altering preparations utilized in any or all these contexts. This becomes
particularly complex within the traditional interrelationships
of religious belief, ritualistic practice, and medicinal use; the
latter is the focus of this chapter. The archeological and historical records offer evidence that *Cannabis* plants provided
a variety of important medicinal uses in antiquity, some of
which persisted into modern times. Ancient medicinal texts
(*materia medica*) of China and India, as well as Egypt, Greece,
and Rome, refer to specific uses for *Cannabis*, and these are
discussed later.

First it is important to recognize that awareness and use of
food as medicine is an ancient tradition in various areas of
the world and continues today (Etkin 2006). Chinese medical
traditions, for example, hold that diet is the most important
determinant of one's health, and there are often no distinct
boundaries between foods and medicine. As Anderson (1988)
explains, "Many things were purely medicines, but medicines
often became food if people learned to like them; many foods
became merely medicines when people stopped relishing
them; and all foods were considered to have medicinal value,

positive or negative, with important effects on health." This
concept of food as medicine applies to the traditional use of
Cannabis in other parts of the world as well. In this chapter
we investigate historical claims of *Cannabis*'s medicinal efficacy as they relate to absorption of cannabinoids (the physiologically active compounds found only in *Cannabis*) into
the bloodstream via inhalation, swallowing, or absorption
through skin or mucous membranes. (Ritual uses of psychoactive marijuana and hashish are discussed in Chapter 7, and
ritual uses of nonpsychoactive stems, fiber, and seed are presented in Chapter 9.) The relationship between mind-altering
and medicinal use in ritual contexts is also often assumed
by the person(s) consuming *Cannabis* or celebrating its symbolic therapeutic potency. References to the use of *Cannabis*
preparations as antibiotic and analgesic medicaments, most
commonly for external use, can be found in folk medicine
traditions and old herbals of the Middle East, East and South
Asia, Europe, Africa, and the New World. Mechoulam (1986)
provided a list of 20 medicinal applications of *Cannabis* by
traditional societies.

Central and East Asia form a huge region to which we attribute the ethnobotanical origins of *Cannabis* use. Medicinal
use of *Cannabis* has a lengthy history there, and much still
remains to be gleaned from literary and potential archaeobotanical records (e.g., see Chapter 7). Benet (1975, citing
Antzyferov 1934) presented some examples. In Tashkent, the
Central Asian capital city of present-day Uzbekistan, a confection consumed to induce a "happy mood," known as *guc-
kand*, was made of *Cannabis* boiled in water, sieved, and then
mashed together with sugar, saffron, and several egg whites.
The paste was formed into small balls and then dried in the

TABLE II

For millennia *Cannabis* has been traditionally
used around the world for relief from a wide variety
of medical conditions.

Medical condition	Medicinal uses substantiated by modern research
Analgesic	X
Anesthetic	
Antiasthmatic	X
Antibiotic	X
Anticonvulsive	X
Antidepressive	
Antidiarrhoeal	X
Antimigraine	
Antiparasitic	X
Antirheumatic	X
Alleviation of memory loss	
Appetite promoter	X
Facilitation of childbirth	
Hypnotic	X
Reduction of fatigue	
Sedative	X

SOURCE: From Mechoulam 1986.

sun. The "candy," which is said to have been "popular among women," was given to children to "keep them from crying" and to boys before circumcision to "reduce pain." A mixture of lamb's fat with *nasha* (hashish) was "recommended for brides to use on their wedding night to reduce the pain of defloration." Apparently, the same formulation worked well for headache "when rubbed into the skin," and it was also eaten "spread on bread."

Early East Asian Medicinal *Cannabis* Use

The utilization of *Cannabis* for medical purposes is ancient in China where origins of therapeutic use can be traced through ancient legends and fables. For example, Abel (1980, citing Doolittle 1966) paraphrased a Chinese story about the mythical emperor Liu Chi-nu that explains connections between *Cannabis* and illnesses:

> One day Liu was out in the fields cutting down some hemp, when he saw a snake. Taking no chances that it might bite him, he shot the serpent with an arrow. The next day he returned to the place and heard the sound of a mortar and pestle. Tracking down the noise he found two boys grinding marijuana leaves. When he asked them what they were doing, the boys told him they were preparing a medicine to give to their master who had been wounded by an arrow shot by Liu Chi-nu. Liu Chi-nu then asked what the boys would do to Liu Chi-nu if they ever found him. Surprisingly, the boys answered that they could not take revenge on him because Liu Chi-nu was destined to become the emperor of China. Liu berated the boys for their foolishness

and they ran away, leaving behind the medicine. Sometime later Liu himself was injured and he applied the crushed marijuana leaves to his wound. The medicine healed him and Liu subsequently announced his discovery to the people of China and they began using it for their injuries.

Another similar folk tale (Abel 1980, citing Wong and Lien-Teh 1936) tells us about a farmer who spotted a snake carrying leaves of *Cannabis* to put on the wound of another serpent. The following day the injured snake was cured. Inquisitive about the efficacy of the hemp remedy, the farmer experimented with *Cannabis* leaves on his own injury and was healed. These and other folk tales provide us with some insight about the notion that *Cannabis* possesses potency.

References to Cannabis in the oldest-known medical text, titled *Shen Nung Pen Ts'ao Ching*, or "Divine Husbandman's Materia Medica," are attributed to the legendary Chinese emperor Shen Nung, who probably lived sometime between 3494 and 2857 BCE (Chang 1962). Shen Nung is traditionally credited with inventing agriculture and introducing medicines to Chinese culture. The first known edition of Shen Nung's *Pen Ts'ao Ching* was published at the end of the Western Han dynasty (ca. 1800 BP) based on manuscripts possibly dating from as early as about 4700 BP. It contains descriptions of 365 medicines from natural sources, including the oldest written record of *Cannabis* use and demonstrates that the Chinese have been knowledgeable about medicinal herbs for several millennia (Li 1974a; Touw 1981). According to legend, Shen Nung had a transparent abdomen and intentionally ingested as many as 70 different plants per day so he could watch their effects and discover their various qualities. He was deified and acclaimed the "Father of Chinese Medicine." Based on his search for efficacious cures derived from prior shamanistic knowledge, and his documented medicinal use of many plants including tea (*Camellia sinensis*) and hemp, Shen Nung is considered the first Chinese medical researcher. Shen Nung is also known as the "Father of Agriculture" and credited with instructing the ancient Chinese people in the cultivation of hemp.

According to the *Pen Ts'ao Ching*, *Cannabis* is included among the drugs of the "first class," which is headed by ginseng. Drugs of the first class were not considered poisonous. In fact, no matter how much they were utilized they were believed to be harmless. Indeed these medicinal plants and their products were usually used to improve breathing and prolong life. *Má fěn* (麻粉), or "hemp powder," a preparation using the flowering tops of female *Cannabis* plants, including the resinous bracts surrounding the seeds, was thought by the ancient Chinese to contain the greatest amount of *yin* energy; *yin* is the receptive female dynamic attribute linked with *yang*, the creative male element in traditional Chinese philosophy and medicine. *Má fěn* was thus prescribed in cases of a loss of *yin*, such as in menstrual fatigue, rheumatism, malaria, beri-beri (Vitamin B$_1$ deficiency), constipation, and absentmindedness (Taylor 1963; Li 1975). On the other hand, the *Pen Ts'ao Ching* indicates that if too many hemp seeds were eaten, they could cause one to see demons, but if taken over a long time, they could enable one to communicate with spirits. Also, Li Shih-Chen's *Pen Ts'ao Kang Mu*, or "Great Herbal," of 1580 tells us that such consumption "makes one see devils" (Li 1974b; Anderson 1988). Both of these ancient references allude to the psychoactive effects of

FIGURE 45. Legendary Chinese Emperor Shen Nung allegedly wrote China's first pharmacopoeia and is said to have personally tested many potentially valuable medicinal herbs including *Cannabis* by swallowing them and observing their effects through his transparent abdomen.

Cannabis seeds, although the seeds contain no cannabinoids, and therefore these early authors were more likely referring to the bracts enveloping the seeds, which can contain large amounts mind-altering THC.

In China, during the second century CE, a new medical use was discovered for *Cannabis*, which is attributed to the famous Chinese surgeon Hua Tuo (lived 110 to 207 CE). According to the *Ho Han Shu*, or "History of the Later Han," the official history of the Eastern Han dynasty (25 to 221 CE) compiled by Fan Ye (died 445 CE), Hua Tuo performed complicated surgical procedures without causing pain, carrying out such amazing operations as incisions into the loin and chest, resectioning of the intestines, and organ grafts. His procedures involved an infusion-like anesthetic called *má yóu* (麻油), literally "hemp oil" but with the modern meaning "sesame oil," which was thought to be made from *Cannabis* resin, *Datura*, and wine (Julien 1894). The character *má* (麻) means "numb" as well as "hemp" (see Chapter 7). A passage in a later work titled *Chêng Lei Pen Ts'ao*, or "Cheng Lei Herbal," written by Tang Shen-wei (1108 CE) and translated by Li (1974b), tells us that *má fěn* has a spicy taste and is used for waste diseases and injuries, clears blood, and cools temperature; it relieves fluxes, undoes rheumatism, and discharges puss.

Wu Pu, a disciple of Hua Tuo, wrote the *Wu Pu Pen T'sao*, or "Wu Pu's Herbal" ca. 200 CE, in which he made a clear distinction between toxic hemp resin *má fěn* and the non-poisonous seeds or kernels (Emboden 1972). Traditional Chinese medicine practitioners still prescribe hemp seed for the relief of digestive problems. This is due in part to their high roughage content and ability to nourish the elderly and others unable to ingest normal amounts of food, as they contain easily digestible edestin-type proteins and high levels of essential fatty acids (EFAs). According to Shou-zhong (1997), hemp seeds are still widely used in modern clinical practice for hastening birth delivery and promoting lactation as well as facilitating urination and defecation; Perry and Metzger (1980) also referred to continued use of hemp seeds in China for aiding birth delivery as well as treating uterine prolapse, the slipping out of place of the uterus.

During the time of the Tang dynasty (618 to 907 CE), pulverized hemp mixed with rice wine was "recommended in various other materia medica against several ailments, ranging from constipation to hair loss" (Dikötter et al. 2004; also see Daihua 2002). The ancient Chinese text *Wu Lei Hsiang Kan Chih*, or "On the Mutual Responses of Things According to Their Categories," is believed to have been written by Lu Tsan-Ning (also known as Su Shih) in about 980 CE. In this text we are told that *Cannabis* is among plants that can drive away mosquitoes when its leaves are burnt, and this could have prevented the spread of mosquito-borne diseases and protected people from discomfort resulting from their bites (Needham et al. 1996).

Trade and communication between China, Korea, and Japan dwindled during the Chinese Qing dynasty (1644 to 1912 CE), the Korean Joseon dynasty (1392 to 1910 CE), and the Japanese Edo and early Meiji periods (1603 to 1912 CE) as each country followed its own introspective path. However, Korea and Japan continued to send scholars and students to learn medicine, agriculture, and science from the Chinese. In Japan, hemp (*tai ma asa*) is considered one of the *sanso*, or "Three Plants," along with red safflower (*benihana, Carthamus tinctorius*) and indigo (*ai, Indigofera tinctoria*), that symbolize long life (Shimamura 1991). Traditional Japanese doctors incorporated much of the ancient Chinese pharmacopoeia and utilized many *Cannabis* preparations (*asashijingan*) for a variety of indications, for example, as a mild laxative, to treat asthma, to relieve poisonous bites, to deworm domestic animals, to counteract skin ailments, and as a general tonic to promote vigor (Olson 1997).

South and Southeast Asian Medical Traditions

There appears to be no doubt that the cannabis plant was believed by the ancient Aryan settlers of India to possess sedative, cooling and febrifuge properties. (CHOPRA AND CHOPRA 1957)

The use of whole *Cannabis* plants for myriad medicinal treatments over the centuries is a testament to the general health value these plants and their products have had for people throughout the history of South Asia. Early references to the medicinal use of *Cannabis* can also be found in South Asian historical records and texts. The *Atharva Veda* (passage 11.6.15), thought to date from well before 500 CE and perhaps as far back as ca. 1600 BCE, refers to the sacred plant *bhanga* with indications that *Cannabis* helps "release us from anxiety" (e.g., see discussions regarding the *Report of the Indian Hemp Drugs Commission* in Campbell 1984 and Kaplan 1969). According to Sharma (1979), several later Vedic texts also make reference to the medicinal use of *Cannabis*. The *Susrita Samhita*, an ancient holy book of the traditional Indian Ayurvedic medical system (written by Punarvasu possibly as early as 800 BCE) listed hemp as a cure for mucus discharge accompanied by diarrhea and as a remedy for biliary fever (Chopra and Chopra 1957). The *Ashtadhyayi* by Panini and the *Vartika* by Katyayana indicate that *Cannabis* was known

FIGURE 46. Glandular trichomes crowded together on the surface of the bracts and small leaflets, shown here in a seedless female *Cannabis* inflorescence, can secrete large amounts of aromatic and psychoactive resin that contains unique, medically valuable compounds (photo ©GrowMedicine.org).

FIGURE 47. Terpenoid compounds, THC, and other medically valuable cannabinoids composing the resin are synthesized in and secreted by the head cells of glandular trichomes perched atop a stalked extension of the green leaf surface. The cluster of secretory cells is reflected inside the shiny surface of the gland head (photo ©Bubbleman).

in India as early as the fourth and third centuries BCE when it had already become part of traditional folk medicine (Dwarakanath 1965). Reliable Ayurvedic literary documents on materia medica such as *Rajanirghanta* (ca. 300 CE), *Dhanwantari nighantu* (eighth century CE), *Sharangadhara Samhita* (thirteenth century CE), *Madanapala nighantu* (1374 CE), *Rajanighantu* (1450 CE), *Dhurtasamagama* (ca. 1500 CE), and *Bhavaprakash* (ca. 1600 CE) describe various aspects of the medicinal use of *Cannabis*. In addition, *Cannabis* has been used as a veterinary therapeutic in India at least since the twelfth to thirteenth centuries CE and continues to this day (e.g., see Chopra and Chopra 1957; Jain 1999). Later Ayurvedic medical works reveal the increasing therapeutic significance of *Cannabis*, which is included in a large number of remedies (Dwarakanath 1965). *Cannabis* has long been prescribed to arouse appetite and as "a source of great staying-power under severe exercise or fatigue" (Nadkarni 1954), as well as for many other purposes including the following: *paphahari* to promote loosening, separation, and the elimination of phlegm; *grahini* to stimulate the retention and binding of the bowels; *pachani* to promote digestion; *ushna* to promote heat; *pitala* to excite the flow of bile; *mada-vardhani* to encourage talkativeness; *moda-vardhani* to promote happiness; *vag-vardhani* to stimulate the digestion; *dipani* to

encourage appetite; *ruchya* to stimulate taste; and *nidraprada* as a hypnotic (Chopra and Chopra 1957).

For centuries, Indian foot bearers transporting goods high into the Himalaya Mountains have relied on *Cannabis* to relieve fatigue. Potions containing juice from the *Cannabis* plant have been used to remove dandruff and vermin from hair, reduce pain from earaches, alleviate bowel complaints such as diarrhea and constipation, and check discharge from gonorrhea (e.g., see Knörzer 2000; Alt et al. 2003). Other significant applications include its use for relieving headaches, acute mania, whooping cough, asthma, and insomnia. In Ayurvedic medical practice, *bhang* has been used to treat fevers, not directly or physically as an ordinary medicine but indirectly or spiritually, by soothing the "angry influences" causing the heats of fever. According to Indian legend, Lord Shiva, enraged by a slight from his father-in-law Daksha, "breathed from his nostrils the eight fevers that wither mankind." If a person stricken with fever poured *bhang* on a Shiva *lingum* (stylized phallus worshiped as a symbol of the god Shiva), "he was pleased, his breath cooled, and the portion of the breath in the body of the sufferer ceased to cause fever" (Campbell 1894; also see Mikuriya 1994). The seminal connections between religious ritual and medical cures strongly influenced traditional therapeutic use of *Cannabis* in India, and its utilization medicinally was not originally separated from religious belief: "The reference in the *Atharva Veda* to overcoming enemies and evil forces may, quite possibly, have included physical as well as spiritual ills. Once medicinal use became increasingly delimited from religious use, it remained distinct from other secular use and could, therefore, be freely and fully explored unhampered by secrecy or disreputability" (Touw 1981). *Cannabis* has long been considered a panacea (cure-all) in India, used commonly in family remedies to treat numerous minor ailments, especially relief from physical pain and mental strain (Chopra and Chopra 1957). Indeed, it is not an overstatement to refer to *Cannabis* as "the penicillin of Ayurvedic medicine" (Sharma 1977a). Even the *Report of the Indian Hemp Drugs Commission* did not oppose the moderate use of *Cannabis* for social, ritual, and medicinal practices in the Indian subcontinent. Indeed, use of the *Cannabis*-based, traditional drink *bhang* was considered "the best of gifts" when consumed in moderation: "*Bhang* is a cordial, a bile absorber, an appetizer, a prolonger of life" (Campbell 1894). Indian Muslims also regard *Cannabis* (*bhang*) as both a holy and medicinal plant, and in Unani-Tibbi, the Arabic-Muslim system of medicine, practitioners use it for treating numerous diseases such as asthma, dandruff, and urinary disorders (e.g., Dwarakanth 1965; Lozano 2006).

Several other traditional medicinal uses for *Cannabis* have been recorded across South Asia. For example, in the uplands of Northwestern Pakistan, among tribal communities in the Hazar Nao forest of the Malakand district, the leaves are used to treat spasms by boiling them and tying them (while they are warm) "over the affected parts of the body" (Murad et al. 2011). East of Pakistan in Rajastan, India, Singh and Pandey (1998) reported that the Meena and Garasia tribal groups apply a paste made of *Cannabis* leaves (*bhang*) on "bleeding and painful" hemorrhoids (also see Bajracharya 1979).

Throughout the Himalayan region a "widespread belief holds that a concoction of young *Cannabis* leaf powder and honey keeps youth, vitality and virility" and it is also applied to the hair to maintain color and texture (Sharma 1977a). In the northwestern Himalayan region, Sarin (1990) reported that a poultice of crushed female flower tops is used to treat prolapse of uterus and hydrocele, the excessive retention of fluid in the testicles that results from obstruction and inflammation of the lymphatic ducts draining the testicular region. In treating prolapse, a warm *Cannabis* poultice is "wrapped in a soft muslin cloth and kept inside the vagina at bed time," and treatment of hydrocele consists of "wrapping the affected testicles with the warm pounded mass at bed time" with a reduction of swelling beginning "within a week or ten days." In a pharmacological appraisal of traditional herbal medicine in far western Nepal, Kunwar et al. (2010) reported that *Cannabis* leaf juice is used to heal wounds, control bleeding, and relieve stomachaches (also see Watanabe et al. 2005), while leaves are smoked and taken internally to relieve discomfort and inflammation (Joshi 2006).

In Nepal, *Cannabis* has a long history of use as a sedative (Bajracharya 1979). According to Fisher (1975) mothers give *Cannabis* mixed with sweets to children as a mild tranquilizer: "by giving her child a small amount of *ganja*" a mother "keeps him less active and less likely to get into trouble while she is occupied in other ways." La Barre (1977) compared use of *Cannabis* as a sedative to use of opium extracts to calm children: "The use of hashish is quite parallel to the use of opiates in soothing syrups given to fretful or teething babies in the nineteenth and twentieth centuries in America."

In answer to the question of whether or not *Cannabis* and opium should be banned as medicines, Dwarakanath (1965) made a case for allowing continued traditional therapeutic use by Ayurvedic Hindu and Unani Muslim practitioners in India. He pointed out that the great majority of people in India reside in a half million villages, almost exclusively attended to by Ayurvedic physicians and Unani *hakims* with significant responsibility for the health of these rural communities. Medical practitioners included *Cannabis* and opium in traditional formulations as important healing agents "especially in the treatment of such conditions as enteritis, dysentery, chronic diarrheas, including sprue syndrome [inability to absorb fats], painful states such as neuralgia, neuritis, rheumatism, insomnia, nervous disorders, etc." Used as pain killers, hypnotics, antispasmodics, and so on, these *Cannabis*-based drugs still continue to play an important role in relieving physical and psychological stress (see Chopra and Chopra 1957 for a list of preparations containing *Cannabis* used in indigenous medicine during the middle of the last century). In sum, the breadth of evidence presented earlier highlights a long history of medicinal *Cannabis* use throughout much of South Asia, within both Hindu and Muslim medical traditions.

We also have evidence of medicinal *Cannabis* use in Southeast Asia, albeit much less than that for South Asia. For example, Martin (1975) studied ethnobotanical aspects of *Cannabis indica* use in this region, mainly in Cambodia and to a lesser extent in Thailand, Laos, and Vietnam. She assumed that the species was probably brought into Southeast Asia within the past 500 years or so from South Asia and pointed out that *ganja*, its Sanskrit name, is the widely used vernacular appellation for this plant in Southeast Asia where it is grown "on a family basis . . . around the house." Besides its widespread use as a pain remedy (similar to opium products) in Southwest Asia, *C. indica* is also used to treat anorexia, dysentery, memory loss, asthma, coughing, lightheadedness, and convulsions, as well as suppressing polyps and calming nerves. It is also used to facilitate digestion and childbirth, stimulate lactation, purify blood, and clear bile, and it is believed to regulate function of the heart, liver, and lungs; eradicate intestinal parasites; induce decongestion; and effectively treat paralysis. A decoction is used in Southeast Asia to mitigate migraine pain and stiffness, and it is also said to be particularly effective in calming nerves when consumed with certain plants prior to bedtime and meals; this soothing effect is acknowledged "both by peasants and by the official pharmacopoeia of these countries."

Martin (1975) pointed out that the medicinal effectiveness of *Cannabis* for many illnesses previously mentioned is not universally recognized through Southeast Asia today because many local folk medicines also include elements of magic. Martin also did refer to the rather broad medical "significance attached by the peasants to a plant which seems never to have been widespread in the region." More than 300 years ago, Georg Eberhard Rumpf (also known as Rumphius), a German doctor serving the Dutch government in Indonesia, listed the local use of *Cannabis* root to treat gonorrhea in his famous botanical work, *Herbarium Amboinense*, first published in 1741. Rumphius also mentioned that "the green leaves of the female plant, cooked in water with nutmeg, [are given] to drink to folks who felt a great oppression in their breasts, along with stabs, as if they had Pleuritis too" (Rumpf and Beekman 1981, quoted in Russo 2002a; also see Rumpf and Beekman 1999).

Egyptian Medicinal *Cannabis* Use

Although it is probable that *Cannabis* and its medicinal use spread into Egypt from areas east of the Mediterranean Sea, the oldest references to such use in the Near Eastern region are associated with ancient Egypt; thus we continue with a discussion of the Egyptian record. Nunn (1996), for example, cited several sources that believe the hieroglyphic word *shemshemet* refers to *Cannabis* and therefore indicates its medicinal use in ancient Egypt. Archeological evidence of ancient hemp fragments were reported from the tomb of Akhenaten (Amenophis IV), dated ca. 1350 BCE (see Manniche 1989), and *Cannabis* pollen was recovered from the tomb of Ramsés II (died 1224 BCE). This evidence supports the contention that *Cannabis* was present and presumably used in ancient Egypt (see Chapters 4 and 5 for further discussion of the history and strength of this conviction).

There are many references to the medical use of *shemshemet* inscribed on ancient Egyptian medical documents made of fibrous sheets of sedge (*Cyperus papyrus*) pith that date from the seventeenth through thirteenth centuries BCE. The earliest identification of the Egyptian word *shemshemet* comes from Fifth dynasty stone carvings dated to ca. 2350 BCE; these list *shemshemet* as a source of cordage. This increases the likelihood that ancient medical references to *shemshemet* in Egypt do refer to *Cannabis* since it is the only common fiber plant also widely employed as medicine. Based on historical research, Russo (2002a, citing Manniche 1989) tells us that in Egypt *Cannabis* "has remained in Egyptian pharmacopoeia since pharaonic times" and "was administered by mouth, rectum, vaginally, on the skin, in the eyes, and by fumigation." Russo (2002b, quoting Ghalioungui 1987) referred to a passage (821) in the ancient *Ebers Papyrus* (dated to ca. 1550 BCE), which he believes is suggestive of nineteenth-century CE use of *Cannabis* to induce contractions during childbirth: "Another: *smsm-t* [*shemshemet*]; ground in honey; introduced into her vagina (*iwf*)." In addition, Russo (2002b) tells us that another passage (618) in the *Ebers Papyrus* "refers to treatment of a toenail with a bandage containing hemp resin."

Mannishe (1989) referred to a relevant passage from *Papyrus Ramesseum III* (dated to ca. 1700 BCE) that describes an eye treatment involving the use of "celery" and "hemp," which were ground up and left overnight in the humid evening air; during the following morning, the patient's eyes were then washed with the mixed celery and hemp potion. Russo (2002b) suggested that this remedy parallels modern use of *Cannabis* in treatment of glaucoma eye disease.

The Egyptian *Fayyum Medical Book* dating back more than 1,800 years and contemporaneous with early Greek medical treatises provides additional references to the use of *Cannabis* to control inflammations such as tumors and acute ear pain. Evidence of *Cannabis* use for medicinal purposes has not been identified for the period from about the third to the ninth or tenth centuries CE; however, by this time "Egyptian medicine had become Islamic medicine" (Russo 2007).

Cannabis in Early Middle Eastern and Later Islamic Medicine

If Thompson (1924, 1949) is correct in his interpretations of ancient manuscripts found in the archive of King Ashurbanipal dated to the seventh century BCE, *Cannabis* was present and used medicinally perhaps as far back as 4,000 years ago among people living in irrigated lowlands between the Tigris and Euphrates Rivers:

The earliest references to cannabis in female medical conditions probably originate in Ancient Mesopotamia. In the 7th century BCE, the Assyrian King Ashurbanipal assembled

a library of manuscripts of vast scale, including Sumerian and Akkadian medical stone tablets dating to 2000 BCE. Specifically according to Thompson, *azallû*, as hemp seeds were known, were mixed with other agents in beer for an unspecified female ailment (Thompson 1924). *Azallû* was also employed for difficult childbirth, and staying the menses when mixed with saffron and mint in beer (Thompson 1949). Usage of cannabis rectally and by fumigation was described for other indications. (RUSSO 2002a; also see RUSSO 2002b)

Other early records from the Middle East indicate medicinal use of *Cannabis* occurred before the rise of Islam. The Zoroastrian scriptures of ancient Persia, related to the Hindu *Vedas*, make reference to *bhang* for inducing miscarriage and producing euphoria. The *Zend-Avesta*, the holy Persian book of Zoroastrianism, which survives only in fragments dating from about the fourth to sixth centuries CE, alludes to the use of "*banga*" in a medical context; this word in the text is identified as *Cannabis* by the translator Darmesteter (1883). In the *XV Fargard* of the *Vendîdâd* (the first book of the *Zend-Avesta*), a compilation of religious laws and mythology, *Cannabis* is referred to as a stimulator of abortion, and thus "the damsel goes to the old woman and applies to her that she may procure her miscarriage, and the old woman brings her some *banga*, or *shalta* or *ghnana*, or *fraspata*, or some other of the drugs that produce miscarriage" (Darmesteter 1883).

The genesis of the ethnomedicinal knowledge and use of *Cannabis* in the Middle East region is not completely clear, but much of Egyptian understanding and application of this medical practice was adopted into Arabic Unani medicine and at least partially integrated into the Greek and Roman heritage. This is manifested in the writings of notable scholars of the time including Dioscorides, a renowned Greek physician and scholar from Cilicia in southern Asia Minor who became the personal physician of the Roman emperor Nero, Galen, a prominent second-century CE Roman physician and surgeon of Greek ancestry who was an accomplished medical researcher, and Pliny the Elder, a famous Roman naturalist and philosopher. Although the works of Dioscorides and Galen were translated into Arabic and had a major influence on development of the early Arabic medical system and to some degree on the use of *Cannabis*, it can be argued that much of this knowledge originated within the Egyptian medical system and that the transfer of medical information was largely from the East to the West. Arab and Greek physicians relied on *Cannabis* for much the same indications, and although many accepted Galen's classification that it had both "desiccating and warming" powers, there were also contrary Arab opinions concerning the roots of its curative nature that may have originated directly from Egyptian concepts.

Awareness of *Cannabis*'s medicinal value shifted from Egyptian through Arabic and then to Greek and Roman medical traditions, but was this knowledge transfer accompanied by availability of *Cannabis* from a more medically effective gene pool? Pliny the Elder, in his *Historia Naturalis* (77 CE), tells us that "*Cannabis*, rather dark and rough in respect to its leaves, first grew in the forests." Butrica (2006) points out that "Pliny's description of the original plant as dark and rough of leaf resembles Dioscorides' description of wild cannabis as having darker and rougher leaves than tame. Perhaps this reflects a belief that tame cannabis had been bred from wild cannabis (Herodotus already distinguishes between cultivated and wild varieties of the plant known to the Scythians); or perhaps—and not for the last time—Pliny confused the two plants or carelessly ignored the distinction." Could it be that the Greek historians were noting botanical phenotypic differences between different biotypes of Cannabis—NLH versus NLD cultivars—rather than between stages in its domestication? Do historical accounts actually differentiate between C. sativa narrow-leaf hemp (NLH), the "tame" cultivated type with which the Greeks and Romans were very familiar, and C. indica narrow-leaf drug (NLD), the "wild" type that was described as being darker and rougher than the "tame." "Darker and rougher" describes accurately the psychoactive and medically potent C. indica biotypes originating to the east in southern Asia, which are characterized as darker green and more highly branched plants. Possibly the ancient Greeks and Romans were describing the differences between foreign and familiar or introduced and local varieties rather than between "wild" and "tame" stages of domestication.

By the first century CE the Greeks and Romans had extensive knowledge of their native NLH, commonly cultivated as a fiber and seed crop, but the content of cannabinoids, particularly psychoactive THC, in NLH biotypes would have been very low. However, if "wild" actually meant introduced NLD varieties from the east, they would have been more curatively potent than the local varieties and therefore worthy of denoting with a separate name, in this case "wild," O'Shaughnessy (1839) had the following to say regarding the difference in medicinal value of European hemp as compared to that of Middle Eastern and South Asian "hemp":

> Much difference of opinion exists on the question, whether the Hemp so abundant in Europe, even in high northern latitudes, is identical in specific characters with the Hemp of Asia Minor and Hindustan. The extraordinary symptoms produced by the latter depend on a resinous secretion with which it abounds and which seems totally absent in the European kind. As the closest physical resemblance or even identity exists between both plants, difference of climate seems to me more than sufficient to account for the absence of the resinous secretion, and consequent want of narcotic [sic] power in that indigenous in colder climates.

The obvious difference in resin content noted by O'Shaughnessy is accurate but cannot be explained simply as a product of differing climate or nurture. Variations in resin content result from a coevolutionary blend of nature (the genetic programming of NLD biotypes to make THC and the adaptation of C. indica to warmer subtropical climates) and nurture (the growing of NLD cultivars at latitudes favoring resin and THC production). Egyptian and Arab physicians lived in much closer proximity to the home of potent NLD Cannabis and were likely familiar with it early on.

A prime example of Cannabis use in Arabic medicine involves hemp seed oil, which was commonly administered externally as ear drops. Arab medical treatises dating from the ninth to the thirteenth centuries CE also recommended using the juice expressed from fresh female inflorescences containing immature seeds in various ways. Recommendations included their use to treat ear discomforts and skin disease, kill intestinal worms, relieve flatulence, purge poisonous humors from many parts of the body, ease uterine contractions and hardening, sooth neurological pain, lower fevers, control vomiting, and kill lice (see Russo 2007; Lozano 2006; Hamarneh 1972; Lozano Cámara 1990).

Abubakr Muhammad Ibn Zakaria Razi (lived 865 to 925 CE), known in the West as Rhazes and regarded as "one of the greatest physicians Islam has ever produced" (Daghestani 1997), provided us with a detailed description of therapeutic Cannabis use. Medicinal use of Cannabis is also referred to in other ancient Arabic and Persian medical works such as Firdo usul-Hikmat and Mujardat Quanan (Dwarakanath 1965; also see Al-Mukhtar by Muhadhahab Al-Deen Al Baghdadi, lived 1117 to 1213 CE). Lozano (2006) in his recent study of Cannabis in Arabic medicine came to the conclusion that Arab scientists were centuries ahead of our current knowledge of its curative powers. By medieval times, medicinal potions containing Cannabis were apparently popular in Arabia, Persia, and Muslim India. According to Benet (1975, citing Dragendorff 1898), medieval Arab physicians referred to hemp as schahdanach, schadabach, or kannab, which they regarded as a "sacred medicine." Cannabis was used in Arabia and Persia during medieval times as a diuretic to increase urine flow, an emmenagogue to stimulate blood flow to the pelvis and uterus, and an anthelmintic to expel intestinal worms (Levey and Al-Khaledy 1967). Medieval Arabs also knew and used Cannabis for its antiemetic, antiepileptic, anti-inflammatory, and pain-killing effects, among others, and for this reason, Lozano (2006) recommended that information regarding medicinal efficacy of Cannabis, which may "be found in Arabic literature could be considered as a possible basis for future research on the therapeutic potential of cannabis and hemp seeds." According to Levey and Al-Khaledy (1967), in more recent Arabic medical practice, Cannabis is still considered to be useful as a retentive, anesthetic, and astringent, all curative powers not alluded to in the Greco-Arabic treatises.

Remarkable physical evidence for ancient medicinal Cannabis use in the Near East was found at the town of Beit Shemesh, Israel, between Jerusalem and Tel Aviv. Archeologists excavating a fourth-century CE late-Roman-period tomb uncovered the skeleton of a girl about 14 years old, which contained in its pelvic area the skeleton of a fetus of a size that would complicate delivery (Zias et al. 1993). A small amount (6.97 grams, or one-quarter ounce) of a dark-colored burnt substance was found in the abdominal area of the skeleton. Initially thought to have been incense, analysis by Israeli police and botanists at Hebrew University determined that the ancient material is a mixture of Cannabis resin (hashish), dried seeds, fruits, and a common reed (Phragmites). Direct evidence of ancient drugs themselves is rare because most organic compounds decay rapidly. However, this substance was preserved because it had been carbonized through burning, and thus the relatively stable cannabinoid Δ^8-THC (Δ^6-THC) could be identified (Zias et al. 1993). Archeologists who uncovered this evidence believe the Cannabis was used as an aid in childbirth since it increases the strength and rate of recurring contractions during labor. Cannabis was commonly used in deliveries well into the nineteenth century, after which newly developed drugs replaced it. However, others have disagreed with the strict medicinal use hypothesis (Prioreschi and Babin 1993), arguing that ritualistic use cannot be discounted. Again, we have evidence that may be explained by both Cannabis's direct medicinal efficacy as well as its use in a ritual context.

It should be noted here that ethnobotanists working in the Rif Mountains, the northern zone of Morocco, have recorded the use of Cannabis in at least two separate herbal preparations associated with abortion applications (Merzouki et al. 2000). In addition to Cannabis, these preparations generally contain known toxic plants. Based on their methodical survey, Merzouki et al. (2000) indicated that such remedies "were commonly used to treat different ailments because many communities and villagers live in remote areas where health facilities are not available." Their ethnomedicinal research in the Rif region since 1992 also documented customary *Cannabis* use to treat various ailments (e.g., see Merzouki et al. 1999; Merzouki and Molero Mesa 1999) but without any record that *Cannabis* has direct abortive effects. However, abortion could be provoked when preparations included one or more variably toxic plants, and since *Cannabis* does not induce uterine contractions, it was likely included for pain relief. Merzouki et al. (2000) also referred to Mathieu and Maneville (1952), who pointed out that in Casablanca a decoction including *Cannabis* leaves was administered orally to produce abortion.

African and South American Medicinal *Cannabis* Use

Although historical references to recreational and ritual *Cannabis* use in Africa are relatively sparse, and those concerning medical use are even more rare, it is likely that awareness of the medicinal values of *Cannabis* developed soon after it arrived in the region, and additional medicinal uses for *Cannabis* were discovered as it was disseminated across Africa and into the New World (see earlier discussion regarding medicinal use of *Cannabis* in Morocco, as well as Chapters 4 and 7). Hewat (1906) referred to a Sotho herbalist in South Africa who eased childbirth by "getting his patient stupefied by much smoking of *dagga* [*Cannabis*]," and Du Toit (1980) described similar facilitation of birthing, saying that "pregnant women were given dagga to make them brave" and avoid pain in the process (also see Watt 1961 and Russo 2002, 2006). According to Watt and Breyer-Branwijk (1932), women of the Sato tribe smoked *Cannabis* to numb themselves during childbirth, and the "Hottentots," more correctly known as the Khoikhoi ethnic group associated with a pastoral culture and language found across southern Africa, made a drink from *Cannabis* leaves to use as a strong laxative and for inducing abortion. *Cannabis* has also been used in South Africa to combat malaria, blackwater fever, blood poisoning, anthrax, and dysentery (cf. Du Toit 1980).

Remedies also developed after *Cannabis* was introduced to several areas in the New World (see Chapter 4). According to Forster (1996), medicinal use of "both hemp and marijuana varieties" of *Cannabis* occurs in Latin America with separate listings in regional ethnopharmacopoeias "indicating that they are considered to be distinct herbal entities" (from our perspective, these "entities" can be grouped into *C. sativa* narrow-leaf hemp, or NLH, and *C. indica* narrow-leaf drug, or NLD). In Chile, Forster (1996) referred to use of hemp (*cañamo*) roots as a purgative, and stems and seeds combined in an infusion used to induce sleep. He also reported that a drink prepared with a relatively small amount of *Cannabis* is used as an all-purpose pain reliever to curtail convulsions, reduce rheumatism, and treat urinary problems. People in Chile apply a plaster of ground up fresh *Cannabis* flowers to body parts experiencing "cold humors" as a blood purgative. The medicinal use of psychoactive *Cannabis* (*hierba, hierba mala, hierba buena*, or *marijuana*) was formally extensive, especially in tinctures produced by "mixing 20 grams [about three-quarter ounce] of dried flowering tops with 90 proof [45 percent] alcohol and soaking for 15 days," after which the liquid was filtered through a cloth or paper. The resulting tincture of *Cannabis* was diluted in water and soaked into

compresses applied to the body to reduce headaches and neuralgia. Because of the strong traditions of herbal healing in Latin America, Forster (1996) also indicated that many Chilean physicians "support its reappearance as an effective and socially acceptable medicine."

According to Kabelik et al. (1960), *Cannabis* has been used in Argentina to treat depression, tetanus, colic, stomachache, swelling of the liver, gonorrhea, sterility, impotency, tuberculosis, and asthma. Colombians also use *Cannabis* for medicinal purposes. For example, Partridge (1975), in his study of a community located at the base of the western slopes of the Sierra Nevada de Santa Marta in Colombia, recorded the use of *Cannabis* soaked in "rum or *aguardiente*" and rubbed into the skin to relieve joint and muscle pain. Partridge also listed *Cannabis* smoking as a part of a "program of health maintenance," use of green hemp leaves "crushed and rubbed on the skin for treatment of pain," and feeding boiled water containing *Cannabis* and raw sugar "to infants for excessive crying."

In northeastern Brazil, an infusion of *Cannabis* leaves and water is consumed "to relieve rheumatism, female troubles, colic and other common complaints," and for toothache, parts of the plant are "packed into and around the aching tooth and left for a period of time, during which it supposedly performs an analgesic function" (Hutchinson 1975; de Pinho 1975). In Jamaica, many people "across socioeconomic lines" have used tonic drinks and pain-relieving salves containing *Cannabis* for "medicinal or prophylactic purposes" (Comitas 1975; see also Rubin and Comitas 1975).

European Medicinal *Cannabis* Use

Ritual Scythian *Cannabis* use was documented by the Greek historian Herodotus in the fifth century BCE, but there is no mention of hemp in the writings of Hippocrates and his followers, indicating that the medicinal use of *Cannabis* had not reached Greece by that time (Butrica 2006; Stefanis et al. 1975). However, by the first century CE, the Greek physician Dioscorides had referred to the medicinal uses of *Cannabis*. Dioscorides marshaled information dealing with medicinal herbs and compiled one of the earliest pharmacopoeias, the *Materia Medica*, published ca. 65 CE. Therapeutic *Cannabis* is among several hundred medicinal plants and their uses he described. Dioscorides listed *Kannabis emeros* (females) and *Kannabis agria* (males) as separate entities, indicating that *Kannabis emeros* could be used to relieve earaches and induce menstrual flow, and that *Kannabis agria* could be used to relieve muscular ailments.

The Roman scholar Pliny the Elder (lived 23 to 79 CE) provided informative accounts of the uses of many plants in his *Historia Naturalis*, completed in 77 CE and dedicated to Emperor Titus. He referred to a number of medicinal uses of *Cannabis*, informing us that the "seed is said to extinguish men's semen," that hemp seed oil "casts out ear-worms and whatever animal has entered," and that the root cooked in water "softens contracted joints, likewise gouts and similar attacks" and "uncooked it is spread on burns" (Butrica 2006; also see Stefanis et al. 1975).

Galen, the famed second-century CE Roman physician of Greek ancestry, mentioned *Cannabis* in at least three passages, which are described and discussed by Arata (2004). In one passage (*De alimentorum facultatibus* 6: 549–50), Galen referred to the use of hemp seed as "difficult to digest and gives pain to the stomach and to the head and spoils humours [certain allegedly important bodily fluids]." Galen also pointed out that seeds were eaten by some people for pleasure during a meal but claimed that eating too many will send heat to the head as well as induce "pharmaceutical fumes." This description of the effects of consuming *Cannabis* seeds is echoed by Oribasius (fourth century CE), a Greek author of medical literature and personal physician of the Roman emperor Julian, as well as by Aëtius Amidenus (late fifth to early sixth centuries CE). The latter was a Byzantine physician who referred to the use of *Cannabis* "fruit," putting less emphasis on the alimentary use of *Cannabis* but adding in this context "that it does not help the formation of gas." Oribasius and Aëtius studied in Alexandria, in what is now Egypt, and both also noted that *Cannabis* "is so desiccating that, if eaten in a rather large quantity, it dries male seed" (Arata 2004; also see Withington 1894), essentially repeating what Galen reported earlier in his *De simplicium medicamentorum temperamentis ac facultatibus*. Indeed Galen repeated the evaluations of Dioscorides and Pliny "that some people, pulling out the juice from it when it is not ripe, use it against ear pains due to an occlusion" (Arata 2004).

Hemp seed oil is expressed from mature seeds and does have medicinal properties of its own, especially in topical and dietary applications due to its high EFA contents. Cannabinoids are not produced in hemp seeds, but a series of ancient Greek and Roman remedies were based on a "juice" or infusion of the immature seeds along with the fresh female inflorescences. The infusion was made by chopping up fresh *Cannabis*, adding it to hot water, wine, and other liquids, and then steeping it for several days (Butrica 2006). This "juice" may have been rich in cannabinoids that could account for the efficacy of early *Cannabis* medicines. According to Arata (2004), the obvious conclusion based on the references to the attributes of "seeds" or "fruits" of *Cannabis* by Galen, Oribasius, and Aetius, especially to their "desiccating power," is its effective use in treating certain ailments. For example, the ancient Greeks and Romans used *Cannabis* "to cure gonorrhea and epistaxis" (the latter being an acute hemorrhage from the nostril, nasal cavity, or nasopharynx). Arata also listed a third condition that Galen attributed to *Cannabis* in his *De Victu Attenuante*, which states that eating too many hemp seeds will be "painful for the head" (*kephalalgès*), in other words, causing "cephalagia." In sum, we suggest that despite the Roman's well-known familiarity with fiber hemp, there is little or no evidence to indicate that early Europeans had medicinal uses for *Cannabis* other than those few learned from the Egyptian medicinal tradition via Greeks and Romans.

Nevertheless, in parts of pre-Christian Europe, *Cannabis* apparently was used therapeutically before and perhaps to some extent after the spread of medicinal knowledge about hemp from early Mediterranean civilizations. Shimwell (2005, citing Rätsch 2001) claimed that *Cannabis* was used in some European regions during this period "for ear ailments, to induce an ecstatic state, for frostbite, herpes, nipple pains, stiffness, swelling and wounds." We can be more certain that *Cannabis* was part of Eastern European medical traditions. Benet (1975), for example, tells us that in Poland, Russia, and Lithuania, toothache was alleviated by inhaling the vapor from hemp seeds thrown on hot stones (see Biegeleisen 1929), much in the same fashion as the Scythians made their "vapor baths" more than two millennia ago. Benet (1975, 1936) provided other traditional uses of medicinal *Cannabis* in Eastern Europe such as its use to treat fevers

in Czechoslovakia, Moravia, and Poland; mixing hemp flowers with olive oil to dress wounds in Poland; and combining hemp flowers with hemp seed oil to treat jaundice and rheumatism in Russia. In addition, Benet (1975) referred to Simon Syrenius (lived 1540 to 1611 CE), a pre-Linnean Polish scholar interested in plants who published a huge botanical atlas (five volumes, 1,540 pages) describing 765 species. Syrenius noted that an ointment containing *Cannabis* resin could be used as an effective remedy for burns and indicated that ailing human joints could be cured with roots of hemp boiled in water. Furthermore, Benet (1975) tells us that during the sixteenth century CE, "Szyman of Lowic" provided a Polish prescription for removing "worms in the teeth." This involved boiling hemp seeds in a fresh pot by adding heated stones and then inhaling the vapors, which is similar to a Ukrainian folk medicine tradition based on belief that fumes of cooked hemp porridge will "intoxicate the worms and cause them to fall out" (see Chapter 9 for a lengthy discussion of the ritualistic traditions of *Cannabis* use, many of which are probably reflections of therapeutic use, especially in Eastern Europe).

Cannabis was probably brought into the British Isles during Roman times (Godwin 1967a/b). Although there is some evidence that it might have been introduced much earlier, hemp only became important as a principal fiber crop after the Romans invaded. In any case, *haenep* (Old English name for hemp) developed into a widely cultivated fiber crop subsequent to the arrival of the Roman legions and their colonial development of Britain. Exactly when medicinal preparations made from *Cannabis* came into use is unclear, but it is said to have become useful for therapeutic purposes in medieval times. For example, it is referred to as an anesthetic in an eleventh-century "Anglo Saxon Herbarium," and it is claimed that parts of the hemp plant, especially roots and seeds, were utilized in medieval Europe to treat pain associated with gout, weight loss, swelling of the head, urinary infections, and birthing problems (Frankhauser 2002; also see Shimwell 2005, citing Pollington 2000; Emerson 2002; Le Strange 1977). The roots of *Cannabis* "have received little research attention in recent decades" (Russo 2007); although we know they do not produce cannabinoids, they do generate terpenoids, sterols, and alkaloids (Sethi et al. 1977; ElSohly et al. 1978), which may explain their past and present use in some medicinal traditions.

In medieval Scotland, hemp cultivation for fiber and medicine was important in some areas, particularly near fishing communities where hemp fiber was used to make nets, ropes, and sails for fishing boats (e.g., see Dingwall 2003). Besides the old place names that reflect the former significance of hemp farming (e.g., Hemphill in Kilmarnock Parish, Aryshire; Hempland in Torthowald, Dumfriesshire; Hempriggs in Wick, Caithness; and Hempy Shot in Oldhanstocks, East Lothian), some record of use of *Cannabis* for fiber or medicine in medieval Scotland can be found in the history, archeology, and archaeobotany associated with the "religious hospitals" and monastic houses of the region (e.g., see Donaldson 1960). It appears that hemp was commonly grown around these ancient institutions for easy access to the plants for fiber and medicinal use (e.g., see Moffat and Fulton 1989 for evidence from the long-running excavation of the hospital at Soutra Mains in Scotland near the border with England and Whittingham and Edwards 1990, who discuss the archaeobotany and history of hemp farming and use of hemp in Scotland).

Among the more remarkable medieval Europeans to discuss the medicinal qualities of *Cannabis* was Hildegard von Bingen (lived 1098 to 1179 CE), the twelfth-century visionary Benedictine abbess of Rhineland, whose contributions to art, literature, linguistics, science, philosophy, poetry, music, herbalism, and medicine are testament to her considerable interest in many fields (von Bingen 2002 [English translation]; also see Hozeski 2001; Throop 1998; Strehlow and Hertzka 1988; Anderson 1977). One of her many works was called "Subtleties of the Diverse Qualities of Created Things," which was produced between 1151 and 1158, and later named *Physica*, or "Medicine" (also known as the "Book of Medicinal Simples"), in the 1533 edition. In *Physica*, von Bingen lists basic qualities of plants and their uses—in other words, are they "hot" or "cold," "dry" or "moist"—and asks if these dualities can be balanced to cure patients. She then expounds on the relative medicinal importance of each and explains how to concoct and apply the medical potion she recommends. Regarding the therapeutic use of *Cannabis* seeds, von Bingen comments at length:

> Hemp (*hanf*) is hot, and it grows where the air is neither very hot nor very cold, and its nature is similar. Its seed is salubrious, and good as food for healthy people. It is gentle and profitable to the stomach, taking away a bit of its mucus. It is easy to digest, diminishes bad humors, and fortifies good humors. Nevertheless, if one who is weak in the head, and has a vacant brain eats hemp, it easily afflicts his head. It does not harm one who has a healthy head and full brain. In one who is very ill, it even afflicts his stomach a bit. Eating it does not harm one who is moderately ill. Let one who has cold stomach cook hemp in water and, when the water has been squeezed out, wrap it in a small cloth, and frequently place it, warm on his stomach. This strengthens and renews that area. Also, a cloth made from hemp is good for binding ulcers and wounds, since the heat it has been tempered. (Translation by THROOP 1998, following the Schott edition based on the 1533 original *Physica*)

The use of medical *Cannabis* was significantly affected following the Papal Bull of Innocent VIII in 1484 in which an association between herbal healers practicing "witchcraft" and *Cannabis* was asserted; this characterized hemp medicine as "an unholy sacrament of satanic rituals" (Frankhauser 2002) and consequently drove its use underground, only to be "resurrected under a pseudonym" in François Rabelais's *Gargantua et Pantagruelion* in the mid-sixteenth century (Booth 2003; Russo 2004).

In his seventeenth-century CE book on plants and their uses, *Theatrum Botanicum*, John Parkinson (1640) described a series of medicinal uses of "hempe," including how the Dutch made an emulsion of the seeds to alleviate several ailments including obstructions of the gall bladder, pains associated with colic, and bowel problems. Parkinson also referred to a decoction of the roots that was said to reduce inflammations in the head or other parts of the body, relieve pain from gout, help cure hard tumors and joint irritations, and lessen swellings of the "sinews" and the hips, and, if mixed with a little oil and butter, it served as a salve for burned skin.

Nicholas Culpeper (lived 1616 to 1654 CE) was a strong advocate of equal access to inexpensive medication. In 1649, Culpeper produced an unauthorized translation of the London College of Physicians' *Pharmacopoeia*, which made accessible a large amount of medical knowledge previously unavailable to the general public, including a good deal of plant lore and herbal medicine. This medical compendium was published as a book in 1653 and later became known as *The Complete Herbal* (Wear 2000; Le Strange 1977), and *Cannabis* was one of several hundred medicinally useful plants that Culpeper listed. Although he did not provide a description of hemp, since he viewed it as "so common a plant, and so well known by almost every inhabitant of this kingdom, that a description of it would be altogether superfluous," his widely available herbal ensured that hemp would have "its place in folk medicine as an antiseptic, anti-inflammatory and anti-spasmodic" (Shimwell 2005). Specifically, Culpeper referred to *Cannabis* as therapeutically useful for inflammations, burns, dry coughs, jaundice, colic, bowel trouble, bleeding, eliminating "worms" and insects in the ear, and relief from many painful ailments.

According to Shimwell (2005, citing Holmstedt 1973), hashish was used not only for its visionary psychoactive powers (e.g., the Parisian "Le Club des Hashichins," or "The Hashish-Eaters' Club") but also as a medicinal product, largely because of a paper presented to the Institute of France in 1809 by Baron Sylvestre de Sacy (lived 1758 to 1838 CE), a notable French scholar of Arabic culture and history and professor at the École des Langues Orientales. Baron Sylvestre de Sacy found previously unknown eleventh- and twelfth-century CE chronicles in the Arabic manuscript collection held at the Bibliothèque Nationale and translated them into French. This work helped stimulate new interest in the psychoactive and medicinal potency of *Cannabis* (Booth 2003).

William Brooke O'Shaughnessy studied medicine in India and played a key role in the introduction of medicinal use of *Cannabis indica* or "Indian hemp" to Europe (Walton 1938; Mikuriya 1969). O'Shaughnessy presented his treatise *On the Preparations of the Indian Hemp, or Gunjah* at the Medical College of Calcutta in 1839, where he reported the efficacy of *Cannabis* extracts in patients suffering from rabies, cholera, tetanus, and infantile convulsions (Chopra and Chopra 1957). However, prior to his return from India with samples of potent *charas* (*Cannabis* resin) and knowledge of its medicinal use, some European physicians were using either wild *Cannabis* or cultivated NLH (*C. sativa*) growing in Europe, yet both of these sources were likely very low in cannabinoids. This could explain why hemp remedies in Europe employed since the Classical Age were quite limited in their scope and emphasized the use of hemp seed and seed oil rather than extracts of *Cannabis* flowers and resin favored in later preparations. Once there was a steady supply of potent NLD *Cannabis indica* from British India, its medicinal use in a wide range of packaged remedies increased significantly.

Previous to its relatively widespread introduction to Western Europe, stimulated by the work of Sylvestre de Sacy in France and O'Shaughnessy in Britain, one of the more common medicinal products of *Cannabis* was a "pharmaceutical preparation, or 'electuary' . . . taken in the form of a greenish paste" (Shimwell 2005 citing Holmstedt 1973) similar to Indian *majoon* (see Chapter 7). In England during the nineteenth century, medicinal *Cannabis* preparations became available in tinctures, pills, and extracts, which were utilized to reduce or soothe pain by lessening the sensitivity of the brain or nervous system, as well as for relief from spasms and inflammations.

In 1899, the well-known British pharmacologist Walter Ernest Dixon (1899) published a paper on "the pharmacology of *Cannabis indica*," which described his extensive research; he concluded that *Cannabis* was pharmacologically useful as

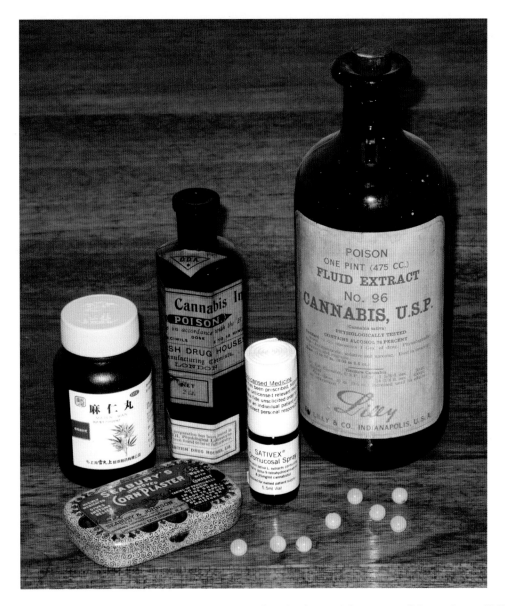

FIGURE 48. Several *Cannabis* varieties were used in popular late nineteenth- and early twentieth-century medicines such as an Eli Lilly tincture made from American-grown narrow-leaf drug (NLD) *Cannabis* (right), a British Drug House tincture of Indian-grown NLD *Cannabis* (center top), and a Seabury's Corn Plaster (bottom left). Present-day prescription *Cannabis* drugs include Sativex sublingual spray (center), containing a blend of THC from a hybrid NLD/BLD drug cultivar and CBD from a narrow-leaf hemp (NLH) cultivar, both grown in the United Kingdom. Marinol capsules (bottom right), known as the "pot pill," contain synthetic THC in sesame oil. Chinese "Ma Ren Wan" pills (left) are made from broad-leaf hemp (BLH) seeds and stimulate digestion (from the collection of David Watson, HortaPharm BV, Netherlands).

a "food accessory." As Russo (2002) has pointed out, Dixon's insightful suggestion was a forerunner of the modern usage of *Cannabis* to alleviate symptoms of weight loss, muscle atrophy, fatigue, weakness, and significant appetite loss associated with cancer chemotherapy and HIV-positive patients. In addition, Dixon refocused attention on smoking medicinal *Cannabis*: "In cases where an immediate effect is desired the drug should be smoked, the fumes being drawn through water. In fits of depression, mental fatigue, nervous headache, and exhaustion a few inhalations produce an almost immediate effect, the sense of depression, headache, feeling of fatigue disappear and the subject is enabled to continue his work, feeling refreshed and soothed. I am further convinced that its results are marvelous in giving staying power and altering the feelings of muscular fatigue which follow hard physical labour [*sic*]" (quoted in Russo 2002b).

During the same year, Shoemaker (1899) reported on a large series of patients who were all effectively treated for painful conditions including "migraine, dental neuralgia, gastralgia, enteralgia, cerebral tumor and herpes zoster" (Russo 2002b).

Although some physicians emphasized the value of medicinal *Cannabis* during the middle of the nineteenth century, it does not appear to have been used extensively. Shimwell (2005) indicates that the use of medicinal *Cannabis* may not always have been looked upon favorably during this period: "Serious flaws, such as unreliability in strength and effect of the drug, the inability to obtain quality and regular supplies of the drug and the stigma that was associated with it were responsible for limited use." Nevertheless, in 1883, two letters to the *British Medical Journal* attested to benefits of an extract of *Cannabis indica* for treating menorrhagia, an abnormally heavy and prolonged menstrual period, with both the associated pain and bleeding successfully reduced with a few doses (Batho 1883; Brown 1883; also Reynolds 1879).

A short, late nineteenth-century editorial titled "Cannabis Indica" from the *Medical and Surgical Reporter*, published in New York, informs us that

> this drug, the most active of non-opiate anodynes or soporifics, which was very popular years ago, although little the fashion at present, is deserving of a large share of professional favor. The principal cause which led to its disuse was fear of its toxic power though there has never been a case of poisoning recorded from its use. Its effect on the system is most marvelous. It causes sleep, overcomes spasms, relieves pain and all nervous irritability, and that too within a few moments after administration. Its soothing and curative effects upon the nervous sympathetic system are great, and there is no one agent that will restore the equilibrium of nerve motion more quickly. The placidity of repose that is produced by this narcotic [*sic*] is rapid and to the point. Further, it does not check secretion or tend to constipation. It does relieve fatigue and arouse vital action, and can and should be given freely until the effect desired is apparent. (KYNETT 1895)

Throughout the latter half of the nineteenth century, a number of prominent physicians in Europe and North America advocated use of extracts of *Cannabis indica* for several ailments, and *Cannabis* was included in the mainstream pharmacopoeias of Britain and America. An example is common use of *Cannabis* in treatment of migraine headaches. In 1915, Sir William Osler, famous English physician and acknowledged father of modern medicine, referred to *Cannabis* as perhaps the best remedy for severe headache treatment. According to Russo (1998), this in turn stimulated physicians in Europe as well as North America to acknowledge efficacy of *Cannabis* in remedies for restlessness, insomnia, and pain, such as that produced by menstrual migraine: "*Cannabis*, or marijuana, has been used for centuries for both symptomatic and prophylactic treatment of migraine. It was highly esteemed as a headache remedy by the most prominent physicians of the age between 1874 and 1942, remaining part of the Western pharmacopoeia for this indication even into the mid-20th century."

Most contemporary doctors are not cognizant of the important former status of *Cannabis* drugs in medical practice; indeed most "remedies fall in and out of favor" with their popular use, rarely exceeding a few decades, and once they are replaced by more popular or easily available remedies, most "fail to re-attain a position of prominence" (Russo 1998). In the case of medicinal *Cannabis*, this remains to be seen. If the recent upsurge of its use is any indication of efficacy, growing acceptance of its valid therapeutic use will rise even more significantly.

Present-Day Western Medicinal Applications of *Cannabis*

To some it's the classic "gateway drug," to others it is a harmless way to relax, or provide relief from pain. (BOOTH 2003)

Cannabis medicines were prescribed for a variety of medical indications throughout the early twentieth century, though on a more limited level than during previous centuries. In 1937, the Marijuana Tax Act was enacted, and in 1941, *Cannabis* preparations were dropped from the United States Pharmacopoeia. Two decades later, in the 1960s, an exponential increase in recreational use of marijuana moved *Cannabis* to the forefront of Western consciousness, eventually reaching a level of notoriety sufficient to discourage almost all legal medicinal use of this age-old and, for countless patients, venerable herb. However, medical research involving *Cannabis* resumed recently, stimulated by "anecdotal reports of patients who serendipitously discovered its benefits" (Russo 1998). In an attempt to dampen newly reinvigorated medicinal interest in herbal *Cannabis*, the US Supreme Court declared medical use of *Cannabis* illegal on the federal level in June 2005, although as of early 2013 its use had been allowed in 18 states and the national capital.

As we outlined earlier, medical use of *Cannabis* has a long association with humans and subjective evidence for its efficacy is plentiful; furthermore, interest in medicinal efficacy of *Cannabis* has been rekindled as commercial interest in plant-based natural medicines increases. During the last decade of the twentieth century public interest in benefits of medicinal *Cannabis* grew exponentially from the knowledge of a limited few to widespread use for largely illicit self-medication in both North America and Europe. As awareness of its medicinal values spread, many patients frustrated by the ineffectiveness of accepted medications tried *Cannabis* for the first time and felt immediate relief from their persistent symptoms. Many more began to realize that their satisfaction with recreational *Cannabis* use lay in part in its ability to mediate long-term psychological and physical conditions. It should be pointed out that although *Cannabis* is a powerful medicine, enabling patients to relieve a wide variety of discomforts and improve their quality of life, it has not as yet been shown to "cure" any medical condition. This situation proves important in the context of the modern pharmaceutical business where single compound "silver bullet cures" are more patentable, profitable, and fundable than whole herbs and plant extracts, although historically plants have proven effective in treating symptoms and improving the quality of life of the sick and dying.

At the federal level in the United States, the only legally available *Cannabis* medicine is synthetic THC called dronabinol in the oral dosage form, and trade-named Marinol. Although approved for sale in 1985, for many people Marinol is not patient-friendly, and it is not easy to titrate dosage as with smoked or vaporized *Cannabis*. Sales of Marinol and Nabilone (a synthetic compound that mimics the action of THC) exceeded US$110 million in 1995 and have increased steadily each year since then. Total Marinol sales in the US were about $190 million in 2008 (Par Pharmaceutical 2008), and worldwide sales of Nabilone in 2009 were nearly $50 million (EvaluatePharma 2013). When humans consume *Cannabis*, Δ^9-THC and its liver metabolite 11-*hydroxy*-THC activate cannabinoid (CB) receptors found throughout the brain and body, resulting in various physiological and psychological responses. For example, when Δ^9-THC is inhaled, it passes from the lungs into the bloodstream, directly to the brain, and then slowly back through the circulatory system to the liver. The liver then converts Δ^9-THC into 11-*hydroxy*-THC. On the other hand, when Δ^9-THC is swallowed, it is absorbed by the gastrointestinal tract and makes a first pass through the liver, where it is converted to 11-*hydroxy*-THC, and then the 11-*hydroxy*-THC enters the bloodstream and is carried to the brain (see Chapter 7). 11-*hydroxy*-THC is considerably more potent and long lasting, and it produces more unpleasant side effects than Δ^9-THC. Since the potency and effects of Δ^9-THC and 11-*hydroxy*-THC differ greatly, so do the clinical effects of the two routes of administration. As a result, the vast majority of medicinal *Cannabis* users in Europe and North America self-medicate by smoking illegally obtained, black market marijuana rather than pursuing legal relief through a pharmacy. In the United States, 17 states as well as Washington, DC, have, over the six years between 1996 and 2012, legalized or decriminalized medical *Cannabis* use. These are Alaska, Arizona, California, Colorado, Delaware, Hawai'i, Maine, Michigan, Montana, Nevada, New Jersey, New Mexico, Oregon, Rhode Island, Vermont, Virginia, and Washington. Several now allow the cultivation of limited amounts of *Cannabis* for medicinal use under a wide variety of localized restrictions.

Since the 1970s, modern North American and European hybrid drug *Cannabis* varieties have been developed, largely by clandestine breeders from crosses between South Asian NLD varieties (that spread early on throughout South and Southeast Asia, Africa, and eventually to the New World), and more recently Central Asian broad-leaf drug (BLD) hashish varieties have been spread widely (see Chapters 4 and 10). In Western societies, recreational *sinsemilla* varieties are commonly used with great efficacy as medicinal *Cannabis*. The primary cannabinoid contained in the vast majority of Western recreational and medicinal *Cannabis* varieties is THC. These varieties usually contain only small amounts of CBD, owing to their predominantly NLD heritage as well as selective breeding for psychoactive potency with increased levels of THC and reduced levels of CBD. In addition to THC and CBD, *Cannabis* produces a myriad of other secondary metabolites, including more than 60 minor cannabinoids and well over 100 terpenoids (Turner et al. 1980). Terpenoids are the primary ingredients in the essential oil of *Cannabis* and are largely responsible for the fragrances associated with different varieties. It is possible that the diverse chemical components in *Cannabis* account for the wide range of psychoactive and therapeutic effects produced by the consumption of various cultivars. In fact, *Cannabis* is considered to be a synergistic medicine by some herbalists, containing primary pharmaceutically active compounds (e.g., THC and CBD) along with many other secondary compounds (e.g., auxiliary cannabinoids and terpenoids) that both enhance the effects of a primary pharmaceutical compound and mitigate its side effects (e.g., see Russo 2011; McPartland and Russo 2001).

Cannabis consumers often associate individual varieties with particular mind and/or body effects as well as differing levels of medical efficacy for certain illnesses. Some medical users often consider varieties with a certain flavor to express similar medicinal effects (e.g., "skunky" smelling varieties are often sedative). These variations can likely be attributed to the auxiliary cannabinoids if present (e.g., CBD and CBN), some of the myriad terpenoids, and possibly other classes of secondary metabolites such as flavonoids. Comparative experiments with THC vaporized and inhaled alone and with the addition of various terpenes indicate that the synergistic effect of each terpene differs (David Watson, personal communication 2007). Several terpenoid and other trace compounds that modify the action of THC and/or CBD may cause the specific medicinal and/or recreational effects associated with each cultivar. Synergized therapeutic effects have also been elucidated using whole *Cannabis* extracts versus THC and THC + CBD versus THC. Cannabinoids act as partial agonists of CB receptors so that they produce more

subtle effects than full agonists and are also less likely to cause tolerance. Medical *Cannabis* users come to realize that they develop tolerance to its minor side effects while developing little if any tolerance to its therapeutic benefits (McPartland and Guy 2004a/b). More biochemical and pharmaceutical research is needed to address these issues.

CBD is nonpsychoactive but produces powerful anticonvulsant and anti-inflammatory effects. CBD has also been shown to attenuate and prolong the effects of THC (Musty 2004). Increased ratios of CBD to THC may prove clinically valuable in reducing anxiety and other unpleasant side effects occasionally experienced with THC while extending the effectiveness of each dose. Terpenoids have been shown to affect CB and other receptors. Terpenoids also alter cannabinoid pharmacokinetics by dilating bronchial capillaries and increasing blood-brain barrier permeability, allowing more THC to reach the brain more quickly (McPartland and Russo 2001; McPartland and Guy 2004a/b). The entire terpenophenolic biosynthesis mechanism (producing both cannabinoids and terpenoids) may have been under positive selection by humans because of two factors: first, terpenoid molecules are required as substrate for cannabinoid biosynthesis, and second, THC is much more effective in the presence of certain associated terpenoid compounds that are biosynthetically similar. Phytocannabinoids stimulate the central nervous system through the human CB1 receptor, but they also stimulate CB2 receptors throughout the body and modulate the immune system. THC also enhances the effects of the opioids, secondary metabolites of the opium poppy—another important medical plant with a long history (Merlin 1984; Russo 2004). Relief of glaucoma symptoms; control of vomiting; and protection of the brain, retina, and nervous system from toxic stresses have also been proven to be effected favorably by cannabinoids.

GW Pharmaceuticals Ltd. in the United Kingdom, continuing the pioneering work of HortaPharm BV in the Netherlands, is developing *Cannabis* varieties that produce only one of the four major cannabinoid compounds (e.g., THC, CBD, CBC, CBG, or their propyl homologs), as well as varieties with consistently uniform mixed cannabinoid and terpenoid profiles. In 1999, GW Pharmaceuticals began clinical trials of a *Cannabis*-based medicine aimed at relieving multiple sclerosis-associated neuropathic pain and spasticity. An oromucosal spray trade-named Sativex with a plant-derived combination of 50 percent THC and 50 percent CBD, has been approved for sale in Canada, the United Kingdom, and Spain and should be approved in the United States in the near future. In addition, plant preparations and extracts delivered through vaporizing or nebulizing (fine spray) devices often offer medicinal efficacy similar to that of smoking *Cannabis* without subjecting the patient to the potentially harmful products of combustion.

Continuing experimentation and medical trials with *Cannabis* extracts and isolated cannabinoid compounds as well as improved delivery systems should return *Cannabis* to the front line of remedies for a variety of indications. In the meantime, patients will largely continue to self-medicate with *Cannabis*, appreciative of the relief they receive, and well aware of the legal risks they often take. For a challenging discussion of the medical value of *Cannabis* in light of its largely illegal status, see Grinspoon and Bakalar (1997). Given the quite lengthy history of *Cannabis* use for pain relief and a wide variety of other medical problems, its future therapeutic utilization seems highly warranted (e.g., see Russo 2001,

FIGURE 49. Resin glands can be collected from dried female flowers and mechanically isolated to make the purest natural medicinal and recreational *Cannabis* preparations. Each of the tiny transparent gland heads contains aromatic essential oil rich in cannabinoid compounds (photos ©Bubbleman).

2002a/b, 2004, 2011; Russo and Grotenhermen 2006; Guy et al. 2004). It will be difficult for modern pharmaceutical companies to improve on nature's tried and true herbal medicine, and until modern cannabinoid medicines become more available patients will continue to self-medicate with natural herbal *Cannabis*. As Dr. Lester Grinspoon (2007), an emeritus professor of psychiatry at Harvard Medical School, explains,

> The pharmaceutical industry is scrambling to isolate cannabinoids and synthesize analogs and to package them in non-smokable forms. In time, companies will almost certainly come up with products and delivery systems that are more useful and less expensive than herbal marijuana. However, the analogs they have produced so far are more expensive than herbal marijuana, and none has shown any improvement over the plant nature gave us to take orally or to smoke. We live in an antismoking environment. But as a method of delivering certain medicinal compounds, smoking marijuana has some real advantages: The effect is almost instantaneous, allowing the patient to fine-tune his or her dose to get the needed relief without intoxication.

It should also be pointed out here that since the mid-twentieth century, hemp seed and its oil also become increasingly recognized as valuable in modern medicinal and nutraceutical applications. For example, in 1955, a Czechoslovakian nutrition study concluded that hemp seed was the "only food that can successfully treat the consumptive disease tuberculosis, in which the nutritive processes are impaired and the body wastes away" (Robinson 1996). Medical uses of hemp seed oil in various regions in the past can serve as models in contemporary societies to successfully treat ear, nose, and throat cases as well as burns and chronic eczema (e.g., see Grigoriev 2002). Modern body care products advertise the emollient effects of hemp seed oil in softening skin. For example, the Fushi Ltd. company based in England, which advertises itself as providing "holistic health and beauty solutions," refers to the polyunsaturated EFAs (linoleic and linolenic acids) in hemp seed oil as balancing dry skin: "It is a non-greasy, emollient and moisturizing compound with excellent anti-ageing and moisture balancing properties" (Fushi Ltd. 2007). EFAs not only help to restore wasting bodies and treat skin

conditions but also may improve damaged immune systems, and researchers are studying them in relation to treating immune system-attacking HIV and reducing the incidence of cancer (e.g., see Leson and Pless 2002).

Zuardi (2006) reviewed the history of medicinal *Cannabis* and pointed out that its relatively widespread use in Western medicine occurred in the middle of the nineteenth century and reached a peak during the last decade of that century with the availability and usage of *Cannabis* extracts or tinctures. Then in the first decades of the twentieth century, use of *Cannabis* in Western medicine decreased significantly mostly because it was difficult to obtain consistent results from plant material that characteristically had variable potencies; this was compounded significantly by the subsequent political and legal pressures against any use of *Cannabis* throughout much of the world. Russo (1998) articulated the situation succinctly: "Problems with quality control and an association with perceived dangerous effects sounded the death knell for *Cannabis* as a recognized Western therapy."

Russo also reminded us that some medicines that potentially produce much more damaging effects than *Cannabis* remain in our pharmacopoeias because of recognized medical efficacy; these include, for example, "opiates for pain control, amphetamines for narcolepsy and attention deficit hyperactivity disorder [ADHD], etc." However, since about 1965, research directed at identifying the chemical structure of *Cannabis*'s components and possibly obtaining its "pure constituents" produced an important boost in *Cannabis* interest among the scientific community. Attention was renewed and greatly accelerated in the late twentieth century with discovery of human cannabinoid receptors and identification of an endogenous cannabinoid system in the brain (e.g., see Devane et al. 1992; McPartland 2004, 2008; McPartland and Guy 2004a).

By 1995 the future of *Cannabis*-based medicines was brightening and cannabinoids were ready to be developed into pharmaceutical rather than political products. Since then there has been a large number of scientific studies to determine the therapeutic benefits and drawbacks of these medicines. Kalant (2001) summarized the history and status of medicinal *Cannabis* use, pointing out that THC and a number of analogs had been shown to offer significant therapeutic benefits in relief of nausea and vomiting, as well as stimulation of appetite in patients with wasting syndrome. He also noted that research "clearly demonstrates analgesic and antispasticity effects that will probably prove to be clinically useful." Kalant argued that "the anticonvulsant effect of cannabidiol [CBD] is sufficiently promising to warrant further properly designed clinical trials" but noted that "a major lack of long-term pharmacokinetic data, and information on drug interactions" remained to be rectified. He emphasized that although "pure cannabinoids, administered orally, rectally, or parenterally [into a vein], have been shown to be effective," smoking medicinal *Cannabis* "might be justified on compassionate grounds in terminally ill patients who are already accustomed to using cannabis in this manner." Kalant called for research that can "yield new synthetic analogs," and these novel products would provide "better separation of therapeutic effects from undesired psychoactivity and other side effects" while having "solubility properties that may permit topical administration in the eye, or aerosol inhalation for rapid systemic effect without the risks associated with smoke inhalation." Five years later, Zuardi (2006) pointed out that a "new and more consistent cycle of the use

TABLE 12

Cannabis produces unique cannabinoid compounds that exhibit a wide range
of potentially beneficial psychological and physiological effects.

Phytocannabinoid	Description, occurrence, and effects
Δ^8-Tetrahydrocannabinol or Δ^9-THC (Δ^1-THC)	• Primary psychotropic ingredient of Cannabis. Primary cannabinoid in marijuana varieties. • Therapeutically used as an antiemetic and to boost appetite in AIDS patients. • A Cannabis-based extract with approx 1:1 ratio of Δ^9-THC and CBD (Sativex) is effective for the symptomatic relief of neuropathic pain in adults with multiple sclerosis and as an adjunctive analgesic treatment for adults with advanced cancer. • Plant-based Δ^9-THC is also used in the generic equivalent of Marinol used in treating AIDS-related anorexia and nausea and vomiting associated with cancer chemotherapy. • Antimicrobial and antifungal. Promising for the treatment of many medical indications.
Δ^8-Tetrahydrocannabinol or Δ^8-THC (Δ^6-THC)	• Artifact resulting from isomerization of Δ^9-THC found only in trace amounts in Cannabis. • Pharmacology similar to Δ^9-THC. • Not used medically.
Cannabinol or CBN	• Product of Δ^9-THC degradative oxidization resulting largely from postharvest storage found only in trace amounts in fresh Cannabis. • Up to 10 percent of the potency of Δ^9-THC. Not used medically.
Cannabidiol or CBD	• Major nonpsychotropic cannabinoid. Commonly produced by hemp and hashish varieties but uncommon in hybrid sinsemilla cultivars. • Anti-inflammatory, analgesic, antioxidant, antispasmodic, antiemetic, antipsychotic, antiepileptic, vasorelaxant, immunosuppressive, and neuroprotective actions. • Effective in anxiety, psychosis, and movement disorders and relieves neuropathic pain in patients with multiple sclerosis (in combination with Δ^9-THC as in Sativex). • Protects against diabetes-induced retinal damage. • Beneficial effects on bone formation and fracture healing. • Antimicrobial and antifungal. • Potential use in the treatment of chemotherapy-induced and anticipatory nausea. • Promising for cancer treatment
Δ^9-Tetrahydrocannabivarin or Δ^9-THCV	• Found in Pakistani hashish varieties. • Antagonizes Δ^9-THC effects in low doses. Antiepileptic and anorectic. • May reduce food intake. • Beneficial effects on bone formation and fracture healing.
Cannabigerol or CBG	• Nonpsychotropic. • Antiproliferative, antimicrobial, antifungal, and antibacterial activity. • Beneficial effects on bone formation and fracture healing. • Potential role in analgesia. Promising for cancer treatment.
Cannabidivarin or CBDV	• Nonpsychotropic. • Found in Pakistani hashish varieties. • Beneficial effects on bone formation and fracture healing. • Physiological effects poorly understood.
Cannabichromene or CBC	• Nonpsychotropic. Along with Δ^9-THC the major cannabinoid in marijuana. • Exerts anti-inflammatory, antimicrobial, antifungal, and modest analgesic activity. • Beneficial effects on bone formation and fracture healing. • Potential role in analgesia. Promising for cancer treatment.
Δ^9-Tetrahydrocannabinolic acid or Δ^9-THCA	• Naturally occurring form of Δ^9-THC in fresh plants. • Exerts antiproliferative and antispasmodic actions. • Potential role in analgesia and in the treatment of prostate carcinoma.
Cannabidiolic acid or CBDA	• Naturally occurring form of CBD in fresh plants. • Exerts antiproliferative action. • Potential role in analgesia, inflammation, and the treatment of prostate carcinoma.

NOTE: Adapted from Izzo et al. 2009.

of *Cannabis* derivatives as medication" had begun with the establishment of effective and safe treatment supported by progressive scientific research. Izzo et al. (2009) reviewed therapeutic opportunities for cannabinoids, concluding that

> recent developments suggest that non-psychotropic phyto-cannabinoids exert a wide range of pharmacological effects, many of which are of potential therapeutic interest. The most studied among these compounds is CBD, the pharmacological effects of which might be explained, at least in part, by a combination of mechanisms of action. CBD has an extremely safe profile in humans, and it has been clinically evaluated (albeit in a preliminary fashion) for the treatment of anxiety, psychosis, and movement disorders. There is good pre-clinical evidence to warrant clinical studies into its use for the treatment of diabetes, ischemia and cancer.

Izzo and his colleagues also referred to a number of other neglected phytocannabinoids such as CBC and CBG, which should be explored for their potential use in pain management.

Russo continues to study the medical history and modern applications of *Cannabis* with some of his most recent research focusing on phytochemicals other than cannabinoids. In his recent review, Russo (2011) explores another stratum of phytotherapeutic *Cannabis* agents—the terpenoids (e.g., limonene, myrcene, a-pinene, linalool, b-caryophyllene, caryophyllene oxide, nerolidol, and phytol). Even though they are regarded as safe by the United States Food and Drug Administration (USFDA) and other regulatory agencies, terpenoids are quite powerful, affecting "animal and even human behavior when inhaled from ambient air at serum levels in the single digits (ng/mL^{-1})." Russo argues the *Cannabis* terpenoids can induce distinctive healing effects that could "contribute meaningfully to the entourage effects of cannabis-based medicinal extracts" and "could produce synergy with respect to treatment of pain, inflammation, depression, anxiety, addiction, epilepsy, cancer, fungal and bacterial infections." He also discusses scientific evidence indicating that noncannabinoid plant components may act as antidotes to intoxicating effects of THC (which might improve its therapeutic value) and suggests a set of experimental methods to investigate the putative "entourage effects" of "phytocannabinoid-terpenoid synergy" in the future. If his thesis is confirmed, it would increase the probability that *Cannabis* will provide a broad array of new healing products: "Selective cross-breeding of high-terpenoid- and high phytocannabinoid-specific chemotypes has thus become a rational target that may lead to novel approaches to such disorders as treatment-resistant depression, anxiety, drug dependency, dementia and a panoply of dermatological disorders, as well as industrial applications as safer pesticides

and antiseptics. A better future via cannabis phytochemistry may be an achievable goal through further research of the entourage effect in this versatile plant that may help it fulfill its promise as a pharmacological treasure trove" (Russo 2011).

Summary and Conclusions

Medicinal *Cannabis* has a long and well documented history across Eurasia reaching back several millennia. *Cannabis* has not been found to cure any illness but is extremely effective in relieving the symptoms of a wide variety of medical conditions. Based largely on anecdotal accounts of its efficacy, *Cannabis* is slowly gaining interest from the pharmaceutical industry.

> The original prohibition of cannabis arose from social pressure as much as safety concerns. It is heartening that the scientific evidence on which a rational reappraisal of cannabis as a prescription medicine can be made has been forthcoming. Patients with intractable disease will welcome this. They are often afflicted in their prime, and availability of an effective and safe prescription medicine will, in turn, lead to an improved quality of life. It is also refreshing to see that the derivation of the first prescription medicine based on whole cannabis represent a return to the roots of science—medicinal plants. (ALAN MACFARLANE, forward in Guy et al. 2004)

Lester Grinspoon (2007) recently summed up the medicinal marijuana situation succinctly: "It is a sad commentary on the state of modern medicine that we still need 'proof' of something that medicine has known for 5000 years. . . . If marijuana were a new discovery rather than a well-known substance carrying cultural and political baggage, it would be hailed as a wonder drug."

While the Pharmaceutical industry fights for patent control of new, politically correct preparations and administration devices, the vast majority of medical use remains self-administered—either smoked or eaten herbal preparations—and illegal in most jurisdictions. Eighteen states in the United States have approved medical *Cannabis* use by popular referendum, yet the federal government has failed to recognize it as an effective medicine. Medical *Cannabis* use is here to stay and will likely become much more popular in the near future. For further, in-depth discussions of historical and modern medicinal uses of *Cannabis* and cannabinoids, see, for example, Russo (2011, 2008, 2006, 2004, 2002, 2001, 1998), Izzo et al. (2009), Guy et al. (2004), Earlywine (2002), Fankhauser (2002), Rätsch (2001), Aldrich (1997), and Mechoulam (1986).

CHAPTER NINE

Nonpsychoactive Ritual Uses of *Cannabis*

Hemp never lost its connection with the cult of the dead.
Even today in Poland and Lithuania, and in former times
also in Russia, on Christmas Eve when it is believed that the
dead visit their families, a soup made of hemp seeds, called
semieniatka, is served for the dead souls to savor.

(RUDENKO 1970, quoted in LA BARRE 1980)

Introduction

A ritual or rite involves speech, singing, and/or other activities that frequently include a symbolic meaning. These activities are performed in a specific order, often during a religious service or a traditional community occasion. Such a set of actions is usually undertaken because of its supposed influence on behavior and its ability to induce emotions in participants. A large body of knowledge exists concerning relations between *Cannabis* and its ritual uses. We explored the psychoactive ritual and recreational uses of *Cannabis* in Chapter 7. Here we will expand upon our theme with supporting evidence for the nonpsychoactive ritual uses of whole plants, hemp fiber, and seed. We have gleaned an extensive sample of this knowledge from ancient ritual texts and treatises, historical accounts, and ethnographic research with a focus on social aspects of traditional hemp growing cultures. Much of this material is somewhat referential, often merely consisting of a passing comment noting the perceived ritual use of hemp fiber or seed. Although we included a considerable amount of speculative interpretation, we have taken a critical approach, being careful not to identify ritual based on its symbolic content alone since many common human actions can be misinterpreted as having symbolic significance (Ashkenazi 1993). Many potential citations have been omitted, yet the case for special ritual significance accorded to *Cannabis* and its nonpsychoactive products is clearly strong. In several geographical locations hemp may have been the only or by far the best available fiber for making cordage, spinning, and weaving and therefore was often the only choice for ritualistic uses calling for rope or cloth. On the other hand, common ritual acts such as tying, binding, and enclosing ritual supplicants can be (and are more commonly) performed with a wide variety of textiles other than hemp. We have taken care to select examples of ritual hemp use where an ancient cultural tradition or ceremony provided a transformative event intended to achieve a certain result (Clothey 1983). In particular, ritual hemp use must include an expectation of a result based on the association of the hemp plant or its products with a particular attribute or power, further supporting the special ritual significance of *Cannabis*.

Plant-derived drugs obviously enhance and expand the transformative nature of ritual. In many traditional rituals that are or were practiced across Eurasia, *Cannabis* has played key roles without any utilization (or possibly even without realization) of its psychoactive potential. *Cannabis* seeds, bark strips, raw fiber, yarn, cordage, and cloth are the plant parts and products most commonly used in hempen rituals. The most common settings for ritual use of hemp across cultures are purification and protection ceremonies, often in the context of healing, and in particular funerals. This chapter also cites many examples of nonpsychoactive use of *Cannabis* in various agricultural, seasonal, and life cycle rituals among traditional hemp-growing cultures in present-day East Asia and Europe and draws comparisons between them.

Hempen Rituals of Major Religions: Shamanic Influences Survive Repression

The role of shamans in traditional Eurasian hunter-gatherer, pastoralist, and agrarian cultures and their uses of psychoactive plants are well documented (e.g., see Schultes and Hofman 1979; also see Chapter 7). Localized, nature-oriented, shamanic traditions are the logical antecedents of organized and widespread religions. Confucianism and other conservative Eastern ways of thought have had a similar effect in East Asia as Christianity has had in the West, largely to disgrace shamanic practitioners and suppress the independent use of drug plants. Schultes (1969b, 1970a) and La Barre (1970) addressed the imbalance between the relative scarcity of plant and fungi drugs traditionally used in Eurasia (primarily *Cannabis* and *Amanita muscaria*) and the plethora of psychoactive species used in New World shamanistic cultures (also see Merlin 2003). All attributed this difference to repression of shamanic and other cultural traditions by changing socioeconomic (especially religious) pressures across Eurasia (such as the demonization and suppression of "witches" or shamans) in comparison with the relatively uninhibited status of traditional naturalistic cultures in the Americas.

Ritual use of nonpsychoactive *Cannabis* is most prevalent in Eastern Europe and the Far East where conservative Christian and Confucian traditions were founded early on and remain popular today. The dearth of ritual traditions incorporating psychoactive *Cannabis* use in these areas cannot be explained by a lack of easy access as moderately psychoactive *Cannabis indica* ssp. *chinensis* (broad-leaf hemp, or BLH) grows in escaped populations across much of temperate East Asia today, and we expect it has been present there and widely traded since ancient times (see Chapters 2 and 4). Present-day European *Cannabis sativa* (narrow-leaf hemp, or NLH) lacks sufficient psychoactive potential to encourage its ritual use in a psychoactive context. However, it is likely that psychoactive *Cannabis* was well known across Central and Eastern Asia, and during the last two millennia, psychoactive ritual use has been suppressed by centralized dominant state religions; all that remains today is the relatively widespread ritual use of nonpsychoactive hemp. *Cannabis* must be cultivated to produce fiber of sufficient quality (fine, straight, and long), and therefore hemp was largely a product of relatively sedentary peoples living in societies with a large labor force, some of which became early city-states encouraging widespread following of a particular favored and approved religion that early on incorporated certain ritual aspects of previously established shamanistic cults. This synthesis has left us with many of the ritual uses of hemp assembled here.

First we address such use in South Asia where there are a number of native subtropical and tropical fiber resources that generally obviate the need for hemp fiber in this region; therefore, *Cannabis* use here has largely focused on its psychoactive potential. In fact, we find a nearly complete absence of hemp cultivation for fiber in South Asia, with the exception of a temperate climate belt across the Himalayan foothills of western Nepal and northeastern India extending into the neighboring Hengduan Mountains of northern Burma and southwestern China (see Chapter 5). Although we identified many uses of the psychoactive narrow-leaf drug (NLD) variety's resin, flowers, and leaves for ritual, recreational, and medicinal purposes in Indian Hindu religion in Chapters 7 and 8, there appears to be little *Cannabis* ritual involving low potency broad-leaf and narrow-leaf hemp (BLH and NLH) in South Asia. However,

some peoples such as the Kumaoni of Uttaranchal state in northern India recognize both cultivated and wild NLD *Cannabis* and employ these plants for both psychoactive and nonpsychoactive ritual uses. Hemp plants are often seen growing around abandoned Kumaoni homes, and thus people, when cursing somebody, say the following: "*Teri kuri bhangau jam jo*," which can be translated as "May hemp grow in your house," meaning, "May your house be ruined and damaged to such an extent that *Cannabis* will grow there" (Shah 1997).

Historically, South Asian, Arabian, and Middle Eastern cultures were more tolerant of ritual and recreational use of psychoactive NLD *Cannabis*, but they have also had their dominant conservative religious factions, and such tolerated use is encountered today only in a few smaller religious sects (e.g., Hindu Naga Babas and Muslim Sufis). There is presently no major world religion that openly promotes the ritual use of psychoactive *Cannabis*. Nevertheless, despite repressive influences of major state religions and many modern-day governments, contemporary ceremonies of many peoples still incorporate elements of shamanic ritual and include hemp in a major role. Nonpsychoactive ritual *Cannabis* use, possibly as a substitute for former psychoactive use, has continued up to the present in contexts common to several traditional cultures across Eurasia, although in many cases the majority of worshippers are unaware of the deep cultural significance of ritual hemp use.

Archeological Remains from Ritual Contexts: Central Asia, China, and Europe

Archeological remains are often interpreted as being indicative of a ritual context. All tomb remains can be considered "ritual" in nature simply because they were recovered from graves and therefore associated with funerary rituals focused on death, interment, and afterlife. In the case of *Cannabis*, ritual context can sometimes be interpreted as psychoactive in nature (e.g., some Central Asian tombs) because remains contain parts of female flowering tissues or seeds, sometimes intentionally charred, or were recovered from a vessel containing remnants of psychoactive cannabinoids (Rudenko 1970; Sherratt 1991; Jiang et al. 2006; Russo et al. 2008; also see Chapter 7).

Some *Cannabis* tomb remains only consist of nonpsychoactive parts of the plant, and thus, ritual significance may have been placed on these artifacts based on factors other than (or in addition to) psychoactivity. In the absence of other plant parts, finds of seeds (although they are produced by the psychoactive female flowers) do not indicate that the plants were utilized for drugs, as seeds are also a historically well-known food source and vital to the continuation of the species. Remains of hemp fibers may present researchers with evidence of nonpsychoactive *Cannabis* use. However, even though found in tomb sites (thereby de facto ritual sites) it is often unclear whether hemp had a specific ritual significance. Interpretation of *Cannabis* ritual depends on whether a ceremonial application was associated with *Cannabis* in particular or whether the choice of hemp represented simply a generic plant product found throughout many regions and ritual contexts (e.g., seeds representing fertility or cordage representing metaphysical connections and delimiting boundaries of souls and spirits).

As chronicled throughout this volume, *Cannabis* has enjoyed a long history in many regions of its native Eurasia and worldwide because it was such an important economic plant utilized for a wide variety of purposes; it was also well adapted to habitats disturbed by humans. Frequent presence of hemp in graves may simply reflect local availability of hemp for cordage and cloth and lack of other sources of plant fibers. medicinal By referring to ancient historical documents and ethnographic reports of more recent cultures, we can gain some insight into how, and most importantly why, *Cannabis* was used in additional ritual settings, and we can then propose a similar ritual use. Today, where hemp is in short supply and is prohibitively expensive, many market substitutes are used in rituals. Lingering ceremonial use of true hemp, often despite its rarity and extreme labor costs reflected in its high market value, indicates ritualistic importance beyond mere availability, thus illustrating the continuing momentum of ancient hempen ritual traditions. When we asked fieldwork informants, "Why do you use hemp in rituals?" the answer was most often, "Because we always have!"

Kurgan pit tomb mounds in several regions of the Central Asian steppe, often attributed to ancient Scythian peoples, have yielded censers, sometimes with charred *Cannabis* seed remains that appear to have been used for purification during funerary rites as described by the ancient Greek historian Herodotus. These finds are commonly cited as early evidence of inhalation of psychoactive vapors from burning *Cannabis* (see Chapters 4 and 7). However, except for the remarkable 2,700-year-old *Cannabis* found in the Yanghai burial (discussed earlier in this chapter), confirmed remains discovered in the steppe tombs are of seeds and no other plant parts. In addition, seeds of the fragrant coriander are also commonly found in association with *kurgan* funerary censers (Rudenko 1970; Clarke 1998a).

Herodotus's *Melpomene*, the fourth book of his "Histories," refers to the Scythian's ritual use of hemp seed in a pungent herbal sweat bath: "When therefore the Scythians have taken some seed of this hemp, they creep under the cloths, and then put the seed on the red hot stones; but this being put on smokes, and produces such a steam, that no Grecian vapour-bath would surpass. The Scythians, transported with the vapour, shout aloud; and this serves them instead of washing, for they never bathe the body in water" (Schleiffer 1979). The presence of *Cannabis* seeds associated with censers recovered from Scythian tombs and little else in the way of edible plant remains or hemp fiber is an obvious indication that they possessed ritual significance. Yet the comments of Herodotus leave us with conflicting views of *Cannabis*'s particular ritual use. On the one hand, his comments about the Scythians being "transported with the vapor" have been interpreted as psychoactive use of *Cannabis* (see Chapter 7); on the other hand, Herodotus's comments can be interpreted as referring simply to burning *Cannabis* as an adjunct to a cleansing and purifying sweat bath ritual associated with funerary rites. (The purifying nature of *Cannabis* as a recurring theme throughout traditional Eurasian hemp cultures is discussed later.) Butrica (2006) provides additional thoughts:

It should be remembered that cannabis seeds were used by the Scythians not recreationally but as a part of their death ritual: instead of a wake, they put the corpse of the deceased into a wagon, and for forty days took it on visits to the homes of friends and kin, where it was served at table along with the other guests. It was at the end of this period of mourning that men resorted to the hemp [and coriander]

baths as a form of cleansing (the head being washed first with soap), while the women pursued a different treatment (they smeared a paste of cypress, cedar, and frankincense on their bodies and allowed it to stand for a day; when removed, it left their skin fragrant, clean and shiny).

Rather than "shouting aloud" in amazement at being "transported by the vapour" with its psychoactive effects, a more accurate translation might be "howling in mourning" as a form of grieving for the departed (Butrica 2006).

Ancient Chinese graves reveal timeless methods used to build tombs and prepare the corpses before interment. Remarkably, excavations of Han dynasty (202 BCE to 220 CE) tombs in Gansu province uncovered "whole specimens of hemp cloth that were used to cover corpses; these wrappers were placed around silk dresses and were bound with hemp ropes, and fibers of hemp were also found as reinforcement of the plaster covering of the brick walls of this ancient crypt" (Kansu Museum 1972). Hemp was used for these same funerary purposes throughout China's history (see Chapter 5) and can be found today among the Han Chinese majority but even more commonly among some of China's minority ethnic groups such as the Miao or Hmong.

Several significant tombs excavated in Europe also contained nonpsychoactive *Cannabis* remains and ritual significance has been inferred from these finds (see Chapter 5). Delaney (1986) refers to a Celtic "princely tomb" dated from the late Hallstatt period (ca. 2550 to 2500 BCE) in Hochdorf, Germany, excavated in 1978 and 1979 by Jörg Biel: "Here the people of prehistoric Hochdorf buried a Celtic chieftain who merited a great mausoleum. He lay on a bronze, high-backed couch embossed with ritual dancing figures and horses pulling a cart. . . . Dr. Biel's fiber analysis revealed, imbedded in the bronze, horsehair, hemp, wool, and the fur of badgers, on which the dead prince had reposed." More evidence of possible ancient ritual use comes from the 570 CE tomb of the Merovingian queen Arnegunde (France-Lanord 1979), whose finely dressed corpse was covered by a hempen shroud (Werner 1964; Godwin 1967b). In these instances, hemp could simply have been utilized because it was locally available. So we must find historical and ethnographical references that may shed light on a special ritual significance of hemp in a particular ceremonial context and aid in the interpretation of ancient *Cannabis* remains.

In 1903, a farmer digging in a large burial mound on his farm, Lille Oseberg in Slagen, Vestfold, Norway, uncovered an impressive Viking ship in which two women were interred around the year 850 CE (see Chapter 5). Among ample nautical equipment recovered within the ship was a small piece of hempen material, perhaps intended for some ritualistic or other uses, which have not yet been determined; this piece of cloth was likely not a scrap of sailcloth as sails during the period were commonly made of coated wool. According to Vindheim (2002), "In Norwegian folklore hemp cloth symbolized the beginning and end, and it was the first as well as the last in which people were swathed in this life. These traditions may be relics from a time when hemp had a religious function in the pre-Christian religion, but the central use of hemp in Norway for the last thousand years has been as a source of fibre [sic]." More compelling perhaps is the recovery of four *Cannabis* seeds from the ancient Viking ship (Holmboe 1927), one found in a small leather bag and now believed to be connected with the woman's priestly functions (Christensen 1992). The esteemed archeologist Anne Stine Ingstad, who was responsible for excavating medieval Norse settlements in Newfoundland, is prominent among many historians who believe the younger of the two buried women—usually called the Oseberg Queen—was a priestess of the great Norse goddess Freya and not only a secular queen as the first excavators thought. Ingstad interprets the presence of the *Cannabis* seed in the (possibly talismanic) pouch as an indication of ritual use of *Cannabis* as a euphoriant in pre-Christian Scandinavia (Fleming and Clarke 1998). Across northern Europe, the love goddess Freya was associated with hemp sowing and harvest Hochzeit, or "high time," erotic rituals conducted in her honor. In feminine *Cannabis* flowers lay the eroticizing and love-generating power of Freya (Neményi 1988). Those who became intoxicated from them experienced the sensual joy and aphrodisiac ecstasies of the love goddess. However, as we shall see repeatedly, cultural reverence for hemp may not always be associated with its psychoactive powers. Worth noting in connection with the Oseberg burial is the lack of ropes and large textiles made from hemp, a strong reason for suggesting a ritual use for the *Cannabis* seeds and even the small piece of hempen cloth. The women in the Oseberg ship wore clothes made from flax, wool, silk, and nettle but not hemp. The ropes were made from lime tree (*Tilia* spp.) fibers despite the superior qualities of hemp rope (Fleming and Clarke 1998).

Interpretations of tomb and other archeological excavations benefit from comparisons with existing ritual traditions. Recent historical and present-day ritual use of hemp can be found in many traditional cultures across Eurasia. In this chapter we cite examples from the Hmong ethnic group with its roots in China, as well as among the Han Chinese and Korean Confucian and Buddhist traditions, Korean shamanism, Japanese Shintō, and European Christian ritual customs. Commonalities of *Cannabis*'s ritual roles throughout these divergent cultural traditions are described and discussed later.

The Hmong: Spirit Travel in Healing, Life Cycle, and Funerary Rituals

My daughters will wear hemp skirts for dressing up, for their weddings and also for their funerals. When my daughters live to be a hundred years old they can wear the hemp clothes to come look for me in the land of the ancestors. Dressed this way they will be able to find me, their mother. (MORGAN AND CULHANE-PERA 1993).

The Hmong people, a subgroup of the "Miao" ethnic minority as they are called in their original Chinese homeland, presently occupy a geographically diverse region stretching from central through southwestern China and spilling over into northern Laos, Thailand, and Vietnam, as well as in overseas communities. Although the Hmong are a large and widely dispersed ethnic group, their ritual uses of hemp are common throughout. The Hmong apparently did not traditionally use *Cannabis* psychoactively, yet they practice many shamanistic rituals associated with divination, healing, and life cycle events, and hemp is of greater present-day importance to the Hmong both as a crop plant and in ritual use than in any other Eurasian culture. Hence we have chosen the Hmong as a starting point in our detailed regional discussion of nonpsychoactive ritual use of *Cannabis*.

Unlike many Asian ethnic groups that tie cotton strings around participants during ceremonies, the Hmong traditionally tied hemp strings around participants during many of their rituals, although today they will also use cotton string in ceremonies (Chindarsi 1976; Lewis and Lewis 1984). In order to "capture good fortune," the Hmong of Laos as well as lowland Lao "tie strings around the neck, wrists and ankles," which are also believed to help "prevent the separation of the [soul] from the body because souls often depart through the extremities" (Catlin et al. 1986). String-tying rituals are performed when there are illnesses, weddings, ceremonies for naming newborn infants, and celebrations before or after long journeys.

Although string tying is a simple and widespread ritual act across Southeast Asia, performed with cordage of many types such as cotton and silk, presently, the Hmong are the only ethnic group to tie hemp string during healing and protection rituals. Hemp strings are most commonly tied around patient's wrists, as noted earlier, to prevent souls escaping from their body or to bring good luck (Lyman 1968) and around the neck to protect against evil spirits entering the body. Hemp twine or crudely joined rough bark strips can also be used to bind a group of persons to the patient (sometimes along with a sacrificial animal during healing rituals) as an act of tying their souls together (Lyman 1968; Chindarsi 1976; Fadiman 1997).

In the course of healing ceremonies the Hmong shaman moves between earthly and spirit worlds to cure illnesses. A hemp string or cloth strip is stretched from the spirit altar centered along the rear wall up and over the rafters to the front door, serving as a bridge or pathway for the shaman's spirit to travel in search of the afflicted person's wandering soul and a route for the spirits of the afflicted and the shaman to follow upon their return (Lemoine 1986). String and cloth bridges are also employed by Hmong shamans during rituals other than healing and funerals. Fadiman (1997) provides insightful background for Hmong ceremonies for barren couples:

If a Hmong couple failed to produce children, they could call in a shaman who was believed to have the ability to enter a trance, summon a posse of helpful familiars, ride a winged horse over the twelve mountains between the earth and the sky, cross an ocean inhabited by dragons, and . . . negotiate for his patients' health with the spirits who live in the realm of the unseen. A [shaman] might be able to cure infertility by asking the couple to sacrifice a dog, a cat, a chicken or a sheep. After the animal's throat was cut, the [shaman] would string a rope bridge from the doorpost to the marriage bed, over which the soul of the couple's future baby, which had been detained by a malevolent spirit called a *dab*, could now freely travel to earth.

Among the Hmong and other traditional hemp-using cultures, hemp could also be interpreted as having little special ritual significance as they have long used hemp for many domestic purposes, and it was their most readily available fiber. Does their ancient dependence upon hemp in daily life reinforce its special importance in ritual, or does hemp's commonality and exclusivity as a source of cordage and cloth diminish or mask its possible ritual significance? These are questions to keep in mind throughout this chapter. It is also possible that hemp's ritual significance is a reflection of its deep cultural history as a vital and highly respected food and fiber plant as well as past psychoactive use by shamans—historically suppressed by Han Chinese influences (Li 1974a; also see Chapter 7).

Possibly the most important achievements in a young Hmong girl's life are mastering the skills of hemp cultivation, processing, spinning, weaving, and sewing her trousseau and other hempen dowry items. Gu (1995) explains why hemp fiber is so important to the Hmong:

Miao [Hmong] girls still learn to spin and weave hemp cloth. If a girl cannot spin, weave, hand-stitch and embroider, she will have a difficult time finding a man willing to marry her. If a housewife cannot spin and weave hemp cloth, she is seen as incompetent. The level and quality of a woman's cloth work is seen as a standard to judge if she is hard working and intelligent. A girl's handicrafts are often the deciding factor in a boy's choice of a mate. Therefore, young girls are eager to learn these skills from their mothers and aunts. They usually begin at the age of six or seven. By the time a young woman is 15 or 16 years old, she can design and produce her own blouses and skirts.

The dowry provides an important part of the Hmong wedding, giving the young bride an opportunity to showcase her vital domestic skills, and the most important element of a bride's dowry is the hempen clothing she has made for herself and her groom. The process from sowing seeds to sewing clothing is lengthy and involved, spanning at least a spring through autumn growing season and a winter of spinning and weaving before cloth is ready for assembly into clothing. Hmong skirts represent the pinnacle of their hemp textile craft, and as Hmong costume specialist Theresa Reilly (1987) tells us, "A pleated skirt, therefore, can take from six months to several years to complete depending on the amount of batik, the size of the embroidered stitches, and the quantity of the appliquéd patterns. In many cases these skirts are part of a woman's trousseau, and a sign not only of her artistic ability, but of the family's affluence. The pleated skirts are often displayed before relatives and friends when the young bride is ready to leave her parents' home for her husband's family."

Another way young women demonstrate their skill is through song. Many young girls begin to practice at the age of six or seven while learning to work hemp, and when young men and women sing love songs, the girl's skill in weaving is commonly a topic her lover inquires about and praises. Here is an example of one such Hmong song called *Stepping Mountain Tune* (translated by Gu 1995):

The boy complains:
    Talking about spinning and weaving,
    I become dumb and helpless.
    Cutting hemp with a sickle,
    I wonder how to fasten them.
    When the hemp dries, I don't know how to peel them.
    Can't twist the peeled hemp,
    I only sweat with anxiety.
    Learning to spin the wheel,
    My feet and hands won't obey me.
    No thread is spun,
    A jumbled mass it has become.
    The loom is set in the parlor,
    But I can't operate it.
    To color a beautiful skirt,
    I have no good motifs in mind.
    Indigo is in the dye vat,
    But I do not know how to dye.
    Looms clicking in the neighbor's home,
    Everyone wears new cloths but me.
    Having no companion to weave,

FIGURE 50. Hmong girls learn to sew their traditional hemp clothing at an early age. This girl belonging to the Flower Hmong ethnic group is embroidering the hem band for a traditional skirt while selling hemp seeds at the market in Sapa, Lao Cai province, Vietnam.

    I have to put on a banana leaf.
The girl replies:
    Spring winds are warmer than autumn's,
    What you've sung is so bitter.
    When the moon gets full on the 15th,
    I will come to your side.
    In the good season of golden autumn,
    Let's harvest the crops together.
    You drive a strong bull,
    plowing plots of terraced lands.
    I sit by the loom,
    Weaving twelve bolts of cloth.
    You take your balance and counterweights out,
    Keeping your mind on doing business.
    I will stay home sewing,
    Making new clothes for you.
    Though it is not silk nor satin,
    Hemp clothes can warm you as well.

A Hmong bride's mother and grandmother sew additional pleated hemp skirts for her dowry. After a woman marries, she is expected to complete a set of traditional hempen clothing for each of her parents, cherished costumes used exclusively to dress the corpse when the parent dies (see Chapter 5 for a lengthy discussion of the historical cross-cultural history of hemp fiber use).

Ritual use of hemp is most highly expressed during Hmong funerary ceremonies. Traditionally, the most important element of Hmong burial clothing is the skirt the bride makes for her mother, which must be made of hemp cloth, the cloth of her ancestors. This skirt is given in exchange for one of the skirts given by the mother to her daughter as part of her dowry. The daughter diligently keeps the dowry skirt to be worn on her own deathbed, so that when she passes away and returns to her cultural homeland in China her mother can recognize her and they will finally be reunited (Mallinson et al. 1988). Referring to the contents of her dowry box an old Hmong woman explains, "At my wedding my mother gave me eight suits, three skirts, five aprons and one headdress. I will wear my mother's clothes and then I will be able to find her in the afterlife" (Morgan and Culhane-Pera 1993). The ritual exchange of traditional hemp clothing between mother and daughter ensures their family link into eternity.

The following is a composite account of Hmong funeral ceremonies, emphasizing the uses of *Cannabis* in nonpsychoactive ritual roles. When a person dies a bier (funeral stretcher) of bamboo poles is lashed together with strips of hemp bark. When an older person dies the children and grandchildren wash the body, dress it in special burial clothing made and kept for that purpose, and lay it on the stretcher facing up (Chindarsi 1976). The body then lies in state on this "spirit-horse" suspended between ceiling and floor so that it hangs horizontally against the back wall of the large central room (Johnson 1985), tied to the rafters with one length of hemp cloth and one length of hemp rope: "Other kinds of rope you cannot use, only hemp. If the sons and daughter-in-laws are hard working then they will have hemp to suspend the dead and shroud the body. If they do not have hemp cloth then the sister will scold them as being very lazy and they may have to pay a fine or obtain hemp cloth from relatives. That is why we still grow hemp" (quoted by Morgan and Culhane-Pera 1993). The Hmong corpse is dressed in strips of hemp cloth wound around each leg and the body is entirely covered with a shroud stitched from hemp cloth strips, leaving the head exposed. The feet are dressed with rough unprocessed hemp bark sandals (Beng 1975). Hemp sandals hold special importance because they are used to walk through the "land of the giant furry caterpillars" on the way to the other world (Chindarsi 1976; Lewis and Lewis 1984; also see later in this chapter).

Each wife sews burial clothing from hemp cloth for her husband and herself, which may differ from that ordinarily worn in daily life and will be worn at death along with dowry skirts and funerary clothing made by their daughters. In addition to trousers for the man and skirts for the woman, there may be three or more upper garments of varying lengths placed on top of the body, each richly embroidered, the outer one being the most elaborate (Chindarsi 1976; Lewis and Lewis 1984). Long, embroidered robes commonly associated with burial attire may also be worn at weddings by the bride and several of her attendants. In both instances, the robe symbolizes a prosperous beginning to a new life and similar ones are frequently worn by elders on special occasions (Catlin et al. 1986). Note that use of hemp fiber and cloth in funeral clothing worn by the deceased and/or mourners is also a common, very long-lasting Confucian tradition among the majority Han Chinese (see Chapter 5 and later in this chapter).

Three balls of hemp yarn are included in the coffin (Chindarsi 1976). The corpse is ritually fed and a chant directing the soul to contact elders in the afterlife is recited (Graham 1978). Hemp cloth and cordage are natural materials that are important in Hmong funerals as they will decompose by the time the dead person arrives in the spirit world. The use of more lasting materials would prevent the soul's ready departure from the body (Catlin et al. 1986). If hemp is not used at a funeral, Hmong believe that the dead person's bones will ache and vengeful spirits will make their children sick, and if Chinese market cloth is worn, then Chinese spirits will steal the dead's clothes and they will have nothing to wear in the afterlife (Morgan and Culhane-Pera 1993). Hemp's organic nature is a recurring theme in its ritual use across Eurasia.

During the most important portion of the Hmong funeral, the shaman recites a long "Opening the Road" or "Showing the Way" chant to the soul of the deceased, describing the route it must take to meet its ancestors in the spirit world. According to Falk (1996), "Death for the Hmong, as for most other non-Han tribal people in southern China, is thought of as a journey—perhaps the penultimate migration—to the sources of life and ultimate rebirth" (also see Falk 2004). Ethnologists have documented several versions of this orally transmitted ritual text in China (Clarke 1911; Graham 1926, 1978; Gu 1995), Thailand (Bernatzik 1970; Beng 1975; Chindarsi 1976; Schworer-Kohl 1984; Cubbs 1986; Tapp 1989; Symonds 1991), Laos (Lemoine 1986; Johnson 1985), and Australia (Falk 1996); these versions are all remarkably similar, especially in their descriptions of the ritual roles of hemp. The following quotations are from different chant versions that pertain to the protective significance of hemp buried with the deceased:

> Now I will lead you to [a] land in heaven to the dancing altar. You must pass in front of [the dragon's] face. By the side of the road he opens his mouth wide. There is also [the tiger] who is lying beside the road. These two want to block the road. At that time you can take a handful of fire hemp [a term often representing medicinal *Cannabis*] and tear it to pieces and cover the dragon's lips. You also take some large hemp [a term usually reserved for fiber hemp] and cover up the tiger's mouth, and after that you can go by unharmed. . . .
>   You also take shoes made of big hemp and put them on, then you go along stepping on the insects, and step on them as you go until you have trodden them to death. (GRAHAM 1978)

> Enter the valley of stinging caterpillars.
> As large as sheep.
> Don't be afraid.
> Just put on your hemp shoes.
> You can pass through.
> Go to the rocky mountain of fierce dragons and tigers.
> They threaten your path.
> Don't be afraid.
> Throw your hemp balls into their mouths.
> Now you can go to meet your grandmother and
>   grandfather.
> (MORGAN AND CULHANE-PERA 1993)

> I shall now show you the way to heaven. On the road there are many creeping things, so you must wear a pair of hemp sandals lest they bite your feet. When you get halfway up, you will see tigers with their mouths wide open waiting to devour you. Carry some hemp on your back, and when a tiger attempts to bite you, let him bite the hemp, and make your escape. When you are halfway up, the sun shines with a burning heat; take this piece of [hemp] calico [cloth] and cover your eyes—you will find it in your breast pocket. (CLARKE 1911)

As we shall see throughout this chapter, protection from various evil spirits is a recurring expectation encountered throughout Eurasian hempen rituals. Hemp cloth also fills a ritual role during funerary divination. Chindarsi (1976) described a Hmong ritual foretelling the future sex of the deceased upon reincarnation but did not identify the type of fabric. Before burying the corpse, a ritual practitioner burned a handkerchief used to clean the dead person's face on the first day after death. If, while the handkerchief is burned they find some black square lines on it resembling the pattern of a skirt, they believe the deceased will be reborn as a woman, but if the entire handkerchief is scorched evenly, they believe the deceased will be reborn as a man. Clarke (1995, personal observation) verified Chindarsi's observations when visiting Hmong living in a remote region of Bao Shan prefecture, Yunnan province, where elderly respondents insisted that the cloth used to wipe the deceased's face and later burned for divination must be made of hemp, as this had always been their funerary tradition.

Indeed hemp is the accepted fiber for use during every aspect of a Hmong funeral, and even today Hmong ritual traditions employ hemp in life cycle ceremonies ranging from birth, through marriage and ill health, to the funeral, and finally into the afterlife. However, common household usage of hemp may at times mask its ritual significance, and it is the ethnographic record that gives us an idea of its significance in Hmong rituals. Reverence for hemp is reflected in its widespread ritual use and high ceremonial status, a strong manifestation of the vital role it plays in the life and cultural identity of the Hmong. Ritual use of hemp and its associations with funeral ceremonies may reflect shared Eurasian traditions with other shamanic belief systems. Although the Hmong did not traditionally practice hempen funerary mourning in ways similar to their Confucian Chinese neighbors, other parallel uses of hemp in funerary and life cycle ceremonies are found in mainstream ritual traditions of China, Korea, and Japan as well as among Eastern European Slavic peoples (e.g., see Mueggler 1998). These regional hemp uses are discussed later.

China: Shamanism, Taoism, and Confucianism

She leaves twisting her hemp, and dances to it through the marketplace. (from the *Shi Jing*, Zhou dynasty, 1122 to 256 BCE)

As we have seen in Chapters 4 through 8, present-day Chinese continue a very ancient and deep relationship with *Cannabis*, having used it for millennia as a source of fiber, paper, food, and medicine. Chinese culture as we know it today began with the Han dynasties more than 2,000 years ago, and by this time diverse ethnic groups occupying East Asia already had extensive contact with and knowledge of *Cannabis*. Early shamanic rituals in China were included in the magico-religious practices of Taoism and some were eventually incorporated into Confucian and Buddhist ritual traditions still extant today. Hemp is used in a few healing rituals, in contexts similar to Hmong shaman healers, but the most important and enduring ritual use of hemp in China is as a vital part of traditional Confucian mourning. This said, the ancient hempen ritual traditions that were once much more widely practiced are at present more frequently observed among many Chinese minority ethnic groups, such as the Hmong, or in the Confucian funerary traditions of Korea.

In ancient China, as elsewhere across Eurasia, healing was attributed to the ceremonial actions and powers of a shaman or other ritual practitioner. In the following account, a relative of the patient performs a healing ritual, most often with the direct participation of a shaman. William Emboden (1972, citing Doolittle 1966) explains,

> When the [*Cannabis*] plant grows under favorable conditions, the main stem becomes highly lignified or woody. It was the practice of the early Chinese to carve this wood into the likeness of a serpent coiled around a rod (not unlike the *caduceus* or Staff of Aesculapius, which had its origin in the Greco-Roman world and is still with us as the traditional symbol of the physician). This image was used in curing rituals; a relative of the patient beat on the sickbed with the snake rod in order to dispel evil spirits.

The association between shamanism and the ritual use of nonpsychoactive *Cannabis* is well established and readily apparent in Hmong as well as Korean and Japanese ritual settings, although apparently less so in Chinese culture. Certainly the Chinese, and likely the Koreans and Japanese, have long had the knowledge that *Cannabis* can be psychoactive. Yet the paucity of references to the ritual use of psychoactive *Cannabis* preparations in China begs the question, did the Chinese use the psychoactive powers of *Cannabis* for ritual and religious purposes? Mia Touw (1981) had the following insight:

> The hallucinogenic use of *Cannabis* seems to have been associated with indigenous Central Asian, shamanistic practices. These were not, nor were they meant to be, shared by the majority of people or openly mentioned in the ancient books. During the Han Dynasty shamanism steadily declined, becoming disreputable as well, and with it, no doubt, the practice of using *Cannabis* as a hallucinogen. By the time Westerners came in contact with the Chinese a millennium and a half later, its psychotropic applications had been entirely forgotten. And since shamanism was rarely spoken of in the old texts, the impression necessarily arose that *Cannabis* was only a fiber and food plant in China.

The ritual use of nonpsychoactive *Cannabis* from prehistory up to the present day can be seen as simply a substitute for psychoactive *Cannabis* originally used in shamanic ritual or as the vestigial associations of nonpsychoactive aspects of hemp with ritual. However, extensive ritual use of hemp by the Hmong, for example, who have no history of using *Cannabis* psychoactively, seems to indicate that the values of *Cannabis* hemp for fiber, food, and medicine (apart from its psychoactive potential) have also imbued this plant and its products with ritual powers.

Confucianism has been the strongest determinant of Chinese ritual practice since its inception and rise to dominance as a social-moralist force during late Han times. Kong Fuzi or Confucius (lived 551 to 479 BCE) was a great thinker and social philosopher who emphasized the importance of funeral ritual as a key element of social cohesion. A series of books beginning with the "Confucian Classics" set forth rules for funerals and especially for the mourning period (see Chapter 4). As Watson (1988) explains, "If anything is central to the creation and maintenance of a unified Chinese culture, it is the standardization of ritual. To be Chinese is to understand, and accept the view, that there is a correct way to perform rites associated with the life cycle, the most important being weddings and funerals. By following accepted ritual routines ordinary citizens participated in the process of cultural unification. . . . What we accept today as 'Chinese' is in large part the product of a centuries-long process of ritual standardization." The most accessible ritual reference book for those of little schooling was the *Chia Li*, or "Rituals for Family Life," attributed to Chu Hsi, a great philosopher of the Song dynasty (960 to 1279 CE), whose popularity led to the *Chia Li* genre of ritual manuals and almanacs. These often included simple drawings and explanations of hempen mourning clothing and ritual paraphernalia, defined the grades of mourning, and described mourning ritual. They are still used today. Based on these manuals of ritual, a thriving business developed to provide the various accoutrements required for Chinese citizens to follow the Confucian edicts, including their hempen mourning costumes (Naquin 1988).

Prior to the spread and popularity of Confucianism, the Chinese spent an inordinate amount of time and resources burying their wealthy citizenry. For example, during the Warring States Period (480 to 221 BCE), the wealthy built huge opulent tombs and held lavish funeral ceremonies. Cristie (1983) quotes from "The Book of Guanzi," a collection of philosophical treatises on statecraft compiled by Liu Xiang in 26 CE but originally attributed to Guan Zhong, who lived during the Spring and Autumn Period (770 to 476 BCE): "Lengthen the period of mourning so as to occupy people's time, and elaborate the funeral so as to spend their money . . . To have large pits for burial is to provide work for poor people; to have magnificent tombs is to provide work for artisans. To have inner and outer coffins is to encourage carpenters, and to have many pieces for the enshrouding is to encourage seamstresses." The Chinese funerary ritual sequence was divided into the burial rites themselves, the rites of the corpse disposal, and finally funerary mourning. These memorial rites began at the moment of death and continued until the encoffined corpse was carried from the home, while rites of disposal included burial or cremation and took on several different regional forms (Watson 1988). Confucian hempen mourning traditions were encouraged throughout the Chinese realm and spread into Korea as well.

We now present a composite synthesis of traditional Han Chinese funeral rituals gleaned from the personal observations of numerous scholars and translations of Chinese accounts ranging from the Song dynasty through the late nineteenth century (e.g., Doolittle 1966; Gray 1878; Wieger 1981; Bogan 1928; Burkhardt 1953; Comber 1963; Lévesque 1969; Tan 1984; Naquin 1988; Rawski 1988; Watson 1988; Ebrey 1991; Lip 1993; Barley 1995; Stafford 1995; Yang 1998). However, it should be remembered that although the number of hempen rituals that took place in any single Chinese funeral varied, all the following ritual aspects of funerary hemp use did appear in one region or another.

When a family member dies their body is ritually bathed by the eldest son or spouse of the deceased (Naquin 1988) and covered with an already prepared hempen shroud. That evening, the sons don their rough hemp mourning clothes and proceed to the village shrine where they report the death to the God of Earth, the spiritual guardian of the village (Yang 1998), and the family mourning lanterns are hung on either side of the main door. These lanterns are traditionally wrapped with one or more strips of hempen mourning cloth, depending on the number of generations of descendants surviving the deceased. The order of the mourning materials has always been the same. The first and highest strip is of the roughest hemp cloth, the second of less rough hemp cloth, and the third of white hemp linen or cotton cloth. This is followed in turn by strips or blue, red, and yellow cotton muslin (Wolf 1970).

Bathing of the corpse is an essential feature of Chinese funerary rites and varies from a full, vigorous scrubbing to a ritualized daubing of the forehead (Watson 1988). Attired in hempen sackcloth, the next of kin hastens to the nearest river or well to collect "River Dragon King" water used to ritually wash the face and body of the deceased (Gray 1878). The water is sometimes "purchased" with a monetary contribution to the deity of a well or stream (Watson 1988).

Ritual bathing is followed by a final clothing of the corpse (Burkhardt 1953). Those of means are dressed in a full set of special "longevity" clothing, deemed to be suitable for the journey to the otherworld and, in anticipation of death, have often been secretly prepared by relatives when a person became elderly or sick (Doolittle 1966). The dead were dressed in their best clothes because Chinese traditionally believed that the spirit would forever appear as the deceased did on his or her day of death (Lip 1993). Longevity clothing could comprise one or more items depending on the deceased's class and could include official robes and gowns of fine cotton or silk or might simply consist of new plain cotton or hemp clothing, soft canvas shoes (Tan 1984), and a black silk hat (Naquin 1988). Outer clothing traditionally consisted of a long coat, usually of indigo blue hemp or silk, depending on the financial status of the family (Bogan 1928). No buttons were used and the garments were fastened with ties of the same fabric (Burkhardt 1953) while the trousers were not tied at the ankles (Wieger 1981). At times the longevity clothing was sewn together on the corpse (Tan 1984). Funeral clothing was often worn in layers, and it was proscribed that two more layers of clothing must be worn on the upper body than the lower (Doolittle 1966). The use of new or special clothing was optional and may not have been common among the poor (Rawski 1988). It was the peasant class, therefore, who were most likely to have been buried in hemp as it was the fabric of their everyday clothing.

The corpse was laid out in a straight posture with hands at the sides (Bogan 1928) and then clothed and furnished with socks and shoes before the feet were bound (Burkhardt 1953) and ankles tied together (Bogan 1928). After the corpse was dressed it was tightly bound with several rolls of hemp cloth torn into strips. After each cloth strip was wound around the corpse it was ritually tied into a special knot considered to be a good omen. Once the body was covered with these auspicious knots (Doolittle 1966) the face was covered with a piece of cloth or paper (Naquin 1988). A screen made of hemp sackcloth was placed before the coffin in a corner of the room (Doolittle 1966), and mourners with uncombed hair, unwashed faces, and wearing no ornaments bowed deeply to the corpse as they wailed, "embracing it, praising the deceased, and lamenting their loss." Encoffining was followed by a ritual involving different quasi-magical procedures for returning the soul back home (Naquin 1988). To mark the end of the funeral ceremony, before the procession began, the chief mourner entered the courtyard with a bowl of water containing medicinal herbs and tossed the contents onto the roof of the house (Stafford 1995). After the funeral ceremonies the family donned mourning clothing: "This is a very formal and important affair. The sons put on garments made of hemp cloth, of the natural color, over their clothing. The grandsons put on garments made of hemp cloth, but of a yellowish tinge. Sons, daughters, and grandchildren, according to strict rules, have braided in their cues threads of hemp, blue or white cotton" (Doolittle 1966). As soon as a death was announced, women neighbors came of their own accord to sew the coffin lining, pillow case, and mourning costumes. Before the mourning garments were put on, they were taken outside and presented to the deities for blessing. Close family members, dressed in hempen garments, also wore a rough straw crown. Sons and daughters carried a mourning staff on which they leaned as a sign of sorrow (Lévesque 1969). Unrelated male friends of the family wore white coats, but of finer hemp material, and each was given a white hemp girdle, "for superstition claimed that the greater the number of girdles worn, in memory of the deceased, the more of his sins would be forgiven." According to ancient traditions, these girdles also enabled friends to assist the sons in pulling the coffin carriage (Bogan 1928).

Outside the house, where the corpse laid at rest, a gathering of friends and relatives formed the funeral procession, along with the officiating priest, his assistants, and possibly a troupe of musicians (Comber 1963). Peasants were simply carried to their graves on a stretcher-like funeral bier. Average citizens were carried to the grave on a cart with white hemp cloth attached to the front by which it was pulled (Tan 1984; Naquin 1988). The fancy wooden coffins of the wealthy were conveyed to their tombs upon an elaborate hearse drawn by pallbearers and accompanied by an entourage several kilometers long, "including dozens of monks and priests, several bands, plaques and colorful spirit-world objects carried by hired men and newly garbed professional mourners" (Naquin 1988). Each funeral procession gave a chance for an affluent family to reveal its wealth to the public, and no expense was spared in making it abundantly clear that the deceased was "popular, prominent, and influential" (Comber 1963). Sometimes large hemp ropes were newly twisted to pull the hearse and coffin in the procession (Ebrey 1991). The hearse was trimmed with crisscrossed strips of white hemp cloth tied in knots in several places, which served as banners of mourning (Doolittle 1966), and pairs of long hemp scrolls with inscriptions eulogizing the dead were hung from the straw awning of the funeral vehicle (Bogan 1928).

Male descendents of the deceased were dressed in hempen mourning clothing and walked immediately in front of the bier or hearse with "the eldest son being held under the arms by two of the other mourners to sustain him in his grief" (Burkhardt 1953). Meanwhile, others carried the hemp ropes feigning that they were pulling the cart (Comber 1963), while remaining mourners wept and wailed as they walked slowly behind the hearse (Doolittle 1966). Pallbearers also wore straw sandals and a white hemp sash and headgear (Lévesque 1969). As Wolf (1970) describes it,

> Seen from a distance, from the top of a building or one of the hills on which most graves are sited, the procession following a Chinese coffin is a colorful site. The mourners wear long robe-like gowns, some of rough dirty-brown sack-cloth, others of gray flax or grass cloth, and still others of unbleached white linen or muslin; scattered among these are blue gowns, red gowns, and, on the rare occasion, a yellow gown. Female mourners cover their heads with a hood that almost hides the face and hangs down the back to the waist; men wear a hempen "helmet" over a short hood or one of two kinds of baglike hats of unbleached or dyed muslin. . . . Sons wear a gown made of *mua po,* a rough, very coarse hempen material ordinarily used in making sacks. . . . *Mua po* hemp sackcloth is rougher and coarser than hemp *tea po,* which is [also] too crude for daily wear. The one exception is the eldest son's eldest son. In the procession to the grave this senior grandson rides in a chair hung in front with a length

of *tea po* covered by a length of *mua po*. The boy himself wears a gown of *tea po* covered by a second gown of *mua po*.

Fine hemp cloth was placed under the coffin to lower it into the grave. This cloth was not pulled out, and any surplus was cut off and discarded (Ebrey 1991). Immediately after the coffin was lowered, the sons of the deceased scattered soil placed in the laps of their hempen mourning garments on top of the coffin (Doolittle 1966).

As we have seen earlier, at various times hemp has played many differing roles in Chinese funeral ritual, but these are only "bit parts" compared to the seminal role hemp played in Confucian funeral mourning practices, which ordained that people in bereavement were to wear rough hemp fabric out of respect for the dead, a custom followed into modern times by millions of Asian people (see also Chapter 5). The specific family relationship to the deceased determined which quality of hemp yarn was used to weave cloth for morning clothing, emphasizing differing degrees of coarseness, and the respective time periods for which it must be worn (Kansu Museum 1972; Li 1975). Filial piety (to care for and respect one's parents), as explained by Confucius, was the natural repayment for the care bestowed by parents and was well articulated as a core Chinese family value. Three years was deemed the appropriate length of mourning because it was the interval in which a child did not leave the arms of its parents (Rawski 1988).

Confucian etiquette and morality were very influential during the formation of Chinese culture. The *Li Chi,* or "Record of Rites," of the second century BCE is considered to be one of Confucius's "Five Classics." It is an etiquette guide containing several revealing references about ritual *Cannabis* use in ancient China. For example, during three years of mourning rites for his parents, when the king went to transact any business, the ropes of his chariot were made of hemp rather than the usual silk (Legge 1885a). When a parent passed away, all members of the household mourned. Sons wore long coats of extremely coarse, undyed hemp material buttoned at the collar but tied with cloth tapes at the side, and the bottom coat hem was frayed to express deep grief. A hemp cloth girdle was worn around the waist and knotted in the front with long hanging ends, which were worn over the left shoulder if the father had died, or worn over the right shoulder if the mother had passed away. This custom originated in the days when the hearse was drawn to the burial ground by the sons, who used their hemp girdles as harnesses to pull it (Bogan 1928). A piece of the hempen corpse shroud was thought to have regenerative powers for women and young children (Naquin 1988; also see parallel Eastern European funerary traditions later in this chapter). After the mourning period had ended, the daughter-in-laws of the deceased absorbed fertility from the dead by converting mourning clothes into baby carriers (Barley 1995), while in other regions mourners would bring their garments to the home of the bereaved to have them burned (Ebrey 1991).

Funeral ceremonies took place either on the day of the removal of the coffin from the home or the day before, and during these rituals hempen mourning clothes had to be worn. Each of the five grades of mourning had its own clothing and accoutrements so that the "status of every mourner was immediately apparent" (Naquin 1988). The mourning period, marked by the wearing of hempen hair ties, began as soon as the corpse was laid out to be ritually bathed and clothed. Immediately following death, "hemp hair binding, hair bands, and hemp hair ties are prepared. Next an oblation [offering] for the dressing of the body is made. Afterward the preliminary laying out is performed. The presiding male and female mourners embrace the body, wail, and beat their breasts. Then they bare their shoulders and put on the hemp hair binding, hair bands, and hemp hair ties in another chamber" (Ebrey 1991). Hair binding was performed by using a hemp cord to pull the hair up into a topknot or by using torn strips of hemp cloth that stretched forward from behind the neck and were tied across the crest of the forehead and then wrapped around the hair and tied, forming a crude topknot (Ebrey 1991).

According to the *Li Chi*, the basic garb worn during the mourning rites of the deceased usually consisted of a hemp sackcloth robe and a headband made of bark strips from the male hemp plant. The *Li Chi* is very specific concerning the identity of *dà má* (大麻) or true *Cannabis* hemp in mourning rituals; it refers to five levels or degrees of mourning dress (*wu fu*) woven of hemp fiber, some of which were made solely of fibers from female plants, while others were woven only using fibers from male plants. Grieving garments were worn by specific people according to their familial relationship to the deceased (Kühn 1987; also see Legge 1885b and Chapter 5), and each involved a specified set of ritual clothing and behavior. Wolf (1970) proposed that the coarsest hemp cloth mourning costumes were worn by the sons and other close relatives as an outward symbol that they were entitled to an inheritance.

First degree or "Unhemmed Sackcloth" mourning lasted up to three years, during which time costumes of the coarsest, naturally colored, unhemmed cloth woven from female hemp plants were worn (Legge 1967) along with a headband of hemp bark strips peeled from old woody (and thus ill-favored) plants, a girdle of rough twisted hemp cord (Steele 1917), straw or hemp bark sandals, and a bamboo mourning staff. First degree mourning clothing was worn by a man for his parents, by a wife for her husband or husband's parents, or by a concubine for her master (Legge 1967): "The wearing of hemp in this unfinished state symbolized the great manifestation of poverty by the son, and represented the traditional idea of giving all one's possessions to the deceased to ensure him a comfortable life in the next world" (Garrett 1987).

Second degree or "Hemmed Sackcloth" mourning for grandparents and great-grandparents lasted one year, and other familial relationships dictated its wearing for a shorter duration. The hemmed mourning costume was sewn from undyed, coarse hemp cloth woven from male plants (Legge 1967) and was worn with a hemp headdress and shoes and also included a bamboo mourning staff (Garrett 1987). The robe, skirt, and hat were the same as those worn for "Unhemmed Sackcloth" mourning, but they were sewn from the next finer grade of coarse, unprocessed hemp cloth with hems along the sides and bottom edge. The second degree mourning hat had a hemp cloth brim and strings, and the headband was made of a very rough "rope" of twisted female hemp plants, with the roots toward the right and the ends wrapped around the head and tied to the roots with hemp cloth strips. The waistband was made of hemp rope and the belt of hemp cloth (Ebrey 1991). The headdress worn with the hemmed sackcloth mourning costume for a deceased father was made from old female plants, because they have an unpleasant appearance

FIGURE 51. Chinese men in "untrimmed" (unhemmed, left) sackcloth and "even" (hemmed, right) sackcloth mourning garments, originally from the clothing section of the *San-ts'ai t'u-hui*, or "Assembled Pictures of the Three Realms" (adapted from Wolf 1970).

and served to show outwardly the internal distress of the mourner, while hemmed sackcloth mourning worn for a deceased mother was fashioned from more desirable male hemp plants (Legge 1967).

Third degree or "Greater Processed Cloth" mourning was usually for brothers and sisters. It consisted of a robe of less coarse hemp or cotton cloth and was worn for nine months (Legge 1967; Garrett 1987). Fourth degree or "Lesser Processed Cloth" mourning for aunts and uncles lasted for five months and consisted of a robe of medium-coarse hemp or cotton cloth. Fifth degree or "Fine Hemp" mourning was for distant relatives, lasted for only three months, and was made of very fine processed hemp linen.

Thus the *Li Chi* specified that people in mourning were to wear undyed hemp fabric out of deep respect for their dead. The specific family relationship to the deceased determined the quality of yarn derived from either male or female hemp plants; this emphasized the differing degrees of coarseness of the clothing worn by the mourners, and the respective bereavement costumes were also worn for varying time periods depending on the relationship to the deceased. The more distant the relative, the less severe the loss and the finer (and more comfortable) the cloth. This custom is followed today by millions of East Asians for whom coarse white cloth woven of hemp or other substitute fibers is still held as a sign of mourning.

It is interesting to note that Benet (1975, citing Klein 1908) referred to "the religious requirement" among ancient Hebrews in the Near East "that the dead be buried in *kaneh* [*Cannabis* hemp] shirts," and it was not until much later that linen (flax) was "substituted for hemp." According to De Groot (1972), the generally prevailing opinion of many Western authors that white is the mourning color of China is completely untrue. More recent white mourning represents the original natural color of hemp and other undyed textiles, and the bleached white color was excluded from traditional deep hempen mourning and was allowed only for lesser mourning. Rough hemp cloth was to the Chinese what "sackcloth" was to the Old Testament Jews: "it is the essence of mourning, expressing sorrow and a complete abnegation of personal comfort" (Wolf 1970).

With the rise of the Qing dynasty in the middle of the seventeenth century, mourning practice was transformed in China. In 1644, the Manchu invaders of northeast China (also known as Manchuria) replaced the ethnic Han Chinese Ming dynasty and expanded from this region to eventually control China proper and its surrounding territories; the Empire of the Great Qing was the last dynasty of China, ruling from 1644 to 1912. During the early Qing, many poems and other written works bemoaned the existing state of funeral observances, expressing a yearning for a return to a more orthodox Confucian tradition of mourning the deceased. In the following section of the first of his 20 "miscellaneous poems" concerning mourning, Zhu Yizun (lived 1629 to 1709) laments the lax adherence to Confucian mourning rites, including the rejection of hemp clothing (Kutcher 1999):

In Luoyang there were substantial customs,
A scholar's conduct was molded in the imperial academy
He invariably gave priority to his parents' personal care,
And thereby made the people know a model to imitate.
At that time they stressed funeral rites,
Close relatives observed each other's mourning periods.
At the correct times they left office,
When in mourning they followed their Confucian
    obligations.
What about the people of today?
Their mothers die and they do not observe mourning.
They reject sackcloth and hempen clothes,
Their faces blush with wine and meat.

One of the foremost scholars of the Qing period, Yan Yuan (lived 1635 to 1704), developed a new path of scholarship that was inspired by his devout Confucianism and experiences during mourning observances. These feelings and experiences were articulated in his "Humble Observations during a Period of Mourning" (*Ju you yu jian*, in *Xizhai jiyu* 10, 1664). Among Yan Yuan's concerns about proper mourning observances involved the fabric of which bereavement clothing was made; Yan referred to mourners of the first and second degree using hemp to make their appropriate gowns and bean-fiber linen (*Pueraria thunbergiana*) to make their belts because these fabrics were cheap and coarse—even though, in fact, these same fabrics were expensive during the time of Yan Yuan because of the rise of the cultivation of inexpensive and widely used cotton: "Perhaps it was because the raw cloth of the ancients was different from our own. Hemp and bean-fiber linen were probably things that a family could easily obtain. Now though they are poor people's luxury clothing. A poor family can rarely obtain them" (Yan Yuan in *Ju you yu jian* 3b, translated by Kutcher 1999). Indeed, the frugal emperor Kangxi (lived 1654 to 1722), who started as a very young supreme ruler, ascending to the throne at the age of 7 and was effectively in control at the age of 15, became an emperor who manifested a minimum of patience for what he believed was unnecessary extravagances such as those involved in the expensive funeral rites of preceding Ming dynasty rulers. For example, Kangxi, who believed that raw silk was the traditional fabric for mourning gowns, declared in an edict to the Board of Rites that his own "mourning clothes should be made of cotton," even though "the traditionally prescribed fiber for three-year mourning was hemp, as it was for mourning worn for a paternal grandmother" (Kutcher 1999).

Regardless of changing fabric availability and attitudes about appropriate bereavement garb, hemp fabric for use in mourning wear has continued. In his description of traditional Chinese mourning rites, Garrett (1987) tells us that by the nineteenth century, sons in mourning of their parents had to take a leave of absence from their public service jobs in order to perform rites only "every seventh day after the death for 49 days," and after the third performance "on the 21st day the hemp [costume] was exchanged for somber clothes." Garrett also tells us that after some time passed "the period of conspicuous mourning became much shorter and there were only two grades of mourning clothes—undyed coarse hemp for close relatives, and undyed fine hemp for others." Relatives of the deceased wore hoods or headbands made of rough hemp sackcloth (Stafford 1995). The width of the headband differed in accordance with the degree of relationship; near relatives wore broad bands, and remote relatives wore narrow ones. Sons of the deceased donned a full hemp cloth cap, and women tied bands around their heads (Wieger 1981). Only sons wore rough hemp during mourning, and all other mourners wore another cloth, such as coarse cotton muslin (Wolf 1970). As the hempen mourning period became shorter and shorter over time, eventually funeral ceremonies were abbreviated, coarse mourning garments were discarded, and the wearing of white clothes alone was considered sufficient to express grief (Comber 1963). By the middle of the twentieth century, traditional Chinese mourning attire and the length of the mourning period had changed considerably; certain elements of traditional ritual were maintained while some modern elements of dress were incorporated. Over time, the ritual role of rough and undyed, natural-colored sackcloth characteristic of hempen mourning has slowly been replaced by any plain weave white cloth, and present-day mourners usually wear white cotton clothing (or even dark-colored business clothing) occasionally decorated with a hemp waist sash, armband, or a small swatch of hemp cloth pinned to their clothing. Therefore, in recent times dress codes for mourning family members were still followed, but in a minimalized ritual display. As Yang (1998) explained, "The sons wear hempen cloth garments and hats when lamenting or conducting funeral ceremonies, but plain white in their daily work. Other members wear garments of coarse white [cotton] calico. . . . Grandchildren wear white jackets and hats at the funeral but are not required to do so at other times. Nephews and nieces wear white only at the ceremonies. A brother wears a white robe and a piece of white cloth bound around his waist, but only in the funeral procession." Although present-day Chinese mourning rituals have become simplified, they still reflect ancient Confucian traditions. The closest relatives, children and their spouses, may wear black sackcloth (Lip 1993) to the funeral or simply attach a small square piece of coarse hemp cloth to their lapel for a week to signify mourning. The recent adoption of black as the mourning color is the result of Western influence.

According to ancient Chinese traditions, hemp seed is also empowered with protective powers, especially for children; such beliefs are also common in Eastern European Slavic folk traditions. The Chinese Lantern Festival or Small New Year is held on the fifteenth day of the first month and is also the birthday of the bed goddess, a deity who watches over children. On that night the streets are crowded with people carrying lanterns as families prepare sacrifices and worship the bed goddess and other deities, making offerings of glutinous rice mixed with hemp seed oil to give thanks for protecting their children (Wei and Coutanceau 1976). It should also be noted here that Doolittle (1966), an American missionary working in southern China in the nineteenth century, referred to consumption of a confection of hemp seed mixed with "molasses" as part of commonly performed ceremony sometimes called "worshiping the measure." This ceremony was held around the middle of the eighth month of the year for the "benefit of children of the family sick or well, *the object being to secure to them longevity and plenty of money*" (italics from the original Doolitle text).

Given the ancient history of *Cannabis* in China, and the active present-day ritual uses of hemp in neighboring Korea and Japan, it is likely that historians will uncover many more traditional ritual uses for hemp. As we describe in the following section, ritual uses in the cultural traditions of Korea and Japan are more widespread today and better documented than they are in China. In fact, to learn more of how hemp may have been used in ancient Chinese ritual it is instructive to investigate neighboring cultures that continue to preserve these traditions.

Korea: Shamanic Funerary Rites, Confucian Mourning, and Ancestor Worship

Ancient shamanic traditions are practiced today throughout Korea much as they have been for millennia. These include the worship of house and village gods as well as the healing and funerary rituals of mansin traditions associated with female shamans. The basic roles of Korean shamans include contacting deities, channeling the spirits of historical personages, or combating evil spirits through mystical ritual techniques. A common impetus to becoming a shaman is a neurological condition such as epilepsy or stroke—a cultural trait shared with Hmong shamans—while others become hereditary shamans. There are many commonalities, as well as lesser regional variations in shamanic rituals practiced on the Korean Peninsula.

Confucian and Buddhist ritual traditions spread extensively in Korea during the ninth century CE and are still widely practiced today, much as they were for centuries in neighboring China. Following the adoption of Chinese moralist doctrines, the ritual activities at Korean funerals came to be largely dictated by the rigid step-by-step procedures of Confucian texts emphasizing ancestor worship and filial piety. Confucian-based funerary rituals continue to be practiced in Korea today and involve hemp in the mourning process in much the same way as in traditional Chinese rites, although they often retain strong links with shamanism, even when performed within a Christian setting. In many cases ritual traditions were better preserved during the recent past in Korea than in China, and when we investigate Korean funerary and mourning rituals it becomes readily apparent that ritual hemp use was more prevalent in the past throughout eastern Asia.

In Korea, shortly after death the corpse is prepared for encoffining and burial. This process involves bathing the corpse in perfumed water, straightening the position of the head, binding the thumbs together, straightening and binding the legs, combing the hair, trimming the fingernails and toenails, and dressing the corpse in special clothes made of silk or fine hemp cloth (Cho 1997). In the past, funeral clothes were prepared for the elders by their daughters and granddaughters, but present-day Koreans more commonly purchase burial clothing from funeral parlors following their parent's death.

Traditional burial clothing is made from handwoven, natural-colored hemp cloth. During the Joseon dynasty (1392 to 1910), great care was taken in preparing burial clothes as death was not considered the end of life, rather its continuation in the next world (Cho 1997). Hemp corpse clothes (*sui*) are only widely used today by well-to-do Koreans who wish to honor the deceased with a "traditional" burial. A full set of *sui* costs at least US$2,000 and can be considerably more expensive. Because of the high price of hemp cloth, less costly silk *sui* are much more popular than hempen *sui*. Excavations of tombs indicate that the former choice of the wealthy was silk, and it was much more likely that commoners were buried in hempen *sui* or simply their own hempen peasant clothing. It is somewhat difficult to understand the present-day popularity of hempen *sui* as there appears to be no ancient archeological context for hemp use as auspicious burial attire. Possibly an honorific display of wealth in the form of what now is expensive burial clothing is considered an auspicious and respectful gesture.

Hemp *sui* for men include inner and outer jackets and trousers, a long white topcoat, and a headpiece. Women's *sui* include inner and outer jackets, bloomers, a skirt, and a topcoat. The remainder of the outfit for both men and women includes wrappings for the face, coverings for the hands and feet, and a small pouch in which fingernail and toenail clippings of the deceased are placed. A cloth to cover the abdomen, two quilts, a sleeping mat, and a pillow are also essential elements for encoffining the corpse. All these loose-fitting hemp clothes have string ties for closures since they are believed to connect the present world with the next (Cho 1997).

Burial costumes in Korea may also combine different fabrics, striking a balance between affordability and people's perception of propriety and tradition. The inner and outer jackets and trousers may be made of silk, while a simple vest may be sewn from rough hemp cloth, normally used for mourning clothes of the closest relatives. Cotton cloth has now replaced hemp in most Korean ceremonies subsequent to the establishment in 1973 of the Family Rites Standards that dictate proper observance of traditional rituals (Dredge 1987) while discouraging the wearing of costly hempen mourning clothing (Kim 1973).

Traditional Korean funerary preparations followed a common format. After the corpse was washed and dressed, it was bound to a board with a long cloth shroud woven of coarse hemp (Bergman 1938). There were two stages to this binding; the first was called "small binding" and the second "large binding" (Kyu 1984). Small binding was performed by laying a long hemp cloth lengthwise under the body and cutting six pairs of opposing slits along each selvage so that the seven resulting strips could be twisted together and carefully knotted over the corpse. Relatives and friends of the deceased slipped folded paper money into the cloth to be used by the deceased in the afterlife. Following the initial binding, the corpse was bound to the "Deity of the Seven Stars" board with a longer hemp cloth wrapped around it (Kyu 1984; Lee 1984) and tied from head to toe seven more times as in the small binding. The seven stars and the seven knots represent Ursa Major, or the Big Bear, the pointer constellation for Polaris or the North Star.

After the corpse was bound, it was placed in the coffin on top of a piece of fine hemp cloth termed the "coverlet of earth," which symbolized a sleeping mat (Dredge 1987) and was then covered with a hemp shroud referred to as the "coverlet of heaven" (Lee 1984). In ancient times, only the finest hemp cloth was used to make funerary shrouds (Hahm 1988). The remaining space within the coffin was filled with old clothes, paper, or straw (Lee 1984; Cho 1997), and the coffin lid was secured with wooden nails while grieving mourners wailed loudly. The coffin was bound with ropes around its upper, middle, and lower parts (Lee 1984), ropes or lengths of cloth were passed under the coffin, and it was lifted and carried slowly, head first, toward the processional bier (Dredge 1987).

Descriptions of traditional Korean funeral processions give us a clear idea of the extent and variety of this ritual application for hemp. Here we provide a composite description of a traditional Korean funerary procession synthesized from various sources and highlighting the ritual uses of hemp.

The traditional Korean funeral procession was led by a funeral bier upon which the coffin was transported to the cemetery. Immediately behind the funeral bier followed the chief mourner (usually the eldest son of the deceased), who wore rough natural-colored hemp clothing and either was carried seated in a "chair" covered with coarse hemp cloth (Hulbert 1969) as at a royal funeral or more commonly walked holding himself up with a staff while crying and wailing (Clark 2000). Mourners from the family, many wearing conical caps of rough hemp cloth, followed behind in decreasing order of relatedness according to the Confucian system (Bergman 1938). The position of the deceased in the traditional social order was proclaimed on a long, narrow hemp cloth banner hung from a long pole at the head of the procession. A second "merit banner" of undyed hemp was also tied to a long pole and carried aloft in the procession; this banner was used at the burial to clean the top of the coffin, becoming one of its coverings in the grave (Dredge 1987). In some villages the coffin was lifted from the bier and lowered into the grave, and in others, the corpse, still tightly bound from head to toe in hemp cloth, was removed from the coffin and placed in a narrow trench specially dug in the bottom of the grave (Clark 2000). On October 8, 1895, the Queen of Korea, who was given the posthumous title of empress, was assassinated, and her funeral took place on November 21 and 22, 1897. According to a firsthand account by Horace Allen (1908), "There were two biers supported on the shoulders of scores of bearers while great ropes extended from the front and back and were manned by hundreds of men, who like the bearers were dressed in [hemp] sackcloth."

Koreans often practice rituals that combine elements of shamanism with Buddhism. The *Chinogi kut* is an elaborate rite of good fortune for the family of the deceased that is performed at funerals by a woman shaman or *mansin*. In its entirety, the *Chinogi kut* consists of a lengthy series of rituals (*kori*) carried out over several days and nights (Lee 1973, 1981). The *Chinogi kut* begins outside the house gate at dusk, where a costumed *mansin* summons the gods and ancestor spirits into the home (Kendall 1985), followed by nine rites performed by the *mansin* on the wooden floor in the middle of the house (Kyu 1984), where rolls of hemp cloth are placed on the memorial table (Yi 1988). In the person of the possessed shaman, the household gods appear throughout the dwelling to be feasted and entertained. Ancestor spirits also arrive, mourn with their living kin, and depart (Kendall 1985). By the following morning, the *kut* shifts from the house into the inner courtyard where eight additional rites for the dead are held (Kyu 1984), with the final *kori* performed back outside the gate (Kendall 1985).

During the second day (and night) of the funerary rites, the memorial offerings are taken outside where the "Gate of Paradise" is erected (Lee 1981). The *mansin* shamaness disappears from view, dresses herself in a yellow hempen mourning robe, ties a cap of rough hemp cloth onto her head with rice straw rope, and winds more rope around her waist. Within her rope belt, the *mansin* tucks a dried fish that is wrapped in a length of hemp cloth or white paper signifying the abode of the departed's spirit. The hemp cloth robe, head covering, rope belt, and wooden staff all represent traditional mourners' apparel. Now begins the final séance with the *mansin* reappearing as the terrifying Saja, or Death Messenger (Yi 1988; Kyu 1984): "The Death Messenger appears at the deathbed to snatch the soul away to judgment in the courts of Hell. In Korean shaman rituals for the dead, the Death Messenger appears in the person of a possessed shaman and stalks the guarded house door, prowling for a new victim" (Kendall 1985).

Once everything essential is cleansed and purified by the *mansin*'s actions, a ritual theme shared by many Eurasian hemp cultures, the *mansin* rings a brass bell and worships the four corners of the universe. Assuming the role of the Abandoned Princess Spirit, the *mansin* leads the deceased's spirit through the dangerous passage to the other world in a dance before the flower-covered "Thornwood Gate of Hell" (Kister 1997). Family members tightly stretch two long cloth bridges at chest height, one made of a length of coarse hemp fabric, known as the "Road out of Hell," and a second made from a length of finer cotton fabric, known as the "Road into Paradise" (Kendall 1985). The hemp cloth symbolizes the unclean bridge that opens the way to the "Ten Kings of Hell," while the cotton symbolizes the clean bridge that opens the way to Buddha or paradise, the "Bridge of the Buddha" (Yi 1988). It is believed that the soul of the dead is released from the spirit of hell and is sent on its way to the "Gate of Paradise" of the Buddha (Lee 1981). During the rite of division of the cloth bridges (known as "Cutting the Gates with Swords"), the *mansin* shamaness (with the fish "soul" bound to her waist in hemp cloth) chants to Buddha and thrusts her body violently, pushing through the cloth bridges with her breast while wielding her ceremonial knives to symbolically remove hindrances on the way; this splits the two cloths in half as the errant soul is led out of hell (Kister 1997; also see Kyu 1984; Covell 1986). The passing of the deceased to the otherworld and the physical separation of the deceased from the world of the living are both dramatized by the cutting and tearing of the cloth bridges (Kyu 1984). When the funerary ceremony ends the cloth bridge is burned before the front gate of the house (Kendall 1985; Yi 1988).

The tearing of hemp cloth is also performed at the healing ceremony called *Kilgallajugi*, immediately following a rite for ancestral deities. If it is determined that the ancestral deities are the cause of illness, the shaman (or patient) runs rapidly forward while quickly tearing a long sheet of yellow hemp cloth into two pieces with his or her own body. The spiritual intent of this process is to "separate the dead ancestor from the living persons without any trouble," emphasizing separation of body and soul as in the *Chinogi kut* (Kim 1973).

A bolt of hemp cloth was provided to each family member when a death occurred in their family so that the women could begin making mourning clothes to be worn on the fourth day after the death (Dredge 1987). Traditionally, the eldest son's bereavement costume consisted of a wide bamboo "mourner" hat, a headband, coat and leggings of coarse hemp cloth, a hemp waist-cord, straw shoes, and a face screen of hemp cloth attached between two sticks held vertically in the hands. Mourning attire for women consisted of wooden hairpins, clothes of coarse hemp, and straw shoes (Hulbert 1969). An early twentieth century Western description of Korean funerary mourning is provided by Allen (1908):

> Mourning it may be inferred, is a serious matter in this erstwhile hermit kingdom. For a parent the family goes into mourning for a period of three years, during which time the sons don sackcloth and a rope girdle, sleep and live in a straw hut erected near their own comfortable dwelling, and indulge in the plainest of food with no luxuries, wine and music, with their accompaniments, being particularly prohibited. Should the mourner be obliged to venture out after a time, he must go clad in unbleached white raiment wearing a hat of fine woven material of about the size of an umbrella, while for greater privacy he wears a shield of sackcloth before his face.

Korean mourning rituals mirror Chinese Confucian traditions from which they are derived. Korean ritual manuals and etiquette books, modeled upon early Joseon dynasty statutes and classical Chinese texts, specified in great detail which relatives should wear certain items of the mourning costume and for what length of time: "Throughout the variations in theory and practice, only one principle remains inviolate; the coarseness of the cloth or the number of items used must correlate with kinship distance" (Janelli and Janelli 1982). We know from historical accounts and museum collections that hemp was the most common fiber in Korean mourning costumes. Although increasingly rare in present-day Korea, hempen mourning traditions are in fact better preserved and more commonly practiced than in China. Korean funeral rituals are described in the *Zhu Zi Jia Li*, or "Book of Family Rites," written by Zhu Xi (lived 1130 to 1200 CE), the Chinese Southern Song dynasty Neo-Confucian scholar, which stipulated 108 different kinds of mourning clothes for the main family, close relatives, and maternal relatives. The customs of a long mourning period date back to the Three Kingdoms Era (57 BCE to 668 CE), but bereavement time was shortened to one hundred days during the Goryeo dynasty (918 to 1392). Later, during the middle of the Joseon dynasty (1392 to 1910), the Confucian code of conduct was forced upon the people by royal command, which decreed that all relatives were expected to wear mourning dress every day. The mourning rites for the deceased father were the most austere of all. The eldest son, clad in coarse hemp sackcloth, fasted, leapt, and wailed for the first three days after the death of his father. Following the burial rites, he was required to wear mourning clothing for 27 months, starving himself and living in a crude mourning hut near the grave. For all but the eldest son, a ceremony on the second anniversary of the death officially ended the mourning period, and only then were the family members allowed to stop wearing their hemp mourning clothes (Cho 1997; Clark 2000). The complexity of the system can be seen in the following table of the traditional grades of Korean mourning, which directly mirror the grades of mourning stipulated by Confucian Chinese texts.

Many present-day Koreans continue to practice Confucian hempen mourning traditions much as their ancestors did before them, albeit in a somewhat abbreviated form. During the period immediately following the funeral, close family members may still wear traditional full mourning costumes made of coarse and loose-woven hemp cloth sewn into loose pantaloons, jackets, and peaked hats (Clark 2000). However,

FIGURE 52. Early twentieth-century Korean man in traditional hempen mourning attire, including of a hemp cloth robe and hat with twisted hemp fiber head and waist bands and a peeled hemp stalk walking stick (from Bergman 1938).

now it is more common to see mourners dressed in Western-style dark suits with a rough hemp or ramie (*Boehmeria nivea*) cloth armband or headband. Men and women often pin bows of hemp cloth to their clothes, which serve as symbols of respect for their traditional Confucian Korean hempen mourning traditions.

Cannabis also plays a role in Korean seasonal rituals, which are usually held in the spring and fall. These rituals are focused on family, especially the women (Lee 1973). One example is the "Dragon Palace Festival" of January 14, during which worshippers offer all kinds of fabrics (e.g., hemp, silk, and cotton) dedicated to the spirits on the altar and pray for blessings and long lives for their children (Lee 1973, 1981). The flying of flags or banners is another ritualistic tradition in Korea. The "Flag of Primary Spirits," unfurled in January to prevent storms and floods, is sewn of hemp cloth. It measures about 180 centimeters long and 120 centimeters wide (six feet by four feet) and bears the image of a dragon spirit. Another example of the ritual use of hemp in Korean flags involves the "Commanding Flag," approximately 10 meters (more than 30 feet) high, which is set up in front of the home for the ascending spirit at ceremonies. It is composed of a large paper sheet covered with a spread of "different colored fabrics and papers, ropes, hemp, rice, coins and other objects" and hemp cloth strips are tied at its top (Lee 1981).

Fiber hemp production is celebrated in other traditional Korean customs such as the *Hangawi* or *Chusok* (Autumn Night) festival that takes place on the fifteenth day of the eighth lunar month. *Chusok* was the most important calendrical celebration in the agrarian countryside largely because fruits and grains ripen at this time: "This is the fairest of the

TABLE 13

In traditional Confucian Korea, as in China, mourning grades for male mourners were determined by the mourner's relationship to the deceased and defined by the required mourning garment and duration of mourning.

Grade	Duration	Mourning garment	Person mourned
Ch'amch'oe samnyon	3 years	Coarse raveled sackcloth with staff	Father
Chaech'oe samnyon	3 years	Coarse hemmed sackcloth with staff	Mother (if father deceased)
Chaech'oe changgi	1 year	Less coarse hemmed sackcloth with staff	Mother (if father still alive) or wife
Chaech'oe kinyon	1 year	Less coarse hemmed sackcloth without staff	Brother, unmarried sister, or grandparents
Chaech'oe sogong	5 months	Less coarse hemmed sackcloth without staff	Great-grandparents
Chaech'oesima	3 months	Less coarse hemmed sackcloth without staff	Great-great-grandparents
Taegong	9 months	Coarse cloth	Patrilineal first cousins
Sogong	5 months	Fine cloth	Patrilineal second cousins or maternal grandparents
Sima	3 months	Fine hemp cloth	Patrilineal third cousins or wife's parents

NOTE: Adapted from Deuchler 1992.

twelve full moons. Families visit the hills for a day of outdoor sports. Wine, cakes, and fresh fruits are enjoyed and offered at family shrines. Graves are visited and contests of wrestling and hemp-spinning are held" (Ha 1958). Today *Chusok* is still one of Korea's major holidays and is loosely regarded as the Korean equivalent of American Thanksgiving Day, although it is more ancient. According to the *Samguk Sagi*, or "Chronicles of the Three Kingdoms," a Korean study of early Silla Kingdom (57 BCE to 935 CE) relations with China compiled by a team of scholars led by Kim Bu Sik and completed in 1145 CE, a king named You Ree selected women from six local towns surrounding his capital, separated them into two groups, and appointed one of his daughters as a leader for each group. Then the King requested that the women make clothing to see who was better at processing and weaving hemp. The two princesses organized their teams and worked diligently to impress their father and prove themselves the better weavers. They also agreed that the losing side would prepare and serve dinner and drinks to the winning side. Following the contest, they enjoyed a sumptuous banquet and various entertaining pastimes. From this ancient fable stems the *Chusok* festival tradition of spinning contests when women amused themselves under the harvest moon (Ha 1968; also see Chapter 5).

The tedious and strenuous work of hemp processing and spinning was traditionally shared among the women in a communal system called *dure sam*, or "processing hemp," which was also recorded in the *Samguk Sagi* chronicles referred to previously. By the fifteenth day of the seventh moon, all the hemp fiber fields had been harvested and the dry stalks were ready to be processed in the farmers' homes. On this night the *dure sam* began as the village women gathered and worked until they finished processing all of that household's hemp. The next evening they convened at another house and on the third moved to another home, and so on, until they completed the tedious processing for the entire village. In a convivial atmosphere of communal spirit each woman performed her tasks with the same careful attention as if working in her own home (Choe 1983). During this arduous work the hemp spinners also found time for lighthearted competitions and harvest celebrations. The harvesting and processing

of hemp followed an annual agricultural cycle, which led to calendrical rites continuing to this day.

In sum, contemporary Korean ritual hemp use stems from a combination of Confucian funerary traditions originating in China and indigenous agricultural rites and shamanic traditions, a fusion of which is still practiced today. Traditional Buddhist ritual uses of *Cannabis* are rare in Korea, although hemp seed offerings were sometimes placed inside Korean temple statues. Eastward in Japan, Confucian traditions have had less influence and ritual hemp use appears primarily within native shamanic Shintō traditions and Buddhist ceremonies.

Japan: Shamanist, Shintō, and Buddhist Hemp Traditions

Shintō, or "the way of the *kami* (gods or spirits)," is the indigenous religion of Japan and has its roots in ancient shamanic ritual traditions. The term "Shintō" was coined in the sixth century CE by combining two Chinese characters *shin* (in Chinese *shan*: unfathomable spiritual power) and *do* or *to* (in Chinese *tao*: way, path or teaching) in order to differentiate the loosely organized native shamanic religious tradition from Buddhism, which was then being introduced into Japan via China (Kitagawa 1987). Shintō is a set of ancient shamanistic beliefs rather than an organized religion and represents the Japanese conception of the cosmos, combining the worship of nature and of their own ancestors (Lowell 1895). Traditionally, Shintō's ritual focus was concerned with purity, fertility, and divination related to the growth and harvesting of crops (Cuyler 1979), and the Shintō priests dressed traditionally in hemp robes. The practice of Shintō is limited to Japanese culture—there are no ultimate gods or sacred scriptures, neither founders nor missionaries. Shintō has become a cultural tradition where followers worship *kami* (spirits or deities residing in every aspect of nature) and celebrates the sacredness of the universe including mountains, rivers, and the natural phenomenon of heaven and earth. Shintōism is not just a religious faith but a state of mind that has shaped the consciousness of the Japanese people.

FIGURE 53. Twenty-first-century Korean mourners dressed in hempen mourning costumes returning from funerary rituals at the grave of a close relative.

FIGURE 55. Hemp *shimenawa* ropes are stretched across the *torii* entrance gates at Shintō shrines in Japan to purify and protect those who pass beneath them as at Ouji Jinja in Omachi village, Nagano prefecture.

FIGURE 54. In 1649, monks sealed an offering of hemp and rice seeds inside the base of a gilded wooden Amitābha Buddha sculpture at Bulguksa temple in Gyeongju province, South Korea. The seeds along with cloth offerings were recovered during restoration of the statue. The *Cannabis* seeds are particularly large, which suggests they came from a broad-leaf hemp (BLH) variety selected for food or oil seed use.

The Ise Jingu, or "Grand Shrine of Ise," in Mie prefecture along the southwestern coast of Honshu Island is the main shrine dedicated to Amaterasu, the sun goddess, celestial weaver, and ultimate ancestor of the Japanese people. According to the *Kojiki*, or "Records of Ancient Matters" (620 CE), after creating Japan, the "primal pair" consulted each other: "We have now produced the great eight-island country, with the mountains, rivers, herbs and trees. Why should we not produce someone who shall be lord of the universe?" (Moore 1913). The pair then begot the founding goddess-figure, Amaterasu Omi, the "Sun Goddess," who is enshrined at Ise. The Emperors of Japan claim to be her direct descendants, and thus Ise is also the main shrine of the Japanese Imperial Family with the emperor himself acting as the high priest of Shintō belief. During the *daijosai*, or "Great Offering Ritual," which is a special version of the annual *shinjosai*, or "First Fruits of the Rice Harvest" ceremony, the new emperor asserts

his authority by personally undertaking the Shintō rituals usually carried out by shrine priests and performs an *oharae*, or "Great Purification" ritual at Ise, to purify the entire nation. The *daijosai oharae* is performed by each new emperor only once near the beginning of his reign and has become associated with his enthronement rites. The *daijosai* has remained the most important Japanese ceremony since it was first performed by the fortieth sovereign emperor Temmu in 672 CE for the worship of Amaterasu, the heavenly deities, and the terrestrial spirits. The rituals of the enthronement ceremonies are explained in detail in the *Engishiki*, or "Codes of Engi," of 927 CE. During the eighth month, Imperial messengers carried symbolic hemp offerings (*onusa*) weighing about half a ton, first to Ise Jingu and then to the official provincial and local shrines of important deities as an announcement of the *daijosai* rituals to be performed at Ise (Bock 1990a). Whenever the "august sovereign" visited shrines he presented this same amount of hemp (*asa*) to the residing *kami* (spirits). When he went to the river for personal purification he always carried a bundle of hemp among other ritual objects, and hemp fiber or cloth (natural color or dyed) was present in a ritual role in the majority of ceremonies surrounding the emperor's accession rites. Furthermore, it was designated that tribute rice associated with the rituals referred to earlier must be distributed in hempen sacks (Bock 1990b).

Hemp cloth has therefore figured importantly in ceremonial offerings associated with the Japanese Imperial Family. The Imperial household traditionally directed the Inbe clan, who were responsible for reciting ritual prayers and supplying *aratae* the sacred hemp tribute cloth, to grow a special hemp crop and prepare sufficient cloth for the accession ceremonies (Bock 1990b). The *Kogoshui*, or "Gleanings from Ancient Stories," written by Hironari of the Inbe clan in 807 CE, claimed that the Inbe's responsibility for providing sacred cloth began with the goddess Ama no Hiwashi, and the clan was charged with the duty of producing cloth from both mulberry and hemp. As fertile ground on which to cultivate these plants, Inbe clan members were given the province of Awa (now Tokushima prefecture) on Shikoku Island (Cort 1989). These same ritual procedures have been followed from the seventh century through to the accession of Emperor Akihito in 1990.

The *Engishiki* prescribed specific numbers of skeins of fiber or pieces of cloth as Shintō ritual offerings. Silk was at this time, more than a thousand years ago, somewhat of a novelty, and therefore, with religious belief being generally conservative, it occupied a less conspicuous place in ritual settings. It was hemp and other bast fibers along with textiles woven from them that were commonly presented as traditional offerings. Plant fiber textiles were more appropriate offerings to be sent to distant shrines than perishable articles such as food, and since cloth was a currency of the day, it made a convenient substitute for rare or inconvenient articles.

Japanese offering cloth was designated by a pair of contrasting terms, *aratae* (rough cloth), a thick sackcloth, versus *nigitae* (smooth cloth), a finely woven fabric made soft and lustrous through fulling—that is, prolonged beating with a wooden mallet. According to the *Kogoshui*, *nigitae* consisted of two types—the inner bark fibers of *kozo*, or paper mulberry (*Broussonetia papyrifera*), and *kaji*, or white mulberry (*Morus alba*), and blue-green *asa* hemp cloth, which were used specifically for sacred garments and rituals. Originally, both mulberry and hemp *nigitae* offerings were made of unwoven strands of fiber, although later they came to be made of cloth or of paper made from the same fibers. The coupled fiber or cloth *aratae* with *nigitae* offerings were an indispensable element in the Shintō altar constructed on any site where a sacred ritual act was performed. A coarse mat was spread over the clean-swept ground and a low table placed upon it. A branch of a glossy-leaved evergreen *sakaki* tree (*Cleyera japonica*) was stood upright on the table, enclosed by a rope tied to posts at the four corners to denote the ritual space and finally draped with cloth *aratae* and *nigitae* offerings (Cort 1989).

The paired mulberry and hemp fiber offerings appear on the archetypal altar described in one of the most important Shintō myths concerning Amaterasu, the sun goddess. In this traditional story, after the unruly male deity Susanoo desecrated the work of his sister, Amaterasu, which included planting the rice fields, performing the harvest festival, and weaving the hempen garments for all the gods, the angry goddess fled to a cave and plunged the world into darkness. The efforts of the other deities to placate her wrath began with setting up a great altar using a five-hundred-branched evergreen *sakaki* tree ornamented with a necklace of sacred jewels, a sacred mirror, and special white mulberry and blue-green hemp cloth offerings (Tsunoda et al. 1958).

Responsibility for producing ritual textile offerings was shared at differing times. According to Cort (1989), "By the time of the accession ceremony held in 927 AD [sic], the members of the ancient Inbe clan had already lost the right to present *asa* [hemp]. . . . Eventually the Inbe clan lost even its right to offer 'rough cloth' of mulberry fiber; the last presentation of *aratae* from the Inbe of Awa is recorded for the *Daijosai* ceremony of 1339. Thereafter all offertory cloth came from the 'Nakatomi (by then known as the Fujiwara) domain.'" By the twentieth century responsibility for producing hemp cloth for the Imperial Family had reverted to the Inbe clan. The Miki family of the Inbe clan of Tokushima prefecture produced hemp *aratae* offering cloth for the accession ceremonies of both Emperor Hirohito in 1928 and the present-day Emperor Akihito in 1990. These offering cloths were extremely fine and of the highest possible quality. However, at least one other group of growers at a Shintō shrine in Gunma prefecture, in concert with a small group of *Mingei* (craft movement) traditional weavers in Nara prefecture, also produce comparable quality hemp cloth for other, possibly secular usage within the Imperial household.

Many of the *daijosai* rituals are held in absolute secrecy within the inner sanctum reserved solely for members of the Japanese Imperial Family and attendants. However, we do know that one key element of the ritual is presentation of "soft and coarse" (silk and hemp respectively) cloth offerings to the deities and spirits (Bock 1990a). These offerings are presented in a special shrine at Ise that is opened for invited guests. The altar within this shrine is flanked on either side with a small *sakaki* tree, each draped with a bundle of long golden hemp fiber strips. Although the Imperial *oharae* purification ritual is also performed in secrecy, we assume that the Imperial rituals and ritual paraphernalia at Ise are reflected in the everyday purification rites of other Shintō shrines.

Apart from the grand *daijosai* ceremony, one of the main rituals at Ise is still called *taima*, the Japanese word for both *Cannabis* hemp and Shintō paper offerings. This ritual is held five times a year (Hiroe 1973) to honor Amaterasu, who is also the goddess of both hemp and rice. These two plants are sacred elements of the rites conducted at Ise, reflections of their ancient roles as the staple products of the original, long-lasting hunting and gathering Jōmon culture and the later rice farming Yayoi culture (Yamada 1995; also see Crawford 2011). Thousands of *taima* amulets tied with a small strip of processed hemp bark are distributed each year at Ise Jingu (Kokugakuin University 1997).

Several examples of hempen ritual paraphernalia may be found at the majority of Shintō shrines. Upon approaching a shrine, one first encounters the purifying *shimenawa* rope across the *torii* entrance arch with several jagged, folded white paper streamers (*shime*) suspended from it. After passing beneath the *shimenawa* and crossing the courtyard to enter the shrine, one immediately sees the thick and tapered *suzunawa* bell rope fashioned from hemp fiber and centrally placed before the altar area. Upon the altar are *gohei* of folded golden paper and a *haraenusa* wand of many white paper *shime*. Both are used in ritual purification and embellished with thin golden-colored hemp bark strips. Small tassels of dyed red and black combed hemp fiber adorn the surrounding bamboo shades, and at major ceremonies worshippers are attended by young women dressed in the same colors, each with a hemp bark strip tied in her hair (Clarke 2006, personal observation). Later we will investigate the religious significance of each of these ritual items.

Central to Shintō tradition is the concept of purity, and the primary means of purification is ritual practice. Since the inception of Japanese culture, hemp fibers have been used by Shintō priests because of their association with purity. Bunce (1960) described the common occurrence of ritual objects associated with hemp, which are still seen today: "Around the precincts could be found straw ropes [*nawa*] from which were suspended small strips of paper [*shime*]. These marked off the sacred spot and were to protect the objects of worship from evil influences. And inside the shrines there were small wands or short sticks with hemp or paper strips inserted at one end [*haraenusa*]. These were symbolic offerings and occasionally

FIGURE 56. Paper *shime* shaped like lightning bolts are tied to *shimenawa* ropes with narrow hemp fiber strips. Japanese hemp fiber has become rare and expensive, so it is often used sparingly to preserve ancient Shintō tradition; photo taken at Yasaka Jinja in Kyoto.

FIGURE 57. Worshippers pull the hemp fiber *suzunawa* bell rope at Shintō shrines and wish for blessings of good fortune; photo taken at Yasaka Jinja in Kyoto.

were regarded as symbols of divinity." During ritual purification, the priest commonly waves and shakes a *haraenusa* (often incorrectly called a *gohei* by Western authors) over the person or thing to be purified (Aston 1905). According to Shintō tradition, evil or impure spirits do not come to places where purity reigns, and so waving the *haraenusa* drives away evil and invites purity (Joya 1951). The simplest traditional *haraenusa* is a strand of undyed hemp fiber attached to the end of a short stick. The modern *haraenusa* is a meter (about three feet) long wooden pole to which are attached many paper *shime* strips hanging down in a rustling bundle from one end and adorned with a single strand of shining hemp bark. Shintō *haraenusa* wands resemble the tasseled horsetail wands of Chinese Buddhist priests and the shaved wooden ritual wands (*inau*) of the indigenous Ainu of Japan's Hokkaido Island (see Hirochika 1988 for suggested resemblance of a *shime* to the *inau* of the Ainu; the *inau* ceremonial prayer wand is among the most sacred symbols in Ainu religion and serves as a messenger and sacrifice to the gods and spirits). Waving the *haraenusa* over the head of a Shintōist supplicant is one of the most common ritual actions in Shintō ceremonies. The *shime* attract the *kami* with their gentle sweeping sound. The *haraenusa* is believed to shake because the *kami* spirit descends into it; the quivering of the *haraenusa* is thought to be the spirit passing into the body of the supplicant. The *haraenusa* therefore serves as a sort of lightning rod to conduct the divine spirit into the human one, and whether the possession takes place in a Shintō or Buddhist rite, the *haraenusa* is placed in the hands of the supplicant to be possessed at the time when the spirit is invited to descend (Lowell 1895). As in Hmong and Korean ceremonies, hemp serves as a pathway guiding spirits to and from the supplicant or patient—acting as a plant of the gods (see Schultes et al. 2001).

As mentioned earlier, the paper *shime* common in Japan today evolved from hemp cloth strips hung on the sacred *sakaki* tree, and a *sakaki* branch decorated with paper *shime* is a ubiquitous feature in the contemporary household *kamidana* (literally "god shelf") shrine, which serves as a family altar. Immediately in front of the *kamidana* there is a small stand for offerings, and there may be candles to either side as well as memorial tablets for deceased relatives hanging just below and perhaps to one side of the altar. Traditionally, a hemp *shimenawa* rope with hemp fiber or paper strips attached to it is hung above the family shrine, but today it is most often twisted of rice straw rather than hemp bark strips. The Shintōist stands before the shrine and purifies himself "by waving before him over each shoulder either a small branch of the sacred *sakaki* tree or an imitation made of hemp and paper cuttings attached to a wand" (Bunce 1960).

Shimenawa also mark the boundary between the sacred ritual space within and the profane world without and purify the sacred shrine preventing entry of impure and malevolent *kami*. According to the *Kojiki*, a *shimenawa* was first used to prevent the sun goddess Amaterasu from reentering her cave and thereby saving the world from eternal night. Although the preponderance of *shimenawa* today are made of rice straw (or polyethylene), major Shintō shrines still have large *shimenawa* made of hemp fiber strips. Today, paper *shime* are usually bleached white and tied to the rice straw *shimenawa* with thin strips of hemp fiber. The continuing use of hemp to tie the *shime* is a remnant of earlier ritual use when hemp was much more common and less expensive and stands as testimony to hemp's continuing importance in Japanese religious traditions (Kuhaulua 1973).

The *gohei* is also a wooden wand but with a single matched pair of large white or gold-covered paper *shime* flattened like

FIGURE 58. The Shintō *haraenusa* wand of many white paper *shime* used in ritual purification is embellished with thin, golden-colored hemp bark strips and takes a prominent place on Shintō altars.

FIGURE 59. Traditional *noren* curtains are hung over the thresholds of homes and businesses to purify those who pass beneath and are commonly seen throughout Japan. Because Japanese hemp has become very expensive, modern *noren* are more likely woven from imported hemp, ramie, or flax.

wings to either side of the wand and stands upon the altar. Lowell (1895) provided an interesting history for the derivation of the paper *shime* used in modern *gohei*:

> They are called *shintai* (god's body), not because they are permanently god, but because they may become his embodiment at any moment. The little we know of the evolution of the *gohei* will help explain what is supposed to take place. Its name signifies cloth, *gohei* meaning august cloth or present; the former meaning having in course of time developed through a whole gamut of gifts in the concrete into the latter meaning in the abstract [e.g., *aratae* and *nigitae* offerings]. For the *gohei* is the direct descendant of the hempen cloth hung on the sacred *sakaki* tree as a present to the gods.

Buddhism arrived to Japan in the eighth century from China via Korea and became intertwined with Shintō and its reverence for hemp. According to Olson (1997), "It is in death that Shintō and Buddhism blend into a common braid. The relatives continue to visit the graves, leaving offerings and praying in the Buddhist way. Yet at home, a family shrine with the departed's picture and memorabilia is tended in the Shintō tradition with hand claps, incense, and worshipping of the *kami* within." The *Obon*, or "All-Souls Feast," is celebrated in July, and although it is considered by most to be a predominately Buddhist ceremony, many *Obon* rituals have grown from Shintō roots. During this time it is believed that the souls of the deceased are permitted to return to their families to be entertained by them. In one room of the house, bamboo canes are erected as a stage. Food and lanterns are placed for the spirits while a Buddhist priest recites ritual texts before them. On the first evening of *Obon* week, worshippers call their ancestor's spirits back to earth and send them away again to their afterlife on the last night. In this invitation for the ancestral spirits to return home, fires of hemp leaves were traditionally ignited before the entrance of the house and incense strewn on the embers (Moore 1913). Today, families normally burn small pieces of the hemp wood that remains after the commercially valuable fibrous bark is peeled off, which serves both to coax the ancestral spirits to earth and to send them back to the afterlife.

Hemp was also used for protection and in purification contexts during *Obon* in the traditional *chinowa* ring. The *Shintō Miomoku* (1699 CE) religious text tells us that the *chinowa* ring represents the round form of the universe and adds that the goal of the ceremony is to avert dangers connected with the change of summer influences to those of autumn. Hemp fiber crops are harvested in midsummer at the same time as the Obon ceremonies, and the traditional *chinowa* was made of hemp stalks bundled into a large ring tall enough to walk through without stooping. Another version of the *chinowa* had hemp plants tied across the top of a ring of reeds so the supplicants walked under hemp as an act of symbolic ritual purification (Aston 1905); in an early twentieth-century form of the ceremony, ritual purification was achieved by passing three times through a large ring made of reeds, holding in the hands hemp leaves, and repeating the verse,

> To the end that
> My impure thoughts
> May be annihilated,
> [With] these hemp leaves,
> Cutting with many a cut,
> I have performed purification.

Ritually purifying acts of Shintō priests waving hemp *haraenusa* over people's heads, or supplicants passing through a *chinowa* beneath hemp stalks, also carry over into Japanese secular life. It is no coincidence that the bast cloth curtains, known as *noren*, hanging above the entrances of numerous Japanese businesses and homes today still function as a means of purifying the entrants and protecting them from harmful forces. Originally *noren* were woven of traditionally appropriate and ritually potent *Cannabis* fibers; contemporary *noren* are usually made of more widely available and less costly ramie fabric.

Apparently hemp is not used in Japanese funerary rituals, which at first may seem puzzling, especially since *Cannabis* fibers play such an important role in the funerary rituals of neighboring China and Korea, as well as among the Hmong peoples of southwestern China and Southeast Asia. This anomaly can be explained by several factors. The funerary traditions of China and Korea are based in Confucian ethics, which had less influence in Japan, especially in terms of funeral rituals. Most importantly, Shintō worship associates death with ritual pollution and divests itself from ritual contact with the dead. Japanese funeral services are generally Buddhist rites and funerary traditions with hemp are largely absent in Buddhist funerary rituals in China and Korea as well.

Japanese people traditionally believed that the *sahe no kami*, a special group of supernatural spirits, could help protect them from evil demons, particularly from those that caused illnesses. *Sahe no kami* were symbolized by phalli, often of gigantic size, set up along highways and especially at crossroads to block any malignant spirits who sought to pass, thereby protecting wayfarers. Travelers prayed to them before setting out on a journey and would make offerings of hemp leaves and rice to each one they passed. These phalli had nothing to do with fertility or reproductive functions, and no peculiar rites were observed in their worship, which in no sense was considered conducive to immorality. By the early twentieth century they were largely removed from major roads "out of regard to the prejudices of Europeans who connected obscene notions with them" (Moore 1913), yet some remain today in out-of-the-way rural areas. Hemp is also used to protect and direct travelers. For example, *seki-mori ishi* are small rocks bound with hemp cordage to show their special significance and placed on stepping stones at forks in the path to show visitors which direction not to follow.

Traditionally in Japan, as well as elsewhere across Eurasia, life cycle events such as birth, coming of age, and marriage ceremonies often incorporated various aspects of hempen ritual, once again most commonly associated with purity and protection. One of Japan's annual rituals is *Shichigosan*, a cleansing rite of passage conducted for children of three, five, and seven years of age occurring around November 15. Held for families and individuals, rather than the community, this simple ritual may be replicated at any ceremony conducted by a Shintō priest. The priest gives the child a *tamagushi*—that is, a sprig of the *sakaki* tree tied with hemp bark strips and paper streamers. Instructed by the priest, who removes himself carefully to a position nearby, the child places the *tamagushi* on a low table before the altar (Ashkenazi 1993).

Japanese traditionally dressed in clothing made of hemp, and hemp fiber was also used to fashion household items such as bedding, matting, furnishing cloths, and netting (see Chapter 5). Later, when cotton and silk fabrics were introduced from China and Korea, hemp fiber continued to be used in life cycle ceremonies because of its association with traditional Japanese culture and Shintō ritual in particular. Originally, hemp and other fibers were spun and woven in their undyed natural color, and as dyed fabrics gained popularity, natural-colored hemp fiber and cloth came to be regarded as symbols of purity. Bundles of natural-colored *asa* hemp fibers were offered to shrines or presented as gifts on occasion of marriages or other joyous celebrations (Joya 1951).

In addition to its use as a symbol of purity at Shintō shrines and festivals, hemp also constituted an important ritual element for marriages. Hemp bark strands were hung on trees as charms to bind lovers, while gifts of hemp cloth were sent as engagement gifts by a man's family to the prospective bride's family as a sign that they were accepting the girl (Joya 1951). *Shiraga* is another name for hemp and also means "white hair." Gifts called *tomoshiraga* are exchanged to signify ties and cooperation in the married life and express wishes for the newly married couple to live a long life together. Hemp fiber strands were also prominently displayed during wedding ceremonies to symbolize the traditional obedience of Japanese wives to their husbands.

In the past, it was believed that the first duty of a Japanese woman was to obey her husband's wishes. Purity and obedience are the basic qualifications of Japanese women for marriage, and they are symbolized by hemp fiber. Hemp fiber can be dyed in any desired color, and thus it came to be regarded as the symbol of obedience. According to a traditional Japanese saying, wives must be willing to be "dyed in any color their husbands may choose" (Joya 1951). The use of hemp fiber in wedding ceremonies expresses hope that the bride will be faithful and will follow the customs of her new family. On the other hand, female shamans are still known from some regions of Japan, such as in Miyage prefecture in the northeast. It is here that we find extant connections between Shintō shaman induction and hemp fiber. For example, Kawamura (1994) reported the traditional use of hemp cords in the *kamitsuke* initiation ritual of divine possession that took place in 1945 for a blind woman 25 years of age.

Tying and binding are important ritual acts in Japan as they are across Eurasia and form key elements of Shintō ritual often utilizing hemp fiber strips and cordage. The Kokugakuin University Museum in Ise displays many ritual offerings presented at special ceremonies such as the enthronement of a new emperor. Many of these parcels are carefully tied in golden, straw-colored natural hemp bark strips.

Hemp is also used in Shintō *muneage* rites performed during the construction of a new home. According to Embree (1946), hemp fiber strips are nailed up on the ceiling along with fans and a dried fish. A packaged kit in the International Hemp Association collection in Amsterdam contains ritual objects used in a ceremony for the occupation of a new Japanese home. The kit contains two strips of individually bundled, processed hemp bark along with a "beard" of white hemp paper, which are hung from the banner standard in front of the home or nailed to the rafters inside the door. Once again these objects are associated with purification and protection, in this case, of the new home.

The *Ogi* Festival (also called *Tofu Matsuri*, or "Bean Curd Festival") is held each year in early February at the Kasuga shrine in Yamagata prefecture on Honshu Island. Three days before the festival, three ritual borders of hemp *shimenawa* are erected; in addition, day lilies (*Hemerocallis* ssp.), pine (*Pinus*

FIGURE 60. Display of Shinto ceremonial offering parcels tied with hemp bark strips.

ssp.) needles, seaweed (*Laminaria* ssp.), and leaves of the native, woody *yuzuri-ha* shrub (*Daphniphyllum macropodum*) are mounted on a 180-centimeter (6 foot) by 60-centimeter (2 foot) rectangular wooden framework. Along the middle *shime* string is placed a pole formed from a chestnut branch 90 centimeters (three feet) long with its bark stripped off, around which paper is wrapped at the top, middle, and bottom. In turn, three places on the middle *shime* string are tied with seven, five, and three hemp cords. The ends of these cords are aligned and cut off evenly; the remaining ends are then frayed like tassels, so that the branch has seven, five, and three ranks of paper offering streamers attached to it (Hirochika 1988; also see Hirochika 2003).

The main function of shrine priests on the occasion of the *Ogi* festival is to add paper *shime* and hemp cloth streamers to the *ogi-sama*, which are paired large ceremonial *gohei* used only during the *Ogi* festival. Each *ogi-sama* is constructed from three cedar staves about seven centimeters (three inches) in diameter and 2.4 meters (eight feet) long. At their tops, a large bundle of five white paper *shime* streamers is attached, and the three poles are joined together by five bands of white hemp cloth and interspaced hemp cords. When the top ends and attached *shime* are spread apart, the *ogi-sama* appears as a large fan-shaped object, thus indicating the association of *ogi* (fan) with its name. The upper and lower guilds of a typical village each possess their own *ogi-sama*, and at all times other than during the festival, the pair of *ogi-sama* are stored inside the shrine's sanctuary with the one belonging to the lower guild pointing vertically and the one of the upper guild laid horizontally (Hirochika 1988).

Early on the first morning of the *Ogi* festival, the *ogi-sama* are taken out of the shrine and new strips of undyed hemp cloth are affixed by the Shintō priests. Each year, 10 new *shime* are added to those already attached to the *ogi-sama*, and the bundle of *shime* on the top of the *ogi-sama* becomes larger year by year. However, every 20 years or so a *mikoromo-gae* or *koisai* (changing of clothes) festival is held and the old accumulation of *shime* are removed. The *ogi-sama* is considered complete when the hempen cloth is attached, as it is the hemp cloth that marks the transformation of the three staffs with attached paper *shime* into a ritual *ogi-sama* (Hirochika 1988). The first cloth wrapped around the *ogi-sama* has been dedicated for the following year, and removing the hemp cloths at the end of the ceremony returns the staffs to their secular state. After the hemp cloth is removed from the *ogi-sama*, it is wrapped around the body of the man chosen to be the next year's host for dressing the *ogi-sama* and is also used to sew his ritual garment (Hirochika 1988).

The *shishimai* is a dance performance (also seen at Chinese New Year celebrations and in Korean mask dances) in which people are dressed to look like lions. Oral traditions tell us that the *shishimai* was first brought to Japan in 612 CE by an immigrant from Korea who had studied the art in China. At first, a kind of masked dance was performed at ancient temples during religious rituals; later these performances spread among the people and took on the various forms seen today. Generally a *shishimai* is performed to celebrate the new spring and serves as a ceremony for exorcising evil influences. The body of the *shishimai* lion costume is made of indigo-dyed hemp with a white crane pattern (Haruo 1988), and the lion's mane is also sometimes formed from strands of hemp fiber.

Taimatsu matsuri, or "fire festivals," are held in several locations in Japan during October or November. *Taimatsu*, originally named after *tai ma* or hemp, are enormous (up to 6 meters, or 20 feet tall) columnar bundles of dried hemp stalks that are stood upright in a clearing and burned during protective rituals to ensure a good harvest. Although *taimatsu* are now more commonly made from rice straw, they were traditionally made from hemp stalks. Young men stood in a circle with their backs to the *taimatsu* and attempted to set fire to it by tossing flaming torches back over their shoulders trying to make them land in the dry leafy crown of the hemp stalk column. Eventually the *taimatsu* caught fire, and as it fell over sending embers into the air and was consumed by the fire, people prayed for a bountiful harvest (Maclagan 1977).

Divination rituals involving hemp are also performed in Shintō ritual contexts. During the *Kudagai* ceremony held every January 14 at the Itakiso shrine in Takayama town of Gifu prefecture, hemp is used to portend prophecies of the year ahead. Hemp stalks are cut into six centimeter (about two and a half inches) sections with diagonally sliced ends, and boiled in rice gruel along with soy and adzuki beans. After the gruel has cooked, four parishioners dressed in hemp *kimono* remove the stem sections and cut them open lengthwise, observing the number and condition of the beans that have made their way inside of the hollow stalks and predicting from these results events of the following year (Ampontan 2008).

Sumo has been Japan's traditional form of wrestling for well over a thousand years. Within the disciplines and rituals of *sumo* there are many manifestations of Shintō thought and worship. *Sumo* was originally practiced as a part of ancient Shintō fertility and divination rituals (Cuyler 1979); it was intended to entertain and appease the deities and ensure a bountiful harvest. Rituals that begin and end each modern-day match are derived from these traditions.

The *sumo* wrestler's basic garment (the *mawashi*) is constructed like a waistband with loincloths and worn for both training and official bouts. For official bouts, the *sagari*, a fringe of starched twisted cotton or silk threads, which is tucked into the front of the belt, is added. The *sagari* fringe symbolizes the hempen *suzunawa* bell ropes hung in front of Shintō shrines and consists of an odd number of strings (usually 11 to 21), as is considered lucky in Shintō custom. Once a wrestler reaches the highest level of grand champion, known as *yokozuna* (great rope), preparations are immediately begun for the making of his first ceremonial *tsuna*. The *tsuna* is the ritual rope belt of twisted hemp bark strips covered with white cotton cloth from which a *yokozuna* takes his name and also indicates his rank. Once again, hemp is chosen because

FIGURE 61. A *sumo* wrestler wearing the hemp fiber *tsuna* belt that distinguishes him as a top-ranking *yokozuna*, or "great rope."

of its association with purity. The use of hemp, the traditional fiber of commoners, elevates the *yokozuna* to the status of a national hero, the highest possible rank in his field, and is thus a symbolic rising of the low to the high. Traditionally, in the elaborate, prebout ceremony, the reigning *sumo* wrestling grand champion *yokozuna* carries a giant hemp rope around his ample girth to purify the ring and exorcise the evil spirits. This continues even today, as when Hawaiian-born champion Akebono wore his special *tsuna* belt, which was also made of hemp, during the ceremony celebrating attainment of the highest *sumo* rank in 1993.

The *tsuna* rope belt is assembled during a vigorous three-day-long ritual that takes place each time a clean one is needed for an important tournament. First, the training area is covered with clean white canvas; then all the wrestlers undergo a purification ritual wearing their cotton practice belt covered with a white towel and white cotton gloves to keep the *tsuna* clean as they work. Three lengths of white cotton cloth are laid on the canvas and golden, straw-colored, processed hemp bark strips, beaten in rice bran until soft, are spread on top of each with the bases overlapping toward the center; this leaves more hemp toward the middle of each strip and less at each end, and when twisted forms a tapered rope largest in the center. Copper wires are inserted in the belts to make stiff cores, and the cotton cloths are rolled up at an angle to form three long tapered bundles bulkier in the middle than at the ends. A smooth wooden pillar is covered with padding and the three strands of cotton-covered hemp fiber are looped around it. While chanting to the accompanying beat of a drum, two groups of wrestlers begin to pull the three bundles from the ends, twisting the strands together into a rope. Another wrestler, with a white cloth over his head to protect the rope from his hair oil, sits on the canvas in front of the pillar underneath the forming rope and

hangs his weight from it to ensure the even distribution of the hemp within the twisted coils (Cuyler 1979). The length of the finished *tsuna* varies based on the girth of the wrestler although through the years they have steadily grown in weight. For example, the *tsuna* of Inazuma, who achieved the *yokozuna* rank in 1830, weighed only 835 grams (almost two pounds), that of Tochinishiki, a star of the 1950s, weighed 5.6 kilograms (almost 13 pounds), and present-day *tsuna* now weigh about 15 kilograms (33 pounds) and are about four meters (14 feet) long (Schilling 1994).

The *tsuna* is tied around the *yokozuna*'s waist by several wrestlers and secured with an intricate knot. Five white zigzag-folded paper *shime* offerings to the deities are hung from it giving the appearance of the *shimenawa* rope seen at all public Shintō shrines. Following ritual purification at a shrine, the *yokozuna* dons his new *tsuna* and mounts the ring to perform the ring-entering ceremonies (Schilling 1994). Kneeling in front of seven *gohei* wands symbolizing the three gods of creation and the four seasons, the chief referee lifts his voice in the manner of a Shintō priest, "Everlasting life to heaven, long life to earth, and may the winds and rains be seasonable" (Cuyler 1979).

In sum, it is clear that Shintō worship is essential to the Japanese culture and hemp forms an integral part of Shintō rites and ceremonies. As in other Eurasian cultures, hemp's primary ritual roles are for protection and purification. These hempen roles are played out in many settings throughout domestic rites and life cycle ceremonies, in sports, and even in the emperor's enthronement. Now we turn our attention to similar Judeo-Christian ritual uses for hemp in Europe and the Middle East.

Europe and the Middle East: Judeo-Christian Hemp Rituals

Throughout much of Europe, especially in southeastern Europe and into parts of the Middle East, nonpsychoactive ritual uses of *Cannabis* seeds, fibers, roots, and whole plants, as well as cordage, cloth and clothing, hemp seed foods, and even the tools of hemp work, are quite common in agricultural, calendrical, and life cycle ceremonies. As we shall see, many expectations of ritual efficacy in Western Judeo-Christian religious contexts, and even the ritual acts themselves, have much in common with those of Eastern cultures and provide strong evidence for the commonality of shamanism at the roots of these diverse traditions. First we will investigate hempen rituals within the Judeo-Christian heartland of the eastern Mediterranean region.

The use of hemp in ancient rites and burials in areas other than East Asia has also been suggested. For example, anthropological and biblical scholar Sula Benet (also known as Sara Benetowa, see Benet 1975) interpreted the word "*kaneh*" to mean *Cannabis* hemp and referred to the "religious requirement" among the early ancient Hebrews in the Near East "that the dead be buried in *kaneh* shirts" (for a different interpretation, see Musselman 2007). In fact, Benet proposed that it was not until hundreds of years later that linen woven from flax was "substituted for hemp" in funeral rites (also see *Encyclopedia Judaica* 1971).

In the Bible, reference is made to one of the ingredients in holy anointing oil. In Hebrew it is *qan:a'bos* or *q'ana bos* (קנה בשם), noticeably similar to the sound of *Cannabis*, and Hebrew dictionaries define *q'ana bos* as hemp. As the Lord said to Moses, "Take the following fine spices: 500 shekels of liquid myrrh, half as much of fragrant cinnamon, 250 shekels of *qaneh-bosm* [*Cannabis*?], 500 shekels of cassia—all according to the sanctuary shekel—and a hind of olive oil. Make these into a sacred anointing oil" (*Exodus* 30: 22–33). So it follows that Jewish rabbis may have poured hemp seed oil (or another oil laced with *Cannabis* products) over the heads of Aaron and the high priests. However, even if the biblical anointing oils did contain psychoactive *Cannabis* compounds, they would have been diluted by other ingredients, and through relatively inefficient transdermal application, they were most likely not potent enough to produce a psychoactive effect (Ohr Somayach International 1998).

The *Talmud* is a collection of rabbinical writings that forms the basis of Orthodox Judaism; this holy book proscribes many taboos concerning combinations of sacred and profane entities. One of these taboos makes casual mention of hemp used in "wine vats" and "olive oil presses" with the following instruction: "The cylinders of twigs [willow] and hemp he must dry" (Neusner 1995). Many Judaic taboos are concerned with the mixing of different classes of plants, animals, and fibers. In Leviticus 19:19 God states the following: "You shall keep my statutes. You shall not let your cattle breed with a different kind; you shall not sow your field with two kinds of seed; nor shall there come upon you a garment of cloth made of two kinds of stuff" (Mendelbaum 1982). An ongoing interpretational dispute among Hebrew scholars concerns whether or not hemp should be considered "diverse kinds" in a vineyard depending on whether it is grown primarily for the sake of its seed or its fiber or whether it is considered an herb or a tree. Also at issue is whether certain types of garments are likely to contain diverse kinds of fibers. Flax and wool are considered to be of "diverse kinds," while a mixture of hemp and wool are not (Mendelbaum 1982).

Folkloric references to hemp use in ritual contexts appear throughout Eastern Orthodox and Western Christian religions, in particular among the Slavic peoples of Eastern Europe and date back to ancient times. As within religious traditions of the Far East, hemp is largely associated with agrarian rituals, life cycle ceremonies, and calendrical rites in the contexts of divination, purification, healing, and protection. It is instructive to note the parallels between ritual hemp use in eastern and western Eurasia spanning divergent cultural and religious traditions. The common denominator of these ritual uses for hemp must date back several millennia to early shamanic traditions shared in part by the major religions today. In the following section we examine many European traditions involving hemp in a nonpsychoactive ritual role.

European folklore casts hemp in both positive and negative lights, possibly due to its diverse uses. Medieval Anglo-Saxons believed that dreaming of hemp betokened misfortune (Folkard 1884; Dyer 1889), yet hemp was also associated with good luck: "In certain areas of Poland, at midnight, a chalk ring is drawn around the [hemp] plant which is then sprinkled with holy water. The person collecting the plant hopes that part of the flower will fall into his boots and give him good fortune" (Benet 1975). Across Europe, hemp was commonly believed to possess protective and healing powers and was also associated with purity. However, the work of boiling hemp yarn was full of regulations and superstitious prohibitions for people's protection, and for this reason hemp was at times considered ritually unclean (Mijatović and Bušetić 1925). Even so, during Eastern European funerals, hemp was traditionally placed in the coffin because it was "clean" in

the ritual sense and could protect the body from vampires (Petrović 1992), and across Central and Eastern Europe the funeral shroud was often woven of hemp (Jovanović 1993; also see Werner 1964 for description of an ancient hemp cloth blanket that was placed over the body of the Merovingian queen Arnegunde when she was interned in a limestone tomb in Paris about 570 CE). On the other hand, in Hungarian funeral rites the use of hemp for covering a corpse was forbidden (Kisgeci 1994). Whether hemp was considered ritually clean or unclean, associated with benevolence or evil, or a harbinger of good fortune or bad, it has traditionally been imbued with magical powers.

The cultivation of hemp and its intense labor are often associated with pain and suffering within the Judeo-Christian realm as they are in eastern Asia. Hemp fiber harvesting and processing involve hard physical work carried out in the heat of summer. Hemp stalks must be dried and collected before being put through a long series of strenuous manual tasks in preparation for spinning and weaving. In Italy, hemp is called *quello delle cento operazioni*, or "the substance of a hundred operations" and has long been regarded as a symbol of suffering (Schaefer 1945). In addition, because of the strong and stupefying smell of the plants, dizziness and violent headaches number among the occasional disagreeable effects of harvesting hemp. Interestingly, the hemp harvest was also traditionally a time of celebration: "The odor of European hemp is stimulating enough to produce euphoria and desire for sociability and gaiety; and harvesting of hemp has always been accompanied by social festivities, dancing, and sometimes even erotic playfulness" (Benet 1975).

Eastern European ethnographers of the nineteenth and early twentieth centuries described numerous peasant ceremonies in folk culture monographs and academic journals that provide us with fascinating firsthand accounts of traditional hemp rituals, and many can be traced back to their origins in much earlier societies. Numerous Eastern European rites and ceremonies have shamanic roots and rely ultimately on belief in hemp's magical powers. For example, hemp plays a role in wedding and funeral customs, as well as rituals intended to promote the fertility of hemp and other crops, cure various illnesses, and protect all members of the household including livestock (Čajkanović 1985). The seasonal tasks of hemp production, from preparing fields and sowing seed through to harvest and preparation of the fiber, were all linked to certain dates in the Roman Catholic and Eastern Orthodox calendars, but these tasks harken back to even more ancient pre-Christian beliefs. Over time, Christian symbolism fused with ancient shamanic traditions to form a new calendrical religious framework for the responsibility and labor of hemp production.

Hemp seeds are also revered in ritual settings. On March 25 Lithuanians traditionally celebrated Stork Day with a feast dedicated to remembrance of Angel Gabriel's announcement that heaven chose Mary to be the Mother of Jesus. In the nineteenth century, Samogitians (a Lithuanian ethnic group) prepared special foods for this feast day, including pastries made of hemp seeds and *krupninkas* (spiced honey liquor made with sweet vodka seasoned by herbal mixtures). Throughout Lithuania eating these ritual pastries and special breads baked from a mixture of many grains, and including hemp seeds, assured an abundant harvest. A Lithuanian tale related to the beginning of Lent concerns the epic battle of Gavenas and Mesinas in the threshing barn at midnight between Shrove Tuesday and Ash Wednesday, 40 days before Easter. The protagonist Gavenas symbolized hemp, while

Mesinas represented bacon or ham. During the Lenten fast, foods were prepared with hemp seed oil since animal fat and meat were forbidden. Thus the beginning of Lent represented the victory of Gavenas and the defeat of Mesinas, or at least his temporary eclipse (Greimas 1992). In another Lithuanian tradition, before Candlemas on February 2, the maidens pooled their money and bought liquor, which they boiled together with poppies, hemp seeds, and honey—roughly symbolizing happiness, well-being, and love. Then they invited the young men who came on horseback and drank all night and into the next day (Greimas 1992). Also, in both Latvia and Ukraine, a dish made of hemp seeds was prepared for Three Kings Day on January 6 (Benet 1975).

Some people in European cultures believed in taboos associating hemp labor and products with ill fortune, especially if they should come in contact with certain other activities at an inauspicious moment. There was, for instance, a widespread superstition that anyone returning home from harvesting hemp must not milk cows, as the milk could not be used for making cheese (Schaefer 1945). In the Piedmont region of Italy, farmers believed that hemp spun on the last day of Carnival (also called Shrovetide) would bring bad luck (Folkard 1884). In Poland, hearse drivers traditionally refrained from whipping their horses for fear that contact with the fiber in the whip would blight the hemp in the fields (Benet 1951). There was a belief in the Cotes-du-nord region of France that contact with hemp would enrage anyone who had been bitten by a dog (Folkard 1884).

Balkan traditions of southeastern Europe (e.g., among Serbs, Macedonians, Bulgarians, and Romanians) dictated that everything associated with hemp products—its seed, fiber, yarn, cordage, and woven fabrics—as well as items related to their manipulation such as tools used to process the plants all held special powers when used in ritual settings. The Balkan region—before the influx of Roman Catholicism and Islam—was largely influenced by Eastern Orthodox Christianity, which incorporated many earlier pagan shamanic traditions shared with Asian cultures. Therefore hemp was used in many life cycle rituals from conception to death. In South Danubian Slavic and Greek folklore there is an important motif manifested in a group of narratives, which focuses special attention on the "passions of the hemp," hemp serving as a metaphor for the "crop power of the field or, more widely, to vegetation in general" (Ionică 1996; see also Ispas 2001). This serves as symbolic representation of more ancient traditional convictions about the visible world, essentially manifesting popular belief in the spiritual relationships among humans and plants throughout much of southeastern Europe, with *Cannabis* playing a key role. According to Slavic traditions, retting hemp in a pond was called "burial." The flowery male hemp was retted at the "Feast of the Assumption" on August 15, and the seeded female hemp was retted after the day of the "Birth of the Blessed Virgin Mary" on September 8. Following retting, the sheaves of stalks were thrown out of the pond, or "resurrected" (Domokos 1930). During the processing of hemp fiber (e.g., cutting, drying, soaking, breaking, and beating) the plants "die" and are transformed from living into dead material. Before spinning begins, raw hemp fiber is in symbolic opposition to the death of the plants during processing, as it is poised on the brink of being "resurrected" as spun fiber and woven cloth (Radenković 1981). The strong links between the powers contained in hemp and its ritual uses stem from its double character as a processed fiber crop plant, first killed by cutting, soaking, and beating but then reborn as a hemp textile. In addition, *Cannabis* is analogous to the human condition because of its dioecious nature and the differentiation of female from male plants. Hemp seeds play a symbolic role in reproductive magic as they are the seminal source of the hemp plant (Mijatović 1985): "The vegetal symbol is constantly present in the rituals performed in southeastern Europe. The motif of suffering ritual plants, wheat and hemp, in legends, fables, and carols, is a significant cultural reference point for the whole area" (also see Ispas 1993).

For example, hemp is often utilized in the Romanian popular tradition of purification rituals in which it is held in very high esteem for its protective role during and after life:

> Hemp is great. It is holy for it dresses man and is so tormented. . . . Hemp is so much work, he who never worked in his life should wear a tow shirt at least on his death [bed] and will find it easier to go there [to Heaven] if he is seen to have carried its burden in this world. In these texts, preserved from old oral tradition, one can easily decode the idea of atonement for sin through passion, not only the passion of the plant, but also that of the person who grows and processes the hemp. At the same time, creation is sacrifice. (ISPAS 2001, quoting NICULIȚĂ-VORONCA 1903)

In the popular literature of Romania there is a legend that a young girl who encounters werewolves or ghosts during a social event can be rescued by telling the "story of the hemp or wheat passions," a legendary tale that focuses on village girls who group together for a hemp spinning social "in a deserted house, out in the field" only to have ghosts appear. The ghosts scare off all the girls except one who had fallen asleep:

> Left to the mercy of the ghosts, this girl spins her hemp and at the same time tells the story of "the agony of the hemp," representing the whole agricultural and domestic process, from sowing to weaving, by which hemp is changed into cloth out of which a shirt is eventually made. This shirt, a variant of "the plague shirt" or "the victory shirt," is also attested in Greek folklore and is produced by an identical technology. When the girl completes her spinning session, the story comes to an end as well; the roosters herald the beginning of a new morning, and the ghosts vanish. (ISPAS 2001)

Hemp's foremost traditional ritual role in southeastern Europe has been associated with its use to make the victory or plague shirt, a hemp garment of great cultural significance among historic, traditional Balkan cultures: "The passion motif of the hemp, as it shows up in certain variants of legends, seems to be the verbalization of an ancient custom by which a community was protected from the spreading of severe epidemics—plague, cholera, typhoid fever, and flu" (Ispas 2001).

In traditional Slavic rituals, a man's hemp shirt was worn for celebrations of births, christenings, Saint George's Day (April 23), the first plowing and sowing, Easter, and Christmas (Petrović 1992). A Ukrainian legend tells of a dragon that lived in Kiev, oppressing the people and demanding tribute. The dragon was killed and the city liberated by a man wearing a hemp shirt (Benet 1975). According to Serbian legend, in 1836, as a defense against a spreading epidemic of cholera, Prince Miloš Obrenović ordered that a special magical *"cuma"* (literally "epidemic") shirt be made. Within the space of one night, nine naked old women sat silently and spun, warped, wove, cut, and sewed a magic shirt of hemp cloth. The prince himself, his family, his escort, and soldiers all squeezed into this large shirt to protect themselves (Djordjević 1938). In Macedonia, in the still of the night three old women would weave a special hempen fabric, which was used for many years in amulets to protect against various illnesses and epidemic disease. The cloth would be sewn into a magic shirt, blessed by a priest in a new church and torn into small pieces to be shared between members of the family who believed in the magical power of each piece (Djordjević 1938; Nikodinovski 1984). This tradition was practiced in many Slavic regions into the 1930s for resistance to various dangerous diseases, and before joining the army a person wore a *cuma* shirt for a night as protection from gunshots (Milošević 1936; see also Filipović and Tomić 1955; Macaj 1966; Zečević 1968/1969; Begović 1986; Trojanović 1990).

This practice also appeared in Greece and involved recitation as the hemp shirt was taken to a designated location: "My dear, from now on mind your own business, spin your hemp, and leave us alone" (Ispas 1993). This special shirt, which could only be fabricated from hemp, was sewn and used following the passing of a sickness whenever any disease affecting many people occurred, and production took place during those same days when hemp labor was prohibited. The most recent record of the production and use of this protective hemp shirt in Romania, according to Ispas (2001), dates to World War I and was carried out as follows: "A number of old women or maids would gather in a house and, during night-time, would spin and weave from hemp a bleached fabric made into a shirt that they would carry ceremoniously to the end of the village or to the nearest crossroads." The protective hempen shirt provided a means of suppressing evil powers so that space and time would be purified, forgiveness would be achieved, and atonement for sins would be obtained via voluntary sacrifice; thus "self-sacrifice was transformed into a preventative mechanism in the battle against harm, and it is these motives of the agony of plants [including *Cannabis*] that have been analyzed and linked to the cult of the dead, [and] as well to agrarian or seasonal ceremonies" (Ispas 2001). The ritualistic association of the purifying hemp shirt and the cult of the dead harken back to the ancient use of *Cannabis* in its ceremonial context of funerals across much of Eurasia (see Chapters 4 and 7 for reference to cult origins and relationships to psychoactive use).

Hemp was considered to be a pure material belonging to the "upper world," and this is why it was used to make garments for priests and mourners in ancient cultures. Furs and wool, however, if they originated from dead animals, were therefore considered impure, and thus their use was forbidden at religious services. On the other hand, people in some regions of Serbia have traditionally believed that during White Week (the last week of Lent, which falls between the third week of March and third week of April), when the souls of the dead are among the living, women must not do any work with hemp, and a funeral shroud must *not* be made of hemp cloth (Čajkanović 1985).

Northern Serbs from the regions of Voivodina and Homolje in northeastern Serbia entertained many taboos concerning hemp processing. The women were not allowed to sow hemp seed, soak hemp stalks, or to do any processing of hemp textiles during the two "wasteful" weeks when the moon wanes; instead, to encourage prosperity they saved hemp work for the two weeks when the moon grows (T. R. Djordjević 1958). After the hemp stalks were cut and soaked in the retting pond, they could not remain there on St. Ilija's Day (August 2) or else they would rot away. In addition, hemp should not be washed on Good Friday (two days before Easter Sunday), and it should not be spun from Christmas Eve on December 24 until St. John's Day on December 31 (Kisgeci 1994). On Transfiguration Day (August 19) when hemp harvesting had commenced, Moravian farmers did no retting work or else the hemp stalks would scatter in the retting pool (T. R. Djordjević 1958). Balkan peasants would not spend the night in a hemp field, and they believed that if cabbage was sown in a place where hemp was previously grown or vice versa, someone in the family would die; hemp also had connections with the underworld and could thus harm pregnant women, children, and eggs (Mijatović and Bušetić 1925).

Many superstitions surrounded the spinning of hemp in the Balkan region. In Croatia, women began spinning when the moon was full so they would be able to produce large amounts of yarn (Lovretić 1902). To protect against vampires, a spinner would not finish her hemp skein in a strange household (Jańćerova 1901). In Herzegovina, it was believed that a girl who spun "black" (female) hemp would find a buried treasure (Bratić 1906). In the Old Szeck Aleksandria region of Moravia, it was said that during a lunar eclipse, "witches spinning hemp yarn in the dark are eating the moon" (D. M. Djordjević 1958). On Saint Triphun Day (February 14) Serbian women did not spin hemp so as to prevent sparrows from eating crops on the field (Mijatović 1907), nor did they comb hemp to prevent wolf attacks (Vukanović 1986; also see Vukanović 1989 for reference to hemp use in protection from witches). Among Croatians, as well as the Serb population in Hungary, all the handwork of women, especially with hemp, was also forbidden on Saint Barbara's Day (December 4), or else Saint Barbara would carry a crucifix, search out the spinner (Medjesi 1978), turn her hemp yarn back into raw hemp fiber, and make her cattle lame (Eperjessy 1975). In Bosnia and Serbia, during the ninth week following Christmas, hemp work was forbidden on "Ninth Tuesday" because children born on this day might be weak and readily perish (Antonić and Zupanc 1999). On Good Friday, the Friday preceding Easter Sunday and commemorating the death of Christ, Croatian women also refrained from washing hemp textiles to prevent rust disease developing on grapes (Ivanišević 1905; Peco 1925). In Romanian folklore, hemp spinning is included in the Great Earth Mother's cult, represented by female deities such as "Jojmerica." People imagined her as a woman who spun hemp and looked for spinners to punish for disobedient spinning on the forbidden days of Wednesday, Friday, and Sunday (Radenković 1981; Mijušković 1985).

Among Serbians, when passing by a weaver, it was considered respectful to express gratitude for her weaving (Mijatović and Bušetić 1925). Within southern Slavic traditions, if one came upon a hemp weaver warping or laying the warp (organizing tensioned yarns before they are placed on the loom), he or she would raise a hand and recite the formula, "For such opened threads," to facilitate her weaving (Djordjević 1938). However, it was also thought that if a person with the evil eye praised a woman's weaving, it would cause the threads to break (Hećimović-Seselja1985). It was considered an ill omen when a priest met a woman winding or warping hemp yarn, which, according to belief, would break the warp (Djordjević 1985). In Herzegovina, people often said it was unlucky to see a woman spinning "black" hemp on her distaff (the vertical stick to which the hemp fiber is attached during spinning), and on such an occasion, a man would cover his forehead, which was believed to be susceptible to evil influences (Bratić 1906). In medieval Montenegro, if a woman was spinning hemp outside her house and met a man, she would hide the hemp from his eyes; according to a statute, upon meeting a man, a woman spinning hemp must stop working and bow to him or move away, and if she did not, she was reprimanded in public (Vukanović 1938; Popović 1953). It was also considered very dangerous for a man to step over the warp yarns on the loom (Mijatović and Bušetić 1925), a superstition shared with the Hmong and other minority peoples in southwestern China. Along the old border area between Austro-Hungary and Turkey if a man was present when a woman finished weaving hemp cloth, people believed he would be killed in war (Begović 1986).

In Serbia, the warping of hemp was also connected with many folk beliefs and superstitions (Mijatović and Bušetić 1925). During warping, as in other areas of southeastern Europe, Serbian women would never cross the warp yarns (Petrović 1948). Warping of hemp yarn was forbidden on several occasions, for instance, between Assumption Day, August 15, and Saint Sava's Day, January 27 (Jovanović and Bjeladinović 1964), or from St. Nicholas Day, December 6, until St. Sava's Day, January 27, with the exception of Christmas Eve (Kisgeci 1994). It was also prohibited to lay a warp for 40 days after the death of a close friend or family member or to stretch a pair of warps (only an odd number of one or three warps were allowed) lest wild animals would harm the household (Mijatović and Bušetić 1925).

After the warp is fully laid out, Serbian women would perform weather rituals. For example, when rain was needed during a drought they would pull the three wooden warping stakes from the ground exposing the "holes" in the earth, and during floods they poured water in the holes and placed ashes or embers inside to stop the rain (Petrović 1948). To promote fast and uniform weaving, women performed rituals of "good magic," such as beating the warp before installing it on the loom to prevent problems with evil spells or throwing a stone as far as possible in the belief that this would encourage the cloth to be long and the weaving fast (Mijatović and Bušetić 1925). Mondays and Thursdays in particular were considered "good days" to begin weaving; this was especially true of early mornings when everybody slept, since the weaving would therefore be protected from evil eyes (Jovanović and Bjeladinović 1964).

Among rope makers in southern Serbia there is a legend concerning the origin of the cordage making craft. Saint Sava (founder of the Serbian Orthodox Church) created a rope of sand to help God protect against the devil. Later, God created hemp, a strong raw material for people to make ropes more suitable for battling the devil (Hadži-Vasiljević 1913). St. Sava is still respected today as the patron saint of rope makers across much of Eastern Europe. Peasants believed that they could stir the "devil" from the bank of the hemp retting pond by calling him and he would come (Milošević 1936).

Slavic hemp farmers obeyed a number of other traditional, superstitious hemp taboos. For example, the open fireplace, where hemp yarn was steamed, was believed to be desecrated. It was forbidden to bake or cook there because of the possibility of contracting a fever, and children were not allowed to approach the hole where the hemp yarn was steamed because it was thought that if they did so they would not grow tall (Kisgeci 1994). During the day of the Autumnal Equinox

(September 21), drying, beating, and breaking hemp were forbidden because it was believed that an early frost would occur and destroy the crops. Between the two Christmas holidays (Catholic, December 25, and Orthodox, January 7) women did not spin flax or especially hemp (Jańčerova 1901).

In Europe, as in Asia, hemp is often incorporated in ritual healing and attributed with beneficial properties. When investigating the ritual use of hemp in healing, it is important to differentiate between practices that may actually have direct medical benefits (e.g., application of poultices, ingestion of plant parts, drinking of infusions, and so on; also see Chapter 8) and those that appear to have only ritual components and psychosomatic efficacy. Brunner (1973) provides a typical example from *De Medicamentis* by the fifth-century Roman author Marcellus Empiricus with the sort of evidence often encountered when searching for healing rituals as opposed to applications of *Cannabis* with an empirically measurable medical effect: "Tie hemp root to your right arm: preferably wrap the entire arm with this root; but if you only have a small amount, then suspend it from your neck by a thread from a loom. To show you how powerful this remedy is: if you tie the root as instructed, the flow of blood will stop immediately. But when you untie and remove the root, the blood will flow again." From our empirical point of view, in this particular case we detect no direct means by which the external application of entire hemp roots could logically be effective in quelling the flow of blood, unless they were being used as a tourniquet. However, there may indeed be some topical or other direct medical effect that we are simply unable to understand and appreciate.

Among many traditional cultures, illness is often attributed to effects of evil spells, and as curative acts, rituals were employed. In these cases it is quite clear that the healing intent gains its power through magical means facilitated by the ritual use of hemp; thus there is little confusion between these strictly ritual acts of healing and more soundly based medicinal applications. As previously noted, hemp yarn and tools associated with hemp processing, spinning, or weaving were also considered to have magical healing powers in parts of Europe (Krstanova 1999).

For example, in some areas, hemp was commonly used in rituals for the relief of pain. From the region of Cerne in Dorset, England, Colley (1899) recorded the following passage quoted from an earlier source: "I hadn't no cause to ax nothing, 'cause he knowed what 'twere all about so soon as ever he sot eyes on I. He told I to get a strand o' new hemp avore 'twere made into ropes, an' a thread or two o' new scarlet silk, an' I were to braid 'em together long enough to bind about the body on the part where the pain were." Serbian Slavs relied on many methods involving hemp to dispel pains from the body caused by an evil curse. Women sorcerers cured patients using ashes of burned raw hemp fiber that were stirred in water; the patient drank this infusion while the healer recited ritual incantations and rubbed ashes on the painful spot (Milićević 1894). Another example involved a young girl's hemp shirt stained with blood from her first menstruation, which was preserved in such condition and later used as medicine by soaking it in water; for instance, a mother would wash her newborn child with the bloody water to prevent infections, or the infusion was used to wash pained eyes (Vukanović 1986). In a cure to relieve pectoral muscle pains resulting from catching a cold caused by sorcery, a patient would wind raw hemp strips around their body, and while holding the hemp they would recite an incantation describing the sequence of steps involved in growing and processing hemp, which would cause the pain to dissipate (Simić 1964). Among southern Slavs, healers collected pieces of hemp yarn from the floor that remained after weaving; then when administering to a patient's pain, the healer would cut up threads and recite from a magic text to drive out the cause of distress and cure the disease, and in Hungary, hemp strings were used for curing various indications such as urinary tract dysfunction or controlling a mother's lactation (Schubert 1984).

Among Slavic peoples hemp was used to ease the discomfort of several other medical conditions. For example, to cure painful mumps or sore throat, one would consume the blood saved from the Christmas pig mixed with raw hemp fiber (Mijatović 1909), and when someone suffered a stomachache from overeating, they would cover the abdominal area with raw hemp fibers (Gundrum-Oriovčanin 1909). Sometimes the magical curative powers of hemp were imbued in other objects. Before a woman began laying the warp, she would place a piece of bread on the spool of hemp yarn, and upon finishing the warp this bread became medicine; if someone with stomach pains tasted a bit of this bread, their pain would stop. And to restore menstruation during menopause, a woman would wash her sexual organs with "black" (female) hemp and recite the following: "In the time when the black hemp blooms, is my menstrual cycle" (Mijatović 1909). After an initial epilepsy attack, the sufferer was laid on the ground while men dug a trench around the patient; a small amount of water was then poured into the trench and given enough time to evaporate before adding raw hemp fibers prepared the previous spring. Then the fibers were set on fire (Milićević 1894), and finally, as the "magic circle" burned, an earthen cup was broken above the patient's head (Raičević 1935). To cure tuberculosis, hemp fiber was boiled in water and used to prepare a special cake for a dog to eat; the aim of this act was to transfer the illness from the person to the dog (Bušetić 1911). In Macedonia, in order to cure thyroid gland problems caused by sorcery, a hemp thread spun by a woman who had since died was tied around the patient's painful neck; the healer then recited magical incantations and the illness would disappear after 40 days (Simić 1964). As we described previously, such ritual tying of hemp during healing ceremonies is common across Eurasia.

In medieval Europe, peasants generally believed that hemp seed should be sown on the days of saints who were known to be tall in order to encourage the crop's growth (Benet 1975), and in each region specific days were chosen for sowing hemp seeds, such as St. Andrei's Day (April 15) in Belarussia (Graves 1955). In the Lovec region of southeastern Bulgaria, hemp seed was sown on Mondays or Wednesdays, while in Serbia, hemp was best sown on Mondays or Saturdays, and in springtime when the moon increases, on either a Tuesday, Thursday, or Saturday (Franić 1935). Women's days for sowing were Tuesday of White Week or Saint Jeremiah's Day, May 1 (Knežević and Jovanović 1958). A mother sowed hemp seed on April 23, the day honoring St. George the patron Saint of Serbia and England as well as the infamous slayer of dragons; this was done so that the mother's children would "grow as fast and high as hemp" (Mijatović 1907) and raised up a holy candle for the crop to grow tall (Krstanova 1999). The ritual first sowing of hemp seed began at sunset since this was when the family had returned after the day's work (Filipović and Tomić 1955). In Germany, hemp seed was sown during the "high" hour, between 11:00 and noon (Benet 1975). Several

other times were ritually avoided when sowing hemp seed. For example, early nineteenth-century Danish hemp-growing traditions considered it bad luck to sow hemp in the week that includes May 15 (Brøndegaard 1979), the Ruthenians of the Voivodina region of Serbia placed a ban on sowing hemp seeds on Easter Thursday and Friday (Medjesi 1978), and in eastern Serbia, hemp was never sown on Tuesday or Friday (Franić 1935).

Before sowing hemp seeds in areas of Serbia, a barefoot woman would raise her seed bag up and shout, "As high as the hemp will grow," and then work silently. After finishing, she would throw the empty bag high in the air to promote a tall hemp crop (Jovanović and Bjeladinović 1964) and recite the words, "May God grant such high hemp" (Kisgeci 1994). In Germany, long paces were taken while broadcast sowing hemp seed, which was thrown high into the air, and cakes baked to stimulate hemp growth were known as *hanfeier* (Benet 1975). A traditional Danish superstition held that if a man sowed hemp during the first year of his marriage, he would be unfaithful to his wife (Brøndegaard 1979).

Following sowing, many more rituals were performed to encourage an abundant harvest of tall, straight plants: "Since the plant was associated with religious ritual and the power of healing, magical practices were connected with its cultivation" (Benet 1975). The custom of dancing or jumping to promote crop growth was known throughout Europe. Polish married women traditionally performed the "hemp dance" on Shrove Tuesday, leaping high into the air (Benet 1975). On Lazarus Saturday (the day before Palm Sunday), Bulgarian masked dance groups would "jump up for hemp" to make the crop grow taller, and they received skeins of hemp yarn as a gift (Krstanova 1999). In the Vosges Mountains of France, "the practice of dancing on the roof was observed to make the hemp grow tall" (Cox 1998). In Saratov, Russia, on Saint John's Day (December 31) a priest or another clergyman, with a few young girls, would roll in the hemp field to increase the crop's fertility, and during the first fasting week of Easter, children jumped high over a bonfire for the hemp to grow tall (Franić 1937). Serbs in Voivodina would also light ritual bonfires, and young people jumped as high as they could over the flames (Bosić 1996).

Hungarian Pentecost (Whitsun Feast) celebrations held on the fiftieth day after Easter Sunday also included ritual means to ensure a tall hemp crop. Near the end of the feast, celebrants lifted the Little Queen, a small girl of five or six, by her arms with the cry, "Let's lift the queen!" Then lifting her veil they say to the lady of the house, "May your hemp grow as high as this!" Raising the Whitsun Queen acted as a magic charm to ensure a good hemp harvest, and conversely, sitting her on the ground would prevent the hemp from growing (Viski 1937). According to Slovak traditions, on Hromnice (February 2) people from the whole village used to sled down a nearby hill to encourage their hemp's growth (Slovak Heritage Live Newsletter 1996). In several regions of Europe, hemp seed would be selected on a clean white hemp tablecloth and sown in the fields from a clean white bag, and the sowers were dressed in white clothing to ensure that the hemp cloth would be as white as possible (Radauš-Ribarić 1988; Krstanova 1999). In Croatia, before leaving home to sow the fields, women would run over the threshold, which had been covered with clean bleached hemp linen carrying hemp seeds (Franić 1935). Traditionally, European men tilled the soil while women sowed the hemp, bringing the seeds to the field in a white hemp apron to ensure future fabrics would be clean and white (Lechner 1954). The Eastern European association of white hemp cloth with cleanliness is similar to the East Asian association of hemp with purity.

Hemp was a well-known symbol of fertility (Viski 1937). In Serbia and Kosovo, a hemp hackle (tool for combing hemp fiber) was placed in the seed sack before sowing to ensure a dense and high quality hemp crop (Veljić 1925). The powerful magic accorded to hemp processing tools was also applied during the sowing of other crops. Before sowing wheat, Serbian peasants put a hemp hackle in the sack of wheat seed along with a silver ring or coin so that the wheat grains would be white like hemp cloth and round like the ring or coin (Mijatović 1907; Vukanović 1986). Eggs were often included in rituals involving hemp seed sowing to encourage their germination and quick growth. Due to their association with fertility, farmers believed that the vital life force from the egg would be magically transmitted to the germinating hemp seed (Lechner 1954). In Moravia, for instance, hemp seed was sown barefooted, and when finished, a hemp hackle and an egg were thrown onto the field for the hemp cloth to be rich and white (D. M. Djordjević 1958). Likewise, in other regions, crop vitality and white cloth were encouraged through local additions and variations to these basic sowing rituals. Most often a boiled egg was buried in the sack of sowing seed (Grbić and Djordjević 1907; Radauš-Ribarić 1988), and in Serbia an Easter egg was kept with stored seeds to prevent hail from damaging the hemp crop (Kisgeci 1994), or a fresh unboiled egg (or the first colored Christmas egg) was dug into the soil to promote good fertility of the hemp crop, and women sowed seeds silently while eating boiled eggs (Jovanović and Bjeladinović 1964). In the Cajnice region of Bosnia, three eggs were dug into the earth and after three days eaten, while the eggshells were thrown into the hemp field (Trojanović 1911; Franić 1935). In Croatia, hemp seeds were prepared by women who ate eggs and ham to ensure rich, clean, and soft hemp cloth, and then the seeds were sown by men (Lechner 1954). The Ruthenians of Voivodina ate scrambled eggs with polenta and rang the church bells during hemp sowing (Medjesi 1978), and the Serbs of Kosovo slaughtered a rooster as a sacrifice (Vukanović 1986). In some Serbian villages, people placed an egg on top of a high stick in the field, or they stole a little hemp from somebody else's field and took it into their own to increase yield (Vladić-Krstić 1997).

Danish traditions tell us that birds will not steal the seeds if the sower walks with three hemp seeds under their tongue (Brøndegaard 1979). After hemp was sown in Croatia and Serbia, people ran three times around the fields with their eyes closed so that birds would not see the seeds and eat them (Lovretić 1902; Kisgeci 1994). Slavic women often took a clod of earth from the home where their ancestors lived, placed it into the bag of hemp seeds to protect the crop, and then sowed the seeds silently so sparrows or poultry would not eat them (Petrović 1948; Hećimović-Seselja 1985; Kisgeci 1994). Along the border area between Austro-Hungary and Turkey, a woman would take bits of soil from each of nine sown hemp fields (Begović 1986) or from under the threshold (Jovanović and Bjeladinović 1964) and spread them in her own field. In Poland, girls removed their shirts, filled their mouths with water, squirted the water upon the hemp seeds, and then ran around the house naked three times (Benet 1975).

Danish traditions also stipulated that when seeds finished ripening, the field must be threshed at full moon, but not on the date of the month it was sown (Brøndegaard 1979).

In Croatia, ripe hemp plants were pulled when the moon was full since this promoted the production of whiter hemp fiber (Lechner 1954). In the Côtes du Nord of France, it was customary to leave the finest stalk of hemp standing in the field after harvest, so that the St. Martin's bird (crow) would be able to rest on it (Folkard 1884). In the Danube Basin of Hungary and in the region of the former Yugoslavia, "white" (male) hemp was gathered around Saint Elias Day (July 20), but "black" (female) hemp was harvested after Assumption Day (August 15), and girls leaving for the fields to pull it were decorated with flowers (Vladić-Krstić 1993). However, the Ruthenians of Voivodina placed a ban on harvesting hemp on Saint Elias' Day (Medjesi 1978). In Moravia, the harvesting of seeded hemp plants commenced quite late on St. John's Day, December 31 (D. M. Djordjević 1958).

Hemp's magical powers were often used to protect farmers' livestock from disease and attack by wild animals. In 1828, near Hanover, Germany, an eyewitness reported the making of a pair of ritual fires utilizing raw hemp fiber and hemp rags and ignited by friction. Then the farmers drove the cattle between the bonfires and blackened each other's faces with the ashes to protect their livestock; Serbians also made a fire by friction using the warp beam of a hemp loom bound with hemp ropes and then burned hemp rags and gave the ashes to their cattle to eat (Trojanović 1990). On the day before St. Martin's Day (November 11), Macedonian cattle breeders drove their cattle between a pair of hemp hackles, and then they snapped the hackles shut, intending to shut the mouths of hungry wolves. The hackles were then hung through the night on the stable door, and the rest of the week they were hidden away (Raičević 1935). On Saint Luke's Day (October 31) women in the Morava Valley of Serbia spun raw hemp fibers and encircled their sheep to protect them from wolves, believing that they had tied the wolves' jaws, and on Saint Marta's Day (March 1) pig bristles were spun together with raw hemp and placed into the opened jaws of hemp hackles, also to protect against wolf attacks (D. M. Djordjević 1958).

Throughout the region of the former Yugoslavia, people generally considered all weasels to be of female gender. When a "lady" weasel made trouble around the hen coop, a little ball of hemp fiber was placed inside so that when the weasel entered, "she" would sense her accustomed feminine calling and spin the hemp rather than harm a fellow female (Viski 1937). In many Slavic regions, the peasants staged an annual fertility rite in August asking the earth goddess to ward off pestilence. They assembled in the fields first turning east, then west and south, and finally north, and poured hemp seed oil onto the land while reciting prayers to subdue evil and cast out demons (Zajaczkowa 2002). On Forty Martyrs Day (March 9), Macedonian women fed hemp seeds to all cattle and poultry in the household to promote the livestock's health and fertility (Vukanović 1986), and in the Morava Valley region of Serbia, people burned hemp threads to keep the livestock fit (Antonijević 1971). In Montenegro, when a horse or bullock collapsed, people would take off their wide hemp underpants, throw them over the animal's head, and run around naked jumping over the fallen creature. When a cow became mad, a healer would take hemp threads from a disused loom and tie them in crosses between the cow's legs, and when the madness disappeared the crosses were cut at the center (Mićović 1952). In Croatia, women have traditionally spun a special hemp thread about two meters (about six feet) long and tied it around a pregnant cow's stomach to ease birth (Vukmanović 1935).

Serbians burned hemp rags and fumigated their households to guard against snakes (Djurić 1934). In Bosnia and Herzegovina, women sowed hemp seeds on St. Jeremiah's Day; before sunrise, they banged together iron dishes and shouted, "Jeremiahs into the field, but snakes into the sea," believing that snakes would never attack their cattle (Lilek 1894). Hemp plants were thought to repel snakes, and therefore hemp fields were considered to be safe havens from snakes as illustrated by the following Polish poem from *Pan Tadeusz*, or "Sir Thaddeus," a Polish historical epic (first published in 1834) quoted in Knab (1995):

> Each bed is girdled with a furrowed border
> Where hemp plants stand on guard in serried order,
> Like cypresses, all silent, green and tall
> Between their leaves no serpent dares to crawl.

In some areas of southeastern Europe, immediately upon arrival of a newborn baby, and indeed throughout a person's entire life, rituals were held to invoke *Cannabis*'s protective powers. For example, Kosovar Serbs placed a pair of hemp hackles along with a skein of hemp yarn next to a woman and her newborn or above the baby's bed for a total of 40 days as protection against supernatural evils (Grbic 1900). Furthermore, the mother and baby were also protected by a magic circle of rope stretched around the birth bed, and the room was fumigated with smoke from burning hemp fibers (Vukanović 1986). Croatian women pulled a skein of yarn across a newborn baby girl's face so she would become a diligent adult (Lovretić 1902). On Saint George's Day in Croatia, mothers bathed their children with various herbs, including hemp, to make them strong and healthy (T. R. Djordjević 1958). When a Croatian child cried persistently, a piece of hempen dish towel was placed under the child's pillow (Djordjević 1985), and old women would set fire to a magic circle made of raw hemp fiber and stand within it with the children to protect them from disease in the New Year (Erdeljanovic 1951; Knežević 1964; Trojanović 1990). To thwart an epidemic, Croatians placed skeins of fiber in the house loft (Grbić and Djordjević 1907). In a southeastern Serbian ritual, family members walked through a fire of hemp ignited by friction and then burned raw hemp fibers for powerful double protection from sickness (Trojanović 1990). It should be noted here that the use of hemp fibers for tinder or other ignition purposes, in loose or cordage form, has ancient roots in human history (e.g., see *Encyclopaedia Britannica* 1911 under "Match," Allen 1900 and Langerman 2009). The Eastern European use of the hempen magic circle provides an example of healing by exorcising evil spells and curses resulting from sorcery. To cure stomach and back pains Slavic healers spun a long hemp thread while reciting a magical incantation about the tortuous work of growing and processing hemp, and a woman healer tied knots along its whole length and fixed it around the patient's hand at bedtime. Then before dawn she made a magic circle of raw hemp fiber around the standing patient in a place where nobody walked, set fire to it, and finally tossed the knotted hemp yarn away to make the illness disappear (Simić 1964; Djordjević 1985). In another variation of the magic circle ritual, a woman healer spun a hemp rope at midnight and placed it in a circle around a female patient; she then recited incantations while burning the hemp, using it to fumigate the patient and cure her suffering from painful sexual organs (Simić 1964).

Serbian peasant women protected themselves from evil spells by wrapping around their waists a hemp rope that had spent the night on the family altar. They also believed strongly in the protective powers of amulets, especially objects fashioned from hemp, such as a piece of a bell tower rope or a hemp string used to tie another powerful amulet (Knežević 1964). In Bosnia, to protect against evil charms a pregnant woman would dress in her husband's hemp shirt and belt (Schubert 1984), or a hemp tablecloth from the holy dining table at a religious feast was torn into pieces and sewn onto the clothes as a protective amulet (Djordjević 1938). In Ukraine, hemp flowers gathered on St. John's Eve (June 23) were thought to counteract witchcraft and protect farm animals from the evil eye (Benet 1975). St. John's Eve and hemp are also associated with love divination.

Jensen (1996) identified certain dates that have been especially important for sowing and harvesting hemp in Belgium. On May 30, hemp seeds were traditionally sown "before May has gone to bed and June has risen," and on July 22, St. Madeline's Day, female hemp stalks were harvested and the rural inhabitants would chant, "Harvest your hemp on St. Madeline's Day. If it's not ready, bale it for another week." In the northeastern French region of Lorraine, which borders three countries (Belgium, Luxembourg, and Germany), women would celebrate their lengthy hemp work on St Agathe's Day (February 5) as an important time to mark the "end of their long nights by the spinning wheel" (see Pezzin 1986).

Hempen rituals were also conducted, although less commonly, in Protestant and Catholic Western Europe. For example, love-associated divination and magic were commonly practiced in nineteenth century Britain. A fictional account of hemp seed divination can be found in Thomas Hardy's novel, *The Woodlanders*, first published in 1887 (Vickery 1995). During the Festival of St. John (December 31), British maidens would practice divination in their love affairs and hemp seed was sown during certain mystic ceremonies (Folkard 1884; Salmon 1902). Many versions of love divination incantations were recited on different nights in several regions across the British Isles, Midsummer Eve (June 21) being the most popular. Additional nights that were traditionally chosen for such activities included St. Valentine's Eve (February 13) in Derbyshire and Devon (Dyer 1889; Wright 1938), St. Mark's Eve (April 24) in parts of East Angolia, All Hallows Eve (October 31) in Scotland, and St. Martin's Eve (November 10) in Norfolk. Two examples of nineteenth-century British incantations involving hemp seed love divination are presented here. As the clock struck midnight, young women desirous of knowing their future husband's identity slipped into a churchyard and ran around the church, scattering hemp seed and repeating all the while,

> I sow hemp seed; hemp-seed I sow;
> He that loves me the best
> Come after me and mow.

> or

> Hemp-seed I sow thee,
> Hemp-seed grow thee.
> And he who will my true-love be
> Come after me and show thee.

It was imagined that the hemp seeds would sprout and grow a crop immediately, and as the maiden ran around the church a third and final time she would look back over her left shoulder for the image of her future husband, who would appear and run after her pulling or mowing the mature hemp plants. If the maiden could not run quickly she believed that this premonition of her future husband might catch her and cut her legs off with his scythe (Parker 1913; Bas 1914; Banks 1939). Folkard (1884) tells us that among the ritual knowledge and skills of an English witch was her ability to "instruct a lass in the mystic rite of hemp-sowing in the churchyard at midnight on St. Valentine's Eve."

On Friday evenings at midnight, Sicilian Catholics employed hemp as a charm to secure the affection of the one they loved, taking hempen yarn and 25 needles full of colored silk and plaiting them together while reciting, *"Chistu è cánnava di Christu, servi pi attaccari a chistu"* (Christ and the *Cannabis* fiber of Christ serves to attach you to Christ). Subsequently, these faithful Sicilians went to church, entering at the moment of their commitment, and then tied three knots in their hemp and silk yarn, adding a little of the hair of their loved one. With this in hand they would "invoke the spirits to entice the person beloved towards the person who craves his or her love." Divine Sicilian rituals associated with hemp sowing and love magic may trace their origins to the use of hemp to make cordage, which is used to bind, attach, or secure an object and would be evocative of both tying the hearts of loved ones together and securing the ties that bind marriage (Skinner 1911; Folkard 1884). In some districts of France, hemp seed was thrown over the bride's head as part of the traditional pre-Christian wedding ceremony, but the Catholic Church interdicted and banned such ritual practice in 1626 (Djordjević 1984).

In Eastern Europe love divination and magic were also associated with ritual hemp use. For example, Benet (1975) tells us that the Eve of Saint Andrew's Day (November 29) was considered a most propitious time in this region for divination about future husbands: "Certain magical spells, using hemp, are believed to advance the date of marriage, perhaps even signal the very day it will occur." In Lithuania, St. Andrew's Day (November 30) marked the beginning of Advent. On this day, girls would use water, hemp seeds, poppies, and other herbs to forecast whether or not they would marry in the following year (Ambrazevicius 1996). Among ethnic Ruthenians of Voivodina in Serbia, young women predicted their future husbands by sowing hemp seeds with nine different young men to determine with whom her next harvest would be (Medjesi 1978). In Hungary, love divination surrounded work in the communal hemp spinnery, during which girls were alone, spending time telling fortunes and making love charms. These were simple games, yet girls believed in them. One of these charms was called *cucorka*, a small hemp lint ball that would rise in the air if set on fire. If it did not fly up, it indicated an unfavorable future, for instance, that her lover was not true to her or possibly there would be no wedding. Sometimes the girls clapped their hands so that the *cucorka* would fly higher (Viski 1937).

In rural villages of the Hungarian Szekely region, families rented a communal house where women and girls shared the tedious task of spinning during long winter evenings. There, with the attendance of young men, they enlivened their work with songs, storytelling, games, and occasional dancing. This arrangement made the spinnery into a pleasant center of village social life where work and play formed a desirable combination and provided the young with an opportunity to meet socially under the watchful eyes of understanding elder women (Kocsis and Kocsis-Hodosi 1998).

Yugoslavian girls applied some fascinating strategies to procure a lover. During the evening before Saint George's Day a girl would roll naked in someone else's hemp field and be overcome by the love of her desired man (Mijušković 1985). When a girl wanted to marry in the near future, she stole the hemp belt of a man who was married three times with virgins and carried it wrapped around her naked skin; it was believed that this would result in her being married the following autumn (Begović 1986).

Hemp also has played a role in Balkan weddings where "the most ancient bride's clothing is the long hemp shirt," worn to ensure that her future hemp cultivation would be successful (Blagojević 1984). In the middle of Lent, when a newly married bride and her mother-in-law visited her mother's home, she would carry a new, unused distaff and a few hemp seeds because of their symbolic fertility (Krstanova 1999). In Voivodina, Serbia, a mother-in-law traditionally gave her daughter-in-law a distaff with hemp fibers, so that in her future household she would spin copious hemp yarn and produce a happy life for her family (Subotić 1904; Karadžić 1867; Blagojević 1984). In Bosnia, the bride approached the bridegroom's house walking across a hemp cloth textile (Karadžić 1867), and in Serbia, the day after the bride's arrival, the family invited the other villagers, receiving a skein of hemp yarn from each as a gift for her. On the second day, wedding guests washed their hands and the bride gave them a hemp towel, which was "paid for" with some small money (Karadžić 1867). The bride would also "steal" the cap of the master of the wedding ceremony and decorate it with objects of ritual meaning, such as a hemp skein, after which the master was required to pay the bride for decorating his cap (Blagojević 1984). Among Ruthenians in Voivodina (Medjesi 1978) and Slavs living along the Adriatic Sea (Zaninović 1964), hemp was a very important element in the bride's trousseau shown publicly on her wedding day.

Young girls in Eastern Europe rarely worked in the hemp fields and spent much of their free time learning to spin, weave, and embroider clothing, bed sheets, pillows, tablecloths, kerchiefs, various domestic textiles, and other hempen dowry items in preparation for their weddings. Bedclothes constituted the majority of a girl's dowry, and as a result, mother and daughter spent years spinning, weaving, and sewing these cloth items, often employing semiprofessional seamstresses to assist them. When the ornate set of bedclothes was complete, it was installed on a bed in the so-called clean room in order to demonstrate to anybody entering that the family had duly prepared their daughter for marriage. The bride's dowry textiles were further exhibited throughout the village on the day before the wedding when they were carried on a wagon before reaching the house of the bridegroom (Fél 1961). As we have previously seen, Eastern European marriage rituals and dowries have direct parallels with the Hmong of southwestern China and Southeast Asia.

In Estonia, the dance "For Hemp's Sake" was traditionally performed at weddings by the young bride and the master of ceremonies (Kolberg 1899). Later, the young bride visited her neighbors in the company of older women asking for gifts, and she was thus "showered" with hemp skeins (Benet 1975). During wedding rituals of the southern Slavs, hemp was considered a symbol of wealth and a talisman for happiness, and when a bride entered her new home after the ceremony, she stroked the four walls with a bunch of hemp fiber and sprinkled hemp seeds to bring good fortune (Benet 1975). Once married, the bride's responsibilities dictated that she must perform the work of mature women. In the Heves region near the Tisza River in eastern Hungary, a newly married girl

FIGURE 62. Loading a Hungarian bride's hempen dowry textiles onto a cart for display to her community. A young woman's skills of weaving, sewing, and embroidering hemp cloth were held as a measure of her suitability for marriage (from Viski 1937).

entering her husband's household was sent to work in the fields or spinnery almost every day if necessary, and she was expected to take part in processing the hemp, spinning yarn, and weaving household textiles (Fél and Hofer 1969).

Marriages could also be terminated with ritual actions involving hemp. According to superstition, in the Balkan Montenegro region, a wife had been unfaithful to her husband if his hemp belt broke suddenly when he danced, and a common way of divorcing involved tearing apart a husband's and wife's hemp belts (Barjaktarević 1951). If a Serb husband in Kosovo exhibited unjustified crudeness toward his wife, she could neutralize his aggression by washing his shirt in the water used to wash raw hemp fiber (Vukanović 1986). Hemp was also commonly used in a ritual context to affect changes of a sexual nature. For example, in Serbia, hemp seed was also considered to be an aphrodisiac, and wearing hemp was thought to increase a man's sexual strength (Benet 1975).

Hemp rituals were also involved in barrenness and conception. The folklore of the Balkan Slavs denoted Wednesdays, Fridays, and Sundays as opportune for conception, since those days were under the influence of their pagan goddess who protected all women's work, including that involving hemp; consequently, during those days all work with textiles was forbidden (Radenković 1981). In Yugoslavia, a barren woman would leave her hemp birth shirt in a church overnight or drink water mixed with ashes of a hangman's hempen rope and then sleep three nights in the bloodied hemp shirt of a woman who had just given birth (Djordjević 1985). In Bosnia, on the eve of Saint George's Day a barren woman would drape an unused hemp shirt on the branches of a fruitful tree, and the next morning, if she saw that an animal had left tracks, the woman believed she would soon conceive (Schubert 1984).

In Eastern Serbia, women who continued to menstruate and could not conceive burned a piece of hempen towel that had been used for washing dishes on Holy Monday (the day after Palm Sunday) and then drank the ashes (Radenković 1996). Barren Serbian women desiring a baby would also find a hemp rope tied around a bundle of hemp stalks, untie it, put it in her bathwater, and bathe with it. Another traditional fertility practice was to hold a mock second wedding ceremony for a childless couple, during which the bride presented a skein of hemp fiber to the priest. Near Prizren in Serbian Kosovo, barren women believed in the fertile power of hemp threads that were wrapped around the church and encircled their own waists with them to become pregnant, and a husband could also carry this thread or a baby's diaper woven of it to aid in conception (Vukanović 1986).

According to Serbian belief, if a woman wished to prevent conception she would tie knots in her hempen belt, and when desiring to make conception possible again she would untie the knots (Schubert 1984). In Slovenian Croatia, as well as in Serbia, magic rituals employed to control birth also included tying knots in a hemp belt or rope, tying a hemp belt around the waist, or tying a hemp rope around a woman's right leg (Lovretić 1902; Jovanović 1993). In order to prevent conception, before the wedding ceremony a bride would tie one knot in her father's hemp belt for each year she wanted to postpone pregnancy, saying, "I am not tying this knot, but my own offspring," and then she would wear it against her skin during the wedding ceremony, and to conceive a baby she would untie all the knots reciting, "I am not untying this knot, but my own offspring" (Grbić and Djordjević 1907). A Serbian bride sometimes wore "white" hemp to encourage impotence of the male gender and prevent pregnancy (Grbić and Djordjević 1907). In Serbia, immediately after birthing and before the first nursing at her breast, the new mother would pierce her hempen birth shirt with an unused spindle to prevent a new pregnancy for a whole year (Grbić and Djordjević 1907; Mijatović 1909). In Risan, on the Bay of Kotor in Montenegro, the bride entered her new husband's house by walking over a carpet, under which was placed a man's hemp belt to encourage conception of a baby boy (Karadžić 1867).

Across Eastern Europe birthing taboos were also important. In the Zepa region of the former Yugoslavia, pregnant women did not walk under a hemp rope because it was believed that this could cause the baby to die during delivery from a tangled umbilical cord (Grbić and Djordjević 1907; Filipović-Fabijanić 1964). Nineteenth-century Russian women gave birth in a steam bath on a table upon which straw was spread beneath a piece of hemp cloth. Only much later, at the turn of the century, did women begin giving birth at home in a more hygienic environment, in a bed covered with coarse hemp cloth, and following the birth, the midwife was presented with a gift of hemp linen (Benet 1970).

Serbian women would bring a newborn girl a piece of hemp fabric to encourage her to become a proficient spinner and weaver (Blagojević 1984). Among Kosovar Serbs, on the occasion of her first visit after the wedding, the bride's mother would push a hemp weaver's distaff through her daughter's shirt as a way of easing the future birth of her daughter's child (Vukanović 1986; Jovanović 1993). In Kosovo and Montenegro, the husband would shake his hempen underpants above his birthing wife and ritually beat her with his hemp belt, which she then donned (Djordjević 1985; Vukanović 1986). Fumigating the mother's sexual organs with smoke from smoldering hemp seeds was also thought to make birth easier (Blagojević 1984; Vukanović 1986). In Germany, hemp sprigs were placed over the pregnant woman's stomach and ankles to prevent convulsions and ease childbirth (Benet 1975).

In Serbia, hemp rituals were employed after giving birth to protect women from attack by evil. In the Aleksinac region of southern Serbia, a circle of rope is placed in a woman's bed for eight days, while in the region surrounding Ohrid Lake, the beds of both the mother and baby are tied with hemp rope, which is also tied around the mother's waist, neck, arms, and legs. In southeastern Serbia near the town of Vranje, after the baby's first bath a woman tied a hemp rope around her waist in the belief that the rope will block disease from entering her body, and a hemp hackle and hemp threads were also placed in the bed for 40 days (Djordjević 1985).

Many examples from Eastern European folklore confirm the idea that hemp could link a spirit to the afterlife (Radenković 1981), and hemp has "never lost its connection with the cult of the dead" (Benet 1975). At funerals it was customary to throw a handful of hemp seeds into a fire as an offering to the dead, echoing Scythians and Pazryk tribes 2,500 years ago. Hempen funerary customs were likely introduced by Scythians or their ancestors during their advance into southeastern Russia and the Caucasus, where they remained for centuries before spreading westward into Eastern Europe (see, for example, Benet 1975; Sherratt 1991, 1997). Old Croatian spinners would burn raw hemp fiber to prevent agony in their dying moments (Trojanović 1911), and Slavic peoples believed that hemp cloth burial garments must not be hemmed as this could bring misfortune upon others in the family (Reljić 1989; compare with Chinese mourning traditions discussed earlier).

Traditional Russian funeral, burial, and mourning attire was especially distinctive. Clothing worn by the deceased was prepared by individuals for themselves during their own lifetimes, and, except for hempen shoes and shrouds, nothing special was made for the deceased by others (Benet 1951). This tradition parallels Hmong culture where family members produce hempen funerary dress for a person around the time of their marriage and make hempen burial sandals following their death. During the early twentieth century, older Russian women were buried in long homespun skirts, their traditional daily costume, as well as a ritual garment saved for death (Benet 1970). The corpse, covered with a coarse, homespun hempen shroud, was laid on a bench with its head under the family's icon, where mourners could pay their last respects. Two hours before the body was taken from the home, it was placed in a coffin lined with homespun hemp fabric (Benet 1970).

At Hungarian funerals, special types of hempen cloths covered the table when a priest performed funerary rituals at the deceased's home. Hempen table cloths were also used as a shroud or canopy for the confinement bed, and the main reason for their tenacious survival may be their ceremonial role: "Hungarian families of the Little Plain region treasure these hemp sheets as precious relics, left to them by grand- or great-grandparents. The sheets are washed in the Danube every spring, though they are only used on the occasion of funerals. At such times, the sheets are spread under the deceased in such a way that the rich embroidery, representing Adam and Eve, or an intricate pattern of birds and carnations, is on the side where the callers—bidding farewell to the dead—can see it" (Fél 1961). In Hungarian tradition, soon after death, the deceased's face and body were washed with soap and a damp ball of hemp fiber, and the corpse was then dressed in special funeral clothing, which for a lifetime

had been "kept . . . with the utmost care" (Mijatović 1909; Stojanović 1968). Funeral clothes were made of hemp linen sewn with hemp thread (Bošković-Matić 1962) and decorated with hemp fringes (Blagojević 1984). A Hungarian bride kept her wedding dress so that years later when she passed away it could become her funeral costume, and women began to embroider their funeral shroud at the age of 40 to ensure that they would enter the afterworld wrapped in their own handiwork (Viski 1937). In Atány village, east of Heves near the Tisza River, every elderly woman, rich or poor, prepared her own funeral costume (Fél and Hofer 1969), and in the Sárkös region, heirloom funeral pillows were laid under the deceased's head after ritual bathing (Fél 1961).

Among Romanians, the hempen shroud was woven in two halves, symbolizing a pair of wings for the dead's soul to let itself free, and an old woman sorcerer would spin a leash of hemp and tie the legs and arms, symbolizing a horse's reins for the ride into the afterlife (Macaj 1966; also see Chapter 4 for a discussion of ancient human, hemp, and horse relationships).

Slavs placed a hemp weaver's spindle and distaff in the deceased woman's coffin, believing that she would spin hemp fibers in her afterlife (Petrović 2000); a little pillow filled with first-gathered raw hemp fiber from the past season was also put in the coffin (Stojanović 1968). The needle with which the burial shroud was sewn was broken and also placed inside the coffin "so that it may not desire to sew another shroud" (Benet 1951). Slav peasants covered the coffin exclusively with a hempen burial shroud, never one of cotton or wool (Petrović 1992). Romanians covered the casket bedding with a hemp linen sheet (Kligman 1988), and after the burial, a candle was wrapped in hemp linen and given to the gravediggers (Macaj 1966). In Bosnia-Herzegovina, after carrying the dead to the cemetery, a piece was cut from the hempen burial shroud and tied around the attic beam in the room where the person died (Dedijer 1908).

The traditional mourning color of Eastern Europe was natural undyed hemp or white, but black became more common during the early twentieth century. However, in most cases mourning clothes were usually "not the rule among the peasants" (Benet 1951), and there was no proscribed period of hempen mourning in Europe as found in the Confucian mourning traditions of East Asia.

Balkan rituals prevented vampires arising from the deceased's grave. For example, the day after burials in the former Yugoslavia, a magic circle of hemp fiber was burned atop the fresh grave, and sorcerers would persuade the dead's spirit to remain in the grave and not wander around the village as a vampire (Djordjević 1985), and Bulgarian women would burn raw hemp fiber alone or mixed with sulfur and gunpowder around the grave (Milićević 1894; Krstanova 1999). Furthermore, before the burial, a skein of hemp yarn was wrapped around the corpse because in the ritual sense hemp was "clean" and could help protect the body from vampires (Prvulović 1982; Djordjević 1985; Petrović 1992). Romanians believed that if the hemp distaff was placed next to the corpse's head, it would prevent the deceased from becoming a vampire, as the spindle would force the vampire's soul to spin around in the grave as thread around a spindle (Macaj 1966). Fumigation with burning hemp was considered the best way to prevent a corpse from becoming a vampire. A naked woman, with unbraided pigtails, walked around the corpse as she burned hemp fiber mixed with bunches of hair taken from deceased's sexual organs and underarms; the

FIGURE 63. A traditional Hungarian woman in white hemp mourning costume; to her left hangs a hand-embroidered hempen death cloth (from Viski 1937).

ashes were then placed in the coffin with the corpse (Zečević 1975; Jovanović 1993).

Cannabis has had many associations with magic, sorcery, and witches. For instance, ancient Finns looked upon the deity Egres as the patron of herbs, peas, turnips, flax, and hemp (Bonser 1928); astrologers assign hemp to the rule of Saturn (Folkard 1884), and among Serbian Slavs, some magical incantations were exclusively performed in the river, especially at hemp retting places (Radenković 1996). Certain hempen "objects of the dead" were used as magical charms, including a hangman's rope, a dead woman's clothes or her distaff, thread from the cover of her coffin, or yarn from a dead woman's unfinished spinning (Radenković 1981, 1996). Along the border area between Austro-Hungary and Turkey, a sorcerer would take these threads and wrap them around a vicious person to cast evil spells on somebody or something (Begović 1986). In Montenegro, because of its connections with hemp, the distaff was believed to possess extraordinary magical power and was often hidden in the house loft so that no outsiders could find it and turn its supernatural force against the household. People would never hit a man with a distaff as he could die, nor hit a child with it because it would limit growth (Vukanović 1938). Croatian women never stored a distaff without leaving at least a few hemp fibers hanging from it ready to spin, or else the spinner and her family would go without clothes that year (Lovretić 1902; Djordjević 1985).

Southern Slavs believed that hemp yarn scraps remaining after warping and weaving were particularly dangerous, and they were used in sorcery to cast magic spells (Radenković 1996). Women never used these threads for making men's clothes as it was believed that they would be shot by hunters or attacked by wild animals (Petrović 1948; Kisgeci 1994). A hemp rope used for suicide or murder was also feared

(Petrović 1948); even though traditionally "it is believed that witches can use the plant to inflict harm, they are not likely to do so in fact, and hemp is often used against persons suspected of witchcraft" (Benet 1975).

Hemp could serve the sorcerer as a powerful tool or play a role in the sorcerer's demise. In November 1636, William Coke and his wife, Alison Dick, were condemned to death for the crime of witchcraft in England, where it was the normal practice for witches to be dressed in rough hemp coats and stuffed inside tarred barrels to facilitate their burning. In this case, the costs of the execution were recorded by the town council (Anon. N.D.; also see *National Archives of Scotland*, Kirkcaldy, CH2/224/1–6):

10 loads of coal	3 pounds, 6 shillings, and 8 pennies
Tar barrel	14 shillings
Hangman's rope	6 shillings
Hemp coats	3 pounds and 10 shillings
Making coats	8 shillings
Expenses of the judge	6 shillings
Executioner—for his pains	8 pounds and 14 shillings
Executioner's expenses	16 shillings and 4 pennies

Interestingly, the rough hemp coats cost as much as 10 loads of coal but less than half as much as hiring the executioner!

Spinning bees were important in Europe as they were in Asia. At the beginning of Lent, women and girls would push their spinning tools down a hill of snow to purify them of their sins before spinning work resumed again at Easter (Benet 1970). Hemp processing was usually performed as collective work and the motto "For me today, for you tomorrow" was uttered as young women gathered at spinning bees to help their elders with hempen chores (Hecimović-Seselja 1985). Girls five or six years old would start their training at spinning bees in preparation for weaving, which was usually carried out in each woman's home. Seniors would joke saying that the girl was "spinning for a gnome's girdle," but she would answer "not, but for my shirt" (Mijatović 1907). Amid hopes of promoting easy and beautiful spinning, young Kosovar Serb girls would throw spun hemp threads into a fire (Vukanović 1986). Along the Austro-Hungarian and Turkish border, "burning the navel" was a common spinning bee amusement, the "navel" being the fibers remaining on the distaff following spinning, and if a girl dozed off, the others would light fire to her "navel" to startle her awake (Begović 1986).

Spinning bees were a common social venue across Eurasia and provided a culturally condoned venue for young women to learn the traditions of their culture and gain fleeting exposure to the male community. Autumn and winter spinning bees were often the only amusement for the young in Serbian countryside villages (Mijatović 1907). The vicinity of the spinneries was where young couples chose each other for marriage (Medjesi 1978), and Bulgarian girls predicted their future husbands by burning dry hemp stalks for each young man and reading the ashes (Krstanova 1999). Sometimes, a young man would steal a skein of hemp fiber from one of the girls and run away, and if she could catch him it was praiseworthy for her. Otherwise, she would become ashamed when the skein of hemp was thrown in front of her house (Damjanović 1985).

In the Gruza region of Serbia, on Saint George's Day Eve, hemp seeds were soaked in water, and the following day the water was used by girls to wash their hair, while the seeds

FIGURE 64. In traditional European cultures spinning bees played an essential role in the social life of rural communities well into the twentieth century (Viski 1937).

were sown in a special plot so women could make hempen wig inserts (Mijatović 1909). To encourage thick hair, girls boiled hemp rope and used the water to wash their hair, and the rope was placed on the threshold of her home where she brushed her hair (Mijatović 1909). In the Gacko region of Bosnia, a girl would brush her hair while sitting on a hemp rope and a shovel, hoping that her hair would become as long as the rope and as strong as the shovel (Lilek 1894).

Hemp fiber and seed also appeared in several ritual roles during the Eastern European Christmas season, and as on many saint's days, work with hemp was proscribed during Christmas week. Between Christmas and the New Year, women did not open a pair of hemp hackles because it could encourage wild animal attacks (Mijatović and Bušetić 1925; Škarić 1939) or cause the cattle to become infected with worms (Vukanović 1986). On Christmas Eve, women would spin hemp (never wool) with a crude wooden stick, instead of their usual distaff, and all members of the household would ritually spin. The Christmas distaff would then take its place near the beehive for the evening, so the bees would stay inside (Djordjević 1985). Along the Austro-Hungarian and Turkish border, a woman would spin a long hemp thread, plait it, and then tie it onto the hearth chains to protect her household (Begović 1986).

Christmas was also a time designated for fertility rituals for the next year's hemp crop. In Zagreb, Croatian women pulled out lengths of straw placed under the tablecloth to foretell the height of the crop (Echkel 1980). On Christmas Eve among Serbs, the host removed the right shoe of the first guest and placed it high on the house ridge to make the following year's hemp crop grow tall. The honored guest would stay for three days of Christmas celebration and be treated to food and drink. Upon leaving he received a skein of hemp, a hemp shirt, or a towel as ritual gifts ensuring that the crop would be tall and of good quality. The first guest's hemp skein was used as medicine for the whole year, and if anyone had a fever, a magic circle of that skein would be burned (Bosić 1996; Milićević 1894; Jovanović and Bjeladinović 1964; Gavazzi 1939).

Farm animals were also involved in Christmas rituals associated with *Cannabis*. Along the border between Austro-Hungary and Turkey, and in Serbia, the head of the household would lead his best ox as the first ritual Christmas guest, and

it was given a ritual cake with hemp fibers on it (Bosić 1996); his wife would also tie a hemp skein around the ox's legs and tail, cover it with a beautiful embroidered hemp sheet, and put a hemp skein on its horns before the ox was taken away to its stall. Three days later, a shepherd would make a whip of the ox's Christmas hemp skein (Djuric 1934; Begović 1986; Vukanović 1986). Early on Christmas morning, people would tie all the children together with the hearth chains and burn a magic circle of raw hemp fiber around them for their good health (Mijatović 1909).

In Russia, Poland, and Lithuania, it was believed that the dead visited their families at Christmas, and hemp seed soup was served for their returning souls (Benet 1975). At Christmas dinner, Ukrainians sprinkled hemp, flax, and poppy seeds under the tablecloth to symbolize fertility; in addition, they presented carolers with hanks of hackled hemp fibers. In Ukrainian tradition, the earliest "gifts" to mankind are still celebrated at Christmas, including the everyday grain foods and fibers for clothing (Tracz 1999). Ruthenians in Voivodina hoping for great hemp fertility in the next season would serve hemp seeds for supper, and seeds, along with the best hemp fiber, were placed atop Christmas breads (Kostić 1975). Serbians broke off pieces of the Christmas cake and held them as high as the next hemp crop should grow (Mijatović 1907; Prvulović 1982), and in Kosovo they would hang a round cake on a hemp skein around the first guest's neck as he departed (Milićević 1894).

There was always plenty of entertainment associated with Christmas. Serbian women and girls, but also even the elderly, would propel themselves very high on swings made of hemp rope shouting wishes for great fertility of their crop (Djordjević 1985; Vukanović 1986). When the head of the family brought the Christmas tree home, his wife tossed hemp seeds over him, wrapped a skein of hemp yarn around the tree, and dressed him in a hemp shirt (Pantelić 1974). Christmas trees were commonly dressed with the youngest male child's (or father's) hemp shirt, and the ritually empowered Christmas tree would then protect the family (Petrović 2000).

Serbian Christmas carolers rode on wooden hemp breaker "horses" (Djordjević 1985) while twirling a woman's distaff for her to make beautiful clothing (Mijatović 1907). Along the border between Austro-Hungary and Turkey, women performed a theatrical ritual called "spinning the thread," and the hemp yarn involved in this ceremony would later be utilized for medicine (Begović 1986). In Voivodina, Serbia, and Kosovo, a special poultry food was mixed on the ground within a ritual circle made of hemp rope that had been used for tying the holy Christmas straw, and a hemp fiber skein was shaken above it so that poultry feeding within the special circle would stay in the yard and lay fertile eggs (Milićević 1894; Mijatović 1907; Djordjević 1985; Petrović 1992, 2000). On Christmas Eve, people tied the legs of the feast table with this rope, which was then placed in the cattle stall for its curative powers (Bosić 1996). Sometimes a Christmas magic circle of hemp rope was covered with hemp bark strips and ignited; then all the family members would huddle together in the center while it burned around them in this protective and purifying rite (Vukanović 1986). Shrovetide is known throughout much of southern Europe as Carnival and marks the beginning of the Lent season, leading up to Easter, which celebrates the resurrection of Christ. Shrovetide is thought to have its origins in pagan spring rituals to chase away evil spirits and that it inherited its sometimes wanton behavior from ancient Roman Saturn festivals. In some

FIGURE 65. Lynch mobs and hemp hanging ropes were common ritual elements of frontier justice. The painting "Cornered at Last" by Stanley Berkeley as it appeared in *The Illustrated London News*, April 2, 1892.

regions it spanned the period from Three Kings Day (January 6) until Shrove Tuesday (falling between February 3 and March 9) and across much of Europe was characterized as a time of excessive indulgence. In some Eastern European regions yarn was spun between the Christmas season and the beginning of Shrovetide when all hemp work became forbidden until the end of Easter season, and following Easter hemp seeds were sown and new crop cycle would begin. Bulgarian women believed their cheese would become worm-eaten if hemp work was carried out during Shrovetide (Djurić 1934). Old Croatian proverbs tell us, "How much you spin until Shrovetide as much you will weave until Easter" and "Hemp dwindles, but yarn appears" (Lechner 1954). In the Piedmont region of Italy, it was believed that hemp should be spun beginning on the last day of Shrovetide. A ceremony would be held to divine what sort of hemp crop might be expected, a bonfire was lighted, and the direction of the flames was attentively watched: "If the flames rose straight upwards, the crop would be good; but if they inclined either way, it would be bad" (Folkard 1884).

Shrovetide ritual swinging to improve the fertility of hemp crops is known from Italy, Greece, Scandinavia, Albania, Bulgaria, Kosovo, and in Serbia as noted earlier. During Shrovetide, Easter, and Pentecost (50 days after Easter), some young Slavs swing on hemp ropes, singing the traditional "Big Hemp" song and shouting, "So high, the highest hemp," expressing their hopes for a tall crop of thin hemp stalks and hopes that the hemp plants would grow tall (Ilijin 1963; Bosić 1985; 1996; Kasuba 1974; Poznanović 1988; Kostić 1989).

In Serbia and Moravia, women would not comb, spin, or do any work with hemp during White Week (the last week of Lent) to prevent their cattle from becoming infested with parasites (Mijatović 1907) and to suppress vermin in general (D. M. Djordjević 1958). Voivodinian Serbs believed that if one wore a shirt woven of hemp sown during White Week, he would contract typhus, and that hemp sown on these days would be stricken by hail stones (Kisgeci 1994). During White Week, after the evening prayer service, old Serbian women would dance "the Peacock," a special ritual folk dance about sowing, growing, and processing hemp for

an abundant harvest (Knežević and Jovanović 1958). These exclusively women's rituals of singing, dancing, and jumping high were all aimed at producing good hemp the next season (Bosić 1985, 1996). On Easter Sunday a hemp rope or hemp threads were burned as a magical means to encourage the fertility of crops, cattle, and people, and the ashes were retained as a medicine for curing diseased limbs (Djordjević 1985).

Hempen rituals were important in many European regions and were employed in a variety of purifying and protective roles. Although found throughout the Christian realm wherever hemp was traditionally grown, these rituals were most popular in the pagan-influenced Eastern Orthodox Balkan and northern Black Sea areas, where aspects of these ancient ritual traditions are still practiced today.

Hangings: Hemp Cordage and Hempen Executions

There is an herbe which light fellowes merily will call Gallowgrasse, Neckeweede, or the Tristrams knot, or Saynt Audres lace, or a basterde brothers badge. (BULLEYN 1562)

Another notable association of hemp with death rituals was its use in the taking of life by hanging. Hemp figured prominently in the history of law enforcement and execution since its fibers were traditionally used to make the hangman's noose in Europe and North America (Shushan 1990; also see Chapter 5). Hemp rope was preferred for hangings because of its strength and relative lack of stretch, although there may also be ritual reasons for choosing *Cannabis*. Since medieval times, the phrase "to stretch hemp" was a euphemism for hanging (Skinner 1911; also see Barr 1891 for a critical discussion of the lack of complete rigidity in hemp rope used in executions). In our overview of hemp rope use in hanging that follows, we focus primarily on such utilization in England and North America.

Throughout its history, execution by hanging was believed to be a more shameful way to die than either death by firing squad or beheading (Laurence 1960): "Historically, hanging was intended as a stamp of indignity emblazoned on the lifeless bodies of criminals" (Huang 1995). Execution by hanging is said to have been first used by Persians and subsequently was spread all the way to the British Isles, most directly from Central Europe with invading Germanic tribes of Angles, Saxons, and Jutes in the fifth century CE, who brought the trait as a significant component of their culture (Gray and Stanley 1989; also see Duff 1953). Hanging "first became an instrument of the state under the reign of Edward III in 1334 and has remained a part of English and American law ever since" (Gray and Stanley; also see Bishop 1965). During the early reign of the English King Henry VIII (lived 1491 to 1547) there was considerable distress among working people as a large number of discharged soldiers were unable to find work. Even though monasteries offered help, England was swarming with vagabonds and thieves as beggars and criminals overran the realm: "Bishop Latimer, a noted preacher of that day, declared that if every farmer should raise two acres of hemp, it would not make rope enough to hang them all. Henry, however, set to work with characteristic vigor and made away, it is said, with great numbers, but without materially abating the evil" (Montgomery 1912; also see Watkins 1824). There are also many colorful literary references to hempen hangings. For example, Pistol, a character in William Shakespeare's *Life of King Henry the Fifth, the Book* (act III, scene VI, "The English Camp in Picardy") refers to the use of hemp rope in a hanging:

> Fortune is Bardolph's foe, and frowns on him;
> For he hath stolen a pax, and hanged must a' be,
> A damned death!
> Let gallows gape for dog, let man go free
> And let not hemp his wind-pipe suffocate.
> But Exeter hath given the doom of death
> For pax of little price.
> Therefore, go speak; the duke will hear thy voice;
> And let not Bardolph's vital thread be cut
> With edge of penny cord and vile reproach:
> Speak, captain, for his life, and I will thee requite.

The playwright John Fletcher (lived 1579 to 1625) presumably collaborated with Shakespeare on some the bard's works such as *Henry VIII* and unfortunately died of the plague; in act 2, scene 2 of his play, coauthored with Francis Beaumont (Beaumont and Fletcher 1718), titled *The Bloody Brother*, the following song appears:

> Three merry boys, and three merry boys,
> And three merry boys are we,
> As ever did sing in a hempen string
> Under the Gallows-Tree.

This passage was minimally altered and used in another old English play, *Rollo, Duke of Normandy* (act III, scene 3), which was first written about 1617, in collaboration by a group of well-known English Renaissance dramatists and rendered here by Rudgley (1998):

> Merry boys are we, as e're did sing.
> In a hempen string, under the gallows tree.

The most famous riverside hanging site in England was the "Execution Dock," a place in London where pirates and smugglers were hanged to death after being sentenced by the British admiralty. It is located adjacent to the Thames at Wapping. Although this "dock" was merely a scaffold, it was used for more than 400 years up until 1834 to execute those who committed "crimes" at sea. Among a procession of dignitaries, the condemned men would be transported in a cart from prison to the scaffold; big crowds of people would line up along the riverbank so they could obtain a clear view of the procession and the hanging that ensued. A shortened rope was deliberately used in these hangings so as to increase the suffering of the convicted that would be strangled to death slowly and painfully. Here they were dispatched or, in the words of the river people, they "danced the hempen jig," as their limbs flailed about in their death struggle (Ackroyd 2007). "Hempen jig" is an old pirate term that was part of an old warning: "If those scalawags catch us, they'll make us dance the hempen jig!"

Because hemp often furnished the means of death, it was sometimes thought to be an evil omen associated with the underworld. However, it was more often considered to be "used for goodly purposes" such as "shutting off the wind of rogues" (Skinner 1911), and therefore hemp was most commonly associated with good fortune. Having "some hemp in your pocket" was to have luck on your side in the most adverse circumstances. The French saying, *Avoir de la corde-de-pendu dans sa poche* ("To have a hanging rope in your pocket") refers to a hempen hanging rope promoting good fortune (Brewer 1898). As noted earlier in this chapter, hemp hangman's ropes were also accorded with special protective powers in Slavic regions. When a Serbian soldier entered the army, his mother sewed a piece of a hangman's rope into his clothes as a protective amulet (Begović 1986). When accused persons went to court in Serbia they would carry a hemp rope with which someone had been hung or a hemp thread from a funeral shroud to protect them from a guilty verdict (Begović 1986; Petrović 1992; Škarić 1939).

Hempen hangings followed colonists to the Americas. Ironically, the backbreaking work of hemp fiber processing in pre-Revolutionary America was often accomplished "through the forced labor of prisoners," with that same hemp fiber being used to "break the necks of some of those same prisoners" (Shushan 1990). A precursor of this ironic "twist" of fate was graphically displayed in the hemp imagery of "A Harlot's Progress" painted in 1732 by the Englishman William Hogarth, which is well described by Abel (1980): "The fourth illustration in the series depicts Mary beating hemp in Bridewell Prison, a house of correction in Tothill Fields, Westminster, for harlots such as herself and other sundry immoral characters. Hogarth portrays her holding a large mallet in her hands while the hemp strands lie in front of her on two tree stumps." Hanging was also a common form of execution in the American "Wild West." Vigilante groups were referred to as "hemp committees," and "sowing hemp" was an epithet for someone "on his way to a rendezvous with the hangman" (Abel 1980). The hangman's noose was often referred to by epithets such as "hempen collar," "hempen necktie," and many others (e.g., see Green 1999), and a person who had been hanged was regularly known as one who had died of "hempen fever." A woman whose husband had been hanged was referred to as a "hempen widow." A horse thief's tombstone epitaph from Rapid City, South Dakota, read, "We're bound to stop this business, or hang you to a man. For we've hemp and hand enough in town, to hang the whole damn clan" (Shushan 1990).

Conclusions

In the course of this study, a number of apparent commonalities between the hempen beliefs and rituals of the Far East and those of Europe have been identified. The ritual use of *Cannabis* seed, fiber, yarn, cordage, and/or cloth was widespread across Eurasia until the late nineteenth century. Hemp is most generally associated with the ritual realms surrounding purification, protection, healing, divination, and the life cycle events of birth, marriage, and death. It is still utilized for a majority of these ritual intents in various cultures including those of the Hmong, Han Chinese, Korean, Japanese, and many European peoples, especially those in the eastern regions. Perhaps most significantly, hemp cordage and cloth commonly served as ritual pathways for souls being directed to and from the afterlife, for ancestor spirits and deities summoned to return and assist in healing ceremonies, and even for the ritual practitioner's spirit traveling to other realms. At other times hemp also has served to contain the soul or bar evil spirits from entry. These themes, and even many of the same ritual actions, reappear time and again in the cultures we have studied.

However, there are a few uses that either are not practiced by certain cultures or were omitted from written records. Apparently the Hmong do not practice hempen mourning except where they have adopted it from their Han Chinese neighbors, and they may not use hemp to purify supplicants or ritual spaces. Han Chinese traditions do not reflect any specific uses of hemp in purification, there is no association with spirit travel, and hemp is also absent in marriages and divination rites. Hemp also seems to be absent from Korean marriage and divination rituals, and Korean shamanic ritual practices do not invoke hemp in a protective role. Only the Japanese omit hemp at funerals, and there is also no record of ritual healing practiced with hemp in Japan. European traditions include hemp in all aspects of ritual intent, except for an absence of spirit travel.

There are inherent pitfalls of circular logic involved in interpreting ritual intent if it places special significance on a very common yet culturally and economically important plant and its products. In each cultural context we must weigh each bit of evidence critically to determine what would give hemp its ritual significance in each situation. Was hemp really ritually important, or was it simply the most commonly available plant from which to procure a specific product for ritual use (e.g., fiber, cordage, cloth, paper, etc.)? What were the ritual intents of events reported in historical documents? As traditional agrarian cultures evolve into modern-day consumer economies, many continue to use hemp in rituals rather than substituting some recently available market alternative. How do present-day ritual practitioners, parishioners, and supplicants explain the persistence of ritual hemp use? Answers to these questions provide us with insights into understanding the significance of ritual hemp use in the past and present.

Skeptics may point out that within the range of each early "hemp culture" we have previously referred to, *the only readily available plant fiber was hemp*, and therefore it was simply by coincidence that these cultures developed ritual systems involving it. It should be noted, however, that the Far East has a long history of ramie (*Boehmeria nivea*), mulberry (*Broussonetia papyrifera*), and kudzu (*Pueraria thunbergiana*) textiles rivaling that of hemp in antiquity, yet considerably more ritual use is accorded to hemp than these other traditional fiber sources. Europe has an ancient record of flax

(*Linum* ssp.), nettle (*Urtica* spp.), and lime tree (*Tilia* ssp.) use as fiber sources, but there are only relatively minor ritual uses for all these plants, while the ritual use of hemp was much more common and widespread. Throughout these cultures hemp is a cultivated crop plant requiring large inputs of time and labor, and hemp fiber and seed were economically valuable. The sacrificial use of hemp during ceremonies indicates that a ritual use was important enough to warrant using a valuable commodity, even though there were other available plant materials growing wild in the immediate surroundings that could have been used instead. *Cannabis* may be accorded higher ritual status than other wild and cultivated fiber plants with long histories of human contact simply because it provides a nutritionally valuable seed and/or psychoactive resins in addition to spinable fiber.

We feel that the shared aims of ritual hemp use across divergent Eurasian cultures rests in their common origins in the shamanic past, dating from the time of first contact between ancient hunter-gatherer bands and stands of wild *Cannabis*. Hemp has a direct link with shamanism because *Cannabis* is psychoactive and served as a valuable ally in shamanic vision and spirit quests, a linkage now manifested in the ritual use by shamans of hemp rope and cloth bridges as passageways for the spirits across much of Eurasia from the British Isles to Japan. Throughout much of Eurasia, we find a widespread belief in the magical powers of *Cannabis* for healing, protection, and purification, and for this it is generally accorded high status. However, the level of ritualized respect for *Cannabis* varies, from the low status of rough hempen sackcloth mourning costumes as an expression of grief and guilt to the highest Shintō reverence for hemp fiber's natural origins and purity. The strongest traditions of ritual hemp seed or fiber use appear to have arisen only across northern temperate Eurasia, within cultures that grew industrial hemp but lacked historical psychoactive *Cannabis* traditions, while hempen ritual beliefs related to seed or fiber use are absent in regions with strong and continuing traditions of psychoactive *Cannabis* use such as South Asia. This dichotomy of ritual emphasis and association with differing ethnobotanical products may once again be perceived as simply an artifact of plant availability and geographical location rather than being culturally determined.

But, we argue, the extreme ritual reverence accorded hemp in historical as well as several contemporary cultures ultimately results from its ancient psychoactive use, probably in ritual settings. Most likely, primordial eastern Asian *C. indica* ssp. *chinensis* or broad-leaf hemp (BLH) contained sufficient amounts of THC to be psychoactive. We assume that ancient humans were sufficiently in tune with their local environments and supplies of plant-based products that they were also well aware of *Cannabis*'s psychoactive potential and at least some of them ingested *Cannabis* in psychoactive ritual contexts (see Chapters 4 and 7). East Asian psychoactive ritual use may have led to the widespread concept of spirit travel facilitated by hemp, while the absence of hempen spirit travel in China is, to a large degree, the result of Confucian proscriptions against certain types of altered consciousness (e.g., see Li 1974a). The lack of ritual concepts of spirit travel in Europe likely reflects the extremely low levels of THC and relative nonpsychoactivity of native European *Cannabis sativa* narrow-leaf hemp (NLH). No matter what the causes of individual regional variations in the nonpsychoactive ritual uses of *Cannabis*, it is readily apparent that ritual hemp use is very widespread and shares an ancient history across Eurasia.

CHAPTER TEN

Recent History of *Cannabis* Breeding

Crops with multiple uses have special significance. The longer a
crop has been in cultivation, everything else being equal, the more
likely it is that various kinds of uses will have been found for it.

(AMES 1939)

Introduction

Only a very limited number of *Cannabis* farmers—whether
they grow it for fiber, seed, or drugs—consciously improve
their crops through selection and breeding. Hemp fiber cultivation and breeding are no longer practiced in the United
States, and presently there are only a handful of industrial
hemp breeding programs worldwide. Clandestine marijuana
breeders secretly work to improve drug types of *Cannabis*, but
the vast majority of marijuana growers practice no selection
at all and either buy seeds of named *sinsemilla* (Spanish for
"without seed") varieties from European or Canadian seed
companies or sow accidentally produced seeds from imported
or domestic marijuana.

Although *Cannabis* produces copious quantities of both
pollen and seed, it is not a particularly straightforward
plant to breed, as its life history presents several obstacles to
improvement by selective breeding. *Cannabis* varieties are
almost always dioecious, with male and female flowers occurring on separate plants, and thus are normally incapable of
selfing (self-fertilizing). Selfing is the most effective sexually
reproductive means of fixing desirable traits, since selected
genes are more likely to be represented in both the male pollen and the female ovule if they come from the same plant. In
Cannabis breeding, the genes controlling a selected trait must
be present in two separate individuals—one male or pollen
parent and one female or seed parent. Female plants supply
most of *Cannabis*'s economically valuable products, including fibers, seeds, or drugs, while male plants merely fertilize
the females and may produce high-quality fiber. This makes
it difficult for plant breeders to recognize potentially favorable traits in a male parent as these traits must ultimately be
expressed in female offspring. All *Cannabis* plants are wind

pollinated, which allows them to intercross freely; therefore,
in order to avoid random fertilization and seed set, selected
female seed parents must be isolated from males until they
are to be fertilized with a selected male pollen parent.

Most importantly, it is highly illegal in most countries for
plant breeders to work with drug *Cannabis*. The illegality and
high visibility of marijuana cultivation makes it preferable
for illicit breeders to limit the size of outdoor gardens and
the frequency of visits to observe their plants. This lowers the
number of potential parent plants breeders have to choose
from; it also limits the amount of time spent observing them
and making effective selections. Modern indoor gardens are
usually small; as a result, clandestine cultivation is focused
on utilizing all available space for the production of vegetative female cuttings and mature flowering plants. Therefore,
little or no space is devoted to males required for breeding.

This chapter documents the twentieth-century breeding of
fiber, seed, medical, and recreational cultivars; in the process,
we outline the cultivars' pedigree lineages, emphasizing the
roles of individual plants from diverse gene pools in the evolution of these cultivars via human selection.

European Hemp Breeding

European hemp breeders, like the vast majority of plant
breeders, started by improving their locally available, native
landraces and eventually made crosses with alien imports to
create improved, higher yielding fiber and seed varieties. It is
evident from their breeding histories and chemo-taxonomic
profiles that considerable mutual genetic relatedness exists
among modern European cultivars (de Meijer 1995; Hillig
2004a). Landraces belonging to the Mediterranean and

Central Russian fiber hemp ecotype groups (Serebriakova 1940) were determined by Hillig (2004a) to be *C. sativa* narrow-leaf hemp (NLH). Crosses between these two ecotype groups formed the basis of present-day European hemp cultivars. Feral populations of *C. sativa* ssp. *spontanea* (NLHA) extant in many continental regions of Europe are most often descendants of previous hemp crops and are genetically related to fiber and/or seed strains once grown in that area, yet they often appear quite different.

Early in the twentieth-century Chinese broad-leaf hemp (BLH) landrace varieties were incorporated into North American hemp varieties and were later used by European breeders. A Chinese strain is presently used in Hungary as a breeding parent to promote heterosis (hybrid vigor) because the Chinese BLH variety is relatively unrelated genetically to the European NLH crossing partners (Bócsa 1993, 1994). The degree of genetic difference between East Asian BLH and European NLH gene pools was illustrated by Hillig (2004a), who placed them in subspecies of different species, *C. indica* ssp. *chinensis* and *C. sativa* ssp. *sativa*, respectively. The heterosis effect seen in the Hungarian cultivar 'Kompolti Hibrid TC' results because the hybrid combines gene pools differing at the species level.

Many European countries produce and market sowing seed of registered industrial hemp varieties, and several are approved for fiber or seed cultivation in the European Union and Canada. Brief summaries of the breeding histories of hemp varieties from various countries are provided later in the chapter, outlining the breeding strategy employed. The emphasis here is on the interactions of differing gene pools and the effects of individual plant selections on the evolution of industrial hemp cultivars. The bulk of this information is adapted from Etienne de Meijer (1995).

French cultivars are monoecious, with male and female flowers appearing on the same plant, and are grown primarily for making specialty paper pulp, animal bedding, and construction materials. Breeding in France today is mainly aimed at maintenance of extant cultivars (conservative breeding) and at further reduction of their THC content. Genetic homogeneity of cultivars must be maintained by careful selection of monoecious parents each year. Seed for sowing is readily available in two types. Crops grown from elite seed consist almost entirely of monoecious plants, while those grown from seed harvested from free-pollinated crops raised from elite seed include 15 to 30 percent dioecious individuals resulting from natural genetic drift in the absence of human selection.

French cultivars are selected directly from 'Fibrimon' a monoecious crossbred cultivar with high fiber content originally developed at the Max Planck Institute in Germany by Reinhold von Sengbusch between 1951 and 1955 (Sengbusch 1956; also see Bredemann et al. 1956, 1961). Parental populations were obtained from several sources and inbred monoecious lines were created from individual monoecious plants occurring spontaneously in 'Havelländische' or 'Schurigs' hemp cultivars, both selected from Central Russian NLH or NLHA gene pools. The German variety 'Fasamo,' registered in the early twenty-first century, was obtained from a cross between 'Schurigs' hemp and 'Bernburger einhäusigen' monoecious hemp bred in Bernburg in the 1940s. Dioecious selections with very high fiber content from Germany (also descendants of Central Russian NLH or NLHA populations) and dioecious late-flowering landraces from Italy and Turkey (NLH or possibly NLH/BLH hybrids) were also used as breeding parents, and pseudomonoecious cultivars were

selected from these cross-progenies. 'Fibrimon' was grown in France and other European countries in the late 1950s, and further hybridization with selected populations began in the 1960s. Monoecious cultivars 'Fibrimon 21,' 'Fibrimon 24,' and 'Fibrimon 56' were selected directly from 'Fibrimon' for differing dates of maturity. 'Férimon 12' is an earlier maturing selection from 'Fibrimon 21,' especially intended for seed production (the higher the numerical appellations of French cultivars, the later they will flower and mature). Pseudomonoecious cultivars were developed by crossing monoecious 'Fibrimon' with an exotic dioecious line and then backcrossing to 'Fibrimon.' 'Fédora 19' resulted from a cross between female plants of the Russian dioecious cultivar 'JUS 9' and monoecious 'Fibrimon 21' individuals, followed by backcrossing of the unisexual female F₁ population with 'Fibrimon 21' plants as pollen donors. The female parent ('JUS 9') comes from crossing 'Yuzhnaya Krasnodarskaya' (originally selected from Italian hemp and therefore possibly an NLH/BLH hybrid) with dwarf northern Russian hemp (probably NLHA). Likewise, 'Félina 34' resulted from a cross between the dioecious female parent 'Kompolti' from Hungary and monoecious 'Fibrimon 24' pollen parents, followed by backcrossing with 'Fibrimon 24.' 'Fédrina 74' and 'Futura 77' both resulted from a cross between a dioecious female parent from the 'Fibridia' cultivar and a 'Fibrimon 24' pollen parent, followed by backcrossing with 'Fibrimon 24.' 'Fibridia' originated from the same German breeding program as 'Fibrimon' and shared the same ancestors, except for the monoecious 'Schurigs' inbreds, and therefore it was dioecious rather than monoecious.

Although the traditional Italian cultivars 'Carmagnola,' 'Fibranova,' 'Eletta Campana,' and 'Superfibra' are practically unavailable today, the Italian gene pool formed an important foundation for European and American hemp breeding. East Asian BLH landrace varieties were brought to Italy early on, and as they represent some of the few introductions of the BLH gene pool to Europe, they were subsequently responsible for much of the heterosis effect (increased vigor resulting from hybridization) sought by European hemp breeders. The present unavailability of Italian cultivars is due largely to legal reasons. After centuries of *Cannabis* farming for fiber, hemp cultivation is now prohibited in Italy unless a cultivar exhibiting an obvious morphological marker (e.g., yellow stems) genetically linked to low THC content is grown, and no such variety is yet available there.

All the Italian cultivars are dioecious. 'Carmagnola' is a northern Italian open-pollinated variety of great antiquity that may have originated from a Chinese BLH landrace and formed the basis of Italy's famed high-quality hemp textile production. 'Carmagnola Selezionata' was selected from 'Carmagnola' in the early 1960s. 'Fibranova' was selected in the 1950s from the progeny of 'Bredemann Eletta' × 'Carmagnola'. The parent 'Bredemann Eletta' was received from the German Max Planck Institute and was one of Gustav Bredemann's high fiber selections also obtained from Northern and/or Central Russian NLH or NLHA landraces used in breeding 'Fibrimon' and 'Bialobrzeskie.' A third variety 'Eletta Campana' resulted from a cross between traditional 'Carmagnola' and high fiber strains of German origin, most likely 'Fibridia' or one of Bredemann's selections.

Most Hungarian cultivars are dioecious and used for production of twine, rope, and heavy weight industrial textiles. 'Kompolti' was selected for high fiber content from 'Fleischmann' hemp and was registered in 1954. 'Fleischmann' hemp was originally selected by Rudolph Fleischmann from an Italian variety, possibly a NLH/BLH hybrid as well. 'Kompolti Sárgaszárú' is a chlorophyll-deficient yellow-stemmed cultivar (registered in 1974) developed as a raw material source for paper mills, where low chlorophyll content in the stalks is of advantage during pulping. This cultivar was obtained from a cross between 'Kompolti' and a spontaneous yellow-stemmed mutant that Helle Stengel-Hoffmann found in Germany in the offspring of a cross between Finnish early maturing (NLH or NLHA) and Italian late-maturing hemp (possibly an NLH/BLH hybrid), which was then repeatedly backcrossed with 'Kompolti.'

Hungary is the only country where intentional heterosis breeding of hemp was implemented under the direction of Dr. Ivan Bócsa of the GATE Research Institute in Kompolt (founded by Rudolph Fleischmann). This breeding for hybrid vigor resulted in several F_1 hybrid cultivars. 'Uniko-B' is a single-cross hybrid cultivar registered in 1969 and is the progeny of 'Kompolti' × 'Fibrimon 21' in which the monoecious 'Fibrimon 21' was the pollen parent. Crops grown from the 'Uniko-B' F_1 hybrid consist of nearly all female plants, resulting in high seed yields. 'Uniko-B' is also used to produce an F_2 commercial sowing seed containing approximately 30 percent males, cultivated for fiber as well as high seed yield. 'Kompolti Hibrid TC' (registered in 1983) is a three-way cross hybrid in which two selections from Chinese origin (both likely BLH), dioecious 'Kinai Kétlaki' and monoecious 'Kinai Egylaki,' were crossed with 'Kompolti.' The first cross ('Kinai Kétlaki' × 'Kinai Egylaki'), where the monoecious line was the pollen parent, produced a unisexual, almost purely female F_1, known as 'Kinai Uniszex.' Unisexual female lines are analogous to male sterile breeding lines in other crops as they produce almost no pollen. The unisex line is subsequently used as the female parent in the crossing 'Kinai Uniszex' × 'Kompolti,' which produces the commercial three-way cross hybrid 'Kompolti Hibrid TC' with a restored 50 percent female to 50 percent male sex ratio. 'Fibriko,' registered in 1989, and 'Fibriko TC,' registered in 2007, are recent Hungarian hybrids. They result from a three-way cross in which 'Kinai Uniszex' is crossed with the yellow-stemmed pollen parent 'Kompolti Sárgaszárú.' However, 'Fibriko' and 'Fibrico TC' are not yellow-stemmed, as the normal green stem color is dominant.

The Polish cultivars 'Bialobrzeskie' and 'Beniko' are monoecious and intended mainly for production of cordage, military fabrics, blended yarns (hemp with wool and cotton), fiber board, and seed oil products. 'Bialobrzeskie' (registered in 1968) resulted from serial crossings of predominately NLH dioecious and monoecious strains ('LKCSD' × 'Kompolti' × 'Bredemann 18' × 'Fibrimon 24') followed by long-term single line selections for high fiber content. The dioecious parent 'LKCSD' was selected from 'Havelländische' or 'Schurigs' hemp (NLH or NLHA) of Central Russian origin as was 'Fibrimon.' Dioecious 'Bredemann 18' is a selection from Germany (originally also Central Russian NLH or NLHA) and is very rich in fiber. 'Beniko' was obtained by individual progeny selection from the cross 'Fibrimon 24' × 'Fibrimon 21' and was registered in Poland in 1985. Continued breeding has resulted in new monoecious cultivars with the tentative names 'W-1,' 'Dolnoslaskie,' and 'D/83.' They are apparently very low in THC and have finer fiber than 'Bialobrzeskie' and 'Beniko.'

Present-day Romanian hemp cultivars are used to produce fine quality yarns and fabric as well as industrial textiles and cordage. 'Fibramulta 151' is a dioecious selection from a single

FIGURE 66. 'Kompolti Sárgaszárú' or Kompolti "yellow stem" is a chlorophyll-deficient cultivar (registered in 1974), developed as a raw material source for paper mills where low chlorophyll content in the stalks is of advantage during pulping. Its pronounced yellow color appears as the crop ripens, making it easy to distinguish from all other *Cannabis* crops lacking the yellow stem trait.

cross, 'ICAR 42-118' × 'Fibridia' (originally from Germany), and was registered in 1965. The parent 'ICAR 42-118' was the progeny of an Italian 'Carmagnola' × 'Bologna' (another Italian landrace) hybrid crossed with the Turkish 'Kastamonu' landrace (NLH). The dioecious 'Lovrin 110' was registered in Romania in 1981 as a replacement for 'Fibramulta 151' and was bred by selection among family groups from the Bulgarian 'Silistrenski' landrace. Monoecious 'Secuieni 1' was registered in 1984 and results from the crossing of 'Dneprovskaya 4' × 'Fibrimon' followed by two semibackcrosses (mating of offspring with members of the parental population) with 'Fibrimon 21' and 'Fibrimon 24,' respectively. The Russian dioecious parent 'Dneprovskaya 4' was selected from 'Yuzhnaya Krasnodarskaya,' which, again, was obtained from Italian hemp (possibly an NLH/BLH hybrid). Another Romanian cultivar named 'Irene' became available in 1995.

Eight cultivars are presently grown in central and southern parts of Ukraine and Russia and are used to produce cordage, core for steel cables, and industrial textiles. Hemp cultivars from the former USSR are classified into maturity groups or geographical types (Serebriakova 1940). Current cultivars belong either to the southern, later-maturing group or to hybrid progenies from the earlier maturing central group crossed with later maturing southern hemp. Hybrid cultivars between the later- and earlier-maturing groups are intended for sowing at higher latitudes in central and northern Europe than those ecologically preadapted to southern Europe. The dioecious southern cultivar 'Kuban' was registered in 1984 and was obtained by 10 cycles of family group selection in the hybrid population from the crossing 'Szegedi 9' × 'Krasnodarskaya 56.' 'Szegedi 9' was selected in Hungary from the native 'Tiborszállási' landrace (NLH or NLHA) crossed with 'Krasnodarskaya 56,' which is probably a selected cross progeny from local Caucasian (NLH or NLHA) and Italian strains (possibly NLH/BLH hybrids). The monoecious southern cultivar 'Dneprovskaya Odnodomnaya 6' was obtained in a similar way by family group selection within the progeny from 'Szegedi 9' × 'Fibrimon 56' and was registered in 1980. 'Zenica' (synonym 'Shenitsa'), a dioecious southern cultivar, was registered in 1990.

The remaining Russian and Ukrainian monoecious cultivars exhibit a southern, later maturing growth pattern but are also cultivated at higher latitudes to promote a longer vegetative growth period, resulting in taller stalks and increased fiber yield. 'USO-11' was bred from three parental populations: 'Dneprovskaya 4' (see 'Secuieni 1'), 'USO-21,' and 'Dneprovskaya Odnodomnaya 6.' 'USO-11' was registered in 1984 and is presently grown in Canada and New Zealand for oil seed production. 'USO-13' was bred from selected progeny of the cross 'USO-16' × 'Dneprovskaya Odnodomnaya 6' and was registered in 1986. 'USO-14' was a further selection from 'USO-1,' which again was bred from selected progeny, this time from the cross 'JUS-6' × 'Odnodomnaya Bernburga' and was registered in 1980. The dioecious parent 'JUS-6' was selected from a cross between 'Yuzhnaya Krasnodarskaya' (originally from Italy and possibly an NLH/BLH hybrid) and "dwarf Northern Russian hemp" (NLHA). 'Odnodomnaya Bernburga' is a monoecious cultivar originally produced in Germany in the 1940s. 'USO-16' was selected directly from the late-maturing French cultivar 'Fibrimon 56' and was registered in 1980. 'USO-31' was selected from the crossing 'Glukhovskaja 10' × 'USO-1' and was registered in 1987. The parental population 'Glukhovskaja 10' came from selections of the central Ukrainian 'Novgorod-Seversk' landrace (NLH or NLHA). 'USO-15' was developed by family group selection among the progeny from the cross 'USO-11' × 'USO-13' and was registered in 1995. Apart from the previous cultivars, the landrace 'Ermakovskaya Mestnaya' is cultivated on a significant scale in Siberia and belongs to the Central Russian maturity group (Serebriakova 1940), but it has low fiber content in comparison to more highly selected cultivars.

The 'Flajsmanova' and 'Novosadska konoplja' cultivars were registered in the former Federal Republic of Yugoslavia where presently hemp is grown in Serbia and Croatia, primarily for cordage. 'Novosadska konoplja' is an improved selection from 'Flajsmanova,' which is the same as 'Fleischmann' hemp of Italian origin (see under Hungarian cultivars) and was bred in the 1950s.

'Finola' is an early maturing Finnish variety used for seed production at more northern latitudes and was registered in the EU in 1999. It is extensively grown in Canada for seed and seed oil production (e.g., see Small and Marcus 2002). 'Finola' was selected from an F_1 hybrid cross between two early ripening northern Russian NLHA landraces (VIR-313 and VIR-315). It was obtained from the Vavilov Research Institute (VIR) *Cannabis* germplasm collection in St. Petersburg, Russia, via the International Hemp Association in Amsterdam. The Canadian seed variety 'Anka' was also selected from VIR accessions. 'Finola' (previously called 'FIN-314') matures early providing both high seed yields (up to 1.7 metric tons per hectare or 1,500 pounds per acre) and short straw (stalks); it can be harvested by combine harvesters for both fiber and seed yields from the same crop (Weightman and Kindred 2005).

European industrial hemp varieties are now also grown in Australia, Canada, New Zealand, and the United Kingdom where they are used for hemp fiber, hurd (woody pith), or seed production. Breeding goals in the recent past were concerned with developing more uniform monoecious varieties and increasing resistance to various pests (e.g., see McPartland et al. 2000). Presently, European fiber hemp breeders are focusing on reducing already negligible levels of psychoactive Δ^9-THC (referred to as THC throughout most of this book) to near zero. The French cultivar 'Santhica' was supposedly selected from a single plant with no THC and high CBG content and may lack the THC synthase gene (B_T allele), but it has not been released commercially (de Meijer 2004).

Most European hemp varieties are still grown commercially within their own countries as well as elsewhere. Research institutes may develop new cultivars for specific uses, while government organizations and/or seed companies continue to reproduce and sell sowing seed of existing varieties. As long as hemp remains an economically viable crop and farmers choose to grow it, these varieties should remain available for the foreseeable future.

North American Hemp Breeding

Landrace varieties of *Cannabis sativa* narrow-leaf hemp (NLH), containing naturally low levels of THC, were first brought to the New World for rope and sail manufacture by British, French, and Spanish colonists during the sixteenth and seventeenth centuries. Early introductions of European NLH eventually became naturalized to the North American climate and were known to hemp breeders as 'Smyrna' hemp, named after the Turkish city of the same name, although the reason for adopting this name remains unclear. 'Smyrna' hemp was able to adapt to a wide range of climatic conditions, as it developed in feral populations, and even though its fiber yield was relatively low, it was well suited for seed production and was the only hemp variety grown in North America until the early 1890s. Then during the late nineteenth and early twentieth centuries, Japanese hemp was introduced into California, and Chinese hemp into Kentucky. At the time, both of these introductions were considered to be varieties of *C. sativa* because they were used in Asia for fiber and seed production rather than drug production. According to recent taxonomic research by Hillig (2004a), early introductions from East Asia were more likely *C. indica* broad-leaf hemp (BLH) varieties, which represent a gene pool that evolved independently from the European NLH varieties. Chinese hemp was acclaimed from the start as a superior variety, no doubt because it had been cultivated and selected for high fiber quality over thousands of years. The Chinese perfected the cultivation and processing of textile hemp very early on and Korean and Japanese weavers still produce hemp cloth nearly as sheer as fine silk (see Chapter 5 for more detail about the lengthy history of Asian fiber hemp). Japanese introductions to California became extinct, while descendants of Chinese introductions into Kentucky became known as 'Kentucky' hemp.

By the turn of the twentieth century several additional varieties had been imported into North America from Europe and China, and the United States Department of Agriculture (USDA) would soon begin to experiment with hemp breeding. Boyce (1900) summarized the situation: "No systematical selection or preservation of any particular variety has been attempted, nor any effort to determine the effects of hybridizing or of climatic conditions. This interesting work remains for the botanist and chemist to elaborate and determine, a very important work." Boyce was unaware, however, that the famed plant breeder Luther Burbank, working in California, would also soon begin his own hybridization experiments (Burbank 1914). The breeding wizard was struck by the potential of hemp as a substitute for wood pulp in the manufacture of paper and thus tried to develop an improved

TABLE 14

European hemp breeders produced many different indigenous, narrow-leaf hemp (NLH) and exotic, broad-leaf hemp (BLH) × NLH hybrid fiber and seed cultivars.

Cultivar	Origin	Year	Sexual type
'Asso'	Italy	2004	Dioecious
'Beniko'	Poland	1985	Monoecious
'Bialobrzeskie'	Poland	1968	Monoecious
'Cannacomp'	Hungary	2004	Dioecious
'Carmono'	Italy	1990s	Monoecious
'Carma'	Italy	1990s	Monoecious
'Chamaeleon'	Netherlands	2002	Monoecious?
'Codimono'	Italy	2004	Monoecious
'CS' ('Carmagnola Selezionata')	Italy	1960s	Dioecious
'Dioica 88'	France	1998	Dioecious
'Dneprovskaya Odnodomnaya'	Ukraine	1980	Monoecious
'Dolnoslaskie'	Poland	Early 1990s	Monoecious
'Eletta Campana'	Italy	1960s	Dioecious
'Epsilon 68'	France	1996	Monoecious
'Ermes'	Italy	1990s	Monoecious
'Fasamo'	Germany	1999	Monoecious
'Fedora 17'	France	1998	Monoecious
'Felina 32'	France	1998	Monoecious
'Felina 34'	France	1974	Monoecious
'Ferimon'	Germany	1981	Monoecious
'Fibramulta 151'	Romania	1965	Dioecious
'Fibranova'	Italy	1950s	Dioecious
'Fibriko'	Hungary	1989	Dioecious
'Fibrol'	Hungary	2006	Monoecious
'Fibriko TC'	Hungary	2007	Dioecious
'Fibrimor'	Italy	2003	Dioecious

(continued)

TABLE 14 *(continued)*

'Fibrimon'	Germany	1950s	Monoecious
'Fibrimon 21'	France	1950s	Monoecious
'Fibrimon 24'	France	1972	Monoecious
'Fibrimon 56'	France	1972	Monoecious
'Finola'	Finland	2003	Dioecious
'Flajsmanova'	Yugoslavia	1950s	Dioecious
'Futura'	France	1981	Monoecious
'Futura 77'	France	1980s	Monoecious
'Futura 75'	France	1998	Monoecious
'Irene'	Romania	1995	Monoecious
'Juso 14'	Ukraine	1980	Monoecious
'Kompolti'	Hungary	1954	Dioecious
'Kompolti Hybrid TC'	Hungary	1983	Dioecious
'Kompolti Sargaszaru'	Hungary	1974	Dioecious
'Kuban'	Ukraine	1984	Dioecious
'Lovrin 110'	Romania	1981	Dioecious
'Monoica'	Hungary	2006	Monoecious
'Novosadska konoplja'	Yugoslavia	1950s	Dioecious
'Red Petiole'	Italy	2002	Dioecious
'Santhica 23'	France	1996	Monoecious
'Santhica 27'	France	2002	Monoecious
'Secuieni 1'	Romania	1984	Monoecious
'Uniko B'	Hungary	1969	Unisex female
'USO 11'	Ukraine	1984	Monoecious
'USO 15'	Ukraine	1995	Monoecious
'USO 13'	Ukraine	1986	Monoecious
'USO 31'	Ukraine	1987	Monoecious
'Zenica'	Ukraine	1990	Dioecious

NOTE: Adapted from de Meijer 1995 and Weightman and Kindred 2005. (See TABLE 1 at the beginning of this book for explanations of Cannabis gene pool acronyms.)

"giant hemp" variety. Burbank's experiments were conducted largely with an improved Chilean NLH variety, but he also incorporated American NLH, BLH, or hybrid lines as well as Japanese and Chinese BLH landraces, and to a limited extent Russian NLH or NLHA landraces into his hybrids. Whether he was successful is unknown, but no varieties attributed to him survived, and they were not incorporated into other varieties.

In 1913 and 1914, Lyster Dewey published results of the USDA's initial hemp breeding experiments begun in 1903. These researchers discovered that the introduced Chinese varieties would "run out" and lose their favorable characteristics of high quality and yield if left to reproduce randomly. In some cases this resulted from indiscriminately crossing the Chinese lines with the less desirable 'Smyrna' landrace variety. However, even if maintained in isolation, the Chinese varieties would decline because the earliest maturing

and highest seed-yielding individual plants, which were also often among the lowest in fiber content and quality, tended to swamp out the later maturing and more desirable phenotypes within a few years. Dewey's report concluded that obviously there would be little improvement of hemp *Cannabis* without continued diligent selection and breeding. As a result, seed production was purposefully restricted to only a few isolated farms under the attention of plant breeders, and consequently the commercial 'Kentucky' Chinese BLH variety became relatively uniform and genetically stable.

In the 1920s, Fritz Knorr of Kentucky bred improved hemp varieties at the Minnesota Agricultural Experiment Station, where he selected for long internodes (the distance between branches, resulting in few branches on each plant), height, stem type, and later maturation. His improved variety was called 'Minnesota No. 8' and proved superior to 'Kentucky'

FIGURE 67. Farmer inspects his experimental fiber hemp field growing in California's Central Valley north of Sacramento (from *Yearbook of Department of Agriculture for 1912*).

hemp in both fiber quality and yield. In 1927, the USDA published its second and final report on hemp breeding. The most significant recorded innovation since the USDA's 1913 and 1914 reports was the notable development of hybrid crosses involving 'Minnesota No. 8' (BLH), 'Ferrara' from Italy (possibly a NLH/BLH hybrid), 'Kentucky' hemp (BLH), and additional Chinese BLH varieties. Selections were made from these complex hybrids for even higher fiber yield. However, not long after the 1927 report, breeding experiments were suspended and hemp was only grown in quantity again as a strategic crop during World War II. By the end of the war softwood pulp dominated publishing and other paper-based industries, and artificial fibers were soon to replace plant fiber packaging, sacking, and cordage. All North American hemp cultivars are extinct today.

Introduction of NLD *Cannabis* to North America

Indian *Cannabis indica* narrow-leaf drug (NLD) varieties have probably been grown throughout the Caribbean and bordering coastal nations from Mexico to Brazil since soon after 1834. This was the year that slavery was abolished in Britain, after which the British began bringing indentured servants to Jamaica from India. Cultivation of drug *Cannabis* in the United States was unreported until 1915, when NLD varieties were first grown as medicinal plants (Stockberger 1915). However, it is likely that NLD *Cannabis* was introduced much earlier during the slave trade, as it was in Brazil (see Chapters 4 and 7). *Cannabis* cultivation was legal in the United States until 1937, but significant importation of NLD marijuana varieties into North America did not commence until the 1960s.

Also during the 1960s, incipient marijuana cultivation began in the United States. Before the 1970s, commercial *sinsemilla* marijuana growers were unknown. Occasionally, seeds found in parcels of imported marijuana were casually

sown by curious smokers and gardeners, but there was no "homegrown" marijuana of any quality or quantity. Nearly all the domestically produced marijuana that lacked seeds was immature, and that which was mature was fully seeded, since the production of *sinsemilla* had not yet caught on. Tropical NLD varieties from Colombia and Thailand rarely matured to the floral stage before northern temperate frosts killed them. Alternately, subtropical Mexican and Jamaican NLD landrace varieties occasionally did mature outdoors across the warmer southern two-thirds of the United States, and some tropical NLD plants survived until maturity in frost-free coastal Florida, southern California, and the Hawai'i where the climate was warm and the growing season long.

Early marijuana growers tried any and all available seeds in their search to find potent plants that would consistently mature before killing frosts. Since most imported marijuana was seeded, the seeds of many different varieties were available in large amounts. Early maturing northern Mexican NLD varieties proved to be the most favored as they most consistently matured in more northern latitudes. The early maturing, domestic NLD varieties of the early and mid-1970s (e.g., 'Pollyanna,' 'Eden Gold,' and 'Haze') resulted from hybrid crosses between Mexican or Jamaican landrace varieties and more potent but later ripening Panamanian, Colombian, and Thai varieties.

Breeding History of NLD Varieties

Traditional drug *Cannabis*-growing cultures in areas of Asia and the New World provided North American marijuana growers with a strong head start by favoring and selecting potent landrace varieties for centuries and through many generations. Home growers in North America combined and recombined imported landraces in multihybrid crosses, and by 1985, the best quality domestic American homegrown *Cannabis* ranked among the world's most potent drug varieties. Domestic varieties were generally adapted to outdoor growing, but some were specially developed for greenhouse or indoor artificial light cultivation; under such conditions, the "season" could be extended to allow later maturing plants to ripen.

In the early 1970s, a handful of marijuana cultivators began to grow *sinsemilla* marijuana. The *sinsemilla* effect is achieved by removing male plants from the fields, leaving only the unfertilized (therefore seedless) female plants to produce mature resin-covered flowers. In lieu of setting seeds in the earliest receptive flowers, as they await fertilization, the female plants continue to mature, producing thousands of additional floral bracts, each covered by a myriad of resin glands. This technique was originally developed in India, and although we are unsure of its history prior to the nineteenth century (Prain 1904), the technique may be quite ancient. The pictorial book *Sinsemilla Marijuana Flowers* (Richardson and Woods 1976) revolutionized illicit marijuana cultivation in the United States. The authors accurately and attractively portrayed the *sinsemilla* technique through informative text and lavish color photographs. They also made the first attempt to describe the proper stages of floral maturity for an optimally potent, pleasant-tasting, and resinous product. Most importantly, understanding how to grow *sinsemilla* allowed many growers to realize that most of the flowers could be cultivated

FIGURE 68. A juvenile narrow-leaf drug (NLD) plant (in the background) is characteristically taller with longer internodes, lax branching, and sparse foliage, while a juvenile broadleaf drug (BLD) plant (in the foreground) is generally shorter with short internodes, more profuse branching, and dense foliage.

FIGURE 69. Ripe inflorescence of a Mexican landrace, 'Eden Gold,' a narrow-leaf drug (NLD) *sinsemilla* variety, growing outdoors in California in the 1970s.

without developing seeds; however, a few branches could be intentionally fertilized with only a tiny amount of select pollen to produce seeds of known parentage. This in turn gave birth to, or at least accelerated, the intentional breeding of potent drug *Cannabis*.

Once growers found potent varieties that would mature under their local climatic conditions, pioneering marijuana breeders continued selecting primarily for stronger potency (high Δ^9-THC content), followed by more aesthetic considerations of flavor, aroma, and color. Modifying adjectives such as "minty," "floral," "spicy," "fruity," "sweet," "purple," "golden," or "red" were often associated with selected varieties—and thus homegrown marijuana connoisseurs were born. Careful inbreeding of the originally favored hybrid crosses resulted in some of the "supersativa" NLD varieties of the 1970s (e.g., 'Original Haze,' 'Purple Haze,' 'Polly,' 'Eden Gold,' 'Three Way,' 'Puna Butter,' 'Maui Wowie,' 'Kona Gold,' and 'Big Sur Holy Weed'). In the late 1970s and early 1980s, growers more commonly began to call these original NLD varieties "sativas" because they more closely resembled NLH varieties in gross phenotype than they did broad-leaf drug (BLD) varieties and to distinguish between the established NLD varieties and recently introduced BLD varieties from Afghanistan, commonly called "indicas." We now know

that all drug varieties, regardless of their origin or gross phenotype, belong to *Cannabis indica* (see Chapter 11).

During the second half of the 1970s, many clandestine marijuana breeders were successfully developing connoisseur NLD varieties; consequently, demand for high-quality drug varieties grew as more potent, aromatic, and colorful flowers brought illicit cultivators greater pride. Purple varieties gained popularity, largely following in the footsteps of the extraordinary 'Purple Haze' grown in Santa Cruz, California. By 1978, commercial *sinsemilla* cultivation was becoming much more common, and many professional growers developed NLD varieties that were both early maturing and high yielding, adding greatly to their profits.

However, police awareness of commercial cultivation was also increasing, especially in the western United States. Small aircraft were routinely deployed in remote terrain to search for larger marijuana fields, while smaller home growers were more often turned over to police by their disapproving families and neighbors. Law enforcement officers knew that marijuana matured late in the autumn and often did not search for it until late in the year. This encouraged marijuana farmers to grow varieties that matured early and therefore could be harvested and out of sight in the drying shed in early autumn. Faced with storage problems as a result of increased seizures, police would often merely count the seized plants and burn the bulk of the confiscated crop immediately, often without weighing it, preserving just enough to be forensically identified and used as evidence in court. The courts considered large numbers of plants, even small seedlings and rooted cuttings, to be evidence of intention to grow large quantities of marijuana, once again encouraging growers to favor large, early maturing NLD varieties. The fewer (yet larger) plants the better, as long as growers continued to realize sufficient yields. Law enforcement efforts had a serious effect on progress in *Cannabis* breeding, especially by raising penalties for commercial-scale growing, which discouraged the cultivation of large populations required for selection of truly superior individual parent plants and expedient variety improvement through intensive selective breeding.

Cannabis responds positively to abundant water, sun, and fertilizer, and huge plants yielding three kilograms (nearly seven pounds) or more of dried flowers were relatively common. Up to a point, the more you feed them, the taller and bushier they get, even when restrained by trellising or pruning. This is a natural tendency for a "camp follower" like *Cannabis*. In the late 1970s, nearly all the *sinsemilla* being grown came from NLD varieties, all originally adapted to South Asian latitudes and therefore later maturing. Imported varieties developed subsequently in Colombia, Mexico, or Thailand average over two and a half meters (eight feet) tall, even when restrained, and can easily reach four meters (13 feet) or more when grown in the most favorable conditions. However, the bigger the plant, the easier it is to spot from an aircraft or over a fence. As marijuana breeders continued to intercross their shortest high-yielding NLD cultivars and prune frequently, their goal was to create something new—a "miracle" variety, short in stature with large, potent flowers, ripe for harvest in late summer.

Introduction of BLD *Cannabis*

Cannabis indica ssp. *afghanica,* broad-leaf drug (BLD) plants, are very distinctive with well-branched, short, and bushy with broad, dark green leaves and many resin glands of high cannabinoid content. They normally mature quite early in northern temperate latitudes from late August through September. BLD varieties often stand only one to two meters (about three to six feet) tall at maturity and produce many large inflorescences, providing an abundant source of highly psychoactive resin traditionally used to make potent Afghan hashish (Clarke 1998a). Seedless BLD flowers smell and taste much like high-grade hashish, and connoisseurs were willing to pay premium prices for this exotic product. These BLD varieties from Afghanistan and Pakistan were introduced to North America and Europe several dozen times during the middle to late 1970s and also during the 1980s when Afghanistan was occupied by the Soviets.

Marijuana breeders intentionally crossed BLD varieties with their sweet-tasting but later-maturing NLD varieties to produce early maturing BLD × NLD (or NLD × BLD) hybrids, commonly called "indica × sativa" or "sativa × indica" hybrids. Initial hybrids spread across North America like wildfire as growers realized they were early maturing, potent, and harder to detect. Hybrid vigor (heterosis) was usually evident in the first hybrid (F₁) crosses and flower yields increased, yet plants rarely exceeded 3 meters (10 feet) in height. Thai × Afghan hybrids were known to be particularly pungent smelling, sweet tasting, and most of all extremely potent. By the early 1980s, the majority of North American marijuana growers began to include at least a few BLD/NLD hybrid plants in their predominantly NLD crops.

By the early to mid-1980s, the vast majority of all illicitly produced *sinsemilla* for sale in North America had probably received some portion of its genome from the BLD gene pool. Therefore it became increasingly difficult to find pre-BLD pure NLD varieties or the pure BLD introductions that had been so popular only a few years earlier. By now, one might be tempted to think that BLD/NLD hybrids were so superior to NLD varieties that they must have been received with total acceptance by all North American and European *sinsemilla*

growers. Although the range of BLD and BLD/NLD hybrid varieties increased throughout the 1980s, owing to their delayed introduction into the central and northern United States, Canada, and Europe, their popularity in pioneering regions of the western United States began to decline.

Although both NLD and BLD varieties can be quite potent, their cannabinoid profiles differ, and hybrids between them can be even more potent. The total cannabinoid content of NLD cultivars reaches as high as 20 percent dry weight of trimmed floral clusters and THC:CBD ratios can exceed 100:1. Total cannabinoid levels in BLD cultivars also reach 20 percent with a THC:CBD ratio ranging from 2:1 to 1:2. Modern BLD/NLD hybrid *sinsemilla* varieties often produce flowers with well over 20 percent THC and negligible CBD (Watson 1993, personal communication).

The major horticultural drawback to BLD/NLD hybrids is their susceptibility to fungal infections. For example, the dense, tightly packed floral clusters of BLD varieties tend to hold moisture and make perfect microclimates for grey mold (*Botrytis cinerea*). BLD/NLD plants have little natural resistance to infection as they are naturally adapted to the arid conditions of Afghanistan where grey mold causes no threat (McPartland et al. 2000). In North America and Europe, grey mold causes significant crop losses in outdoor crops, often averaging 20 percent or more, and occasionally resulting in total crop failure. Such fungal attacks were almost unheard of when only NLD varieties were grown.

Since *Cannabis* is wind pollinated, and *sinsemilla* is usually grown in small, crowded gardens, accidental pollination often results in undesirable, profuse seed production. During the 1970s and 1980s, accidental seeds were far more common than intentionally produced seeds because they were widely dispersed as a constituent in commercial marijuana. Under normal conditions, intentionally bred seeds were only passed along from one diligent marijuana grower or breeder to another, and therefore their distribution was much more limited. Accidentally produced seeds containing varying proportions of the introduced BLD gene pool were grown and indiscriminately crossed again and again, producing a multihybrid condition in which suites of favorable traits were more rarely observed and the quality became uniformly lower. Almost all the offspring were dissimilar to their parents or siblings in appearance and potency, their gene pools representing many randomly collected genetic scraps handed down from their assorted predecessors, and as the randomly mixed and unselected gene pools were reproduced, they often manifested many undesirable characteristics. Accidental recombination into complex multihybrids brought out some of the less desirable traits of BLD varieties that had previously been suppressed through meticulous selection. When production expands to meet market demands every seed is sown and no selection is exercised. This is particularly true of the majority of *Cannabis* originating in Mexico since the late twentieth century. In the absence of careful selection and breeding, drug *Cannabis* also begins to turn weedy, and as natural selection takes over, it loses its vigor, palatability, and potency. This is what has happened to *Cannabis* in many regions formerly known for fine quality products (e.g., Kerala state in southern India, Laos in Southeast Asia, and Colombia).

Consumption characteristics such as a slow, flat, dreary "high," along with an unattractive, acrid aroma and sour, harsh taste quickly became associated with many BLD/NLD hybrid plants. This was often what the average *sinsemilla* consumer experienced in the 1990s and is still common today;

FIGURE 70. Widely varying traditional drug *Cannabis* landraces from several geographical regions were used to breed modern *sinsemilla* cultivars. Early maturing South African (A) narrow-leaf drug (NLD) and Afghan (B) broad-leaf drug (BLD) plants were often crossed with later maturing Thai (C) NLD landraces.

FIGURE 71. The broad leaflets and dark purple-green color of a female broad-leaf drug (BLD) inflorescence (left side) contrasts with the narrow leaflets and golden-green color of a narrow-leaf drug (NLD) inflorescence (right side); the artwork shown here was designed for a late 1970s Sacred Seeds Company seed package.

FIGURE 72. Ripe terminal inflorescence of a 'Skunk No. 1' open pollinated hybrid *sinsemilla* cultivar from California. 'Skunk No. 1' has been incorporated into more modern *sinsemilla* cultivars than any other artificially selected variety.

therefore, *sinsemilla* connoisseurs often feel that the introduction of BLD varieties did little to improve the general quality of *sinsemilla*—only its quantity. However, the points of view of average commercial or home cultivators, whose overriding considerations are yield and potency, are quite different (Clarke personal observation 2008). The hardy growth, rapid maturation, and tolerance to cold, which BLD varieties offered illicit growers, allowed *sinsemilla* to be grown outdoors in the United States across the northern tier of states, from Washington to Maine and even farther north into southern Canada. This revolutionized the domestic marijuana market geographically (and to some degree culturally) by making potent "homegrown" a reality for those living at northern latitudes, as well as widening the scope and intensity of *sinsemilla* cultivation across North America and Europe. Domestic marijuana production spread rapidly from early epicenters along the Pacific Coast and Hawai'i, and presently at least some *sinsemilla* is grown outdoors, in glasshouses, or indoors under artificial light in all regions of the United States, Canada, and Europe and continues to spread throughout Asia and South America.

The introduction of BLD varieties has also had more subtle but possibly longer-lasting effects on *sinsemilla* breeders and breeding than the NLD varieties. The relatively high demand of consumers for exotic purple *sinsemilla* NLD varieties created the short-lived "Purple Haze Craze" of the mid to late 1970s. Subsequently, BLD/NLD growers discovered that their plants would frequently turn purple if they were left standing in the field through a frost. During the early 1980s, many illicit *Cannabis* smokers paid considerably more for purple

BLD/NLD flowers as purple coloration had become a sign of quality and potency in late-maturing NLD cultivars like 'Purple Haze.' However, when early maturing BLD/NLD (erroneously called "indica") varieties are left in the field through a frost, they become overripe and lose much of their potency. This realization abruptly ended the "Purple Indica Craze," and more important, it suggested to more insightful and diligent *Cannabis* breeders that many traits prove to be advantageous only under certain conditions in particular varieties and that conscious breeders must be extremely selective when experimenting with new introductions. Although consumers and commercial growers of the late 1970s largely accepted BLD varieties, serious connoisseur breeders of the late 1980s began to view BLD/NLD hybrids with more skepticism.

On the other hand, the futures of domestic *sinsemilla* cultivation and breeding are not necessarily hopeless as a result of BLD contamination and the spread of BLD/NLD varieties in North America and Europe. Although BLD varieties have fallen out of grace with some clandestine growers and connoisseurs, BLD/NLD hybrids have certainly provided significant enhancement of THC content for the average Western *sinsemilla* smoker, and their short stature and high yield have endeared them to indoor artificial light growers. Dedicated *sinsemilla* breeders continue to produce additional potent and desirable BLD/NLD hybrids, many of which favor their original NLD parentage. *Sinsemilla* breeders form a close-knit group, although connected by no tangible ties beyond their outlaw life-style, common interests, and mutual respect. Open exchanges of information, seeds, and cuttings are commonplace, and connoisseur marijuana breeding will continue to develop during the twenty-first century regardless of its illegality.

Prior to the increased popularity of seed sales in the Netherlands throughout the 1990s, most inexperienced growers still did not recognize the importance of purchasing high-quality, intentionally bred seeds; these cultivators most likely started with accidentally produced seeds collected from randomly available, illicit commercial marijuana. Since 1985, seeds of BLD/NLD hybrid drug varieties emanating from *Cannabis* breeders in the Netherlands have been widely available from distributors in Canada and across Western Europe. Accompanying the availability of specially selected, hybrid drug cultivars was relatively easy access to the technology to produce uniform crops from vegetative cuttings indoors under artificial grow lights. This combination has revolutionized *sinsemilla* production and heavily influenced the selection criteria of marijuana breeders.

Recent Trends in *Cannabis* Breeding

Further advances in industrial hemp breeding are expected in the near future. Economical fiber production relies on increasing the scale of production and reducing production costs and could be achieved in part by breeding for increased yield of biomass along with increased fiber yield and quality, reducing waste such as dust, and improving fiber color. Experiments have shown that hemp has the potential to produce much higher stalk and fiber yields than are presently achieved. Improved yields of both fiber and seed varieties could be achieved by continued selection for optimal flowering date, improved resistance to pests and diseases, optimizing canopy architecture and improving fiber size, aspect ratio, strength, stiffness, density, surface characteristics, adhesive properties, lignification, and light color for specific end uses carried out in concert with development of optimized agronomic regimes and specific processing strategies for woven, nonwoven, and composite applications. Ideal hemp fiber varieties would have a high proportion of fine, easily extracted fibers that enhance rapid processing and lower wastage (Weightman and Kindred 2005).

Cannabinoid production is controlled by a limited number of genes, and heritability is extremely high. High variability and heritability are also characteristic of both flowering and maturity times; in addition, reduced sensitivity to day length could improve the versatility of hemp varieties making breeding and seed multiplication easier. Where it is environmentally feasible, late flowering varieties could significantly improve biomass yield by lengthening the vegetative period and encouraging stalk growth. Fiber quality and ease of retting are controlled by a complex set of genes that interact with agronomic and processing conditions, which can result in low heritability of fiber traits and difficulties for breeders; on the other hand, fatty acid and protein synthesis are controlled by fewer genes encoding for specific products making selection and breeding more straightforward. Essential oils are made up of many terpenoid and other aromatic compounds, each controlled by interrelated genes, and therefore breeding is also problematic. In general, breeding would be aided by legislative tolerance of higher THC levels (Weightman and Kindred 2005).

Law enforcement pressure on marijuana growers in North America forced the thrust of progress in *Cannabis* breeding to the Netherlands where the political climate was less

FIGURE 73. 'Original Haze' from Santa Cruz, California, is a late-maturing narrow-leaf drug (NLD) hybrid cultivar of Thai, Indian, and Colombian ancestry, and hybrid 'Haze' varieties often turn many shades of purple as they ripen.

threatening and cultivation of *Cannabis* for seed production, as well as seed sales, were tolerated. Beginning in 1985, marijuana seed companies selling varieties originally developed in California, as well as landraces from international marijuana exporting regions, began to appear in the Netherlands, and many small companies continue to breed, reproduce, and sell seeds of exotic high-quality *sinsemilla* varieties to mostly North American and European growers (Clarke 2001).

The basic North American/Dutch gene pool continues to be recombined by seed companies, and as novel individuals become founders of new populations, selection moves steadily onward. Some serious connoisseur *sinsemilla* breeders have returned to some of their older breeding stock of pure NLD and BLD varieties and crossed them into their more highly inbred BLD/NLD hybrid lines to enhance their flavor and potency. Also, ever-curious breeders continually search for new sources of exotic, imported landrace seeds. Pure BLD varieties are still highly prized breeding stock, and new BLD introductions from Central Asia, Afghanistan, and Pakistan are occasionally used in hybrid crosses. In the 1990s, NLD varieties from South Africa gained favor, as they matured early, but did not suffer from grey mold and other shortcomings of BLD varieties. South African varieties, since they come from far south of the equator, mature in late summer

THE COSMIC BOOGIE — 1976

In clandestine locations sparsely scattered throughout California, a few dedicated farmers cultivate the herb cannabis as a way of life. Anyone can grow weeds and see the virgin greens, but to cultivate the exotic queens beyond your dreams, you should have dedication, sensitivity to plant life, consistency, a little basic knowledge, Good Karma, and a great desire to find harmony with Mama Nature.

The basis for any strong, healthy plant is the soil it grows in. Resin-producing herbs grow best in soil that is neutral to semi-sweet. The larger the bush or tree, whatever it may be, the larger the root structure needed to support it.

Roots develop most freely in a loose, rich soil. Mix one yard of soil composed of 30% mushroom compost, 30% redwood sawdust, 40% sandy loam topsoil. Mix organic nutrients: three pounds of hoof & horn meal, five pounds of rock phosphate, four pounds of kelp, one pound natural compost, one pound soil iron, five pounds of gypsum, and five pounds of dolomite.

Mix soil and minerals thoroughly and let sit for at least one week, then check the soil with a soil-testing kit for pH and add agricultural lime to bring the pH to 7.5 — 8.5 range, semi-sweet for the connoisseurs' treat.

Spring arrives, and with it comes new life, fresh spirit, and great hope for the new crop. Plant your favorite seeds on the 6th or 7th of April; the moon is in Cancer, best sign for planting above-ground herbs. In May, the 3rd, 4th, and 5th are good. If you're still not together, try June 8th and 9th while the moon is in Scorpio.

When you irrigate, make sure that the moon is in one of these signs (when possible): Cancer is the most excellent time to irrigate; Taurus, Scorpio, and Pisces are also good for irrigation. Pruning should be done only in the third quarter.

Patience is the _real_ secret and the true test. Try to make it until December 10th or 11th, when the moon is in Leo and harvest conditions are perfect. So, if you can dig it . . .

In Nature We Trust
Spirit of '76

Copyright © 1975 R.L.

FIGURE 74. The 'Original Haze' poster from 1976, commemorating fine _sinsemilla_ marijuana and the two hundredth anniversary of American independence.

TABLE 15

Recent trends in the evolution of hybrid drug cultivars resulting from intentional selection and breeding.

Taxon	Quantitative vs. qualitative traits	Before selection	Following selection
BLD X NLD	Quantitative	Moderate to high THC content	Very high THC content
		Moderate to low CBD content	Very low CBD content
		Low bract to leaf ratio	High bract to leaf ratio
		Low flower yield	High flower yield
	Qualitative	Moderately branched	Profusely branched
		Medium internodes	Short internodes
		Later maturing	Earlier maturing

(See TABLE 1 at the beginning of this book for explanations of Cannabis gene pool acronyms.)

and are short and moderately potent but relatively low yielding. However, hybrids between South African varieties and established inbred NLD, BLD, or hybrid varieties are usually vigorous, early maturing, and express many of the desirable NLD and/or BLD traits (e.g., high potency, strong aroma, and high yield).

NLD landrace varieties have also been imported from India, Kashmir, Nepal, Indonesia, Korea, West and Central Africa, and Southeast Asia. These are occasionally incorporated into hybrids to impart particular flavors to the smoke or to enhance potency but have become increasingly difficult to find in recent years. Since commercial shipments do not often originate in these regions, seeds are usually collected in small numbers by travelers and are as such relatively rare compared to seeds from the major marijuana producing regions of Colombia, Jamaica, Mexico, Thailand, and more recently North America. Australia and New Zealand also have long-established clandestine *sinsemilla* growers who tend their own unique varieties. Australian and Polynesian outdoor varieties are more often based on Southeast Asian landraces introduced in the 1970s and until recently were less affected by BLD/NLH introductions.

Prior to 1980, a few drug *Cannabis* breeders worked with varieties of weedy plants from Central Europe, which they called "ruderalis." We consider these to be either escapes from NLH fiber cultivation or possibly the putative NLHA (see Chapters 2 and 12). At temperate latitudes, these weedy varieties begin to mature in July or early August, making them seemingly desirable parents in hybrid drug variety development aimed at hastening maturity. However, since they are almost entirely devoid of THC, production of THC is lowered in their hybrid offspring, and subsequent backcross selections must be made to restore a higher level of potency. In addition, adaptation to a weedy life cycle has suppressed the tendency for an evenly timed ripening of flowers and seeds. A dependable ripening time is much desired in cultivated *Cannabis* crops because it facilitates uniform harvesting. Wild and weedy ruderal varieties gain an ecologically adaptive survival advantage from seeds ripening gradually throughout the warm summer months as they continue to make new flowers until killed by frost. NLD, BLD, and BLD/NLD hybrid crosses with feral NLHA landraces will likely prove of value only to outdoor drug *Cannabis* cultivators living in very high latitudes where no other drug varieties will mature.

Because *Cannabis* is a difficult plant in which to fix traits through selective breeding and only female plants are of economic importance for psychoactive or medicinal use, it is often advantageous to reproduce exceptional female plants by rooting vegetative cuttings (cloning). In this way, practically unlimited numbers of identical, select female plants can be grown. Cloning produces uniform crops of female plants in one generation without using seeds and thereby circumvents the vagaries of genetic recombination resulting from dioecious sexuality. Removal of male plants is not required to produce *sinsemilla*, as no male plants are grown. All the flowers in vegetatively produced monocrops mature at the same time and the entire crop can be harvested at once. These are obvious advantages in commercial *sinsemilla* cultivation, and BLD/NLD hybrid clones have proven to be well adapted to indoor cultivation. Maturing quickly, a grow room can be harvested three to four times per year and can yield more than 450 grams (one pound) of dry flowers per square meter (10 square feet) per harvest. NLD varieties are sparse and tall and often take too long to mature. The tops of the tall plants, near the lights, often are stressed by the hot lights and shade the poorly lit bottom branches, preventing them from producing many flowers; this makes NLD varieties less desirable for indoor growing. Clones of male plants valued as pollen sources in breeding programs can also be preserved in a vegetative "library" under long photoperiod and induced to flower when pollen is required.

Diversity in the drug *Cannabis* gene pool will become further narrowed by gradual culling through human selection of economically unfavorable clonal lines. If a grower has several clones, in all likelihood the one or two that are higher yielding, more potent, or otherwise more desirable and commercially valuable will be propagated; the remaining clones will be destroyed. However there are some potential longer-term drawbacks to the clonal cultivation strategy. When the drug *Cannabis* gene pool reaches a genetic bottleneck caused by the preponderance of asexual, vegetatively reproduced commercial populations, genetic diversity will decrease, and susceptibility to agricultural pests and diseases will increase. Lowered genetic diversity and a lack of sexual recombination will result in lowered potential for evolving resistance. Should pathogenic organisms such as viruses, fungi, mites, and aphids attack a genetically uniform, asexually reproduced clonal population crop, loss would be extensive. Table and wine grapes (*Vitis vinifera*) have traditionally been propagated vegetatively much the way modern *Cannabis* varieties are now propagated. A recent genetic survey of grape accessions in the United States Department of Agriculture collection by Myles et al. (2010) came to the following conclusion: "We propose that the adoption of vegetative propagation was a double-edged sword: Although it provided a benefit by ensuring true breeding cultivars, it also discouraged the generation of unique cultivars through crosses. The grape currently faces severe pathogen pressures, and the long-term sustainability of the grape and wine industries will rely on the exploitation of the grape's tremendous natural genetic diversity." Research by Arrigo and Arnold (2007) comparing the ecological traits and genetic diversity of self-sowing naturalized *Vitis* rootstocks originating in agriculture against wild rootstocks of *Vitis vinifera* ssp. *silvestris* determined that the genetic diversity of naturalized rootstocks exceeded that of wild populations; in addition, introduced rootstock cultivars were invasively displacing endangered wild populations considered vital for grape breeding. Similar situations likely occurred when cultivated *Cannabis* varieties were introduced to regions where *Cannabis* grew wild (see Chapter 2).

The final blow to genetic diversity will result from commercialization of a small number of select varieties that satisfy limited consumer preferences. Under the assumption of eventual legalization, only a few varieties will pass the consumer filter, and only these will be proliferated. Given the present-day political climate and increasing trend toward asexual reproduction, in both indoor and outdoor crops, this scenario seems likely. The outcome will certainly be narrowing of the gene pool to only the commercially viable and therefore preserved gene pools much as with grape selections. In the past, as early agriculturalists spread across new geographical frontiers, they encountered novel environments that challenged not only their own survival but also the survival and proliferation of the economic plants that traveled with them, and in the case of *Cannabis*, they simultaneously exposed these plants to new and evolutionarily challenging environments. The result was the tremendous genetic diversity of geographically isolated wild or feral populations and traditional landraces—produced by sexual reproduction and genetic recombination—which today present novel phenotypes for natural and human selection.

Indoor clonal environments offer no chance for evolutionary change as sex is prevented. As varieties fade away through custodial or economic neglect, they become extinct. This has happened to many heirloom varieties of fruits and vegetables, which are now gone forever. Should there remain an indigenous population maintaining localized varieties, they will survive in situ. This will preserve the genetic base since landrace populations reproduce sexually, constantly allowing for genetic recombination, mutation, and evolution under human selection. However, the ranges of in situ traditional agricultural societies are presently diminishing at a more or less constant rate worldwide. This does not bode well for the future of the *Cannabis* gene pool, since the extinction of any branch of the genetic diversity is detrimental for the ultimate survival of its unique genes and allelic combinations.

Female seeds (those bred to give rise only to female plants) are marketed by several Canadian and European companies. Female seeds preclude the need for growers to remove male plants, eliminate any chance of accidental pollinations, and confer some of the advantages of vegetative cuttings, all in the convenient packaging of a seed. This allows them to be used clandestinely outdoors in remote areas where it is difficult for growers to visit their gardens often enough to remove all male plants. Sterile cuttings of female varieties, or seed varieties that are infertile, would guarantee that the female plants will be seedless; this could offer an additional advantage to illicit *sinsemilla* growers.

On the other hand, increasing use of legal medical marijuana adds new parameters to drug *Cannabis* breeding based on feedback from medical *Cannabis* users; consequently breeders are endeavoring to develop cultivars with specific medicinal effects. *Cannabis* breeders continually search for new sources of exotic seeds, explore variations in objective and subjective effects between different clones, and develop varieties with enhanced medical efficacy for particular indications. Pure BLD varieties are still highly prized by medicinal *Cannabis* breeders—largely because they contain CBD, largely bred out of recreational drug varieties—and new accessions from Central Asia are occasionally introduced. NLD varieties from Mexico, South Africa, and Korea have regained favor with breeders, as they mature early but express fewer of the undesirable traits of many BLD varieties.

Breeding projects aimed at using *Cannabis* as a source of pharmaceutical compounds strive to produce cultivars with a high yield of dry biomass, a high proportion of flowers as opposed to leaves and stems, a high total cannabinoid content in the flowers, and a high proportion of the target cannabinoid (e.g., CBD, THC, etc.; see Chapter 8); these are objectives largely consistent with the breeding of recreational *Cannabis* selected for high THC content. Hybrid crossing, followed by serial inbreeding of select individuals from the F_1 hybrid population, as well as a subsequent restoration of hybrid vigor by making crosses between favorable lines, has proven effective in achieving these goals (de Meijer 2004). Two medical varieties, 'Medisins' registered by HortaPharm in the Netherlands in 1998 and 'Grace' registered by GW Pharmaceuticals in the United Kingdom in 2004, have so far been awarded plant breeders rights (Weightman and Kindred 2005).

The renewed but still generally illicit popularity of hashish production has led to many new technological developments now commonly put into practice by Western connoisseurs (Clarke 1998a). *Sinsemilla* breeders are keen to develop varieties that are better suited for modern hashish production by concentrating on selections for traits such as large and easily removed resin glands, dry rather than sticky or oily resin texture, and terpenoid compositions that can withstand water or organic solvent extraction so aromas and flavors are preserved.

Commercial breeding programs may also try to protect the intellectual property of novel cultivars by incorporating particular biological traits. These could include the following: triploid chromosome number resulting in sterility, selection of a high level of heterozygosity in F_1 hybrids so their offspring will not be true to type, recessive morphological markers such as abnormal pigmentation (e.g., yellow stem, purple flowers, or red leaves), abnormal leaf shapes (e.g., hooked serrations or webbed leaves), or a unique chemical profiles (e.g., traces of a minor unique aromatic terpenes or aldehydes) that would make the cultivar phenotypically distinct and readily identifiable. Genetic fingerprinting of DNA sequences could also be used to identify cultivars; in addition, molecular markers for suites of genes associated with biosynthesis of individual target compounds and agronomic traits would

FIGURE 75. Modern hybrid NLD × BLD *sinsemilla* cultivars have been bred for high yields and exotic appearance, and consequently they barely resemble the landraces from which they originated (photo ©Mojave Richmond).

allow mass screening of juvenile populations and could be of great advantage to breeders (de Meijer 2004).

Biotechnology in the form of genetic modification has also reached *Cannabis*. An Irish research group successfully transferred genes for grey mold resistance to an industrial hemp variety (Clarke and Watson 2007). Grey mold is one of the leading pests of *Cannabis*, causing crop loss and contaminating medical supplies, and the transfer of resistance into medical varieties would therefore be of great value. Yukihiro Shoyama and colleagues at Fukuoka University in Japan have transferred the THC-synthase gene from *Cannabis* to tobacco (*Nicotiana tobaccum*) and induced it to convert CBG (cannabigerol, the precursor molecule to THC) into THC (Shoyama et al. 2001). Other agronomically valuable traits may also be transferred to *Cannabis* such as enhanced pest resistance, increased yields of medically valuable compounds, tolerance of environmental extremes, and sexual sterility.

The reaction to development and release of genetically modified (GM) organisms has been guarded. *Cannabis* presents a particularly high risk of transmitting GM genes to industrial hemp crops and weedy *Cannabis* because it is wind pollinated. The EU has installed strict regulations to prevent the accidental release of GM genes, and therefore production of GM *Cannabis* in the EU may prove impractical. However, nonfood industrial fiber and pharmaceutical cultivars may not receive as much resistance from consumers and environmentalists as food crops. For example, genes coding for cannabinoid biosynthesis might also be transferred from *Cannabis* to less politically sensitive organisms than tobacco. However, expression of THC synthesis in a widely distributed plant may also present a new set of obstacles to law enforcement. On the other hand, transferring cannabinoid synthesis systems from *Cannabis* into other genera of plants, fungi, and bacteria opens up the possibility of producing medically valuable cannabinoids in industrial fermenters and circumventing *Cannabis* growing altogether.

Summary and Conclusions

Presently, few if any imported *Cannabis* shipments are as psychoactively potent as the best *sinsemilla* grown in North America and Europe. However, seeds of improved hybrid cultivars have spread to many *Cannabis*-producing nations, and the potency of *Cannabis* imported from select locations is steadily increasing. Due to continuing international establishment pressure against marijuana growers, and the highly inflated price of *sinsemilla* (in the United States from $5,000 to $10,000 per kilogram, or $2,000 to $4,000 per pound), indoor cultivation will continue to gain popularity worldwide in any location with a reliable electrical grid. Relevant information is readily available on the Internet. Metal halide and high pressure sodium lighting systems are easily set up in attics, bedrooms, or basements, but space is often limited. Under these circumstances there is no room for nonproductive plants, and the single most productive female clone is usually selected for all future cultivation. The use of a single clone improves grow room performance but precludes the possibility of seed production. Breeding is no longer possible and variety improvement ceases entirely. Presently, due to the rapid spread of indoor cultivation, growers are increasingly likely to exchange vegetative cuttings of proven clones rather than carefully bred seeds, and consequently, the genetic diversity in drug *Cannabis* cultivars has decreased in the past two decades. We expect that cloning will have a lasting effect on *sinsemilla* production and that it will further slow the evolution of drug *Cannabis*. In addition, North American/Dutch varieties are presently used in many of the traditional growing areas (e.g., Jamaica, Mexico, and Southeast Asia), where they are, in effect, genetically contaminating local landraces and will eventually displace them. This makes the collection of traditional landrace varieties increasingly difficult and their reproduction more important than ever.

Two predominant trends of *Cannabis* breeding have developed in modern times. On one hand, *Cannabis* has been modified by industrial hemp breeders who have lowered THC content and raised fiber content while developing uniform, high-yielding cultivars. On the other hand, *sinsemilla* breeders have also altered the genetic makeup of drug varieties effecting quantitative traits such as raising THC content, flower to leaf ratio in the inflorescences, and yield of flowers. Qualitative traits have also been altered such as increasing branching, shortening internodes, selecting for rapid maturation, and early floral response to inductive photoperiods. Breeders are expected to make further advances on all fronts as *Cannabis* cultivation becomes more widespread for a variety of uses.

CHAPTER ELEVEN

Classical and Molecular Taxonomy of *Cannabis*

In this age of rapidly expanding information, we have reached a period where the most important contributions to our knowledge are syntheses of the disparate information in diverse disciplines.

(WILLIAM EMBODEN, in RUDGLEY 1995)

Introduction

In this chapter, we focus on the scientific classification of *Cannabis* in light of its evolutionary biology. First, we will briefly review our knowledge of *Cannabis*'s evolution to date, then summarize the various nineteenth- and twentieth-century taxonomic systems for *Cannabis*, and finally investigate the genetic evidence for *Cannabis* evolution as illustrated by the disciplines of chemotaxonomy (classification based on variability of secondary metabolites) and molecular taxonomy (classification based on variability of nucleotide sequences). Data provided by several academic disciplines have proven useful in building our case for the evolution of *Cannabis* under natural and artificial selection. Studies of the botany and ecology of *Cannabis* provide us with an understanding of its environmental requirements for successful growth, as well as insights into its reproductive strategies, both key aspects of plant evolution. Geography gives us an understanding of the earth's environments, landscapes, and climates and their influences on the natural selection and dispersal of *Cannabis*. Archeology and archaeobotany supply us with material evidence for the ancient occurrence of *Cannabis* and provide guidelines for establishing its migration routes, while history gives us written records of human dispersal and thus offers additional information regarding the anthropogenic influences on the dispersal and distribution of *Cannabis*.

Studies of human culture teach us about societal determinants affecting preferences for traditional *Cannabis* products including fiber, food, medicines, and psychoactive drugs, as well as how these use incentives have affected the biodiversity of the genus.

Cannabis grows abundantly in many temperate, subtropical and tropical regions of the world, in some areas both spontaneously (truly wild and feral) and cultivated. Today, cultivated *Cannabis* is utilized over a remarkably wide geographical range, most of it the result of human-related movement. The many associations of *Cannabis* and humans began thousands of years ago. As humans began to migrate into and across Eurasia, they encountered *Cannabis* very early on, either in its putative pre-Pleistocene home in Central Asia or during the early Holocene in East Asia, South Asia, or Europe (see Chapters 2 and 4). Over many millennia, natural and human selection have directed the evolution of *Cannabis* into both a multiple use crop plant valuable for fiber, seed, and drug production, as well as a pernicious and costly weed (see Chapters 3 and 4). Genus *Cannabis* encompasses a wide range of plant varieties including spontaneous wild, escaped, or weedy ones that have been mostly selected naturally, as well as cultivated types with phenotypes that are largely reflective of human selections favoring one *Cannabis* product over others. Natural environment and culture work hand-in-hand determining human needs and preferences for varying plant products (see

311

Chapters 5 through 9). It is our contention that through dissemination and selection, human intervention has been the major force in the evolution of *Cannabis*, at least during the Holocene, and has led to most of the diversity observed in *Cannabis* today.

Cannabis is an interesting, if somewhat complex, plant for evolutionary study. Despite its lengthy association with humans, until very recently it had not been fully domesticated and continues to undergo rapid evolution under intense human selection (see Chapter 10). If we accept that *Cannabis*'s recent evolution has largely been a consequence of human intervention, we may focus on defining the relationships between distinct gene pools produced by isolation and hybridization prior to and during cultivation and/or following escape. This investigative plan assigns taxonomic groupings to gene pools originating in divergent geographical locations within varying cultures. This avoids the improbable assumption that *Cannabis* simply followed random human dispersals and was subject only to natural selection. Although natural selection, gene flow, genetic drift, and mutation rates are difficult to measure, they are undoubtedly important in the evolution of domesticated plants, but human influences on these natural factors is paramount and in many cases are well documented.

Operating at a population level, natural selection rarely chooses the fittest phenotype but rather the most successful average phenotype and accounts for little actual directed evolution. On the other hand, sexual reproduction does tend to produce unique genetic combinations that, if they are favored by humans and reproduce, lead to more rapid evolutionary change than can be accounted for by natural selection alone. Cultivation and human selection favor genetically unique individuals, protecting them from hostile natural environments, which, through natural selection, often tend to eliminate offspring with rare traits in favor of an average phenotype. Humans recognize rare traits when they are of apparent advantage ethnobotanically and through artificial selection shelter the genes determining those traits, giving them an even better chance of being carried throughout subsequent generations with increasing frequency. Present and future biotechnology notwithstanding, humans do not invent beneficial mutations, we merely select naturally occurring mutations that are attractive for various purposes and attempt to recombine them with other beneficial traits until they are expressed regularly.

One, Two, or Three Species?

Cannabis is a botanical genus that includes a group of flowering herbs that vary in their form (morphology) and function (physiology). Genus *Cannabis* is assigned as the type genus in the relatively small family Cannabaceae, which presently consists of just two accepted genera: *Cannabis* and *Humulus*. Cannabaceae has been variously included in a number of families but most commonly Moraceae (mulberry or fig family) and Urticaceae (nettle family), with which it shares some characteristics. Many members of Moraceae are arborescent (trees) and produce milky latex, while Cannabaceae and Urticaceae have clear vascular fluids. In Moraceae the lobes of the calyx are usually in fours but are often reduced or absent, while in Cannabaceae the five-lobed calyx is freely lobed in male flowers and reduced to near absence as the seed coat in females. Moraceae does not contain any commonly used, long-fiber bast plants but does include a common paper pulp plant, paper mulberry (*Broussonetia papyrifera*), which, like some fig trees (*Ficus* spp.), has been used traditionally as a bast (bark) fiber source to make paper and pounded cloth. Urticaceae, on the other hand, contains several plants utilized for their long fibers, including common nettle (*Urtica dioeca*), giant nettle (*Girardinia diversifolia*), and ramie (*Boehmeria nivea*). McPartland and Nicholson (2003) investigated the relationships between obligate *Cannabis* parasites and obligate parasites of Urticaceae and Moraceae and discovered that while *Cannabis* shares seven obligate parasites with members of Urticaceae it shares none with Moraceae. Still, overall evidence has been mixed as to relatedness between Cannabaceae and Urticaceae or Moraceae. It should be noted here that recent research on the structural organization of nuclear ribosomal DNA (rDNA) has shown a high degree of sequence similarity in the rDNA coding regions of *Cannabis sativa* and *Humulus lupulus*, thus supporting the lumping of *Cannabis* and *Humulus* into one family, Cannabaceae (Pillay and Kenny 2006).

Even though some recent phylogenetic research suggested the reclassification of *Cannabis* into Celtidaceae based on chloroplast restriction sites maps (Weigreffe et al. 1998) and chloroplast *mat* K gene sequences (Song et al. 2001), newer research by Sytsma et al. (2002) using genetic analyses of plastid DNA has shown that Moraceae and Urticaceae are indeed the closest familial relatives of the Cannabaceae. However, Cannabaceae appears to have diverged from the Urticales group of order Rosales before the divergence of the Moraceae and Urticaceae and is an evolutionarily older plant family, explaining its similarities to both families. In addition, Sytsma et al. proposed that the genera *Aphananthe*, *Lozanella*, *Parasponia*, and *Pteroceltis* should have their own subfamily grouping, or more correctly be included in the Cannabaceae, as they are all derived from the Celtidaceae clade (2n = 20) presently assigned to the less related Ulmaceae. Although Cannabaceae and Celtidaceae are genetically and floristically similar, they differ greatly in growth form—*Cannabis* is an erect annual and *Humulus* a twinning annual with overwintering roots, while the four genera of the Celtidaceae grow as small perennial, woody trees, or shrubs. *Pteroceltis* appears more closely related to *Cannabis* and *Humulus* and, along with *Celtis*, *Aphananthe*, *Lozanella*, and *Parasponia*, are presently held to be members of the Cannabaceae (Stevens 2008). It is interesting that *Cannabis*'s sister genus *Humulus* also includes three species (see Chapter 2). One of these is *Humulus lupulus*, the common hop plant, the dried inflorescences of which (known as hop cones or hops) are used for flavoring in the brewing of beer, a very popular psychoactive beverage of great antiquity!

The most well-known, and commonly used, scientific name of hemp is *Cannabis sativa* L. The "L." indicates that this Latin binomial, or species name, was provided by the great Swedish naturalist and taxonomist Carolus Linnaeus (Carl von Linné) when he published his seminal *Species Plantarum* in 1753. Linnaeus listed several "varietas" of the taxon, assuming that all *Cannabis* plants belonged to the single species he first recorded, with *Cannabis* as the genus and *sativa* as the species. In 1785, famous French biologist Lamarck recognized two distinct species, "*Cannabis sativa*" as a taller, more fibrous plant and "*Cannabis indica*" as a shorter, more psychoactive plant. The nomenclature proposed by Linnaeus and Lamarck has recently been supported by Hillig (2005a/b) and Hillig and Mahlberg (2004).

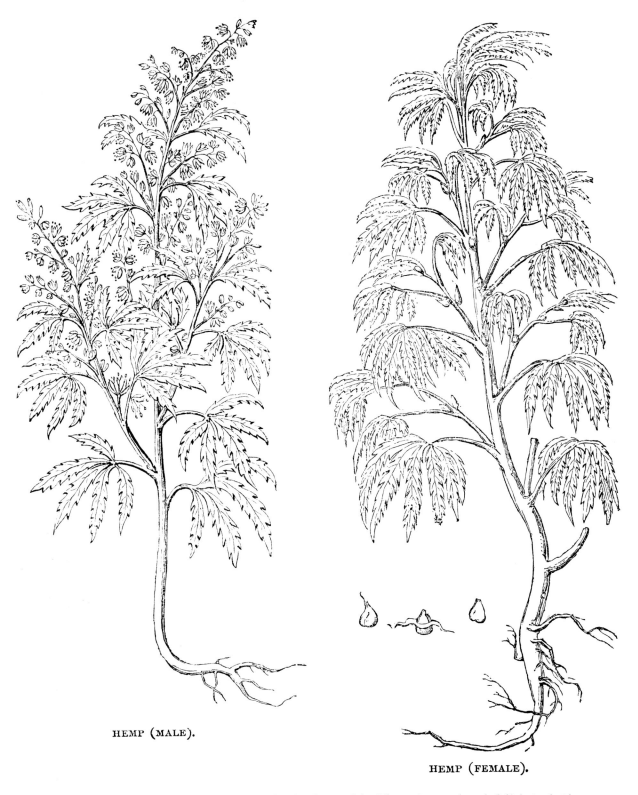

HEMP (MALE).

HEMP (FEMALE).

FIGURE 76. *Cannabis* often appeared in early botanical encyclopedias that noted the difference in sexes, the male (left) depicted with more apparent flowers than the female (right). From "Cultivation and Management of Hemp" in *The Illustrated London News*, December 14, 1850.

This polytypic (more than one species) approach prevailed for almost two centuries, as a series of new "species" names for a wide variety of *Cannabis* plants were published in the literature. However, eventually a majority of botanists came to accept the position that all *Cannabis* plants belonged to the single species of *C. sativa*. This more or less orthodox monotypic concept lasted for decades until it was reexamined in the 1970s by a series of scientists and scholars. Some, such as Schultes et al. (1974) and Emboden (1974), proposed that genus *Cannabis* is indeed polytypic (including three species: *C. sativa*, *C. indica*, and *C. ruderalis*). According to those arguing for this polytypic classification, the three species concept is based on their own field experience as well as other studies, including more modern Russian sources long overlooked in the literature (Vavilov 1926, 1931; Vavilov and Bukinich 1929; Serebriakova 1940; Zhukovskii 1950). Russian studies also indicated that the psychoactive chemical content of *C. indica* is higher than that of *C. sativa* or *C. ruderalis*.

After further study, Ernest Small (1979; also see Small and Cronquist 1976) lumped all *Cannabis* plants back into *C. sativa*, with a number of subspecies and varieties. Clarke (1981) provided additional insight into this confusing taxonomic situation, contending that all classifications to date are "more semantic than scientific and data can be interpreted differently by various taxonomists." Clarke argued that the uses of the plants "are our most important keys to distinguishing individual varieties, strains or species." The biological form and function of an individual organism, such as a *Cannabis* plant, are determined by its genetic inheritance, the environment in which it lives, and the effects of natural and artificial selection. Although the environment plays its role in shaping the life history of an organism, genetic inheritance is the key to differentiating one species from another. Thus Clarke (1981) pointed out that "a Himalayan *Cannabis indica* plant and a Kentucky fiber *Cannabis sativa* plant grown under the same conditions year after year show morphological differences suggesting that at least two separate species should be defined." However, based on East Asian use for fiber rather than drugs Clarke (1981) and McPartland et al. (2000) considered Chinese, Korean, and Japanese cultivars to be members of *C. sativa* rather than *C. indica* as we accept today.

Whatever position one takes on this monotypic versus polytypic debate involving *Cannabis*, it is a striking example of the classic controversy of "lumping" versus "splitting" in biological classification. For those readers with botanical, ecological, legal, financial, or other interests in this challenging botanical problem, a more in-depth discussion of *Cannabis* taxonomy is presented in the next section.

History of *Cannabis* Taxonomy

Taxonomic decisions concerning cultivated plants have been heavily influenced by the compulsions of experimental botanists to discover and define external "natural" (nonhuman) causes for extremes in phenotypic variability. Semantics and law have also played significant roles in confusing species disputes. As a result, taxonomists as well as lay persons occasionally fail to consider and acknowledge, and sometimes overemphasize, the ultimate role humans play in the evolution of *Cannabis*. Despite decades of committed debate over the species status of *Cannabis*, the question of species assignment has only recently been approached from genetic, chemotaxonomic, and molecular taxonomic perspectives (Gilmore, Peakall, and Robertson 2003; Hillig 2004a, 2005a/b; Hillig and Mahlberg 2004; Datwyler and Weiblen 2006). This has produced a more thorough characterization of modern *Cannabis* taxa, as well as more accurate interpretations of their evolutionary interrelationships.

Genus *Cannabis* has been variously characterized by modern taxonomists. As noted earlier, some favored lumping all *Cannabis* taxa together into the single species *C. sativa*, circumscribing two subspecies (hemp and drug), each further divided into two varieties, wild or feral and cultivated (Small and Cronquist 1976). Others split *Cannabis* into three species: *C. sativa*, *C. indica*, and *C. ruderalis*, each species

FIGURE 77. Harvard professor Dr. Richard Evans Schultes with small female (left) and male (right) *C. indica* ssp. *afghanica* broadleaf drug (BLD) plants in a harvested field in Kandahar, Afghanistan, in December 1971 (photo courtesy of Neil Schultes).

circumscribing its own varieties (Schultes et al. 1974). Still others did not recognize *C. ruderalis* but preserved the remaining two species (Vavilov and Bukinich 1929; Serebriakova 1940; Zhukovskii 1950). In all the popular systems of nomenclature preceding the twenty-first century, *C. sativa* was considered the most varied and geographically diverse taxon and included the majority of fiber, seed, and drug varieties.

Narrow-leaf hemp (NLH), *C. sativa* L., has been characterized by its tall stature, less developed branching (especially in fiber varieties), light to medium green foliage, and spicy or sweet aroma. Narrow-leaf drug (NLD) varieties of *C. indica* subspecies *indica*, which were formerly assigned to *C. sativa*, secrete resins of high cannabinoid content, predominately psychoactive THC, while NLH fiber varieties of *C. sativa* secrete resin that is much lower in total cannabinoids, and particularly low in THC. According to Zhukovskii (1950), *C. sativa* grows wild in river basins and on slopes in the Transvolga and islands of the Volga Delta as well as in the Himalaya, Hindu Kush, Tian Shan, and Altai Mountains. The vast majority of the spontaneous and cultivated varieties of *Cannabis* have also previously been classified as members of *C. sativa*. More recent genetic and chemotaxonomic studies demonstrate that this conclusion is erroneous.

The female-type specimen of *C. indica* chosen by Lamarck in 1783 is an NLD variety with a relatively elongated and sparse inflorescence. The epithet "indica" literally means "from India," and this is exactly what Lamarck intended when he named the second species. By the onset of the nineteenth century, the name "*Cannabis indica*" denoted pharmaceutical *Cannabis* imported from India and widely used in popular medications.

Vavilov and Bukinich (1929) first reported the occurrence of the Afghan *C. indica* broad-leaf drug (BLD) biotype recognized by Hillig (2005a). Zhukovskii (1950) assigned *C. indica* to a wild range of Pakistan and Afghanistan and a cultivated range of India, Iran, Turkey, Syria, and North Africa. Although Vavilov never reported observing NLD varieties in South Asia, he included two BLD varieties from Afghanistan and Pakistan as subspecies of *C. indica*. Because the Afghan varieties differed greatly in gross phenotype and biochemistry from familiar European NLH varieties of *C. sativa* but were native to the Indian subcontinent, Vavilov assigned them to the second established taxa, *C. indica,* rather than to *C. sativa.* Vavilov did not mention if the Afghan varieties were drug or fiber types, although based on later observations from Afghanistan (Schultes et al. 1974) and neighboring regions, we suggest that they were used almost exclusively for hashish production. Although Vavilov included BLD varieties from Afghanistan within *C. indica,* he assigned them to a subspecies rank of *afghanica* (Vavilov and Bukinich 1929) since they were from Afghanistan, not India, and were dissimilar to Lamarck's NLD-type specimen in gross phenotype, leaf shape, and inflorescence morphology. Vavilov and Bukinich's nomenclature is also supported by Hillig (2004, 2005a/b) and Hillig and Mahlberg (2004). In 1972, Richard Evans Schultes of Harvard University traveled to Afghanistan specifically to obtain specimens of then little-known Afghan hashish cultivars. Following Vavilov, Schultes accepted a broad interpretation of *C. indica* to include Afghan varieties, as well as Indian drug types and their new world descendants, much as we presently accept. Therefore, we now include the short, broadleaved, acrid-smelling Afghan *Cannabis* in the BLD biotype (*C. indica* ssp. *afghanica*), as described by Vavilov and Bukinich (1929) and Schultes et al. (1974) and presently referred to as the "indica" or "hashish plant" by domestic marijuana growers.

The Indian biotypes described by Lamarck either were included in *C. indica* by some subsequent researchers (Schultes et al. 1974) or were considered drug varieties of *Cannabis sativa* indigenous to South Asia (Small and Cronquist 1976); we now classify Indian biotypes as narrow-leaf drug (NLD) varieties of *C. indica* ssp. *indica* based on Hillig (2005a/b). Lamarck was correct in naming the distinct NLD biotypes *C. indica,* although later researchers such as Small and Cronquist (1976) would include Indian *Cannabis* and its NLD descendants worldwide as varieties of *C. sativa.* As interpreted by Schultes et al. (1974), Lamarck's *C. indica* was the progenitor of the vast majority of both Old and New World NLD varieties, and we believe this to be accurate. Neither Linnaeus nor Lamarck was familiar with the BLD hashish varieties from Afghanistan, and both Vavilov and Schultes placed the Afghan BLD biotypes with the Indian NLD biotypes, both circumscribed by *C. indica.* However, Schultes et al. (1974) chose and described a short and profusely branched Afghan hashish variety with broad dark green leaves, a typical BLD variety, for their type specimen of *C. indica.* Consequently, BLD *Cannabis* from Afghanistan has come to typify *C. indica,* especially in the minds of marijuana growers, who use the term "indica" to differentiate Afghan BLD varieties from NLD varieties or "sativas" (Clarke 1981). Based on their familiarity with a wide range of drug type *Cannabis* varieties, North American marijuana growers were correct in differentiating taxonomically Afghan BLD *Cannabis* from all other NLD *Cannabis* varieties. However, following Schultes's lead and choosing the name "indica" to represent Afghan *Cannabis* alone leaves out the remainder of NLD varieties that should also properly be included in *C. indica.*

The problem arose when Lamarck's valid name *C. indica* was first used by Schultes, albeit correctly, to differentiate the distinctive Afghan varieties. Apparently Lamarck was completely unfamiliar with Afghan *Cannabis* as he never reported it, and therefore it is inappropriate to use his species name without subspecies modifiers; Lamarck intended *C. indica* to represent another taxon of *Cannabis* entirely—one

phenotypically distinct from European NLH and specifically native to India but not Afghanistan. Small and Cronquist (1976) applied the term "indica" incorrectly when they used it as a subspecies name to indicate the high-THC varieties of *C. sativa.* Additionally, Small and Cronquist were unable to recognize Afghan BLD *Cannabis* as a separate taxon, as it was not among their accessions. Clarke (personal observations) surveyed the *Cannabis* collection in the New York Botanical Garden herbarium in 1986, which includes those specimens collected by Small and Cronquist among many others, and did not observe any accessions from Afghanistan or any resembling Afghan BLD varieties in leaf shape or floral morphology. Schultes's Afghan specimens of BLD hashish varieties are now accessioned in the Harvard University herbarium. The herbarium of the Vavilov Research Institute in St. Petersburg, Russia, contains hundreds of *Cannabis* specimens viewed by Schultes, yet he assigned only a few of the broad-leaf accessions to *C. indica* (Clarke personal observations 1994, 1995).

Vavilov was the first to document the Afghan BLD varieties, and consequently he should receive recognition for their discovery. McPartland et al. (2000) erroneously chose to use Vavilov's varietal name "afghanica" as the species name for the morphologically distinct Afghan BLD varieties, and hence they were included as varieties of *C. afghanica,* although they have been shown to be genetically distinct enough to warrant species rank (Hillig 2005a). Lamarck's term "indica" was preserved and also erroneously applied by McPartland et al. (2000) as the subspecies name for the NLD varieties of *C. sativa* originating from India. Their interpretation followed Lamarck's recognition of NLD varieties, although the NLD drug varieties of India were considered by McPartland et al. (2000) to be varieties of *C. sativa* distinct from the NLH *C. sativa* of northern Eurasia rather than members of *C. indica,* and thereby agreed, albeit incorrectly, with Small and Cronquist (1976). McPartland et al. (2000) characterized the BLD varieties by their shorter stature, well-developed branching, thick brittle stem, dark green foliage, broad coarsely serrated leaves, dark gray shiny seeds, acrid aroma, copious production of easily abscised resin, and susceptibility to *Botrytis* gray mold, and they respectfully, but mistakenly according to the rules of nomenclature, called them *C. afghanica.*

Female floral morphology varies markedly between BLD and NLD varieties, the NLD plants having evolved lax inflorescences with more flowers and fewer leaves, while BLD biotypes evolved compact inflorescences with fewer flowers and more leaves. This characteristic of BLD biotypes, along with their origin in an arid climate devoid of pathogenic fungi, increases their susceptibility to fungal blights (McPartland et al. 2000), especially when they are cultivated in areas beyond their native range. The seeds of most NLH and NLD varieties are borne in racemes with medium to long internodes and fewer leaflets. The hundreds of bracts, each surrounding a single seed, are superior to many of the larger leaflets, such that they are exposed and shatter more easily, readily dispersing the seed. However, cultivated varieties tend to have more persistent seeds than feral or wild varieties as a result of unintentional human selection during harvest (see Chapter 3). The fewer seeds of Afghan BLD varieties are borne in bracts, nestled deeply between the leaflets of the floral clusters near the central axis, where the leaflets tightly surround the bracts and natural dispersal of seed is extremely limited. BLD *Cannabis* has been utilized almost exclusively for the production of sieved hashish (see Chapter 7) and does not appear to have escaped as a weed in areas beyond its origin and traditional cultivation regions. Persistent fruits and low seed production are likely characteristics derived from human selection, indicating that BLD varieties depend more on humans as their primary agent of dispersal than NLH or NLD varieties do. Vavilov and Buchinich (1929) reported "wild" populations along the Kunar River in eastern Afghanistan with freely shattering seeds, which they named *C. indica* var. *kafiristanica*. Hillig (2005a) suggests that *C. indica* ssp. *kafiristanica* may represent the putative narrow-leaf drug ancestor (NLDA) of the *C. indica* NLD and BLD biotypes.

The third putative species, *C. ruderalis*, was first described by Janischevsky (1924) as a very small weedy variety that was not cultivated, and it produced oils in glands near the base of each seed that attracted beetles, which then facilitated its dispersal, this being interpreted as a "wild" characteristic. However, no other researchers have identified this curious biological dispersal. Purported *C. ruderalis* has a wild range throughout the Eurasian Steppes and occurs as a weed in the Volga Basins, northern Siberia, Central Asia, and into Europe (Zhukovskii 1950); Vavilov (1931) proposed that *C. ruderalis* is the primordial putative ancestor (PA) of both *C. sativa* and *C. indica*. Janischevsky's juvenile-type specimens for *C. ruderalis* have reduced leaves resembling those of BLD varieties in shape. However, ruderal *Cannabis* from Hungary resembles dwarfed *C. sativa* with freely shattering seeds, limited branching, and narrow leaflets (Clarke personal observation 1992; also see Benécs-Bárdi 2002) and therefore may more accurately be classified as *C. sativa* ssp. *spontanea*, while "wild" *Cannabis* from eastern China may be more accurately classified as *C. indica* ssp. *chinensis* (see Hillig 2005a). Serebriakova (1940) did not recognize *C. ruderalis*, although she described several small feral biotypes of *C. sativa* from central and southern Russia and classified them as *C. sativa* ssp. *spontanea*. Chrtek (1981) reported hybridization between southern populations of *C. ruderalis* and more northern populations of *C. sativa* in Czechoslovakia and named these hybrids *C. × intersita*. This is the first documentation of hybridization between the proposed PA and its possible narrow-leaf hemp ancestor (NLHA) descendants, which would lead to the genetic degradation and modification of the original PA *C. ruderalis* gene pool. Because of crosses between the PA and feral NLH varieties (NLHA), there probably are no true PA *C. ruderalis* populations existing today, although the ruderal Central Asian varieties may represent our closest genetic link to the now extinct PA.

Vavilov (1922) was the first to describe *C. sativa* var. *spontanea* as a "true wild plant," found throughout the northern Caucasus, Ural, and Volga regions, as well as the Altai Mountains and Central Asian Kashgaria in areas where cultivation of *Cannabis* was unknown. However, there is some confusion between *C. sativa* var. *spontanea* and *C. ruderalis* in the later writings of Vavilov (1931). Both are weedy, spontaneously growing plants with freely shattering inflorescences containing small mosaic-patterned fruits characterized by a horseshoe-like structure (caruncle) at the base. Vavilov (1931) described the characteristics of *Cannabis* in its wild and cultivated state throughout a broad area from European Russia through Central Asia to the Altai Mountains. The most prominent forms of spontaneous *Cannabis* he described as *C. sativa* var. *spontanea*, a medium height (60 to 150 centimeters, or two to five feet), well-branched plant bearing freely shattering small dark mosaic-patterned seeds with the characteristic "horseshoe" at the base; this form was encountered

by Vavilov all across Asia. According to Janischevsky (1924), *C. ruderalis* shares these same seed characteristics and appears to be a shorter and less branched relative of *C. sativa* var. *spontanea*. In the species key of Schultes et al. (1974), *C. ruderalis* is more closely allied to *C. indica* than *C. sativa*. *C. ruderalis* can also be interpreted as a *C. sativa* × *C. indica* hybrid or even the common progenitor of both *C. sativa* and *C. indica*. Hillig (2005a/b) also tentatively recognizes *C. ruderalis* and proposes it as a PA of the genus *Cannabis*.

Serebriakova (1940) divided *Cannabis* into two species based solely on morphological characters. She characterized *C. sativa* as a tall, sparsely branched plant with large leaves and large, dull, and gray-brown seeds and *C. indica* as a short, well-branched plant with small leaves and small, shiny, and dark-colored seeds. She further subdivided *C. sativa* into two subspecies. Subspecies *spontanea* circumscribed the feral and possibly wild varieties of *C. sativa*, which she characterized as short, well-branched plants with small leaves of narrow leaflets and small, elongated seeds, dark in color with a well-defined pattern and obvious horseshoe-shaped base; essentially this description was similar to that which Janischevshy (1924) provided for *C. ruderalis*. According to Serebriakova (1940), subspecies *culta* encompassed the cultivated varieties of *C. sativa*, which were characterized as tall and sparsely branched with medium to large leaves and wide leaflets. The seeds were described as medium or large, oval-circular in shape, and gray-brown or dark colored with or without pattern and lacking a horseshoe at the base. Fiber and seed populations from European Russia were characterized by Serebriakova (1940) as short to medium in height (60 to 150 centimeters or two to five feet) and unbranched with larger light-colored fruits that did not shatter readily. Some northern strains had darker gray fruits. These were primarily cultivated forms with a few escaped weedy types. Serebriakova also described a "dwarfish and early-ripening" type (*C. sativa* variety *praecox*) used to produce oil but not utilized for fiber (Smekalova 2008). Later taxonomists may have confused this variety with *C. ruderalis*.

Serebriakova was well acquainted with the northern temperate varieties of *Cannabis*, but she lacked exposure to more southerly temperate, subtropical and tropical varieties. This research bias resulted from both her living in temperate northern Eurasia and the Soviet interest in studying only plants with culturally established, high economic values. The study of fiber or seed varieties of *Cannabis* was much more important to Soviet officials than the study of drug varieties. Even some varieties designated by Serebriakova, which by their names are related to drug varieties (e.g., subnarcotica), are not described as such in her key, and psychoactivity was not used as a taxonomic character. Comparisons of the descriptions in her taxonomic key with her annotations on the NLH *C. sativa* accessions contained in the Vavilov Research Institute herbarium in St. Petersberg (Clarke personal observations 1994, 1995) revealed that her characterizations of NLH varieties match the specimens well, and therefore we accept the remainder of her descriptions as likely being accurate. In deference to Serebriakova (1940), Hillig (2005a) elevated the taxon *C. sativa* var. *spontanea* to include several additional spontaneous varieties and renamed it *C. sativa* ssp. *spontanea* or the putative narrow-leaf ancestor (NLHA) of narrow-leaf European hemp (NLH), *C. sativa* ssp. *sativa*.

The classification systems proposed by all these leading taxonomists make logical sense within their own research setting and do not significantly contradict each other. However, these studies included only a limited part of *Cannabis*'s diversity, as they all lacked complete *Cannabis* germplasm collections that ideally should have included accessions common to all the other researchers. As a result, each was only able to describe a part of the total range of diversity within the genus, and they did so quite accurately. By reconciling only a few small differences we were able to fit them together into the taxonomic scheme presented in this chapter. Based on firsthand descriptions of foreign populations and domestically grown *Cannabis* accessions (Clarke personal observations), we have added to Serebriakova's key by including additional feral varieties and the drug cultivars of which she lacked voucher specimens. Most recently, we have incorporated the chemotaxonomic data of Hillig (2004a, 2005a/b) to add his biotype designations to each proposed taxon.

Recent Advances in *Cannabis* Taxonomy

As noted earlier, McPartland et al. (2000) described four biotypes representing three established species, *C. ruderalis*, *C. sativa*, *C. indica*, and added a fourth species, incorrectly named *C. afghanica*, to circumscribe the unique Central Asian hashish varieties first reported by Vavilov and Bukinich (1929). Recent doctoral research by Karl Hillig at Indiana University correlated morphological characteristics of 157 *Cannabis* seed accessions of diverse geographical origins grown in a common greenhouse with restriction mapping of various proteins as well as cannabinoid and terpenoid contents. His research provides us with five key papers (Hillig 2004a/b, 2005a/b; Hillig and Mahlberg 2004) further elucidating the taxonomic structure of genus *Cannabis* and providing additional clues to its evolution. Hillig concluded that *Cannabis* should be divided into seven taxonomically distinct biotypes of three species with six putative subspecies levels. We have adopted and modified Hillig's biotype designations (e.g., NLH, BLH, NLD, BLD, NLHA, NLDA, and PA) and used them throughout this book.

Much of Hillig's work relied on detecting patterns of allozyme variation between different populations corroborated by morphological and chemical evidence. "Allozymes are enzyme variants that have arisen through the process of DNA mutation. The genetic markers (allozymes) that are commonly assayed are part of a plant's primary metabolic pathways, and presumed neutral to the effects of human selection. Through allozyme analysis, it is possible to discern underlying patterns of variation that have been outwardly obscured by the process of domestication" (Hillig 2005b). Hillig (2005a/b) tentatively identified *C. ruderalis* based on only five accessions. He considered *C. ruderalis* to be "wild" and that it may represent the putative ancestor (PA) of domesticated *Cannabis*, a primitive biotype possibly unmodified by human selection and domestication. By definition, if *C. ruderalis* is truly the PA, it must never have been cultivated, although some populations considered "wild" were occasionally utilized in the twentieth century for fiber (Vavilov 1957) or drugs (Clarke 1998a). *Cannabis* probably did evolve in Central Asia millennia ago, but it was likely reintroduced into this region during the early Holocene from more temperate Pleistocene refugia following glaciations (see Chapters 2 and 12). Truly "wild" *Cannabis* is very difficult to differentiate from feral populations. This problem can be compared with that involving spontaneous and cultivated opium poppies (*Papaver somniferum*), for which no known truly wild plants have been documented. The hypothetical range of *C. ruderalis* in

Central Asia lies centrally within the range of feral *Cannabis* spreading west into Europe, east into Siberia and China, and south along the Himalayan foothills of South Asia. However, what we today classify tentatively as the PA might more correctly be viewed as an unselected, degenerate multihybrid between various escapes from cultivation (see Chapter 3). It is unlikely that *C. ruderalis* could have maintained its genomic integrity during the domestication process; for example, if domestication occurred within the natural range of *C. ruderalis*, windblown pollen would have facilitated the introgression of genes back into the truly wild *C. ruderalis* gene pool from the selected semidomesticates. Plants that appear to be *C. ruderalis* may be more accurately described as primordial examples of *C. sativa* ssp. *spontanea*, the NLHA, or *C. indica* ssp. *kafiristanica*, the NLDA with a higher frequency of predomestication gene combinations and a lower frequency of postdomestication gene combinations. Gilmore et al. (2007) felt there was insufficient genetic evidence to support *C. ruderalis* as a separate taxon and that the "wild" accessions more likely represent feral European hemp. We suggest that backcrossing to ancestral populations was common and the existence of genetically distinct PA populations is unlikely (for further discussion of the evolutionary and ecological effects of hybridization in *Cannabis*, see Chapters 2 and 3). *C. ruderalis* may have been a valid taxon in the distant past, but it may be impossible to find accessions that have remained truly isolated from human intervention.

Hillig (2005a) divided *C. sativa* into two biotypes or putative subspecies: *C. sativa* ssp. *sativa* or narrow-leaf hemp (NLH, which includes European industrial hemp cultivars and has been selected for centuries for both fiber and seed production; see Chapters 5 and 6), and *C. sativa* ssp. *spontanea* or the narrow-leaf hemp ancestor (NLHA, which includes spontaneously growing populations from eastern Europe and western Asia and exhibits characteristics shared by both the proposed PA, *C. ruderalis*, and NLH). The NLHA biotype may result either from interbreeding of NLH escapes from cultivation with the truly wild PA or from the atavistic (spontaneously appearing from the past) emergence of ancestral traits resembling those of the PA within escaped NLH populations following decades of evolving under natural selection alone. Spontaneously growing populations found throughout Eurasia are all made up of relatively well-branched plants with reduced leaflets and readily dispersed seeds (Clarke personal observations), a condition that may echo the wild phenotype of the original PA. Total cannabinoid levels in purported *C. ruderalis* and the NLHA rarely exceed 1 percent dry weight with the predominate cannabinoid being CBD. The only other accessions with a preponderance of CBD are the European NLH cultivars artificially selected for an abnormally low THC content. European law stipulates that THC content must be less than 0.2 percent dry weight and that the ratio of CBD to THC must exceed 3:1 (Hennink 1997).

Hillig (2005a) divided *C. indica* into four biotypes or putative subspecies. These biotypes include the following: (1) *C. indica* ssp. *kafiristanica*, the proposed narrow-leaf drug ancestor (NLDA), which circumscribes spontaneous populations from Nepal and northern India, based on similarities in both enzyme banding patterns and cannabinoid profiles—this biotype may represent either the ancestor of NLD *Cannabis* or, more likely, feral escaped populations of NLD varieties rather than a common taxon from which all three other *C. indica* subspecies evolved; (2) *C. indica* ssp. *indica*, which circumscribes cultivated narrow-leaf drug or NLD biotypes native to the Indian subcontinent; (3) *C. indica* ssp. *afghanica*, which circumscribes broad-leaf drug or BLD hashish cultivars originating in Afghanistan or western Turkestan—it is characterized by its short stature and broad and dark green leaflets and is the taxon erroneously designated as *C. afghanica* by McPartland et al. (2000); and (4) *C. indica* ssp. *chinensis*, which includes the East Asian broad-leaf hemp (BLH) landraces traditionally used for fiber and seed production.

Interestingly, Chinese, Japanese, and Korean BLH landraces more closely resemble South Asian NLD and Central Asian BLD biotypes, both phenotypically and in terms of cannabinoid production, than they resemble European NLH. This suggests that East Asian landraces, although more recently utilized only for fiber and seed, were selected in the past also or exclusively for their psychoactivity (see Chapters 4, 5, 6, and 7). This also indicates that BLH followed a divergent evolutionary path from NLH and its NLHA early on. Did BLH evolve from BLD or NLD or did BLD and NLD evolve from BLH, or perhaps more plausibly, did NLDA, BLD, NLD, and NLH all evolve from a common drug producing putative drug ancestor or PDA population? Possibly BLH came to East Asia at a later date from a South Asian or southern East Asian Pleistocene refuge rather than directly from Central Asia (e.g., see Olson 2002b and Chapter 2).

Hillig (2005a) convincingly presented a new and modern taxonomy for *Cannabis* incorporating previously established nomenclature to circumscribe and accurately represent discrete biotypes. Hillig's taxonomic groupings correspond closely with those of Serebriakova and Vavilov. However, there are some exceptions. Unlike Serebriakova, Hillig preserves the taxon *C. ruderalis*, and, unlike Vavilov, he subdivides *C. sativa* into *C. sativa* ssp. *sativa* (called *C. sativa* ssp. *culta* by Serebriakova and commonly referred to as "European" hemp) and *C. sativa* ssp. *spontanea* (adopting Serebriakova's nomenclature and commonly referred to as "feral" or "wild" hemp).

More recently, Gilmore et al. (2007) performed polymerase chain reaction assays on 76 populations of *Cannabis* representing crop usage for either "fiber" or "drugs" as well as purportedly "wild" populations. They established six organelle DNA haplotypes (six sets of closely linked genes that tend to be inherited together), which they assigned to three haplotype groups. All three haplotype groups (A, B, and C) support the chemotaxonomic conclusions and taxonomic proposals of Hillig (2005a/b), which are based on earlier isozyme and secondary product research that we use throughout this book.

Haplotype Group A, encompassing haplotypes I and II, is representative of NLH and NLHA taxa delimited by Hillig (2005a/b). Haplotype I in Group A includes all the cultivated and escaped *Cannabis* of Europe and North America (21 accessions) and represents what we classify as either NLH or NLHA biotypes. Six apparent "drug" types were also included within haplotype I. Two accessions were of "*nederweit*," which was the traditional balcony and garden *Cannabis* grown in the Netherlands before glasshouse and indoor *sinsemilla* cultivation began. These *Cannabis* plants were derived from commercial bird seed and plants that matured and produced seeds outdoors so far north that they were of very low potency; thus these Dutch "drug" accessions are more likely European hemp varieties. The two Australian "drug" accessions are likely derived from the accidental purchase of European hemp seed from unscrupulous drug-seed sellers who occasionally sell inexpensive hemp seed for high prices

and which they claim is seed for *sinsemilla* cultivation. This could also explain the Lebanese and South African "drug" accessions in haplotype I; in addition, there may be some other reason that hemp seed was being grown for purported drug purposes in these regions. Corroborating cannabinoid analysis could resolve this question by demonstrating if these "drug" accessions were actually very low in THC. Both Small and Cronquist (1976) and Gilmore et al. (2007) chose to assign crop-use characteristics to each accession, often relying on police accounts of intended usage, and in both cases this has led to confusion in the interpretation of the results. As we have mentioned previously, it is very difficult to determine whether a population is truly "wild," and sometimes people also grow "fiber" hemp accidentally when they are trying to grow "drug" marijuana. Haplotype II in Group A includes only two accessions, both "fiber" types from Korea, and their unique haplotype may result from a founder effect during introduction of *Cannabis* from China. It would be interesting to determine if *Cannabis* from Japan also belongs to haplotype II.

Haplotype Group B *Cannabis* includes an assortment of *C. indica* biotypes of "wild" and cultivated "drug" varieties from Afghanistan, Australia, China, Mexico, Nepal, Netherlands, and Turkey; Gilmore et al. (2007) refer to Group B varieties (in marijuana breeder's taxonomy) as "indica." All in Group B are circumscribed by haplotype III with the single exception of an accession from China, which was assigned to haplotype IV. The accessions from Afghanistan we classify as BLD types. Traditional accessions from Mexico, Nepal, and Turkey are NLD biotypes. *Sinsemilla* varieties from the Netherlands and Australia are typically hybrids between BLD and NLD biotypes. All three groups are commonly used for drug production. We classify the Chinese accessions as BLH biotypes, many of which also have the genetic potential to produce at least moderate amounts of THC even if they have been traditionally grown for fiber and seed for thousands of years. Eight supposedly "fiber" types were also included in haplotype III. Four are from Turkey where landrace varieties have traditionally been used for both fiber and drug production (see Hakki et al. 2007 for ISSR analyses and Pinarkara et al. 2009 for RFLP analyses of Turkish *Cannabis*). Two accessions are of the French fiber cultivar 'Fibrimon,' which has Turkish ancestors in its pedigree (see Chapter 10). The remaining two are Russian and Czech hemp cultivars and may also share Turkish ancestry.

Haplotype Group C includes an assortment of *C. indica* biotypes of "wild" and cultivated "drug" varieties from Africa, India, Jamaica, Mexico, Nepal, the Netherlands, and Thailand, all representing what breeders refer to as "sativa." One supposedly "fiber" accession was from the northwestern Himalayas and the "wild" type was from Nepal, both drug-producing regions where drug types are also used for fiber production (Clarke 2007). All are represented by haplotypes V and VI and are members of *C. indica* ssp. *indica* or the NLD biotype. Gilmore et al. (2007) also note that only accessions from the Netherlands included all three haplotype groups. This results from the Netherlands position as supplier of the majority of "drug" *Cannabis* seeds grown worldwide.

Additional genomic and chemical analyses also provide data that further support the taxonomy we present here. This may also allow us to understand the early evolution of differing *Cannabis* taxa. Faeti et al. (1996) used RAPD analysis of industrial hemp cultivars to identify three distinct gene pools: one originating in Italy, another from Hungary, and a third from outside of Europe (Korea), groupings that reflect their their breeding histories. Traditional Italian landrace hemp varieties are likely derived solely from European NLH, while many Hungarian cultivars were developed from hybrid crosses between European NLH and Chinese BLH landrace varieties (see Chapter 10 for a discussion of industrial hemp breeding); the Korean accession probably represents an unhybridized East Asian BLH landrace variety. Therefore, the groupings based on DNA sequences identified by Faeti et al. (1996) indicate origins within differing taxa as described by Hillig (2005b).

A variety of molecular genetic techniques have been applied to *Cannabis* taxonomy. Carboni et al. (2000) and Forapani et al. (2001) utilized variations in Restriction Fragment Length Polymorphisms (RFLP) molecular markers to survey industrial hemp varieties. Gilmore et al. (2003) used Short Tandem Repeat (STR) DNA markers or microsatellites to determine relationships and degree of genetic diversity. Datwyler and Weiblen (2006) explored genetic variation using Amplified Fragment Length Polymorphism (AFLP) molecular markers. Hakki et al. (2007) utilized Inter Simple Sequence Repeats (ISSR), and Pinarkara et al. (2009) used Random Amplification of Polymorphic DNA (RAPD) to differentiate between industrial hemp and drug populations for forensic purposes. Most recently, van Bakel et al. (2011) established a draft genome and transcriptome (complete set of RNA molecules) for cannabinoid biosynthesis using gene sequencing techniques. Their research showed that two NLD/BLD hybrid drug cultivars ('Purple Kush' and 'Chemdawg') expressed an active THCA-synthase gene, while a NLH seed variety ('Finola') and a NLH fiber variety ('USO-31') expressed an active CBDA-synthase gene. These genetic studies all support biosynthetic divergence early in the evolution of *Cannabis* leading to drug producing *C. indica* (NLDA, NLD, BLD, and BLH) and nondrug producing *C. sativa* (NLHA and NLH). This research unanimously supports Hillig's essential taxonomy.

A large quantity of amazingly well-preserved, 2,700-year-old *Cannabis* remains was recently recovered from an ancient Yanghai tomb in Xinjiang province, China, that was probably used to inter a "shaman" (Jiang et al. 2006; Russo et al. 2008). This discovery provides a wealth of genomic data of taxonomic and evolutionary significance, as well as its cultural importance (see Chapters 4, 6, and 7). Mukherjee et al. (2008) used polymerase chain reaction (PCR) techniques to determine organelle DNA sequences and compared them to the DNA sequences of feral *Cannabis* growing near the site, as well as a purported *C. indica* accession and the Japanese cultivar 'Tochigi Shiro.' Two primary groups within the *Cannabis* accessions were identified, one included the archeological Yanghai sample and its weedy relative, and a second circumscribed the purported *C. indica* accession and the Japanese industrial hemp cultivar. This second group follows the taxonomy proposed by Hillig (2005b) as East Asian BLH varieties, also members of *C. indica*. The modern-day Chinese sample matches the archeological material closely, as we might expect, but most importantly the two Yanghai samples form a group separate from the *C. indica* accessions indicating that they do not share a close relationship with East Asian *Cannabis* and could belong to another taxon. Mukherjee et al. (2008) proposed that the Yanghai *Cannabis* may be of European-Siberian origin. The Yanghai *Cannabis* may be more closely related to the proposed narrow-leaf

hemp ancestor (NLHA), *C. sativa* ssp. *spontanea* or *C. ruderalis* the PA of both *C. sativa* and *C. indica* (Hillig 2005b).

In their interdisciplinary study of the ancient *Cannabis* macrofossils found at Yanghai, Russo et al. (2008) employed chromatographic techniques in concert with molecular PCR analysis to identify degradation products of THC (CBN), THCV (CBNV), CBC (CBL), and CBD (CBE) as well as gene sequences coding for the production of THC synthases (enzymes that catalyse a synthesis process); from the perspective of our study, the most relevant results of their research are as follows:

> Using high performance liquid chromatography (HPLC), the largest cannabinoid peak was cannabinol (CBN) at 7.4 min, but concentration levels were very low, averaging 0.007% w/w. CBN is an oxidative breakdown product THC, generated non-enzymatically, with increasing age. . . . In the 36.3–40.5 min region, the known THC degradant cannabitriol (CBO) . . . was seen, as well as a series of peaks with spectral similarities to CBN, three of which are tentatively identified by the NIST database as either *hydroxyl-* or *oxo-* CBN . . . There was a very small peak detected at the correct retention time in the sample for THC, but the spectra could not confirm its identity. (RUSSO ET AL. 2008; also see BRENNEISEN 2007)

The large amount of *Cannabis* flowers (789 grams) recovered from the tomb probably represents tissues from a population of several to many plants, and therefore all the constituent plants could have been of the high-THC type. Although there is no indication of actual potency (THC percentage) these flowers were most likely used for mind-altering purposes and are in essence equivalent in cannabinoid profile to modern-day medicinal and recreational marijuana. Furthermore, plants within feral and cultivated Eurasian BLH and BLD populations (evolving by natural selection for cannabinoid profile) usually display a bell curve of THC:CBD ratios ranging from a few plants with high THC:CBD ratios, through the majority of individuals with roughly equal amounts of THC and CBD, to a few plants with very low THC:CBD ratios, resulting in an average THC:CBD ratio of approximately 1:1 (de Meijer et al. 2003; Russo 2007). However, the average cannabinoid ratio is easily altered through selective breeding and within a few generations a population can produce almost all THC or almost all CBD (Clarke personal observations). The chances that a population would naturally evolve to be high in THC and low in CBD are very low. Therefore, the high-THC Yanghai materials are very likely the result of human intervention by selective breeding for psychoactive potency.

Information collected from the Yanghai remains confirm that by 2,700 years ago humans had discovered the psychoactive nature of *Cannabis* and had selectively bred drug varieties. The occurrence of a novel gene sequence coding for a second THC synthase protein (see Russo et al. 2008; Russo 2011) that has not yet been discovered in present-day *Cannabis* may indicate that several biosynthetic pathways for THC synthesis may once have existed; apparently only one of these has been favored through natural and human selection, and therefore this is the single pathway that survives today. Genomic evidence also points to the possibility that the Yanghai *Cannabis* is more closely related to a taxon other than present-day *C. indica*; if so, then either the NLHA *C. sativa* ssp. *spontanea* or the PA *C. ruderalis* may have been preadapted to produce THC, possessing the appropriate genetic programming that only required simple human selection to become psychoactive populations, although psychoactivity is not characteristic of these taxa today. A more likely choice may be NLD *Cannabis* imported from farther south nearer the Himalaya or Pamir Mountains.

"The excellent preservation of the *Cannabis* from [the 2,700 year old Yanghai] tomb allowed an unprecedented level of modern botanical investigation through biochemistry and genetics to conclude that the plant was cultivated for psychoactive purposes" (Russo et al. 2008). We surmise that the *Cannabis* in question was highly revered and that is why it was placed in quantity with the shaman in his final resting place.

Recent taxonomic research enhances our study by allowing us to propose a model for the evolution of *Cannabis* gene pools (see the following section). It also allows us to apply our knowledge of these gene pools to understanding early selection as a crop plant (see Chapter 3), human migrations (see Chapter 4), and modern cultivar breeding (see Chapter 10), especially in terms of their impact on *Cannabis*'s evolution.

Genetic and Historical Model for the Evolution of *Cannabis* Biotypes

In Chapters 2 and 12, we present a hypothetical model for the early evolution of *Cannabis* based largely on chronological, environmental, and geographical constraints associated with Pleistocene glaciations and the early Holocene epoch. Biochemical and physiological changes have also had major influences on the evolution of *Cannabis* biotypes. Many millennia ago, the primordial *Cannabis* gene pool experienced a momentous genetic event—a mutation of the cannabinoid biosynthesis gene, suddenly coded for the production of psychoactive THC—altering both the course of its own evolution and that of human culture. At some point following this mutation, nomadic bands of early humans encountered wild populations of the PA, or its direct descendants the PDAs, that produced noticeable amounts of THC. During the following centuries, *Cannabis*, with the ability to produce at least nominal amounts of THC, was dispersed across much of southern and eastern Eurasia. Following its introduction into new environments, humans further selected *Cannabis* for differing products, Europe and East Asia eventually favoring fiber and seed uses and South Asia favoring drug uses. Hemp fiber and seed use was largely a northern temperate phenomenon, and drug use was traditionally a subtropical and tropical phenomenon. Hashish production and use have usually been at higher latitude than marijuana use and ranges much farther north into temperate latitudes.

The major remaining question focuses on the role of people in the evolution of drug *Cannabis*: did this great cannabinoid divide between *C. sativa* NLH and *C. indica* NLD, BLD, and BLH result from natural selection or from artificial selection by humans? As you might expect by now, the answer is both. Small and Cronquist (1976) proposed that the split occurred as a result of human selection alone. Although humans would later be important facilitators, the biosynthetic and taxonomic split between CBD-producing populations and THC-producing populations must have occurred well before human contact with *Cannabis*. Following initial dispersal out of Central Asia and/or later diffusion from Pleistocene refugia, humans in some regions positively selected for THC production, while those in other regions did not. Peoples living at northern latitudes needed hemp as a fiber source,

as there were few native fiber plants (e.g., flax, nettles, elm, lime, and hemp), whereas peoples living at southern latitudes possessed a myriad of fiber plants to choose from. However, we would expect that both northern and southern peoples explored the use of *Cannabis* for its euphoric potential. Why does traditional drug use of *Cannabis* appear restricted to mostly subtropical latitudes? Is this only a relatively recent situation? Was the use of psychoactive *Cannabis* much more widespread in the past?

There is another hypothesis that may answer some of these questions. Although the initial mutation allowing the biosynthesis of THC likely occurred many millennia ago, it is possible that this mutation only occurred in the Asian PDA population after the European putative hemp ancestor (PHA) populations and East Asian (or South Asian) PDA populations had split into different gene pools in geographically isolated refugia during earlier glaciations. If this did occur, with one refuge for the proto-*sativa* PHA in Europe and the other for proto-*indica* PDA in southeastern Eurasia, these ancestral *Cannabis* populations would have eventually evolved into separate geographically isolated taxa. Possibly only the proto-*indica* PDA carried the B_T allele for THC biosynthesis. Did humans bring the two primordial gene pools (PHA and PDA) together at some relatively recent time during the Holocene? Did this account for what is perceived by some today as *C. ruderalis*, the "original" PA, possessing both the B_D allele for CBD synthesis *and* the B_T allele for THC synthesis?

Presently, the allele that facilitates THC biosynthesis (B_T) and the one that results in CBD synthesis (B_D) are found in both *C. sativa* and in *C. indica*. However, the B_T allele is weakly expressed in *C. sativa* and the B_D allele is more strongly expressed, while in *C. indica*, although the B_D allele is also expressed, the B_T allele is always expressed at a higher level than it is in *C. sativa*, even in East Asian BLH varieties. Possibly the B_T allele did not exist in the proto-*sativa* or earliest *C. sativa* gene pools before the introduction of drug varieties from the East. This could explain why we have found no confirmed evidence of early European peoples using *Cannabis* for drug purposes—because European NLHA and NLH did not contain any psychoactive THC until much later in history (ca. 4000 to 3000 BP), when drug varieties were introduced from western Asia. Over time, introgression may have led to the B_T allele appearance in the European NLHA and NLH gene pools, as well as in the purported *C. ruderalis* PA, which is recorded at low frequencies in taxonomic studies.

Introgression between BLH and NLH is indicated by the low frequency appearance of additional common BLH alleles in southern NLH populations from Spain, Italy, the Balkans, and Turkey, and reverse gene flow is detected in BLH populations that possess an NLH allele. According to Hillig (2004b), geographically associated cannabinoid variation provides evidence of relationships that are progenitor derived: "The low frequency of B_T and the low levels of propyl cannabinoids (THCV, etc.) in accessions assigned to *C. ruderalis* suggest that this putative taxon could be the progenitor of *C. sativa*, but not of *C. indica*." NLHA populations comprising feral escapes from cultivated NLH populations introgressed with naturalized populations of putative *C. ruderalis* as Vavilov (1926) suggested. Wide variation in cannabinoid contents within feral NLDA suggests that some may represent "wild" progenitor populations. It is less likely that NLDA populations represent feral escapes from NLD populations, since NLD populations do not exhibit the cannabinoid diversity of NLDAs. Frequent occurrence of propyl cannabinoids in *C. indica* and their near absence in *C. sativa* indicates that *C. indica* was not the predecessor of *C. sativa*. Additionally, the high frequency of the B_T allele in East Asian BLH "suggests the possibility that one or both drug biotypes [NLDA and NLD] could have been secondarily derived from this taxon's gene pool" (Hillig and Mahlberg 2004). We suggest that NLD, BLD, and BLH are all descendants of a common PDA that evolved within the Hengduan-Yungui region of present-day southwestern China.

The BLH varieties in China today still possess the genetic potential to produce THC yet have not been selected recently for THC production; indeed the balance of the cannabinoid synthase alleles in these varieties favors the B_D allele coding for CBD synthesis rather than the B_T allele coding for THC synthesis (see de Meijer et al. 2003). Preferred use and consequent selection in East Asia during the past two millennia have favored fiber and seed rather than drugs, and as a result the THC/CBD ratio has become lower due to lack of positive selection for THC. When puritanical religions and moral codes became pervasive social forces—Christianity in Europe and Confucianism in China—shamanism fell into disfavor. *Cannabis* was still needed for its fiber and seed, but entheogenic (religiously inspired) use began to fade away and there were fewer and fewer positive selective pressures encouraging proliferation of the B_T allele and THC synthesis. (See Chapter 9 for a discussion of the traditional and contemporary ritual use of nonpsychoactive *Cannabis* across northern temperate latitudes of Eurasia.)

We now accept, based on Hillig's research, that the vast majority of *Cannabis* worldwide should be classified as *C. indica*, which possesses the capacity to produce both CBD and THC, the latter often in great quantity. *C. sativa* has a much smaller range in Europe and the Americas and exhibits reduced potential for cannabinoid biosynthesis and low THC levels. The PA, NLHA, and NLH biotypes of *Cannabis* all express the B_D allele. All four subspecies of *C. indica* also express the B_D allele and, at least to some degree, the B_T allele. BLD and BLH biotypes are likely descendants from the same gene pool of broad-leaf *C. indica* since they are both broad leafed with THC:CBD ratios ranging from 1:2 to 2:1, whereas NLD varieties produce predominantly THC and almost no CBD in ratios as high as 100:1. As we have suggested, BLD and BLH biotypes originate from a common ancestor evolved within the Hengduan-Yungui refugium, with BLD evolving in isolation in the Hindu Kush Mountains of Afghanistan and more recently used for hashish production and BLH evolving as it spread eastward through the valleys of central China, possibly surviving the Last Glacial Maximum (LGM) in refugia located in eastern China, the Korean Peninsula, and the Japan Archipelago. NLD populations also originated in southwestern China and evolved their high THC content as they spread westward along the Himalayan foothills throughout South Asia where they were primarily utilized for marijuana production.

It appears that the NLD, BLD, and BLH accessions all evolved from a common protodrug gene pool (possibly NLDA but now tentatively called the PDA), and following the same train of thought, NLH evolved from a proto-NLH gene pool (possibly NLHA but tentatively called the PHA). Both these gene pools must have evolved from the Central Asian PA, *C. ruderalis*, or more likely a now extinct common ancestor. At some ancient time there may have been two biochemically distinct gene pools undergoing selection—southern European PHA populations and East Asian PDA populations. East Asian *Cannabis* certainly provided fiber and seed for ancient cultures but probably at the same time was selected for high THC content for shamanic ritual use. Some Korean "hemp" varieties produce significant levels of THC and have even been used in modern drug *Cannabis* breeding (Clarke 2001; see also Chapter 10), and several accessions from China have also been shown to produce THC (Shao and Clarke 1996). *C. sativa* was only selected for fiber and seed (NLH), while *C. indica* was selected for fiber and seed use (BLH) as well as drug content (NLD and BLD). There are no genetic linkages that prevent hemp fiber and seed varieties from producing THC in addition to fiber and seed if they are selected in such a fashion. In fact, traditional Nepalese farmers still collect fiber, seed, and psychoactive resin from a single plant (Clarke 2007). The potential to produce THC has been present in *Cannabis* since before human contact and has always been there for humans to utilize, if and when they chose to do so. *Cannabis* was either selected for higher and higher THC content or not, according to the preferences of individual societies during various periods in history.

The genetic variation of *Cannabis* as a whole approximates values estimated for other crop plants. Generally, within dioecious and wind-pollinated species, genetic diversity is higher on the whole than in self-pollinating weedy species but lower at the individual plant level within the species population. *C. sativa* circumscribes hemp landraces from Europe as well as Southwest and Central Asia and is more genetically diverse than *C. indica* at the species level, while it is also more genetically homogenous than *C. indica* at the individual plant level (Hillig 2005a); in other words, *C. sativa* possesses a greater variety of alleles (diversity) spread throughout the majority of individuals (homogeneity), while *C. indica* possesses a lesser variety of alleles, with much of this genetic diversity restricted to individual, geographically isolated subspecies. This may indicate that *C. sativa* descended directly from its hypothetical PHA ancestor population with little differentiation into geographical races and varieties (Hillig 2005a) and spreading easily from its putative Pleistocene refugia in the Balkan and Caucasus Mountains and radiating across the relatively small and geographically uniform early Holocene European and Central Asian landscape. Geographical isolation appears to have been limited and therefore gene exchange was more frequent, encouraging homogeneity (an even distribution of alleles), while limited human selection, primarily for hemp fiber and seed, encouraged diversity (survival of more alleles). Increased homogeneity may also have developed much more recently due to germplasm sharing in twentieth-century European hemp breeding programs and because of the selections made following World War II for extremely low THC content and subsequent introgression between NLHA and NLH populations (see the section on European hemp breeding in Chapter 10). Europe is also much smaller than Asia, with fewer obstacles to human migration; this allowed for easier dispersal and exchange of *Cannabis* germplasm. *C. indica*, on the other hand, exhibits much phenotypic diversity across its four putative subspecies but is less genetically diverse yet more heterogeneous than *C. sativa*; in other words, it has a smaller variety of alleles unevenly distributed among individual populations. This may have resulted from (or was at least encouraged by) adaptive radiation into diverse geographical zones and diversification into myriad geographical races and cultivars (fiber, seed, and drug). Hillig (2005a) asserts that "the alleles that differentiate *C. indica* from *C. sativa* [are] common in the *C. sativa* gene pool and uncommon in the *C. indica* gene pool, which suggests that a founder event may have narrowed the genetic base of *C. indica*. However, a considerable number of mutations appear to have subsequently accumulated in both gene pools, indicating that the *indica/sativa* split may be quite ancient."

Was the proliferation of individuals with the B_T allele that controls biosynthesis of THC the key founder effect for *C. indica*? Within *C. indica*, BLH exhibits the greatest genetic diversity and NLD the least, and within *C. sativa* NLH and NLHA exhibit almost identical diversity. In contrast to Europe and Central Asia, the immense, culturally heterogeneous and geographically diverse, divided landscapes of South and East Asia, combined with successive introductions, resultant founder effects, and long-term cultural selection for divergent uses, have decreased homogeneity while lowering genetic diversity within *C. indica*. Reduced genetic diversity results from genetic bottlenecks caused by founder effects and is balanced by a plethora of unique genetic combinations adapted to a wide range of habitats and human selective pressures. Genetic diversity generally decreases as populations move away from the source population both in geographical distance and in terms of human selection. Increased diversity can arise in areas of refuge and rapid expansion or in areas of hybridization where two or more dispersal routes coincide. In general, NLD populations exhibit more evidence of inbreeding than NLH and BLH landraces. Narrowed diversity results from selection almost entirely for drug content; this reduced genetic diversity is intensified because NLD crops are propagated from a small amount of seed collected from a few select plants, and most of the male plants are culled from the population.

According to this scenario, *C. indica* radiated across eastern Eurasia from a separate Asian Pleistocene refuge for the putative drug ancestor (PDA) or proto-*indica* population—in present-day southwestern China—under a steadily warming Holocene climate. Climate change in the Himalaya Mountains and Qinghai-Xizang Plateau separating northern Eurasia from South Asia, and the formidable physical barrier this terrain presents to human and avian migrations even today, led to a diversity of ecological conditions and effective reproductive isolation on different sides of this great mountainous divide. Prehistorically, the primordial *Cannabis* population was probably split by natural events or human intervention into two geographically isolated populations: (1) the putative hemp ancestor (PHA), the proto-*sativa* population, which evolved into low-THC, *C. sativa* NLH, and (2) the PDA proto-*indica* population, which evolved into high-THC, *C. indica* NLD and BLD, and lower THC *C. indica* BLH.

In addition, gene flow between these two isolated populations, resulting from human dispersal, could have been limited to infrequent individual founding introductions into southern Asia and along the Himalayas. These founder populations could have then evolved in isolated locations such as Afghanistan and along the Himalayan foothills resulting in genetic bottlenecks. Hillig (2005a) points out that these perceived genetic bottlenecks may have developed because of the way drug *Cannabis* is grown—cultivated fields are often established from a handful of seeds collected from only a few preferred plants. In the case of modern Western *sinsemilla* cultivars, this narrowing effect is even more pronounced as seeds of commerce are usually produced on a single vegetatively propagated female parent pollinated by a single vegetatively propagated male parent, thereby further narrowing the parental gene pool to only two plants. In any case, under the circumstance described earlier, the evolution of each subspecies then proceeded on its own individual evolutionary path by both natural and human selection.

There also may have been significantly more recent gene flow from the East Asian BLH gene pool into the European NLH gene pool due to the importation of Chinese hemp varieties (BLH) into Europe and North America and their use in breeding projects. During the early 1960s, Korean breeders produced hybrid crosses with either the Korean landrace from Gangneung province or the improved Japanese landrace 'Tochigi' and the Italian cultivar 'Eletta Campana' (Ree 1966), although these varieties may no longer be grown. Although there has been little historically documented reciprocal gene flow from European NLH into East Asian BLH, it should be noted that Hillig (2005b) was able to detect introgression of NLH into BLH.

Human selection and protection of unique allelic combinations and the accompanying converse that other alleles are *not* selected should narrow genetic diversity when compared to the ancestral population. The self-sown feral or ancestral taxa NLHA and NLDA are both less genetically diverse than their cultivated subspecies NLH and NLD, indicating that they are more likely descendant from, rather than ancestral to, the cultivated varieties; therefore, they may be feral escapes from cultivation that are closely related to cultivated varieties but not necessarily ancestors. We propose the following three important steps in the early natural and cultural evolution of *Cannabis*: (1) a PA, possibly resembling *C. ruderalis*, evolved first into proto-*sativa* (PHA) and proto-*indica* (PDA) gene pools in the Hengduan-Yungui region of southeastern China, (2) *C. sativa* and *C. indica* evolved independently from their respective ancestral gene pools, and (3) these PHA and PDA gene pools were extensively modified by human selection and, along with the PA, no longer exist in the wild.

One of the most interesting implications of our evolutionary theory for *Cannabis* is that Chinese BLH varieties have evolved from the same primeval gene pool as the South and Central Asian NLD and BLD varieties rather than descending more directly from the Central Asian PA, proto-*sativa* (PHA), or European *C. sativa* gene pools. Genetic evidence implies a dispersal path during the evolution of BLH, proceeding not directly from Central Asia into contiguous northern China but either via an intermediate PDA gene pool in southwestern East Asia that gave rise to BLD and NLD as well as BLH varieties or, less likely, via migration into South Asia and then either back across the Himalaya Mountains into Central Asia or from Southeast Asia into southwestern China (e.g., see Shah 1997). The higher levels of THC found in Chinese hemp landraces compared to European hemp cultivars are due to recent selection in Europe for particularly low levels of THC in industrial hemp cultivars compared to relatively less selection in Chinese hemp landraces, and/or more likely because Chinese hemp evolved from the PDA gene pool that had already evolved higher THC content. NLH *C. sativa* evolved from the proto-*sativa* PHA originating in the Pleistocene refuge of southern Europe and expressed the B_T allele at a much lower frequency.

The world's *Cannabis* biotypes, so far sampled, can be placed into one of four phenotypic categories: (1) NLD marijuana varieties originating in South Asia, (2) NLH fiber and seed varieties from Europe, (3) BLD hashish varieties from Afghanistan, and (4) BLH fiber and seed varieties from East Asia (Hillig 2005a/b; Hillig and Mahlberg 2004). Previously, either the NLD, NLH, and BLH varieties, as well as their escaped or ancestral relatives, were all classified as *C. sativa* (Small and Cronquist 1976; McPartland et al. 2000) or, alternately, the low-THC, NLH varieties were classified as *C. sativa* and the high-THC, NLD varieties were classified as *C. indica*

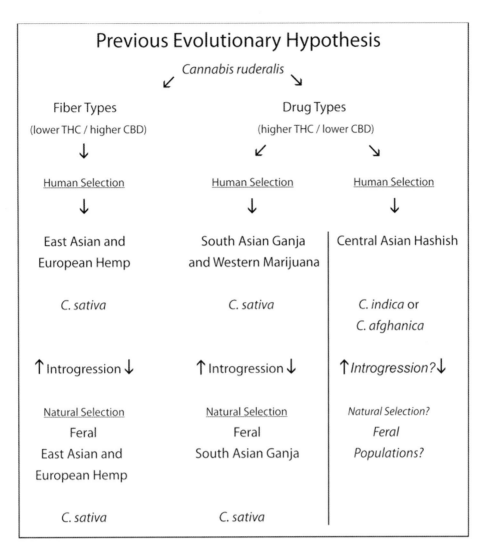

FIGURE 78. Previous hypotheses concerning *Cannabis* evolution based on twentieth-century research. We no longer accept that the majority of *Cannabis* biotypes (NLHA, NLH, NLDA, NLD, and BLH) be assigned to *C. sativa* and only the Afghan hashish varieties (BLD) be assigned to *C. indica* (adapted in part from Vavilov and Bukinich 1929; Serebriakova 1940; Schultes et al. 1974; Small and Cronquist 1976; McPartland et al. 2000).

(Schultes et al. 1974) and the markedly different BLD accessions were classified as either *C. indica* (Schultes et al. 1974) or *C. afghanica* (McPartland et al. 2000). *C. ruderalis* has been considered a primeval PA of the other two species (Hillig 2004a, 2005a/b; Hillig and Mahlberg 2004).

One apparent contradiction in the preceding evolutionary tree is the placement of BLH at the terminus of the NLH lineage. The placement of BLH varieties within an evolutionary scheme based on the research by Hillig (2005a) presents only a small problem because he has named the biotypes based primarily on leaf shape. Leaf morphology is considered to be a weak character for taxonomic placement as it has little genetic correlation with other taxonomic traits, and leaf shape varies considerably due to environmental effects. The basic taxonomic division in *Cannabis* is the ability or inability to produce THC, in other words, expression of the B_T allele.

Upon reviewing the systematic research of Hillig (2005a/b) and Gilmore et al. (2007), it becomes apparent that East Asian BLH varieties are more closely related to *C. indica* than *C. sativa*. Hillig also suggested that *C. sativa* ssp. *spontanea* may be ancestral to *C. sativa* ssp. *sativa* and that *C. indica* ssp. *kafiristanica* may be ancestral to *C. indica* ssp. *indica*. Conversely,

Gilmore et al. (2007) suggested these taxa may be made up mostly of feral escapes from cultivation, a scenario we feel is much more likely. In fact, we believe it is extremely unlikely that ancestral populations still exist because of the outcrossing nature of *Cannabis*, its long history of association with humans, and its tendency to escape cultivation and become feral. However, a thorough search of Central Asia to collect possible ancestral populations should be carried out and the accessions analyzed for similarities and differences in gene nucleotide sequences.

Recent Geographical Distributions of *Cannabis* Biotypes

That grand subject, that almost keystone of the laws of creation, geographical distribution. (CHARLES DARWIN in a letter to Joseph Dalton Hooker in 1845).

As Richard Evans Schultes (1969a) informed us, "*Cannabis* is probably the most widely disseminated hallucinogenic plant, now known in virtually all inhabited parts of the world,

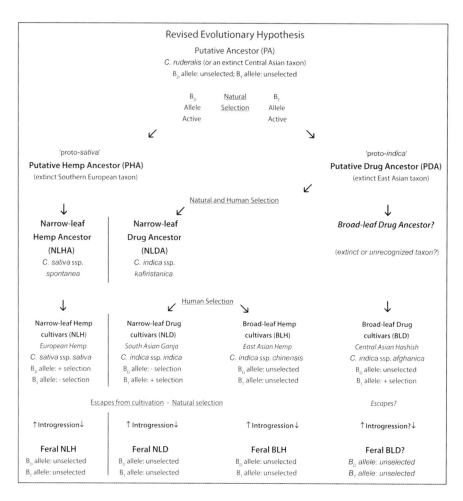

FIGURE 79. Revised hypothesis and family tree for the evolution of *Cannabis*. We argue that *C. indica* now circumscribes the vast majority of biotypes (NLDA, NLD, BLD, and BLH) and *C. sativa* circumscribes only European hemp (NLHA and NLH; adapted from de Meijer et al. 2003; Hillig 2005a/b).

escaping easily from cultivation and growing spontaneously." Although almost every country in the United Nations forbids the possession and use of *Cannabis* for all but a few legally sanctioned, industrial purposes, its illegal use is well established on a worldwide scale and continues to grow during the early part of the twenty-first century (e.g., see Global Commission on Drug Policy 2011). This widely adapted genus is found growing in either wild, feral, or cultivated populations almost everywhere humans have lived. It is generally a hardy plant that thrives in such diverse regions as continental Manchuria, the tropical wet-dry portions of Central America, arid northwestern and temperate southern Africa, the Ganges River plains and hill tracts of monsoon India, and the grasslands of North America.

EUROPE AND THE FORMER SOVIET UNION

Cannabis occurs both spontaneously and cultivated across much of Europe and western Asia. All spontaneous (NLHA) populations and cultivated (NLH) varieties are classified as *C. sativa*, and the "wild" populations are often classified as *C. ruderalis* the PA of all *Cannabis* species (Hillig 2005a). All are characterized as low-THC types and the cultivated varieties are used predominantly for fiber or oil seed production.

General trends in morphology and maturation can be discerned by comparing the descriptions of Russian populations from various latitudes. As you move south from Archangelsk above the Arctic Circle through central European Russia and the Volga River basin, as well as southeast through the Central Asian Altai Mountains, *Cannabis* varieties become taller and more branched, have larger leaves with more leaflets, and mature later. This trend can be followed in a series of intergrading steps through these populations and is most likely attributable to variations in length of growing season, photoperiod, light intensity, and additional latitude-determined parameters (Serebriakova 1940, see also Chapter 2).

Differences in seed size, shape, and coloration do not appear in intergrading steps. Vavilov (1931) reported variations between cultivated and escaped populations of European hemp NLH varieties (*C. sativa* ssp. *sativa*) and the remainder of spontaneous Russian and Central Asian varieties (*C. sativa* ssp. *spontanea* or NLHA). He characterized European varieties with larger, lighter colored seeds as products of artificial selection by hemp breeders. Serebriakova (1940) included both NLH and BLH varieties as part of *C. sativa* based on large plant size, light seed color, and useage for fiber and seed.

We postulate that the trait of larger light-colored seed was contributed by Chinese fiber varieties imported into Europe during the nineteenth century, and prior to this, native European populations and derived fiber cultivars likely had smaller

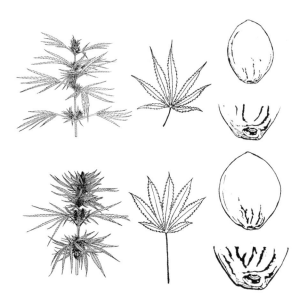

FIGURE 80. Inflorescences, seeds, and leaves of *C. sativa* ssp. *spontanea* (top), the naturally growing putative narrow-leaf hemp ancestor (NLHA), are smaller than *C. sativa* ssp. *sativa* (bottom), cultivated narrow-leaf hemp (NLH; adapted in part from Small and Cronquist 1976 and Clarke 1981).

FIGURE 81. East Asian *C. indica* ssp. *chinensis* broadleaf hemp (BLH) is grown for both fiber and seed production. Although some individual plants possess modest levels of psychoactive THC, it was not traditionally grown for drug production.

darker mottled seeds resembling members of *C. sativa* ssp. *spontanea* (NLHA). It would be valuable to compare European herbarium specimens of *Cannabis* from the early nineteenth century with those from the late nineteenth century. The wide-ranging populations of self-sown NLHA are likely among the most primitive populations of *Cannabis* living today, and although they likely do not represent direct descendants of the original wild populations of primordial *C. ruderalis,* they may echo the appearance of the primordial PA due to natural selection in the absence of human selection and the atavistic reappearance of primordial traits (see Chapter 3).

CHINA

In earlier chapters we presented archeological, linguistic, and historical evidence supporting Central Asia as the original pre-Pleistocene home of *Cannabis.* Subsequent to its evolutionary origins in Central Asia, *Cannabis* was displaced into refugia in both the Balkan and Caucasus regions of southeastern Europe and southwestern Asia, where *C. sativa* evolved, and in the Hengduan Mountain and Yungui Plateau regions of southwestern East Asia, where *C. indica* evolved. We have also argued that *Cannabis* and its use came to China very early on. Hemp has been cultivated in China for at least 4,500 years and probably much longer. We believe that wild *Cannabis* plants have been used for both fiber and seed since early Neolithic times, perhaps even well before the Holocene (see Chapters 4, 5, and 6). *Cannabis* was domesticated independently as a fiber and seed crop in Europe (NLH) and northern China (BLH). The ancient Chinese were also familiar with *Cannabis's* use as a medicinal and sacramental drug plant. In early China *Cannabis* was employed ritually for its spiritual, psychological, and healing powers, at least until Confucian, Buddhist, and to some extent its Taoist philosophies suppressed shamanism (see Chapter 7). These ancient changes

in Chinese cultural and religious preferences account for the paucity of drug *Cannabis* use in recent times; this model also explains the lack of ritual psychoactive, shamanistic use in Korea and Japan in more recent times. According to one historical theory, *Cannabis* originated in Central Asia and was brought to Western China by the Zhou culture around 3700 BP (Lacouperie 1893), although archeological evidence indicates that *Cannabis* was found in China long before this time. *Cannabis* served as one of the most important edible grain and oil seeds in northern China for centuries if not millennia until it was supplanted by the introduction of upland rice about 3000 BP (see Chapter 6).

Chinese *Cannabis* includes diverse phenotypes and many regional varieties have evolved. Variation most likely resulted from its early and wide dispersal into different latitudes and climate zones of China, concurrent with early Chinese interest in *Cannabis* as a multipurpose plant and selection for a variety of uses. Some fiber varieties in continental East Asia reach heights of more than 5 meters (16 feet) and large-seeded food varieties have also been reported (Clarke 1998b). Chinese *Cannabis* is characterized by large broad leaves and large, light-colored seeds and is most similar in appearance to some BLD varieties. No *Cannabis* drug production has been reported in any region of China outside of Chinese Turkestan (Xinjiang province), and no Chinese varieties containing potent levels of THC (> 5 percent) have been documented. However, all possess the B_T allele and the genetic capability to synthesize THC. Because the Chinese varieties have not been selected for drug content in recent times, they tend to synthesize both CBD and THC in only moderate amounts. Chinese BLH varieties were imported by European and New World hemp growers and breeders during the late 1800s.

FIGURE 82. Broadleaf drug (BLD) hashish plants growing in a mixed garden in Afghanistan in the early 1970s.

Hybridization of acclimatized NLH landraces with the Chinese BLH gene pool had a very strong impact on the breeding of fiber and seed varieties in Europe and North America (see Chapter 10).

CENTRAL ASIA, AFGHANISTAN, AND TURKESTAN

Herodotus (ca. 2500 BP) was the first to describe the Scythian practice of inhaling *Cannabis* smoke to induce altered consciousness, and Scythian tombs containing *Cannabis* seeds and smoking equipment dated to around 2500–2300 BP were found frozen in the Altai Mountains (Artamanov 1965; Rudenko 1970). During the thirteenth century CE the cult of the Haschischin assassins was active in northern Persia , and Marco Polo recounted famous journeys of his time and mentioned hemp growing around the oases of Kashgar and Khotan in eastern Turkestan. *Cannabis* has occasionally supplied fiber for crude cordage throughout Central and South Asian history, but it has rarely been utilized extensively for fiber or seed production.

"Hashish" is the proper Arabic term for the agglutinated *Cannabis* resin product originally produced in Central Asia. Chopra and Chopra (1957) described the technique used by Muslim residents of the Chinese Turkestanian Yarkand region (presently in Xinjiang province, China) to produce hashish: "The female flower heads are first dried, then broken and crushed between the hands into a powder which is passed through sieves so that it obtains the fineness and consistency

of sand or sawdust." The Sanskrit word *"charas"* is traditionally used in Hindu India for *Cannabis* resin. *Charas* refers more accurately to resin collected by hand-rubbing the flowers of living or freshly harvested plants. Sieving or rubbing isolates the resinous glandular trichome heads containing psychoactive THC (see Chapter 7 for more detailed descriptions of the preparation of psychoactive *Cannabis* products). Generally, modern Asian commercial sieved hashish is produced predominantly from BLD varieties and hand-rubbed *charas* from NLD varieties, although in more recent years NLD varieties have also been used for making sieved hashish in Morocco and Nepal. Indian and Nepalese cultivated *ganja* and hand-rubbed *charas* were, and still are, products of NLD varieties. However, it is more difficult to determine which *Cannabis* varieties were used by Turkestanians for the manufacture of sieved hashish. When Vavilov visited Afghanistan in 1924, he discovered two varieties of what he called *C. indica*, growing wild along the Kunar River in eastern Afghanistan and northern Pakistan. Vavilov and Bukinich (1929) described *C. indica* var. *kafiristanica* as characterized by its "short habit, profuse branching from the very first internode [node], and by short internodes . . . small leaves with obovate [egg-shaped with the pointed end toward the base] leaflets of narrow shape. Achenes [seeds] extremely small, mosaic, dark gray with pattern." Hillig (2005a) named this biotype *C. indica* ssp. *kafiristanica* and considered it to possibly be the narrow-leaf drug ancestor, or NLDA. *Cannabis indica* var. *afghanica* was described as having seeds with a "colorless involucre (perianth or seed coat)" and broad leaves and was apparently similar to *C. indica* ssp. *kafirstanica* in its other morphological characteristics. Hillig (2005a) calls this biotype *C. indica* ssp. *afghanica* or broadleaf drug (BLD) *Cannabis*. However, there were probably additional, unnoticed *Cannabis* biotypes growing in Afghanistan at the time Vavilov and others made their observations.

Vavilov and Bukinich (1929) reported that BLD and NLDA varieties, what they called *C. indica* var. *afghana* and var. *kafiristanica*, were not cultivated along the Kunar River and that "ordinary *C. sativa* L. is chiefly cultivated for hashish around Herat and Faizabad in northern Afghanistan." The seeds they collected were large, light-colored, and unpatterned, resembling those of *C. sativa* NLH grown for fiber in central European Russia. However, it is very unlikely that NLH was used for hashish production as it typically expresses a very low THC content. It is more likely that what Vavilov saw were NLD cultivars that appear much more like NLH plants than BLD plants. Although BLD varieties may not have been cultivated for hashish when Vavilov visited Afghanistan in 1924, or when he reported that "*C. sativa*" was used for hashish production in northern Afghanistan in 1929, BLD varieties have certainly been cultivated there since at least 1965, and probably much longer. In recent years, BLD plants have been used exclusively for hashish production in Afghanistan, northern Pakistan, and Kashmir (Clarke 1998a). It is not known whether the Central Asian hashish imported into India from Bukhara (in present-day Uzbekistan) and Yarkand (in present-day Xinjiang) during the nineteenth century was prepared from NLD or BLD cultivars. However, the extensive, spontaneous range of NLD populations makes it a likely choice for domestication in Turkestan at that time. BLD varieties apparently had an extremely limited range only within Afghanistan and possibly neighboring Pakistan and Kashmir prior to the late twentieth century when they became incorporated into the North American domestic *sinsemilla* gene pool.

FIGURE 83. Naturally growing, putative narrow-leaf drug ancestor (NLDA), *C. indica* ssp. *kafiristanica* (top), has much smaller inflorescences than cultivated broadleaf drug (BLD), *C. indica* ssp. *afghanica* (bottom), otherwise both biotypes are similar in appearance (adapted in part from Small and Cronquist 1976 and Clarke 1981).

Vavilov must have overlooked the cultivation of *Cannabis*, as well as hashish making, when he visited Yarkand in 1929. He reported his observation in Yarkand of "*C. sativa* var. *spontanea*," which he described as being similar to the *Cannabis* plants of the Tian Shan and Altai Mountains; Hillig (2005a) refers to this biotype as *C. sativa* ssp. *spontanea*, the NLHA. Vavilov observed these plants growing spontaneously on "waste lots" everywhere; however, much of this *Cannabis* may actually have been cultivated rather than self-sown. During 1929, as they had for many years, the farmers of the Yarkand region of Chinese Turkestan (Xinjiang province) exported more than 90 metric tons of hashish south into the Northwest Frontier and Punjab regions of India, and during 1937 and 1938 this traffic represented 42 percent of the total value of Xinjiang's exports to India (Lattimore 1950). This legal trade was recorded and taxed by the colonial British government (United Nations Office on Drugs and Crime 1953). Uncalculated amounts of hashish were also illicitly smuggled into India each year, and the total production in Yarkand must also account for local consumption within Turkestan. Hashish may have been collected from spontaneous plants, but it is more likely that extensive fields of irrigated and cultivated *Cannabis* would be required to produce such huge volumes of hashish and to maintain such substantial exports of hashish year after year (Clarke 1998a). Besides, it is difficult to imagine "waste lots" everywhere in an arid oasis, where arable land was at a premium. Since these plants were not described as *C. indica*, which Vavilov and Bukinich (1929) earlier reported from Afghanistan, and because they resembled European NLH, they must have been NLD cultivars. Russian authorities had forced the hashish makers from Bukhara in western (Russian) Turkestan into eastern (Chinese) Turkestan before the end of the nineteenth century. If Vavilov did not realize that the fields of *Cannabis* were cultivated, it may have been because he was a Russian visitor and therefore was not told about the clandestine hashish production being carried on there.

Could Vavilov also have overlooked evidence of cultivated *C. indica* ssp. *afghanica* on his earlier visits to the Kunar River valley in Afghanistan and Pakistan? It is possible that this area was not used for *Cannabis* cultivation in 1924, even though it lies along the trade routes for Turkestanian hashish. BLD varieties were cultivated for hashish here from at least 1970 to 1980. Perhaps BLD varieties were not regularly or intensively utilized for hashish until after 1935 when production became illegal in Chinese Turkestan. *Cannabis indica* var. *afghanica* (*C. indica* ssp. *afghanica* or BLD), with light-colored seeds as reported by Vavilov and Bukinich (1929), may have been the cultivated variety, and *C. afghanica* var. *kafiristanica* (*C. indica* ssp. *kafiristanica* or NLDA), with dark-colored seeds, may have been the wild or feral type.

If farmers of the older hashish producing areas of northern Afghanistan grew NLD varieties, then they did not prevail, as BLD varieties are favored for hashish production there today. We feel it is possible that prior to the shift in production from Turkestan into Afghanistan, Pakistan, and Kashmir, hashish was manufactured primarily from NLD varieties. Later, farmers began growing their NLD hashish cultivars brought from Turkestan; in addition, they hybridized with feral NLDA populations and cultivated BLD varieties native to their new home. This could account for the vigor and morphological variation characteristic of BLD hashish cultivars collected in Afghanistan in the late twentieth century. Hashish production by the sieved method is certainly ancient. However, this does not mean that BLD varieties were used to make hashish in ancient times. Available historical evidence indicates that BLD cultivars were possibly domesticated for hashish production quite recently. During the latter half of the twentieth century, BLD types were found in the foothills to both the north and south of the Hindu Kush Mountains of eastern Afghanistan and northern Pakistan, especially along the Kunar River valley, but as far west as Kandahar and as far north as Mazar-i-Sharif, and their southern limit lies along the Afghanistan-Pakistan border in northern Pakistan. This region is occupied predominantly by Muslim Turkestanians and Afghans of Iranic heritage and has been under their religious influence for centuries. BLD cultivars used for sieved hashish manufacture in this region provide the distinctive aroma, flavor, and psychoactivity of Afghan hashish (Clarke 1998a).

INDIA AND NEPAL

The most complete and concise references to *ganja* (marijuana) and *charas* (hashish) cultivation and trade within India and Nepal are found in the nineteenth-century "Report of the Cultivation of and Trade in Ganja in Bengal" (Kerr 1877). Although this detailed report described the legal trade in *Cannabis* products during the nineteenth century, it also indicates that trade in *Cannabis* drugs had been carried on in much the same manner for centuries. Kerr described *Cannabis* cultivated in Bengal in eastern India for *ganja* as "rarely rising above six feet . . . slender, delicate, regularly pyramidal assuming more the character of the cypress, though not quite so fastigiate [branches not so upright and clustered, but more open growth]. . . . The stem is erect, six or eight inches in circumference, simple when crowded, but when growing apart branched even from the bottom. . . . The leaves are opposite or alternate, on long weak petioles . . . leaflets five to seven in number, narrow lanceolate [narrow and tapering to a point at

FIGURE 84. In Nepal, *Cannabis indica* appears in different growth forms depending on where and why it grows. Putatively wild-growing *C. indica ssp. kafiristanica* narrow-leaf drug ancestor (NLDA) plants are highly branched, low growing, and compact with sparse inflorescences (A). Cultivated *C. indica ssp. indica* narrow-leaf drug (NLD) plants are grown tall and thin to encourage fiber growth as well as flower development (B). Spontaneously growing Himalayan NLD *Cannabis* escapes from nearby cultivation, and when growing in a favorable human-disturbed niche will develop long branches and can reach a great size (C).

the apex] and sharply serrated." Kerr also quoted a proficient botanist named Benjamin Clarke, who had made careful observations concerning the isolation of wild and cultivated *Cannabis* populations in Bengal:

> The whole country abounds (especially in waste spots around villages) with wild hemp, *Cannabis sativa* . . . I have not been able to discover anything like a hybrid between the wild hemp and *ganja*, which favors the supposition that the two are very distinct varieties, if not to be reckoned as species. The *ganja* plant differs from the wild hemp in its woody, thick, straight stem, its bushy pyramidal habit, the crowded female flowers, the shape of the calyx [*sic*] and bracts of the female flowers, and the presence of the viscid, *ganja*-bearing glandular hair on the calyx [*sic*] and the bracts.

The cultivated *ganja* plants described by Clarke were quite different from either European NLH varieties or feral NLDA biotypes and were likely Indian NLD cultivars. In written accounts of cultivated Bengali *ganja,* Prain (1904) described tall and laxly branched plants with loosely clustered inflorescences and narrow lanceolate leaves. These plants may appear similar to *C. sativa,* which was the binomial chosen by both Kerr and Prain, although presently we follow

Lamarck and call them *C. indica* ssp. *indica* or NLD biotypes (Hillig 2005a).

According to the *Report of the India Hemp Drugs Commission* (1893–94; see Kaplan 1969), throughout the nineteenth century limited amounts of reportedly very high quality *charas* (hand-rubbed hashish) were imported from the Himalayan foothills of Nepal into Lucknow in northern India. The cultivated and feral Himalayan Nepalese drug varieties are characteristically very tall and laxly branched with thin, loose inflorescences and narrow, lanceolate leaflets and, like Indian *Cannabis,* are members of NLD *C. indica* ssp. *indica.* Some naturally growing NLDA populations may be truly wild, growing profusely in the lowland Terai region of Nepal and the Gangetic Plain of India. Feral plants are characteristically diminutive and bushy with small, narrow leaflets; very small inflorescences; and small, dark seeds. Interestingly, Terai populations also mature much later (December and January) than the Himalayan varieties (October and November; Clarke personal observations 2006). Apparently, they are only rarely utilized, and then only by mendicant Sadhus in search of sacramental *Cannabis* (see Chapter 7 for a discussion of the ritual uses of *Cannabis* in South Asia).

The range of NLD varieties encompasses the Indian subcontinent from the southern foothills of the Himalaya

Mountains in northern India and Nepal extending from Arunachal Pradesh, India; west through Bhutan, Sikkim, and Nepal; into northern India and Kashmir; and south across the Gangetic plain, with additional cultivation in various regions across the remainder of India (e.g., Kerala). This region is peopled primarily by Hindu and Buddhist religious groups of Tibeto-Burman or Indic heritage. *Charas* throughout this region has been traditionally produced by rubbing the fresh resin glands from the pistillate inflorescences by hand, subsequently collecting the rubbings from the palms and pressing them into a shaped ball, rod, or patty. Both cultivated and spontaneously growing NLD plants are used for *charas* rubbing throughout this region; these factors account for the distinctive aroma, flavor, and psychoactivity of Nepalese *charas* when compared with Afghan sieved BLD hashish.

NLDA populations range through Nepal and northern India and may represent relic wild populations. Rare spontaneously growing populations across southern India are more likely descendants of escapes from cultivation. NLDA populations are occasionally used for the production of seeded *ganja* or hand-rubbed *charas*, although of low quality.

Cannabis of the Himalaya and Hindu Kush Mountains can be described as two distinct allopatric (nonoverlapping) taxa in geographically isolated regions. There is only a small region of possible sympatry (overlap), located roughly between the BLD region of Afghanistan and the NLD region of Nepal and neighboring northern India, extending northwestward into northern Kashmir and northeastern Pakistan. This region of sympatry is occupied primarily by peoples of Dardic heritage (an Indic linguistic group from the upper Indus Valley and Kashmir). It would be informative to include seed accessions from throughout the Himalaya and Hindu Kush foothill regions in future taxonomic studies. The current Asian range of BLD biotypes within Afghanistan, northern Pakistan, and Kashmir is not very extensive, whereas NLD biotypes are much more widely spread across most of South and Southeast Asia.

Hashish and *charas* production within Kashmir involves techniques used respectively in neighboring regions east and west. Both cultivated NLD and BLD biotypes are utilized, although BLD landraces are usually preferred for sieved hashish production and NLD landraces are usually utilized for hand-rubbing *charas*; in addition, they may be blended together in the *Cannabis* trade. This is an excellent contemporary example of the effects of geographical isolation and cultural preference on preserving the genetic integrity of *Cannabis* varieties. Although these farmers may be aware of other *Cannabis* varieties and methods of making hashish, they prefer to grow the same varieties and use the same methods as their cultural tradition dictates. However, in regions of sympatry such as Kashmir, populations used for sieved hashish production exhibit a wide range of phenotypes intermediate between the BLD and NLD types, and limited hybridization is suspected (Watson personal communication 1997).

South and Central Asia host a wide diversity of *Cannabis indica* drug varieties reflecting early introduction and domestication of *Cannabis* for drug use within these regions. However, *C. indica* is not as genetically diverse as *C. sativa*, possibly indicating a genetic bottleneck caused by limited introductions of the founding germplasm and subsequent intense selection by South Asian cultures for high drug content at the expense of other characteristics.

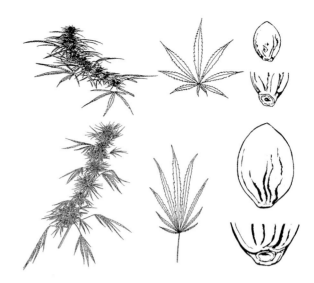

FIGURE 85. *C. indica* ssp. *indica* narrow-leaf drug (NLD) biotypes from India (top) and Thailand (bottom; adapted in part from Small and Cronquist 1976 and Clarke 1981).

SOUTHEAST ASIA

Southeast Asian varieties are also members of *C. indica* ssp. *indica* and resemble South Asian varieties in their tall growth and sparse inflorescences, although the limbs are often even longer and more meandering. The leaves are also usually larger than those of South Asian NLD varieties, and the leaflets are more coarsely serrated. In addition, the leaflets are often arranged in a narrow, handlike—rather than the more usual radial—array. Female inflorescences are longer and thinner with longer internodes. Bracts are large and readily distinguished, although there are often not as many within each inflorescence as in Indian *ganja* cultivars. Fruits are usually large and light brown with a few darker brown stripes at or near the base. Floral maturation is very late, with plants ripening in November through January. The Southeast Asian biotypes are found in cultivation and occasionally as escaped populations throughout southwestern China (Clarke 1995 through 2003, personal observations), Southeast Asia, and Sumatra (Watson personal communication 1979). Southeast Asian varieties bear a strong phenotypic resemblance to Chinese BLH varieties grown for fiber and seed (Clarke 1996 through 2003, personal observations). Earlier, we proposed the Hengduan-Yungui region within mountainous Southwest China as a Pleistocene refuge for *Cannabis*, which may have evolved within more northern temperate latitudes during much earlier and warmer interglacial periods (see Chapters 2 and 4), and many diverse phenotypes resembling BLH, BLD, and NLD biotypes are found there.

Thai *ganja* cultivars are historically the best known members of what we call *C. indica* ssp. *indica* var. *indochinensis* (following the nomenclature of Serebriakova 1940), as their seeds were collected by Westerners from illicit shipments of marijuana smuggled from Thailand during the middle 1970s and early 1980s. Interest in Thai varieties was high at the time, since they were some of the most potent drug varieties available in Europe and North America. Several illicit samples were analyzed at 12 percent THC (Baker, Bagon, and Gough 1980; Baker, Gough, and Taylor 1980). However, late maturation raised problems with var. *indochinensis* when it was grown in temperate latitudes. The only way to ensure complete floral maturation was to protect the plants from autumn weather in a glasshouse or grow them indoors under lights. To overcome these problems, while preserving their potency, Thai varieties were commonly used in hybrid crosses with earlier-maturing NLD and BLD varieties (see Chapter 10).

Thai varieties are often partially monoecious with a preponderance of female individuals developing male flowers. Up to 90 percent of the population will initially appear to be female, and then, later in flowering, shoots of male flowers will form along female limbs. Fertilization by pollen from these late appearing male flowers is often quite effective and many mature seeds form. As cosexual *Cannabis* is relatively rare, we suspect the monoecious condition has evolved in response to persistent selection for female plants in seedless *ganja* production with male flowers on female plants overlooked by farmers serving as the majority pollen donors and thereby perpetuating the monoecious female trait.

AFRICA AND THE MIDDLE EAST

The African continent is divided by the Sahara Desert. Lying north is the predominantly Muslim region bordered by the Mediterranean Sea and extending southward into the desert itself. The region south of the Sahara is divided into a tropical, equatorial region and a temperate region farther to the south. Sub-Saharan Africa has historically been more isolated than the Islamic north, and cultural influences have largely come from its own endemic tribal religions and limited contact with Arab sea traders.

The northern region is characterized by the production of sieved hashish and, occasionally, seedless marijuana (*kif* or *banga*). Egypt has a long history of hashish production and use, dating from at least 1200 CE, and possibly much earlier (Abel 1980). The only remaining large-scale, commercial cultivation of *Cannabis* in North Africa is at high altitude farms in the Rif Mountains of Morocco where production of hashish is de facto legal. The preparation of *kif* may be quite ancient, but the sieved method of hashish making was brought to Morocco during the 1960s (Clarke 1998a).

Drug cultivars from North Africa are presently classified as NLD varieties, although they express strong morphological similarities to European NLH cultivars, and introgression between NLD and NLH is suspected. We have named them *C. indica* ssp. *indica* var. *mediterraneana* following the nomenclature of Serebriakova (1940). In Morocco, hashish fields are sown at 100 or more seeds per square meter; plants do not branch at all unless they grow alone, and then they are only weakly branched. There are no self-sowing or weedy *Cannabis* populations in northern Africa. Variety *mediterraneana* is also cultivated for sieved hashish production throughout the Middle East. Turkey, Lebanon, and Syria are well known for hashish production and illicit *Cannabis* cultivation continues in these regions today. Varieties from sub-Saharan Africa strongly resemble those from India, reflecting the dissemination of *Cannabis* by Arab seafarers down the east coast of southern Africa from India (Du Toit 1980). Drug varieties were not introduced into western equatorial and Sahelian Africa until after World War II, making these areas among the last to receive drug *Cannabis* prior to its modern expansion beginning in the 1970s

FIGURE 86. *C. indica* ssp. *indica* narrow-leaf drug (NLD) biotypes from sub-Saharan Africa (forma *equatoria*, top left, and *forma sudafricana* top right; drawings by Pam Elias in Clarke 1981) and general phenotypic characteristics (bottom; adapted in part from Small and Cronquist 1976).

(see Global Commission on Drug Use 2011). Escapes from cultivation now grow spontaneously in many areas across semitropical Africa.

Southern African varieties are of medium height with spreading limbs. The inflorescences are usually relatively dense with many bracts and resemble other NLD cultivars. The leaves are often made up of leaflets in a handlike array and resemble those of *C. indica* ssp. *indica* var. *indochinensis*. We group the sub-Saharan varieties together in *C. indica* ssp. *indica* var. *africana*, again following the nomenclature of Serebriakova (1940). We have further divided var. *africana* into two forms based on differences in floral maturation. Forma *equatoria* circumscribes more equatorial populations, which mature at temperate latitudes from early October until early December. Forma *sudafricana* circumscribes earlier maturing, more southern populations, which mature from late August to late September at temperate latitudes.

The African continent presents an interesting situation for studies of *Cannabis* taxonomy. Two different, historically documented paths of dispersal have resulted in two morphologically divergent populations culturally and geographically isolated by the Sahara Desert. In addition, the coastal Mediterranean regions are of particularly interest because European NLH varieties grown for fiber or seed hybridized with South Asian NLD varieties grown for hashish production. Although the acquisition of Moroccan *Cannabis* seed accessions would be relatively straightforward, *Cannabis* from this region has been omitted from taxonomic studies and certainly should be included in the future.

NEW WORLD

New World NLH and NLD biotypes conform largely to a pattern of distribution much as they do in Asia, which is linked to the distance from the equator. European NLH was disseminated to the New World during the sixteenth century by European colonists. East Asian BLH was brought to

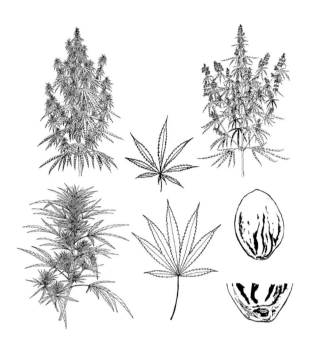

FIGURE 87. *C. indica* ssp. *indica* narrow-leaf drug (NLD) variety *mexicana* biotypes from Mexico, adapted in part from Small and Cronquist (1976) and drawings by Pam Elias (top row) in Clarke (1981).

North America from China during the early twentieth century by plant explorers, such as Frank Meyer of the United States Department of Agriculture, who were interested in collecting seed accessions for improving Western industrial hemp varieties (see Chapter 10). NLH is generally limited to temperate regions north of 30 degrees north latitude and south of 30 degrees south latitude, where hemp was historically cultivated. All feral populations of NLH, such as those commonly found as weeds throughout the Midwest of the United States, are descendants of escapes from hemp cultivation dating as far back as the colonial period. North American self-sowing hemp populations are herein classified as *C. sativa* ssp. *sativa* var. *americana*, once again following the nomenclature of Serebriakova (1940). NLH cultivars were also imported from Europe into the Santiago region of Chile during the nineteenth century for fiber production (Forster 1996; see Chapter 5); if feral populations exist in this region, seed accessions would be of great interest in future taxonomic studies.

All indigenous New World NLD populations are cultivated and no feral populations have been reported. We divide New World NLD populations into three types. Variety *colombiana* is of medium height and strongly erect, with many branches, and flowers mature from late October to late December or later. Variety *jamaicana* is slightly taller and matures from the middle of October to the middle of December. Variety *mexicana* plants grow strongly erect, often with large with long upright branches, and its maturation usually occurs from early October through the middle of November.

Many NLD and BLD varieties coming from diverse regions via many introductions have contributed to the modern drug *Cannabis* gene pool. Hybrids between NLD marijuana and BLD hashish cultivars are grown extensively, albeit illicitly, across North America. Since the 1960s, seeds from nearly all the drug *Cannabis* gene pools have come to the West in illicit shipments of marijuana and hashish. Many were grown to fruition and have become incorporated in the North American hybrid domestic *sinsemilla* gene pool. More recently they have become popular in Europe, Australia, and New Zealand.

Summary and Conclusions

The predecessors of modern-day *Cannabis* evolved for untold millennia in Eurasia, probably first in Central Asia. Then at some point in evolutionary time, long before the arrival of humans in the region, a single offspring experienced a genetic mutation and began to produce THC. The mutation survived, and subsequent offspring perpetuated this unique biosynthetic capability. This single mutation creating the B_T allele has proven to be the most evolutionarily important milestone along *Cannabis*'s evolutionary path—favored by some cultures and shunned by others. THC has often been a focus of forensic debate and its geographical distribution has proven to be a salient point in modern *Cannabis* taxonomy. The *Cannabis* gene pool was initially divided along genetic, geographical, and usage lines, descendant from a hypothetical medicinal and/or entheogenic gene pool (the PDA "proto-*indica*" gene pool) or from a hypothetical fiber and seed use gene pool (the PHA "proto-*sativa*" gene pool). The present-day genetic distinction is essentially "dope versus rope" or NLD and BLD versus NLH. Drug biotypes have subsequently been selected for subsidiary fiber and seed use (BLH), while NLH has been selected for lower and lower drug content. Hypothetical ancestors of *C. sativa* (NLHA) and *C. indica* (NLDA) also commonly occur in a "wild" state, but most are more likely feral escapes from cultivation and simply represent the initial stages of domestication when semidomesticated landraces continued to hybridize with their progenitor populations.

Basic questions concerning the early evolution of genus *Cannabis*—especially the evolutionary relationships between the two proposed cultivated species, their subspecies, feral relatives, and putative ancestral populations—warrant further research including both traditional landraces and feral populations. It will be most important to collect and genetically characterize purported *C. ruderalis* germplasm from Central Asia to determine its genetic and evolutionary relationships with proposed ancestral subspecies *C. sativa* ssp. *spontanea* (NLHA) and *C. indica* ssp. *kafiristanica* (NLDA), and in turn their cultivated relatives. Seed accessions from regions of possible hybridization such as the Mediterranean, Kashmir, and Southeast Asia will also be of importance in unraveling the evolution of *Cannabis*. We hope that future molecular biological, biochemical, morphological, and physiological analyses will, in combination, provide further evidence to support or modify our conclusions.

CHAPTER TWELVE

Hypotheses Concerning the Early Evolution of *Cannabis*

Climate changes have caused the main modifications on the spatial distribution of organisms. Over the past millennia, human activity has also played a significant role in the distribution patterns of plants. Both factors, climatic and human, are considered fundamental to understanding the present distribution of plants.

(PANAREDA AND BOCCIO 2009)

Introduction

Where did *Cannabis* originate? Where and how did it survive the great upheavals of the glacial expansions and retreats of the Pleistocene? And where and how did humans significantly affect the evolution of *Cannabis*?

During our search for ice age refugia, the determination of prehistoric vegetation ranges and climate zones has helped us present plausible environmental settings in which *Cannabis* survived ice ages, evolved into multiple species, and expanded to colonize most of Eurasia. Pleistocene ice ages and early Holocene warming greatly affected early human dispersals. Studies of the human genome allow us to establish a time frame for early human dispersals and to determine how they could have influenced the diffusion of *Cannabis* and the origins of agriculture. It is also important to understand the similarities and differences among the members of Cannabaceae (the hemp and hop family), including their suite of reproductive strategies that give us clues to their origins. Based on these geographical and biological parameters, we offer hypotheses concerning the origin and early evolution of *Cannabis* and propose a hypothetical Cannabaceae ancestor.

Prehistoric Climate Change and Plant Distribution

Viewed from a geological perspective, present-day animal and plant communities in many parts of the world have a remarkably short history. The environmental revolution at the end of the Pleistocene, a mere 10,000 years ago, triggered major shifts in the ranges of species and hence composition of communities. Present-day communities in the taiga and temperate zones assembled at this time by combining species that survived the northern environment of the [last glacial period] with those returning from more temperate refugia. (STEWART AND LISTER 2001)

Prior to widespread human population expansion, climate had the largest impact on the world's vegetation distributions. We begin by exploring ancient climate change and its effect on the early evolution of Cannabaceae when the family was completely under the control of natural factors, long before human contact. Climate change also influenced human dispersal and colonization, which in turn affected plant dispersal and eventually the domestication of crop species. Later, we present theories concerning the evolution of flowering plants and especially traits shared by hemp and hop. This will help us determine the processes under which *Cannabis* species may have evolved, as well as how and where they survived glacial periods and by which routes they expanded following the last glacial maximum (LGM).

During its lengthy geological history spanning more than 4.5 billion years, the earth passed through many eras before plant life first appeared on land around 420 million years ago with the evolution of pollen-producing and seed-bearing plants. Around 140 million years ago (or perhaps much earlier), the first true flowering plants appeared, eventually diversifying and becoming widespread by 100 million years ago when they began to replace conifers as the world's dominant vegetation type. The Tertiary period, beginning about 65 million years ago, was characterized by climate fluctuations increasing in amplitude and eventually leading to the ice ages of the Quaternary period, which began around 2.6 million years ago. The Quaternary is divided into the prehistoric Pleistocene Epoch and the relatively recent, present-day Holocene Epoch that started around 12,000 years before the present (BP). The great factors determining the Quaternary distribution of terrestrial plant communities are plate tectonic activity and associated movements of continental masses, as well as more recent fluctuations of the polar ice cap (Hu et al. 2009). Quaternary climatic cycles of differing amplitudes and durations included long cycles of gradual mean climate change lasting up to 100,000 years, with approximately 10,000 to 20,000 year periods of warmer and moister conditions punctuated by much shorter and more striking changes occurring over as little as a thousand years before returning to the norm. The LGM before the beginning of the Holocene lasted from about 22,000 to 18,000 BP. During glacial periods much of the higher latitude northern hemisphere was covered by a widely expanded polar ice cap; localized glaciations also occurred in lower latitude mountain ranges such as the Alps and the Himalayas (Comes and Kadereit 1998), and detailed regional studies indicate that there were considerable variations in the extent and timing of glaciation from one area to the next (Owen et al. 2008). Climatic fluctuations affected the length of time taxa were confined to refugia (regions of persistent favorable habitat during extreme climate fluctuation) and how extensively they could expand as the climate ameliorated. Profound changes in climate are thought to have been a major driving force for population divergence in temperate species: "The Quaternary has witnessed the most recent, and possibly the most fruitful, of a series of evolutionary episodes which have occurred over geological time" (Knapp and Mallet 2003).

Scientists cannot directly measure past climates, so insights into prehistoric climate changes, plant distributions, and possible refugial events are derived largely from two disciplines—paleontology and genetics. Paleoecological data derived from fossils recovered from datable geological contexts reflect ancient plant distributions, and often times the relative abundance and density may be estimated; these data, in turn, are used to infer rates of extinction and expansion in response to past climate change, as well as to interpret present-day patterns of biodiversity. Early paleoclimates are reconstructed based on proxy data of the ancient ranges of various plants determined by radiometric dating of pollen microfossil and seed macrofossil remains. These floral fossils are often recovered from core samples drilled from dry sea beds, swamps, bogs, lakes, and riverbeds. The recovery of a single pollen grain can indicate local presence of a particular species; however, the absence of pollen data does not preclude the possibility that a plant species occurred in that region in low abundance and that its pollen has not been recovered. Ideally, increased pollen frequency brings better resolution to paleoclimate and vegetation studies. Although fossil pollen and seeds can provide accurate data, a paucity of fossil evidence for small local populations frequently makes estimates of dispersal routes and refugia difficult (Hu et al. 2009). Moreover, pollen of wind-pollinated species and macrofossils of wetland and peatland species are commonly overrepresented in the fossil record (Comes and Kadereit 1998; Roberts 1998) as abundant and widely dispersed windborne pollen grains are more likely to be recovered from core samples, and anerobic bog environments favor preservation of plant remains. In addition, long-distance pollen transport can contaminate local samples and certain types of pollen are often difficult to classify accurately (Krebs et al. 2004; e.g., *Cannabis* vs. *Humulus* [hop] pollen, see Chapter 4). Seed, fiber, and charcoal analysis may provide more accurate data as identification can often be carried out to the genus or even species level.

Genetic data are derived from molecular studies of genome (hereditary information) diversity in extant plant and animal populations including humans. This information can be analyzed following the assumption that populations exhibiting the greatest genetic diversity are ancestral and their present range represents the region of origin prior to divergence of less diverse populations. Genetic data can be used to extrapolate the timing of events such as bottlenecks and founder effects (loss of genetic diversity when populations are established by only a few individuals) and population expansion (see later in this chapter). Molecular markers may allow improved reconstruction, at or below species level, of past distributions, refugial population ranges, recolonization dynamics, and genomic divergence (Hewitt 2001). Although ancient pollen and seed remains help us explain these expansions at a generic or specific rank, hereditary data are necessary to identify and comprehend subspecific events involving *Cannabis* biotype and ecotype evolution. Molecular markers indicate close resemblance between populations of a single genome expansion and thus help us trace probable refugia: "Global climate has fluctuated greatly during the past three million years, leading to the recent major ice ages. An inescapable consequence for most living organisms is great changes in their distribution, which are expressed differently in boreal [taiga], temperate and tropical zones. Such range changes can be expected to have genetic consequences, and the advent of DNA technology provides the most suitable markers to examine these" (Hewitt 2000). Although molecular markers can be very useful in the determination of purely natural distributions over time, molecular phytogeographic approaches are less accurate when humans influence a species' dispersal and distribution (Krebs et al. 2004); hence, these approaches are more difficult to apply to *Cannabis*, which has an extraordinarily lengthy interaction with humans.

Modern biogeographical reconstructions include data from many geological and paleobiological indicators such as moraine boulder analysis, lake sedimentation studies, oxygen isotope analysis of limestone cave stalactites and glacial ice cores, historical records of changes in glacial size and flow, variations in radiolarian (microscopic organisms that form elaborate mineral skeletons) populations, fossil wood and tree ring dating, and population simulations. Sufficient data from several disciplines allows accurate estimation of a taxon's ancient ancestry, abundance, and distribution as well as when migration, isolation, and diversification events may have occurred. Each type of proxy evidence has its strengths and limitations with respect to dating, resolution, and accuracy of interpretation.

Glacial-interglacial (colder-warmer) cycles throughout geologic history have been driven by the path of the earth's orbit around the sun (therefore major cycles occurred at regular intervals), and organisms have undergone changes in abundance and range as a result: "It is apparent that these major climatic shifts were felt differently across the globe owing to regional differences in landform, ocean currents and latitude. Furthermore, species responded individually, and their range changes were particular to local geography and climate" (Hewitt 2000). As temperature decreased and ice sheets grew during each glacial period, species (1) became extinct over much of their range, (2) dispersed to new locations, or (3) survived in refugia of reduced range to expand only when the climate returned to more favorable conditions during the next interglacial cycle. When the polar ice sheets and mountain massif glaciers were most expansive, vegetation zones were compressed toward the equator and world sea level was lowered by 120 meters (almost 400 feet). This exposed several important land bridges such as those between eastern Siberia and northern Japan or southeastern Asia and the Indo-Malay archipelago. During glacial periods, most temperate forest species survived south of the continental ice sheets and/or in smaller refugia scattered along glacial margins and ice-free niches at higher latitudes. During times of ameliorated climate plant communities expanded producing landscapes such as those of the mid-Holocene (Hu et al. 2009).

The term "refugia" was developed from paleoecological studies (predominately the occurrence of tree pollen) and has been applied commonly to regions where modern floral and faunal taxa survived glacial periods in reduced distribution and abundance (smaller range and population). A biological refugium may be defined as a region, such as a steep mountain range with deep gorges, where a distinct genetic lineage persisted through serial climatic changes (Médail and Diadema 2009). In a genetic context, refugia represent areas where more widespread modern populations originated (Bennett and Provan 2008). Accordingly, plant recolonization during the interglacial warming periods started out from these core sites (refugia), and disappearance of a species (extinction) implies that it failed to survive in a refugium (Krebs et al. 2004).

Presently, Cannabis has a wide natural or naturalized range existing as feral and possibly wild populations across much of temperate Eurasia, having evolved into two extant species encompassing four cultivated and two feral subspecies (described in more detail later and in Chapter 11). Since our main focus is the relationship between humans and Cannabis, it is important for us to investigate the climatic conditions during Pleistocene glaciations at the beginning of the Holocene, determine where both Cannabis and early humans would have survived the LGM, and propose how they expanded into their current ranges. Central Asia has frequently been suggested as the original home of C. ruderalis, the putative ancestor (PA) of all Cannabis plants, which diverged from a common ancestor with hop (Humulus), and from whence Cannabis evolved into two species and the wide diversity of biotypes that colonized Eurasia and eventually much of the world. Europe is presently the home of cultivated C. sativa ssp. sativa, or narrow-leaf hemp (NLH), and its feral relative, subspecies spontanea (NLHA), while much of the remainder of Eurasia is home to four subspecies of C. indica. These include ssp. indica, or narrow-leaf drug (NLD); ssp. afghanica, or broad-leaf drug (BLD); ssp. chinensis, or broad-leaf hemp (BLH) cultivated varieties; and their feral common relative, ssp. kafiristanica (narrow-leaf drug ancestor, or NLDA). Early human encounters with each Cannabis biotype probably took place within a subset of their present-day ranges. Therefore, it is instructive to reconstruct climates for each region during these crucial periods: (1) the late LGM around 18,000 years

ago when much of northern Eurasia was covered with ice, (2) at the start of the Holocene about 12,000 years ago as the earth warmed and humans began to spread into nearly every biome (large regions of similar climate, flora, and fauna) on the planet, and (3) approximately 5,000 years ago as humans began to bring about large-scale environmental change through clearing land for agriculture. Evidence for the ancient climates of Eurasia varies regionally. Consequently, we will consider the following regions separately as each may have a bearing on the evolution and dispersal of Cannabis. These areas include Europe and Central, East, and South Asia, as well as the Qinghai-Xizang (Tibetan) Plateau, the Himalaya and Hengduan Mountains, northeastern China, the Korean Peninsula, and the Japanese Archipelago. It is important at each time, and in each region, to imagine whether humans and/or Cannabis would have perished, survived, or flourished and how the evolution and dispersal of each affected the other.

The Pleistocene

During the late Pleistocene, climate changes across Eurasia followed general trends, yet regional differences and exceptions to the norms also arose based on localized geography and microclimates. Around 150,000 years ago Eurasia was colder and drier than at present; then during the Eemian interglacial period, beginning 130,000 years ago, a warm and relatively moist climate lasted about 15,000 years. The warmest phase of the Eemian, the interglacial preceding the present-day Holocene, occurred from about 130,000 to 125,000 years ago. For instance, north of the Alps, summers during this period were more moist and warmer than today, and winters were also warmer. Under such relatively mild climates, Cannabis as well as many woodland species could have thrived. Beginning about 115,000 BP, several thousand years of cooler and moister summers and milder winters than today allowed relatively frost-sensitive temperate species to spread across northern Europe, Central Asia, and much of Siberia before a rapid cooling event around 110,000 years ago caused the renewed growth of ice sheets and resulted in a fall in sea level lasting several millennia (Adams and Faure 1998). It is around 120,000 and 100,000 BP in northern Central Asia that we find the earliest pollen evidence for Cannabaceae (either Cannabis or Humulus; see Chapter 4). Climate during the last hundred millennia of the Pleistocene was highly unstable across Eurasia (yet generally cooler and drier than today) as the extent of the ice sheets waxed and waned, reaching maximum size during prolonged stadial or cooling periods. During the coldest phases, steppe (semiarid grassy plain) and mixed steppe and tundra (treeless region between polar ice and treeline) were the predominant vegetation types, while forests were restricted to localized ranges in the mountains of southern Europe, eastern Turkey, and eastern Siberia where Cannabis may also have found refuge. Also, during the previous 100,000 years several dozen major interstadials (warmer phases during glacial periods) brought prolonged climatic conditions about as warm as the present, lasting from a few centuries to around 2,000 years, before rapid cooling returned the climate to glacial conditions. In southern Europe and eastern Asia, temperate woodlands expanded northward out of localized refugia during each interstadial but shrunk again when cold dry glacial conditions returned. Cannabis would also have followed warming trends by extending its range along with other temperate species and then would have experienced reduced distribution as the earth cooled once again.

Beginning around 39,000 BP a mild interstadial (though still much colder and drier than today) lasted several thousand years. During this time forest extended across most of southern Europe, while steppe-tundra and open woodland predominated across northern Europe forming isolated clumps or pockets of fragmented woodland rather than continuous closed forest cover (Adams and Faure 1998). This would have offered perfect conditions for *Cannabis* to proliferate as it does in regions with similar climates and vegetation cover today, although as yet we have no physical evidence for the presence of *Cannabis* in Europe during this period. By 30,000 BP, semiarid conditions prevailed and much of southern Europe may also have become dry forest-steppe with only a sparse element of trees (Adams and Faure 1998); under these conditions *Cannabis* would not flourish but could possibly survive. This is also when anatomically modern humans (*Homo sapiens sapiens*, or AMHs) in small bands began to colonize the warming Eurasian landmass only to be reduced to refugia as the earth chilled once again.

The last glacial period lasted from about 24,000 to 15,000 years ago when very cold and dry conditions persisted throughout Europe, and by the time of the LGM, northern Eurasia, as well as the Alps, Pyrenees, and Carpathian Mountains, was covered by ice. Sparse grassland or semidesert covered most of southern Europe, while a mixture of dry, open "steppe tundra" and polar desert covered parts of northern Europe not occupied by ice sheets (Kamruzzahan 2003; Naydenov et al. 2007). Throughout most of the Mediterranean zone, summer and winter temperatures were 8 to 10 degrees lower than at present and permafrost extended as far south as central France and Hungary.

These are not the climatic conditions and vegetation types that support naturally growing (wild or feral) *Cannabis* populations today. Open forest and woodland were almost nonexistent, except for isolated pockets of woody vegetation on the Iberian, Italian, and Balkan Peninsulas (Adams and Faure 1998). The Caucasus Mountain region was also only locally glaciated during the Quaternary, ice cover never merging with the eastern European ice sheet. In this mountainous region, a refuge formed for many temperate and subtropical species, where *Cannabis* and early humans may also have survived the LGM. Occasional findings of *C. sativa* pollen throughout the early Holocene indicates the local but very rare presence of this species, in contrast to the conventional opinion that *C. sativa* was introduced into Western Europe much later during the Roman period (Hofstetter et al. 2006).

A rapid warming and moistening of climate all across Europe occurred 14,500 years ago, and for about 500 years conditions as warm or warmer than at present began rapid deglaciation, filling river systems and lakes with meltwater and raising sea levels. Across most of Europe herbaceous communities shifted from dry and cold climate steppe-tundra toward mixed steppe, and open woodland returned to cover much of Europe, although northwestern Europe remained essentially treeless 13,000 years ago as the 200-year "Older Dryas" cooling phase encouraged a resurgence of cold climate tundra vegetation. Along the Mediterranean Sea steppe, vegetation gave way to northern woodlands spreading from southern France and the Pyrenees, and dense forest was already present in some areas (Adams and Faure 1998).

Warming climate following the LGM caused coniferous forests in the Caucasus to retreat northwards and westward. These woodlands were replaced by mixed deciduous broadleaf and evergreen pine forests, providing likely habitats for *Cannabis* to recolonize from southerly glacial refugia and avoid colder regions to the northwest during dispersal toward Central Asia. Today this region supports widespread naturally growing *Cannabis* populations, some of which have been classified as *C. ruderalis*. By the LGM, central and eastern Anatolia were covered by arid steppe-like vegetation and post-LGM conditions became moister but were still drier than today. The climate across southern Asia generally seems to have been much more arid than at present as the summer monsoon was much weaker during the LGM (Adams and Faure 1998). Arid conditions are unfavorable for *Cannabis*, which requires summer rain and cool winters; this severely limited the regions where *Cannabis* could survive the LGM.

Commencing approximately 12,800 BP, the intensely cold 1,300-year "Younger Dryas" or "Big Freeze" produced climate conditions much drier than present over much of Europe and the Middle East, and once again temperate woodlands that previously extended over much of Europe, both north and south, disappeared and were replaced by dry steppe and steppe-tundra. During this time *Cannabis* populations, following prior expansion into newly favorable regions, would have once again shrunk into refugia. This probably resulted in geographical isolation of *Cannabis* populations across northwestern Europe, where mixed patches of trees and grassland favorable for its growth were once widespread and may have persisted as short-lived cryptic refugia. The Younger Dryas ended rather suddenly over a few decades, returning to a relatively warm and moist climate (although possibly still slightly cooler generally on a global scale than it is now). This marked the beginning of the Holocene interglacial in which we presently live (Adams and Faure 1998), and relatively soon after, around 10,000 years ago, humans and agriculture began to spread from the Near East, and perhaps from parts of East Asia, as the climate once again warmed (Hewitt 2000).

Eastern Eurasia also experienced climatic diversity at the LGM. As in Europe, the climate was generally colder and drier than today (Zhuo et al. 1998), with steppe and desert vegetation extending to the coast of present-day eastern China. In the northeast, a semidesert steppe-tundra predominated (Wang and Sun 1994), grading into forest-steppe farther south. Presently, these areas are where *Cannabis* grows naturally, but during the LGM *Cannabis* could not have survived except in coastal areas as inland it was too cold and dry. In southern China, conditions cooler than today but fairly moist prevailed at least part of the time. Broadleaved evergreen and warm mixed forests had retreated toward tropical latitudes, while taiga (evergreen coniferous forest) extended southward (Adams and Faure 1998). As in Europe, the range of *Cannabis* was likely restricted to southern latitudes of East Asia during the LGM as its previous area of distribution across temperate Eurasia became too dry and cold supporting only arid steppe and tundra.

Within present-day warm temperate and subtropical latitudes of southern China there seems to have been substantial forest regression during the LGM; even the vast majority of southern lowland China would have been covered in steppe, with upland cool temperate forest and open woodlands only in the east, where even medium and low elevation mountains may have been locally glaciated (Shuang et al. 2008).

For example, on Taiwan Island and the Leizhou Peninsula in southernmost China, which are currently affected by a tropical rainforest climate too hot and humid for natural *Cannabis* growth, warm temperate mixed pine forest and oak woodland with incomplete canopy cover was present during the LGM (Adams and Faure 1998). On Jeju Island near the Korean Peninsula grassland with patches of cool temperate, deciduous, broadleaved forest developed under cold dry conditions, but from 14,400 to 11,800 years ago, woodland expanded and grasslands retreated (Chung 2007), once again providing suitable *Cannabis* habitats near where feral *Cannabis* populations are found today.

Across northern and eastern China before the LGM, the climate was colder and drier than today (areas north of 37 to 40 degrees north experienced permafrost), and steppe and desert biomes expanded southward and eastward. In addition, conifer forests expanded and cool mixed forests shifted about 1,000 kilometers (over 600 miles) eastward into the lowlands. As a result, broadleaved evergreen and warm mixed forests were displaced southward by about the same distance, replaced by herbaceous plants composed mainly of drought tolerant ragweeds (*Artemisia*: Asteraceae or sunflower family) and chenopods (Chenopodiaceae or goosefoot family; Zhang et al. 2007). The climate had generally become too cold and dry for *Cannabis* to flourish, although some populations could have survived in local microclimate refugia. Cool temperate forest grew in a belt crossing southern China, well to the south of its present distribution, which may also have supported *Cannabis* populations (Adams and Faure 1998).

On Hokkaido, the northernmost island in the Japanese Archipelago, feral *Cannabis* still grows today. During the LGM, this island was connected to the Siberian mainland via the Sakhalin Peninsula (now also an island). In both the LGM and the subsequent Younger Dryas periods, subarctic forest cover began to return to Hokkaido (Adams and Faure 1998), and both taiga- and cool-mixed forests covered the lowlands (Takahara et al. 2000). Although these environmental conditions would have been to the detriment of any resident *Cannabis* populations, cryptic microclimates for successful refugial populations may have existed. The cool dry climate of the southern part of the Japanese archipelago hosted open cool temperate or subarctic woodland vegetation in most areas (Adams and Faure 1998), which would have favored *Cannabis*'s range expansion southward were it extant there at this early date. Following the LGM, in the present cool temperate latitudes of Japan, large areas of grassland or forest-steppe developed in combination with open cool temperate and deciduous woodland cover. This vegetation response graded northward into open woodland on Hokkaido (Adams and Faure 1998), presenting widely favorable habitat for the expansion of *Cannabis*. Pollen and seed evidence from Japan and Korea are among the earliest Holocene *Cannabis* remains, indicating the possibility of dispersal from refugia in this region soon after the Younger Dryas.

Throughout the rainforest zone of Southeast Asia, during the LGM, aridity greatly reduced the amount of forest cover, and lowering sea levels would have linked most islands of the present-day Indo-Malay Archipelago; however, the equatorial tropics in the region remained sufficiently humid, and consequently, there was almost no change in vegetation type (Adams and Faure 1998). Tropical regions are not where *Cannabis* flourishes today, and feral populations are absent. Although *Cannabis* could have survived the LGM in a cryptic temperate microclimate within the tropical zone, there is no

paleoecological evidence to support this scenario. In present-day tropical, semitropical, and temperate southern Asia, conditions were also much drier and colder than now, with regression of forests and expansion of deserts.

Despite many cycles of climate change, by the mid-Pleistocene, temperate forest and woodland habitats with modern plant families and genera were distributed across southwestern China much as they are today. The rising Himalaya Mountains blocked penetration of the southeasterly monsoon causing a rain shadow and increasing aridity on the Qinghai-Xizang (Tibetan) Plateau. From around 36,000 to 20,000 BP the Yungui (Yunnan-Guizhou) Plateau to the southeast experienced slightly cooler and wetter winters than today; however, by 17,000 to 16,000 years ago, the still moist climate had cooled substantially. By 10,000 BP, corresponding to the end of Younger Dryas in Europe, Holocene climate was firmly established, and the Yungui Plateau was dominated by tropical air and was only somewhat colder to the north and wetter to the south especially in winter (Walker 1986).

Where did Asian *Cannabis* find a favorable habitat to survive the LGM and Younger Dryas? The Qinghai-Xizang Plateau, northern Eurasia, and many mountain ranges were covered in ice. Subtropical climates where *Cannabis* thrives as a plant today were likely far too arid during the LGM, while the tropics were far too hot and humid except at higher elevation and lacked suitable open vegetation types and climatic seasonality. However, temperate woodland vegetation occurred within a narrow band across the Hengduan Mountains and Yungui Plateau of present-day Yunnan and Guizhou provinces, as well as localized upland areas of eastern China, the Korean Peninsula, and insular Japan, presenting favorable microclimates for *Cannabis* refugia.

General climatic trends and events of the Quaternary were shared throughout Eurasia, yet differences in topography and proximity to oceans created localized climatic conditions that would have impinged on survival of endemic flora and fauna. The Hengduan Mountain and Yungui Plateau region of southwestern China, the Caucasus Mountains of southwestern Asia, and the northern Mediterranean peninsulas were only locally glaciated during the LGM and may have supported refugial *Cannabis* populations. By the end of the Younger Dryas the stage was set for *Cannabis* to spread from its LGM refugia and recolonize most of Eurasia. Human relationships with *Cannabis* intensified with the start of the Holocene Epoch—12,000 years ago—when our planet began its most recent warming trend as humans (AMHs) began to intensively colonize an increasingly hospitable world.

The Holocene

Early Holocene conditions across central and northern Europe were slightly warmer and often moister than at present with temperatures peaking around 9000 BP. Following the end of the Younger Dryas, the transition to peak Holocene levels of tree cover in the eastern Mediterranean was rapid, but it seems to have taken another millennium before forest cover had returned to most of Europe. Even 8,000 years ago, in many parts of Europe the forest cover was more open than at present with more herbaceous glades, a preferred *Cannabis* habitat. About 7,500 years ago, rising Mediterranean Sea waters broke through the Bosporus River valley, and in a very short time "the Black Sea became brackish and rose several hundred feet, inundating former shores and river valleys

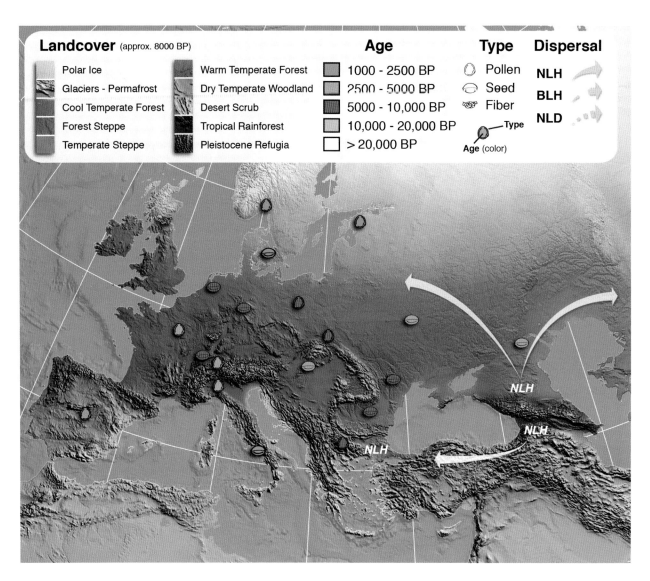

MAP 16. Proposed European glacial refugium and postglacial expansion routes of *Cannabis*, with earliest archeological evidence shown in each region. Climate warmed in the early Holocene and *C. sativa* narrow-leaf hemp (NLH) spread from the Caucasus Mountain region across the Eurasian Steppe and Anatolia first into southern Europe and later into northern Europe. Holocene vegetation reconstruction ca. 8000 BP (cartography by Matt Barbee). (See TABLE 1 at the beginning of this book for explanations of Cannabis gene pool acronyms.)

deep into the interior" with water rising 15 centimeters (6 inches) a day and thus encroaching upon the land at about a kilometer (0.6 miles) daily, eventually increasing the size of the Black Sea by 30 percent (Adams 1997). By about 7000 BP forests had developed closed canopies in Eastern Europe, with conifers more abundant than at present, and by 6000 BP closed oak forest prevailed in the Sierra Nevada of southern Spain while oak and pine savanna (subtropical grassland) vegetation covered many parts of Greece. Only southern Crete seems to have had true Mediterranean vegetation with abundant steppe, savanna, oak scrub, and some xerophytic (drought tolerant) vegetation during the early Holocene (9000–8000 BP; Adams 1997), but some northern trees were also present. Across the Mediterranean and Near East, forests were most extensive between 7000 and 5000 BP during the warmest part of the Holocene, with abundant woodlands or steppe in drier places and closed deciduous forest in moister regions (Adams and Faure 1998). Southern England, southern Scandinavia, and the eastern Baltic were slightly drier than

today. However, between 5550 and 5300 BP western central Europe's climate became cooler coinciding with advancing glaciers and lowering treeline in the Alps (Magny and Haas 2004). With the closing of forest canopies and drying out of steppe regions, *Cannabis* could have found favorable microclimates spread over its once more continuous range. However, within this large Eurasian range, it would have been restricted to geographically isolated niches in the last remaining open woodlands, along the savanna margins of closed forests, near water courses, clearings opened by fires, and disturbed areas near and often associated with human settlements. At the same time forest canopy naturally closed, agriculture spread to most parts of Europe, and *Cannabis* survived on lands opened by human activities.

For about 4,500 years, European climate has been fairly similar to the present, and vegetation changes are attributed primarily to human influence. For example, on the Italian Peninsula between 5200 and 4350 BP a dense evergreen oak forest dominated the landscape, and then the forest steadily opened

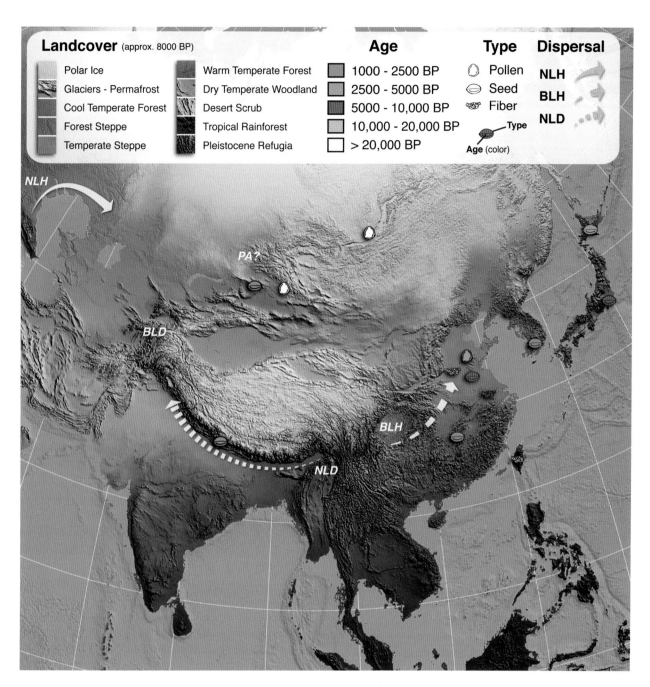

Landcover (approx. 8000 BP)

- Polar Ice
- Glaciers - Permafrost
- Cool Temperate Forest
- Forest Steppe
- Temperate Steppe
- Warm Temperate Forest
- Dry Temperate Woodland
- Desert Scrub
- Tropical Rainforest
- Pleistocene Refugia

Age

- 1000 - 2500 BP
- 2500 - 5000 BP
- 5000 - 10,000 BP
- 10,000 - 20,000 BP
- \> 20,000 BP

Type

- Pollen
- Seed
- Fiber

Type
Age (color)

Dispersal

- NLH
- BLH
- NLD

MAP 17. Proposed Asian glacial refugia and postglacial expansion routes of *Cannabis*, with earliest archeological evidence shown in each region. Very early pollen evidence may represent the earliest ancestors of *Cannabis* that evolved in Central Asia. Climate warmed in the early Holocene and *C. indica* narrow-leaf drug (NLD) Cannabis spread westward along the Himalayan foothills from the Hengduan Mountain region and *C. indica* broad-leaf hemp (BLH) *Cannabis* spread eastward across Asia. Holocene vegetation reconstruction ca. 8000 BP (cartography by Matt Barbee). (See TABLE 1 at the beginning of this book for explanations of Cannabis gene pool acronyms.)

until around 3,900 years ago. From 3900 to 2100 BP evergreen forest expanded in this area as Mediterranean climate conditions prevailed and human disturbance increased. Roman impact on the Italian Peninsula from 150 BCE to 450 CE resulted in significant opening of the forest and increases of salt tolerant plants and managed wild olive trees. By 1,500 years ago human impact caused a further decrease of natural woodland in favor of an extraordinary expansion of cultivated olive trees (Di Rita and Magri 2009). On a broader scale, deforestation in Europe became significant after about 5,000 years ago and was most rapid around 3,500 to 2,000 years ago (Adams and

Faure 1998; also see Kaplan et al. 2009; Dearing 2006; Brewer et al. 2008). Since early Holocene climates seem to have been slightly moister in southern Europe, expansion of arid zone vegetation is believed to have taken place more recently due to human intervention in cutting and burning forests, resulting in thinner soils and sparse, more drought-susceptible vegetation rather than a real effect of unaltered (natural) climate patterns. In much of Europe, however, agriculture was probably still not significant as a modifier of landscape except on a very local scale, even though agriculturalists were present at low densities through most of Europe by 4,000 years ago. By about

3000 BP, agriculture exerted significant effects on landscape in most parts of Europe, although before about 2500 BP large areas of forest in central and northern Europe were probably still untouched by farming. Before 3,000 years ago, most archaeobotanical evidence of agriculture and related deforestation comes from the Balkan and Baltic regions, with widespread occurrence across much of Western Europe soon thereafter.

In the Colchis region between the Black Sea and Caucasus Mountains, Late Holocene pollen cores are associated with charcoal deposits and reductions in tree species along with increases in herbaceous pollens such as *Cannabis*-type reflecting the spread of agriculture. Strabo, a Greek citizen of Anatolia, a historian, a philosopher, and the world's first published geographer who traveled extensively in western Eurasia 2,000 years ago, informs us in volume 11 of *Geographica*, his huge encyclopedia of geographical knowledge, that the Colchis country "is excellent both in respect to its produce . . . and in respect to everything that pertains to ship building; for it not only produces quantities of timber but also brings it down on rivers. And the people make linen in quantities, and hemp, wax, and pitch." Wheat (*Triticum* spp.) did not tolerate the humid climate and millet and hemp were possibly the only "grain" crops grown, and even then only on raised and drained lands or lands farmed using swidden (slash and burn) agriculture (Connor et al. 2007).

As in Europe, Holocene warming stimulated extensive and repeated human expansion across Asia and greatly accelerated human-*Cannabis* contact and interaction. Regional variations in early Holocene climate determined where different species colonized and how they adapted to local ecological changes. Asia covers a vast territory with diverse microclimates and ecological niches offering many chances for genome divergence and speciation.

Intensity of East Asian and Indian summer monsoons has been weakening since about 9000 BP (Maher 2008). In northern Russia (including Siberia), taiga or boreal forest expansion commenced by 10,000 BP, and from 7000 to 6000 BP Siberia was warmer than today (Adams and Faure 1998). Over most of Russia, forest advanced to near the current arctic coastline between 9000 and 7000 BP and retreated to its present position by between 4000 and 3000 BP (MacDonald et al. 2000). At the beginning of the Holocene on Jeju Island at the tip of the Korean Peninsula, warm temperate evergreen and deciduous broadleaved forests with a fern understory were supported by warm humid conditions similar to the modern climate (Chung 2007), and Japan's climate appears to have been much the same as today since 5,000 years ago.

In eastern China around 9000 to 6000 BP temperatures are estimated to have been slightly warmer than today, causing glaciers to melt and disappear, and the consequent rise of sea level resulted in great changes in the coastal environments. Permafrost was located far to the north of its present-day southern limit. Desert regions and loess (sediment formed by the accumulation of windblown silt) deposition were greatly reduced and the highest lake levels occurred mostly during this time. Conifer forests grew to the north of their present locations, and many east Asian steppe areas were covered with forests, the northern limit of temperate deciduous forest shifting about 800 kilometers (500 miles) north relative to today (Yu et al. 2000). *Cannabis*'s range may have been somewhat reduced toward the south, except at higher elevations due to arid conditions, but huge tracts of the north would have been available for its rapid proliferation, especially as it was dispersed by humans expanding into newly opened environments.

Climatic events strongly affected early Yangzi and Huanghe (Yellow River) cultures. The western part of the Chinese Loess Plateau along the Huanghe River valley experienced a series of environmental changes during the early Holocene (11,000 to 7400 BP) when late glacial desert-steppe vegetation was succeeded by pine-dominated forest-steppe. During the mid-Holocene (7400 to 4700 BP), pine forest was succeeded by dense and diverse deciduous forest before reverting to pine-dominated forest-steppe. During the late Holocene (4700 BP to the present), forest-steppe and steppe vegetation dominated until further desiccation resulted in replacement by desert-steppe alternating with steppe vegetation; this process of change lasted until around 1,000 years ago (Feng et al. 2006). Since the mid-Holocene, around the time agriculture became well established, none of the naturally occurring vegetation types were conducive to the natural growth of *Cannabis*, although it was utilized and likely grown widely as an agricultural crop in ancient China. *Cannabis*'s camp following tendencies and early adoption as an agricultural plant ensured its survival in regions that did not readily support its natural growth. The earliest *Cannabis* seed evidence from the Huanghe basin dates from 5500 to 4500 BP, pollen to 4500 BP, and fibers to 4000 BP (also see Chapters 4 and 5).

In the lower Yangzi River drainage basin in eastern China, a mixed evergreen and deciduous broadleaved oak forest predominated from around 10,500 BP and became fully developed between 8250 and 7550 BP. Trees declined and herbs increased between 7550 and 3750 BP, as a result of progressive removal of the forest by humans through cutting and burning. By approximately 3750 BP the broadleaved forest of the lower Yangzi basin had been largely replaced by terrestrial herbs. Pine populations continued to rise alongside the majority of herbaceous vegetation between 3750 and 2000 BP and then also declined after about 2,000 years ago. Again, disappearance of broadleaved forest after 3750 BP indicates increased cultural impact on the natural environment primarily due to habitat alteration through increasing human population and farming (Chen et al. 2009).

Farther east, in the Huanghe River delta region, significant human-induced vegetation change also began around 4,000 years ago as a result of deforestation and agriculture (Yi et al. 2002). Prior to 5,000 years ago, the climate of the Jiaodong (Shandong) Peninsula between the mouths of the Huanghe and Yangzi Rivers was warm and humid, accommodating southern plant species. After 5000 BP, vegetation shifted to mostly mixed coniferous-broadleaf forest, followed by dominance of herbaceous plants and shrubs resulting from a shift toward a dry, temperate climate. The earliest evidence of *Cannabis* seed from the Yangzi River basin comes from Changsha and is dated to only 2000 BP, while fiber evidence dated to 5000 BP was recovered from the lower Yangzi. Earlier evidence for the Huanghe basin and coastal regions argues for dispersal of *Cannabis* following the LGM from coastal northeastern China, Japan, or Korea rather than from southern China.

In almost all regions of China north of the Yangzi River the early Holocene from 10,000 to 6000 BP is characterized by an expansion in forest cover reaching a maximum and declining thereafter, with temperate deciduous forest eventually experiencing a remarkable reduction largely as a consequence of human activity. By the early Neolithic, around 7000 BP, the lower Yangzi River provides evidence of deforestation and possibly food production by early agriculturalists. After 6000 to 5000 BP, forest cover along the middle and lower Huanghe River began to decline, and during the mid- to late

Holocene it continued to decline in most regions north of the Yangzi River. Presently, forest cover has decreased by about 92 percent in the middle and lower reaches of the Huanghe, 64 percent in the easternmost part of the Qinghai-Xizang Plateau, and 37 percent between the Yangzi and Huanghe Rivers.

In sum, changes in forest cover prior to 6,000 years ago were probably caused mostly by natural climate change, but the decline in forest cover since that time was induced predominantly by humans. Anthropogenic disturbance was of overwhelming importance as ancient agriculture and high-density settlement expanded outward from the middle and lower Huanghe regions (Ren 2007; Ren and Beug 2002). As *Cannabis* does not tolerate shade, the cutting of forests offered opportunities for its rapid expansion, especially into areas near human settlements. Humans likely also cleared some land to cultivate hemp. During the mid-Holocene period, *Cannabis* expanded into most of the open habitats and associated niches where it is found today throughout much of China. These areas are largely those created in association with human activities—that is, anthropomorphic landscapes that allowed diverse opportunities for *Cannabis* populations to proliferate.

In the Hengduan Mountains of southwestern China vegetation began to respond very soon after the LGM as deglaciation made terrain available. The course of vegetational changes was often complex, although less so where ecological differences from present conditions were smallest; in any case, most natural major vegetation shifts were over by around 10,000 years ago (Walker 1986). After about 11,750 BP, in the area around Erhai Lake in northwestern Yunnan province where the Hengduan Mountains meet the Yungui Plateau, a warming climate coupled with enhanced summer monsoon precipitation resulted in expansion of hemlock (*Tsuga*: Pinaceae or pine family) and evergreen broadleaved trees. An increase in evergreen oaks and dry-tolerant species after 10,300 BP suggests a greater seasonality in rainfall, reflecting a southward shift in the winter frontal system. This trend of increasing temperature and seasonality continued into the mid-Holocene.

Archeological and historical records indicate a steady increase in human expansion throughout the late Holocene; this data, as well as an apparent decline in trees coupled with increased levels of grasses and other disturbance-adapted species, provide the first evidence for human impact in southeastern China at around 6400 BP: "This early phase of forest clearance led to the collapse of the natural altitudinal vegetation gradient that existed in the area from the last glacial period. The subsequent expansion of secondary pine forest suggests that these early clearances were part of a sustained period of shifting agriculture" (Yang et al. 2005). Early deforestation resulted in increased surface runoff and erosion. Forest clearance increased in intensity from 2140 BP through the end of the twentieth century due to agricultural intensification and urbanization (Shen et al. 2006).

The Holocene of the Xinjiang and Inner Mongolian Plateau (Western China/Central Asia) regions north of the Oinghai-Xizang Plateau is divided into three phases: a warming and dry early stage from 11,000 to 8700 BP, a warm and wet middle stage from 8700 to 4500 BP (about a millennium later in Mongolia), and a fluctuating cool and dry late stage continuing into the present (Zhang et al. 2007). As the Holocene commenced much of Central Asia provided areas where *Cannabis* could grow wild or as an escape, yet many of the available transitory microclimates may soon have disappeared and the resident populations perished. Today *Cannabis* grows as a feral escape in favorable locations throughout northern Central Asia.

The Himalaya Mountain range stretches across northern India to southwestern China and forms the southern boundary of the Qinghai-Xizang Plateau. Generally, during cold periods Himalayan glaciers expanded to well beyond their current extent pushing temperate flora southward to lower elevations. The extreme altitudinal gradient of the Himalayas provided diverse habitats and niches for species survival within a relatively compact region; these conditions could have also accelerated evolution of species through geographical isolation. Today *Cannabis* grows both naturally and cultivated throughout this region. In the section that follows, we present paleoclimate data for different Himalayan regions, moving from west to east.

In Kashmir, periglacial (glacial margin) conditions predominated prior to 15,000 years ago, followed by pioneer meadows with sparse tree cover, including pine-oak *(Pinus-Quercus)* woodlands with some thermophilous (warmth-loving) trees becoming established around 12,000 to 10,000 years ago (Walker 1986). Between 9500 and 6125 BP the climate was warm and humid supporting mixed oak-broadleaf and pine forest, with a dominance of pine stimulated by the onset of cooler and drier climate attributed to decreased monsoon rainfall. Between 6125 and 4330 BP the climate turned moderately more humid and warm, presumably due to enhanced monsoonal effect, and as a result, broadleaf forests expanded. A short-lived pluvial (relatively wet and/or rainy) environment formed between 4330 and 4000 BP before a cool and dry climate prevailed until 2100 BP, which was marked by a sharp decline in oak forests (Trivedi and Chauhan 2009). In the northwestern Himalayan region, the climate was warm and moist during most of the Holocene with short phases of colder and drier climate (Bhattacharyya and Shah 2008). Most of these climate fluctuations would not seriously limit *Cannabis* growth, and some may have proven favorable; *Cannabis* would have found many suitable niches for survival without moving far.

Early Human Dispersals

By 65,000 years ago, modern humans had made their way along the shorelines of southern Asia to Australia. They moved north into eastern Asia within the past 40,000 years. In northern Asia, people from the East eventually mixed with groups that had moved from the Middle East into Siberia. More recently, people from East and West have mixed in the oases and steppes of Central Asia. (OLSON 2002b)

Based on climate reconstructions and related vegetation responses summarized earlier, it is readily apparent that about 18,000 years ago during the LGM there were only two relatively small regions of Eurasia with suitable climates to support the natural growth of *Cannabis*. These were southern Europe and southern East Asia. It is likely that early forms of *Cannabis* survived Pleistocene glaciations, including the LGM, in temperate refugia much farther south from their putative origin in Central Asia, which was by then covered by ice. By this time, *Humulus* and *Cannabis* had long since diverged along their separate evolutionary pathways. The PA populations of *Cannabis* would have been easily recognizable with their resemblance to modern species in this genus—with one having already evolved the capability to biosynthesize THC. If *Cannabis*'s evolution is largely the story of its interaction with humans, then what evidence do we have for the presence of early humans in its Eurasian homeland?

Fossil finds, corroborated by genetic data (derived from studies of the frequency of Y-chromosome mutations found in various genetically isolated human populations), suggest that the first humans (*Homo sapiens*) appeared in Africa around 500,000 years ago, and AMHs (*Homo sapiens sapiens*) evolved there around 200,000 to 100,000 years ago. Pleistocene climatic fluctuations imposed range restrictions on early human populations as well as their host biomes. When the earth's climate cooled, ranges of thermophilous plants and humans were reduced to southern refugia, although, as mobile mammals, early humans were much better able to adapt and/or move than were sedentary plants.

About 71,000 BP the climate suddenly chilled coincident with the massive volcanic eruption of Mount Toba in Sumatra (the earth's largest eruption over the past 450 million years). During the subsequent millennium, it is likely that AMH populations were reduced to as few as 10,000 individuals, living in refugia in equatorial, semitropical Africa, which created a bottleneck limiting genetic diversity. Genetic "bottleneck" describes the genomic and especially evolutionary consequences of reduced population size followed by population increase (Hawks et al. 2000). Fewer than 60,000 years ago, AMHs migrated from tropical refugia to northeastern Africa, entered the European Levant about 41,000 BP, displacing Neanderthals (*Homo neanderthalensis* who became extinct around 28,000 BP), and reached Central Asia by 35,000 BP (Wells 2002, see also Chapter 4). During the Upper Paleolithic, from 40,000 to 15,000 years ago, the climate of northern Asia became steadily colder, and early humans dispersed southward into southern Europe, South Asia, and southern East Asia. This set the stage for increasing human expansion and colonization spreading very far north after the LGM. Around 20,000 BP, during the last glacial period, human ranges were restricted to southern refugia north of the Mediterranean and in southern East Asia near the South China Sea. Immediately following the last glaciations lower sea levels exposed extensive land bridges between Siberia and the Japanese Archipelago and across Beringia from Siberia to Alaska (Adams and Faure 1998); these terrestrial pathways facilitated human colonization (or recolonization) and some plant dispersal. AMHs would have radiated northward from southern refugia as the earth warmed anew around 16,000 years ago, possibly carrying favored plants with them.

As the Holocene Epoch, in which we live in today, began and earth's atmosphere heated up once again, the climate across northern Eurasia generally became more warm and moist. Precipitation in South Asia began to increase, although vegetation recovery was slow. Tree cover increased until most of northern Asia was covered by open woodlands of birch and conifers (Adams and Faure 1998), setting the stage for the spread of *Cannabis* into favored microclimates throughout much of northern temperate Eurasia. Shrub tundra vegetation invaded polar desert latitudes, to be succeeded by open steppe-woodlands, and ancestors of today's temperate flora began to recolonize earth from their southerly Pleistocene refugia.

Renewed human dispersals and recolonization of northern latitudes offered repeated chances for humans to contact *Cannabis*. The rate of population expansion was determined largely by favorable changes in climate, and as climate change was quite rapid, likely so was human recolonization (Hewitt 2001, 2004). Following the setback of the later glacial advances, *Cannabis* surviving in southern refugia was transported northward by humans into eastern China and westward across the temperate foothill regions of the Himalaya Mountains to which it was climatically preadapted. Holocene warming also stimulated the "Neolithic agricultural revolution" in several regions of Eurasia promoting dramatic human population growth, fueling colonization, and further spreading early crop plants such as *Cannabis*.

By the middle Holocene, agriculture spread to much of Eurasia, and *Cannabis* found many natural habitats and anthropogenic (human created) niches where it survived and flourished. Several expanding forest-steppe and open woodland vegetation zones offered natural conditions favorable for *Cannabis* and at times formed a nearly continuous east-west corridor for its natural dispersal and colonization. Northward extension of warm temperate forest belts in northeastern China, Korea, and Japan favored *Cannabis* growth and productive cultivation as long as suitable open terrain was available amid forest cover. Humans began to move widely across Eurasia, which further accommodated *Cannabis*'s spread. By this time early humans had reached many favorable places to live, but there were important climatic and cultural events about to unfold. *Cannabis*'s many attributes became apparent to many different peoples who utilized (and selected) it for several uses such as food, fiber, medicine, and ritual purposes leading eventually to the diversity of twentieth century cultivated varieties.

Plant Speciation and Colonization

The places where species persist during glaciations have generally been described as refugia. Isolation within such glacial refugia, and the timing and mode of expansion from them, have become topics of increasing importance in our understanding of evolutionary processes such as adaptation, speciation and extinction. (STEWART et al. 2010)

Researchers have favored several different taxonomic schemes for *Cannabis* ranging from a single species to three, and we recognize two extant and one ancestral species (see Chapter 11). We know approximately when and where *Cannabis* populations spread during warm periods and then contracted and found refuge during cold periods, but we have not hypothesized how the evolutionary process of speciation (differentiation into separate species) may have occurred. It is important that we understand some basic concepts of plant diversification and speciation before trying to understand the evolution of family Cannabaceae and genus *Cannabis* in particular. Speciation is a consequence of diverging evolution resulting from imposition of obstacles to genetic exchange between previously interbreeding populations and associated genetic changes (i.e., reproductive isolation; see Rieseberg and Willis 2007). There are two basic scenarios proposed for plant speciation—between allopatric (geographically isolated) populations or within sympatric (shared range) populations—accompanied by several evolutionary strategies followed independently or in concert.

If a plant's range becomes divided by climatic or other changes into geographically disjunct populations separated by more than the effective distance for pollen travel, the subpopulations become genetically isolated. Geographic isolation allows allopatric speciation to proceed via mutation (random change in genome sequence), natural selection (consistent effects upon survival and reproduction), and genetic drift (random change in gene frequencies), promoting genetic diversity (Yuan et al. 2008) accompanied by ecological adaptation of novel phenotypes to new niches. This results in a divergent population that no longer interbreeds with its progenitor population and proceeds along its own evolutionary path. In addition to geographical separation, reproductive barriers appearing both before and after mating enforce genetic isolation. Ecological, seasonal, and genetic incompatibilities often prevent successful mating between populations, and even following successful hybridization embryos and seedlings may die, or living offspring may be ecologically unfit or sterile, producing no offspring and further promoting genetic isolation (Moulton 2004).

In plants, barriers preventing mating or fertilization are more common than influences such as lowered fitness of hybrid offspring (Rieseberg and Willis 2007). Narrow ecological tolerance and limited seed dispersal curtail expansion of some plants into new geographically isolated territories, and under such conditions new species must evolve sympatrically within the parental population. Sympatric speciation results from a mutation (usually polyploidy or chromosome set doubling) that causes infertility between offspring and parents, yet offspring inheriting the same mutation remain interfertile, and reproductive isolation becomes genomic rather than geographical (Moulton 2004). There is evidence for aneuploidy in Cannabaceae, which consists of the diploid genera *Cannabis* (2n = 20) and *Humulus* (2n = 20 and 16), but only in the latter genus. Both *Cannabis* and *Humulus* include three putative species: *C. ruderalis*, *C. sativa*, and *C. indica* and *H. lupulus*, *H. yunnanensis*, and *H. scandens* (synonymous with *H. japonicus*). Diversification within a species is driven by genetic drift, which, if it is directed by selection and reinforced by isolation barriers sufficient to restrict gene transfer between a new population and its ancestral population, can accumulate genetic differences leading to speciation. Ecological isolation is the most important reproductive barrier in closely related plant species, subspecies or biotypes resulting from restricted habitat preferences within a shared geographical range. Adaptation of interfertile subspecies to different local habitats restricts gene flow. Isolation need not be complete as any reduction in the transfer of genes between populations facilitates divergence, eventually reducing gene flow to near zero. Reproductive barriers can arise quickly, especially when created by humans changing the local environment (Rieseberg and Willis 2007) or transporting propagules (seed or other plant parts that can give rise to another plant) to isolated niches, although these conditions are more likely to lead to smaller changes on the subspecies or variety level.

Incomplete isolation of interfertile populations can lead to hybridization. Hybridization is viewed by many as "a potent evolutionary force that creates opportunities for adaptive evolution and speciation" where "increased genetic variation and new gene combinations resulting from hybridization promote the development and acquisition of novel adaptations," or it is dismissed by a few as insignificant "evolutionary noise." We argue that presently there is insufficient evidence to confirm either opinion. However, it is generally accepted that hybrid genotypes may become established through "diploid hybrid speciation," whereby a reproductive barrier such as ecological divergence or spatial separation arises in the hybrid lineage limiting gene flow from parental populations (Rieseberg et al. 2003). This can ultimately result in hybrid or "recombinational" speciation (Ungerer et al. 1998). Following the LGM, speciation via hybridization probably occurred in regions of population overlap where populations lived sympatrically and hybridized. This concept also applies to habitats recently created through human activity (Hewitt 2001). Hybridization between taxa (species, subspecies, and ecotypes) leading to introgression (exchange of genes between populations) is common even between widely distributed and geographically isolated populations and may constitute a "powerful and creative evolutionary process" acting upon diversification and speciation (Cronn and Wendel 2003). According to Lexer and Widmer (2008), "Even strong and genetically complex isolating barriers are unlikely to prevent widespread introgression."

Hybridization infers greater adaptive potential than genetic drift because genetic changes are achieved at many loci (gene locations) simultaneously rather than at fewer loci spread over a greater time span, creating increased opportunity for a more successful phenotype to appear. Furthermore, new hybrid recombinations are between alleles (pairs of alternative gene forms) already bestowing fitness, making hybrid offspring somewhat preadapted to local conditions and increasing their survival rate. Many hybrid species diverged ecologically from their parental lineages and often occur in more ecologically extreme conditions, although there is no proof that hybridization provided the fitness and genetic divergence required for speciation or if new species arose by an alternate pathway (Rieseberg et al. 2003). Only a few cases of hybrid speciation have been convincingly documented.

One group of campions (*Silene*: Asteraceae or daisy family) is similar to Cannabaceae in that it contains anemophilous, dioecious species with heteromorphic (X and Y) sex chromosomes. At least two species in this group of campions, differing "most prominently in floral traits and habitat preferences," are "able to hybridize when they come into contact, yet maintain morphological integrity at the boundaries of such hybrid zones." Gene flow between these campion species occurs via backcrosses with "hybrids acting as bridges to gene flow" (Lexer and Widmer 2008), and this continuing gene flow between parental populations prevents the establishment of a hybrid species. Research with sunflower (*Helianthus*), poplar (*Populus*) and iris (*Iris*) populations has shown that introgression between species is significant even across substantial geographical barriers and that beneficial alleles may be transferred great distances in outcrossing (fertilization between two different individuals) taxa such as those in Cannabaceae to overcome geographical isolation barriers, encourage gene sharing, and discourage species formation (Lexer and Widmer 2008). On the other hand, ancient hybrid sunflower species occupy the most ecologically extreme habitats and hybrid gene combinations facilitated ecological speciation in new niches. Independent hybridization events occurring 60,000 to 200,000 years ago between two widespread and often sympatric sunflower species that favor moist clay and sand soils led to three different hybrid species restricted to divergent habitats with extreme desert, dune, and brackish marsh conditions (Rieseberg et al. 2003).

In ragworts (*Senecio*: Asteraceae or daisy family), rapid homoploid (diploid as opposed to polyploid) speciation has been documented within the past 300 years; this accelerated evolution has been attributed to natural hybridization followed by geographical isolation via human dispersal into new habitats. Oxford ragwort (*Senecio squalidus*), a morphologically distinct diploid hybrid species of the British Isles, evolved allopatrically following long-distance dispersal of hybrid material from a natural hybrid zone between higher altitude *S. aethnensis* and lower altitude *S. chrysanthemifolius* on Mount Etna in Sicily, Italy, during the early eighteenth century to the Oxford Botanical Gardens from whence it began to spread in its new insular environment (James and Abbot 2005). Ecological speciation via homoploid hybrids, as in ragwort and sunflower, is relatively rare, having been documented in only a few species, likely because "fit hybrid genotypes that reproduce by out-crossing must escape from hybrid zones" and become reproductively isolated (geographically or ecologically) from parental populations or the hybrid genome will constantly be eroded by backcrossing. Spatial isolation from the parental populations also encourages establishment, as there is less competition for favorable ecological niches (James and Abbot 2005). The most convincing demonstration of restricted gene flow (introgression) between *Cannabis* taxa comes from Hillig and Mahlberg (2004). Synthesis of THC, the primary psychoactive cannabinoid compound in *Cannabis*, is modified by a pair of alleles at one locus, B_T, which control the THC production characteristic of *C. indica* populations; similarly, the synthesis of CBD (a nonpsychoactive chemical component in *Cannabis* resin), which is a characteristic of *C. sativa* populations, is controlled by paired alleles at the BD locus. *Cannabis indica* expresses both B_T and B_D alleles, and while B_T alleles are rare in *C. sativa* they are present in low frequency, indicating the possibility that "introgression from *C. indica* may have played a role in the evolution of *C. sativa*" with alleles migrating via wind-borne pollen. However, within *C. sativa* B_T allele frequency is greatest in southern populations near regions where *C. indica* was introduced for hashish production (Hillig and Mahlberg 2004), supporting introgression during recent historical times, long after the evolution of the two species (see Chapter 11).

"When new species arise as a result of divergent natural selection mandated by abiotic or biotic environmental forces, the process may be referred to as ecological speciation" (Lewin 2006). Early on in the speciation of *Cannabis*, ecological diversification, especially in disturbed habitats, led to a wide divergence in *Cannabis* subspecies, biotypes, and landraces (traditional cultivated varieties). Ecological speciation is belived to be common in unsaturated floras such as the leading edge of an expanding population where populations can reap the benefit of "ecological opportunities" including comparatively small intensities of "competition, herbivory, and disease." In saturated floras such as dense woodlands, ecological disturbance (often induced by human activities) is a key factor for the colonization and expansion of many invasive alien species. Disturbed areas allow newly immigrant species a chance to become established where they would otherwise be excluded and to become established in sufficient abundance to compete with the dominant plant community. In a new open niche, natural selection for greater adaptation and fitness for habitat tolerance, as well as resource utilization, can lead to species divergence. Ecological speciation is well documented in several members of the herbaceous Asteraceae or sunflower family. For example, in the Asteraceae (daisy and sunflower family) New World thistles (*Circium*) species are restricted to very specific habitats, diverging rapidly and filling a wide range of open ecological niches including "prairies, montane meadows, rocky desert canyons, forests, lacustrine and coastal dunes, seeps and streamsides, and brackish marshes," and in eastern Africa, different giant groundsel (*Dendrosenecio*)

species occupy ecologically diverse montane forest, heather, or meadow habitats ranging from 2,500 to 4,600 meters altitude on the same mountain slope (Levin 2004).

Cannabis often becomes quickly established in disturbed areas, especially when disturbances open up the tree or brush canopy and allow direct sunlight to reach the ground. Being preadapted to open human-induced microclimates, *Cannabis* became an invasive agricultural weed in temperate continental Europe and later in North America. Mature plant communities are usually saturated with each niche filled by well-adapted species that are effective competitors, and accordingly, less fit species are most often excluded from a saturated community rather than occupying suboptimal niches. *Cannabis* may be an exception to this "rule" in that populations will survive under extremely marginal conditions in a much diminished stature and population size and still produce enough seed to establish a breeding population the next year. However, this environmental tolerance may only be displayed for a short time, and as other vegetation types encroach, *Cannabis* may become better established (with larger populations) in a more favorable niche nearby. Disturbed areas open up space for effective wind pollination, and opening up closed canopy forest habitat encourages *Cannabis*'s gene flow. Some regions such as the Himalaya Mountains are prone to natural disturbances such as wind falls and erosional slips brought on by storms; furthermore, these massive mountains have long been tectonically active, and even today seismically induced landslides on steep slopes also open up new areas for colonization. Rapid diversification (or reduced extinction) in plants is correlated with new niches created by major geological changes such as the uplift of the Himalayas and Qinghai-Xizang Plateau (Rieseberg and Willis 2007). The ultimate evolutionary response of Cannabaceae to forest encroachment may have been the development of the vining habit of *Humulus*.

Cannabis species are interfertile and there is evidence for hemp and hop crop introgression into wild populations (Jarvis and Hodgkin 1999), so it is possible that alleles from wild *Cannabis* populations may introgress into cultivated seed hemp crops. However, there was no plausible time during the early evolution and expansion of *Cannabis* where two species, subspecies, or biotypes lived sympatrically and could have given rise to hybrid populations. Rather, two documented species—*C. sativa* in Europe and *C. indica* in Asia—diverged early on from a common ancestor via geographical isolation, with multiple subspecies and biotypes of *C. indica* evolving later during population expansion from Pleistocene refugia following the LGM, which involved adaptive radiation and ecological diversification. Only much later did *Cannabis* populations hybridize but without species formation. Although both diploid hybrid divergence and genome introgression (which are particularly applicable to outcrossing taxa) are plausible evolutionary scenarios, they probably became important only after humans began to manipulate the *Cannabis* gene pool, at first unintentionally through human dispersals (see Chapter 4) and later on by artificial selection (see Chapter 10). Hybridization and introgression were likely much more important in generating diversity at the biotype and cultivar level than during speciation. Throughout the Holocene, *Cannabis* has been in a state of constant expansion, diversification, and ecological adaptation and, along with human assistance, recolonized most of Eurasia as well as the New World in recent centuries. Only in the twentieth century did the diverse genomes become hybridized, largely for the purposes of developing *Cannabis* cultivars. *Cannabis* species are interfertile and cross easily, producing viable offspring; therefore, intentional hybridization has been a common modern breeding tool. However, hybridization and introgression played only limited roles in *Cannabis* evolution prior to the twentieth century. Allopatric speciation, supported by geographical isolation within the highly diversified topography of montane refugia, intensified during postglacial range expansion into new territories, offering the most likely scenario for *Cannabis*'s evolution. This applies as well to the evolution of cultivated *Cannabis* varieties under human selection where geographical isolation was encouraged by long-distance human dispersal of seeds to novel habitats.

Pleistocene Refugia

During the Quaternary period, the northern hemisphere has been subjected to multiple cold climate cycles. These have been characterized by advancing ice sheets and changing weather patterns that resulted in expanding deserts (both hot and cold) and shrinking temperate regions. These colder cycles have been alternating with warming climate conditions such as the Holocene Epoch. These in turn have caused the glaciers to melt and recede, equatorial areas to become more humid, and tropical and temperate forest, woodland, and steppe vegetation to spread across Eurasia. The ranges of all temperate organisms, including *Cannabis*, were greatly reduced during glacial periods and many became extinct with loss of habitat. Temperate-adapted taxa were confined to refugia when their ranges shrank during glacial periods, while colder-adapted taxa were confined to refugia during warm interglacials like the Holocene when they became restricted to high latitude polar regions and high altitude mountainous regions. Quaternary cold periods reduced woodland diversity, both in number of plant taxa and in genetic diversity within each taxon (Krebs et al. 2004). In the Northern Hemisphere, survival during glacial periods in southerly refugia followed by northward postglacial recolonization has been observed for a wide range of temperate taxa: "It was necessary to move, adapt or go extinct, and present lineages had the ability and luck to survive such shifts. Each time they would colonize new territory, face new environments and meet new neighbours [*sic*]. These challenges would cause genomes to diverge, both through selection and chance, and ultimately speciate" (Hewitt 2000). Latitudinal and elevational range shifts occurred over many generations and provided many opportunities for extinction or adaptation, leading to variations in genome structure. For example, populations became extinct, alleles were lost in genetic bottlenecks and then expanded during founder events, and mutations occurred, which, if selected, were spread as potential range expanded (Hewitt 2004).

An individual species' physiology largely determines its reaction to cooling climate, although survival is also linked to the surrounding plant community and, for many plant species, their animal pollinators. Cold-sensitive plants such as *Cannabis* suffer irreversible damage as temperatures approach freezing. Species that are tolerant to freezing conditions, such as northern temperate and arctic trees, can withstand sustained temperatures below freezing, but no plant communities survive total ice cover. Cold tolerant plants live closer to glacial ice (higher elevation and/or latitude) than cold sensitive plants such as *Cannabis*, which live farther from the ice

along an increasing temperature gradient in temperate or subtropical climates. As ice sheets advance and retreat, the gradient of vegetation types along glacial margins shifts as well; this distribution is evident today in regions of natural plant growth.

In the Northern Hemisphere, it is generally accepted that refugia for temperate species were relatively smaller and likely recurrent. These refugia usually corresponded with the southern part of their ranges during interglacial periods like the present. During glacial periods species also survived within smaller and more temporary periglacial "cryptic" refugia often located at higher latitudes than southern refugia. These more diminutive refugia were probably smaller because they were surrounded by unfavorable frozen habitats: "Increasing evidence suggests that the well-studied southern and eastern refugia for thermophilous animal and plant taxa were supplemented by cryptic refugia in northern Europe during the Late Pleistocene" (Stewart and Lister 2001). Archeological *Cannabis* finds from early dates (9000 to 5000 BP) come from the area north of the Baltic Sea, the Balkan Peninsula, northeastern China, and Japan, reflecting more exhaustive research in these regions and also possible cryptic refugia (see Chapter 4). Smaller fleeting refugia may have been most important during short-term climatic changes than during more persistent glaciations.

Could *Cannabis* have remained pan-Eurasian in its distribution throughout the LGM and previous glaciations, albeit at lower abundance, and then simply become more abundant and expansive during interglacials? If the cryptic refugia concept applies to *Cannabis*, then it would have survived across a broader range rather than, or in addition to, discreet southerly refugia, but at low abundance and density. Cryptic northern refugia for temperate taxa such as *Cannabis* would have been sheltered habitats at low elevation in deep valleys or along coastlines where environments were ameliorated by maritime conditions; such refugia would have provided microclimates that allowed for survival at latitudes where surrounding populations perished. Since *Cannabis* is dioecious and anemophilous, it requires a large population size and high density to persist in wild populations. Interbreeding population sizes must be adequately large to avoid inbreeding depression (Stewart et al. 2010). Inbreeding increases the frequency of homozygotes (both dominant and recessive), which facilitates the expression of beneficial as well as deleterious recessive traits and provides a wide range of new mutations for selection. Inbreeding depression from repeated inbreeding causes offspring to accumulate and express deleterious recessive mutations, resulting in reduced fertility and decreased survival rates. The death of lethal inbred recessive recombinants purges many deleterious genes from the genome, but ultimately the fixing of mildly deleterious mutations leads to slow deterioration of the genome and eventual extinction (Charlesworth 2006). This is particularly critical for outcrossing plants like *Cannabis* in which population sizes must be larger than other plants with different reproductive systems (i.e., self-pollination) to avoid inbreeding depression and eventual extinction. Therefore, suitable refugia for *Cannabis* would have had to have been larger and survival in smaller cryptic refugia is thus a less likely scenario.

As the glacial periods were longer in duration than interglacials, temperate taxa spent protracted periods of time in their diminished refugial range. It is likely that shorter term climate shifts also caused ecological disturbances and gave rise to contractions and expansions of ranges on a shorter time scale. Many populations die off and species become extinct as their geographical ranges contract. Populations occupying long-term refugia and still extant today are descendants of individuals that were already living there during the postglacial expansion phase (Stewart et al. 2010).

Recurring postglacial colonization and range expansion, natural selection and adaptation to differing niches led to genome reorganization and evolution of local biotype populations suited to differing photoperiod and temperature conditions. This occurred across a wide latitudinal range: "The current evidence indicates that although genomic divergence proceeds over millions of years, morphological, physiological and behavioural [sic] changes may produce subspecies and species at almost any time" (Hewitt 2000). This is probably how the subspecies and ecotypes of *Cannabis* evolved, although they have not yet diverged sufficiently to warrant species level status.

Glacial phases forced populations to split into isolated refugia, leading to genetic variation and speciation, followed by radiation into an expanded range often shared by both new species. Several regions in southern Europe including the Iberian, Italian, and Balkan Peninsulas have been identified as glacial refugia for many plant and animal species. It is possible for a region to have been the area where a species evolved (its endemic homeland) as well as the subset of its full range that served as a refuge during climate shifts (Stewart et al. 2010). Regions exhibiting the greatest genetic diversity within the range of a temperate taxon can be interpreted either as long-term refugia where a taxon survived repeated glaciations or as centers of origin. Pleistocene refugia "played a key role in structuring biodiversity, notably species richness and endemism, but also genetic diversity" by providing favorable niches during unfavorable climatic cycles, preventing extinction and favoring diversification (Hampe and Petit 2010).

There are three possible evolutionary results for a species when its widespread range is reduced in size into localized glacial refugia—species extinction due to competition, continuation without adaptive change, or evolution into new taxa (Knapp and Mallet 2003). Ice ages are long, and therefore temperate species can spend from 10 thousand to several hundred thousand years in a lower latitude refugium, and a region may serve as a refuge during more than one glaciation (Stewart et al. 2010). When populations are isolated in refugia, they may diverge evolutionarily. Speciation results from the establishment of reproductive barriers such as geographic features that isolate a portion of the population, which then evolves in genetic isolation. Refugia theory implies breaks in species distribution as range decreases, imposing geographical isolation (Knapp and Mallet 2003). Southern refugia in the northern hemisphere have been located nearer the equator and were often separated by great distances, setting the stage for evolutionary change. Populations living in differing plant communities are exposed to new ecological associations and interactions during rapid climate change and therefore undergo intense natural selection and adaptation; given 100,000 years or less of isolation, these populations can diverge into different species (Stewart et al. 2010).

Mountain ranges have played a key role as persistent, cumulative refugia during repeated cycles of glaciation and deglaciation; therefore, they have been regions of diversification and speciation. Populations from higher latitude regions thrived at lower latitudes by living at higher elevations, thereby finding an appropriate temperate climate

and open habitat suitable to their life cycles; they must have always done so as climate changed. Mountainous regions are characterized by steep altitudinal gradients, climatic variations between north- and south-facing slopes, weather exposure, and rain shadows; thus they offer a wide range of ecological niches within a relatively small area (Bennett and Provan 2008). Varied montane topography provides diverse ecological niches and possible refugia within a narrow altitudinal band allowing survival without great geographical displacement. The diversity of montane habitats also reinforces genetic isolation barriers leading to divergence over short distances and serves as an in situ "source and sink of genetic diversity" (Médail and Diadema 2009). Fragmented topography promotes division of species into isolated populations that may evolve independently until later contact during postglacial expansion (Hewitt 2000). In mountain refugia, bottlenecks may be reduced because altitudinal range shifts are of shorter distance and duration than long-distance range shifts, allowing more contact between progenitor and descendant populations (Comes and Kadereit 1998; Roberts 1998). This may be particularly important in the case of wind-pollinated plants such as *Cannabis* where the male genetic component may be dispersed over great distances, but the likelihood of effective pollination is greatly increased by proximity. Refugia within the Balkan, Caucasus, and Hengduan Mountain regions would have provided just such a favorable altitudinal gradient. It is in one or more of these refugia that *Cannabis* survived during the lengthy ice ages.

The Mediterranean Basin is one of the world's major biodiversity hotspots, resulting largely from the local presence of glacial refugia at the nexus of the European, Saharan, and Iranian biogeographical regions. The Mediterranean bioclimatic region, spread out both north and south of the Mediterranean Sea across southern Europe and northern Africa from Gibraltar to Turkey, hosts more than 30,000 plant species (Diadema and Médail 2009); this region has 52 likely refugia proposed as evolutionary hotspots, 25 within the Iberian, Italian, and Balkan Peninsulas and 33 in mountainous regions (Médail 2009; Médail and Diadema 2009). Several types of Mediterranean refugia have been characterized, including three that would have allowed temperate mesophilous and thermophilous (moderate and warm temperature) plants, including *Cannabis*, to survive glacial periods: (1) geographically widespread middle elevation (400 to 800 meters above sea level) moist regions; (2) more geographically limited gorges and sheltered river valleys with warm humid rocky and streamside microclimates; and (3) lower elevation valley bottom, coastal plain, and wetland areas (Médail and Diadema 2009).

In addition to the Iberian and Italian Peninsulas, the Balkans provided refugia for both Western and Eastern European species and "a place for further diversification and formation of new species further enhancing its richness in endemic[s]" (Bardy et al. 2008). While the icy Alps, and to a lesser degree the Pyrenees, presented barriers to northward expansion, the Carpathian Mountains form a discontinuous island system, separated by large forested and subalpine areas and, unlike the Alps, were only locally glaciated with lower slopes remaining ice-free offering favorable refugia for a wide variety of temperate plants (Ronikier and Intrabiodiv Consortium 2008). The Carpathian foothills along the Black Sea coast may have provided a corridor for post-LGM expansion of *Cannabis* from Balkan refugia.

In many ways, the life histories and evolution of perennial, long-lived trees are not analogous to annual herbaceous plants because their generation times are much longer, providing less chance for genetic drift. However, forests provide habitat for herbaceous species such as *Cannabis*, and the spread of forests is key to the understanding of dispersals of herbaceous plants having similar climatic requirements. Several tree species with life cycle and ecological preferences similar to *Cannabis* provide examples of adaptation to Mediterranean Pleistocene refugia.

Examples dicussed later include beech, oak, and chestnut. Paleobotanical and genetic data for European beech (*Fagus sylvatica*), a common thermophilous tree, indicate that southeastern France, Slovenia, and possibly southern Moravia, along with the Italian, Iberian, and Balkan Peninsulas, served as refugia during the LGM. Furthermore, evidence suggests that these refugia participated differentially in post-LGM recolonization of Europe, which was facilitated by mountain chains, and that diversity within present-day beech genomes was affected by multiple glacial-interglacial cycles since the mid-Pleistocene (Magri et al. 2006). Closely related Oriental beech (*F. orientalis*) ranges from northwest Turkey east to the Caucasus and Alborz Mountains of Iran. In the regions where *F. sylvatica* and *F. orientalis* overlap in Anatolia they hybridize freely (Comes and Kadereit 1998), much as *Cannabis* species do.

European deciduous oaks (*Quercus*) found glacial refuge in southern parts of the Iberian, Italian, and the Balkan Peninsulas (Comes and Kadereit 1998), first spreading from these refugia northward into the mountains of central Europe. By 6,000 years ago, as Holocene climate stabilized, they had expanded northwest across much of Europe (Brewer et al. 2002), while Asian oaks were restricted to survival on Taiwan Island (Cheng et al. 2005).

European sweet chestnut (*Castanea sativa*), a thermophilous and moisture requiring tree, has an original range and distribution that is masked by lengthy human impact, yet chestnut refugia have been identified in the Transcaucasian region, northwestern Anatolia, along the Apennine range in southern Italy, and the Iberian Peninsula. One common ecological feature links all presumed chestnut and oak refugia; all are located in low- to midelevation regions at the margins of mountain ranges.

These are regions that provided sufficient water and favorable microclimates where thermophilous plants could survive even during generally cool and dry glacial periods (Krebs et al. 2004). These regions also served as refugia for many other species during the LGM and previous glacial periods and may well have done so for *Cannabis sativa* NLH populations.

In addition to the main Mediterranean peninsulas (Iberian, Italian, and Balkan) with their mountainous topography, the Caucasus Mountains also provided Pleistocene refugia within diverse and isolated montane habitats "spanning large temperature and moisture gradients," thus promoting ecological diversity and resulting in about "6350 plant species . . . of which 1600 are endemic" (Russian Nature 2010). The Caucasus experienced more favorable Pleistocene and early Holocene climate and environment than most of Eurasia and has a long history of human occupation commencing in the middle Pleistocene. Agricultural settlements based on the cultivation of wheat, barley, oats, millet, vegetables, and stock breeding were established in Transcaucasia (south of Caucasia and northeast of Turkey) by 8,000 years ago and then spread into Caucasia (stretching from the Black Sea eastward to the Caspian Sea including the Caucasus Mountains); this led to early alteration of the natural landscape through deforestation and irrigation. By 5,000 years ago, the climate had become more arid, and pastoralism replaced agriculture in many areas. Prolonged human impact on soil cover, vegetation, and wildlife has been extensive, and much of the vegetation we see today "is in fact secondary successions formed over centuries of human activity" and is not indicative of earlier vegetation types (Russian Nature 2010). Sweet chestnut survived repeated northern glaciations in southern refugia, although present-day chestnut forests have a very restricted distribution in mixed deciduous forests, "confined to the sides of steep valleys in the foothills ringing the lowlands" (Connor et al. 2007).

While Eurasia was largely covered in steppe and tundra vegetation during the LGM, mixed coniferous and temperate deciduous forests survived in the Colchis region of lowland present-day western Georgia, below the southern slopes of the Caucasus Mountains facing the Black Sea. Colchis is primarily a vast alluvial lowland, largely below 100 meters elevation. It is broadly triangular in shape, spreading from the northwestern part of the Caucasus to northeastern Anatolia in Turkey and is bordered on the north by the Caucasus Mountains. Toward the east, it narrows and reaches the watershed between the Black and Caspian Sea basins, while the southern extent reaches Anatolia, and to the west lies the Black Sea. Colchic climate is oceanic with mild winters and warm summers, and Black Sea storms bring rains year-round. Water level changes in the Black Sea from the time of a proposed early Holocene deluge 7,550 years ago until recently have greatly influenced the Colchis and caused the formation of "extensive alder swamps and peat bogs." Colchis forms part of the Caucasian refugium and is perhaps the most important Pleistocene refuge for thermophilous plants in Eurasia, hosting 450 endemic plants including many Tertiary relicts and combining elements of many of today's extant floras. Highly dissected topography with numerous isolated gorges, altitudinal range from sea level to snowline, and mild climate contribute to the biodiversity of the Colchis flora (Kikvidze and Ohsawa 2001; Connor et al. 2007). Mid-Holocene Colchis forests were predominately chestnut with an understory of grape, along with hop and possibly hemp as indicated by *Cannabis*-type pollen grains (Connor et al. 2007).

Members of the walnut genus *Juglans* are thermophilous, light-demanding, and monoecious trees that may have survived in multiple refugia in southern Europe, Syria, Kyrgyzstan, China, and the Himalayas (Beer et al. 2008). Pollen evidence indicates the cultivation of walnut along with other species including hemp, hop, and flax. Alder buckthorn (*Frangula alnus*, also known as *Rhamnus frangula*) is a bird-dispersed shrub or small tree distributed over most of Europe and West Asia and is present in three major European refugia: the Balkan and Iberian Peninsulas and Anatolia. "Bird-mediated seed dispersal has apparently allowed not only a very rapid postglacial expansion of *F. alnus* but also subsequent regular seed exchanges between populations of the largely continuous species range in temperate Europe. In contrast, the disjunct *F. alnus* populations persisting in Mediterranean mountain ranges seem to have experienced little gene flow and have therefore accumulated a high degree of differentiation, even at short distances" (Hampe et al. 2003). This situation may be analogous to *Cannabis*—rapid expansion and regular seed exchanges facilitated by humans in addition to birds, with isolated pioneer populations exhibiting a high degree of differentiation and divergence.

Herbaceous plants also offer examples that pertain to the identification of refugia and early evolution of *Cannabis*. Campions (*Silene* spp.) provide an example of relatively recent divergence and speciation. *S. latifolia* and *S. dioica* arose when they were limited to isolated glacial refugia—*S. latifolia* on the Balkan or Iberian Peninsulas and *S. dioica* in the Balkan region or Caucasus Mountains (Harper 2009). Their common ancestor arose around 900,000 years ago and the two species diverged about 500,000 years ago consistent with Pleistocene glacial cycles. Following the LGM, deciduous forest communities spread northward from the Iberian and Balkan Peninsulas and the Caucasus allowing many herbaceous woodland species to follow into favorable niches for which they were preadapted. The two *Silene* species would have expanded from their separate refugia, *S. dioica* "spreading rapidly with the expansion of deciduous forests" into northern Europe, and *S. latifolia* following "considerably later with the spread of agriculture" 8,000 to 6,000 years ago, allowing secondary contact and hybridization although with little introgression. As agriculture spread into Europe, humans carried the seeds of plants that were unable to breach mountain or ocean barriers on their own to reach new natural habitats and human-disturbed niches preferred by *Silene* species (Harper 2009); this scenario also applies to *Cannabis*.

Several regions in Asia are also considered possible glacial refugia for northern temperate plants, and some of these may have supported *Cannabis* populations during glacial periods. For example, Western Hubei, adjacent portions of Sichuan and Chongqing in central China, Yunnan, southern Guizhou, as well as western Guangxi in southwestern China and contiguous mountainous regions of northern Myanmar, Thailand, and Vietnam, produced many endemic temperate genera (Yan et al. 2007).

The Himalayan range swings south toward its eastern end and, together with a number of parallel ridges extending southward from the Qinghai-Xizang Plateau, forms the Hengduan ("transverse" or "transecting") Mountain range. This highland region has a maximum elevation exceeding 6,000 meters (more than 9,000 feet), sloping steeply southward toward lower ground on the Yungui Plateau with heights above 2,000 meters at its southern extremity, separating the Bay of Bengal from the South China Sea. The Gaoligong Mountains compose the southwestern section of the Hengduan range including contiguous 4,000- to 5,000-meter ridges straddling 500 kilometers (310 miles) along the border between China's Yunnan province and Myanmar (Burma) and forming the drainage divide between the Nujiang River in China and the Irrawaddy River in Myanmar. The Hengduan Mountains were formed before the uplift of the Qinghai-Xizang Plateau and were again uplifted during formation of the Himalayas; today they remain a geological work in progress. Deep, narrow river gorges associated with active seismic faults—oriented north to south, with few south- or north-facing slopes—ameliorate the climate compared to other regions at similar latitude, and until relatively recently, the steep terrain dissuaded large-scale human habitation, thereby preserving native biodiversity (Chaplin 2005). As the mountain ranges tail off, they merge with the "less dissected country further east," forming the predominately karst limestone Yungui Plateau with smaller rivers flowing in relatively shallow valleys. The western half of the mountain range is characterized by larger rivers running north to south cutting mountain gorges more than 1,000 meters deep, and an extreme north to south slope gradient results in a pronounced temperature gradient "from permanent snow to temperatures perennially above 10° C [50° F]." The diverse topography of the Hengduan Mountains and Yungui Plateau forms a natural division between subtropical and temperate zones. As a result these areas were less affected by cold air masses from Siberia and therefore experienced a relatively stable Pleistocene climate. Much of this highly convoluted terrain remained warm and moist during glacial advances offering a key refuge for temperate plant communities with many chances for speciation; indeed, this region is considered a present-day biodiversity hotspot (Yan et al. 2007). Contemporary Yunnan province in China straddles these descending ridges, valleys, and plateaus at subtropical latitudes between about 21 degrees and 29 degrees north and ranges in altitude from less than 80 meters (260 feet) along the Hong (Red) River at the Vietnam border to mountain peaks more than 6,700 meters (22,000 feet) on the edge of the Qinghai-Xizang Plateau. Yunnan supports at least 15,000 plant species (more than half of all the vascular plants in China), including several ancient taxa (Shen et al. 2005) and 37 percent of China's endemic species (Li and Walker 1986). The Hengduan Mountains alone support more than 8,590 seed plant species, 32 percent of which are endemic (Zhang et al. 2009). Monsoon-driven summer rains, falling from May through October, account for 85 to 90 percent of precipitation, while the winter is clear, dry, and cool supporting widespread agriculture. Plants normally assigned to temperate Asian, Mediterranean, and Central Asian floras together make up only a small component (2.4 percent) of Yunnan's flora, and most are only important locally in nature, although wheat, peas (*Pisum*), and *Cannabis* are widely cultivated (Li and Walker 1986).

A wide range of substrate, elevation, latitude, and climate conditions gave "rise to considerable soil diversity, hence contributing to the region's biodiversity" (Chaplin 2005). High genetic diversity is attributed to massive geological uplifts during the development of the Himalayan and Qinghai-Xizang Plateau, which produced a dynamic flux of diverse topographies, microclimates, and soil types, and it is in this region that a number of ancient species were preserved during glacial periods (Wang et al. 2008b). The Hengduan Mountain floristic region borders the Yungui Plateau region to the south and the Tangut region to the north. To the south a wider variety of habitats combined with increased solar energy levels and limited ice coverage resulted in the evolution of a greater number of species (many of relatively recent origin) that relied on this region as a refuge during climate changes (Zhang et al. 2009). The region south of 29 degrees north occupies about 40 percent of the Hengduan range but contains more than 80 percent of all the seed plants (Zhang and Sun 2010). In addition, plant populations of the Qinghai-Xizang Plateau are genetically restricted compared to sister populations in the Hengduan Mountains, indicating that these species survived glaciations within the Hengduan Mountains, recolonized the Qinghai-Xizang Plateau to the north following the LGM, and were influenced by strong founder effects (Wang 2009).

North-south orientation of the Hengduan ranges facilitates the dispersal of plant species between higher and lower elevations and latitudes within the region along river valley corridors. Simultaneously, steep fractured topography discourages dispersal across mountain ridges as well as the development of broad rivers in an easterly or westerly direction; such conditions may have restricted the interglacial expansion of species, restricting populations within refugial boundaries (Chaplin 2005; Wang et al. 2008b). The main rivers are the Yangzi (Jinsha), flowing eastward to the Yellow Sea, the Mekong (Lancang), flowing southward to the South China Sea, and the Salween (Nujiang), flowing southwestward to

FIGURE 88. The Hengduan Mountain and Yungui Plateau biodiversity regions of southwestern China host a wide range of phenotypically variant feral and cultivated *Cannabis*. Plants of reduced foliage and stature (A) appear to grow wild in open areas. Feral populations (B) surround the Songzanlin Tibetan Buddhist monastery in northern Yunnan province. Seed fields (C) growing near Kunming exhibit great phenotypic diversity even within the same crop, and favored spontaneously growing plants (D) are often seen along village streets.

the Bay of Bengal (Li and Walker 1986). These huge rivers supply water to several major lowland cultures, facilitating agriculture and plant movement in and out of the Hengduan region whenever climate allowed.

As in Europe, the bulk of Asian paleoclimate data comes from tree pollen. Multiple refugia for plant species in southern China have been suggested by several studies, and long-term isolation within these refugia may have resulted in high genetic differentiation in several woody species that were more resistant to cold than *Cannabis*, although several thermophilous trees have also been studied (Shuang et al. 2008). One example is *Eurycorymbus cavaleriei* (Sapindaceae: soapberry family), an endangered, deciduous canopy tree growing to heights of about 20 meters (66 feet) in broad-leaved forests ranging in elevation from 300 to 1,400 meters (984 to 4593 feet) across southern China from Yunnan to Taiwan; this tree exhibits high genetic differentiation between populations resulting from "long-term range fragmentation and limited seed dispersal for the species," suggesting "multiple potential refugia across subtropical China" (Wang et al. 2009). Another example is *Dipentodon sinicus*, a small, deciduous tree that is the only generally recognized species in the family Dipentodontaceae that grows primarily in the southeastern Qinghai-Xizang Plateau, the Hengduan Mountains, and much of the Yungui Plateau as well as adjacent northern Myanmar (Burma), northeastern India, and northern Vietnam from 800 to 2,800 meters (2,625 to 9,186 feet) above sea level. Restricted genetic diversity in this species is attributed to inefficient and localized seed dispersal, isolation resulting from rapid mountain building, subsequent genetic drift, and ecological adaptation to local environments. *Dipentodon* exhibits greater genetic diversity in the more uplifted southeastern Qinghai-Xizang Plateau than the less uplifted Yungui Plateau, indicating that uplift has encouraged diversification (Yuan et al. 2008). We suggest that *Cannabis* followed a similar evolutionary path within this region of southern China.

Chinkapin (*Castanopsis hystrix*) is a subtropical broad-leaved evergreen, wind-pollinated, and monoecious tree found on moist, acid to neutral, medium (loamy) to heavy (clay) soils in open woodlands, spread across the eastern Himalayas into southern China ranging up to 2,400 meters (7,874 feet) elevation. Much as with walnut in Europe (discussed earlier), human activities played an important role in shaping the present distribution pattern of chinkapin trees (Li et al. 2007), as well as with dioecious, wind-pollinated "living fossil" ginko (*Ginkgo biloba*; see Gong et al. 2008); this conclusion is supported by the small populations of nonindigenous haplotypes appearing at great geographical distance from regions of high haplotype frequency. Chinkapin were originally planted for lumber production and were often preserved for ritual "*fengshui*" geomantic reasons as were ginkgo trees, although the glacial refugia of both species probably included southwestern China.

Coniferous trees also found glacial refuge in southwestern China. For example, Himalayan hemlock (*Tsuga dumosa*: Pinaceae or pine family) is a monoecious, wind-pollinated lumber tree native to the eastern Himalaya and Hengduan regions with an adaptation to cold climate and high rainfall and humidity at an elevation of 2,000 to 3,600 meters (6,562 to 11,811 feet) above sea level. Before the LGM, hemlocks recolonized the Himalaya range dispersing westward from the Hengduan Mountains. The fossil record indicates that this was accompanied by serial genetic bottlenecks with strong founder effects, which preceded waves of rapid population expansion, resulting in decreasing genetic diversity along the Himalaya Mountains from east to west. Hemlock probably survived the LGM along the southern slope of the Himalayas, possibly in the Kathmandu Valley. The range of hemlocks before the Quaternary glaciations may have been approximately the same as today, indicating that the influence of earlier glaciations on population histories could be much stronger than that of the LGM (Cun and Wang 2010). *Cannabis* may have been widespread across Eurasia before the Pleistocene with some of its taxa now extinct. The case of such trees provides some insight, as we discuss later.

Larches (*Larix*: Pinaceae or pine family) form a genus of deciduous conifers and are dominant members of northern Eurasian subarctic forests. Two Eurasian sections of this genus were isolated early on and evolved independently in two refugia—southern European and eastern Himalayan—followed by rapid expansion with accompanying founder effects, recolonization, and morphological divergence. Some of these larch species are endemic to northwestern Yunnan and southeastern Tibet, while others are confined to the eastern Himalayas, southern and eastern Tibet, or Sichuan, and, in some cases, these species have evolved local biotypes or subspecies that, in regions of overlap, grow in the same stands and may hybridize. Populations expanding from southern European refugia met with more persistent ice cover, while those expanding into partially glaciated northern Siberia and the Russian Far East were less restricted (Wei and Wang 2004). The evolution of *Cannabis* has resulted in a similar situation to that of larch trees, with two Eurasian taxa subdivided into several independently evolved subtaxa. *Cannabis*'s evolutionary path may have many parallels with larches, such as two (or more) independent refugia, rapid recolonization accompanied by founder effects, ecological adaptation, and morphological divergence.

The life cycles and ecological preferences of several herbaceous plants also give us insight into the establishment and preservation of refugial populations and their expansion during warm periods. For example, thale cress (*Arabidopsis thaliana*: Brassicaceae or mustard family) is a European and Central Asian native that is highly successful at colonizing human-disturbed landscapes and consequently is naturalized worldwide. It has been hypothesized that the Yangzi River populations of thale cress originated from a common ancestor and may have dispersed eastward from the Himalayas approximately 90,000 years ago, with Altai Mountain populations surviving in a local refugium during late Pleistocene glaciations (Yin et al. 2010). Present-day European populations of *A. thaliana* are descendants of populations that expanded by natural colonization from an Asian refuge around 10,000 years ago with a westward spread of almost a kilometer (0.62 mile) per year, and more fit populations may have replaced previously established ones. Dispersal of *A. thaliana* into Europe was most likely facilitated by the spread of agriculture (François et al. 2008). The genus *Capsella* (Brassicaceae) is closely related to *Arabidopsis* and contains two self-incompatible diploid species, *Capsella grandiflora* and *C. rubella*. *C. rubella* diverged from *C. grandiflora* 50,000 to 30,000 years ago via an extreme bottleneck, and self-incompatibility broke down at the same time, possibly via a single self-compatible individual. The new species of *Capsella* probably originated in Greece and was dispersed around the Mediterranean Basin concurrent with the spread of agriculture (Guo et al. 2008).

During the Pleistocene, several herbaceous genera found refuge in the Hengduan Mountain region and radiated from there during interglacial periods such as the Holocene, for example, *Pedicularis* (Orobanchaceae), moisture-loving root parasites with many Asian species (Yang et al. 2008); *Aconitum* (Ranunculaceae), perennials favoring mountain meadows; snow lotus (*Saussurea*), widespread in the Himalayas; *Gentiana* (Gentianaceae), a cosmopolitan group favoring alpine temperate climates and full sun (Wei and Wang 2004); and Primula (*Primulaceae*), which prefers filtered light and are adapted to subalpine climates (Wang et al. 2008a/b).

Across much of the Hengduan-Yungui region, forests of subtropical evergreen broadleaved and pine trees occur in limited areas at from 1,000 to 2,800 meters (3,280 to 9,190 feet) above sea level. It is in this region that habitats suitable for a wide diversity of present-day *Cannabis* populations can be found within remnant patches of spontaneous vegetation, including subtropical evergreen, deciduous broad-leaved and mixed deciduous-evergreen broad-leaved/conifer forests, as well as evergreen and mixed deciduous-evergreen shrublands, and lower-altitude montane conifer forests and shrublands (Li and Walker 1986). The Hengduan Mountain and Yungui Plateau region likely hosted early Pleistocene *Cannabis* refugia, and as the climate warmed, *Cannabis* expanded from this region westward along the Himalayan foothills and eastward into the East Asian lowland valleys. Today Yunnan harbors many phenotypically diverse *Cannabis* populations growing naturally at 1,000 to 3,000 meters (3,280 to 9,840 feet) elevation. Gross phenotypes range from narrow-leaf biotypes of short stature and early maturation in the north to broadleaf central biotypes of immense stature resembling fiber and seed landraces. Cannabinoid (compounds unique to *Cannabis*) production exhibits a continuous range from almost entirely CBD with little THC to almost entirely THC and little if any CBD (although absolute cannabinoid percentages are relatively low compared to selected drug varieties). Terpene content also varies widely producing most of the characteristic *Cannabis indica* NLD, BLD, and BLH aromas (Clarke 1993–2004 and Watson 2003–2005, personal observations). Such a relatively broad diversity of phenotypes supports designation of the Hengduan-Yungui region as a Pleistocene refuge and area of broad ecological adaptation.

Not all temperate forest populations became extinct at northern latitudes during the LGM; indeed, there is evidence that temperate glacial forest refugia were present on the Shandong Peninsula in northeastern China, the Korean Peninsula, and Japan Archipelago. The temperate-deciduous Manchurian walnut tree (*Juglans mandshurica*) grows in northern and northeastern China, Japan, and Korea. Genomic data indicates that it survived the LGM in two refugia in northwestern China and that it did not move south of 30 degrees north latitude as predicted for many temperate trees (Bai et al. 2010). Archeological Cannabaceae (*Cannabis* or *Humulus*) pollen finds from northeastern Asia dated from 6,000 to 9,000 years ago indicate that Cannabaceae may have survived the LGM there, and possibly following the LGM, BLH *Cannabis* populations spread across eastern Asia from more northern refugia rather than from the Hengduan-Yungui refugium. *C. indica* ssp. *chinensis* BLH ecotypes may have evolved from ancestors radiating from a northeastern Asian refugium. In this case we have three possible Pleistocene refugia—the Caucasus region for NLH, the Hengduan Mountains and Yungui Plateau for NLD, and perhaps northeastern Asia for BLH, while BLD evolved later from an NLD population isolated in the Hindu Kush Mountains.

Postglacial Population Expansion

As we have discussed earlier, postglacial warming often proceeded quickly, leading to rapid expansion of populations along the northern limits of refugia into large areas of favorable habitat. This occurred via long-distance dispersal agents setting up new colonies and rapidly expanding before the arrival of competing taxa, with only a limited number of pioneer plants giving rise to new populations: "This would be repeated many times over a long colonizing route, and these founding events would lead to loss of alleles and

homozygosity" (Hewitt 2000). Variables such as differences in species dispersal capacities, variation in the geographic locations of refugia or the existence of barriers, and the complex processes of genetic structuring occurring during population expansions make it "difficult to generalize about the mode and tempo of recolonization of genetic lineages" (Médail and Diadema 2009). Range shifts, expansions, and contractions occurred "at various times, at different rates, and in various directions, and therefore plant assemblages were formed repeatedly and subsequently disassembled. Successful population expansion is linked to effective rate of migration [dispersal] based on biological parameters such as reproduction mode, ability to compete with other species, and resistance to extinction" (Comes and Kadereit 1998). Alleles trapped behind the leading edge of expansion result in two possible outcomes, either the genome remains more diverse within refugia than in derived populations or, if the refugium climate changes and becomes unfavorable, the refugial population shrinks, becomes divided, and genetic diversity is severely reduced (Hewitt 2000). Different biotypes exhibited differing patterns of population expansion during interglacial periods, and some became extinct (Bennett and Provan 2008). Expansive radiation of geographically restricted gene pools began anew as climate warmed, and deserted refugia may have been recolonized during a later glacial period. As a result of glacial-interglacial range shifts, exponential population expansion, and founder effects, northern populations near the limits of northward recolonization are usually less genetically diverse, while southern populations often survived several glaciations in persistent refugia, offering ample time for further genetic divergence leading ultimately to speciation (Hewitt 2000).

Although climate has been the main force driving changes in distribution and abundance during glacial–interglacial transitions, some plants such as *Cannabis* have reached maximum range as a result of human activity. Human transport of seeds overcame geographical barriers to plant dispersal, and increasingly widespread anthropogenic landscapes provided newly disturbed niches for colonization without much competition from other species and thereby promoted rapid population expansion. Some plants (predominately herbaceous taxa) appear in the late Pleistocene fossil record and disappear during the early Holocene, only to reappear in larger numbers in the later Holocene: "These are all examples of taxa that have been present but scarce throughout the Quaternary, but have increased to be common after human activity caused reduction of forests and modified upland habitats" (Bennett and Provan 2008). *Cannabis* fits this scenario well as it also has a scanty fossil record before the middle Holocene, and its abundance increased rapidly as human activity increased the frequency of disturbed habitats favorable for its proliferation.

Successful invasion of new habitats and niches progresses in two phases. Initially, a new habitat is colonized predominately by marginally adapted individuals. The minorities that are preadapted to the new ecological conditions will survive, reproduce, and form a persistent population. Subsequently, populations adapt under strong selection pressure, "allowing them to persist where progenitors struggled, and prosper in the face of a somewhat different package of competitors, herbivores, pathogens, and mutualists than their progenitors encountered" (Levin 2004). Populations that fail to adapt to change diminish and eventually perish. Adaptive change occurs over many generations, and speciation is completed "once the new suite of adaptations has been assembled, resulting in the ecological isolation of the new entity from its progenitor." Newly evolved and highly adapted, divergent populations are small, and longevity is only assured once they multiply (Levin 2004).

As long as expanding animal-pollinated plant populations are accompanied by their obligate pollinators, they can spread more efficiently with an initially small population size. The same holds true for self-pollinators. Wind-pollinated, dioecious *Cannabis* requires a relatively larger population size to simply reproduce successfully and produce enough seed and plants for the next generation. However, because *Cannabis* produces copious pollen and seed, fewer individuals are required to ensure population stability and promote expansion. Large population size is most important to curtail inbreeding and ensure genetic diversity in subsequent generations.

Speciation Rate

When and how quickly did new species and subspecies arise within *Cannabis*? Genetic divergence within a species, which first leads to phenotypically distinct biotypes, then to subspecies, and eventually new species, can occur at any time during its evolution; however, many estimates of divergence times (largely ecotype divergence rather than speciation events) fall within the Quaternary, especially clustered "near the Pleistocene-Holocene boundary" (Comes and Kadereit 1998) around 12,000 years ago when abrupt climatic changes were of considerable magnitude.

But timing of proposed events leading to divergence of lineages varies widely. For example, northern and southern allozyme lineages (variant forms of an enzyme coded by different alleles at the same locus) of the tulip tree (*Liriodendron tulipifera*), which is native to eastern North America, diverged about 1.5 million years ago; in Europe, separate genetic lineages of sessile oak (*Quercus petraea*) diverged approximately 85,000 BP, although morphological differences remain small. Coniferous European spruce (*Picea abies*) lineages are thought to have diverged approximately 40,000 BP, while "northern and southern lineages of Rocky Mountain lodgepole pine (*Pinus contorta* ssp. *latifolia*) from western North America have differentiated only during the last 12,000 years" (Comes and Kadereit 1998).

Recent evolutionary divergence has greater probability among short-lived herbaceous plants such as *Cannabis* that pass through more reproductive cycles in a given period of time than longer-lived species, thus allowing for more recombination and increased genetic drift. Evidence for rapid speciation is provided by maritime goldfields (*Lasthenia maritima*), an annual herbaceous, flowering species that diverged from coastal goldfields (*L. minor*) 15,000 to 10,000 years ago; this occurred when islands and offshore rocks inhabited by seabirds and *L. maritima* were cut off from the mainland following a rise in sea level as Holocene warming began. *L. maritima* differs from *L. minor* in various characters, including breeding system, edaphic tolerance, and fruit morphology, all of which evolved in adaptation to the seabird rocks (Comes and Kadereit 1998).

Cannabis has remained diploid, and polyploidy therefore does not account for speciation. Speciation in *Cannabis* likely occurred in the Pleistocene during glacial periods when temperate populations were restricted to more southerly refugia and then expanded as climate warmed. With

FIGURE 89. Wild hop (*Humunlus scandens*) and feral hemp (*Cannabis* sp.) occupy the same environmental niche in temperate continental Shandong province, China.

rising temperatures, vast tracts of land became suitable for *Cannabis*, and it expanded rapidly into new niches in suitable habitats. Then founder effects combined with ecological pressures induced speciation to form the two or three species we recognize as well as others that have become extinct. *C. ruderalis* may be the PA of the narrow-leaf hemp ancestor (NLHA) and NLH, although it is more likely that it should be classified as NLHA. It appears that *C. ruderalis* is not the ancestor of NLD, BLD, or BLH biotypes, and the PA of present-day *Cannabis* species is most likely extinct. By the last glacial period, leading up to the LGM, the range of NLHA and NLH biotypes of *Cannabis sativa* were reduced to refugia in the Caucasus Mountains and possibly also the Balkan Peninsula. *Cannabis indica* biotypes found refuge in the Himalayan and Hengduan-Yungui region. BLD and NLD biotypes moved westward along the Himalayan foothills, with BLD biotypes adapting to the arid and cold Hindu Kush Mountains of present-day Afghanistan, while the NLD biotypes spread across the Himalayan foothills from present-day northern Myanmar to northern India. BLH biotypes likely remained in the Hengduan-Yungui refugium during the LGM and may also have survived in favorable microclimates near the coast on the Shandong Peninsula in eastern China, the Korean Peninsula, and/or the Japanese Archipelago.

FIGURE 90. Wild *Cannabis* plants are often very short and compact when young and resemble young *Humulus* plants. In this example from Tai Shan Mountain in Shandong province, China, the branches are longer than the plant is tall.

As Holocene warming commenced, populations spread from refugia and quickly recolonized northern territories. It is during this Holocene expansion that *Cannabis* passed through founder effects and adapted to a wide range of habitats and niches, evolving into many *naturally* evolved ecotypes. Although humans may have played a role in dispersing *Cannabis* during the Pleistocene, we have no supporting evidence. On the other hand, human expansion also began anew during the early Holocene with human population growing and agriculture eventually spreading to many new regions after its initial development. We believe that it was during this early Holocene period that humans transported *Cannabis* to several parts of Eurasia—to areas that had favorable environmental conditions where people began to select these plants for a wide range of uses. In the next section we will explore close relatives within Cannabaceae, investigate evolution of their shared traits and offer hypotheses concerning their early evolution.

Early Evolution of Cannabaceae: The Hemp and Hop Family

Cannabis sativa L. and *Humulus* [*scandens*] are both annual, widespread, and aggressive weeds, and although variable, are on the whole relatively poorly differentiated on a geographical basis. There are pronounced similarities in the propagules of these two species, with the achenes [seeds] being enveloped by relatively small bracts (shed at maturity) which apparently have a protective rather than a dispersal function. Although these achenes have no special adaptations for long distance dispersal, they nevertheless do appear to be easily transported by water, rodents, birds and man. *Humulus yunnanensis* is a localized endemic, whose longevity has not been established. *Humulus lupulus* is a perennial, propagates extensively in situ by underground rhizomes, [and] is not particularly aggressive as a weed . . . Thus it would appear that *C. sativa* and *H.* [*scandens*] have "general purpose genotypes", flexibly adapted to colonization and cosmopolitanism, and tend to localized habitation and differentiation. (SMALL 1978)

Cannabaceae consists of two closely related genera—*Cannabis* or hemp with two extant species, *C. sativa* and *C. indica* and one putative species *C. ruderalis* (all with diploid chromosome number or 2n = 20) while *Humulus* or hop has three species, *H. lupulus* (2n = 20), *H. scandens* (2n = 16), and *H. yunnanensis* (2n = unknown). Understanding the evolution of Cannabaceae as a whole, and hop species in particular, sheds light on the evolution and spread of hop's closest relative, hemp. For example, *Cannabis* and *Humulus* are annual plants, and both produce sufficient seed under proper conditions, but *Humulus* varies significantly in that it overwinters by persistent roots and buds in the spring to produce new vines. Ancient humans seized upon this characteristic and were able to spread novel individuals easily by vegetative propagation.

All Cannabaceae species are heliotropic, dioecious, and anemophilous. As a result, *Cannabis* and *Humulus* favor open sunny habitats where wind pollination is facilitated. *Cannabis* likely evolved its strongly upright growth habit, extensive branching and reduced foliage in similar open habitats, most likely mixed open woodland and woodland steppe vegetation zones. *Humulus* vines require wind pollination as well but achieve exposure to wind (both for pollen dispersal and for trapping pollen) by vining up neighboring vegetation. As *Humulus* is also shade intolerant, it likely evolved along sunny exposed woodland fringes, in disturbed areas and along waterways, in exactly the same sort of open niche favored by self-sown *Cannabis* today.

Closely related *Cannabis* and *Humulus* are genetically quite similar and obvious morphological differences such as growth habit (erect vs. vining), leaf form (hand-shaped vs. webbed or heart-shaped), and female inflorescence form (compact elongated raceme flower cluster vs. cone-shaped strobilus), as well as divergent secondary metabolite chemistry (oxygen-containing aromatic cannabinoids vs. bitter alpha-acids), are genetically controlled by simple allelic inheritance. These traits most likely evolved through genetic drift in an ecological niche favoring mutant phenotypes. *Cannabis* occasionally exhibits atavistic (ancient ancestral) Cannabaceae traits such as webbed leaves like hop and female inflorescences with nested bracts and a very high bract to leaflet ratio resembling hop cones (Clarke, personal observations). Apart from these simple differences *Cannabis* and *Humulus* generally share relatively rare evolutionary states in plants but common to all Cannabaceae—in addition to being anemophilous and dioecious, both have X and Y sexual determination.

All three *Humulus* species are herbaceous vines with hooked climbing hairs. As they climb, the majority of vining plant species form right-handed (counterclockwise) helicies, but a handful, including hop and yam (*Dioscorea* spp.), are left-handed. *Humulus* employs hooked trichomes (eglandular or nonsecretory hairs) to attach to substrate as it climbs using its paired stipules at the base of each leaf petiole to create opposing tension between stem and climbed surface, further securing the vine to its support (Isnard and Silk 2009); this is a highly specialized adaptation. Since both *Cannabis* and *Humulus* likely evolved in sunny open environments, erect habit probably appeared first, and vining evolved as a later adaptation to a more specialized niche—woodland fringe. Possibly vining (or *Humulus* itself) evolved as a result of competition with *Cannabis*, a closely related and aggressive colonizer of their mutually favorable habitat.

Female *Humulus* inflorescences form a strobilus or cone, and *H. lupulus* cones grown seedless in the absence of male pollinators are the commercial hops used as a bittering agent and preservative in beer. Yunnan hop, or *H. yunnanensis*, is a dioecious perennial that also reproduces by seed and is native to the Hengduan Mountains and Yungui Plateau of

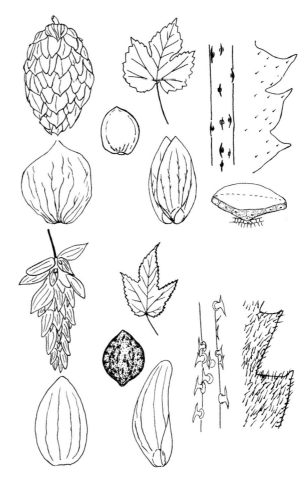

FIGURE 91. Inflorescence, bract, leaf, seed, and gland anatomy of *Cannabis* (2n = 20, top) and its closest relative East Asian *Humulus scandens* (2n = 16, bottom; adapted in part from Small 1978).

FIGURE 92. Highly derived inflorescence, bract, leaf, seed, and gland anatomy of European and North American brewer's hop *Humulus lupulus* (2n = 20, top) and wild Chinese *H. yunnanensis* (2n = ?, bottom; adapted in part from Small 1978).

southwestern China. Although the leaves of *H. yunnanensis* are less lobed and cones larger than those of *H. scandens*, in general, they resemble those of *H. scandens* more than those of *H. lupulus*. *H. yunnanensis* is distinguished from the other *Humulus* species by "the stiffness of the climbing hairs, pubescent leaf surfaces, small leaf glands, distribution of the trichomes and large pollen grains" and, since *H. yunnanensis* produces only a few secretory glands on the cones and low alpha-acid levels, it is not used in brewing. Japanese hop, or *H. scandens*, is a dioecious, annual native to Asia where it is now widespread as an aggressive, invasive weed. It is characterized by large, seven-lobed leaves, long internodes, extremely strong hooked climbing hairs, and small cones with few glands, so it is also not used in brewing (NCGR 2010). *H. scandens*, which seems to be a more primitive relative of the cultivated *H. lupulus*, has smaller female bracts that surround the fruits much more closely as in *Cannabis*, yet its seeds fall easily from the cones and are freely dispersed. *H. scandens* has a general purpose genotype that adapts easily to a wide variety of environmental conditions, whereas *H. yunnanensis* and *H. lupulus* are more narrowly adapted through propagation by rhizomes and less durable seeds that limit their dispersal to locations relatively near the mother plant (Small 1978). Grudzinskaya (1988) classified *H. scandens* as *Humulopsis scandens* and proposed it as the ancestor genus of *Humulus*, thereby designating *H. scandens* (*Humulopsis scandens*) as the ancestor of both *H. yunnanensis* and *H. lupulus*. Mukherjee et al. (2008) point out that Chinese *Cannabis* and East Asian wild *H. scandens* are closely related based on DNA studies and graft compatibility indicating gene transfer between them in ancient times, while *H. lupulus* appears to result from a differing lineage. Brewers hop, or *H. lupulus*, is a perennial that under cultivation regrows each spring from underground rhizomes and

in natural conditions also spreads by seed. Mature cones of female *H. lupulus* range in length from about 1 to 10 centimeters (two-fifths inch to 4 inches) and are made up of papery green bracts covered with tiny glands that secrete sticky yellow resin containing bitter alpha-acid compounds valued in brewing beer. Only *H. lupulus* is cultivated, and *H. lupulus* cultivars are limited in their morphological variation in comparison with *Cannabis* cultivars and have a narrower genetic base. This may result from historically early asexual propagation of only a few phenotypes favored for medicinal value and eventually brewing, contrasted with the hybrid ancestry of most *Cannabis* cultivars, which naturally reproduce sexually, resulting in greater diversity within the genus (see Chapter 10). Since cultivated *H. lupulus* was traditionally propagated vegetatively, seeds were not so important in its evolution through artificial selection.

Support for the evolutionary origin of *H. lupulus* (indeed *Humulus* as a whole) in present-day China is provided by the presence of the two other hop species *H. scandens* and *H. yunnanensis* in eastern Asia but not Europe and North America. American and Asian hops began to diverge between 700,000 to 500,000 years ago, and long-distance dispersal of *H. lupulus* into North America (rather than range restriction caused by a vicariance event such as glaciation leading to geographical isolation) is a more likely explanation of its wide range.

FIGURE 93. Wild *Humulus scandens* male flowers (left) look much the same as *Cannabis*, while the female bracts form compact, leafless, cone-like strobili (right).

Present-day northern Chinese feral *H. lupulus* genomes reflect hybridization between Chinese and European lineages resulting from the import of female hop clones from Europe. "Introgression between these introduced cultivars and a wild native population may have produced a haplotype (a genetic combination inherited as a unit) having a mixture of European nuclear and Chinese cytoplasmic DNA." Wide distribution of European hop from Portugal to the Altai region of Central Asia "suggests that migration, or gene flow, occurred across Eurasia at a high rate, similar to those for European trees." *H. lupulus* populations in the Caucasus region are more genetically diverse than European populations; therefore, the Caucasus region may have served as a refugium during glaciations. After the LGM, *H. lupulus* spread from its southern refugia, and as no founder effect can be identified, population expansion "may have occurred naturally rather than through human association." Hop cultivars share alleles with native European *Humulus* populations, indicating that "traditional cultivars were simply selected from natural wild populations, presumably because of their superior qualities for beer production" and were thereafter vegetatively propagated (Murakami et al. 2006).

Faeli et al. (1996) used RAPD analysis of industrial hemp cultivars to identify three distinct gene pools, one originating in Italy, another from Hungary, and a third from outside of Europe (Korea). These groupings reflect their breeding histories. Traditional Italian landrace hemp varieties are likely derived solely from European NLH, while many Hungarian cultivars were developed from hybrid crosses between European NLH and Asian (Chinese) BLH landrace varieties, and the Korean accession likely represents an unhybridized East Asian BLH landrace variety. Therefore, the base pair sequences identified by Faeli et al. (1996) reflect the accession's origins within differing taxa as described by Hillig (2005b).

Remarkably well-preserved, 2,700-year-old *Cannabis* remains recovered from one of the ancient Yanghai tombs in Xinjiang province, China, have provided a wealth of genomic data of taxonomic and evolutionary significance (see Chapter 4, 6, and 10 for more detailed discussion regarding the ethnobotanical and taxonomic significance of this archeological discovery). Suffice to note here again that information collected from the Yangtai tomb remains confirm that humans had discovered the psychoactive nature of *Cannabis* and had selectively

bred drug varieties at least by the first millennium BCE (see Jiang et al. 2006; Russo et al. 2008); we believe these human-Cannabis relationships began much earlier in Holocene, if not in Pleistocene times.

Both *Cannabis* and *Humulus* evolved within independent refugia. This led to separate Asian and European species and was followed by post-LGM expansions and establishment of local biotypes. More recently both *Cannabis* and *Humulus* were subjected to hybridization and selection by plant breeders. *H. yunnanensis* may have remained within the Hengduan-Yungui region as we have hypothesized for *C. indica* drug biotypes (narrow-leaf drug ancestor or NLDA, NLD, and BLD), while *H. scandens* may have found refuge in northeastern China, Korea, and Japan. In that case, *H. scandens* may have spread from the Hengduan-Yungui region following the LGM as we propose for *C. indica* BLH biotypes, and *H. lupulus* could have spread across Europe and Central Asia from its refugium in the Caucasus Mountains as we propose for *C. sativa* NLHA and NLH biotypes.

Hop plant breeders pursued differing goals in response to diverse brewing styles; as a result, *H. lupulus* cultivars form two taxonomic groups reflecting geographical origin and usage—one of purely "European origin representing the aroma pool" and another with "European germplasm infiltrated by wild American genes." These two taxonomic groups of *H. lupulus* are further subdivided into five varieties based on morphological characters. North American varieties have "less aroma quality, but a higher bittering potential" (Seefelder et al. 2000), and early English hop breeders noted their high resin content, describing them as "rich with glands and buttery to touch" (NCGR 2010). Cultivated hop followed much the same evolutionary path we propose for *Cannabis*—initial rapid genome expansion via seed followed by asexual clonal reproduction—although *Cannabis* only reached the asexual reproduction stage in the late twentieth century as marijuana production moved indoors (see Chapter 10).

In sum, Cannabaceae is a small plant family made up of two important genera, *Cannabis* and *Humulus*, both with very long ethnobotanical associations. Based on evolutionary and taxonomic principles we assume that *Cannabis* and *Humulus* share a common ancestor. At some time before contact with early humans, *Cannabis* evolved into a genus

FIGURE 94. Cultivated *Humulus lupulus* female strobili or brewer's hop cones are much larger and more abundant than those of wild *H. scandens*.

with several traits that distinguished it from its closest relative *Humulus*. Unfortunately, early Cannabaceae ancestors are not yet found in the fossil record, so we have neither physical nor historical evidence of any of the intermediate evolutionary stages leading to what we today know as *Cannabis*. However, some of the general hypotheses of angiosperm (flowering plant) evolution, and an investigation of plant families with similar reproductive strategies and ecological requirements, help us determine in which geographical region *Cannabis* was likely to have evolved and what the progenitor of modern *Cannabis* may have looked like. A key issue involves the determination of when, where, and how the two *Cannabis* species *sativa* and *indica* diverged and evolved.

Breeding Systems and Reproductive Strategies as Clues to Geographical Origin

Breeding systems are important, and often neglected, aspects of the natural biology of organisms, affecting homozygosity and thus many aspects of their biology, including levels and patterns of genetic diversity and genome evolution. (CHARLESWORTH 2006)

In addition to matching paleoclimate reconstructions with *Cannabis*'s ecological requirements, and drawing comparisons

with the evolution of similar species, understanding breeding systems and reproductive strategies also gives us insight into where *Cannabis* originated and clues to its early evolution. Angiospermy (being a flowering plant), annuality (flowering once a year), anemophily (wind pollination), and dioecy (producing unisexual male and female plants) all have natural environmental limitations for maximum effectiveness. Correlations between reproductive strategies and geographical distributions are proposed for many plant taxa and prove useful when speculating about origins of less studied plant groups, such as *Cannabis*. As basic reproductive traits likely evolved very early on, it is important for us to study evolution of Cannabaceae as a whole.

ANGIOSPERMY

Although angiosperms are often reported as originating around 140 to 130 million years ago (Axelrod 1974; Stebbins 1976), more recent research suggests they began to evolve much earlier, perhaps in the Early to Middle Jurassic, 179 to 158 million years ago (Wikstrom et al. 2001), or they may even have begun to evolve as early as 400 million years ago (Benton and Ayala 2003). Disagreements have developed as a result of how researchers define "flowering" plants versus other "seed" plants. Nevertheless, evolutionary taxonomists generally argue in favor of an arid upland habitat as the site of origin for angiosperms. Such an upland area with its highly dissected topography would have offered steep environmental selection gradients; they also would have allowed opportunities for geographic isolation by drought, flood, or other climatic or geological factors. Seasonal variations in temperature and rainfall may have helped select for the annual life cycle of many angiosperms. Takhtajan (1987) suggested that angiosperms originated in Southeast Asia and radiated from there; indeed the earliest purported angiosperm fossils have been recovered from East Asia (Raven et al. 2005).

Four evolutionary trends are shared by a majority of angiosperm families: flower parts became reduced and less variable in number, the number of whorls decreased and the axis shortened as floral parts fused, ovaries became inferior in position to stamens (in hermaphrodite flowers with both sexes) while the perianth differentiated into calyx and corolla, and radial symmetry replaced bilaterally symmetric floral forms (Raven et al. 2005). In Cannabaceae (both hemp and hop), flower parts became quite reduced and fused in pistillate (female or seed) flowers (single ovule enveloped by a subtending bilaterally symmetrical green bract), the perianth is reduced to fruit or "seed" coat, and there is no elaborate calyx nor corolla; on the other hand, staminate (male or pollen) flowers have reduced green corolla and preserve their ancient radial symmetry. In some respects, Cannabaceae appear to be both evolutionarily advanced (reduced and fused flower parts) and primitive (bilaterally symmetrical pistillate flowers, no elaboration of perianth), possibly remaining anemophilous since early evolution rather than coevolving biotic (animal-mediated) pollination relationships and then returning to anemophily as is hypothesized for other wind-pollinated plants. Flowers of Cannabaceae remained simple without showy perianths as there was no need to attract pollinators. Obviously this reproductive strategy has proven effective as Cannabaceae thrive in naturally open and generally disturbed agricultural settings worldwide.

Abiotic pollination proved evolutionarily successful in mosses, ferns, gymnosperms, and other plants long before flowers evolved. The earliest seed-bearing plants produced unisexual male and female organs on separate structures and reproduced passively and abiotically, releasing large amounts of pollen into the wind to reach receptive ovules by random chance. Ovules were borne on leaves (similar to floral bracts in *Cannabis*) or in cones (similar to strobili in *Humulus*) and exuded a sticky substance to trap pollen grains. Animals began to feed on reproductive organs and inadvertently transferred pollen to other flowers. Biotic pollination evolved independently many times, and this is explained by increased efficiency (fewer pollen grains actively directed to ovules) offered by animal-mediated pollination (Raven et al. 2005), although wind pollination may be equally efficient.

ANNUALITY

As latitude increases toward the poles, day length becomes more variable, the tropics having summer and winter days and nights of nearly the same length, while in the Arctic summer days and winter nights last 24 hours. In temperate or higher latitude regions, winters are colder and temperate zone annual plants like *Cannabis* must finish their life cycle each autumn. The tropics are consistently warmer than temperate regions and seasonality imposes less selective pressure. As plants moved northward from tropical latitudes with little annual variation in day length into temperate latitudes with increasingly wide seasonal day length variations, they evolved an annual life cycle with short-day flowering and autumn seed set to circumvent cold winters. Closer to the equator where climate is more constant, annuality is of little adaptive significance.

It is most likely that annuality and short-day photoperiodism originally evolved from perennial day-neutral characteristics as the range of the first angiosperms spread northward from tropical to temperate regions, and these traits would have evolved in Cannabaceae ancestors long before Cannabaceae evolved. Cannabaceae are not cold tolerant plants and often have difficulty surviving frosts. This trait was not a problem as long as they could find refuge from the advancing glacial ice by shifting their range southward. It is during these recurrent range changes that *Cannabis* and *Humulus* diverged and later evolved into separate species within each genus. In its original home in Central Asia, *Cannabis* had already evolved into an autumn-flowering, short-day plant in adaptive response to shortened seasons at higher temperate latitudes. Thus it was preadapted for rapid recolonization following the LGM as ancestral populations spread northward and across Eurasia (see Chapter 4). An example of flowering time adaptation is provided by goat grasses (*Aegilops*: Poaceae or grass family). These grass species found refuge within the temperate desert vegetation zone during the LGM, and when the climate became warmer and more moist, they evolved various flowering times (photoperiod responses) as they radiated northward. Due to the weedy growth habit of goat grasses, human activity likely contributed to their expansion following the advent of agriculture about 10,000 years ago (Matsuoka et al. 2008). *Cannabis* flowering response would have evolved in the same way as it spread northward from southern refugia and consequently the maximum day length that allows flowering increased from 12 to 15 hours. This enabled vegetative growth until after the summer solstice (June 21–22), when day length decreases sufficiently to promote flowering. If *Cannabis* flowered only at very short day length (less than 12 to 13 hours) at temperate latitudes, either it would flower in the spring without reaching large size, producing only a few seeds, or it would flower too late in the autumn, and many seeds would be destroyed by frost before they ripened. Through natural selection, *Cannabis* and many other plant groups adapted from tropical to temperate habitats by evolving annuality and short-day photoperiodism reproductive strategies. Though the progenitors of Cannabaceae may have lived closer to the equator, annuality and short-day photoperiodism must have evolved during expansion further north, where increased variation in natural light cycles and shorter growing seasons would select for an annual reproductive strategy.

ANEMOPHILY

Anemophily (wind pollination) is characteristic of both *Cannabis* and *Humulus*. However, this form of pollination evolved independently in only 10 percent of angiosperm species spread over 20 percent of angiosperm families, being most common in three related families—Cannabaceae, Moraceae or mulberry family, and Urticaceae or nettle family. *Cannabis* and *Humulus* were included in either Urticaceae or Moraceae until assigned to their own family Cannabaceae, and recent genome evidence strongly suggests subsuming Cannabaceae with Celtidaceae or hackberry family (Culley et al. 2002; also see Chapter 11). *Cannabis* and *Humulus* express many traits common to wind pollinated plants that may have encouraged the evolution of anemophily; these traits include growing in dense populations spread across open habitats of temperate climates, producing many greenish or whitish unscented unisexual flowers with small or absent petals, and synchronous flowering—females with protruding feathery styles and a single ovule and males producing large amounts of simple dry pollen grains (Friedman and Barrett 2009). Cannabaceae are well adapted to wind pollination, producing copious small pollen grains in nearly leafless terminal inflorescences. A recent study has documented the long distance wind dispersal of *Cannabis* pollen across the Mediterranean Sea from North Africa to southern Spain (Simons 1995). In female *Cannabis* flowers, pollen is readily trapped by exposed paired stigmas. Reproductive similarity between the only two genera of Cannabaceae supports the hypothesis that the ancestor of this family originated in northern temperate latitudes. It is generally believed that wind pollination evolved under selection pressures such as climate change in which biotic pollination became less advantageous (Culley et al. 2002; Friedman and Barrett 2009). The common ancestor of *Cannabis* and *Humulus* may have evolved its reproductive strategy of wind pollination in response to great distances between its adaptive habitats, such as along more or less isolated streams and lakes in the relatively arid, vast continental temperate region of Central Asia.

Wind pollination may seem to be a random reproductive strategy wasteful of male gametes, especially compared to highly directed biotic pollination, yet research shows that "pollen-transfer efficiencies in wind-pollinated species are not substantially lower than in animal-pollinated taxa" (Friedman and Barret 2009). Males produce copious amounts

TABLE 16

Anemophilous or wind-pollinated plants differ from animal-pollinated plants in many key aspects of their evolution.

C = Cannabis; H = Humulus; and *C/H = Cannabis* and *Humulus*

	Wind-pollinated	Animal-pollinated
	Floral traits	
Fragrance (floral)	Absent (*C/H*) or reduced	Present
Perianth	Absent or reduced (*C/H*)	Showy
Flower type	Usually unisexual (*C/H*)	Usually bisexual
Ovule number	Usually single	Single or multiple
Inflorescence structure	Pendulous (male *C/H*, female *H*), catkin-like, often condensed (female *C/H*)	Variable, sometimes simple and diffuse
Inflorescence position	Held away from vegetation (male *C/H*; female *H*)	Variable
Pollen	Small, light, and dry individual grains with simple surface (*C/H*)	Sticky clumped grains or grains with complex surface architecture
	Habitat variables	
Optimum wind speed	Low to moderate	Zero to low
Humidity	Low (during pollen dehiscence)	Medium to high
Precipitation	Infrequent (during pollen dehiscence)	Infrequent to common
Surrounding vegetation	Open (steppe and open woodland)	Open to closed
Plant population density	Moderate to high (*C/H*)	Low to high

NOTE: Adapted from Culley, Weller, and Sakai 2002.

of pollen because male fitness is equated with pollen quantity, not because wind pollination is inherently inefficient. Self-pollination provides reproductive assurance but at the cost of inbreeding consequences, while dioecy, in concert with anemophily, results in obligate outcrossing with wide genetic variation. This encourages adaptive success in colonizing populations without reliance on local biotic pollinators and thus engenders wind-pollinated dioecious plants with wide-ranging ability to colonize new habitats. The genomes of outcrossers such as Cannabaceae are more heterozygous than inbreeders, and allelic dominance, rather than homozygous pairs of alleles, largely controls expression of traits. Today, anemophily in angiosperms is correlated to middle elevations and temperate latitudes and generally is predominant in northern hardwood forests, especially in open blow downs or light gaps within the forest ecosystem. In contrast, dense tropical forests are made up predominately of animal-pollinated (bat, bird, or insect) species. Upland or exposed temperate environments offer an open windy situation favoring pollen travel, while dense tropical forests discourage pollen travel by reducing air circulation and presenting competing pollen-catching surfaces. Wind-pollinated species are more abundant in open temperate habitats, as these conditions favored both evolution of anemophily and its persistence. However, the thriving existence of many anemophilous plants including *Cannabis* today in higher latitude regions with seasonally arid temperate climates and open vegetation does not prove that anemophily evolved under these conditions or whether it is simply more successful today within these parameters. And if data are adjusted for the preponderance

of wind-pollinated Gymnospermae (pines, firs, etc.), Poaceae (grasses, sedges, rushes, etc.), and Fagaceae (beeches, oaks, etc.), then there is no statistical correlation between wind pollination and geographical distribution (Friedman and Barrett 2009).

Anemophilous taxa such as Cannabaceae share certain traits that differ greatly from animal pollinated plants. Small green flowers with single ovules and no nectar are characteristic of anemophilous species, yet it is not certain that wind pollination "evolves more frequently in lineages that have small inconspicuous flowers and no nectar, or if these traits are lost after the evolution of wind pollination because of energetic reasons associated with a loss of function" (Friedman and Barrett 2009). It seems equally plausible that Cannabaceae never evolved biotic pollinator relationships as abiotic pollination provided robust fitness without further reproductive system evolution. Wind pollination differs in dynamics and range from biotic pollination mediated by mammals, birds, and insects and in certain conditions may provide an advantage.

DIOECY

Approximately 30 percent of genera with anemophilous species are also dioecious (Culley et al. 2002), and dioecy may have evolved either before or after anemophily. If dioecy evolved after the development of biotic pollination of hermaphrodite flowers and before anemophily, then biotic pollinators attracted to pollen rather than nectar may have shunned female flowers lacking pollen; if this is so, effective biotic pollination could have become restricted leading to an advantage for wind pollination. If, on the other hand, dioecy evolved *after* anemophily, then inbreeding between spatially near and self-compatible stamens and pistils on the same plant could have directed evolution toward dioecy by partitioning male and female flowers to separate unisexual plants and thus circumvented inbreeding depression. It is possible that a primarily animal-pollinated Cannabaceae ancestor lost its obligate pollinator during postglacial range expansion, and hermaphrodite or unisexual flowers that could pollinate abiotically, even with low fertility, had a distinct evolutionary advantage (Friedman and Barrett 2009).

Resource allocation (relatively ephemeral males devoting energy to copious pollen production and longer lived females to seed production), absence of biotic pollinator coevolution, seed dispersal mechanics (seeds form on ends of branches), and selective predation (flower versus seed) by herbivores and pathogens are also important factors influencing the evolution of dioecy (Traveset 1999). In dioecious species, herbivory is almost always biased toward both flowers and leaves of male (pollen) plants; therefore, they incur more damage than female (seed) plants (Ashman 2002). Obligate outcrossing of dioecious species prevents competition for resources between male and female floral organs within an individual plant and may allow female flowers to evolve better fruit quality and better seed productivity, which may in turn promote dispersal and increase survival rate (Thompson and Brunet 1990). This is especially important for annual plants.

The origin of dioecy is difficult to reconstruct in *Cannabis*. The vast majority of plant species are cosexual, either monoecious (male and female flowers on the same individual) or hermaphroditic (male and female organs in the same flower). Dioecy is more common in wind-pollinated species such as *Cannabis* and *Humulus*, perennial vines such as *Humulus* and wild grape (*Vitis*: Vitaceae or grape family), and plants with greenish flowers such as *Cannabis* and *Humulus* (see Traveset 1999; Negrutiu et al. 2001). Evolution of abiotic pollination from biotic pollination, and colonization of marginal habitats are also associated with dioecy (Ashman 2002), which also may have imparted enhanced fitness during postglacial expansion. Dioecy has independently evolved on many separate occasions in several cultivated taxa such as papaya (*Carica papaya*) and spinach (*Spinacea oleracea*). However, the majority of dioecious plants, in contrast to *Cannabis*, are usually large perennial woody species. Perhaps dioecy is most closely correlated with anemophily because the separation of sexes offers a favorable situation for wind pollination and outcrossing.

SEX DETERMINATION

Cannabaceae share another relatively uncommon trait linked with dioecy: an X and Y chromosome genetic sex determination system more prevalent in mammals, which has independently if only rarely evolved in several plant taxa. Among approximately 160 families with dioecious species (Charlesworth 2002) heteromorphic (different-shaped) sex chromosomes have been discovered in only four (Ming et al. 2007). Cosexual hermaphrodite and monoecious plants do not require a sex determination system as each plant contains both sexes. The chromosomal sex determination system in Cannabaceae likely arose as an evolutionary consequence of natural selection in favor of dioecy. Sex chromosomes evolve from autosomes (nonsex chromosomes containing sex

determination genes) that have lost the ability to recombine; the former homologs evolve separately and develop specialized features such as reduced Y chromosome size and function, resulting in heteromorphic sex chromosomes. Two basic sex chromosome systems are found in plants. The first is the XY system similar to that of mammals in which females have two copies of the same chromosome (XX) and males have one copy of the X chromosome as well as one copy of an active Y chromosome, the inheritance of which determines male sex (XY). In the other X and Y sex determination system, the highly reduced and genetically inactive Y chromosome has no function in sex determination other than not being an X and is expendable. These two sex chromosome systems may represent different stages of sex chromosome evolution—the inactive Y system of Cannabaceae being the more highly evolved (Ming et al. 2007).

Plant chromosomal sex determination appears to have evolved relatively recently. An age of less than 20 million years has been suggested for evolution of the heteromorphic X and Y sexual system in docks and sorrels (*Rumex*: Polygonaceae or buckwheat family; Negrutiu et al. 2001). In campions (*Silene*: Caryophyllaceae or carnation family), there are both hermaphrodite as well as dioecious species with an X and Y chromosome sex determination system; divergence of X and Y sex chromosomes began in campions even more recently, around 10 million years ago, while animal sexual systems evolved more than 300 million years ago (Nicolas et al. 2005). Both hemp and hop share the same highly evolved sex determination system, which likely appeared in their common ancestor before divergence. Does the highly evolved nature of the X and inactive Y sex determination system indicate that more evolutionary time was required to reach this level? Or does this imply that the ancestor of Cannabaceae evolved X and Y sexual determination relatively late in evolutionary time yet before *Cannabis* and *Humulus* diverged?

Since the 1920s, the dioecious condition of *Cannabis* has enticed researchers to decipher its means of sexual determination. These researchers have provided many models ranging from simple male heterogamy of the Y chromosome to complex sex determination resulting from the "interaction between individual hereditary potencies and environmental factors," and along with *Humulus* and *Silene*, *Cannabis* is considered to have one of the more complicated sex determination systems in dioecious plants (Truţă et al. 2007). Several forms other than unisexual male and female plants have been observed, both in Indian NLD *ganja* fields (Prain 1904) where female plants developed male flowers and partial hermaphrodites and in monoecious NLHA and NLH individuals employed in twentieth-century European hemp breeding projects (see Chapter 10). Environmental stress such as short photoperiod and low temperature can also cause sex reversals. Plant breeders employ exogenous applications of plant hormones to reverse the sexual development of *Cannabis* flowers—thereby producing male flowers on female plants or female flowers on male plants—without changing the plant's genetically determined sex: "*Cannabis sativa* L. has a very complex genetic constitution and heredity which explains the dioecy, amplitude of phenotype variability, polymorphism and the great biological plasticity of this species. The flexibility of *Cannabis* sexual phenotype often leads to the differentiation of hermaphrodite flowers or monoecious phenotype. Therefore, the genetical component of sex determination and of sex control displays, by phenotype expression, a great variability in hemp" (Truţă et al. 2007). We suggest that the dioecious characteristic evolved relatively early in Cannabaceae as the trait is shared by *Cannabis*, *Humulus*, and several members of Celtideae. The dioecious condition in *Cannabis* may be explained best by X and Y chromosomal control, while phenotypic expression of sexuality is strongly influenced by several environmental factors, possibly via autosome and/or organelle nucleic acid (DNA and RNA) regulation. The combination of these factors may result in varying phenotypic expression of monoecy and more rarely hermaphrodism, although *Cannabis* may possess a unique sex determination system not yet understood (Truţă et al. 2007).

Reconstruction of a Cannabaceae Ancestor

Using evidence from its breeding system evolution, we can begin to reconstruct a hypothetical ancestor to Cannabaceae. If we assume that before *Cannabis* and *Humulus* diverged, their common ancestor was already wind pollinated as they both are, and we follow the evolutionary convention that wind pollination is a condition derived from biotic pollination when pollinators are absent (Guo et al. 2008), then most likely the ancestor to Cannabaceae was hermaphrodite or monoecious and visited by animal pollinators. However, evolution may have been more economically direct, with Cannabaceae arising directly from wind-pollinated ancestors without evolution and extinction of an intermediate animal-pollinated stage.

Fossil remains of putative ancestors are helpful in reconstructing early species ranges, as well as possible evolutionary scenarios and hypothetical ancestors, but Cannabaceae-like finds are quite rare. Zhilin (2006) reported the identification of remarkable leaf fossils from central Kazakhstan discovered on a sloped bank of the Zhaman-Kaindy River and dated to the end of the late Eocene, 38 million years ago. According to Zhilin, these particular fossils are "extremely similar to *Cannabis*." Zhilin referred to another Eocene *Cannabis*-like fossil, *Cannbis oligocenica*, which has been described from Germany, but suggested that Kazakhstan Eocene specimens deserve being erected as a separate species. Based on the fossil flora "composed of species with small narrow, leaves, mostly dentate, with tightly condensed venation," among which this *Cannabis*-like specimen was found, Zhilin characterized the vegetation as "quite xeromorphic." Although this evidence needs further confirmation, by the end of the Eocene the earth's climate had cooled and the continental interiors were drying, with forests in retreat in some areas. Evolving grasses were restricted mainly to riverbanks and lakesides and had not yet expanded into open areas such as plains and savannas, and this likely applied to ecologically related plants such as *Cannabis*.

Prior to the evolution of flowering plants, more primitive liverworts, mosses, ferns, and gymnosperms proliferated for millions of years relying primarily on the vectors of water and wind to mobilize their gametes. However, in most flowering plant families anemophily is considered a derived state because some ancestors of wind-pollinated species are animal pollinated, and pollination of many early angiosperms appears to have been via beetles and flies. Although bees collect pollen from male *Cannabis* plants, the female plants offer no attractive lure so bees rarely alight on them, and these or any other insects transfer little, if any, pollen. However, during their foraging, bees dislodge large amounts of pollen that is carried away by air currents. Why waste evolutionary time and energy to first coevolve biotic pollination systems when abiotic systems already existed in angiosperm ancestors?

FIGURE 95. Female *Cannabis* inflorescences with pistillate bracts and apical male anthers occasionally appear, likely an atavistic intersex trait, manifesting characteristics of early Cannabaceae ancestors.

Because coevolution may drive speciation—in this case evolution of a pollinator paralleled by evolution of a flower requiring pollination—it could help explain the evolution of biotic pollination, followed in some cases by a reversion to abiotic pollination. Likewise, in the absence of coevolution of biotic pollinators, there is little reason to evolve and then lose a biotic pollination mechanism; wind pollination may have proven adequate since the earliest evolution of Cannabaceae.

Conditions favoring anemophily are more narrowly proscribed than those favoring biotic pollination, and a combination of factors likely directed the evolution of wind pollination. These must have included a preadapted ancestor population (e.g., upright growth, pendulous exerted stamens with small dry pollen, exposed pistils with reduced corolla, and gregarious growth habit) without biotic pollinators (e.g., lack of floral fragrance) growing in an open vegetation zone with a relatively arid and windy climate (e.g., open steppe and woodland). Effective wind pollination requires a relatively constant physical environment within seasons and across many years (Culley et al. 2002). Therefore, given that natural and human-induced climate and vegetation change has proceeded rapidly since the dawn of the Holocene, the natural range of *Cannabis* must have repeatedly shifted to open environments favorable for wind pollination. Wind pollination may have been encouraged in *Cannabis* and *Humulus* because they moved in and out of refugia repeatedly and were less likely to coevolve with a reliable suite of animal pollinators.

An early Cannabaceae ancestor could have resembled Chinese *Archaefructus sinensis*, the earliest flowering plant

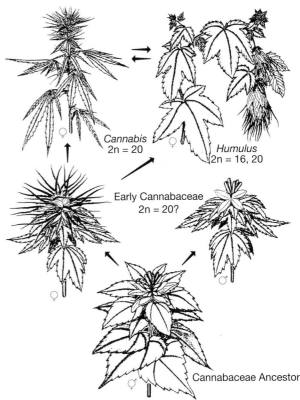

FIGURE 97. Hypothetical primeval ancestors of the Cannabaceae family. The dioecious Cannabaceae likely evolved from a monoecious Moraceaean ancestor about 60 million years ago. Later, about 22 million years ago *Cannabis* and *Humulus* diverged into separate genera.

FIGURE 96. Nested bracts in female drug *Cannabis* inflorescences appear much like clusters of tiny hop cones and may be an atavistic trait indicating evolutionarily links between *Cannabis* and *Humulus*.

fossil so far identified (Raven et al. 2005; Russell et al. 2011), which lived about 125,000 years ago. The sexual system of *A. sinensis* was inferred from fossil evidence indicating it may have been pollinated by an abiotic mechanism. Open single- or double-ovule carpels appear at the branch tips above clusters of stamens along the lower part of the branch in a monoecious configuration allowing pollination by a combination of wind or water and gravity. Increasing the number of unisexual flowers and compressing the floral axis during evolution would result in a male inflorescence with female flowers at the tip. An inverse configuration, inflorescences with staminate organs arising at the terminal meristem late in development (probably an atavistic expression of an ancestral trait), is well documented in *Cannabis* (Prain 1904). Also, female Cannabaceae inflorescences are formed from many single ovules each subtended by a green leaf-like bract, and in *Humulus* the inflorescense forms a pendant cone-like strobilus; these traits resemble ancient pre-angiosperm seed plants. Although an ancient ancestor to Cannabaceae may have had an animal pollinator, and wind pollination reevolved more recently, we propose that Cannabaceae did not require hermaphrodite flowers or biotic pollination for evolutionary fitness and success and may have remained unisexual and anemophilous throughout its evolution.

Both *Cannabis* and *Humulus* are anemophilous, dioecious, short-day flowering, herbaceous annuals. The male inflorescences are semipendulous in both *Cannabis* and *Humulus*; indeed they are remarkably similar. Both *Cannabis* and *Humulus* possess smooth, medium-sized seeds enveloped by a bract covered in protective eglandular and glandular trichomes. Female inflorescences of these two closely related genera differ morphologically by only one key trait. In *Cannabis* the female flowers are individual, produced in the axils of leaves, and surrounded by an enlarged, erect bract. A terminal inflorescence forms from many bracts, but a reduced leaflet usually subtends each pair of single flowers. In *Humulus* the floral bracts are greatly enlarged and gathered in a pendulous, cone-shaped terminal strobilus, but a reduced leaflet does not subtend each bract.

While *Cannabis* prefers an open habitat and develops a strongly erect profile, *Humulus* prefers the fringes of clearings and low brush where support can be found for its lax, vining branches. Although *Humulus* evolved into a vine that flowers and produces seed annually, it has maintained its more primitive perennial root system. *Cannabis* has been propagated traditionally from seed and only recently asexually, while *Humulus* has traditionally been vegetatively propagated from persistent rootstocks as well as by seed. *Cannabis* remains a strict annual, at least without human intervention (see Chapter 10). Much like their modern relatives, primordeal Cannabaceae most likely preferred an open, sunny habitat and plenty of water. Many species of Moraceae and Urticaceae are also early successional herbs that thrive in open sites of various sizes. Anderson (1956) characterized post-Pleistocene angiosperm evolution as largely an elaboration of weed-like herbaceous species. These species typically inhabit open sunny areas including those disturbed by increased drainage associated with the melting of glacial ice after the LGM and increased human alteration of the environment. Following Sauer (1952, 1967) and Anderson (1967), Merlin (1972) predicted that pluvial environments in river valleys at the end of the most recent Pleistocene glaciation should prove to be ecological settings in which relationships between humans and *Cannabis*, if not the plant species itself, first evolved.

Summary and Conclusions

Pre-Pleistocene conditions in the northern hemisphere were sometimes warmer and more humid than today. Ecological requirements and reproductive traits of *Cannabis* indicate that it originated in what was then a temperate region somewhere within northern Eurasia. The southern slopes of the Tian Shan Mountains or river valleys to their north may have provided an ideal location for the early evolution of the ancestors of annual, wind-pollinated, and dioecious *Cannabis*. From the beginning of the final Pleistocene glaciation around 24,000 years ago until the beginning of the Holocene there was no widespread midlatitude, temperate climatic zone in northern Eurasia where thermophilous plants such as *Cannabis* could persist. As ice sheets advanced, frost-tolerant plants largely replaced plants that thrived at northern latitudes prior to the Pleistocene glaciations, and warmth-loving populations became restricted to temperate refugia.

Present-day Cannabaceae evolved two relatively rare reproductive strategies, wind pollination and dioecy with associated (and even rarer) X and Y sex determination, and these likely characterized their common ancestor as well. Polyploidy and introgression, both profound evolutionary forces, are not recognizable in the early evolution of Cannabaceae. *Cannabis* is neither animal pollinated nor polyploid, so speciation did not occur through pollinator-flower coevolution or sudden genetic isolation due to chromosomal incompatibility. *Cannabis* evolved into separate species by ecological adaptation during climate changes and diversified during post-LGM expansion into local ecotypes. *Cannabis* species, subspecies, and biotypes are interfertile and rely on no natural reproductive barriers other than latitudinal photoperiod response—temperate biotypes flower under longer day length than tropical biotypes—but this does not present a reproductive barrier within localized populations. Hybridization has led to much variation in twentieth-century cultivars, yet it is unlikely that hybridization played a role in *Cannabis* speciation.

How does our knowledge of ancient climate change, Pleistocene refugia, flowering plant evolution, and genetics impinge on theories of *Cannabis*'s origin, evolution, and dispersal? *Cannabis* is often considered a native plant of Central Asia with several researchers suggesting it as the original evolutionary home of *Cannabis*. Much evidence supports this scenario. However, if *C. ruderalis* is recognized as a valid species and is truly a descendant from ancient PA populations, when, where, and how did *C. sativa* and *C. indica* evolve? Moreover, how did *Cannabis* radiate into the subspecies and ecotypes we recognize today?

Annual life cycle, short-day flowering response, anemophily, dioecious sexuality, and X and Y sex chromosomes are all *Cannabis* characteristics providing clues to its ancient origins. Although in many flowering plant families wind pollination is considered a derived condition resulting from a breakdown in coevolved biotic pollination systems, in Cannabaceae anemophily is most likely the original pollination mechanism. Dioecious sexual reproduction derived from an earlier monoecious condition, with male and female flowers borne on the same plant.

Repeatedly throughout the Quaternary, and possibly earlier geological periods, climatic conditions approximated those favorable for *Cannabis* today, and its ancient range was determined by environmental and biotic factors but without human influence. Fewer than 10 million years ago and possibly much more recently *Cannabis* diverged from its Cannabaceae ancestor and began to evolve independently. *Humulus* originated from the same or a closely related ancestor around this time or slightly thereafter. Some *Cannabis* populations divided, with geographical separation providing genetic isolation in concert with ecological adaptation leading to successful radiation and speciation, while other populations of the genus failed to adapt and became extinct.

What environmental factors would be required for *Cannabis* to survive and what was its refugial range during the Pleistocene? *Cannabis* requires a warm and moist temperate climate to flourish but can survive in suboptimal and even marginal circumstances. Natural populations endure snow-covered winters in subalpine and subarctic regions as seeds, but beginning each spring they must receive sufficient annual warmth, moisture, and light to grow and reproduce during a short growing season. *Cannabis* need not be abundant throughout its range to be successful, but each population must be of sufficient size and stand density to ensure adequate pollen and seed production, promote outcrossing (limit inbreeding), and preserve genetic diversity; if not, fitness and survival are compromised. These conditions surely would have been even more significant when *Cannabis* populations were restricted to reduced refugial ranges. *Cannabis* is a survivor, a plant for which favorable niches remained even during the most severe glaciations. Hypothetical changes in *Cannabis*'s distribution and abundance largely fit classical models of "glacial refugia," but additional evolutionary scenarios may better explain certain phases in its evolution. During the LGM, tundra dominated a large area of northern Eurasia along the leading edge of the Scandinavian ice sheet. The Tian-Shan Mountains and northern Mongolia hosted steppe and taiga, the vegetation types predominant there today (Tarasov et al. 2000). Steppe-like vegetation dominated the latitudinal band from western Ukraine, where temperate deciduous forests grow today, to western Siberia, where taiga and cold deciduous forests currently predominate. Feral populations now thrive in favorable microclimates within generally inhospitable Central Asian steppe-land regions, and *Cannabis* may also have survived the LGM (or earlier glaciations) in cryptic refugia spread across the steppe vegetation band; surely *Cannabis* would have recolonized this favorable climatic region early on during interglacial warming periods. Broadleaved trees were confined to small refugia, such as along the eastern coast of the Black Sea and western Caucasus Mountains where forests thrived more than 1,000 meters lower than today, across the Hengduan Mountain-Yungui Plateau region where diverse feral *Cannabis* populations abound, and along coastal East Asia where the oldest archeological evidence of *Cannabis* has been found. All these regions may have harbored *Cannabis* populations during recurring cold cycles.

If *Cannabis* persisted through the LGM within several refugia, then speciation must have occurred during or before this time. As the Holocene began, *C. sativa* radiated from southern European putative hemp ancestor (PHA) refugia while *C. indica* radiated from Asian putative drug ancestor (PDA) refugia, leading to three *C. indica* subspecies—*indica* or NLD (and possibly also BLH) in the Hengduan-Yungui region, *chinensis* or BLH in East Asia, and *afghanica* or BLD in the Hindu Kush Mountains—while other subspecies level taxa, if they existed at the time, have since become extinct. Biotype differentiation began as humans carried *Cannabis* to nearly every region on earth—divergence, isolation, and evolution taking place in agricultural niches and other human-disturbed environments. Speciation in *Cannabis* most likely proceeded via dispersal to new isolated environments, and in new habitats ecological adaptation accompanied by genetic drift eventually led to diploid speciation without polyploidy. Hybridization and introgression were likely of little importance during early evolution but weighed heavily in twentieth-century cultivar development.

CHAPTER THIRTEEN

Cannabis and Homo sapiens

Present Position and Future Directions

> It is only through the interdisciplinary approach that such discoveries can be made. In fact, the unprecedented strides achieved in the study of hallucinogens in the past 30 or 40 years owe their spectacular success to interdisciplinary studies and consequent integration of data gleaned from many seemingly unrelated fields of investigation: anthropology, botany, ethnobotany, chemistry, history, linguistics, medicine, pharmacognosy, pharmacology and psychology.
>
> (SCHULTES 1969b)

Introduction

In this summary chapter, we briefly review the long and complex history of *Cannabis* and its human use for many purposes, with special focus on the potential recreational, medical, agricultural, and industrial applications of this resource-rich group of plants. We also examine the evidence for coevolution between humans and *Cannabis* to support our model for *Cannabis*'s evolution and offer food for thought for future researchers as well as the general public.

Cannabis, the ancient multiuse genus, is one of *Homo sapiens*' most favored yet controversial plant allies. According to the independent Global Commission on Drugs (2011), an estimated 160 million people consumed psychoactive *Cannabis* in 2008, nearly all of those illicitly! There exists a synergistic relationship between the influence of *Cannabis* on the development of human society and the parallel effects of humans on the evolution of *Cannabis* as an economic plant. These two essential elements form the core of the "*Cannabis* Phenomenon." It can be argued that very few other plants have had such profound effects on human cultural evolution. On the other hand, some say none has felt the harsh hand of humanity more than *Cannabis*. So how will the human-*Cannabis* relationship continue to evolve during and beyond the twenty-first century? Will governments continue to condemn *Cannabis*, or will the virtues of our ancient ally once again be appreciated and more fully utilized for the benefit of humanity?

Cannabis was among our earliest cultivated plants, and for centuries it ranked as one of our most important agricultural crops. For more than 7,000 years, *Cannabis* was utilized to produce necessities of life such as fiber, fabric, food, lighting oil, and medicine, and it has become one of the most widely cultivated plants in the world. *Cannabis* hemp produces one of the strongest and most durable natural fibers, versatile enough to be used for the manufacture of fine clothes as well as tough canvas, cordage, and various paper products. Until the introduction of the cotton gin, hemp was the most widely used textile fiber. The invention of petrochemical fibers and acid-process pulp paper, and the popularity of petroleum products such as kerosene lighting oil, lessened the importance of hemp. *Cannabis* has been recommended in the treatment of diverse medical conditions for more than 3,000 years, and *Cannabis* extracts were the analgesics most widely prescribed by Western doctors during the nineteenth and early twentieth centuries. *Cannabis* offers great promise as a medicine for the future, even though presently its use as a prescription drug is limited to a few countries.

Unfortunately, the prohibition of *Cannabis* drugs has led to the prohibition of *Cannabis* cultivation in general, and some important past uses of *Cannabis* have been, more or less, forgotten—lost largely because of unresolved controversy or, as some have argued, because of conspiratorial corporate pressures to eliminate hemp from competition with more profitable resources and products (e.g., see Herer 1992). *Cannabis* has a long history as an economic plant and has participated significantly in shaping world history. *Cannabis* has been utilized for its profound psychoactive powers for thousands of years and has influenced the creative thought of humans, thereby making an additional impact on culture through religion, music, art, and literature. More recently, governmental efforts to eradicate the international drug trade continue to have significant effects on both the evolution of *Cannabis* and the cultural development of twenty-first-century societies.

Documenting and analyzing the influence of *Cannabis* on human cultures has been the task of many authors who have largely concentrated on seeking evidence for either the beneficial or detrimental ways in which *Cannabis* affects humans and our societies. Few have explored the ways in which humans have affected the evolution of *Cannabis* as a cultivated economic plant and perhaps how *Cannabis* in turn has affected us (e.g., see Pollan 2001). When fragments of evidence illustrating the history of human association with *Cannabis* are collected from several disciplines and collated into a historical framework, a fascinating story unfolds. We propose that *Cannabis* and humans have coevolved during their lengthy association, each participating in shaping the future of the other in significant and often beneficial ways. Presently the morality, legality, and practicality of using *Cannabis* for psychoactive purposes are all quite controversial. Is *Cannabis* a benefit or curse for humanity? It seems quite foolish to pass judgment on such matters without at least attempting to understand how, when, why, and with what effects the plant was used in the past.

Two famous scientists who have focused on psychoactive plants, Richard Evans Schultes, the "father of ethnobotany," and Albert Hofmann, the world famous chemist who, in 1943, first recognized the intensively powerful mind-altering power of semisynthetic LSD (lysergic acid diethylamide), have given us guidance on the issue of *Cannabis*. In their book dealing with the "Plants of the Gods" (1992), Schultes and Hofmann beseeched us to "consider the role of *Cannabis* in [our] past and learn what lessons it can teach us . . . for it appears that it will be with us for a long time."

Most people are unaware of the many diverse uses of *Cannabis* and their antiquity, and we hope that our research sheds additional light on the subject. Perhaps this work will serve as encouragement for others to study the fascinating, puzzling, and what *we believe* to be important questions concerning the biological and cultural ramifications of human and *Cannabis* coevolution.

The Long-Term Relationship

By this point in the text it should be clear to readers that humans have had a very lengthy, important, and multifaceted association with *Cannabis*, yet we are not sure exactly when and where these diverse human-*Cannabis* relationships began. Was it the Mesolithic peoples who first developed a conscious association with *Cannabis* plants during the beginning of the Holocene Epoch about 10,000 years ago, or did this occur much earlier? The initial people to utilize *Cannabis* extensively probably did so near fresh water sources where both the plant and early humans would have found favorable niches. The resilient fibers, edible fruits, and psychoactive drugs produced by *Cannabis* have attracted humans at least since Neolithic times, and we propose that the roots of its use by people may be much more ancient in certain regions of Eurasia. Among the many applications of *Cannabis*, use of its water-resistant fiber for making nets to catch fish and hunt game and consuming its seed as a rich food source

were undoubtedly important. It is also quite possible that early human association with *Cannabis* was motivated by religious or euphoric interests with people learning that *Cannabis* could be used for fiber, food, seed oil, and medicines only after parts of the plant were first knowingly ingested to induce spiritual or psychoactive effects. *Cannabis* obviously had a variety of resources to attract human attention and interaction, and people expanded the favored habitat of *Cannabis* as they extended the distribution of this plant and further affected the well-watered and nutrient-rich areas near fresh water.

We believe that *Cannabis* had its biological origin in Central Asia or somewhere in the European steppes. Then following its course of natural evolution for millions of years, it was eventually dispersed by humans over many millennia to nearly all the temperate and subtropical regions of the world. Prehistoric environmental reconstructions support the contention that *Cannabis* spread from Central Asia into the remainder of Eurasia, possibly with movements of early humans, and then found refuge during the last glaciations leading up to the Holocene in the Caucasus and Hengduan Mountain-Yungui Plateau regions. More recently, historical sources suggest that the cultural diffusion of *Cannabis* across Europe, Southwest Asia, India, and possibly even China was further facilitated by wide-ranging, nomadic migrations that followed the perfection of chariot warfare in the second millennium BCE. More recently it was transported by people from its huge range within Eurasia to most of the habitable areas of the world. Taken as a whole, *Cannabis* presently grows in a wider variety of natural and human-created niches than probably any other plant.

However, much remains to be understood about when and how *Cannabis* was first used. For what purposes and because of what motivations did people begin their associations with these plants? Although *Cannabis* has been cultivated for several thousand years, it remains incompletely domesticated, and weedy escapes from cultivation are relatively common, especially in some temperate environments that are similar ecologically to its original homeland. Wild ancestral populations may still be found in remote regions of Central Asia although they may well be products of interbreeding with cultivated varieties. *Cannabis* can also become a pernicious weed where it grows spontaneously; however, it did not evolve into a cultivated plant from a weed present in other crops. Humans have had the greatest influence on the evolution of *Cannabis* under domestication. Extreme geographical isolation of populations in regions where humans have introduced it, in concert with human selection of cultivated varieties for varying purposes, have resulted in the extreme diversity manifested in modern *Cannabis*. Humans have selected naturally adapted local varieties with their own morphological and physiological traits specifically for fiber, seed, and various drug preparations; consequently each extant landrace variety is morphologically and physiologically distinct. Most recently, controlled hybridization by plant breeders has played the dominant role in the evolution of *Cannabis* cultivars.

The most important single mutational step in the evolution of *Cannabis* involved the development of the ability to synthesize THC. This mutation occurred long before first human contact and then attracted the attention of early humans. Humans have spread *Cannabis* over a huge area of our planet largely because of *Cannabis*'s ability to produce psychoactive THC. Recent research by McPartland and Guy

(2004a/b), exploring possible genetic coevolution of the THC synthase gene in *Cannabis* and the human cannabinoid (CB or CNR) receptors (specific protein binding sites often controlling cellular function), suggests that there may have been a causal relationship between human consumption of THC and increased mutation rate of our own cannabinoid receptors. Increased understanding of the coevolution of humans and *Cannabis* will come as we accumulate more data concerning evolution of the endocannabinoid gene system in other primates such as chimpanzees.

More accurate analysis of the origin and early evolution of *Cannabis* can only be made after more detailed anatomical, chemotaxonomic, and phylogenetic data have been obtained, utilizing accessions from both the proposed regions of origin in Central Asia and along routes of historically traceable dispersal. We hope that more researchers will collect meaningful data using rigorous protocols to test some of our hypotheses.

Summary of *Cannabis*'s Evolution

Hypotheses of origin and dispersal based on archeological and historical data are corroborated by botanical observations of *Cannabis*'s gross phenotypes and seed morphologies. Until the late nineteenth century, *Cannabis* gene pools largely evolved independent of one another as a result of geographical, human cultural, and/or agricultural isolation. During the twentieth century, diverse *Cannabis* gene pools were introduced into Europe and North America. Subsequently, much well documented hybridization was carried out by plant breeders, providing us with additional data concerning the evolution of *Cannabis*. Biochemical records such as cannabinoid and terpenoid profiles, protein banding patterns, and DNA mapping add to our ability to characterize and classify different gene pools.

Following from taxonomic work by Karl Hillig (2004a/b, 2005a/b), we now classify modern-day cultivated *Cannabis* into four biotypes circumscribed by two species: (1) *C. sativa* ssp. *sativa* includes the European narrow-leaf hemp (NLH) landraces, hemp cultivars, and their descendants, and (2) *C. indica*, by far the more morphologically diverse species, circumscribes the Asian subspecies *chinensis* broad-leaf hemp (BLH) biotypes, as well as subspecies *indica* narrow-leaf drug (NLD) and subspecies *afghanica* broad-leaf drug (BLD) biotypes and their descendants. A third species, tentatively accepted by some, is the putative ancestor (PA) *C. ruderalis* but more likely also represents feral escapes (see Chapter 11).

Based on data provided by diverse scientific disciplines, assertions related to the evolution of *Cannabis* gene pools can be made with relative certainty, and thus we propose a genetic and taxonomic model for the following chronological stages in the evolution of *Cannabis*:

1. *Cannabis* and *Humulus* (hemp and hop) likely share common ancestry among the Urticacean branch of the Rosales, diverging into separate genera more than 20 million years ago. Reproductive strategies indicate that *Cannabis* (and *Humulus*) evolved in northern temperate latitudes during warmer Pleistocene interglacial periods.

2. The division of the primordial *Cannabis* gene pool into a nondrug proto-*C. sativa* putative hemp ancestor (PHA) gene pool and a psychoactive THC producing proto-*C. indica* putative drug ancestor (PDA) gene pool occurred before human contact and resulted from a single gene mutation (B_T allele) coding for the production of the THC-synthase enzyme and THC biosynthesis.

3. The *C. sativa* gene pool originated in western Eurasia and may have evolved via subspecies *spontanea*, the narrow-leaf hemp ancestor (NLHA), and then into subspecies *sativa*, circumscribing the European narrow-leaf hemp (NLH) varieties. Both are characterized by low THC content. Alternatively, the *spontanea* biotypes may represent escapes from cultivation and are more likely products of natural selection beyond the selective pressures of humans combined with interbreeding with cultivated populations (NLHA ←X→ NLH) rather than representing relict ancestral populations.

4. The *C. indica* gene pool originated in eastern or southern Asia and may have first evolved into subspecies *kafiristanica*, the putative narrow-leaf drug ancestor (NLDA), although as in the case of *C. sativa* ssp. *spontanea*, subspecies *kafiristanica* may represent descendants from escaped feral populations or hybrids with cultivated populations rather than ancestors. *C. indica* evolved into three biotypes that may represent subspecies, including *indica*, the narrow-leaf drug (NLD) varieties originating in South Asia; *afghanica*, broad-leaf drug (BLD) varieties used for hashish production in southern Central Asia; and *chinensis*, the broad-leaf hemp (BLH) of East Asia.

5. The main natural traits of cultural and economic importance that initially attracted human interest and encouraged continued human selections were fiber quality and quantity, high yields of nutritious seeds, and medicinal and psychoactive potency of the female flowers and their associated resin.

6. The four major divisions of *Cannabis* extant today are represented by four cultivated biotypes or subspecies, which evolved from two early and independent series of selections for fiber and/or seed usage (NLH in Europe and BLH in East Asia) and two early and independent series of selections for increased THC content (NLD in South and Southeast Asia and BLD in Central Asia).

7. Fiber, seed, and drug landrace varieties and cultivars extant today are primarily the result of human selective pressures over a wide range of environments for specific plant products.

8. *Cannabis sativa* has only been selected for hemp fiber and seed usage (NLH), while *C. indica* has been selected for psychoactive and medicinal use (NLD and BLD) as well as fiber and seed (BLH).

9. The two drug *Cannabis* gene pools (NLD and BLD) have separate ancient origins, apparently from the primary *C. indica* putative drug ancestor (PDA) gene pool as indicated by their extreme morphological divergence. The NLD gene pool is characterized by tall stature; supple stems with long internodes; lax, uncrowded inflorescences with a higher bract to leaf ratio; medium to large, finely serrated green leaflets; and light brown and often striped seeds. The BLD gene pool is characterized by plants of short stature; brittle stems with short internodes; crowded leafy

inflorescences; wide coarsely serrated, dark green leaflets; and shiny, dark, mottled grey or brown seeds.

10. Southeast Asian drug varieties may have arisen from hybridization between large-seeded Chinese BLH varieties and South Asian NLD varieties or may have arisen independently from a common PDA population originating in Pleistocene refugia located in southeastern Asia.

11. BLD varieties were most likely domesticated by Muslim farmers in Afghanistan, possibly as late as the nineteenth century, for the preparation of sieved hashish.

12. Chinese BLH fiber and seed cultivars were dispersed into Europe and the New World during the nineteenth and twentieth centuries and used for industrial hemp breeding. Chinese BLH cultivars are significant genetic contributors to the modern European and North American NLH cultivars and may be the origin of the B_T allele and trace THC production in NLH cultivars.

13. In the twentieth century, the vast majority of the cultivated drug *Cannabis* in the world was descendant from the South Asian NLD gene pool.

14. Prior to 1960, all New World NLD varieties were descendants of South Asian NLD varieties dispersed to the Caribbean Islands by indentured laborers brought from British India in the middle to late nineteenth century or possibly earlier via varieties brought directly by slaves or traders from Africa to New World areas in South America, the Caribbean, or Mexico.

15. Up until the 1970s, the range of BLD varieties was restricted only to contiguous parts of Afghanistan, Pakistan, and Kashmir. Since the mid-1970s, many hybrid crosses between more recently introduced BLD and previously introduced NLD varieties were made and dispersed across North America and Western Europe. Throughout the 1990s, hybrid NLD × BLD cultivars were spread across the globe by indoor hobbyists as well as commercial growers, resulting recently in widely expanded international distribution for the once isolated BLD gene pool, and hybrids between BLD and NLD landrace varieties now make up the vast majority of the Western marijuana drug *Cannabis* gene pool. (See Chapter 10 for a detailed discussion of modern *Cannabis* breeding.)

Clearly the human drive to ensure regular access to the benefits of *Cannabis* by carrying it to new environments and selecting it for specific products has affected the evolution of the individual *Cannabis* gene pools in profound ways. All four of the extant gene pools are exploited by humans at an increasing rate, although the three *C. indica* biotypes (NLD, BLD, and BLH) are spread more widely and exhibit far more localized morphological and biochemical variation than *C. sativa* NLH. Largely owing to their ability to synthesize THC, the three cultivated *C. indica* subspecies attracted the attention of humans who broadened their ranges around the world. Human selection of both fiber and drug cultivars continues at many locations, and it will be interesting to see how *Cannabis*'s evolution is influenced by the future needs of human cultures for textile, paper, food, and feed, as well as for medicinal, ritualistic-religious, and recreational purposes.

Cannabis's Influence on the Evolution of Human Culture

Throughout this volume we have established that humans had a marked effect on the evolution of the *Cannabis* genome during the past 10,000 or more years, as we have with most domesticated species. These changes have resulted from humans disseminating *Cannabis* throughout the world, introducing it to diverse new habitats and selecting it for various uses. The broadening of *Cannabis*'s range has been advantageous to the survival and proliferation of its gene pool. But how has *Cannabis* influenced the evolution of human culture? *Cannabis* provided raw materials, tools, and seminal lessons in survival for early humans, thereby playing a significant role in molding the early cultural traits of Eurasian societies. *Cannabis* is an adept camp follower with readily apparent resources and products for people to utilize. *Cannabis* was preadapted to early human's primal concepts of agriculture (e.g., annual life cycle, affinity for disturbed niches, easily stored seed, etc.), and thus became one of our earliest cultivated plants, certainly in its native temperate environments of Eurasia if not beyond. *Cannabis* has provided many varied products of strong regional economic and strategic importance, and each has had its profound effect on human culture. In addition to providing material to weave early clothing, the fiber of *Cannabis* was used even earlier to make cordage for nets, snares, and tethers to capture, consume, or tame wild animals. Later hemp was used to lay the lines, weave the sails, and seal leaky hulls that empowered fleets of ships, which changed the face of world conquest and history. In addition, hemp seed grains provided an important food, nourishing early societies that dispersed across the northern temperate latitudes of Eurasia.

Today, *Cannabis* still offers humans a relatively environment-friendly crop plant, producing durable fiber, nutritious seed, and a host of medicinally active compounds. Hemp fiber gradually decreased in importance throughout the late twentieth century as more profitable natural and synthetic fibers were developed or discovered. However, because of hemp's traditional and counterculture popularity and its qualities of durability, ease of dyeing, and modern environmentally friendly processing and manufacture, *Cannabis* remains a significant natural fiber source in international as well as local markets. Although the popularity of natural fiber textiles and garments is steadily increasing, it is still quite uncertain whether hemp will once again play a major role in the now highly competitive and technological textile industry. A much more cost effective and technologically appropriate use for the fibrous biomass of industrial hemp is pulp paper production, especially the strengthening of postconsumer recycled paper products. Another use for *Cannabis* is in the building materials industry where parts of its stalks can be used to make nonwoven insulations and reinforce concrete. In Europe, especially France, where there is still a significant hemp industry, the hurds (the leftover fragments of the stalk once all the fibers have been removed) are commonly combined with powdered limestone to form mineralized hemp blocks used for housing construction or poured between the walls in refurbished older homes as insulation. Hemp hurds are also used in stables as dust-free animal bedding. Furthermore, hemp fiber and hurds can be used to press light and medium density fiber board, and the international automobile industry has recently begun to use biodegradable interior molded panels utilizing hemp and other natural fibers in accordance with more stringent recycling standards in the automotive industry.

Another key value of *Cannabis* lies in its seed. Hemp seed contains high levels of easily digestible protein and essential fatty acids (EFAs). As a whole grain food, its EFA profile and protein composition are perfectly adapted to human and domesticated animal nutrition (DeFerne and Pate 1996; also see Khan et al. 2010). Humans, and our intensively fed livestock, have largely relied on fish and other marine products for supplies of EFAs. In the face of declining world fisheries, alternative supplies of dietary EFAs will be required, and *Cannabis* offers one of the most cost effective renewable sources. The EFA content of hemp seed oil also has an emollient effect on the skin, and its popularity in body care products is increasing (Pate and DeFerne 1996). Hemp seed oil can also be burned for fuel or used in industrial products such as paint and soap. The predominate terpenoids found in the fragrant essential oil produced by the resin glands of female inflorescences and extracted with steam distillation has found limited use as a scenting and flavoring agent for cosmetic products, candy, and soft drinks.

Medical *Cannabis* research has made great advances in recent years. Relying on the receptor-agonist-antagonist model of cannabinoid activity, and using it as a springboard, cannabinoid researchers have discovered and isolated human cannabinoid receptors (e.g., CB1 and CB2), endocannabinoids produced by the human body (e.g., anandamide [AEA] and 2-AG), and additional natural plant products (e.g., cannabinoids, terpenoids, and flavonoids). Scientists have also developed synthetic compounds (e.g., nabilone) that mediate cannabinoid receptor activity. Privately owned pharmaceutical companies continue clinical trials to investigate the efficacy and safety of *Cannabis* preparations and cannabinoids (both natural compounds and synthetic analogs) for treatment of ailments such as multiple sclerosis, pain management, blood pressure regulation, glaucoma, migraine, asthma, appetite stimulation, weight management, nausea control, and so on. Oral dosage forms of cannabinoid medicines (e.g., the "pot pill" Marinol) have proven less effective (and much less popular) than smoked *Cannabis*. Coupled with investigations into the efficacy of various *Cannabis* preparations and cannabinoids is the development of more effective delivery systems and dosage forms ranging from vaporizers of whole plant parts or extracts to nebulizers, atomizers, skin patches, injectables, suppositories, sublingual drops, and sprays of single or blended cannabinoids. Because of *Cannabis*'s strong effects on the basic functions of human brain and body, it is expected that *Cannabis* medicines will target a large number of disorders in the muscular, nervous, circulatory, and gastrointestinal systems, as well as mental health problems. Simultaneously, *Cannabis* is demonized fervently by prohibitionists as an evil drug plant and a threat to society (for a more sympathetic position regarding its medicinal value, see, e.g., Grinspoon and Bakalar 1997; Rätsch 2001).

A Case for the Social Benefits of *Cannabis*'s Psychoactivity

It can be argued that through the ages *Cannabis* drugs have heightened the senses; soothed human emotions; facilitated communication across time and space; and provided creative inspiration for innumerable inventors, artists, musicians, and writers. For example, the eating of hashish confections kindled fantasies in nineteenth-century Western writers, and smoking marijuana had strong effects on "all that jazz" and inspired legions of rock and roll musicians. The use of *Cannabis* and other psychoactive drug plants has altered the way many humans think and, through their altered consciousness, has determined, in part, our cultural evolution.

Hochman (1972) described marijuana as a "facilitator" that allows the breaking of "gestalt" or automatic responses to situations and promotes more independent thoughts, which have a direct effect on cultural evolution. In reference to the early marijuana culture of the United States during the 1960s and early 1970s he reflected upon his research:

> After several years of observation, one becomes convinced that there is an intrinsic aspect of the experience of marijuana which seems to facilitate change in the personal outlook of chronic marijuana users. The existence of the relationship between a drug and the form of the culture in which it is used is not unique. History reveals that the use of coca leaves, opiates and hallucinogenic plants have all been important aspects of the cultures in which they were used, shaping patterns of worship, self-perception, and social interrelationships. Marijuana, too, has been an important factor in numerous cultures. But contemporarily, the novelty of the current situation seems to be that the use of the drug is involved in rapidly changing the social forms of history's most technologically advanced and complex society [the United States].

Hochman went on to muse further about the impact of *Cannabis* use in modern American culture: "The increasing self-awareness and pleasure orientation of a whole segment of our society is becoming institutionalized in alternative social structures of work, education, and community. Marijuana has been, and continues to be, an essentially important catalyst to this evolution." In his bestselling book, *The Botany of Desire* (2001), Michael Pollan presented his hypothetical, visionary point of view about the philosophical and perceptual impact of *Cannabis* consumption: "What if these plant toxins [in this case THC] function as a kind of cultural mutagen, not unlike the effect of radiation on the genome? They are, after all, chemicals with the power to alter mental constructs—to propose new metaphors, new ways of looking at things, and, occasionally, whole new mental constructs. Anyone who uses them knows they also generate plenty of mental errors; most such mistakes are useless or worse, but a few inevitably turn out to be the germs of new insights and metaphors." Certain psychological and physiological effects induced by *Cannabis* may have positive social and evolutionary consequences. As we have shown throughout this book, many cultures have perceived *Cannabis* use as both beneficial and harmful, often at alternating periods in their social and political evolution. Although some of the effects explained later may be counterproductive in certain settings, in the long run they may affect our personal growth beneficially and therefore enhance our chances of survival.

The lowering of aggression could have been a great disadvantage in the early days of human history, during the hunter-gatherer phase of our evolution when we needed to fight to survive, especially during the hunt. But as family clans grew into tribes and eventually began more sedentary agricultural life during the early Holocene, population densities rose (and continue to rise) exponentially. In certain circumstances the moderation of aggressive behavior allowed early humans to coexist more easily with their fellow humans. Diminished aggression may have been of great advantage in surviving the confrontational social situations that humans have been increasingly exposed to during the evolution of contemporary societies. Our highly evolved human brain allows introspection into our lives, encourages us to assess our actions, and stimulates positive decisions concerning our future activities. *Cannabis* can enhance introspective thought, while allowing the user to experience differing realities and weigh differing points of view. Anything that promotes a human's abilities to get along with one another will certainly be of benefit in our overpopulated future.

A well-known side effect of *Cannabis* use is short-term memory loss. It may be aggravating to forget a person's phone number or a name told to you moments earlier, or even more so to become overly preoccupied with washing dishes and then smell smoke, realizing that you forgot the toast you started; however, forgetting may have distinct evolutionary advantages. Our minds have a gating mechanism in the form of filters that essentially screen our sensory input and decide what we need to retain in our active memory. Many of the little things that would simply load up our memory space pass right through and we remember only enough to maintain continuity in our thoughts. Short-term memory loss caused by *Cannabis* could be of evolutionary advantage by helping forget pain and suffering as well as less important mind-cluttering thoughts or allowing humans more time to focus on major survival or other challenges, such as finding shelter and procuring food.

Another common consequence of *Cannabis* consumption is the profound enhancement of appetite commonly referred to as the "munchies." Feeding is one of the most vitally important human endeavors, although an active appetite for food is presently of little value for most people in our over-fed modern Western societies. In fact, an increased appetite could lead to an even higher obesity rate and its attendant health problems. On the other hand, the stimulation of appetite may well have presented an advantage to our physically active ancestors who regularly faced monotonous diets with little flavor.

Cannabis has long been thought to increase the free association of ideas and boost intellectual creativity. Problem solving is another important aspect of human evolution. *Cannabis* is particularly effective in facilitating lateral thinking or "thinking outside the box" and thereby enhances creativity. Only with creative thought is problem solving possible, and without being able to solve problems no human progress would have been made. Generally speaking, promoters of conscious thought have been highly regarded by traditional societies.

Normally, *Cannabis* smoking is quite relaxing, allowing many users to feel euphoric, content, and untroubled by stressful surroundings. As Benjamin (2006) explains it, "Always the same world—yet one has patience." As our world becomes increasingly complex and troubling, *Cannabis* will become an increasingly valuable tool allowing one to step away for a brief time and see situations from a more objective viewpoint. *Cannabis* also serves as an excellent sleep aid for many users. Stress is increasingly recognized as one of our leading contemporary causes of disease, and for many people, *Cannabis* relieves stress and its unpleasant consequences effectively. *Cannabis* can also activate the pleasure centers and enhances cerebral as well as sensual activities such as the creation and enjoyment of musical and visual entertainment, writing and reading, food and drink, and of course

sexual and other fantasies. These are viewed by many to be additional, healthful benefits of the psychoactive effects of *Cannabis* (e.g., see Rätsch 2001; Guy et al. 2004; Russo and Grotenhermen 2006).

Human Influence on *Cannabis*'s Evolution

The widespread geographical distribution of *Cannabis* has obviously benefited from its relationship with humans. We greatly expanded its range by carrying it from its native homes, through a complex history of dispersal, to most regions of the earth (the spread of *Cannabis* due to human agency is detailed in Chapter 4). In addition, the morphological and physiological divergence of *Cannabis* has been expanded by human selection with its concurrent domestication for a wide range of uses (see Chapters 5 through 10). Human selection has improved fiber quality and yield, increased seed yield and oil content, altered chemical profiles to either maximize or minimize psychoactivity, and by increasing its THC level has made *Cannabis* one of the world's most widely used plants. Suites of characteristics have become modified to achieve each of these ends, based largely on the preferential selection pressures of humans. We have explored the evolution of *Cannabis* from its hypothetical prehistoric beginnings through its lengthy and at times intimate relationship with humans, and where possible we have highlighted the individual factors leading to the diversity observed in *Cannabis* today.

As this book has demonstrated, *Homo sapiens* and *Cannabis* have shared a long relationship spanning several millennia—at times fruitful and at times fitful—each constantly adapting to changes invoked by the other and evolving new strategies for continued coexistence. Humans have entertained fancies with many plants and have exercised varying levels of control over their evolutions, but few species have experienced such constant attention from humans as *Cannabis*. The ups and downs of this love-and-hate relationship have forged tenuous yet continuous links among human history, the evolution of *Cannabis*, and our shared futures. Humans have had lasting effects on *Cannabis* through changes in both cultural preferences for various plant products and societal pressures of acceptance or eradication impinging on the growth and reproduction of *Cannabis* populations.

Cannabis landraces and cultivars were selected essentially for fiber, food, or psychoactive drug content based on human needs and preferences. This has resulted in a generally dichotomous evolution of populations adapted for either industrial hemp or drug use, with few intermediate phenotypes. In the past three decades, *Cannabis* cultivation has enjoyed a global renaissance due initially to the international fascination with *Cannabis* as a recreational drug and more recently with the increased awareness of the values of medical *Cannabis* and industrial hemp. Small amounts of seed of both fiber and seed varieties, as well as drug varieties, have been disseminated widely in recent years although generally under clandestine circumstances. Nevertheless, geographically and culturally isolated gene pools were blended in unprecedented combinations to develop both industrial hemp and recreational/medicinal *Cannabis* cultivars that will be productive when grown in environments to which they are adapted.

Human selection has increased seed and fiber yield and quality, as well as altered cannabinoid profiles. The production of THC has been enhanced profoundly by human selection for potency and secondarily by natural selective pressures such as latitude, temperature, or ultraviolet light levels, which were facilitated by human introduction of *Cannabis* into new geographical areas where it thrived (e.g., see Pate 1983; Lydon et al. 1987). In addition, population size has influenced the rate of evolution, and selections for varying plant products have also altered the sexual expression of different populations.

During the early twentieth century, Western *Cannabis* breeders concentrated on improving fiber hemp varieties. Varietal improvement continues today, but with a largely covert focus on developing recreational and medicinal drug varieties. Currently, *sinsemilla* (seedless) marijuana growers in North America and Europe are breeding more potent drug varieties, many of which also find medical applications, but in general the breeding of new drug cultivars by growers has decreased due to the late twentieth- and early twenty-first-century trend of cultivating all-female crops from vegetative cuttings indoors using artificial lights. Under such conditions seeds are neither sown nor produced, and accelerated evolution due to human selection ceases.

Industrial hemp is once again spreading across the northern temperate regions of Asia, Europe, and North America as the popularity of hemp fiber textiles and paper, as well as hemp seed foods and body care products, grows. Industrial hemp varieties approved for cultivation in Europe and Canada were all artificially selected to produce very low levels of cannabinoids in general and THC in particular. Industrial hemp varieties are presently being developed for a wide range of individual purposes by following specific selection criteria. These include low pigment and lignin contents for paper pulp production, early or late maturation adapted to new growing regions from the Arctic Circle to the tropics, high quality fiber for spinning yarn and weaving fabrics, enhanced essential fatty acid (EFA) and protein contents for food and nutraceutical uses, and single component or blended chemical profiles for medical use (e.g., see Small and Marcus 2002 and Chapter 10). As additional specialized uses for hemp fiber, seed, and resin constituents are identified, the incentive to breed new varieties with enhanced traits to fulfill these needs will increase.

Long ago the quest for more potent *Cannabis* drugs promoted the technique of cultivating seedless *sinsemilla* marijuana, the current mainstay of Western smokers both for illicit recreational purposes and for legally approved medicinal applications. Due to unrelenting pressure from law enforcement, *sinsemilla* production has most recently moved indoors. Marijuana crops of several dozen or more plants were traditionally grown from seeds outdoors in the soil under natural light and at their most uniform were somewhat variable in composition. Indoor crops grown under electric lights in artificial growing media usually consist of rooted cuttings from a single female plant, so all the individuals in the crop are genetically identical. This results in an entirely different set of selection criteria due to the shift from an outdoor environment with sunlight and sexual reproduction via gametes (male pollen and female ovules) to an indoor environment with artificial light and asexual reproduction (only rooted female cuttings). Outdoor crops grown from seed exhibit much wider morphological and physiological variation than indoor clonal crops; they are the products of random sexual recombination, and therefore natural selection continues to play a background role to selection by farmers. Indoor, vegetatively propagated cultivars are strictly selected for early ripening, high potency, and large, dense inflorescences;

generally they produce very high yields of THC per area per year. Because growing space is normally limited and costly for illicit indoor cultivators, economic parameters weigh heavily in breeding selections. Consequently it is only those female plants that perform favorably under artificial conditions that are asexually multiplied. Under such circumstances, the female genetic base normally becomes very narrow (sometimes only one genotype) and there are no male plants grown. The absence of males precludes seed production as well as both natural and human selection of new genotypes, thus curtailing evolution at the crop level.

A recent breeding development (and another genetic dead end) is the production of uniform all-female seeds, essentially a female clone in a more convenient package—the seed. Storable and transportable, all-female seeds obviate the need for time wasting and space consuming planting and cutting operations of the vegetative mother. In addition, reproductively sterile female cultivars would allow the production of seedless marijuana no matter how much pollen is floating through the air. Breeders will always be trying to make further improvements and create new varieties, especially those that improve the quality and quantiy of their chosen *Cannabis* product, in other words, seeds, fiber or psychoactive resin.

Although a number of new cultivars have been developed over the past 50 years, the overall biodiversity of the gene pools of both wild-growing and cultivated *Cannabis* have been greatly reduced. This reduction can be attributed to a number of causes, including the acculturation of traditional *Cannabis* cultures, natural and agricultural habitat degradation and loss, failure to value and preserve genetic diversity, commercial exploitation, the spread of single cultivars, and outright eradication by law enforcement Vegetatively propagated NLD × BLD cultivars have evolved to the level of full domestication and complete dependence on humans for their reproduction—as with the majority of houseplants. Without the nurturing of humans (e.g., when threatened by an interruption in water or energy supply) these modern drug cultivars could quickly die and become extinct. *Sinsemilla Cannabis* clones have become exotic pets, like highly bred tropical aquarium fish or caged songbirds, unable to escape their confines and reproduce naturally.

Industrial hemp cultivation will continue to enlarge and may eventually come to dominate outdoor *Cannabis* production from temperate through tropical latitudes, and as long as it remains illegal, ever increasing amounts of the drug varieties will be grown indoors. Having become highly adapted to artificial lighting and hydroponic agriculture, as well as developing a reliance on the electric grid, *Cannabis* is strategically poised to continue its evolution in the artificial environments of grow rooms. No other drug-producing organism can be so readily cultivated in such a wide variety of natural as well as artificial environments—not the opium poppy, coca bush, tobacco plant, nor even psychoactive fungi.

Biotechnologies in the form of genome mapping and gene regulation have many potential applications that could be applied to *Cannabis*. Researchers have already identified unique gene sequences associated with a number of traits, including the sex of the plant, production of various fiber constituents and cannabinoids, regulation of EFA biosynthesis, and resistance to pathogens. The ability to regulate targeted regions of the genome could result in cultivars enriched in various cannabinoids for medical or recreational use, devoid of cannabinoids for use as a placebo in medical trials, enriched in EFA and protein content, or more resistant

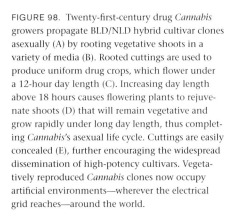

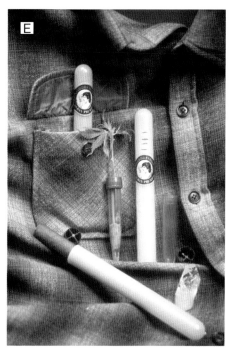

FIGURE 98. Twenty-first-century drug *Cannabis* growers propagate BLD/NLD hybrid cultivar clones asexually (A) by rooting vegetative shoots in a variety of media (B). Rooted cuttings are used to produce uniform drug crops, which flower under a 12-hour day length (C). Increasing day length above 18 hours causes flowering plants to rejuvenate shoots (D) that will remain vegetative and grow rapidly under long day length, thus completing *Cannabis*'s asexual life cycle. Cuttings are easily concealed (E), further encouraging the widespread dissemination of high-potency cultivars. Vegetatively reproduced *Cannabis* clones now occupy artificial environments—wherever the electrical grid reaches—around the world.

to attack by a specific pest or disease. Genome mapping may allow the accurate identification of the genetic background of *Cannabis* accessions for plant breeding; such mapping may also enhance taxonomic and forensic studies (e.g., see Jagadish et al. 1996; Gilmore et al. 2007).

Environmental Impact of the Human-*Cannabis* Relationship

Will *Cannabis* play a role in saving our environment? Or will it turn out to be just another water-grabbing and soil-degrading crop plant? Some environmentalists predict that potable water shortages will be one of the major environmental crises of the immediate future, so the feasibility of continuing to grow thirsty fiber crops comes into question, especially when artificial fibers are so diverse, functional, and affordable, as long as petrochemical supplies last. Also, intensive fertilizer use and certain fiber processing methods can lead to water pollution. Overall, *Cannabis* has had a somewhat checkered relationship with the environment. Although touted as an environmentally friendly crop for the future—primarily because it requires less pesticide application than other food and fiber crops such as cotton—*Cannabis* has shared in extensive environmental damage on several fronts. *Cannabis* crops generally require large amounts of water for proper growth and productivity, while industrial hemp fiber processing also uses large amounts of water for both plant growth and the processing of the desired product; indeed, many times throughout Eurasian history the retting of hemp stems in water caused severe water pollution (eutrophication), leading to fish kills as well as strongly offensive odors (see Chapter 5). In an environmentally controlled and water-rationed future, natural fibers may eventually have to yield to synthetic ones. On the other hand, if natural fibers remain in demand and water supplies allow production, hemp still promises to be a relatively environment-friendly fiber crop when compared to cotton. Cotton also requires plenty of water and nutrients, and more significantly it accounts for nearly 50 percent of all pesticide use internationally. Hemp has few pests of major economic impact even when grown in large monocrops (McPartland et al. 2000).

As *Cannabis* cultivation increased during the late twentieth century and law enforcement pressure forced growers farther into the countryside, more and more rural and wilderness land came to be appropriated for clandestine cultivation. Greedy marijuana growers have degraded natural environments worldwide by slashing and burning natural vegetation and polluting waterways with chemical fertilizers (e.g., in Northern California and Oregon, British Columbia, Morocco, Nepal, India, Colombia, Jamaica, and Mexico). Indoor growers also dump concentrated nutrient solutions and pesticide residues into urban sewage systems.

The American-inspired "War on Drugs" causes its own environmental and social damage through the aerial spraying of herbicides such as Paraquat to destroy marijuana fields and adjacent farm fields, resulting in crop losses and serious health problems for rural residents (e.g., Colombia, Jamaica, Mexico, etc.). Actions of this type have only served to increase the level of animosity between local citizens and advisors representing the United States government. Western nations also continue to experiment with soil-borne fungal pathogens for use as biological controls for marijuana crops

(McPartland et al. 2000). This is an obvious threat to the environment in general and legal industrial hemp cultivation in particular by cross-infection of nontarget organisms (e.g., industrial hemp or hop crops). Moreover, the release of a biological weapon into the environment is prohibited by international treaty.

Coevolution of *Cannabis* and Humans: Fresh Concepts

In this book we set out to investigate the evolutionary relationships and ethnobotanical history of humans and *Cannabis*. We have focused primarily on the impact of artificial selections that humans imposed on *Cannabis* during its evolution as a cultivated crop plant (Chapter 3) and the influences *Cannabis* has had on human history and cultural evolution (Chapters 3, 5 through 9). Repercussions upon the evolution of *Homo sapiens* resulting from human-*Cannabis* relationships have not been the main theme of our book and such effects were usually only considered in situations where changes in the human condition may have brought about subsequent changes in the evolution of *Cannabis* (Chapter 4). Whenever possible, we have tried to weigh all evidence from the "plant's perspectives" as well as from our human vantage point.

Cannabis has been used in shamanic rituals for a very long period of human history and vestiges of these ancient practices persist today within the major religious disciplines. While most of the dominant present-day religions have maintained some shamanic traditions, they have certainly expunged many more that were not in keeping with centralized coordination between church and state. Nevertheless, hemp smoke is still used among some cultures to fumigate and purify, hemp cordage is used by some to tie and protect souls, hemp cloth provides a bridge or gateway for the spirits during some healing practices, and hemp clothing is still used as the final attire of the corpse and worn by mourning relatives as a reminder of their filial piety in some areas of the world (see Chapters 5 and 9). Are these nonpsychoactive ritual uses relics of past ritual use of psychoactive *Cannabis*?

Had *Cannabis* not contained THC and therefore not altered human consciousness in such a profound way, would it be of much interest to us today? Indeed, would we have even written this book? Probably not, even though the historical contributions of *Cannabis* fiber and seed use have proven quite significant over time. It is undoubtedly the unique, some would say magical compound THC that has captured the affirmative interest of many millions of people in more modern times, with strongly contrary opinions by countless others who have come to consider it an abomination to human society. Without THC, *Cannabis* would just be another flax—a fine fiber and oil seed plant—but with limited geographical range and cultural usage. *Cannabis* resin glands—by secreting not only THC but also terpenes that impart the characteristic *Cannabis* aroma and modify and enhance THC's effects—are often believed to repel foraging insects or to protect the plant from the sun (e.g., see Russo 2011). Although at times these adaptive factors may have played a part in *Cannabis*'s evolutionary success, there is no denying that THC has attracted humans for millennia and has been a primary factor leading to its dissemination worldwide.

Since the onset of research into isolating the active ingredients of *Cannabis* in the 1960s, thousands of scientific and scholarly papers have been published concerning cannabinoid chemistry and use (e.g., for reviews see Mechoulam and Hanus 2000; Mechoulam et al. 2007). During the late 1980s and 1990s, scientists discovered the receptor sites activated by THC within the body and brain of humans as well as other animals from hydra and zebra finches to rats and primates (e.g., Devane et al. 1988; De Petrocellis et al. 1999; Soderstrom and Johnson 2001; Justinová et al. 2011). In this period, the first endocannabinoid (arachidonoylethanolamide), a substance produced by the human brain that binds to the same receptor as THC, was also discovered and named "anandamide" in honor of the Sanskrit word *ananda*, or "bliss." Synthetic antagonists were soon developed to block the action of THC, as well as other cannabinoids and anandamide. With this "Holy Trinity" of pharmacological research—the receptors (CB1, CB2, etc.), agonist ligands (molecules that bind to a receptor protein and activate the receptor, e.g., THC, anandamide, 2-AG, etc.), and the antagonist ligands (molecules that bind to a receptor protein and block the receptor, e.g., SR141716A)—researchers have begun to reveal the secrets of *Cannabis*'s effects on human consciousness and physiology (see Chapters 7 and 8). Endocannabinoids play a role in memory, sleep, pain perception, inner ear orientation, regulation of blood pressure, and immune system response. Based on the effects of endocannabinoids discovered so far, it seems likely that THC and other phytocannabinoids will find many additional medicinal uses.

But what have been the effects of *Cannabis* on the evolution of our species, *Homo sapiens*? Humans have obviously modified the course of *Cannabis*'s evolution, but has *Cannabis* significantly affected human evolution as well? And if so, do these mutual changes provide evidence for reciprocal coevolution between a simple plant, no matter how useful it has proven itself, and *Homo sapiens*—the species often perceived as the pinnacle of mammalian evolution and certainly the species that has created by far the most environmental change on earth?

Twenty-first-century research has brought us interesting insights into the evolution of *Cannabis* and has opened a discussion focusing on the possibility of human-*Cannabis* coevolution (e.g., see Pollan 2001; McPartland and Guy 2004a/b). *Cannabis* produces a number of secondary metabolites such as cannabinoids and terpenoids. The varying levels of these compounds have been proposed as possible taxonomic markers and indicators of geographical origin (Baker, Bagon, and Gough 1980; Baker, Gough, and Taylor 1980; Turner et al. 1979; see also Chapter 11). More recently, studies have utilized the banding patterns of proteins (Hillig 2005b)—the direct allelic products of genes—rather than the secondary products of gene-mediated biosynthesis, such as cannabinoids and terpenoids (Hillig and Mahlberg 2004; Hillig 2004a/b), to determine taxonomic classification and biogeography (Hillig and Mahlberg 2004); in addition, DNA banding patterns have been used to establish provenance for forensic purposes (Gilmore et al. 2007). Each of these chemotaxonomic parameters has proven to be of great value in understanding the evolution and breeding of *Cannabis* (see Chapters 3 and 10). As noted earlier, it was not until late in the twentieth century that researchers discovered the human cannabinoid (CB) receptors and their natural human endocannabinoid ligands, and these discoveries have helped us begin to unravel the mysteries of THC's effects on the mind and body. At least two different cannabinoid receptors are found in humans. CB1 (discovered in 1988) is found in the central nervous system

and is one of the most common receptors in the brain, while CB2 (discovered in 1993) is found widely in immune cells and tissues (Russo 2004). CB1 and CB2 have an equal affinity for THC while CB2 has approximately three times the affinity for CBD and CBN (minor physiologically active *Cannabis* constituents) than for THC. The varying affinity of the cerebral CB1 receptor and corporeal CB2 receptor may explain in part the variations in mind versus body effects of the various strains of *Cannabis*, which have diverse origins and chemical contents (McPartland and Guy 2004a/b).

The meta-analytical research of John McPartland and his coauthors (McPartland and Pruit 2002; McPartland 2004; McPartland and Guy 2004a/b; McPartland et al. 2006) draws on prior investigations in many disciplines to compare the evolution of cannabinoid profiles in *Cannabis* with the distribution of cannabinoid receptors in a wide range of organisms (e.g., fish, amphibians, birds, rodents, monkeys, and humans). This research also focused on relative rates of evolution of the CB1 receptor and proposes a case for biological coevolution between humans and *Cannabis*, casting a seminal role for *Cannabis* in the lengthy drama of human evolution.

Mutualism and coevolution are closely associated concepts. Mutualism is the condition whereby two species benefit reciprocally from their symbiotic interaction. Throughout this book we have described a wide variety of relationships between humans and *Cannabis* that can be interpreted as mutually beneficial, for instance, the widespread dissemination of hemp as a fiber crop. Biological coevolution is the expansion of a mutually beneficial relationship through enhancement of each other's fitness for survival and the eventual alteration of both species' genomes through natural selection, and in many cases of mutual relationships between humans and other species, this has involved artificial selection. Coevolution can and often does occur between differing taxonomic groups such as flowering plants (e.g., *Cannabis*) and mammals (e.g., *Homo sapiens*). The essence of coevolution is not simply the phenotypic change that we can perceive and measure but lies more fundamentally in the allelic changes in the genes themselves. The evolutionary interaction between *Cannabis* and humans investigated by McPartland and Guy (2004a/b) rests at the gene level, through the interaction of two gene products—THC synthase enzymes in *Cannabis* and cannabinoid receptor proteins in humans.

The human body produces several different fatty acid metabolites of arachidonic acid (a polyunsaturated essential fatty acid), which act as endogenous ligands of the cannabinoid receptors (endocannabinoids) and are the primary compounds that coevolved along with the cannabinoid receptors over millions of years. The best known of these endogenous ligands is anandamide, discovered in 1992 (Devane et al. 1992; Mechoulam and Hanus 2000). A complete understanding of the basic functions of the cannabinoid receptor system awaits further study, but it probably influences such essential mental realms as pain, personality, mood, creative thought, and concentration while participating in maintaining the basic homeostatic equilibrium of life itself. Zimmer et al. (1999) demonstrated that rats lacking CB receptors soon die. Furthermore, Kittler et al. (2000) showed that phytocannabinoids effect the functioning of at least 50 rat genes shared by humans.

Functionally similar but structurally different CB receptors occur in many animal life forms; CB1 is found in a broad range of organisms, ranging from the evolutionarily primitive hydra (polyps related to coral) through to complex mammals, and CB2 is found in all vertebrates (McPartland et al. 2006).

The amino acid sequences of these receptor proteins can be compared for homology of nucleotide base pairs, and the number of accumulated base substitutions can be measured and used to construct a phylogenetic tree for the CB receptor gene to establish the timing of evolutionary events. At the base of the theoretical evolutionary tree lies an ancestral CB receptor gene that gave rise to CB1 and CB2 genes during a duplication event that occurred at least 590 million years ago (McPartland 2004). Flowering plants may have begun evolving as long as 400 million years ago (Benton and Ayala 2003). Cannabaceae diverged from its Urticales lineage before Moraceae (Sytsma et al. 2002), more than 90 million years ago, when early Moreacean dioecious pollination systems first evolved (Datwyler and Weiblen 2004). *Cannabis* and *Humulus* must therefore have diverged and become separate genera even more recently around 22 million years ago (Zerega et al. 2005). *Homo sapiens* did not evolve until 400,000 to 250,000 years ago, and anatomically modern humans (AMHs) did not enter central Eurasia until about 50,000 years ago (Wells 2002; also see Olson 2002b; Forster 2004). Binding with phytocannabinoids such as THC, produced by a relatively recently evolved flowering plant such as *Cannabis*, is obviously not the original evolutionary role of cannabinoid receptors in animals. In terms of evolutionary scale, exposure of human CB receptors to phytocannabinoids is a very recent occurrence, theoretically less than one ten-thousandth of the time since the CB1 receptor evolved. The CB receptor system and other gene systems such as cytochrome P450, which protects us from plant toxins, were inherited by us from other animals because they were beneficial for survival, and millions of years have allowed considerable adaptation and evolutionary change in these systems (Sullivan et al. 2008).

CB1 receptors in humans are found in regions of the forebrain associated with development of language and music skills and could also be responsible for at least some of the psychological effects of *Cannabis*. High CB1 levels are also found in the thalamus, which contributes to development of the human personality and provides access to pain pathways, thus allowing THC-induced analgesia. Lack of CB1 receptors in the part of the brain controlling the cardiorespiratory functions may explain THC's low toxicity.

From an adaptive, mechanistic point of view, humans must have benefited from possessing the endocannabinoid system; therefore, our CB receptor genes have been preserved, perpetuated, and even proliferated over time. As early humans, we probably also benefited from our early association with *Cannabis*. For example, *Cannabis* use has been shown to improve night vision (Russo et al. 2004). Hunter-gatherer groups would have gained an advantage from heightened sensitivity to smells, color, sound, and subtle plays of light and shadow induced by *Cannabis*, which in turn enhanced our evolutionary fitness and survival as a species (e.g., see Sagan 1977). *Cannabis* use may also have precipitated other adaptive evolutionary changes in the human condition. Ethan Russo (2001) described an "endocannabinoid deficiency syndrome" (EDS) caused by abnormally low levels of anandamide or reduced sensitivity to its effects. EDS can result from dietary deficiencies of essential fatty acids (EFAs), resulting in lowered levels of anandamide synthesis. Genetic mutations could also cause deficits of CB receptors and/or decreased sensitivity of CB receptors to anandamide. Hemp seeds are one of nature's richest sources of *omega*-3 (linolenic) and *omega*-6 (linoleic) EFAs and provided a valuable food for early humans. Human consumption of THC and other phytocannabinoids may also help balance endocannabinoid deficiencies. Annual harvests of hemp seeds and THC-laden inflorescences, both products with at least a one-year storage life, could have provided ancient humans with a continual source of food rich in brain-enriching EFAs as well as medicines high in exogenous cannabinoids—both valuable during times of endocannabinoid deficiency. THC stimulates creative thought, improves lateral problem solving, and increases appetite, all advantageous to survival, particularly in times of environmental change. Both schizophrenia and Attention deficit disorder (ADD) may be associated with decreased CB1 receptivity to endocannabinoids, and many sufferers have received relief through self-medicating with exogenous cannabinoids in the form of smoked *Cannabis*.

Terrance McKenna (1992) proposed that the use of *Cannabis* by Paleolithic humans may have stimulated the emergence of language, a major key to the "great leap forward" in human evolution about 50,000 years ago, accompanied by the crafting of specialized tools from new materials (e.g., rope, nets, and cloth from hemp fiber), the utilization of new food plants (e.g., hemp seed), the altering of consciousness (e.g., ingestion of *Cannabis* drugs), the development of art and music, and the inception of social systems. Sudden acceleration in the evolution of human mental powers was preceded by an ancient random mutation resulting in the CB1 receptor and then much later may have been positively selected through prolonged contact with THC, beginning in Central Asia as early as 50,000 years ago during a warm interglacial period of the Pleistocene (see Chapter 4). Not only was *Cannabis* among the first fibers to be woven (see Chapter 5) and an early provider of a seed rich in EFAs (see Chapter 6), but it may also have promoted the inception of the very cognitive processes required to invent such complex crafts as weaving.

Cannabis also benefited simultaneously through range expansion, and the previous examples in this chapter provide strong evidence of mutualistic relationships between humans and *Cannabis*. However, proof of biological coevolution depends on establishing that changes in *Cannabis*'s genome during domestication were reciprocated by changes in humans simultaneously using *Cannabis*. Do human CB receptors exhibit an accelerated rate of evolution, and does this result from relatively recent exposure (beginning less than 50,000 and possibly only 15,000 to 10,000 years ago) of the ancient endocannabinoid system to the "new" ligands Δ^9-THC and 11-*hydroxy*-THC? We are measuring a mere moment in our recent genetic history, the last 50,000 years since AMHs evolved and began to move into Central Asia; on the other hand, the evolutionary process to develop and perpetuate the CB receptor system in the animal kingdom has spanned about 590 million years. Throughout human history we have been increasingly exposed to *Cannabis* drugs, yet on an evolutionary time scale, we are still making relatively early contacts with the naturally occurring phytocannabinoids. As a result of this contact with plant-based exocannabinoids over the past 10 to 50 years, evolution of the endocannabinoid receptor system in humans may have progressed to a new level. The rate of evolution may have been accelerated through contact with novel plant ligands that bind to the CB receptors and mimic the essential effects of the endocannabinoids. Elucidation of chimpanzee CB1 mutation rates will be very informative as we would expect chimps to have a lower rate of CB receptor mutation than humans as they have not been exposed to exogenous cannabinoid ligands.

Human genes associated with nervous system development evolve more than 50 percent faster than those in rats and mice, and receptor protein evolution is particularly accelerated in the lineage leading from macaques to humans; therefore, increased rate of gene mutation may provide underlying reasons for rapid and advanced evolution of the human brain (Dorus et al. 2004). By comparing amino acid sequences of humans, rats, and mice, McPartland and Pruit (2002) were able to determine that the CB1 receptor in humans evolved at a rate 3.25 times greater than rats, whose receptors in general usually undergo mutational changes at twice the rate of humans. Based on this accelerated rate of receptor evolution, it has been proposed that CB1 is undergoing positive selection in humans (McPartland and Guy 2004a/b). However, subsequent research by McPartland et al. (2007a) showed that the mean rate of evolution of 18 endocannabinoid-associated genes (including CB1 and CB2) in rats and mice was almost three times that in humans and also indicated strong purifying selection (few surviving mutations) with evolutionarily adaptive, positive selection in only four genes, which did not include those affecting CB1 or CB2. However, if the CB1 receptor does evolve at an accelerated rate in humans, could this be driven by exposure to phytocannabinoids?

Prolonged contact between humans and *Cannabis* was possible as much as 50,000 years ago when modern humans began to disperse into Eurasia while Central Asia was experiencing a warmer and more humid period before the final Pleistocene glaciation. Our view of the time frame involved is merely a brief window into a very long evolutionary process and raises many questions concerning the coevolution of *Cannabis* and *Homo sapiens*. Contact with early human relatives could have occurred as much as 400,000 years ago, long before the appearance of modern humans, but what if first contact was only about 12,000 years ago at the beginning of the Holocene? Voight et al. (2006) reported several recent, positively evolving gene systems related to changes in human plant and animal usage patterns brought on by the spread of agriculture during the early Holocene (e.g., lactose tolerance and the spread of cattle and gluten tolerance in Europeans where wheat has become the dietary staple) fewer than 11,000 years ago. Could accelerated evolution of the cannabinoid receptor system have occurred and still be occurring in such a short time span of 10 thousand rather than hundreds of thousands of years? Do all humans exhibit this accelerated CB1 evolution? Are there geographically variant levels of cannabinoid receptor gene mutation? Are CB1 receptors in more ancient populations in Central Asia or populations only recently exposed to THC as in the Americas evolving at different paces than those of African, Indian, and European descent? Obviously all humans have cannabinoid receptors and anandamide, but if we argue that exogenous THC stimulated accelerated receptor evolution, then we should demonstrate that groups unexposed to THC exhibit a different rate of mutation. Alternatively we could assume that the accelerated rate of CB receptor evolution occurred before humans spread out of Africa, before contact with *Cannabis*, and that presently cannabinoid receptor evolution may not be accelerating.

Cannabis evolved the B_D and B_T alleles for CBD and THC synthesis well before humans first encountered it and spread it across Eurasia. Humans have simply selected (consciously and unconsciously) for a higher frequency of one allele or the other according to which use of *Cannabis* they favored. Selection for drug use raised the frequency of the B_T allele and in many cases also lowered the frequency of the B_D allele. Selection for fiber and seed use lowered the frequency of the B_T allele and raised the frequency of the B_D allele. In this way humans have impinged directly on the biosynthesis of the two most prevalent exogenous phytocannabinoid ligands—CBD and THC.

Accelerated rates of nucleotide substitutions (mutations) in the human CB1 receptor gene could result from the presence of exogenous stimuli such as THC and the resultant selective pressure for the survival of CB receptor mutants. Could accelerated mutation rate also result from increased requirements for endocannabinoids in modern urban humans, leading to increased survival of CB1 receptor mutations by internal selective pressures apart from any requirement for cannabinoids as exogenous ligands? McPartland et al. (2007b) presented evidence for the coevolution of endocannabinoid ligands and cannabinoid receptor proteins. Have these two channels—exocannabinoid selective pressures from consuming *Cannabis* and endocannabinoid selective pressures imposed by rapid cultural and environmental change—worked together in accelerating the evolution of the CB1 receptor system?

Human CB1 receptors are much more sensitive to THC than CB1 receptors in rats. More actively evolving CB1 receptors might explain the 50-fold increase in sensitivity to THC of humans over laboratory rats reported by Varvel et al. (2001). Human CB1 gene mutants may have been positively selected both for greater sensitivity to, and resulting efficacy of, cannabinoids. This would only be of evolutionary advantage if modern *Homo sapiens* require more of the beneficial effects of cannabinoids. What McPartland and colleagues may have discovered is that the increasingly stressful changes of modern life have accelerated cannabinoid receptor evolution.

THC was an important plant compound in human evolution, and its early appeal may have led the way for humans to entertain symbiotic relationships with *Cannabis* based on the suite of benefits that *Cannabis* offers—fiber, food, medicine, and recreation. Because there are no reported fibers, seed protein, or EFA receptors, we are only able to genetically investigate the coevolutionary relationship with the cannabinoids and the rate of cannabinoid receptor evolution. Accelerated CB1 receptor evolution may result not just from THC's positive influence but also from other medicinally valuable compounds (e.g., EFAs, other cannabinoids, flavonoids, and terpenoids) that may also have been important in human evolution. *Cannabis* induces many physical and mental reactions, several of which may be simultaneously beneficial; therefore, CB1 mutations are even more likely to be preserved through positive selection. It is possible that recent contact with *Cannabis* has stimulated other receptor systems to evolve more rapidly; even the adaptive survival advantages of eating hemp seed and wearing hemp clothing may have exerted timely influences on human fitness for survival and may also have indirectly affected the evolution of modern *Homo sapiens* (e.g., see Gilligan 2007).

Rodents do not consume cannabinoids except as an aside to foraging for edible seeds; however, *Cannabis* seeds are a favored food of rats and mice, and their relationship with orally consumed THC could be very much more ancient than that of humans. Attraction to the aroma of *Cannabis* resin early on may have allowed rats to benefit from ingestion of nutritious hemp seed, which possess essential EFAs. Humans have utilized *Cannabis* for many purposes, the most significant for this discussion being its psychoactive use to reduce pain and anxiety, which can have positive consequences for survival. So how did *Cannabis* exert selective pressures on the human gene pool and the cannabinoid receptor system in particular? The cannabinoids are not mutagenic compounds and, thus, do not have the power to directly alter gene function or evolution by direct physical action. If cannabinoids are not directly biochemically responsible for invoking mutational change in genes, then how could they have accelerated the rate of mutational change in humans?

Nucleotide base pair substitutions occur spontaneously, and when they survive they give rise to genes with a differing nucleotide composition and function. These alterations of nucleotide sequences are called "mutations." The vast majority of receptor protein mutations are not beneficial to the survival of the organism, either being detrimental by binding to novel ligands that cause unfavorable reactions or simply being neutral and having no effect on the organism at all; these new mutated genes either are not passed on to subsequent generations or are passed on passively with no selection. However, a small minority of accidental and random, spontaneous mutations are beneficial to the organism, and if they are passed on to subsequent generations, they may confer an evolutionary advantage to the species as a whole. Rates of random nucleotide substitutions generally appear fixed across the animal kingdom, as detrimental changes in nucleotides are dropped from the gene pool and cannot be measured. The vast majority of surviving nucleotide substitutions are functionally neutral (conferring no evolutionary advantage or disadvantage) and, along with a very few beneficial mutations, are preserved in the genome, and can be measured.

The cannabinoid receptor system continues to be perpetuated and improved because it is of fundamental importance to the successful functioning of nearly all branches of the animal kingdom. If we accept that nucleotide substitutions occur spontaneously at a rate based on the average number of random substitutions in a given time period, then an accelerated evolution rate for CB receptors indicates that more total substitutions are being positively selected than the average for related organisms and genomes. Under ambient selective pressures the mutation rate is constant, and when exposed to new selective pressures the mutation rate increases. This could occur because an additional set of exogenous ligands offers more opportunities for survival of random mutations of the CB receptor gene. When there are new exogenous ligands added to the ancient suite of endocannabinoid ligands, there are more types of molecules present to potentially bind to a mutant CB receptor. This can create a beneficial endocannabimimetic effect; for example, some mutant receptors that allow bonding by exogenous cannabinoid ligands may provide a physiological advantage (increased fitness) and would have a higher chance of being perpetuated in subsequent generations. We assume that early humans consumed Cannabis soon after they made first contact with this multipurpose plant. This would have immediately supplied more ligands for mutants of the CB receptors, thereby providing possible beneficial outcomes (e.g., hunting and foraging success, language development, improved health, pain relief, and increased happiness) and thus promoting the survival of the mutant gene in the human gene pool. The process of base pair substitution followed by death (purifying selection) or survival (adaptive selection) of mutant genes occurs constantly in all organisms. The survival of CB1 receptor gene mutants in Homo sapiens may have been enhanced by coevolution of cannabinoid receptors with endocannabinoid ligands, the exogenous application of phytocannabinoids, and the suite of benefits Cannabis provides, thereby increasing the rate of evolution of the cannabinoid receptor system.

In Cannabis the THC synthase gene (B_T) has been positively selected by humans for millennia, and on the human side of this coevolutionary partnership is the CB1 receptor, which may have undergone accelerated evolutionary change under exposure to THC. In other words, there has been a reciprocal relationship driven by both partners in the equation.

Certainly Cannabis has been attractive to humans for many uses and we have indeed disseminated it globally. Humans have benefited from the relationship by receiving a seminally important food, fiber, fodder, and fuel plant, as well as the benefits of medicines and psychoactive stimulation. In turn, humans selected varieties with increased seed and fiber production as well as enhanced medicinal efficacy with a minimum of side effects. Homo sapiens and Cannabis are intimately involved in what Geertz (1973) described as a "reciprocally creative relationship" between human culture and biology, which provides an accurate way to characterize the human-Cannabis relationship without having to prove actual biological coevolution. Whether or not the case for coevolution is proven, there obviously exists a strong mutualistic relationship between humans and Cannabis, and this symbiosis has certainly benefited both. Indeed, it is difficult to imagine the evolution of human culture and human nature without the interplay of powerful plants such as Cannabis (see Pollan 2001).

Humans have likely used Cannabis drugs for millennia, spanning innumerable human generations, yet the worldwide psychoactive use of Cannabis has been on the increase for only a few centuries. The spread of Cannabis use is apparently unstoppable, possessing an internal momentum that disregards each and every legal system that attempts to suppress it. Are our cannabinoid receptor genes mutating faster than ever, as we apply more and more exogenous THC? Does our collective consciousness as Cannabis users (in an ecologically adaptive sense) guide us to consume Cannabis at this time in our evolution? Is that why the will of Cannabis users has remained so strong, despite the increasing threat imposed by law enforcement? The endocannabinoid system is present throughout much of the animal kingdom and appears to be of fundamental importance in human physiological and psychological control systems. We humans, as members of the experimenting species, Homo sapiens, have encountered many beneficial plants and incorporated them into our traditional pharmacopeia. The widespread popularity of recreational psychoactive Cannabis use is entirely the result of endocannabinoid receptors of the human brain being activated by THC. Could Cannabis play a key role in adaptively modifying our behavior and assist human evolution in a new and challenging industrialized environment? Will Cannabis provide a voice of reason and help us reflect inwardly and imagine a new and appropriate path toward our shared future? This point of view has, of course, many detractors at all levels of governments, not only as the past century has shown, but with a much deeper history also bearing witness to varying amounts of acceptance and repression of Cannabis use for mind-altering purposes (see Chapters 7 and 9).

Present Position of the Human-Cannabis Relationship

Cannabis is certainly one of the most widespread groups of plants on the planet. Cannabis owes its nearly universal presence and immense popularity to its ancient history of use for fiber, food, oil, and drugs but most importantly today to its its psychoactive effects. *Cannabis* is now found in diverse environments on every continent, ranging from marijuana grow rooms wherever the electric grid has penetrated, through traditional temperate and semitropical agricultural settings to steamy tropical jungles. The human-*Cannabis* relationship is unfolding now more rapidly than at any previous time. In the United States, staunch anti-*Cannabis* attitudes are largely held by government institutions and private corporations that feel threatened by the widespread introduction and use of an economic plant with so many advantages and so few proven faults (e.g., see Grinspoon and Bakalar 1997; Joy et al. 1999; Pollan 2001; Russo and Grotenhermen 2006). Recreational and medical *Cannabis* use has caused a radical polarization of thought. In the continuing debate over the legal status of *Cannabis*, there seems to be little middle ground, and favoring either side places one in certain opposition to those favoring the other. The freedom to use *Cannabis* industrially, nutritionally, medically, or recreationally is viewed by many as a basic human rights issue and has become a rallying point for civil rights and state's rights activists. Others believe psychoactive use of *Cannabis* is a "sin" and a "gateway drug" to physically addictive drugs, or they may be convinced that there is no place for another legally sanctioned mind-altering drug—rationalizing that we already have enough societal problems associated with alcohol and tobacco misuse and addiction. *Cannabis* is perceived either as a beneficial plant ally of real and potential value to us and to our environment or as a potentially dangerous adversary promoting widespread freedom of thought and creating major sociocultural problems.

Such moral and environmental controversies have precipitated an avalanche of strong public opinion in certain circles. However, the most serious obstacle to the success of marijuana decriminalization movements is widespread political apathy, fed by the general social tolerance of *Cannabis* use by a majority of Western societies (with the obvious exception of most active-duty law officials). Bringing an end to *Cannabis* prohibition has largely become a nonissue in the eyes of most citizens, even more so as medical *Cannabis* gains acceptance on the one hand and legal cultivation loopholes are increasingly abused on the other (e.g., in California). There is certainly no shortage of marijuana in most of the Western world—where legislators have tried their hardest to eradicate it—further encouraging general apathy on the part of users. The most immediately important political question for those who favor legalization of *Cannabis* use for all its myriad applications is the following: will the collective consciousness of *Cannabis* consumers mature to form a unified voice in defense of the cultivation and general use of a presently illegal plant?

Politicians (most often intentionally) and the general public (due to lack of education) still confuse marijuana and industrial hemp. It is unreasonable to fear social repercussions from the rampant smoking of hemp fiber varieties, since they nearly always contain an insufficient amount of THC to cause any psychological effect (less than 1 percent THC) and often contain higher amounts of CBD as well. It is unfortunate that harmless and extremely useful plants such as fiber and seed hemp are confused with their psychoactively potent relatives and that the cultivation of industrial hemp is restricted by laws intended to control illicit marijuana cultivation. Various groups representing American farmers have had no success in gaining approval for industrial hemp cultivation, while Canadian farmers have grown commercial hemp seed crops since the 1990s and supply the vast majority of hemp seed oil products consumed in the United States. American legislators continue to confuse fiber and seed varieties of *Cannabis* with drug varieties, and industrial hemp will probably not be grown again in the United States until marijuana is legalized, even though it is legally cultivated in Canada and throughout Eurasia.

Almost every country in the United Nations forbids the possession and use of *Cannabis* for all but a few legally sanctioned industrial purposes, yet we know that illegal recreational and medical use is widespread on a global scale. *Cannabis* grows prolifically in diverse climates, even indoors, when provided with good soil and enough water and light. In response to law enforcement pressure and the limited availability of high quality marijuana, growing your own, whether for personal pleasure, commercial profit, or self-medication, is a trend rapidly spreading in North America and Europe. There are an increasing number of locations in the world where casual, discrete *Cannabis* cultivation and use by individuals is generally overlooked by law enforcement, although every nation has strict laws against *Cannabis* smuggling and sales. Traditional drug *Cannabis* production areas (e.g., parts of India, Nepal, and Morocco) remain more tolerant of users, and many areas of South and Southeast Asia, as well as Central and South America, are also increasingly lenient in terms of personal possession and use. In some places, the political climate surrounding medical *Cannabis* legislation has become more informed, compassionate, and moderate. The localized approval of medical *Cannabis* use in many regions has blurred the line between recreational enjoyment and medical need, and marijuana and hashish use are de facto legal in parts of North America and Europe. However, in most of the world, recreational and medical *Cannabis* is only safely grown and used in the privacy of one's own home. The Western world (led primarily by the United States) hounds "drug-producing nations" around the world to eradicate *Cannabis* production and exports claiming that it is world supply that drives the American appetite for drugs. Now the United States has become a major marijuana supplier and use is higher than ever, while foreign governments search for differing legal solutions to their own situations.

Cannabis cultivation for personal medical use will eventually be legalized or tolerated in many additional jurisdictions, resulting from increasing governmental awareness of the conflict inherent in attempting to prohibit a popular and effective medicine. Pharmaceutical research companies are developing new natural cannabinoid formulations and delivery systems that will meet government regulatory requirements and be used to relieve a growing number of medical indications. As clinical trials prove successful, and the understanding of *Cannabis*'s efficacy and safety as a modern medicine spreads, patients can look forward to a steady flow of new *Cannabis* medicines providing welcome relief.

Increasing public awareness is also illustrated by examples from fashion and the media. The prevalence of *Cannabis* leaf imagery despite of—or perhaps in reaction to—increased repression by law enforcement is a growing phenomenon. The outline of the *Cannabis* leaf is among the most widely recognized and rapidly spreading popular images worldwide, associated with mainstream personal adornment products such as clothing, apparel, and jewelry as well as body art. More and more films and television programs portray the casual recreational use of *Cannabis* in decreasingly judgmental settings. Users may still run afoul of the law as always, but they are much less likely to be ostracized by their neighbors.

On the one hand, marijuana smoking seems condoned and at times even accepted by the media, while on the other hand, thousands more marijuana users are imprisoned worldwide each year. Consumers will continue to seek illicit marijuana, and a clandestine market with its associated social ills will persist until *Cannabis* use becomes legal.

Does humanity have the right or the reason to continue waging an already long, expensive, socially disrupting, and perhaps intractable war against *Cannabis*, especially before we have allowed sufficient time to understand its many potential benefits? The high societal expenses resulting from the enforcement of anti-*Cannabis* laws, in terms of both the direct costs of law enforcement and the taxes required to support it, as well as the impact of incarceration on those found guilty and their families, have become significant burdens on modern societies. The United States is experiencing the highest incarceration rates ever for simple possession and cultivation of *Cannabis* for personal use. The overreaction by Western (especially US) lawmakers to the perceived threat of recreational *Cannabis* use and the pressuring of beleaguered governments to adopt political and economic anti-*Cannabis* campaigns have created an international "drug war" consciousness resulting in thousands of *Cannabis* cultivators, sellers, and users languishing in prisons worldwide. In a recent study, specifically drawing attention to the "war on marijuana" and how it transformed the overall war on drugs, King and Mauer (2006) argued that law enforcement resources were not being allocated effectively to those offenses that are most costly to society: "The financial and personnel investment in marijuana offenses, at all points in the criminal justice system, diverts funds away from other crime types, thereby representing a questionable policy choice; by 2002, marijuana arrests in the United States had reached nearly 700,000 annually and are still increasing today." The US Federal Bureau of Investigation (FBI) reports that police prosecuted 858,408 persons for marijuana violations in 2009, an increase over 2008, and the second highest level ever, 2007 having been the highest on record (NORML 2010). The FBI reported the arrest of 757,969 people in the US in 2011 for marijuana-related offenses (NORML 2012),

In the absence of outside influences, many traditional cultures understand the uses of *Cannabis* and are comfortable with them. Martin (1975) contends that in Southeast Asia: "Distrust regarding hemp appears among individuals having cultural and social attitudes patterned after those of the West. As for the peasants, they experiment with everything that belongs to their universe, often have complete knowledge of all of the elements that compose it, and know how to use them in moderation. There is thus nothing surprising in the fact that they consider *Cannabis* to be a plant that is socially beneficial. This is an example of application of knowledge regulated by folk understanding." This simple, time-tested control mechanism has proven effective for generations and may be trusted to moderate *Cannabis* use today. Will greater acceptance and use of either industrial or drug *Cannabis* really have a lasting effect on world civilization? *Cannabis* alone certainly cannot save our planet, yet clearly, the inequities of today's world are largely unacceptable to those who favor *Cannabis* use for medicine, recreation, or both. Therefore, *Cannabis* users present an open challenge to the societal and cultural status quo even under the stigma of official illegality. We believe that marijuana consumption and cultivation for personal use will eventually be tolerated or possibly legalized in a majority of nations; this could occur through increasingly favorable public attitudes toward legalization as societies become more

aware of the injustice inherent in marijuana prohibition. Suppression of personal choice continues, but widespread tolerance is probably inevitable. We are presently living in times of more rapid change in all aspects of our lives and societies than ever before in history, and the human-*Cannabis* relationship is not insulated from the rapid pace of environmental and social changes. We can foresee—in the relatively near future, at least within our youngest reader's lifetimes—that humans will come to peace with *Cannabis* and both may benefit more fully from our intimate relationship.

Remaining Questions and Future Directions

There are several predictions that can be proposed for the near future of the human-*Cannabis* relationship. These are based on a number of factors including the current role of *Cannabis* in modern society as a provider of medical relief, its continuing status as one of the most popular recreational drugs over a huge geographic range, its survival through countless economic and legal conflicts, and increasing awareness of hemp as an environmentally friendly crop.

First, we predict that *Cannabis* will continue to provide efficacious medicines for an increasing array of medical conditions, and these will be accepted by a widening range of people. The medical *Cannabis* movement is founded on the real life medical experiences of citizens living in wealthier Western countries who have found profound relief from their suffering. It is a movement based on compassion for the medically desperate who can find no other medicine as effective as *Cannabis*. Medical researchers and pharmaceutical companies, as well as patient's rights groups, civil libertarians, and other activists, are making steady progress in the public reevaluation of *Cannabis* as an effective medicine. It is very likely that in the near future the medical *Cannabis* movement will be successful in removing penalties for those in real medical need. It is also hoped that governments will support systems for providing *Cannabis* medicines to patients—those who are unable to work, have few resources, and/or may be physically or mentally challenged. Pilot programs started by concerned activists, and largely operating illegally, have continued to supply medical *Cannabis* to their patient groups. Many have become well integrated into local society and are likely to proliferate across the United States and Canada unless governments intervene.

Second, the "drug war" will end: either by the more amenable (yet less likely) scenario of governments simply declaring victory—"OK, we successfully lowered drug use and now we can get on with core issues like drug education"—or the multinational efforts to illegitimatize the recreational use of *Cannabis* will dwindle or possibly even cease through the eventual implosion of the unrealistic war against drugs; this would occur due to blowback or extensive collateral damage associated with the "drug war," which will no longer be deemed acceptable by the international community. This "war" has had its greatest impact in the United States. Michael Pollan (2001) pointed out that "by the end of the 20th century this plant and its taboo had appreciably changed American life not once but twice: the first time rather mildly, with marijuana's widespread popularity beginning in the sixties, and then again, perhaps more profoundly, in its role as *casus belli* [justification] in the war against drugs." Unfortunately, prolongation of the "drug war" through at least the near future now seems unavoidable as illicit drugs of all kinds, including

Cannabis, are under scrutiny as potential funding sources for suspected "terrorist" activities worldwide.

Third, the popularity of hemp products will continue to grow as new uses arise in the fiber and food industries. Hemp products constitute a strong, if presently relatively minor, presence in international markets, and the hemp industry continues to grow, driven by the realization of Western consumers that hemp provides sensible alternatives to mainstream products. Since 1990, industrial hemp production has increased in many parts of the world resulting from both the rediscovery of traditional uses and the invention of new ones.

While imagining the future of *Cannabis* during the remainder of our lifetimes, two main questions must be addressed. From the human perspective, which of its many uses will we deem appropriate in aiding our cultural progress? First, let us look at what we likely may or may not require of *Cannabis* in the future. Because human cultural requirements have largely determined the course of *Cannabis*'s evolution up to now, it makes sense to also present the plant's perspective: how will humans continue to impinge on the evolution of *Cannabis* as an industrial crop?

It should be clear to the reader that *Cannabis* has proven a useful ally to many of the world's ancient as well as presently dominant cultures. It is also evident that the utilization of *Cannabis* products has varied among peoples living in diverse geographical regions at differing times in their cultural development. Our knowledge of when, where, and why past cultures chose to adopt *Cannabis*, and what are its most popular uses today, should shed some light on its possible appropriate future uses. The gifts that *Cannabis* has provided for humans can be grouped most simply as fiber, food, and drugs.

Hemp fiber was obviously of great value in temperate regions throughout history until only very recently in the human cultural timeline. This claim is supported by the extensive historical records of its use for textiles and paper-making (see Chapter 5). The invention of joining, twisting, and weaving plant fibers had a profound influence on late Mesolithic and early Neolithic cultures, allowing much greater choice of clothing and domestic fabrics. This eventually stimulated a growing human dependence on the farming of fiber crops, of which *Cannabis* was among the first and in some regions the foremost. *Cannabis* provided some of the earliest plant fibers to be manufactured into pulp paper and fueled early attempts at industrial mass production. Although there should remain little doubt as to the value of hemp fiber in our past, what role might hemp fiber play in our future?

Hemp fiber was of much greater value to preindustrial agrarian societies, falling steadily into disuse during the increasingly industrialized times of the late twentieth century. There were several reasons for this decline in popularity including competition from the rampant felling of softwood trees for paper pulp, the widespread growing of cotton, and the wholesale adoption of petrochemical-based and other synthetic and natural fibers. Although these strategies may not be sustainable for the long run, so far hemp has not been accepted generally as a viable economic alternative. Presently only a handful of small companies produce hemp fiber or seed. Hemp fiber does remain economically valuable to a small fraction of rural peasant farmers, although they also aspire to a higher economic status. Therefore they are rapidly giving up their traditional subsistence life-styles while turning with increasing frequency to selling other farm produce such as vegetables and purchasing industrially manufactured goods. No matter how appropriate hemp fiber production may once have been to subsistence farmers and cottage industries, there is no reason, short of total economic collapse, that will encourage most farmers to return to subsistence agriculture.

Cannabis can grow spontaneously in a range of environments disturbed by humans, and its inherently weedy nature and ability to survive harsh conditions as well as a lack of human attention have led to a common misconception of its requirements as a crop plant. Economically feasible production of hemp fiber or seed requires large tracts of arable farmland. *Cannabis* crops also require large amounts of natural rainfall or irrigation, well-drained alluvial soils, and significant amounts of nutrients. *Cannabis* crops cannot tolerate marginal field conditions if they are to thrive and be productive. As a consequence of global warming, in the relatively near future much of our planet's arable regions will become more arid, and as unpolluted water supplies become scarce, traditional field agriculture itself may be threatened. Farm fields are very inefficient users of water, yet we depend almost entirely on agriculture for the sustenance of our growing population, and as potable and agricultural water supplies dry up, less lucrative industrial crops will begin to lose out to food crops. Hemp fiber utilization is also very labor intensive and textile production costs are much higher than for cotton and synthetics. Plant fibers are at the lowest end of the range for profitability for farmers, below most food crops and well below drug crops. Food, in the form of highly nutritious seeds, is another important natural gift from *Cannabis* that has served humans well for millennia and may offer a brighter future than fiber. It should also be remembered that hemp seeds have been a food source of varying significance for domestic animals (see Chapter 6).

The psychoactive properties of *Cannabis* have certainly provided its most enduring and controversial legacy, at least when measured by its increasingly widespread medicinal and recreational use. In spite of the continuing heated disagreement concerning its reputed societal harms and medical benefits, it has become quite obvious that *Cannabis* drugs are here to stay. The acceptance of the medical benefits of *Cannabis* and the concurrent failure of the "drug war" could ultimately stimulate the legalization of *Cannabis* cultivation and possession for personal medicinal and recreational use. The growing realization that moderate *Cannabis* use constitutes one of the least threatening of our contemporary social problems should eventually lead to the general acceptance of *Cannabis*, for its benefits as well as its shortcomings. The use of hemp in familiar medical and household products will show *Cannabis* in a new light as a safe and beneficial plant. A concerted effort to properly educate youth and reeducate adults could change the prejudice against *Cannabis* into a positive force promoting it as a beneficial ally. Widespread acceptance of *Cannabis* use for many purposes will open more doors to meaningful scientific research—from farmer's fields to the laboratories of molecular biologists.

In many ways, despite centuries of familiarity and study, *Cannabis* is still a mysterious plant and although we have developed several evolutionary and ethnobotanical hypotheses, many important questions germane to our more complete understanding remain only partially answered. Where was the original home of *Cannabis*? Where and when did it evolve? Which scientific methods will improve our understanding of its early evolution? Where and when were the first widespread uses of *Cannabis* for drug, food, or fiber?

Several research disciplines could contribute much more to our understanding of these issues. Ethnobotanical studies of the remaining cultures that cultivate and use *Cannabis* would allow informed conjecture that could shed light on earlier traditional uses. The discovery of more detailed historical references from ancient eras would enlighten us with accounts of earlier *Cannabis* use. Additional archaeobotanical discoveries of ancient *Cannabis* remains will provide us with more accurate dates for the early range and first uses of *Cannabis* while also corroborating historical accounts and biochemical data. It is hoped that more DNA studies of *Cannabis* to determine the genetic affinities of its different geographical races will give us a more fundamental understanding of prehistoric evolution, dispersal, and early domestication of the genus.

Although recent advances in *Cannabis* research have led us to a better understanding of *Cannabis* taxonomy and evolution (Hillig 2004a/b; Hillig and Mahlberg 2004), continued study could clarify remaining questions. *C. sativa* is more genetically diverse than either *C. ruderalis* or *C. indica*, which may result from its lengthy association with humans (Hillig 2004a). Conversely, does this mean that because it is less genetically diverse *C. indica* has a shorter history of contact with humans? Is European fiber and seed use more ancient than Asian fiber and drug use? Future *Cannabis* evolutionary studies should move from protein electrophoresis directly to analysis of homologous DNA gene sequences. Additional seed accessions from evolutionarily interesting regions such as areas of possible origin, early evolution, and suspected hybridization of the primary gene pools omitted from previous taxonomic studies should be surveyed and compared (e.g., Central Asia, southern China, the Himalayan Foothills, North Africa, the Balkans, the Caucasus Mountains, and Anatolia).

Social questions will only be answered after the passing of time. Given evidence that we have presented, it appears that human-*Cannabis* relationships much more often than not have been beneficial to humans with few if any disadvantages. Why then is *Cannabis* prohibited today? Cynical scrutiny directs us to determine who benefits from prohibition. Many suffer due to the illegality of *Cannabis*, but who benefits? How will *Cannabis* weigh in the future of human evolution? Will industrial hemp be of major economic value and importance and what will it take to make it so? Will the majority of humans overcome their deep-seated prejudices against *Cannabis*, which are constantly reaffirmed through the intentional or misguided presentation of misinformation? Will we allow *Cannabis* to provide us with the beneficial gifts with which it is naturally endowed and we are naturally entitled?

A major report by the Global Commission on Drug Policy (2011) recently concluded that international criminalization of selected psychoactive drugs, including those affecting the use of *Cannabis*, have failed with devastating consequences. This report was produced by a 19-member commission that included former United Nations secretary-general Kofi Annan, former presidents of Mexico, Brazil, and Colombia, and other leading members of civil society. They summarized their conclusions regarding the unsuccessful, planetary "war on drugs" and all the repressive measures directed toward it over the past half century:

> The United Nations Single Convention on Narcotic Drugs came into being 50 years ago, and when President Nixon launched the US government's war on drugs 40 years ago, policymakers believed that harsh law enforcement action against those involved in drug production, distribution and use would lead to an ever-diminishing market in controlled drugs such as heroin, cocaine and cannabis, and the eventual achievement of a "drug free world". In practice, the global scale of illegal drug markets—largely controlled by organized crime—has grown dramatically over this period. While accurate estimates of global consumption across the entire 50 year period are not available, an analysis of the last 10 years alone shows a large and growing market.

The Global Commission on Drug Policy compared the United Nations' estimates of annual drug consumption over a 10-year period. In 1998 an estimated 147.4 million people consumed psychoactive *Cannabis*; that amount increased to 160 million by 2008, an increase of 8.5 percent (also see the recent report by the United Nations Office on Drugs and Crime 2011). The Global Commission on Drug Policy summarized their main recommendations as follows:

> End the criminalization, marginalization and stigmatization of people who use drugs but who do no harm to others. Challenge rather than reinforce common misconceptions about drug markets, drug use and drug dependence. Encourage experimentation by governments with models of legal regulation of drugs to undermine the power of organized crime and safeguard the health and security of their citizens. *This recommendation applies especially to cannabis*, but we also encourage other experiments in decriminalization and legal regulation that can accomplish these objectives and provide models for others. (Italics added by the present authors)

In the course of our extensive study of *Cannabis*, we hope that we have adequately described and explained much of the ethnobotanical and evolutionary history of a long-term, multipurpose plant resource. Nevertheless, a great amount of this lengthy history needs to be elucidated, and we expect the next few years to provide us with additional insights. It is the opinion of G. K. Sharma (1977b), renowned Himalayan ethnobotanist and a student of Richard Evans Schultes, that "a combination of folklore and modern tools of research in biology, biochemistry, and anthropology will surely result in a wealth of information. Investigation on *Cannabis* and biological problems associated with it have not yet received the support and attention warranted by their potentialities for human welfare. International and national efforts must be made to understand basically this plant of such great significance." Many questions do remain, and much of this investigative research suggests paths for further inquiry. We strongly urge emerging scientists to apply creative scientific techniques in their search for answers. The parallel evolutions of *Homo sapiens* and *Cannabis* continue, and it will be fascinating to watch where our human-*Cannabis* relationships take us. In closing, another quotation from our inspirational mentor and friend Richard Evans Schultes (1973) seems even more pertinent than it did about 40 years ago:

> There can be no doubt that a plant that has been in partnership with man since the beginnings of agricultural efforts, that has served man in so many ways, and that, under the searchlight of modern chemical study, has yielded many new and interesting compounds will continue to be a part of man's economy. It would be a luxury that we could ill afford if we allowed prejudices, resulting from the abuse of *Cannabis*, to deter scientists from learning as much as possible about this ancient and mysterious plant.

Cannabis in History

12,000 BCE	The earth begins to warm as the Holocene Epoch begins, and plants and animals begin to recolonize Eurasia from glacial refugia.
8000 BCE	Antecedents of Japanese Jōmon culture already using hemp seed and leaving remains.
5000 BCE	Earliest European hemp seed remains deposited in Germany.
5000–4000 BCE	*Cannabis* seed imprints in pottery, Dniester-Prut region, Moldova
4000 BCE	Ancient Egyptians build the first sailing ships.
3300 BCE	Indus Valley Civilization comes to an end, and introduced drug *Cannabis* biotypes eventually become widely used in many areas of South Asia.
3000 BCE	Hemp seed remains appear in the Baltic region.
2800 BCE	Earliest hemp seed remains from China, and the first assumed written record of *Cannabis* use for medicine is in the pharmacopoeia of Emperor Shen Nung, the legendary father of Chinese medicine.
2700 BCE	Remarkably well-preserved *Cannabis* flowers, seeds, stems, and leaves are left in a Yanghai burial tomb of a shaman in western China.
2200 BCE	Ancient Yellow River civilization begins to consolidate power in northern China, and *Cannabis* is an important multipurpose, cultivated plant.
2000 BCE	First hemp seed evidence from the Balkan region.
800 BCE	Yangzi River cultures become powerful in southern China.
600 BCE	Phoenicians pioneer the first sea trade routes in the eastern Mediterranean and Red Seas.

500 BCE	*Cannabis* is described in the Persian Zoroastrian *Avesta* sacred text.
500 BCE	Earliest hemp seed remains from the Korean Peninsula.
440 BCE	Greek historian Herodotus describes the Scythians of the Eurasian steppes who breath in fumes of burning *Cannabis*.
420 BCE	Hemp seed offerings left in Scythian kurgan tombs in Central Asia.
325 BCE	Greek geographer and astronomer Pytheas makes first recorded sojourn to England and Scandinavia by sail.
100 BCE	Chinese make first paper from *Cannabis* and mulberry.
70 CE	Roman physician Dioscorides records *Cannabis*'s medical properties.
600 CE	Papermaking spreads to Korea.
640 CE	The Koran, Islam's central religious text, tolerates *Cannabis* use but forbids alcohol.
900 CE	Viking expeditions begin reaching Iceland, Greenland, and Newfoundland.
950 CE	Muslim Moors introduce papermaking with *Cannabis* to Spain from North Africa.
1000 CE	The English word "hempe" first listed in a dictionary.
1023	Chinese Song dynasty issues the first paper money.
1149	Oxford University is founded in Oxford, England.
1160	Hildegard von Bingen writes *Physica* describing the medicinal use of *Cannabis*.
1200s	The magnetic compass commonly used on Chinese oceangoing ships.
1206	Genghis Khan leads the Mongol armies and conquers much of Eurasia.

(continued)

1241	Gunpowder introduced to Europe by the Mongols.
1250	European sailors begin to use the magnetic compass.
1275	Marco Polo starts on his alleged 20-year trip to China and reports use of hemp fiber for paper making, caulking of Chinese ships, and cultivation near oases in eastern Turkestan.
1315	The Great Famine begins in Europe.
1346	The bubonic plague starts in China and spreads westward through Europe killing at least one-quarter of Europe's population.
1405	Chinese send forth a huge fleet of 300 ships under the command of Admiral Zheng He, exploring as far as the Middle East.
1453	Christian Constantinople conquered by Turkic invaders.
1440s	German inventor Johann Gutenberg revolutionizes knowledge transfer by combining the printing press, movable metal type, and an oil-based ink.
1484	Pope Innocent VIII vilifies *Cannabis* as an unholy sacrament of Satan.
1492	Spanish explorer Christopher Columbus lands in the Bahamas, leading ultimately to the colonization of the New World and introduction of several Old World plants including *Cannabis*.
1495	First paper mill in England started.
1498	Portuguese explorer Vasco da Gama sails to India via southern Africa and the Cape of Good Hope.
1500s	Tobacco and smoking reach Europe from the New World, eventually this leads to the smoking of psychoactive products of *Cannabis* most often combined with tobacco.
1514	Copernicus argues that the earth revolves around the sun.
1519 to 1521	Ferdinand Magellan's fleet circumnavigates the globe in relatively small, tubby, hemp-rigged ships.
1533	King Henry VIII issues his first royal decree ordering each farmer to set aside a quarter acre of land for every 60 acres he controlled to cultivate hemp as a strategic crop.
1545	Spanish bring hemp cultivation to Chile for cordage and cloth.
1550–1850	The Little Ice Age strikes Europe; crops fail and many starve.
1563	Queen Elizabeth I decrees that land owners must grow hemp or pay a £5 fine.
1564	King Philip orders hemp to be grown throughout the Spanish Empire from Argentina to Oregon.

1569	Mercator publishes his cylindrical projection map of the earth.
1585	Thomas Hariot first writes about inhaling an amazing herbal remedy called tobacco.
1602	Dutch United East India Company (VOC) founded.
1606	British begin to grow hemp in Canada for maritime use.
1611	British begin to grow hemp in Virginia colony.
1619	Virginia becomes first American colony to make hemp growing mandatory.
1630s	Hemp traded throughout the American Colonies.
1735	Carolus Linnaeus introduces his taxonomic system for naming species.
1750s to 1790s	George Washington and Thomas Jefferson experiment with growing hemp on their farms.
1753	*Cannabis sativa* described and classified by Linneaus.
1778	After visiting Australia, James Cook is the first European to travel to Hawai'i.
1783	*Cannabis indica* described and classified by Lamarck.
1791	President Washington imposes import duties on hemp to encourage domestic industry, and Thomas Jefferson urges farmers to grow hemp instead of tobacco.
1800s	Narrow-leaf hemp (NLH) *Cannabis* sativa is widely grown in the United States and used to make rope, sails, and clothes. Narrow-leaf drug (NLD) *Cannabis indica* becomes a common ingredient in popular medicines.
1807	Napoleon signs treaty with Russia severing all legal Russian hemp trade with Britain. American sailors commence illegal trade in Russian Hemp.
1812	Napoleon invades Russia hoping to control the supply of hemp.
1841	Scotsman William O'Shaughnessy learns of the medical use of *Cannabis* in India.
1845	Frenchman Jacques-Joseph Moreau de Tours documents the medical benefits of *Cannabis*.
1844 to 1849	The most active years for the Club de Hashichins in Paris, which attracted many literary and intellectual elite in the city including, among others Moreau, Gautier, Baudelaire, Delacroix, and Dumas.
1857	Fitz Hugh Ludlow's *The Hasheesh Eater* is published.

(continued)

1872–76   Scientific Challenger expedition makes many discoveries and established oceanography; the expedition's, mother vessel, HMS Challenger, was supplied with 291 km (181 miles) of Italian hemp for depth sounding.

1859   Charles *Darwin* publishes his classic *The Origin of Species* describing evolution by natural selection and opens the ongoing debate of "evolution versus creationism."

1860   Ohio State Medical society conducts first governmental study of *Cannabis* use and health.

1860s   Augustinian friar Gregor Mendel lays the foundation for modern genetics.

1869   University of California established.

1870   *Cannabis* is listed in the United States Pharmacopoeia as a treatment for various ailments.

1881   Charles and Sir Francis Darwin publish *The Power of Movement in Plants*, investigating fundamental aspects of plant growth.

1890   Queen Victoria's personal physician, Sir Russell Reynolds, prescribes *Cannabis* for menstrual cramps and claims that when pure preparations of *Cannabis* are administered carefully, it is a most valuable medicine.

1894   The *Indian Hemp Drugs Commission Report* concludes that *Cannabis* has medical uses, no addictive properties, and a number of positive emotional and social benefits.

1910s   The song "La Cucaracha" about marijuana smoking becomes popular during the Mexican Revolution.

1910s-1920s   Following the Mexican Revolution, Mexicans immigrated to the United States and introduced recreational *Cannabis* use.

1910   African American "reefer" use reported in New Orleans jazz clubs, and Mexicans reported to be smoking marijuana in Texas.

1911   Hindus reportedly using "*gunjah*" in San Francisco. South Africa begins to outlaw *Cannabis*.

1912   Controlling *Cannabis* use is discussed at the first International Opium Conference in the Netherlands.

1913   California outlaws marijuana possession and use.

1916   United States Department of Agriculture calls for expansion of hemp acreage to replace timber use by the paper pulp industry.

1919   Texas outlaws marijuana possession and use.

1920   Alcohol prohibition begins in the United States.

1923   League of Nations delegates call for international *Cannabis* controls. Britain insists on further research before controls are imposed.

1924   Nikolai Ivanovich Vavilov famed Russian botanist, geneticist, and researcher of crop plant origins visits Afghanistan and describes and names *C. indica* ssp. *afghanica* broad-leaf drug (BLD) hashish plants.

Russian botanist Janischewsky describes and names *Cannabis ruderalis*.

The second International Opiates Conference declares *Cannabis* a narcotic and recommends strict international controls.

1925   The International Opium Convention in Geneva bans the use of *Cannabis* except for authorized medical and scientific purposes.

1925   The *Panama Canal Zone Report* concludes that there is no evidence that *Cannabis* use is habit-forming or deleterious and recommends that no action be taken to prevent its use.

1928   The Dangerous Drugs Act of 1925 becomes law and *Cannabis* becomes illegal in Great Britain.

1930   The Federal Bureau of Narcotics is formed with Harry J. Anslinger appointed as its first commissioner. American jazz icon Louis Armstrong is arrested in Los Angeles for possession of marijuana.

1930s   Marijuana use steadily becomes regulated, state by state, across the United States through the Uniform State Narcotic Drug Act.

1933   Alcohol prohibition ends in the United States.

1936   Antimarijuana movie drama *Reefer Madness* depicts lurid behavior and insanity resulting from smoking marijuana.

1937   The United States Marihuana Tax Act makes possession or transfer of *Cannabis* illegal except for medical and industrial uses, which remain highly regulated. Prohibitively high taxes destroy the American hemp industry.

Samuel Caldwell sentenced to four years in Leavenworth prison for selling two "joints" and becomes the first casualty of the Marijuana Tax Stamp Act.

(continued)

1938   *Popular Mechanics* magazine publishes an article written before the Marijuana Transfer Tax was passed extolling the virtues of "Hemp—the New Billion Dollar Crop."

1941   *Cannabis* dropped from the American Pharmacopoeia. *Popular Mechanics* magazine reveals Henry Ford's plastic car made using hemp fiber reinforcement and fueled with hemp seed oil.

1942   United States Department of Agriculture releases the movie *Hemp for Victory*, encouraging American farmers to resume hemp cultivation to support the war effort.

1944   New York mayor LaGuardia's marijuana commission reports that marijuana use causes no violence at all and cites other positive results. Anslinger responds by threatening doctors with prison sentences if they carry out independent *Cannabis* research.

1945   *Newsweek* reports that more than 100,000 Americans smoke marijuana.

1948   Anslinger, who previously claimed that marijuana smoking incited extreme violence, now states that it causes users to become pacifistic.

1951   The United Nations *Bulletin on Narcotic Drugs* estimates there are 200 million *Cannabis* users worldwide.

1952   First *Cannabis* arrest in the United Kingdom at the Number 11 Club, Soho.

The American Geographical Society publishes University of California, Berkeley, professor Carl O. Sauer's *Agricultural Origins and Dispersals*, his early treatise on the evolution of crop plants.

1956   The United States Narcotics Control Act makes first-time *Cannabis* possession offense punishable by a minimum of 2 to 10 years in prison and a fine of up to $20,000.

1961   The United Nations Single Convention on Narcotic Drugs limits production and trade of drugs exclusively to medical and scientific purposes and establishes international cooperation to prevent illegal drug trafficking. International restrictions are placed on *Cannabis* aiming to eliminate its use within 25 years.

1962   President John F. Kennedy retires Anslinger as Bureau of Narcotics commissioner.

1964   The active ingredient of *Cannabis*, delta 9-tetrahydrocannabinol or Δ^9-THC, isolated by Yechiel Gaoni and Raphael Mechoulam at Hebrew University in Jerusalem, Israel.

The first "head shop" opens in the United States.

1967   In New York on Valentines Day, Abbie Hoffman and the Yippies mail out 3,000 marijuana cigarettes to addresses chosen at random from the phonebook.

Rolling Stones' Keith Richards and Mick Jagger sentenced to prison for marijuana possession, but their convictions are overturned on appeal.

More than 3,000 people hold a mass "Smoke-In" in Hyde Park in London.

University of California Press publishes Carl O. Sauer's "Environment and Culture during the last Deglaciation" in John Leighly's *Land and Life* and Edgar Anderson's *Plants, Man, and Life* key resources in the study of biogeography and human-plant interactions.

1968   The United States Bureau of Narcotics and Dangerous Drugs (BNDD) formed by combining the Bureau of Narcotics and the Bureau of Drug Abuse Control into one agency.

1968   A British Home Office select committee concludes that *Cannabis* is no more harmful than tobacco or alcohol and recommends that penalties be reduced.

1969   England's Misuse of Drugs Act prescribes a maximum five years' imprisonment for *Cannabis* possession.

1970s   Broad-leaf drug varieties (*C. indica* ssp. *afghanica*) introduced to Europe and the United States from Afghanistan. Hybrids with narrow-leaf drug varieties (*C. indica* ssp. *indica*) from Colombia, Mexico, and Thailand form the basis of modern drug *Cannabis*.

1970   United States Controlled Substances Act regulates the prescribing and dispensing of psychoactive drugs, including stimulants, depressants, and hallucinogens. *Cannabis* is classified as having high potential for abuse and no medical use.

United States Congress repeals mandatory penalties for *Cannabis* offenses.

Canadian *Le Dain Report* claims that the debate on the nonmedical use of *Cannabis* "has all too often been based on hearsay, myth and ill-informed opinion about the effects of the drug."

Marijuana Transfer Tax of 1937 declared unconstitutional by the United States Supreme Court.

England's' Misuse of Drugs Act lists *Cannabis* as a Class B drug and bans its medical use.

(continued)

United States President Nixon declares drugs "America's public enemy No. 1."

Harvard University Press publishes *Marihuana Reconsidered* by Harvard psychiatrist Dr. Lester Grinspoon.

1972   The United States passes a $1 billion antidrug bill and the government's *Shafer Report* voices concern at the 4-year 10-fold increase in the level of spending used to stop illicit drug use.

The California Marijuana Initiative (CMI) to remove penalties for adults using, possessing, growing, processing, or transporting marijuana for personal use fails by a two-thirds majority.

Mark D. Merlin releases *Man and Marijuana*, the first scholarly attempt at understanding the diverse aspects of their ancient relationship.

1973   The United States Drug Enforcement Administration (DEA) created by President Richard Nixon through an Executive Order with a mandate to wage a global war on drugs.

President Nixon declares, "We have turned the corner on drug addiction in America."

1975   Doctors call on United States government to instigate further research on *Cannabis*.

Supreme Court of Alaska declares that a citizen's "right of privacy" protects *Cannabis* possession in the home. Limit for public possession is set at one ounce.

1976   President Gerald Ford's administration bans government funding of medical *Cannabis* research. Pharmaceutical companies allowed to research synthetic cannabinoid analogues.

The Netherlands adopts policy of tolerance toward *Cannabis* users.

Robert Randal becomes first American to receive medical marijuana from federal government supplies under an Investigational New Drug program.

Robert Dupont, Ford's chief advisor on drugs, declares that *Cannabis* is less harmful than alcohol or tobacco and urges its decriminalization.

1976   Sacred Seeds established as the first drug *Cannabis* seed company.

Beautifully illustrated *Sinsemilla Marijuana Flowers* by Jim Richardson and Arik Woods revolutionizes seedless marijuana growing and presents the possibility of making select seeds.

1977   Robert Connell Clarke publishes *The Botany and Ecology of Cannabis*, exploring the natural parameters of *Cannabis* growth and development.

1978   New Mexico makes *Cannabis* available for medical use.

1980s   The ready availability of seed in the Netherlands since the 1980s accelerated the spread of drug *Cannabis* in many regions.

1980   The Beatles' Paul McCartney spends 10 days in a Japanese prison for *Cannabis* possession.

1981   *Marijuana Botany* by Robert Connell Clarke explores the propagation and breeding of distinctive *Cannabis* and presents techniques for vegetative propagation (cloning).

1983   United Kingdom convictions for *Cannabis* possession exceed 20,000, having risen from just under 15,000 in 1980.

1985   Unimed releases Marinol, the "pot pill" containing synthetic THC called dronabinol in sesame seed oil.

Charles B. Heiser of Indiana University publishes his popular *Of Plants and People*, an introduction to the study of economic plants.

1988   In Washington, DC, judge Francis Young concludes that "marijuana in its natural form is one of the safest therapeutically active substances known to man" and recommends that medical use of marijuana should be allowed. The Drug Enforcement Administration rejects the ruling, and the US Senate adds $2.6 billion to federal antidrug funding.

Allyn Howlett and William Devane characterize the existence of a cannabinoid receptor in rat brains.

1989   Outgoing US president Ronald Reagan declares victory in the War on Drugs as a major achievement of his administration, while Secretary of State James Baker reports that the global war on narcotics production "is clearly not being won."

1990   Science journal *Nature* reports the mapping of a cannabinoid receptor system in the human brain by Miles Herkenham and his team.

1991   In the United Kingdom, 42,209 people are convicted of *Cannabis* offences and 19,583 are issued cautions.

Hortapharm BV founded in Amsterdam, Netherlands, begins research and development of single cannabinoid varieties for use by pharmaceutical companies and produce cannabinoids from natural plant sources.

(continued)

1992 Anandamide, the endogenous cannabinoid compound produced in the human body, is discovered by Czech analytical chemist Lumír Hanuš and American molecular pharmacologist William Devane and synthesized by organic chemist Aviva Breuer in the laboratory of Raphael Mechoulam at Hebrew University in Jerusalem, Israel.

International Cannabinoid Research Society holds their first official meeting at Keystone, Colorado.

HortaPharm BV established in the Netherlands to breed medical *Cannabis* cultivars and produce cannabinoids from natural plant sources.

1993 Hemcore becomes the first British company to be granted a license to grow industrial hemp as the Home Office lifts restrictions on cultivation.

Winifred Rosen and Andrew Weil write *From Chocolate to Morphine*, an easily accessible guide to mind-altering drugs.

Yale University Press publishes *Marihuana the Forbidden Medicine* by Lester Grinspoon and James B. Bakalar.

1994 Britain's Home Secretary Michael Howard increases maximum fines for possession from £500 to £2,500.

Hempflax reestablishes hemp cultivation in the Netherlands.

The United Kingdom's Association of Cannabis Therapeutics talks to the Department of Health about possibility of legalizing *Cannabis* for medical use.

Germany becomes the first European country since the Netherlands to decriminalize possession of small quantities of cannabis for occasional use.

1995 Ten millionth *Cannabis* arrest in the United States.

1996 Compassionate Use Act passes in California allowing the use and cultivation of medical *Cannabis*.

1997 The British newspaper *The Independent on Sunday* launches a "Decriminalise Cannabis" campaign.

1998 'Medisins' high-THC medical cultivar granted Plant Breeders Rights in the Netherlands.

Robert Connell Clarke's *HASHISH!* explores hashish history, cultures, ingredients, recreational and medicinal use, and production techniques.

2000 After four years of *Cannabis* repression under the Labour Party administration of Tony Blair, the climate of British public opinion begins to change and the crackdown on *Cannabis* loses support.

2001 The US Supreme Court rules that federal antidrug laws do not permit an exception for medical *Cannabis*.

The British government sets up a Select Committee to look at drugs policy, and Home Secretary David Blunkett announces his intention to move *Cannabis* from Class B to Class C, making possession a nonarrestable offence.

Michael Pollan's best-selling *The Botany of Desire: A Plant's-Eye View of the World* investigates the mutually beneficial relationships between humans and *Cannabis*.

2004 GW Pharmaceuticals granted Plant Breeders Rights for 'Grace' cultivar in the United Kingdom.

2005 The United States Supreme Court rules that the Commerce Clause of the United States Constitution allows the federal government to ban the use of *Cannabis*, including medical use.

2008 The Global Commission on Drug Policy compared the United Nations' estimates of annual drug consumption over a 10-year Period; in 1998 an estimated 147.4 million people consumed psychoactive *Cannabis*, which increased 8.5 percent to 160 million by 2008.

2009 *Cannabis* is restored to Class B of the British Misuse of Drugs Act.

2011 In the United States, 16 states and Washington, DC, have approved medical *Cannabis* programs.

2012 Sativex made from *Cannabis* grown in the United Kingdom by GW Pharmaceuticals from natural *Cannabis* extracts is now approved in the United Kingdom, Spain, Germany, Denmark, New Zealand, Canada, and Austria for the treatment of spasticity due to multiple sclerosis, cancer pain, and neuropathic pain of various origins.

In the United States, 18 states and Washington, DC, have approved medical *Cannabis* programs, while the FBI reported the total arrest of 757,969 people in the US in 2011 for marijuana-related offenses.

REFERENCES

Aalto, M., and H. Heinäjoki-Majander. 1997. "Archaeobotany and Palaeoenvironment of the Viking Age Town of Staraya Ladoga, Russia." *Pact* 52:13–30.

Abbo, S., A. Gopher, B. Rubin, and S. Lev-Yadun. 2005. "On the Origin of Near Eastern Founder Crops and the 'Dump-heap Hypothesis.'" *Genetic Resources and Crop Evolution* 52 (5): 491–95.

Abel, E. L. 1980. *Marijuana: The First Twelve Thousand Years.* New York: Plenum.

Academia Turfanica. 2006. *Selected treasures of the Turfan relics.* Turpan, China: Academia Turfanica.

Accorsi, C. A., M. Bandini Mazzanti, and A. M. Mercuri. 1998. *Evidence of the cultivation of* Cannabis *in Roman Times in the Holocene Pollen Diagrams of Albano and Nemi Lakes (Central Italy).* Proceedings of the *VII International Congress of Ecology—INTECOL*, Florence, Italy, July.

Accorsi, C. A., M. Bandini Mazzanti, A. M. Mercuri, C. Rivalenti, P. Torri, C. Balista, and P. Bellintani. 1998a. "Analisi pollinica di saggio per l'insediamento palafitticolo di Cànar-Rovigo, 680–700ms.l.m. (AnticaEt`a del Bronzo)." In *Cànar di San Pietro Polesine. Ricerche archeo-ambientali sul sito palafitticolo. Padusa Quaderni, 2,* edited by C. Balista and P. Bellintani, 131–49. Rovigo, Italy: Centro Polesano di Studi Storici Archeologici ed Etnografici [in Italian].

Accorsi, C. A., M. Bandini Mazzanti, L. Forlani, N. Giordani, M. Marchesini, S. Marvelli, and G. Bosi. 1998b. "Archaeobotany of the Cogneto Hiding Well (Modena; Northern Italy; 34 m a.s.l.; 44°38'12" N 10°35'2"E; Late Roman-Modern Age)." *Proceedings of the 1st International Congress on: "Science and Technology for the Safeguard of Cultural Heritage in the Mediterranean Basin,"* November 27–December 2, 1995, Catania, Siracusa-Italy. 1: 1537–44.

Ackroyd, P. 2007. *Thames: The Biography.* New York: Nan A. Talese/Doubleday.

Adams, J. M. 1997. *Global Land Environments Since the Last Interglacial: Europe During the Last 150,000 Years.* Oak Ridge, TN: Oak Ridge National Laboratory. http://www.esd.ornl.gov/projects/qen/nercEUROPE.html.

Adams, J. M., and H. Faure, eds. 1998. "Preliminary Land Ecosystem Maps of the World Since the Last Glacial Maximum—Review and Atlas of Palaeovegetation." *Quaternary Environments Network (QEN).* http://www.esd.ornl.gov/projects/qen.

Adovasio, J. M., O. Soffer, and B. Kléma. 1996. "Upper Palaeolithic Fibre Technology: Interlaced Woven Finds from Pavlov I, Czech Republic, c. 26,000 Years Ago." *Antiquity* 70 (269): 526–34.

Adovasio, J. M., O. Soffer, and J. Page. 2007. *The Invisible Sex: Uncovering the True Roles of Women in Prehistory.* New York: Smithsonian Books.

Agadjanian, A. K. 2006. "The Dynamics of Bioresources and Activity of the Paleolithic Man, Using the Example of Northwestern Altai Mountains." *Paleontological Journal* 40 (5): 482–93.

Agadjanian, A. K., and N. V. Serdyuk. 2005. "The History of Mammalian Communities and Paleogeography of the Altai Mountains in the Paleolithic." *Paleontological Journal* 39 (6): 645–821.

Ahn, S. M. 2004. "The Beginning of Agriculture and Sedentary Life and their Relation to Social Changes in Korea." In *Cultural Diversity and the Archeology of the 21st Century, The Society of Archeological Studies 50th Anniversary Symposium, Okayama, Japan,* 40–52. Okayama, Japan: The Society of Archeological Studies.

Ahokas, H. 2002. "Cultivation of *Brassica* Species and *Cannabis* by Ancient Finnic Peoples, Traced by Linguistic, Historical and Ethnologicala Data; Revision of *Brassica Napus* as *B. Radice-Rapi.*" *Acta Botanica Fennica* 172:1–32.

———. 2003. *Major Discharge and Floods of the Kymi River (Kymijoki) over the Salpausselkä I esker to Coastal South Finland Traced by* 14C *datings, Unequal Land Upheaval, Oral Tradition and Onomastics.* Helsinki, Finland: Hannu Ahokas.

Aikens, C. M., and T. Higuchi. 1982. *Prehistory of Japan.* New York: Academic.

Alberta Government. 2007. "Industrial Hemp Production in Canada." *Agriculture and Food.* http://www1.agric.gov.ab.ca/$department/deptdocs.nsf/all/econ9631.

Aldrich, M. R. 1971. *A Brief Legal History of Marijuana.* Paper presented at the Western Institute of Drug Problems, Marijuana Conference, August 7, Portland, Oregon.

———. 1977. "Tantric Cannabis Use in India." *Journal of Psychedelic Drugs* 9:227–33.

———. 1997. "History of Therapeutic Cannabis." In *Cannabis in Medical Practice,* edited by M. L. Mathre. 35–55. Jefferson, NC: McFarland.

Alekseev, A. Y., N. A. Bokovenko, Y. Boltrik, K. A. Chugunov, G. Cook, V. A. Dergachev, N. Kovalyukh, G. Possnert, E. M. Scott, A. Sementsov, V. Skripkin, V. Vasiliev, and G. Zaitseva. 2001. "A Chronology of the Scythian Antiquities of Eurasia Based on New Archeological and 14C Data." *Radiocarbon* 43:1085–107.

———. 2002. "Some Problems in the Study of the Chronology of the Ancient Nomadic Cultures in Eurasia (9th–3rd Centuries BC)." *Geochronometria: Journal on Methods and Applications of Absolute Chronology* 21:143–50.

Alenius, T. 2007. "Environmental Change and Anthropogenic Impact on Lake Sediments During the Holocene in the Finnish – Karelian Inland Area." PhD diss., University of Helsinki, Faculty of Science.

Alexander, D. 2000. "The Geography of Italian Pasta." *The Professional Geographer* 52 (3): 553–66.

al-Hassan, A. Y., and D. R. Hill. 1986. *Islamic Technology: An Illustrated History.* Cambridge: Cambridge University.

Allen, H. N. 1908. *Things Korean.* New York: Fleming H. Revell.

Allen, J. L. 1900. *The Reign of Law.* New York: Macmillan.

Allison, E. P., A. R. Hall, H. K. Kenward, W. J. B. McKenna, C. M. Nicholson, and T. P. O'Connor. 1996. "Further Excavation at the Dominican Priory, Beverley, 1986–89." In *Sheffield Excavation Reports* 4, edited by M. Foreman, 195–212. Sheffield, SD: Academic Press.

Alonso, N., and J. Juan-Tresserras. 1994. "Fibras de Lino en las Piletas del Poblado Ibérico del Coll del Moro (Gandesa, Terra Alta): Estudio Paleoetnobotánico." *Trabajos de Prehistoria*, 51 (2): 137–42.

Alt, K. W., J. Burger, A. Simons, W. Schön, G. Grupe, S. Hummel, B. Grosskopf, W. Vach, C. Buitrage Téllez, C. H. Fischer, S. Möller-Wiering, S. S. Shrestha, S. L. Pichler, and A. von den Driesch. 2003. "Climbing into the Past—First Himalayan Mummies Discovered in Nepal." *Journal of Archeological Science* 20:1529–35.

Altisent, A. 1967–68. "Una Societat Mercantile a Catalunya a les Darreries del Segle XII." *Boletin de la Real Academia de Buenas Letras de Barcelona* 32: 45–65.

———. 1970. "Comerç Maritim Icapitalisme Incipient. Episodis de la Vida Econòmica d'un Matrimony Tarragoni (1191–1203)." In *Micellánia Histórica Catalana*, edited by R. Saladrigues, 161–80. Tarragona, Spain: Abadia de Poblet.

Ambrazevicius, R., ed. 1996. "Lithuanian Roots: An Overview of Lithuanian Traditional Culture." *Lithuanian Folk Culture Center*. http://thelithuanians.com/booklithuanianroots/node55.html.

American Heritage Dictionary: Dictionary of the English Language. 2000. 4th ed. Boston: Houghton Mifflin.

Ames, F. 1958. "A Clinical and Metabolic Study of Acute Intoxication with *Cannabis Sativa*." *Journal of Mental Science* 104:972–99.

Ames, O. 1939. *Economic Annuals and Human Culture*. Cambridge, MA: Harvard University.

Ampontan [Bill Sakovich]. 2008. "Hemp Stalk Prophecies in Japan," January 19. http://ampontan.wordpress.com/2008/01/19/hemp-stalk-prophesies-in-japan/.

Anderson, E. 1948. "Hybridization of the Habitat." *Evolution* 2:1–9.

———. 1956. "Man as a Maker of New Plants and New Plant Communities." In *Man's Role in Changing the Face of the Earth*, edited by W. L. Thomas, 763–77. Chicago: University of Chicago.

———. (1952) 1967. *Plants, Man, and Life*. Berkeley, CA: University of California.

Anderson, E. N. 1988. *The Food of China*. New Haven, CT: Yale University.

Anderson, F. J. 1977. *The Physica of Hildegard of Bingen*. In *An Illustrated History of Herbals*, edited by F. J. Anderson, 51–58. New York: Columbia University.

Andersson, J. G. 1923. *An Early Chinese Culture*. Reprint from *Bulletin of Geological Survey of China*, no.5. Peking: Ministry of Agriculture and Commerce.

Andrews, F. H. 1935. *Descriptive Catalogue of Antiquities Recovered by Sir Aurel Stein, K.C.I.E., Ph.D., D.Litt., D.Sc., Archeological Survey of India (Retd.), Fellow of the British Academy, During his Exploration in Central Asia, Kansu and Eastern Iran*. New Delhi, India: Central Asian Antiquities Museum.

Ansorge, J., K. Igel, H. Schäfer, and J. Wiethold. 2003. "Ein Holzschacht aus der Baderstraße 1a in Greifswald. Aus der Materiellen Alltagskultur Einer Hansestadt in der Zweiten Hälfte des 14. Jahrhunderts." *Bodendenkmalpflege Mecklenburg-Vorpommern Jahrb* 2002 (50): 119–57.

Anthony, D. W. 1986. "The 'Kurgan culture,' Indo-European Origins, and the Domestication of the Horse: A Reconsideration." *Current Anthropology* 27 (4): 291–313.

———. 1991. "The Domestication of the Horse." In *Equids in the Ancient World*, edited by R. H. Meadow and H. P. Uerpmann, 250–77. Weisbaden: Ludwig Reichert.

———. 1998. "The Opening of the Eurasian Steppe at 2000 BCE." In *The Bronze Age and Early Iron Age Peoples of Eastern Central Asia*, vol. 1 (Journal of Indo-European Studies Monograph 26), edited by V. H. Mair. Washington, DC: The Institute for the Study of Man.

———. 2007. *The Horse, the Wheel, and Language: How Bronze-Age Riders from the Eurasian Steppes Shaped the Modern World*. Princeton, NJ: Princeton University.

Anthony, D. W., and D. R. Brown. 1991. "The Origins of Horseback Riding." *Antiquity* 65:22–38.

Antonić, D., and M. Zupanc. 1999. *Srpski Narodni Kalendar za 2000 Godinu* [Serbian Folk Calendar for 2000]. [In Serbian.] Beograd, Serbia: Vukova zaduzbina.

Antonijević, D. 1971. *Aleksinacko Pomoravlje*. In *Srpski etnografski zbornik LXXXIII*, 111–20. [In Serbian.] Beograd: Srpska akademija nauka i umetnosti.

Antzyferov, L. V. 1934. "Hashish in Central Asia." *Journal of Socialist Health Care in Uzbekistan* 65:259–82.

Arata, L. 2004. Nepenthes and Cannabis in Ancient Greece. *Janus Head* 7 (1): 34–49.

Ardelean, I. I. 1893. *Monographia comunii Chítichaz. Kétegyháza község monográfiája* [Monograph of the Chitighaz village]. Arad, Hungary.

Arensberg, I. 2006. *Ostindiefararen Götheborg seglar igen: The Swedish Ship Götheborg sails again*. Bilingual Swedish/English ed. Sävedalen, Sweden: Warne förlag.

Argant, J., J. A. López-Sáez, and P. Bintz. 2006. "Exploring the Ancient Occupation of a High Altitude Site (Lake Lauzon, France): Comparison between Pollen and Non-Pollen Palynomorphs." *Review of Palaeobotany and Palynology* 141 (1–2): 151–63.

Arobba, D. 2001. "Macroresti botanici rinvenuti nei livelli tardoantichi e medievali del Battistero della Cattedrale di Ventimiglia." *Rivista Studi Liguri* 66:197–212.

Arobba, D., R. Caramiello, and P. Palazzi. 2003. "Ricerche archeobotaniche nell'abitato medievale di Finalborgo (Savona): primi risultati." *Archeologia Medievale* 30:247–58.

Arrigo, N., and C. Arnold. 2007. "Naturalised *Vitis* Rootstocks in Europe and Consequences to Native Wild Grapevine." *PLoS ONE* 2 (6): e521. http://www.plosone.org.

Artamanov, M. I. 1965. "Frozen Tombs of the Scythians." *Scientific American* 212:101–9.

Ash, A. L. 1948. Hemp—Production and Utilization. *Economic Botany* 2:158–69.

Ashkenazi, M. 1993. *Matsuri: Festivals of a Japanese Town*. Honolulu: University of Hawaii.

Ashman, T. L. 2002. "The Role of Herbivores in the Evolution of Separate Sexes from Hermaphroditism." *Ecology* 83(5): 1175–84 (Concepts and Synthesis section).

Asplund, H. 2008. *Kymittae: Sites, Centrality and Long-Term Settlement Change in the Kemiönsaari Region in SW Finland*. Turku, Finland: Turun Yliopisto.

Assis Cintra, F. 1934. *Os Escândalos de Carlota Joaquina*. [In Portuguese.] Rio de Janeiro: Civilização Brasileira.

Aston, W. G. 1905. *Shinto (The Way of the Gods)*. London: Longmans, Green.

Athenaeus of Naucratis. 1854. *The Deipnosophists, or Banquet of the Learned of Athenæus*. Edited by C. D. Yonge. London: H. G. Bohn.

Axelrod, D. 1974. "Revolutions in the Plant World." *Geophytology* 4:1–6.

Bahn, P. G. 1996. *Tombs, Graves and Mummies*. London: Weidenfeld & Nicholson.

Bai, W. N., W. J. Liao, and D. Y. Zhang. 2010. "Nuclear and Chloroplast DNA Phylogeography Reveal Two Refuge Areas with Asymmetrical Gene Flow in a Temperate Walnut Tree from East Asia." *New Phytologist* 10:892–901.

Bailey, J. 2001. *John Bailey's Complete Guide to Freshwater Fishing*. London: New Holland (UK).

Bajracharya, M. B. 1979. *Ayurvedic Medicinal Plants and General Treatment*. Kathmandu, Nepal: Piyusavarsi Ausadhalaya.

Bakels, C. C. 1991. "Western Continental Europe." In *Progress in Old World Palaeoethnobotany*, edited by W. Van Zeist, K. Wasylikowa, and K. E. Behre, 279–98. Rotterdam, The Netherlands: Balkema.

———. 2003. "The Contents of Ceramic Vessels in the Bactria-Margiana Archeological Complex, Turkmenistan." *Electronic Journal of Vedic Studies* 9:IC. www.ejvs. laurasianacademy.com.

———. 2005. "Crops Produced in the Southern Netherlands and Northern France during the Early Medieval Period: A Comparison." *Vegetation History and Archaeobotany* 14 (4): 394–99.

Bakels, C., R. Kok, L. I. Kooistra, and C. Vermeeren. 2000. "The Plant Remains from Gouda-Oostpolder, a Twelfth Century Farm in the Peatlands of Holland." *Vegetation History and Archaeobotany* 9 (3): 133–87.

Baker, P. B., K. R. Bagon, and T. A. Gough. 1980. "Variation in the THC Content of Illicitly Imported *Cannabis* Products—Part I." *Bulletin on Narcotics* 32 (4): 47–54.

Baker, P. B., T. A. Gough, and B. J. Taylor. 1980. "Illicitly Imported *Cannabis* Products: Some Physical and Chemical Features Indicative of their Origin." *Bulletin on Narcotics* 32 (2): 31–40.

Balabanova, S., F. Parsche, and W. Pirsig. 1992. "First Identification of Drugs in Egyptian Mummies." *Naturwissenschaften* 79:358.

Balassa, I., and G. Ortutay. 1979. *Magyar neprajz*. Hungarian ed. Budapest: Corvina Kiadó. Published online as *Hungarian Ethnography and Folklore*, translated by Maria and Kenneth Bales, http://mek.niif.hu/02700/02790/html/index.html.

Ball, M. V. 1910. "The Effects of Haschisch Not Due to *Cannabis Indica*." *Therapeutic Gazette* 34:777–80.

Balodis, F. 1956. *Aleksandrs Senā Latvija*. [In Latvian.] Chicago: Dzimtā zeme.

Bancroft, H. H. 1886. *History of California*. San Francisco: The History Company. Reprinted 1963, Santa Barbara, CA, by Wallace Herberd, as *The Works of Hubert Howe Bancroft*.

Bandini Mazzanti, M., A. M. Mercuri, G. Trevisan Grandi, M. Barbi, and C. A. Accorsi. 1999. "Il fossato di Argenta (Ferrara) e la sua bonifica in età medievale: contributo alla ricostruzione della storia del sito in base ai semi e frutti del riempimento." In Il Tardo Medioevo ad Argenta—Lo scarico di via Vinarola-Aleotti. Quaderni di archeologia dell'Emilia Romagna 2. *Edizioni All'Insegna del Giglio*, edited by C. Guarnieri, 219–37. Firenze, Tuscany.

Bandini Mazzanti, M., G. Bosi, A. M. Mercuri, C. A. Accorsi, and C. Guarnieri. 2005. "Plant Use in a City in Northern Italy during the Late Mediaeval and Renaissance Periods: Results of the Archaeobotanical Investigation of 'The Mirror Pit' (14th–15th century AD) in Ferrara." *Vegetation History and Archaeobotany* 14 (4): 442–52.

Banks, M. M. 1939. "Scottish Lore of Earth, Its Fruits, and the Plough." *Folklore* 50:12–32.

Banpo Museum. 1982. *Xian Banpo. Neolithic Site at Banpo Near Xian*. Beijing: Wenwu.

Barber, E. J. W. 1991. *Prehistoric Textiles: The Development of Cloth in the Neolithic and Bronze Ages with Special Reference to the Aegean*. Princeton, NJ: Princeton University.

———. 1994. *Women's Work: The First 20,000 Years: Women, Cloth and Society in Early Times*. New York: Norton.

———. 1999. *The Mummies of Ürümchi*. London: Pan Books.

———. 2007. *New Evidence for Early Trans-Eurasian Connections: The Xinjiang Mummies and the Horsemen of the Steppes*. Abstract of speech presented at the University of Hawaii at Manoa, November 8, 2007.

Barber, K. E., and S. N. Twigger. 1987. "Late Quarternary Palaeoecology of the Severn Basin." In *Palaeohydrology in Practice*, edited by K. J. Gregory, J. Lewin, and J. B. Thornes, 217–50. Chichester, England: Wiley.

Bard, K. A., and S. B. Shubert. 1999. *Encyclopedia of the Archeology of Ancient Egypt*. London: Routledge.

Bardy, K., P. Schönswetter, D. C. Albach, and M. A. Fischer. 2008. "Phylogeography on the Balkan Peninsula—Examples from *Veronica (Plantaginaceae)*." In *Xth Symposium of the International Organization of Plant Biosystematists, Book of Abstracts* 2–4 (July 2008): Symposium Secretariat, 5. Vysoké Tatry, Slovakia.

Barfield, T. J. 2001. "The Shadow Empires: Imperial State Formation along the Chinese-Nomad Frontier." In *Empires: Perspectives from Archeology and History*, edited by S. E. Alcock, T. N. D'Altroy, K. D. Morrison, and C. M. Sinopoli, 10–41. Cambridge: Cambridge University.

Barjaktarević, M. 1951. *Obicajni razvod braka u nasem narodu* [A popular way of divorcing in former times]. [In Serbian.] Beograd, Serbia: Zbornik Etnografskog muzeja u Beogradu.

Barkmann, U. B. 2002. "Qara Qorum (Karakorum)—Fragmente zur Geschichte einer vergessenen Reichshauptstadt." In "Qara Qorum City (Mongolei) I. Preliminary Report of the Excavations 2000/2001," edited by H. R. Roth, U. Erdenebat, E. Nagel, and E. Pohl. *Bonn Contributions to Asian Archeology* 1:7–20.

Barley, N. 1995. *Dancing on the Grave—Encounters with Death*. London: John Murray.

Barnes, G., and M. Okida. 1999. "Japanese Archeology in the 1990s." *Journal of Archaeological Research* 7 (4): 349–95.

Barnett, L. D. 1914. *Antiquities of India: An Account of the History and Culture of Ancient Hindustan*. New York: G. P. Putnam's Sons.

Barr, J. 1891. "Judicial Executions." *British Medical Journal* 2: 822–23.

Barrett, T. 1983. *Japanese Papermaking, Traditions, Tools, and Techniques*. Tokyo: Weatherhill.

Bartley, D. D., and A. V. Morgan. 1990. "The Palynological Record of the King's Pool, Stafford, England." *New Phytologist* 116 (1): 177–94.

Bar-Yosef, O. 1990. "The Last Glacial Maximum in the Mediterranean Levant." In *The World at 18,000 BP*, vol. 1, edited by O. Soffer and C. Gamble, 58–68. London: Unwin Hyman.

Basham, A. L. 1959. *The Wonder that was India: A Survey of the Culture of the Indian Sub-Continent before the Coming of the Muslims*. New York: Grove.

Batho, R. 1883. "Cannabis Indica." *British Medical Journal* 1 (May 26): 1002.

Beales, P. W. 1980. "The Late Devensian and Flandrian Vegetational History of Crose Mere, Shropshire." *New Phytologist* 85:133–61.

Beaumont, F., and J. Fletcher. 1718. *The Bloody Brother; or, Rollo. A Tragedy*. London: Black Swan.

Becker, C. J. 1941. "Fund af Ruser fra Danmarks Stenalder." *Aarbøger for Nordisk Oldkyndighed og Historie* 1:131–48.

Beckwith, C. I. 2009. *Empires of the Silk Road: A History of Central Eurasia from the Bronze Age to the Present*. Princeton, NJ: Princeton University.

Beer, R., F. Kaiser, K. Schmidt, B. Ammann, G. Carraro, E. Grisa, and W. Tinner. 2008. "Vegetation History of the Walnut Forests in Kyrgyzstan (Central Asia): Natural or Anthropogenic Origin?" *Quaternary Science Reviews* 27 (5–6): 621–32.

Begović, N. 1986. *Zivot Srba granicara u XIX veku* [Way of living the life among Serbs in the border area between Austro-Hungary and Turkey in the XIX century]. [In Serbian.] Beograd, Serbia.

Behre, K.-E. 2007. "Evidence for Mesolithic Agriculture in and around Central Europe?" *Vegetation History and Archaeobotany* 16:203–19.

Bellinger, L. 1962. "Textiles from Gordian." *Bulletin of the Needle and Bobbin Club* 46:5–33.

Bellwood, P. 2005. *First Farmers: The Origins of Agricultural Societies*. Oxford: Blackwell.

Benécs-Bárdi, G. 2002. "Taxonomy and Morphology of Uncultivated Hemp (*Cannabis sativa* L.) as Weed in Hungary." *Acta Botanica Hungarica* 44 (1–2): 31–47.

Beneš, J. 1996. "Archeologický a archeobotanický výzkum pozdněstředověkého vodovodního dila z Prachatic" [Archeological and archaeobotanical studies of late medieval water pipe works at Prachatice]. *Zlatá stezka* 3:158–81.

Benet, S. [S. Benatowa]. 1934. "Le chanvre dans les croyances et les coutumes populaires." *Comtes Rendus de Séances de la Société des Sciences et des Lettres de Varsovic* 27(2):51–55.

———. 1951. *Song, Dance and Customs of Peasant Poland*. London: Dennis Dobson.

———, ed. and trans. (1958) 1970. *The Village of Viriatino*. Garden City, NY: Anchor Books.

———. 1975. "Early Diffusion and Folk Uses of Hemp." In *Cannabis and Culture*, edited by V. Rubin, 39–49. Paris: Mouton.

Beng, T. C. 1975. "Meo death customs." In *Farmers in the Hills—Ethnographic Notes on the Upland Peoples of North Thailand*, Data Papers in Social Anthropology, edited by Anthony R. Walker, 81–84. Pulau Pinang, Malaysia: Penerbit Universiti Sains Malaysia.

Bengtsson, S. 1975. "The Sails of Wasa." *International Journal of Nautical Archeology* 4 (1): 27–41.

Benjamin, W. 2006. *On Hashish*. Cambridge, MA: Belknap of Harvard University.

Benjelloun-Laroui, L. 1990. *Les biblithèques au Maroc*. Paris: Maisonneuve et Larose.

Bennett, C. 1999. "Hemp Seed, the Royal Grain." *Cannabis Culture Magazine* January. http://www.cannabisculture.com.

———. 2011. "Early/Ancient History." In *The Pot Book: A Complete Guide to Cannabis, Its Role in Medicine, Politics, Science, and Culture*, edited by J. Holland, 17–26. Rochester, VT: Park Street.

Bennett, K. D., and J. Provan. 2008. "What Do We Mean by 'Refugia'?" *Quaternary Science Reviews* 27:2449–55.

Benoît, P., and F. Micheau. 1995. "The Arab Intermediary." In *A History of Scientific Thought: Elements of a History of Science*, edited by M. Serres, 160–90. Cambridge, MA: Blackwell.

Benton, M. J., and F. J. Ayala. 2003. "Dating the Tree of Life." *Science* 300 (5626): 1698–1700.

Bergfjord, C., S. Kar, A. Rast-Eicher, M. L. Nosch, U. Mannering, R. G. Allaby, B. M. Murph, and B. Holst. 2010. "Comment on '30,000-Year-Old Wild Flax Fibers.'" *Science* 328 (5986): 1634.

Bergman, J., D. Hammarlund, G. Hannon, L. Barnekow, and B. Wohlfarth. 2005. "Deglacial Vegetation Succession and Holocene Tree-Limit Dynamics in the Scandes Mountains, West-Central

Sweden: Stratigraphic Data Compared to Megafossil Evidence." *Review of Palaeobotany and Palynology* 134:129–51.

Bergman, S. 1938. *Korean Wilds and Villages*. Translated by Frederic White. London: John Gifford Ltd.

Bernabó Brea, M., A. Cardarelli, and M. Cremaschi. 1997. "Il crollo del sistema Terramaricolo." In *Le Terramare. La più antica civiltà padana*, edited by Bernabó Brea M., A. Cardarelli, and M. Cremaschi, 745–53. Milano, Italy: Electa.

Bernatzik, H. A. 1970. *Akha and Miao—Problems of applied ethnography in farther India*. New Haven, CT: Human Relations Area Files.

Bērziņš, V. 2008. *Sērnate: Living by a Coastal Lake during the East Baltic Neolithic*. PhD diss., Faculty of Humanities of the University of Oulu. Oulu, Finland: Oulu University.

Betts, E. M., ed. 1953. *Thomas Jefferson's Farm Book*. Princeton, NJ: Princeton University.

Beug, H.-J. 2004. *Leitfaden der Pollenbestimmung für Mitteleuropa und angrenzende Gebiete*. München, Germany: Verlag Dr. Friedrich Pfeil.

Beutler, J. A. and A. H. der Marderosian. 1978. "Chemotaxonomy of *Cannabis* I. Crossbreeding between *Cannabis Sativa* and *C. Ruderalis*, with Analysis of Cannabinoid Content." *Economic Botany* 32 (4): 387–94.

Bezusko, L. G., S. L. Mosyakin, and A. G. Bezusko. 2009. "Flora and Vegetation of the Ovruch Ridge (Northern Ukraine) in the Early Middle Ages according to Palynological Evidence." *Quaternary International* 203 (1–2): 120–28.

Bhakuni, D. S., M. L. Dhar, M. M. Dhar, B. N. Dhawan, and B. N. Mehrotra. 1969. "Screening of Indian Plants for Biological Activity: Part II." *Indian Journal of Experimental Biology* 7:250–62.

Bharati, A. 1965. *The Tantric Tradition*. London: Rider and Company.

Bhattacharyya, A., and S. K. Shah. 2008. "Spatio-Temporal Variation of Vegetation During Holocene in the Himalayan Region." *Himalyan Journal of Sciences* 5 (7): 31. Special Issue: Extended Abstracts, 23rd Himalayan-Karakoram-Tibet Workshop, August 11, 2008, Ladakh, India

Biegeleisen, H. 1929. *Lecznictwo ludu Polskiego* [Polish folk medicine]. [In Polish.] Kraków, Poland: Polska Adademia Umiejes.

Biel, J. 1981. "The Late Hallstatt Chieftain's Grave at Hochdoft." *Antiquity* 55 (213): 16–18.

Bird, I. 1898. *Korea and her Neighbours: A Narrative of Travel, with an Account of the Vicissitudes and Position of the Country*. London: John Murray.

Birdwood, G. C. M. 1865. *Catalogue of the Vegetable Products of the Presidency of Bombay*. Bombay, India: Government Central Museum.

Bishop, G. V. 1965. *Executions: The Legal Ways of Death*. Los Angeles: Sherbourne.

Bishop, J. L. 1966. *A History of American Manufactures*. New York: Augustus M. Kelly.

Björkman, L. 1999. "The Establishment of Fagus Sylvatica at the Stand-Scale in Southern Sweden." *Holocene* 9:237–45.

Björkman, L., and R. Bradshaw. 1996. "The Immigration of *Fagus Sylvatica* L. and *Picea Abies* (L.) Karst. into a Natural Forest Stand in Southern Sweden during the Last 2000 Years." *Journal of Biogeography* 23 (2): 235–44.

Blagojević, N. 1984. "Običaji u vezi sa rođenjem, ženidbom i smrću u titovoužičkom, požeškom i kosjerskom kraju." [In Serbian.] *Glasnik Etnografskog muzeja u Beogradu* 48:209–310.

Blagoveshchensky, G., V. Popovtsev, L. Shevtsova, V. Romanenkov, and L. Komarov. 2002. "Country Pasture/Forage Resource Profiles, Russian Federation." *FAO Website*. Edited by J. M. Suttie and S. G. Reynolds. May, http://www.fao.org/ag/AGP/AGPC/doc/counprof/russia.htm.

Blanford, N. 2007. "A Comeback for Lebanon's Hashish." *Time World*, October 16, 2007. http://www.time.com/time/world/article/0,8599,1672244,00.html.

Bloom, J. M. 1999. "Revolution by the Ream—A History of Paper." *ARAMCO World Magazine* 50 (3): 26–39.

———. 2001. *Paper before Print: The History and Impact of Paper in the Islamic World*. New Haven, CT: Yale University.

Blum, A. 1934. *On The Origin of Paper*. New York: R. R. Bowker.

Blumler, M. 1996. "Ecology, Evolutionary Theory and Agricultural Origins." In *The Origins and Spread of Agriculture and Pastoralism in Eurasia*, edited by D. Harris, 25–50, Washington, DC: Smithsonian Institution.

Bock, F. G. 1990a. "The Great Feast of Enthronement." *Monumenta Nipponica* 45 (1): 27–38.

———. 1990b. "The Enthronement Rites: The Text of Engishiki, 927." *Monumenta Nipponica* 45 (3): 307–37.

Bocquet-Appel, J. P. 2011. "When the World's Population Took Off: The Springboard of the Neolithic Demographic Transition." *Science* 333 (6042): 560–61.

Bócsa, I. 1994. "Professor Dr. Ivan Bocsa, the Breeder of Kompolti Hemp (Interview by the JIHA)." *Journal of the International Hemp Association* 1 (2): 61–62.

———. 1998. "Genetic Improvement: Convential Approaches." In *Advances in Hemp Research*, edited by P. Ranalli, 153–84. Binghampton, NY: Haworth.

Boere, G., A. Colin, D. S. Galbraith, and L. K. Bridge. 2006. *Waterbirds around the World: A Global Overview of the Conservation, Management and Research of the World's Waterbird Flyways*. Edinburgh: The Stationery Office.

Bogan, M. L. C. 1928. *Manchu Customs and Superstitions*. Peking: China Booksellers, Ltd.

Boileau, G. 2002. "Wu and Shaman." *Bulletin of the School of Oriental and African Studies* 5:350–78.

Bolens, L. 1981. *Agronomes andalous du Moyen-Age*. Geneva, Switzerland: Droz.

———. 1992. "The Use of Plants for Dyeing and Clothing: Cotton and Woad in al-Andalus: A Thriving Agricultural Sector (5th/11th–7th/13th Centuries)." In *The Legacy of Muslim Spain*, edited by S. K. Jayyusi, 1000–1015. Leiden, The Netherlands: Brill.

Bolikhovskaya, N. S., and A. N. Molodkov. 2006. "East European Loess–Palaeosol Sequences: Palynology, Stratigraphy and Correlation." *Quaternary International* 149:24–36.

Bolikhovskaya, N. S., A. P. Derevyanko, and M. V. Shun'kov. 2006. "The Fossil Palynoflora, Geological Age, and Climatostratigraphy of the Earliest Deposits of the Karama Site (Early Paleolithic, Altai Mountains)." *Paleontological Journal* 40 (5): 558–66.

Bolikhovskaya, N. S., M. Kaitamba, A. Porotov, and E. Fouache. 2004. "Environmental Changes of the Northeastern Black Sea's Coastal Region during the Middle to Late Holocene." In *Impact of the Environment on Human Migration in Eurasia*, edited by E. M. Scott, A. Y. Alekseev, and G. Zaitseva, 209–23. Dordrecht, The Netherlands: Kluwer Academic.

Booth, M. 2003. *Cannabis: A History*. New York: St. Martin's.

Bonser, W. 1928. "The Mythology of the Kalevala, with Notes on Bear-Worship among the Finns." *Folklore* 39:344–58.

Bos, J. A. A., and R. Urz. 2003. "Late Glacial and Early Holocene Environment in the Middle Lahn River Valley (Hessen, Central-West Germany) and the Local Impact of Early Mesolithic People—Pollen and Macrofossil Evidence." *Vegetation History Archaeobotany* 12:19–36.

Bosi, G. 2000. "Flora e ambiente vegetale a Ferrara tra il X e il XV secolo attraverso i reperti carpologici dello scavo di Corso Porta Reno—Via Vaspergolo nell'attuale centro storico." [In Italian.] PhD diss., Università degli Studi di Firenze.

Bosi, G., M. Bandini Mazzanti, A. Florenzano, I. M. N'siala, A. Pederzoli, R. Rinaldi, P. Torri, and A. M. Mercuri. 2011. "Seeds/Fruits, Pollen and Parasite Remains as Evidence of Site Function: Piazza Garibaldi—Parma (N Italy) in Roman and Mediaeval Times." *Journal of Archeological Science* 38 (7): 1621–33.

Bosić, M. 1985. "Običaji i verovanja Srba u Vojvodini sa osvrtom na okolinu Sombora" [Customs and beliefs among Serbs in Vojvodina focused on Sombor surrounding Northwest of Vojvodina]. [In Serbian.] In *Zbornik radova XXXII, kongresa Saveza udruženja folklorista Jugoslavije odrzanog u Somboru 1985*, 22–23, 90–91 Novi Sad, Serbia: Vuk Karadžić.

———. 1996. *Godisnji običaji Srba u Vojvodini* (*Annual customs among Serbs in Vojvodina*. [In Serbian.] Novi Sad, Serbia.

Bošković-Matić, M. 1962. "Narodni običaji. Svadbeni običaji." [In Serbian.] *Glasnik Etnografskog muzeja u Beogradu* 25:17–194.

Bouby, L. 2002. "Le chanvre (*Cannabis sativa* L.): Une plante cultivée à la fin de l'âge du Fer en France du Sud-Ouest?" [In French with English abstract.] *Comptes Rendus Palevol* 1 (2): 89–95.

Bourguignon, E., ed. 1973. "Religion, Altered States of Consciousness, and Social Change." Columbus, OH: Ohio University.

Bower, G. H., and E. R. Hilgard. 1997. *Theories of Learning*. New York: Prentice Hall.

Bowman, J. N. 1943. *Notes on Hemp Culture in Provincial California.* Unpublished typewritten manuscript. English Archival Material 1 p. l., 14 no. 1. Berkeley, CA: University of California at Berkeley, Bancroft Library.

Boyce, S. S. 1900. *Hemp (Cannabis sativa).* New York: Orange Judd.

Boyer, R., and A. Encart. 1996. "Étude d'une carde en chanvre." In "Toiture et restes carbonisés d'une maison incendiée dans l'habitat de Lattes au 4e s. av. n. è.," edited by R. Buxó, L. Chabal, and J.-C. Roux. *Lattara* 9: 378–79.

Bradshaw, E. G., P. Rasmussen, and B. V. Odgaard. 2005. "Mid- to Late-Holocene Land-Use Change and Lake Development at Dallund Sø, Denmark: Synthesis of Multiproxy Data, Linking Land and Lake." *The Holocene* 15 (8): 1152–62.

Bradshaw, E. G., P. Rasmussen, and J. Anderson. 2000. *Long-Term Lake Responses to Changing Landscape: Evidence from Dallund Sø, Funen, Denmark.* LUNDQUA Report 37. Lund, Sweden: Lund University, Department of Quaternary Geology.

Bradshaw, R. H. W., P. Coxon, J. R. A. Greig, and A. R. Hall. 1981. "New Fossil Evidence for the Past Cultivation and Processing of Hemp (*Cannabis sativa* L.) in Eastern England." *The New Phytologist* 89:503–10.

Bramanti, B., M. G. Thomas, W. Haak, M. Unterlaender, P. Jores, K. Tambets, I. Antanaitis-Jacobs, M. N. Haidle, R. Jankauskas, C.-J. Kind, F. Lueth, T. Terberger, J. Hiller, S. Matsumura, P. Forster, and J. Burger. 2009. *Ancient DNA Says Europe's First Farmers Came From Afar. Science* 4 September 325 (5945): 1189. DOI: 10.1126/science.325_1189.

Brande, A. 1985. "Mittelalterlich-neuzeitliche Vegetationsentwicklung am Krummen Fenn in Berlin-Zehlendorf." *Verhandlungen des Berliner Botanischen Vereins* 4:3–65.

Bratić, A. T. 1906. "Pripravljanje predje, snovanje, navijanje i tkanje u Hercegovini" [Processing yarn, warping, winding the warp on the warp beam, and weaving in Herzegovina]. [In Serbo-Croation.] *Glasnik Zemaljskog muzeja u Bosni i Hercegovini* 18, 391.

Bray, F. 1984. "Agriculture." In *Science and Civilization in China,* edited by J. Needham, Volume 6: *Biology and Biological Technology,* Part 2. London: Cambridge University.

Bredemann, G., F. R. Schwanitz, and R. von Sengbusch. 1956. "Problems of Modern Hemp Breeding with Particular Reference to the Breeding of Varieties of Hemp Containing Little or No Hashish." *Bulletin in Narcotics* 8 (3): 31–35.

Bredemann, G., K. Garber, W. Huhnke, and R. von Sengbusch. 1961. "Die Züchtung von monözischen und diözischen, faserertragreichen Hanfsorten (Fibrimon und Fibridia)." *Zeitschrift für Pflanzenzüchtung* 46 (3): 235–45.

Bremmer, J. N. 2001. *The Rise and Fall of the Afterlife: The 1995 Read-Tuckwell Lectures at the University of Bristol.* London: Routlege.

Brenneisen R. 2007. "Chemistry and Analysis of Phytocannabinoids and Other Cannabis Constituents." In *Marijuana and the Cannabinoids,* edited by M. Elsohly, 17–49. Totowa, NY: Humana.

Bretschneider, E. 1870. *On the Study and Value of Chinese Botanical Works with Notes on the History of Plants and Geographical Botany from Chinese Sources.* Fuzhou, China: Rozario, Marcal.

———. 1895. *Botanicon Sinicum. Notes on Chinese Botany from Native and Western Sources. Part 3. Botanical Investigations into the Materia Medica of the Ancient Chinese.* London: Kelly & Walsh.

Brewer, E. C. 1898. *Dictionary of Phrase and Fable.* Philadelphia, PA: Henry Altemus.

Brewer, S., J. W. Williams, P. Tarasov, and B. A. S. Davis. 2008. "Changes in European Tree Cover during the Holocene based on Pollen and AVHRR Data." *Terra Nostra* 2:37–38.

Brewer, S., R. Cheddadi, J. L. de Beaulieu, M. Reille, and data contributors. 2002. "The Spread of Deciduous *Quercus* throughout Europe since the Last Glacial Period." *Forest Ecology and Management* 156:27–48.

Brinkkemper, O., and R. de Man. 1999. "5 Archeobotanisch onderzoek van beerput 1 (15e eeuw)." In *De opgraving van het St. Agneslooster in Oldenzaal, Rapportage Archeol Monumentenzorg 50,* edited by Ostkamp, 50: 52–59. Amersfoort, The Netherlands.

Brinkley, F. and D. Kikuchi. 1915. *A History of the Japanese People from the Earliest Times to the End of the Meiji Era.* New York: The Encyclopaedia Britannica.

British Broadcasting Corporation. 2008. "Hemp Catching On with NW Farmers." *BBC News.* May 20, 2008. http://news.bbc.co.uk/2/hi/uk_news/northern_ireland/foyle_and_west/7409103.stm.

"British Museum Compass Collections Online." 2000. http://www.thebritishmuseum.ac.uk/compass.

Brockington, J. 1998. *The Sanskrit Epics.* Leiden, The Netherlands: Brill Academic.

Brombacher, C. 1998. "Les macrorestes botaniques (graines et fruits) de Develier/Courtételle-Étude 1997." In "L'habitat du haut Moyen âge à Develier-Courtelele (JU, Suisse)," edited by M. Federici-Schenardi and R. Fellner, 91–104. Porrentruy, Switzerland: République et Canton du Jura.

———. 1999. "La Neuvevill. L'histoire du paysage medieval révélée par l'étude des macrorestes végétaux." *Archäologie im Kanton Bern* 4B:277–84.

Brombacher, C., and A. M. Rachoud-Schneider. 1999. "Develier-Courtételle (Jura). Paysage et plantes cultivées." *Helvetia Archeologica* 30 (118/119): 95–103.

Brombacher, C., and A. Rehazek. 1999. "Ein Beitrag zum Speisezettel des Mittelalters. Archäobiologische Untersuchungen von Latrinen am Beispiel der Stadt Schaffhausen." *Jörg Archäologie der Schweiz* 22 (1): 44–48.

Brombacher, C., and M. Klee. 2007. "Environment, Agriculture and Environment from the Bronze Age to the Medieval Period in the Delemont Basin (Jura, Switzerland)." *Program and Abstracts for the 14th Symposium of the International Work Group for Palaeoethnobotany,* June 13–23, Krakow, Poland.

Brøndegaard, V. J. 1979. *Folk og Flora: Dansk Etnobotanik.* [In Danish.] Tønder, Denmark: Rosenkild og Bagger.

Brook, J. S., C. Zhang, and D. W. Brook. 2011. "Developmental Trajectories of Marijuana Use from Adolescence to Adulthood: Personal Predictors." *Archives of Paediatric & Adolescence Medicine* 165 (1): 55–60.

Brough, J. 1971. "Soma and *Amanita muscaria.*" *Bulletin of the School of Oriental and African Studies* 34 (2): 331–62.

Brown, J. 1883. "Cannabis Indica: A Valuable Remedy in Menorrhagia." *British Medical Journal* 1 (May 26): 1002.

Brunner, T. F. 1973. "Marijuana in Ancient Greece and Rome? The Literary Evidence." *Bulletin of the History of Medicine* 47 (4): 344–55.

———. 1977 Marijuana in Ancient Greece and Rome? The Literary Evidence. *Journal of Psychedelic Drugs* 9:221–25.

Bryant, A. T. 1949. *The Zulu People: As They Were Before White Man Came.* Pietermaritzburg, South Africa: Shuter & Shooter.

Buisseret, D. 1998. *Envisioning the City: Six Studies in Urban Cartography.* Chicago: University Of Chicago.

Bulleyn, W. [W. Bullein]. 1562. *Bulleins Bulwarke of Defence against All Sicknesse, Soarenesse, and Vvoundes that Doe Dayly Assaulte Mankinde: Which Bulwarke is Kept with Hilarius the Gardener, [and] Health the Phisicion, with the Chirurgian, to Helpe the Wounded Soldiours. Gathered and Practised from the Most Worthy Learned, Both Olde and New: To the Great Comfort of Mankinde.* London: Thomas Marshe. Reprinted as facsimile edition 1971, New York by Da Capo.

Bunce, W. K. 1960. *Religions in Japan.* Reprint, Rutland, VT: Charles E Tuttle.

Burbank, L. 1914. *Luther Burbank: His Methods and Discoveries and their Practical Application.* Vol. 8. New York: Luther Burbank.

Burkert, W. 1972. *Lore and Science in Ancient Pythagoreanism.* English ed. Translated by E. L. Minar Jr. Cambridge, MA: Harvard University.

Burkhardt, V. R. 1953. *Chinese Creeds and Customs.* Hong Kong: South China Morning Post.

Burkill, I. H. 1962. "Habits of Man and the Origins of the Cultivated Plants of the Old World." In *Readings in Cultural Geography,* edited by P. Wagner and M. Mikesell, 274–75. Chicago: University of Chicago.

Burney, D. A.1987a. "Pre-Settlement Vegetation Changes at Lake Tritrivakely, Madagascar." *Palaeoecology of Africa* 18:357–81.

———. 1987b. "Late Holocene Vegetational Change in central Madagascar." *Quaternary Research* 40:98–106.

———. 1988. "Modern Pollen Spectra from Madagascar." *Paleogeography, Palaeoclimatology, Palaeoecology* 66:63–75.

———. 1997. "Tropical Islands as Paleoecological Laboratories: Gauging the Consequences of Human Arrival." *Human Ecology* 25 (3): 437–57.

Burney, D. A., G. A. Brook, and J. B. Cowart. 1994. "A Holocene Pollen Record for the Kalahari Desert of Botswana from a U-Series Dated Speleothem." *The Holocene* 4:225–32.

Burns, R. I. 1981. "The Paper Revolution in Europe: Crusader Valencia's Paper Industry—A Technological and Behavioral Breakthrough." *Pacific Historical Review.* 50 (1): 1–30.

———. 1985. *Society and Documentation in Crusader Valencia. Diplomatarium of the Crusader Kingdom of Valencia: The Registered Charters of Its Conqueror Jaume I, 1257–1276.* Princeton, NJ: Princeton University.

———. 1996. "Paper Comes to the West, 800–1400." In *Europäische Technik im Mittelalter. 800 bis 1400. Tradition und Innovation* (4th ed.), edited by U. Lindgren, 413–22. Berlin: Gebr. Mann Verlag.

Buštić, T. M. 1911. "Narodna medicina Srba seljaka u Levcu i temniću." [In Serbian.] *Srpski etnografski zbornik* XVII: 529–95. Beograd, Serbia: Srpska kraljevska akademija.

Butrica, J. L. 2006. "The Medicinal Use of *Cannabis* among the Greeks and Romans." In *Handbook of Cannabis Therapeutics: From Bench to Bedside*, edited by Russo, Ethan B. and Franjo Grotenhermen, 23–42. New York: Haworth.

Čajkanović, V. 1985. *Rečnik srpskih narodnih verovanja o biljkama* [Dictionary on Serb folks' beliefs about herbs]. [In Serbian]. Beograd, Serbia: Srpska knjizevna zadruga, Srpska akademija nauka i umetnosti.

Callaway, J. C. 2004. "Hempseed as a Nutritional Resource: An Overview." *Euphytica* 140:65–72.

Callaway, J. C., and A. M. Hemmilä. 1996. "Cultivation of *Cannabis* Fiber Varieties in Central Finland." *Journal of the International Hemp Association* 3 (1): 29–31.

Campbell, J. M. 1894. "The Religion of Hemp." *Report of the Indian Hemp Drugs Commission 1893–1894*, vol. 3, edited by W. M. Young, 250–52. Simla, India: Government Printing Office.

Caramiello, R., A. Zeme, P. Gemello, M. Barra, and C. Preacco. 1992. "Palynological Findings in the Centocamare and Marasà Sud Sites (Locri Epizephyrii) and their Relation to the Historical and Archeological Hypotheses." *Alloionia* 31:7–19.

Carboni, A., C. Paoletti, V. M. Cristiana Moliterni, P. Ranalli, and G. Mandolino. 2000. "Molecular Markers as Genetic Tools for Hemp Characterization." Proceedings of Bioresource Hemp 2000 Symposium, September 13–16, Wolfsburg, Germany.

Cârciumaru, M. 1996. *Paleoetnobotanica. Studii în preistoriaşi protoistoria României. Istoria agriculturai din România.* [In Romanian.] Iaşi, Romania: Ed Glasul Bucovinei Helios.

Carmichael, M. 2007. "Beyond Stones & Bones." *Newsweek*, March 19, 53–58.

Carozza, J.-M., L. Carozza, and L. Bouby. 2002. "Le bassin versant du Boulou (Lot, France) au cours de la seconde moitié de l'Holocène: stabilité, rupture et rythme d'évolution d'un petit hydrosystème." In *Équilibres et ruptures dans les écosystèmes durant les 20 derniers millénaires en Europe de l'Ouest*, edited by H. Richard and A. Vignot, 239–53. Besançon, France: Presses Univesitaires Franc-comtoises.

Carter, T. F. 1925. *The Invention of Printing in China and Its Spread Westward.* New York: Columbia University. Second edition published 1955.

Casteel, R. W. 1976. *Fish Remains in Archeology.* New York: Academic.

Castelletti, L., E. Castiglioni, and M. Rottoli. 2001. "L'agricoltura dell'Italia settentrionale dal Neolitico al Medioevo." In *Le piante coltivate e la loro storia. Dalle origini al transito in Lombardia nel centenario della riscoperta della genetica di Mendel*, edited by O. Failla and G. Forni, 33–84. Milano, Italy: Franco Angeli.

Catlin, A., J. Cubbs and T. Dunnigan. 1986. *Hmong Art—Tradition and Change.* Sheboygan, WI: John Michael Kohler Arts Center.

Cato, M. P., and M. T. Varro. 1935. *Marcus Porcius Cato, on Agriculture; Marcus Terentius Varro. On agriculture [De re Rustica].* With an English translation by W. D. Hooper. Revised by H. B. Ash. Cambridge, MA: Harvard University.

Cederlund, C. O. 2006. *Vasa I, the Archeology of a Swedish Warship of 1628.* With a contribution by G. Hafstrom. Series edited by F. Hocker. Stockholm: Statens Maritima Museer.

Central Museum of Hannam University. 2003. "Daecheon-ri Neolithic Site at Okcheon-gun." [In Korean]. In *Archeological Research Report No. 16.* Hannam, Korea: Central Museum of Hannam University.

Cerina, A., L. Kalnina, and G. Grube. 2007. "Plant Macroremains and Pollen Analyses as a Source of Information about the Stone Age Human Diet in Luban Plain, Eastern Latvia." Paper presented at 14th Symposium of the International Work Group for Palaeoethnobotany—Program and Abstracts, June 13–23, Krakow, Poland.

Chaix, R., F. Austerlitz, T. Hegay, L. Quintana-Murci, and E. Heyer. 2008. "Genetic Traces of East-to-West Human Expansion Waves in Eurasia." *American Journal of Physical Anthropology* 136 (3): 309–17.

Chang, C. 1962. *Chinese History of Fifty Centuries.* Vol. 1. Taipei: Institute for Advanced Chinese Studies.

Chang, C., N. Benecke, F. P. Grigoriev, A. M. Rosen, and P. A. Tourtellotte. 2003. "Iron Age Society and Chronology in South-East Kazakhstan." *Antiquity* 77 (296): 298–312.

Chang, K. C. 1963. *The Archeology of Ancient China.* New Haven and London: Yale University.

———. 1968. *The Archeology of Ancient China.* 2nd ed. Revised and enlarged, London: Yale University.

———. 1977. 3rd ed. *The Archeology of Ancient China.* Revised and enlarged, London: Yale University.

———. 1979. *Food in Chinese Culture: Anthropological and Historical Perspectives.* New Haven, CT: Yale University.

———. 1986. 4th ed. *The Archeology of Ancient China.* Revised, London: Yale University.

Chang, T. T. 1970. *Der Kultder Shang-Dynastieirn Spiegleder Orakelinschriften. Eine paldogrpahische Studie zur Religion i marchaischen China.* Wiesbaden, Germany: Otto Harrassowitz.

———. 1976. "Rice." In *Evolution of Crop Plants*, edited by N. W. Simmonds, 98–104. London: Longman.

———. 1988. "Indo-European Vocabulary in Old Chinese: A New Thesis on the Emergence of Chinese Language and Civilization in the Late Neolithic Age." *Sino-Platonic Papers* 7:1–54.

Changpei, O. n.d. "Ancient Paper Making." *Travel China.* Xiamen, China. http://www.chinavista.com/experience/paper_make/paper_make.html.

Chaplin, G. 2005. "Physical Geography of the Gaoligong Shan Area of Southwest China in Relation to Biodiversity." *Proceedings of the California Academy of Sciences, Fourth Series* 56 (28): 527–56.

Charlesworth, D. 2002. "Plant Sex Determination and Sex Chromosomes." *Heredity* 88:94–101.

———. 2006. "Evolution of Plant Breeding Systems Review." *Current Biology* 16:R726–R735.

Chauhan, M. S., A. K. Pokharia, and I. B. Singh. 2005. "Preliminary Results on the Palaeovegetation during Holocene from Lahuradewa Lake, District sant Kabir Nagar, Uttar Pradesh." *Pragdhara* 15:33–40.

Chen, C., and J. W. Olsen. 1990. "China at the Last Glacial Maximum." In *The World at 18,000 BP* vol. 1, edited by Olga Soffer and Clive Gamble, 276–95. London: Unwin Hyman.

Chen, G., K. S. Katsumata, and M. Inaba. 2003. "Traditional Chinese Papers, their Properties and Permanence." *Restaurator* 24 (3): 135–44.

Chen, T., Y. Wu, Y. Zhang, B. Wang, Y. Hu, C. Wang, and H. Jiang. 2012. "Archaeobotanical Study of Ancient Food and Cereal Remains at the Astana Cemeteries, Xinjiang, China." *PLoS ONE* 7 (9): 1–9, e45137, www.plosone.org.

Chen, W. 1984. "Agricultural Science and Technology in Ancient China." In *Recent Discoveries in Chinese Archeology*, edited by Z. Boyang, 49–55. Beijing: Foreign Languages.

Chen, W., W.-M. Wang, and X.-R. Dai. 2009. "Holocene Vegetation History with Implications of Human Impact in the Lake Chaohu Area, Anhui Province, East China." *Vegetation History and Archaeobotany* 18 (2): 137–46.

Cheng, T.-K. 1959. *Archeology in China, Volume One—Prehistoric China.* Cambridge: W. Heffer & Sons Ltd.

———. 1963. *Archeology in China, Volume Three—Chou China.* Cambridge: W. Heffer & Sons Ltd.

———. 1966. *Archeology in China, Supplement to Volume One: New Light on Prehistoric China.* Cambridge: W. Heffer & Sons.

———. 1982. *Studies in Chinese Archeology.* Hong Kong: Chinese University.

Cheng, Y. P., S. Y. Hwang, and T. P. Lin. 2005. "Potential Refugia in Taiwan Revealed by the Phylogeographical Study of *Castanopsis Carlesii* Hayata (Fagaceae)." *Molecular Ecology* 14:2075–85.

Chessa, B., F. Pereira, F. Arnaud, A. Amorim, F. Goyache, I. Mainland, R. R. Kao, et al. 2009. "Revealing the History of Sheep Domestication Using Retrovirus Integrations." *Science* 324 (5926): 532–36.

Childe, V. G. 1936. *Man Makes Himself.* London: Watts.

China Heritage Newsletter. 2005. "2,000-Year-Old Hairpiece Unearthed in Sichuan." *China Heritage Newsletter*, no. 4 (December). Canberra, Australia: Australian National University, China Heritage Project.

Chindarsi, N. 1976. *The Religion of the Hmong Njua.* Bangkok: The Siam Society.

Chłodnicki, M., and L. Krzyżaniak, eds. 1998. *Catalogue for Pipeline of the Archeological Treasures.* Translated to English by P. M. Barford

and B. Gostyńska. Poznań, Poland: AN Studio. http://web.archive
.org/web/20110707003051/http://www.muzarp.poznan.pl/
archweb/gazociag/contents.htm.

Choe, S.-S. 1983. *Annual Customs of Korea*. Seoul: Seomun-dang.

Choi, H. 1971. *The Economic History of Korea*. Seoul: The Freedom
Library.

Choi, J.-H., ed. n.d. "Korea Folk Village Museum." http://www
.koreanfolk.co.kr/folk/english.

Chopra, I. C., and R. N. Chopra. 1957. "The Use of *Cannabis* Drugs in
India." *Bulletin on Narcotics* 9:4–29.

Chopra, R. N. 1958. *Indigenous Drugs of India*. 2nd ed. Calcutta, India:
U. N. Dhur and Sons.

Chopra, R. N., and G. S. Chopra. 1939. "The Present Position of
Hemp-Drug Addiction in India." *Indian Medical Research Memoirs*.
Calcutta, India: Thacker, Spink.

Chōsen iseki ibutsu zukan (Illustrated book of ruins and relics of
Korea). 1990. Seoul: Tonggwang Ch'ulp'ansa.

Choson Iibo. 2004. "The Choson Iibo, National/Politics, English.
Choson.com." Seoul: Chosen Ilbo [English Edition]. http://
english.chosum.com/w21data/html/news/200408/200409310022
.html.

Chou, K. 1963. *Analysis of Pollen from the Neolithic Site at Pan-p'o. near
Sian. K'ao-ku 9:520–22*. Peking: China [in Chinese].

Christensen, I. M. 1992. *Osebergdronningens grav—vaar arkeologiske
nasjonalsskatt i nytt lys*. [In Norwegian.] Oslo: Schibsted.

Chrtek, J. 1981. "Remarks on the Distribution of the Species *Cannabis
Ruderalis* in Bohemia Czechoslovakia." *Casopsis Narodniho Musea v
Praze* 150 (1–2): 21–24.

CHTA/ACCC. 2004. "Canadian Hemp: A Plant With Opportunity."
Canadian Hemp Trade Alliance. http://www.hemptrade.ca

Chung, C.-H. 2007. "Vegetation Response to Climate Change on Jeju
Island, South Korea, during the Last Deglaciation based on Pollen
Record." *Geosciences Journal* 11 (2): 147–55.

Ciaraldi, M. 2000. "Drug Preparation in Evidence? An Unusual Plant
and Bone Assemblage from the Pompeian Countryside, Italy."
Vegetation History and Archaeobotany 9:91–98.

Clapper, J. R., R. A. Mangieri, and D. Piomelli. 2009. "The Endocannabinoid System as a Target for the Treatment of Cannabis Dependence." *Neuropharmacology* 56 (supplement 1): 235.

Clark, D. N. 2000. *Culture and Customs of Korea*. Westport, CT: Greenwood.

Clark, V. S. 1929. *History of Manufacture in the United States*. New York:
McGraw Hill.

Clarke, R. C. 1977. *The Botany and Ecology of Cannabis*. Ben Lomond,
CA: Pods.

———. 1981. *Marijuana Botany*. Berkeley, CA: And/Or.

———. 1995. "Scythian *Cannabis* Verification Project." *Journal of the
International Hemp Association* 2 (2): 104.

———. 1998a. *Hashish!* Los Angeles: Red Eye.

———. 1998b. "Botany of the Genus *Cannabis*." In *Advances in
Hemp Research*, edited by P. Ranalli, 1–20. Binghampton, NY:
Haworth.

———. 2001. "Sinsemilla Heritage: What's in a Name?" In
The Cannabible, edited by J. King, 1–24. Berkeley, CA: Ten Speed
Press.

———. 2006a. "Hemp (*Cannabis*) Cultivation and Use in the Republic of Korea." *Journal of Industrial Hemp* 11 (1): 51–86.

———. 2006b. "Searching for Hempen Treasures—Field Identification of Hemp Fiber in Markets, Museums and Private Collections."
Journal of Industrial Hemp 11 (2): 73–90.

———. 2007. "Traditional *Cannabis* Cultivation in Darchula District,
Nepal—Seed, Resin and Textiles." *Journal of Industrial Hemp* 12 (2):
19–42.

———. 2008. "Four Generations of Sani Hemp Satchels." *Journal of
Industrial Hemp* 13 (1): 58–72.

———.2010a. "Traditional Fiber Hemp (*Cannabis*) Production,
Processing, Yarn Making, and Weaving Strategies—Functional
Constraints and Regional Responses. Part 1." *Journal of Natural
Fibers* 14 (2): 118–53.

———. 2010b. "Traditional Fiber Hemp (*Cannabis*) Production,
Processing, Yarn Making, and Weaving Strategies—Functional
Constraints and Regional Responses. Part 2." *Journal of Natural
Fibers* 14 (3): 229–50.

Clarke, R. C. and D. P. Watson. 2007. "Cannabis and Natural Cannabis Medicines." In *Marijuana and the Cannabinoids*, edited by
M. A. ElSohly, 1–17. Totowa, New Jersey: Humana Press.

Clarke, R. C., and W. Gu. 1998. "Survey of Hemp (*Cannabis sativa* L.)
Use by the Hmong (Miao) of the China/Vietnam Border Region."
Journal of the International Hemp Association 5 (1): 1, 4–9.

Clarke, S. R. 1911. *Among the Tribes in South-west China*. London:
China Inland Mission, Morgan and Scott.

Clothey, F. W. 1983. *Rhythm and Intent: Ritual Studies from South India*.
Bombay: Blackie and Son.

Clottes, J. 1996. "Thematic Changes in Upper Palaeolithic Art: A
View from the Grotte Chauvet." *Antiquity* 70 (268): 276–88.

Clutton-Brock, J. 1992. *Horse Power: A History of the Horse and the Don-
key in Human Societies*. Boston: Harvard University.

Colley, M. H. 1899. "Micellanea: Dorset Folklore Collected in 1879."
Folklore 10:478–89.

Columella, L. J. M. 1941–55. *On Agriculture with a Recension of the Text
and an English Translation*. 3 vols, vol. 1 translated by H. B. Ash,
vol. 2–3 translated by E. S. Forster and E. H. Heffner. Cambridge,
MA: Harvard University.

Comber, L. 1963. *Chinese Ancestor Worship in Malaya*. Singapore:
Eastern Universities.

Comes, H. P., and J. W. Kadereit. 1998. "The Effect of Quaternary
Climatic Changes on Plant Distribution and Evolution." *Trends in
Plant Science* 3 (11): 432–38.

Comitas, L. 1975. "The Social Nexus of Ganja in Jamaica." In *Cannabis and Culture*, edited by V. Rubin, 119–32. Paris: Mouton.

Comşa, E. 1996. Viaţa oamenilor din spaţiul carpato-danubiano-
pontic în mileniile 7–4 î. [Life of humans in the Danube—
Carpathians Region in millennia 7–4 BP]. Hr. Bucureşti [Bucha-
rest, Romania]: Editura. Didactică şi pedagogică [in Romanian].

Connor, S. E., I. Thomas, and E. V. Kvavadze. 2007. "A 5600-yr History of Changing Vegetation, Sea Levels and Human Impacts from
the Black Sea Coast of Georgia." *The Holocene* 17:25–36.

Constantine, P. 1992. *Japanese Street Slang*. Tokyo: Tengu Books.

Cook, B., C. Chirstiansen, and L. Hammarlund. 2002. "Viking Woollen Square-Sails and Fabric Cover Factor." *The International Journal
of Nautical Archeology* 31 (2): 202–21.

Cooremans, B. 1995–96a. "2 Plantenresten [Plant remains]." In "De
voedselvoorziening in de Sint-Salvatorsabdij te Ename (Oude-
naarde, prov. Oost-Vlaanderen. 4. Een beer—en afvalput uit het
gastenkwartier (1350–1450 A.D.)," edited by A. Ervynck, B. Cooremans, and V. Van Neer. *Archeologie in Vlaanderen* 5:303–15.

———. 1995–96b. "Plantenresten [Plant remains]." In "Granaatap-
pels, een zeeëngel en rugstreeppadden. Een greep uit de inhoud
van een baksteen beerput uit het 15de-eeuwse Raversijde (Oostende, prov. West-Vlaanderen)," edited by M. Pieters, F. Bouchet,
B. Cooremans, K. Desender, A. Ervynck, and V. Van Neer. *Archeologie in Vlaanderen* 5:193–224.

Corillion, R., and N. Planchais. 1963. "Recherches sur la vegetation
actuelle et passee d'une lande torbeuse Armoricaine: Malingue
(Mayenne)." *Pollen Spores* 5:73–286.

Corominas, J. 1959. "Tarifa dels corridors de Barcelona l'any 1271."
In *Hispanic Studies in Honour of I. González Lluber*, edited by
F. Pierce, 117–27. Oxford: Oxford University.

Cort, L. A. 1989. "The Changing Fortunes of Three Archaic Japanese
Textiles." In *Cloth and the Human Experience*, edited by A. Weiner
and J. Schneider, 377–415. Washington, DC: Smithsonian Institution.

Costain, T. B. 1954. *The White and the Gold: The French Regime in
Canada*. New York: Doubleday.

Cotterell, Maurice. 2004. *The Terracotta Warriors: The Secret Codes of
the Emperor's Army*. Rochester, NY: Bear.

Covell, A. C. 1986. *Folk Art and Magic: Shamanism in Korea*. Seoul:
Hollym International Corporation.

Covington, J. 1874. *The Satires of A. Persiu Flaccus with a Translation
and Commentary*. Edited by H. Nettleship. 2nd ed. Revised, Oxford:
Clarendon.

Coward, H. G., R. W. Neufeldt, and E. K Neumaier-Dargyay. 2007.
Readings in Eastern Religions. Waterloo, ON: Wilfrid Laurier University.

Cox, G. W. 2004. *Alien Species and Evolution: The evolutionary Ecology
of Exotic Plants, Animals, Microbes and Interacting Native Species*.
Covelo, CA: Island.

Cox, M., J. Chandler, C. Cox, J. Jones, and H. Tinsley. 2000. "Early
Medieval Hemp-Retting at Glasson Moss, Cumbria in the Context
of the Use of *Cannabis Sativa* during the Historic Period." *Transactions of the Cumberland & Westmoreland Antiquarian & Archeological
Society* 100:131–50.

Cox, W. 1998. *The Origins of Christmas and Easter*. No. 235. 3rd ed. Woden, Australia: Christian Churches of God. http://www.ccg.org/english/s/p235.html

Crawford, G. W. 1992. "Prehistoric Plant Domestication in East Asia." In *The Origins of Agriculture: An International Perspective*, edited by C. W. Cowan and P. J. Watson, 17–38. Washington, DC: Smithsonian Institution.

———. 2009. "Agricultural Origins in North China Pushed Back to the Pleistocene." *Proceedings of the National Academy of Sciences* 106 (18): 7271–72.

———. 2011. "Advances in Understanding Early Agriculture in Japan." *Current Anthropology* 52(S4): S331–S345.

Crawford, G. W., and G. Lee. 2003. "Agricultural Origins in the Korean Peninsula." *Antiquity* 77 (295): 87–95.

Crawford, G. W., and H. Takamiya. 1990. "The Origins and Implications of Late Prehistoric Plant Husbandry in Northern Japan." *Antiquity* 64 (245): 889–911.

Crawford, G. W., and M. Yoshizaki. 1987. "Ainu Ancestors and Prehistoric Asian Agriculture." *Journal of Archeological Science* 14:201–13.

Creel, H. G. 1937. *The Birth of China*. New York: F. Ungar.

Cristie, A. 1983. *Chinese Mythology*. Feltham, England: Newnes Books.

Cronn, R., and J. F. Wendel. 2003. "Cryptic Trysts, Genomic Mergers, and Plant Speciation." *New Phytologist* 161:133–42.

Crosby, A. W. 1965. *America, Russia, Hemp, and Napoleon: American Trade with Russia and the Baltic, 1783–1812*. Columbus: Ohio State University.

———. 1973. *The Columbian Exchange: Biological and Cultural Consequences of 1492*. Westport, CT: Greenwood.

Crumlin-Pedersen, O., and O. Olsen, eds. 2002. *The Skuldelev Ships I: Topography, History, Conservation and Display. Ships and Boats of the North 4.1*. Roskilde, Denmark: The Viking Ship Museum in Roskilde and Centre for Maritime Archeology of the National Museum of Denmark.

Cubbs, J. 1986. *Hmong Art: Tradition and Change*. New York: John Michael Kohler Arts Center.

Culley, T. M., S. G. Weller, and A. K. Sakai. 2002. "The Evolution of Wind Pollination in Angiosperms." *Trends in Ecology and Evolution* 17 (8): 361–69.

Cun, Y. Z. and X.-Q. Wang. 2010. "Plant Recolonization in the Himalaya from the Southeastern Qinghai-Tibetan Plateau: Geographical Isolation Contributed to High Population Differentiation." *Molecular Phylogenetics and Evolution* 56:972–82.

Cuyler, P. L. 1979. *Sumo from Rite to Sport*. New York: Weatherhill.

Cyprien, A. L. and L. Visset. 2001. "Paleoenvironmental Study of the Carquefou Site (Massif Armoricain, France) from the End of the Sub-Boreal." *Vegetation History and Archaeobotany* 10:139–49.

Daghestani, A. N. 1997. "al-Razi (Rhazes), 865–925." *American Journal of Psychiatry* 154:1602.

Daihua, T. 2002. *Zhiwuming shitukao changbian* [An illustrated encyclopedia of plants]. [In Chinese.] Vol. 1. Beijing: Renmin weisheng chubanshe.

Dálnoki, O., and S. Jacomet. 2002. "Some Aspects of Late Iron Age Agriculture Based on the First Results of an Archaeobotanical Investigation at Corvin tér, Budapest, Hungary." *Vegetation History and Archaeobotany* 11:9–15.

Damania, A. B. 1998. "Diversity of Major Cultivated Plants Domesticated in the Near East." In *The Origins of Agriculture and Crop Domestication*, edited by A. B. Damania, J. Valkoun, G. Willcox, and C. O. Qualset, 51–64. Proceedings of the Harlan Symposium, May 10–14, 1997, Aleppo, Syria.

Damjanović, G. 1985. "Kordunasko prelo u Conoplja." [In Yugoslavian.] In *Zbornik radova XXXII, kongresa Saveza udruženja folklorista Jugoslavije odrzanog u Somboru*, 38–39. Novi Sad: Udruženje folklorista SAP Vojvodine.

D'Andrea, A. C. 1992. *Palaeoethnobotany of Later Jomon and Yayoi Cultures of Northeastern Japan: Northeastern Aomori and Southwestern Hokkaido*. PhD diss., University of Toronto, Department of Anthropology.

———. 1999. "The Dispersal of Domesticated Plants into Northeastern Japan." In *Prehistory of Food: Appetites for Change*, edited by C. Gosden, and J. Hather, 166–83. London: Routledge.

D'Andrea, A. C., G. W. Crawford, M. Yoshizaki, and T. Kudo. 1995. "Late Jomon Cultigens in Northeastern Japan." *Antiquity* 69: 146–52.

Dannaway, F. 2009. "Thunder among the Pines: Defining a Pan-Asian Soma." *Journal of Psychoactive Drugs* 41 (1): 67–84.

Dark, P. 2005. "Mid- to Late-Holocene Vegetational and Land-Use Change in the Hadrian's Wall Region: A Radiocarbon-Dated Pollen Sequence from Crag Lough, Northumberland, England." *Journal of Archeological Science*, 32 (4): 601–18.

Darlington, C. D. 1969. *The Evolution of Man and Society*. New York: Simon and Schuster.

Darmesteter, J. 1883. *The Zend-Avesta, Part I, The Vendîdâd*. 2nd ed. London: Oxford University.

Darwin, C. 1881. *The Power of Movement in Plants*. New York: Da Capo.

Dash, M. 2002. *Batavia's Graveyard: The True Story of the Mad Heretic who Led History's Bloodiest Mutiny*. London: Weidenfeld & Nicolson.

Datwyler, S. L., and G. D. Weiblen. 2004. "On the Origin of the Fig: Phylogenetic Relationships of Moraceae from *ndh*F Sequences." *American Journal of Botany* 91:767–77.

———. 2006. "Genetic Variation in Hemp and Marijuana (*Cannabis Sativa* L.) according to Amplified Fragment Length Polymorphisms." *Journal of Forensic Science* 51 (2): 371–75.

Davidyan, G. G., 1972. "Hemp (Biology and Initial Materials for Breeding)." *Works on Applied Botany, Genetics and Breeding*. VIR [Saint Petersburg, Russia: N. I. Vavilov Institute for Plant Industry] 48 (3): 10–11.

Davis, A. 2003. "The Plant Remains." In "Urban Development in North-West Southwark. Excavations 1974–90," edited by C. Cowan, 182–91. *MoLAS Monograph 16*. London: MoLAS.

Davis, T. W. M., C. G. Farmillo, and M. Osadchuk. 1963. "Identification and Origin Determinations of *Cannabis* by Gas and Paper Chromatography." *Analytical Chemistry* 35 (6): 751.

Davis-Kimball, J. 1997. *Ice Mummies: Siberian Ice Maiden*. Boston: PBS. Nova/WGBH co-production. Airdate: Nov. 24, 1998. Videocassette (VHS) Video WG2517, 56 minutes.

Dearing, J. A., 2006. "Climate-Human-Environment Interactions: Resolving our Past." *Climate of the Past* 2:187–203.

de Bonneville, F. 1994. *The Book of Fine Linen*. Paris: Flammarion.

de Candolle, A. 1967. *Origin of Cultivated Plants*. New York: Hafner. First edition published 1882 in French.

Dedijer, J. 1908. "Vrste nepokretne svojine u Hercegovini" [Kinds of real estate in Herzegovina]. [In Yugoslavian.] *Glasnik Zemaljskog muzeja u Bosni i Hercegovini* (Sarajevo) 20:387–402.

Deferne, J.-L., and D. W. Pate. 1996. "Hemp Seed Oil: A Source of Valuable Essential Fatty Acids." *Journal of the International Hemp Association* 3 (1): 1, 4–7.

de Groot, J. J. M. 1972. *The Religious System of China*. 6 vols. Taipei, China: Ch'eng Wen. First published ca. 1892–1910, Leiden, The Netherlands, by E. J. Brill in 3 volumes.

de Hingh, A., and C. Bakels. 1996. "Palaeobotanical Evidence for Social Difference? The Example of the Early Medieval Domain of Serris-Les Ruelles, France." *Vegetation History and Archaeobotany* 5:117–20.

de Jong, R., S. Björck, L. Björkman, and L. B. Clemmensen. 2006. "Storminess Variation during the Last 6500 Years as Reconstructed from an Ombrotrophic Peat Bog in Halland, Southwest Sweden." *Journal of Quaternary Science* 21 (8): 905–19.

de Lacouperie, T. 1893. "On Hemp from Central Asia to Ancient China, 1700 B.C." *Babylonian and Oriental Record* 6:247–53.

Delaney, F. 1986. *The Celts*. London: BBC Publication.

Delusina, I. V. 1991. "The Holocene Pollen Stratigraphy of Lake Ladoga and the Vegetational History of the Surroundings." *Annales Academiae Scientiarum Fennicae* 153 (3): 1–66.

de Man, R. 1996. 1996. De botanische inhoud van twee Andennepotten uit een 12/13e eeuwse waterput te Lieshout-Nieuwenhof. Interne Rapporten Archeobotanie ROB 1996/18.

Dembinska, M. 1999. *Food and Drink in Medieval Poland*. Translated by M. Thomas with revision by W. W. Weaver. Philadelphia: University of Pennsylvania. First published 1963 in Polish by the Polish Academy of Sciences.

de Meijer, E. P. M. 1995. "Fibre Hemp Cultivars: A Survey of Origin, Ancestry, Availability and Brief Agronomic Characteristics." *Journal of the International Hemp Association* 2 (2): 66–73.

———. 2004. "The Breeding of *Cannabis* Cultivars for Pharmaceutical End Uses." In *The Medicinal Uses of Cannabis and Cannabinoids*, edited by G. W. Guy, B. A. Whittle, and P. J. Robson, 55–69. London: Pharmaceutical.

de Meijer, E. P. M., M. Bagatta, A. Carboni, P. M. Crucitti, V. M. Cristiana, P. Ranalli, and G. Mandolino. 2003. "The Inheritance of Chemical Phenotype in *Cannabis sativa* L." *Genetics* 163:335–46.

Dempsey, J. M. 1975. *Fiber Crops*. Gainesville, FL: University of Florida.

Deng, W., B. Shi, X. He, Z. Zhang, J. Xu, B. Li, J. Yang, L. Ling, C. Dai, B. Qiang, Y. Shen, and R. Chen. 2004. "Evolution and Migration History of the Chinese Population Inferred from Chinese Y-Chromosome Evidence." *Journal of Human Genetics* 49 (7): 339–48.

Denisova, R. 1997. "The Most Ancient Populations of Latvia." *Humanities and Social Sciences: Latvia* 3 (16): 5–18.

De Petrocellis, L., D. Melck, T. Bisogno, A. Milone, and V. Marzo. 1999. "Finding the Endocannabinoid Signaling System in Hydra, a Very Primitive Organism: Possible Role in the Feeding Response." *Neuroscience* 92:377–87.

de Pinho, A. R. 1975. "Social and Medical Aspects of the Use of Cannabis in Brazil." In *Cannabis and Culture*, edited by V. Rubin, 293–301. Paris: Mouton.

Derevianko, A. P. 2001. "The Middle to Upper Paleolithic Transition in the Altai (Mongolia and Siberia)." *Archeology, Ethnology & Anthropology of Eurasia* 2 (3): 70–103.

Derevianko, A. P., A. V. Postnov, E. P. Rybin, Y. V. Kuzmin, and S. G. Keates. 2005. "The Pleistocene Peopling of Siberia: A Review of the Environmental and Behavioural Aspects." *Indo-Pacific Prehistory Association Bulletin* 25:57–68. Taipei Papers, vol. 3.

Derham, B. 2004. "Archeological and Ethnographic Toxins in Museum Collections." In *Impact of the Environment on Human Migration in Eurasia*, edited by E. M. Scott, A. Y. Alekseev, and G. I. Zaitseva, 185–98. Dordrecht, The Netherlands: Kluwer Academic.

Devane, W. A., F. A. Dysarz 3rd, M. R. Johnson, L. S. Melvin, and A. C. Howlett. 1988. "Determination and Characterization of a Cannabinoid Receptor in Rat Brain." *Molecular Pharmacology* 34 (5): 605–13.

Devane, W. A., H. L. Breuer, A. Pertwee, R. G. Stevenson, L. A. Griffin, G. Gibson, D. Mandelbaum, A. Etinger, and R. Mechoulam. 1992. "Isolation and Structure of a Brain Constituent that Binds to the Cannabinoid Receptor." *Science* 258:1946–49.

Devlet, E. 2001. "Rock Art and the Material Culture of Siberia and Central Asian Shamanism." In *The Archeology of Shamanism*, edited by N. Price, 43–55. London: Routledge.

———. 1914. "Hemp." In *Yearbook of the United States Department of Agriculture for 1913*, 283–346. Washington, DC: United States Government Printing Office.

———. 1920. *USDA Bureau of Plant Industry, Report of the Chief: No. 26*. Washington, DC: United States Government Printing Office.

———. 1931. "Hemp Fiber Losing Ground, Despite Its Valuable Qualities." In *USDA Yearbook of Agriculture*, 284. Washington, DC: United States Government Printing Office.

Dewey, L. H., and J. L. Merril. 1916. "Hemp Hurds as Paper-Making Material." *United States Department of Agriculture Bulletin*. No. 404. Washington, DC, 1–7, 9.

Diadema, K., and F. Médail. 2009. "Mediterranean Mountains: Plant Biodiversity and Phylogeographical Hotspots." In *Abstracts of the The 45th International SISV & FIP Congress on Biodiversity Hotspots in the Mediterranean Area: Species, Communities and Landscape Level* (June 22–24). Sardinia, Italy: University of Cagliari.

Diakonova, V. P. 1994. "Shamans in Traditional Tuvinian Society." In *Ancient Tradtiions: Shamanism in Central Asia and the Americas*, edited by G. Seaman and J. S. Day, 245–56. Boulder: University Press of Colorado.

Diamond, J. 1998. "The Japanese Roots (Part II)." *Discover Magazine* 19:86–94.

Dickson, C., and J. H. Dickson. 2000. *Plants and People in Ancient Scotland*. Gloucestershire, England: Stroud.

Dickson, C. A., and G. F. Mitchell. 1984. "Appendix VI Botanical Report." In "Excavations at Shop Street, Drogheda," edited by P. D. Sweetman. *Proceedings of the Royal Irish Academy* C84:5, 219–22.

Dikötter, F., L. Laamann, and Z. Xun. 2004. *Narcotic Culture: A History of Drugs in China*. Chicago: University of Chicago.

Dimitrijević, S. 1958. *Običaji u licnom i porodicnom zivotu kod Banatskih Hera* [Customs in personal and family life]. In *Banatske Here* [The Hera in Banat], edited by M. S. Filipović. [In Serbian.] Novi Sad, Serbia: Vojvodanski Muzej. Posebna izdanja, I.

Dinacauze, D. F. 2000. *Environmental Archeology, Principles and Practice*. Cambridge: Cambridge University.

Dingwall, H. M. 2003. *A History of Scottish Medicine: Themes and Influences*. Edinburgh: Edinburgh University.

Dioscorides, P. 1968. *The Greek Herbal of Dioscorides*. Translated by J. Goodyer and R. W. T. Gunther. London: Hafner.

Di Rita, F., and D. Magri. 2009. "Holocene Drought, Deforestation and Evergreen Vegetation Development in the Central Mediterranean: A 5500 Year Record from Lago Alimini Piccolo, Apulia, Southeast Italy." *The Holocene* 19 (2): 295–306.

Dirksen, V. G. 2000. "Modern Treeless Pollen Spectra Studying for Paleogeographical Reconstructions." *Palaeonotogical Journal* 34 (2): 221–26.

Dirksen, V. G., and B. van Geel. 2004. "Mid to Late Holocene Climate Change and Its Influence on Cultural Development in South Central Siberia." In *Impact of the Environment on Human Migration in Eurasia*, edited by E. M. Scott, A. Y. Alekseev, and G. Zaitseva, 291–307. Dordrecht, The Netherlands: Kluwer Academic.

Dixon, W. E. 1899. "The Pharmacology of *Cannabis Indica*." *British Medical Journal* 2:1354–57.

Djordjević, D. M. 1958. "Lazarice u Leskovackoj Moravi" [Processions on Saint Lazarus Day in the valley of Morava near Leskovac, south of Serbia]. [In Serbian.] *RCE Kongresa folklorist jugoslavije na Bjelasnici*, 117–24.

———. 1985. *Život i običaji narodni u leskovačkom kraju*. [In Serbian.] Vol. 35. Leskovac, Serbia: Narodni muzej.

Djordjević, T. R. 1938. *Zle oči u verovanju Južnih Slovena* [Beliefs about bewitching eyes among Southern Slavs]. [In Serbian.] Beograd, Serbia: Srpska Kraljevska Akademija. Second edition published 1985, Beograd, Serbia, by Prosveta.

———. 1958. "Priroda u verovanju i predanju našega naroda" [Nature in the folk beliefs and tradition]. [In Serbian.] *Srpska akademija nauka, Srpski etnografski zbornik* 1–2:71–72.

———. 1984. *Naš narodni život I* [Our National Culture I]. [In Serbian.] Reprint, Beograd, Serbia: Prosveta. First published 1923 by Štamparija "Rodoljub."

Djurić, S. 1934. "Srpski narodni običaji u Gornjoj Krajini." [Serbian folk customs in the Upper district in the Krajina mountain range of the Adriatic coast]. [In Serbian.] Vol. 50. Beograd, Serbia: Srpska kraljevska akademija, Srpski etnografski zbornik.Dodds, E. R. 1951. *The Greeks and the Irrational*. Berkeley, CA: University of California.

Dodge, C. R. 1896. "A Report on the Culture of Hemp and Jute in the United States." *USDA Office of Fiber Investigations Report*, no. 8:7.

———. 1897. *A Descriptive Catalogue of Useful Fiber Plants of the World*. Washington, DC: United States Government Printing Office.

Dolukhanov, P. M. 1986. "Foragers and Farmers in West-Central Asia." In *Hunters in Transition: Mesolithic Societies of Temperate Eursia and their Transition to Farming*, edited by M. Zvelebil, 121–32. Cambridge: Cambridge University.

———. 2004. "Prehistoric Environment, Human Migrations and Origin of Pastoralism in Northern Eurasia." In *Impact of the Environment on Human Migration in Eurasia: Proceeding of the Nato Advanced Research Workshop, Held in St. Petersburg, 15–18 November 2003*, edited by E. M. Scott, A. Y. Alekseev, and G. Zaitseva, 225–42. Dordrecht, The Netherlands: Kluwer Academic.

Dolukhanov, P., A. Shukurov, D. Gronenborn, D. Sokoloff, V. Timofeev, and G. Zaitseva. 2005. "The Chronology of Neolithic Dispersal in Central and Eastern Europe." *Journal of Archeological Science* 32:1441–58.

Dombrowski, J. C. 1971. "Excavations in Ethiopia: Lalibela and Natchabiet Caves, Begemeder Province." PhD diss., Boston University.

Domokos, P. P. 1930. "A kender feldolgozása és eszközei Menaságon" [The way and tools of hemp processing in Menaság]. [In Hungarian.] *Né* 22:145–49.

Donaldson, G. 1960. "Sources for Scottish Agrarian History before the Eighteenth Century." *The Agricultural History Review*. 8 (2): 82–90

Doolittle, J. 1966. *Social Life of the Chinese: Religious, Governmental, Educational, and Business Customs and Opinion*. 2 vols. Taipei, China: Ch'eng Wen. First published 1865 by Harper Brothers.

Dörfler, W. 1990. "Die geschichte des hanfanbaus in Mitteleuropa aufgrund palynologischer untersuchungen und von Großrestnachweisen." *Praehistorische Zeitschrift* 65:218–44.

Doria, R. 1986. "Os fumadores de maconha: efeitos e males do vicio." In *Diamba Sarabamba*, edited by A. Henman and O. Pessoa Jr., 19–38. São Paulo, Brazil: Ground.

Dorus, S., E. J. Vallender, P. D. Evans, J. R. Anderson, S. L. Gilbert, M. Mahowald, J. Wyckoff, C. Malcom, and B. T. Lahn. 2004. "Accelerated Evolution of Nervous System Genes in the Origin of Homo Sapiens." *Cell* 119 (7): 1027–40.

Dragendorff, G. 1898. *Die Heilpflanzen der verschiedenen Völker und Zeiten*. Stuttgart, Germany: F. Enke.

Drake-Brockman, H. (1963) 1995. *Voyage to Disaster*. 2nd ed. Nedlands, Australia: University of Western Australia.

Dredge, C. P. 1987. "Korean Funerals: Ritual as Process." In *Religion and Ritual in Korean Society*, edited by L. Kendall and G. Dix, 71–92. Institute of East Asian Studies. Berkeley, CA: University of California.

Drekmeier, C. 1962. *Kinship and Community in Ancient India*. Palo Alto, CA: Stanford University.

Drews, R. 2004. *Early riders: the beginning of mounted warfare in Asia and Europe*. New York: Routledge.

Duff, C. 1953. *A New Handbook on Hanging* (2nd ed.). Chicago: Henry Regnery.

Dumayne-Peaty, L. 1999. "Late Holocene Human Impact on the Vegetation of Southeastern Scotland: A Pollen Diagram from Dogden Moss, Berwickshire." *Review of Palaeobotany and Palynology* 105 (3–4): 121–41.

Du Toit, B. M. 1975a. "Dagga: The History and Ethnographic Setting of *Cannabis Sativa* in Southern Africa." In *Cannabis and Culture*, edited by V. Rubin, 81–116. Paris: Mouton.

———. 1975b. "Continuity and Change in *Cannabis* Use by Africans in South Africa." *Journal of Asian and African Studies* 11 (3–4): 203–8.

———. 1977. "Historical Factors Influencing Cannabis Use among Indians in South Africa." *Journal of Psychedelic Drugs* 9:235–46.

———. 1980. *Cannabis in Africa: A Survey of Its Distribution in Africa, and a Study of Cannabis Use and Users in Multi-Ethnic South Africa*. Rotterdam, The Netherlands: A. A. Balkema.

Dutt, U. C. 1900. *The Materia Medica of the Hindus*. Calcutta, India: Dwarkanath Mukerjee.

Dwarakanath, C. 1965. "Use of Opium and Cannabis in the Traditional Systems of Medicine in India." *Bulletin on Narcotics* 17:15–19.

Dyer, T. F. T. 1889. *The Folklore of Plants*. London: Catto and Windus.

Earleywine, M. 2002. *Understanding Marijuana: A New Look at the Scientific Evidence*. Oxford: Oxford University.

Eaton, C. 1966. *A History of the Old South*. New York: Macmillan.

Eberhard, W., comp. and trans. 1938. *Chinese Fairy Tales and Folk Tales*. New York: E. P. Dutton.

Eberland, W. 1968. *The Local Cultures of South and East China*. Leiden, The Netherolands: E. J. Brill.

Ebrey, P. B. 1991. *Chu Hsi's Family Rituals—A Twelfth-Century Chinese Manual for the Performance of Cappings, Weddings, Funerals, and Ancestral Rites*. Princeton, NJ: Princeton University.

Ebrey, P. B. n.d. "A Visual Sourcebook of Chinese Civilization." *University of Washington*. http://depts.washington.edu/chinaciv/tg/tmiltech.pdf.

Echkel, N. 1980. "Tekstilno rukotvorstvo uže okolici Zagreba" [Traditional textile handwork in the vicinity of Zagreb]. [In Serbo-Croatian.] *Etnološka tribina* 10 (3): 27–38.

Ecsedy, I. 1979. *The People of the Pit-Grave Kurgans in Eastern Hungary*. Budapest: Adadémia Kiadó.

Edwards, K. J. and G. Whittington. 1992. "Male and Female Plant Selection in the Cultivation Hemp, and Variations in Fossil Pollen Representation." *The Holocene* 2:85–87.

Egerton, F. 2002. "A History of the Ecological Sciences Part 7. Arabic Language Science: Botany, Geography, and Decline." *Bulletin of the Ecological Society of America* (October): 261–66.

Eglītis, J. 1956. "Rūpnieciskās zvejas tehnika" [Commercial fishing methods]. Riga: *Latvijas PSR Republikāniskā zvejnieku kolchozu savienība* [Latvian SSR Republican fishermen kolchozu Union, Riga]. [In Latvian.] 1 (1).

Ehrensing, D. T. 1998. *Feasibility of Industrial Hemp Production in the United States Pacific Northwest*, May, State Bulletin 681. Corvallis, OR: Oregon State University Extension Service.

Ekwall, E. 1960. *The Concise Oxford Dictionary of English Place-Names*, 4th ed. Oxford: Clarendon.

Eliade, M. 1972. *Zalmoxis, the Vanishing God*. Translated from French by W. R. Trask. Chicago: University of Chicago.

Eliade, M., and W. R. Trask. 1972. "Zalmoxis." *History of Religions* 11 (3): 257–302.

Ellstrand, N. C., and K. A. Schierenbeck. 2000. "Hybridization as a Stimulus for the Evolution of Invasiveness in Plants." *Proceedings of the National Academy of Sciences* 97 (13): 7043–50.

Ellstrand, N. C., H. C. Prentice, and J. F. Hancock. 1999. "Gene Flow and Introgression for Domesticated Plants into their Wild Relatives." *Annual Review of Ecology and Systematics* 30:39–563.

ElSohly, M. A., C. E. Turner, C. H. Phoebe Jr., J. E. Knapp, P. L. Schiff Jr., and D. J. Slatkin. 1978. "Anhydrocannabisativine, a New Alkaloid from *Cannabis Sativa* L." *Journal of Pharmacological Science* 67 (1): 124.

Elwin, V. 1949. *The Myths of Middle India*. Oxford: Oxford University.

Emboden, W. 1972. "Ritual Use of Cannabis Sativa L: A Historical-Ethnographic Survey." In *Flesh of the Gods; the Ritual Use of Hallucinogens*, edited by P. T. Furst, 214–36. New York: Praeger.

———. 1974. "Cannabis—a Polytypic Genus." *Economic Botany* 28 (3): 304—10.

———. 1995. "Art and Artifact as Ethnobotanical Tools in the Ancient Near East with Emphasis on Psychoactive Plants." In *Ethnobotany, Evolution of a Discipline*, edited by R. E. Schultes and S. von Reis, 93–107. Portland, OR: Dioscorides.

Embree, J. F. 1946. *A Japanese Village—Suye Mura*. London: Routledge.

Emerson, G. 2002. *Sin City London in Pursuit of Pleasure*. London: Granada.

Enami, K., S. Sakamoto, Y. Okada, K. Masuda, and M. Khono. 2010. "Origin of the Difference in Papermaking Technologies between Those Transferred to the East and the West from the Motherland China." *Journal of the International Association of Paper Historians* 14 (2): 11–22. www.paperhistory.org.

Encyclopaedia Britannica: A Dictionary of Arts, Sciences, Literature and General Information. 1911. 11th ed. Vol. 15. Cambridge: Cambridge University.

Encyclopedia Judaica. 1971. Vol. 8. New York: Macmillan.

Englebrecht, T. 1916. "Über die Entstehung einiger feldmässig angebauter Kulturp-flanzen." *Geographische Zeitschrift* 22 (6): 328–34.

Enters, D., A. Lücke, and B. Zolitschka. 2006. "Effects of Land-Use Change on Deposition and Composition of Organic Matter in Frickenhauser See, Northern Bavaria, Germany." *Science of the Total Environment* 369 (1–3): 178–87.

Enters, D., W. Dörfler, and B. Zolitschka. 2008. "Historical Soil Erosion and Land-Use Change during the Last Two Millenia Recorded in Lake Sediments of Frickenhauser See, Northern Bavaria, Central Germany." *The Holocene* 18 (2): 243–54.

Eperjessy, E. 1975. "Verovanja i običaji u vezi sa Barbarinim danom" [Folk customs and beliefs connected with the Saint Barbara's day]. [In Serbian.] *Etnografija Juznih Slavena u Mađjarskoj* (Budapest, Hungary) 9: 55.

Erasmus, U. 1993. *Fats that Heal, Fats that Kill*. Burnaby, BC: Alive Books.

Erdeljanović, J. 1951. "Etnološka građa o Šumadincima" [Ethnological material on Šumadinci]. [In Serbian.] *Srpski etnografski zbornik* (Beograd, Serbia) 64:2. Rasprave i građa knj [Discussions and structure] 2. Erdtman, G. 1943. *An Introduction to Pollen Analysis*. Waltham, MA: Chronica Botanica.

Etkin, N. 2006. *Edible Medicines: An Ethnopharmacology of Food*. Tucson, AZ: University of Arizona.

EvaluatePharma. 2013. "Nabilone: Worldwide Sales 2009/10." http://www.evaluatepharma.com/Universal/View.aspx?type=Entity&entityType=Product&IType=modData&id=3995&componentID=1003.

Evans, J. and T. O'Connor. 2001. *Environmental Archeology, Principles and Method*. Sutton, England: Stroud.

Evans, L. 1871. *The Satires of Juvenal, Persiu, Sulpicia, and Lucilius, Literally Translated into English Prose with Notes, Chronological Tables, Arcuments, etc*. London: Bell & Daldy.

Fadiman, A. 1997. *The Spirit Catches You and You Fall Down*. New York: Farrar, Straus and Giroux.

Faeti V., G. Mandolino, and P. Ranalli. 1996. "Genetic Diversity of Cannabis Sativa Germplasm Based on RAPD markers." *Plant Breeding* 115 (5): 367–70.

Fagan, B., and D. W. Phillipson. 1965. "Sebanzi: The Iron Age Sequence at Lochinvar and the Tonga." *The Journal of the Royal Anthropological Institute* 95 (2): 253–94.

Falk, C. 1996. "Upon Meeting the Ancestors: The Hmong Funeral Ritual in Asia and Australia." *Hmong Studies Journal* 1 (1): 1–15.

———. 2004. "Hmong Instructions to the Dead: What the Mouth Organ *Qeej* Says (Part One)." *Asian Folklore Studies* 63:1–29.

Falk, H. 1919. "Altwestnordische Kleiderkunde. Mit besonderer Berücksichtigung der Terminologie." In *Videnskabs selskapets skrifter II, Hist.-Fil. Klasse no. 3*, 141–47. Kristiania, Norway: In Kommission bei J. Dybwad.

Fang, L. 1935. *Taiping yulan* [an imperial encyclopedia compiled between 977 and 983 CE that was titled *Taiping leibian*, "Anthology of the Taiping Era"]. SBCK ed. Shanghai: Commercial.

Fankhauser, M. 2002. "History of Cannabis in Western Medicine." In *Cannabis and Cannabinoids: Pharmacology, Toxicology and Therapeutic Potential*, edited by F. Grotenhermen and R. Russo, 37–51. Binghampton, NY: Haworth.

FAO. 2009. "Hemp's future in Chinese Fabrics: 2009 International Year of Natural Fibres." http://www.naturalfibres2009.org/en/stories/hemp.html.

Farnsworth, N. R. 1968. "Hallucinogenic Plants." *Science* 162:1086

Farrington, I., and J. Urry. 1985. "Food and the Early History of Cultivation." *Journal of Ethnobiology* 5:143–57.

Farris, W. W. 1998. *Sacred Texts and Buried Treasures: Issues in the Historical Archeology of Ancient Japan*. Honolulu: University of Hawaii.

Fél, E. 1961. *Hungarian Peasant Embroidery*. Translated from Hungarian by A. Barat and L. Halapy. London: B. T. Batsford Ltd.

Fél, E., and T. Hofer. 1969. *Proper Peasants: Traditional Life in a Hungarian Village*. Budapest: Corvina.

Feng, Z.-D., L. Y. Tang, H. B. Wang, Y. Z. Ma, and K.-B. Liu. 2006. "Holocene Vegetation Variations and the Associated Environmental Changes in the Western Part of the Chinese Loess Plateau." *Palaeogeography, Palaeoclimatology, and Palaeoecology* 241:440–56.

Fetterman, P. S., E. S. Keith, C. W. Waller, O. Guerrero, N. J. Doorenbos, and M. W. Quimby. 1971. "Mississippi-Grown *Cannabis Sativa* L.: Preliminary Observations on Chemical Definition of Phenotype and Variations in Tetrahydrocannabinol Content versus Age, Sex, and Plant Part." *Journal of Pharmaceutical Sciences* 60 (8): 1246–49.

Feurdean, A. 2005. "Holocene Forest Dynamics in Northwestern Romania." *The Holocene* 15 (3): 435–46.

Feurdean, A., and C. Astaloş. 2005. "The Impact of Human Activities in the Gutâiului Mountains, Romania." *Studia Universitatis Babeş-Bolyai, Geologia*, 50 (1–2): 63–72.

Filipović, M., and P. Tomić. 1955. "Gornja Pčinja: Rasprave i građa 3" [Upper Pcinja river district in South Serbia: Dissertations and scientific material 3]. [In Serbian.] *Srpski etnografski zbornik* [Serbian ethnographic proceedings] 68:92. Filipović-Fabijanić, R. 1964. "Glasnik Zemaljskog muzeja u Sarajevu." [In Yugoslavian.] *Etnologija, Nova serija* 19:212, 220.

Filippi, M. L., E. Arpenti, N. Angeli, F. Corradini, S. Frisia, and K. Van der Borg. 2005. "Human Presence and Its Impact as Recorded in Lake Lavarone (Dolomites, NE Italy) Sedimentrecord." Paper Presented at Paleoclimate, Environmental Sustainability and our Future 2nd Open Science Meeting, August 10–12, Beijing, China.

Finlayson, C. 2005. "Biogeography and Evolution of the Genus *Homo*." *Trends in Ecology and Evolution* 20 (8): 457–63.

———. 2009. *The Humans Who Went Extinct: Why Neanderthals Died Out and We Survived*. Oxford: Oxford University Press.

Fisher, J. 1975. "*Cannabis* in Nepal: An Overview." In *Cannabis and Culture*, edited by V. Rubin, 247–56. Paris: Mouton.

Flannery, K. 1986. "The Problem and the Model." In *Guila Naquitz*, edited by K. V. Flannery, 1–18. Orlando, FL: Academic.

Flannery, K. V. 1969. "Origins and Ecological Effects of Early Domestication in Iran and the Near East." In *The Domestication and Exploitation of Plants and Animals*, edited by P. J. Ucko and G. W. Dimbleby, 73–100. London: Duckworth.

Flattery, D. S., and M. Schwartz. 1989. *Haoma and Harmaline: The Botanical Identity of the Indo-Iranian Sacred Hallucinogen "Soma" and Its Legacy in Religion, Language, and Middle Eastern Folklore*. Berkeley, CA: University of California.

Fleming, M. P., and R. C. Clarke. 1998. "Physical Evidence for the Antiquity of *Cannabis sativa* L." *Journal of the International Hemp Association* 5 (2): 80–93.

Folkard, R. 1884. *Plant Lore, Legends and Lyrics*. London: Sampson Low, Marston, Searle and Rivington.

Forapani, S., A. Carboni, C. Paoletti, V. M. C. Moliterni, P. Ranalli, and G. Mandolino. 2001. "Comparison of Hemp Varieties Using Random Amplified Polymorphic DNA Markers." *Crop Science* 41:1682–89.

Forbes, R. J. 1964. *Studies in Ancient Technology*. Vol. 4. 2nd. ed. Revised, Leiden, The Netherlands: E. J. Brill.

Forster, E. 1996. "History of Hemp in Chile." *Journal of the International Hemp Association* 3 (2): 72.

Forster, P. 2004. "Ice Ages and the Mitochondrial DNA Chronology of Human Dispersals: A Review." *Philosphical Transactions of the Royal Society of London* B 359: 255–64.

Fortenbcry, T. R., and M. Bennett. 2004. "Opportunities for Commercial Hemp Production." *Review of Agricultural Economics* 26 (1): 97–117.

Foster-Carter, A. 2009. "North Korea–South Korea Relations: Things Can Only Get Better? Money for New Rope: First Major Southern Joint Venture Opens in Pyongyang." *Comparative Connections: A Quarterly E-Journal on East Asian Bilateral Relations*. Leeds, England: University of Leeds. January 2009. http://csis.org/files/media/csis/pubs/0804qnk_sk.pdf.

France-Lanord, A. 1979. "La fouille en laboratoire: Méthodes et résultats." *Dossiers de l'Archéologie* 32:66–91.

François, O., M. G. B. Blum, M. Jakobsson, and N. A. Rosenberg. 2008. "Demographic History of European Populations of *Arabidopsis Thaliana*." *PLoS Genetics* 4 (5): e1000075. doi:10.1371/journal.pgen.1000075.

Franić, T. I. 1935. "Narodni običaji i obredi uz prvo oranje i sijanje u srezu slavonsko-pozeskom" [Folk customs and rituals connected with first ploughing and sowing in Slavonija (Croatia)]. [In Yugoslavian.] *Glasnik Etnografskog muzeja u Beogradu* 10: 33–41.Frank, T. 1959. *An Economic Survey of Ancient Rome*. Vol. 4. Patterson, NJ: Pageant Books.

Frankhauser, M. 2002. "History of Cannabis in Western Medicine." In *Cannabis and Cannabinoids: Pharmacology, Toxicology, and Therapeutic Potential*, edited by Grotenherman, F. and E. Russo, 37–49. New York: Haworth.

French, C. N., and P. D. Moore. 1986. "Deforestation, *Cannabis* Cultivation, and Schwingmoor Formation at Cors Llyn (Llyn Mire), Central Wales." *New Phytologist* 102:469–82.

Freyre, G. 1946. *The Masters and the Slaves*. 1st American ed. Translated from Portuguese (of 4th and definitive Brazilian ed.) by S. Putnam. New York: Knopf.

———. 1967. *A Brazilian Tale*. Translated from Portuguese by B. Shelby. New York: Knopf.

———. 1981. *Sobrados e mucambos*. Rio de Janeiro: José Olympio.

Friedman, J., and S. C. H. Barrett 2009. "Wind of Change: New Insights on the Ecology and Evolution of Pollination and Mating in Wind-Pollinated Plants." *Annals of Botany* 103:1515–27.

Fries, M. 1962. "Studies of the Sediments and the Vegetational History in the Osbysjo Basin, North of Stockholm." *Oikos* 13: 76–96.

Fujiwara, H., M. R. Mughal, A. Sasaki, and T. Matano. 1992. "Rice and Ragi at Harappa: Preliminary Results by Opal Plant Phytoliths." *Pakistan Archeology* 27:129–42.

Fuller, D. Q. 2002. "Fifty Years of Archaeobotanical Studies in India: Laying a Solid Foundation." In *Indian Archeology in Retrospect*, vol. 3. *Archeology and Interactive Disciplines*, edited by S. Settar and R. Korisettar, 247–64. New Delhi: Manohar.

———. 2006. "Agricultural Origins and Frontiers in South Asia: A Working Synthesis." *Journal of World Prehistory* 20:1–86.

———. 2008. "The Spread of Textile Production and Textile Crops in India Beyond the Harappan Zone: An Aspect of the Emergence of Craft Specialization and Systematic Trade." In *Linguistics, Archeology and the Human Past*, edited by T. Osada and A. Eesugi, Occasional Paper 3, 1–25. Kyoto, Japan: Indus Project Research for Humanity and Nature.

———. 2011. "Finding Plant Domestication in the Indian Subcontinent: The Origins of Agriculture: New Data, New Ideas." *Current Anthropology* 52 (S4): S347–S362.

Fuller, D. Q., and M. Madella. 2001. "Issues in Harappan Archaeobotany: Retrospect and Prospect." In *Indian Archeology in Retrospect*, vol. 2. *Protohistory. Archeology of the Harappan civilization*, edited by S. Settar and R. Korisettar, 317–90. New Delhi: Manohar.

Fuller, D. Q., L. Qin, and E. Harvey. 2007. "A Critical Assessment of Early Agriculture in East Asia, with Emphasis on Lower Yangzte Rice Domestication." For publication in a special issue of *Pradghara* (Journal of the Uttar Pradesh State Archeology Department), edited by Rakesh Tewari, relating to "First Farmers in Global Perspective" seminar of Uttar Pradesh State, January 18–20, 2006, Lucknow, India, Department of Archeology.

Fuller, D. Q., R. Korisettar, P. C. Venkatasubbaiah, and M. K. Jones. 2004. "Early Plant Domestications in Southern India: Some Preliminary Archaeobotanical Results." *Vegetation History and Archaeobotany,* 13 (2): 115–29.

Fuminori, S. 1998. "Expansion of Chinese Paddy Rice to the Yunnan-Guizhou Plateau." [In Japanese.] *Agricultural Archeology* 1:255–62.

Furst, P. T., ed. 1972. *Flesh of the Gods: The Ritual Use of Hallucinogens.* New York: Praeger.

Fushi Ltd. 2007. "Fushi: Holistic Health and Beauty Solutions. Hempseed (Cannabis Sativa[*sic*])." http://www.fushi.co.uk/Hempseed-Cannabis-Sativa_591.aspx.

Gaillard, M.-J., and B. Berglund. 1988. "Land-Use History During the Last 2700 Years in the Area of Bjaresjo, Southern Sweden." In *The Cultural Landscape—Past, Present and Future,* edited by H. H. Birks, H. J. B. Birks, P. E. Kaland, and D. Moe, 409–28. Cambridge: Cambridge University.

Gale, E. M. 1931. *Discourses on Salt and Iron. A Debate of Commerce and Industry in Ancient China. Chapters 1–19.* Lieden, The Netherlands: Brill. Reprinted 1973, Taipei, as *Sinica Leidensia Edidit Institutum Sinologicum Lugduno-Batavum, II* by Ch'eng-Wen.

Galen, C. 2003. *On the Properties of Foodstuffs/De Alimentorum Facultatibus.* Translated by O. Powell. With introduction and commentary written by O. Powell. Cambridge: Cambridge University.

Garrett, V. M. 1987. *Traditional Chinese Clothing in Hong Kong and South China, 1840–1980.* Hong Kong: Oxford University.

Garside, P., and P. Wyeth. 2006. "Identification of Cellulosic Fibres by FTIR Spectroscopy: Differentiation of Flax and Hemp by Polarized ATR FTIR." *Studies in Conservation* 51 (3): 205–11.

Gates, P. W. 1965. *Agriculture and the Civil War.* New York: Knopf.

Gavazzi, M. 1939. *Godinu dana hrvatskih narodnih obicaja, II oko Bozica* [A year of Croatian folk customs, volume II, Christmas time]. [In Serbo-Croatian.] Zagreb [Croatia]: Izdanje Matice Hrvatske, 51.

Gavriljuk, N. A. 2005. "Fishery in the Life of the Nomadic Population of Northern Black Sea Area in the Early Iron Age." In *Ancient Fishing and Fish Processing in the Black Sea Region,* edited by T. Bekker-Nielsen, 105–13. Aarhus, Denmark: Aarhus University. Proceedings of an interdisciplinary workshop on marine resources and trade in fish products in the Black Sea region in antiquity, April 4–5, 2003, University of Southern Denmark, Esbjerg.

Gearey, B. R., A. R. Hall, H. Kenward, M. J. Bunting, M. C. Lillie and J. Carrott. 2005. "Recent Palaeoenvironmental Evidence for the Processing of Hemp (*Cannabis sativa* L.) in Eastern England during the Medieval Period." *Medieval Archeology* 49:317–22.

Geertz, C. 1973. *The Interpretation of Cultures: Selected Essays.* New York: Basic.

Geijer, A. 1938. *Die Textilfunde aus den Gräben. Birka: Untersuchungen und Studien, III.* Stockholm: Kungliga Vitterhets Historie och Antikvitets Akademien [in German].

———. 1983. "The Textile Finds from Birka." In *Cloth and Clothing in Medieval Europe,* edited by N. B. Harte and K. G. Ponting, 80–99. London: Heinemann.

Gelling, M. 1984. *Place-Names in the Landscape.* London: J. M. Dent & Sons.

Gelorini. V., L. Meersschaert, and J.-P. Van Roeyen 2003. "Archeobotanisch onderzoek van enkele laat—en postmiddeleeuwse archeologische contexten uit de onderzoekszone Verrebroekdok (Beveren, prov. Oost-Vlaanderen)." *Archeol Vlaanderen* 7 (1999/2000): 201–24.

Gepts, P. 2004. "Crop Domestication as a Long-Term Selection Experiment." *Plant Breeding Reviews* 24 (2): 1–44.

Gerard, J. 1633. *Herball or General Historie of Plantes.* Edited by Thomas Johnson. London: Adam Islip, Joice Norton & Richard Whitakers, 708–9.

German Press Agency (Deutsche Presse-Agentur). 2006. "Queen Praises Latvian Hemp's Role in Battle of Trafalgar," October 18.

Ghalioungui, P. 1987. *The Ebers Papyrus: A New English Translation, Commentaries, and Glossaries.* Cairo: Academy of Scientific Research and Technology.

Ghildiyal, R. M. 2005. *Garhwali Cuisine, Savvy Cook Book,* April. http://rushina-mushaw-ghildiyal.blogspot.com/2006/03/garhwali-cuisine-savvy-cook-book-april.html.

Giele, E. 1998. "Early Chinese Manuscripts: Including Addenda and Corrigenda to New Sources of Early Chinese History: An Introduction to the Reading of Inscriptions and Manuscripts." *Early China* 23–24 (1998–99): 247–337.

Gilligan, I. 2007. "Clothing and Farming Origins: The Indo-Pacific Evidence." *Bulletin of the Indo-Pacific Prehistory Association* 27:12–21.

Gilmore, S., R. Peakall, and J. Robertson. 2003. "Short Tandem Repeat (STR) DNA Markers are Hypervariable and Informative in *Cannabis sativa*: Implications for Forensic Investigations." *Forensic Science International* 131:65–74.

———. 2007. "Organelle DNA Haplotypes Reflect Crop-Use Characteristics and Geographic Origins of *Cannabis sativa.*" *Forensic Science International* 172:179–90.

Gimbutas, M. 1956. *The Prehistory of Eastern Europe.* Cambridge, MA: Peabody Museum.

Ginzburg, C. 2012. *Threads and Traces.* Translated by A. C. Tedschi and J. Tedeschi. Los Angeles: University of California.

Glick, T. F. 1979. *Islamic and Christian Spain in the Early Middle Ages.* Princeton, NJ: Princeton University.

———. 2005. *Islamic and Christian Spain in the Early Middle Ages.* Revised, Leiden, The Netherlands: Brill.

Global Commission on Drug Policy. 2011. *Report of the Global Commission on Drug Policy.* www.globalcommissionondrugs.org.

Godwin, H. 1967a. "Pollen-Analytic Evidence for the Cultivation of *Cannabis* in England." *Review of Palaeobotany and Palynology* 4:71–80.

———. 1967b. "The Ancient Cultivation of Hemp." *Antiquity* 41 (161): 42–49.

Gong, W., Z. Zeng, Y. Y. Chen, C. Chen, Y. X. Qiu, and C. X. Fu. 2008. "Glacial Refugia of Ginkgo Biloba and Human Impact on Its Genetic Diversity: Evidence from Chloroplast DNA." *Journal of Integrated Plant Biology* 50 (3): 368–74.

Gongliang, Z., and D. Du. 1988. "Wujing zongyao [General Military]." [In Chinese.] Late Ming Dynasty (Wanli Period) ed. In *Zhongguo bingshu jicheng [Chinese book on military strategy integration],* vols. 3–5, edited by the Editing Committee. Beijing: Jiefangjun chubanshe and Liaoshen shushe.

Good, I. 2001. "Archeological Textiles: A Review of Current Research." *Annual Review of Anthropology* 30:209–26.

Goodrich, S. G. 1856. *History of All Nations from the Earliest Periods to the Present Time; or, Universal History: In Which the History of Every Nation, Ancient and Modern, is Seapraetly Given.* New York: Miller, Orton & Mulligan.

Graham, D. C. 1926. "The Chuan Miao of West China." *Journal of Religion* 6:302–7.

———. 1978. *The Tribal Songs and Tales of the Ch'uan Miao. Asian Folklore and Social Life Monographs, vol. 102.* Taipei: Orient Cultural Service.

Granoszewski, W., D. Demske, M. Nita, G. Heumann, and A. A. Andreev. 2005. "Vegetation and Climate Variability during the Last Interglacial Evidenced in the Pollen Record from Lake Baikal." *Global Planetary Change* 46 (1–4): 187–98.

Grassi, F., F. De Mattia, G. Zecca, F. Sala, and L. Massimo. 2008. "Historical Isolation and Range Expansion of Wild Grape." *Biological Journal of the Linnean Society* 95 (3): 611–19.

Graves, R. 1955. *The Greek Myths.* London: Penguin.

Gray, I., and M. Stanley. 1989. *A Punishment in Search of a Crime: Americans Speak Out against the Death Penalty.* Dresden, TN: Avon Books.

Gray, J. H. 1878. *China—A history of the Laws, Manners, and Customs of the People.* Vol. 1. London: Macmillan.

Gray, L. 2002. "The Botanical Remains." In "Settlement in Roman Southwark Archeological excavations (1991–8) for the London Underground Limited Jubilee Line Extension Project," edited by J. Drummond-Murray, P. Thompson, and C. Cowan. *MoLAS* [Museum of London Archeological Service] 12:242–59.

Gray, L. C. 1958. *History of Agriculture in the Southern United States to 1860.* Vol. 1. Gloucester, MA: Peter Smith.

Grbić, S. M., and T. R. Djordjević. 1907 "Običaji naroda srpskoga" [Traditional customs among Serbs]. *Srpski etnografski zbornik* 14. [In Croatian.] Beogradu: Stampano v Državnoj štampariji kraljevine Srbije.

Green, F. J. 1979a. "Collection and Interpretation of Botanical Information from Medieval Urban Excavations in Southern England." In *Veröffentlicht mit Mittelen des Landschaftsverbandes Rheinland,* edited by M. Ludwig, 39–55. Bonn, Germany: Rheinisches Landesmuseum.

———. 1979b. "The Plant Remains." In "Excavations at 1 Westgate Street. Goucester, 1975," edited by C. M. Heighway, A. P. Garrod, and A. G. Vince. *Medieval Archeology* 23:186–90.

———. 1986. "The Archaeobotanical Evidence from the Castle Ditch." In "Excavations at Southampton Castle. Southamption City Museums," edited by J. Oxley. *Southampton Archeogical Monographs* 3:45–46.

Green, J. 1999. *The Cassell Dictionary of Slang*. Reprint, London: Cassell.

Greig, J. R. A. 1988. "Plant Remains." In S. Ward. (ed.). Excavations at Chester: 12 Watergate Street 1985. Roman headquarters to medieval row. *Grosvenor Museum Archeological Excavation and Survey Report* 5. Chester: City Council, 59–69.

Greimas, A. J. 1992. *Of Gods and Men—Studies in Lithuanian Mythology*. Bloomington, IN: Indiana University.

Grierson, G. A. 1894. "The Hemp Plant in Sanskrit and Hindi Literature." *The Indian Antiquary* 23:260–62.

Griffin, K., and P. U. Sandvik. 1991. "Plant Remains from Medieval Trondheim, Norway." *Acta interdisciplinaria archeologica* 7: 111–15.

Griffiths, R. R., W. A. Richards, U. McCann, and R. Jesse. 2006. "Psilocybin Can Occasion Mystical-Type Experiences Having Substantial and Sustained Personal Meaning and Spiritual Significance." *Psychopharmacology* 187:268–83.

Grigoriev, O. V. 2002. "Application of Hempseed (*Cannabis Sativa* L.) Oil in the Treatment of Ear, Nose and Throat (ENT) Disorders." *Journal of Industrial Hemp* 7 (2): 5–16.

Grigoryev, S. 2007. *Hemp (Cannabis Sativa L.) Genetic Resources at the VIR: From the Collection of Seeds, through the Collection of Sources, towards the Collection of Donors of Traits*. http://www.vir.nw.ru/hemp/hemp1.htm.

Grinspoon, L., and J. Bakalar. (1993) 1997. *Marihuana, the Forbidden Medicine*. New Haven, CT: Yale University.

———. 2007. "Marijuana, the Wonder Drug." *The Boston Globe*, March 1.

Grinter, P., and J. P. Huntley. 2002. "The Plant Remains." In L. Truman (ed.). Excavations at Stockbridge, Newcastle upon Tyne, 1995. *Archeologia Aeliana* 29:210–17.

Grlic, L. 1968. "A Combined Spectrophotometic Differentiation of Samples of *Cannabis*." *Bulletin on Narcotics* 20:25.

Gronenborn, D. 1999. "A Variation on Basic Theme: The Transition to Farming in Southern Central Europe." *Journal of World Prehistory* 13:123–210.

———. 2003. "Migration, Acculturation and Culture Change in Temperate Europe and Eurasia, 6500–5000 cal BC." In "The Neolithization of Eurasia: Reflections in Archeology and Archaeogenetics," edited by M. Budja. *Documenta Praehistorica* (Lubljana, Slovenia) 30:79–91.

Grönlund, E., H. Simola, and P. Huttunen. 1986. "Paleolimnological Reflections of Fiber-Plant Retting in the Sediment of a Small Clear-Water Lake." *Hydrobiologia* 143 (1): 425–31.

Grudzinskaya, I. A. 1988. "On the Taxonomy of Cannabaceae." *Botanicheskii Zhurnal* (St. Petersburg, Russia) 73 (4): 589–93.

Grun, R., and C. Stringer. 2000. "Tabun Revisited: Revised ESR Chronology and New ESR and U-Series Analyses of Dental Material from Tabun C1." *Journal of Human Evolution* 39, 601–12.

Gu, W. 1995. *Weavers of Ethic Culture: The Miaos*. Kunming, China: Yunnan Education.

Guerra Doce, E. 2006. "Evidencias del consumo de drogas en Europa durante la Prehistoria." *Trastornos Adictivos* 8 (1): 53–61.

Guerra Doce, E., and J. A. López Sáez. 2006. "El registro arqueobotánico de plantas psicoactivas en la prehistoria de la Península Ibérica. Una aproximación etnobotánica y fitoquímica a la interpretación de la evidencia." *Complutum* 17:7–24.

Gundrum-Oriovčanin, F. S. 1909. "Luiceva lekarusa" [Medicine prescriptions by Lujic]. In *Zbornik za narodni život i običaje Južnih Slavena XIV*, vol. 14. Zagreb, Croatia: Jugoslavenska akademija znanosti i umjetnosti [in Yugoslavian].

Gunther, R. T., ed. (1655) 1959. *The Greek Herbal of Dioscorides*. Translated by John Goodyer. New York: Hafner.

Guo, Z. T., B. Sun, Z. S. Zhang, S. Z. Peng, G. Q. Xiao, J. Y. Ge, Q. Z. Hao, Y. S. Qiao, M. Y. Liang, J. F. Liu, Q. Z. Yin, and J. J. Wei. 2008. "A Major Reorganization of Asian Climate Regime by the Early Miocene." *Climate of the Past—Discussions* 4:535–84.

Guy, G., W. Whittle, B. Anthony, and P. Robson, eds. 2004. *The Medicinal Uses of Cannabis and Cannabinoids*. London: Pharmaceutical.

Ha, T.-H. 1968. *Guide to Korean Culture*. Seoul: Yonsei University.

———. 1958. *Folk Customs and Family Life*. Seoul: Yonsei.

Haak, W., O. Balanovsky, J. J. Sanchez, S. Koshel., and V. Zaporozhchenko, et al. 2010. "Ancient DNA from European Early Neolithic Farmers Reveals Their Near Eastern Affinities." *PLoS Biol* 8 (11): e1000536. www.plosone.org.

Habu, J. 2004. *Ancient Jomon of Japan*. Berkely, Ca: University of California.

Hadži-Vasiljević, J. 1913. "Presevska oblast" [The area of Presevo in present Macedonia]. In *Južna Stara Srbija II,—istorijska, etnografska i politička istraživanja*. Beograd, Serbia: Davidović [in Serbian].

Hafsten, U. 1956. "Pollen Analysis Investigations on the Late-Quaternary Development in the Inner Oslo Fjord Area." *Arbok. Univ. Bergen, Mat. Naturv* 8:1–162.

Hahm, P. C. 1988. "Shamanism and the Korean World-View: Family Life-Cycle, Society and Social Life." In *Shamanism: The Spirit World of Korea*, edited by R. Uisso and C-S. Yu, 60–97. Berkeley, CA: Asian Humanities.

Hai, H., and R. Rippchen. 1994. *Das Hanf-Handbuch*. Löhrbach: Werner Pieper Medienexperimenten.

Hajnalová, E. 1993. "Praveké osidlenie lokality Šarišské Michal'any. Dokumentované rastlinnými zvyškami" (Original occupation at the Sarisské Michal'any site shown by plant remains). *Východoslovenský Pravek (Košice)* 4:49–65 [in Serbian].

Hakki, E. E., S. A. Kayis, E. Pinarkara, and A. Sag. 2007. "Inter Simple Sequence Repeats Separate Efficiently Hemp from Marijuana (*Cannabis Sativa* L.)." *Electronic Journal of Biotechnology* 10 (4): 570–81.

Hald, M. 1980. *Ancient Danish Textiles from Bogs and Burials: A Comparative Study of Costume and Iron Age Textiles*. Translated by J. Olsen. Copenhagen: Danish National Museum.

Hall, A., H. Kenward, and J. Carrott. 2004. "Technical Report: Plant and Invertebrate Remains from Medieval Deposits at Various Sites in Aberdeen." *Reports from the Centre for Human Palaeoecology, University of York*. Report 2004/06.

Hall, A. R., and D. Williams. 1983. "The Plant Remains." In "Environment and Living Conditions at Two Anglo-Scandinavian Sites," edited by A. R. Hall, H. K. Kenward, D. Williams, and J. R. A. Greig. *The Archeology of York* 14(4): 157–225. London: Council for British Archeology.

Hall, A. R., and H. K. Kenward. 1990. "Environmental Evidence from the Colonia: General Accident and Rougier Street." In *The Archeology of York* 14(6): 289–434+ plates II–IX and fiche 2–11. London: Council for British Archeology.

Hall, A. R., H. K. Kenward, and D. Williams. 1980. "Environmental Evidence from Roman Deposits in Skeldergate." In *The Archeology of York* 14(3):101–56. London: Council for British Archeology.

Hall, A. R., H. K. Kenward, D. Williams, and J. R. A. Greig. 1983. "Environment and Living Conditions at Two Anglo-Scandinavian Sites." *The Archeology of York* 14 (4): 157–240 and fiche 1. London: Council for British Archeology.

Hall, A. R., H. K. Kenward, and J. M. McComish. 2003. "Pattern in Thinly-Distributed Plant and Invertebrate Macrofossils Revealed by Extensive Analysis of Occupation Deposits at Low Fisher Gate, Doncaster, U.K." *Environmental Archaeolology* 8:129–44.

Hall, A. R., and J. P. Huntley. 2007. *A Review for Macrofossil Plant Remains from Archeological Deposits in Northern England*. Ser. 87-2007. London: English Heritage Research Department Report.

Hall, W., and L. Degenhardt. 2007. "Prevalence and Correlates of Cannabis Use in Developed and Developing Countries." *Current Opinion in Psychiatry* 20:393–97.

Hamarneh, S. 1972. "Pharmacy in Medieval Islam and the History of Drug Addiction." *Medical History* 16 (2): 226–37.

Hammer, E. 2001. *The Arts of Korea: A Resource for Educators*. New York: The Metropolitan Museum of Art.

Hampe, A., J. Arroyo, P. Jordano, and R. J. Petit. 2003. "Rangewide Phylogeography of a Bird-Dispersed Eurasian Shrub: Contrasting Mediterranean and Temperate Glacial Refugia." *Molecular Ecology* 12:3415–26.

Hampe, A., and R. J. Petit. 2010. "Cryptic Forest Refugia on the 'Roof of the World.'" *New Phytologist* 185:5–7.

Haney, A., and B. B. Kutscheid. 1973. "Quantitative Variation in the Chemical Constituents of Marijuana from Stands of Naturalized *Cannabis Sativa* in East-Central Illinois." *Economic Botany* 27:193–203.

Hardy, K. 2007. "Where would We Be without String? Evidence for the Use, Manufacture and Role of String in the Upper Palaeolithic and Mesotithic of Northern Europe." In *Plant Processing from a Prehistoric and Ethnographic Perspective* (Proceedings of a workshop at Ghent University, November 28, 2006, Belgium), edited by V. Beugriier and P. Crombier, British Archeogical Reports International Series 1718:9–22. Oxford: John & Erica Hedges.

———. 2008. "Prehistoric String Theory. How Twisted Fibres Helped to Shape the World." *Antiquity* 82 (316): 271–80.

Harlan, J. R. 1965. "The Possible Role of Weed Races in the Evolution of Cultivated Plants." *Euphytica* 14:173–76.

———. 1971. "Agricultural Origins: Centers and Non Centers." *Science* 174:468–74.

———. 1975. "Geographic Patterns of Variation in Some Cultivated Plants." *Journal of Heredity* 66 (4): 182 91.

———. 1992. *Crops and Man*, 2nd ed. Madison, WI: American Society of Agronomy-Crop Science Society.

———. 1995. *The Living Fields: Our Agricultural Heritage*. Cambridge: Cambridge University.

Harper, A. L. 2009. "Population Genetics and Speciation in the Plant Genus *Silene* (Section *Elisanthe*)." PhD diss., University of Birmingham, School of Biosciences.

Harris, D. R. 1967. "New Light on Plant Ddomestication and the Origins of Agriculture." *Geographical Review* 57:90–107.

———. 1996. "The Origins and Spread of Agriculture and Pastoralism in Eurasia: An Overview." In *The Origins and Spread of Agriculture and Pastoralism in Eurasia*, edited by D. R. Harris, 552–73. Washington, DC: Smithsonian Institution.

Harrison, R. G., ed. 1993. *Hybrid Zones and the Evolutionary Process*. Oxford: Oxford University.

Hartley, D. 1979. *Lost Country Life*. New York: Pantheon Books.

Hartyányi, B. P., and G. Nováki. 1975. "Samen-und Fruchtfunde in Ungarn von der Neusteinzeit bis zum 18. Jahrhundert." In *Agrártörténeti Szemle* (Historia Rerum Rusticarum, Budapest) 17:1–88.

Haruo, S. 1988. "The Symbolism of the *Shishi* Performance as a Community Ritual: The *Okashira* Shinji in Ise." *Japanese Journal of Religious Studies* 15 (2–3): 137–53.

Haskins, C. H. 1927. *Studies in the History of Medieval Science*, 2nd ed. Cambridge, MA: Harvard University.

Hassan, F. 1988. "The Predynastic of Egypt." *Journal of World Prehistory* 2 (2): 135–85.

Hastorf, C. A. 1998. "The Cultural Life of Early Domestic Plant Use." *Antiquity* 72:773–82.

———. 1999. "Recent Research in Paleoethnobotany." *Journal of Archeological Research* 7 (1): 55–103.

Hawkes, D. 1959. *Ch'u Tz'u—The Songs of the South, an Ancient Chinese Anthology*. Boston: Beacon.

Hawkes, J. 1964. *The Achievements of Paleolithic Man before History*. In *Man before History*, edited by C. Gabel, 21–35. Englewood Cliffs, NJ: Prentice Hall.

———. 1969. "The Ecological Background of Plant Domestication." In *The Domestication and Exploitation of Plants and Animals*, edited by P. J. Ucko and G. W. Dimbleby, 18–29. London: Duckworth.

———. 1983. *The Diversity of Crop Plants*. Cambridge, MA: Harvard University.

Hawks, J., K. Hunley, L. Sang-Hee, and M. Wolpoff. 2000. "Population Bottlenecks and Pleistocene Human Evolution." *Molecular Biology and Evolution* 17 (1): 2–22.

Hayashi, Y. 1975. *Ainu no Noko Bunka [Ainu plant cultivation]*. Tokyo: Kokominsosho [in Japanese].

Hayatghaibi, H., and I. Karimi. 2007. "Hypercholesterolemic Effect of Drug-Type *Cannabis Sativa* L. Seed (Marijuana Seed) in Guinea Pig." *Pakistan Journal of Nutrition* 6 (1): 59–62.

Hayden, B. 1990. "Nimrods, Piscators, Pluckers and Planters: The Emergence of Food Production." *Journal of Anthropological Archeology* 9:31–69.

Hećimović-Seselja, M. 1985. *Tradicijski život i kultura ličkoga sela Ivčević Kosa (Traditional life and culture of the village Ivcevic kosa in the Lika region, Croatia)*. Zagreb, Croatia: Mladen Seselja [in Croatian].

Hegi, G. 1957. *Illustrierte flora von Mittel-Europa*. Vol. 3. München, Germany: Lehmann's Verlag.

Hehn, V. 1885. *The Wandering of Plants and Animals from their First Home*. London: S. Sonnenschein.

Heinäjoki, H., and M. Aalto. 1997. "Zur Geschichte der Vegetation beim Burgwall der Ladogaburg." In *Archeologiya, istoriya, kul'tura. Doklady Rossiisko-Finlyadskogo simpoziuma po voprosam arkheologii*, edited by A. N. Kirpichnikov and S. Ryabinin, 158–73. St. Petersburg, Russia: Dmitrii Bulanin [in German].

Heiser, C. B. 1973. "Introgression Re-Examined." *The Botanical Review* 39 (4): 347–66.

Hellwig, M. 1997. "Plant Remains from Two Cesspits (15th and 16th Century) and a Pond (13th Century) from Göttingen, Southern Lower Saxony, Germany." *Vegetation History Archaeobotany* 6:105–16.

Hennink, S. 1997. "EU Rregulations on Hemp Cultivation." *Journal of the International Hemp Association* 4 (1): 38–39.

Henriksen, P. 2003. "Rye Cultivation in the Danish Iron Age—Some New Evidence from Iron-Smelting Furnaces." *Vegetation History and Archaeobotany* 12 (3): 177–85.

Herer, J. 1992. *The Emperor Wears No Clothes: The Authoritative Historical Record of the Cannabis Plant, Marijuana Prohibition, and How Hemp Can Still Save the World*. Van Nuys, CA: Access Unlimited.

Hermione-La Fayette Association. 2011. "Arsenal Maritime, Rochefort, France: 'Caulking on the Orlop Deck.'" http://www.hermione.com/en/the-construction/2008/585-caulking-on-the-orlop-deck.html.

Herndon, G. M. 1959. "The Story of Hemp in Colonial Virginia." PhD diss., University of Virginia.

———. 1966. "A War-Inspired Industry." *Virginia Magazine of History and Biography* 74:301–11.

Herodotus. 1921. *The Histories*. Vol. II, Book IV. Loeb Classical Library. Translated by A. D. Godley. Cambridge, MA: Harvard University.

Hewat, M. L. 1906. *Bantu Folk Lore*. Capetown, South Africa: Maskew Miller.

Hewitt, G. M. 2000. "The Genetic Legacy of the Quaternary Ice Ages." *Nature* 405 (6789): 907–13.

———. 2001. "Speciation, Hybrid Zones and Phylogeography: Or Seeing Genes in Space and Time." *Molecular Ecology* 10:537–49.

———. 2004. "Genetic Consequences of Climatic Oscillations in the Quaternary." *Philosophical Transactions Royal Society London B* 359:183–95.

Hey, J., W. M. Fitch, and F. J. Ayala. 2005. *Systematics and the Origin of Species: on Ernst Mayr's 100th anniversary*. Washington, DC: National Academies.

Hiebert, F. 1994. *Origins of the Bronze Age Oasis Civilization in Central Asia*. Cambridge, MA: Cambridge University.

Hiie, S., and K. Kihno. 2008. "Evidence of Pollen and Plantmacro-Remains from the Sediments of Suburban Area of Medieval Tartu." *Estonian Journal of Archeology* 12 (1): 30–50.

Hiie, S., K. Kihno, and Ü. Sillasoo. 2007. "Archaeobotanical Investigations in the Suburban Area of Medieval Tartu (Estonia)." In *Programme and Abstracts: 14th Symposium of the International Work Group for Palaeoethnobotany, Krakov, Poland, June 17–23*, 137. Krakow, Poland: Polish Academy of Sciences.

Hill, J. E., trans. 2003. "The Western Regions according to the *Hou Hanshu*." The *Xiyu juan* "Chapter on the Western Regions" from *Hou Hanshu* 88. Extensively revised with additional notes and appendices. http://depts.washington.edu/silkroad/texts/hhshu/hou_han_shu.html.

———. 2009. *Through the Jade Gate to Rome: A Study of the Silk Routes during the Later Han Dynasty, First to Second Centuries CE*. Charleston, SC: BookSurge.

Hillig, K. W. 2004a. A Chemotaxonomic Analysis of Terpenoid Variation in Cannabis." *Biochemical Systematics and Ecology* 32: 875–91.

———. 2004b. "A Multivariate Analysis of Allozyme Variation in 93 *Cannabis* accessions from the VIR Germplasm Collection." *Journal of Industrial Hemp* 9 (2): 5–22.

———. 2005a. "A Systematic Investigation of Cannabis." PhD diss., Indiana University.

———. 2005b. "Genetic Evidence for Speciation in *Cannabis* (Cannabaceae)." *Genetic Research and Crop Evolution* 52 (2): 161–80.

Hillig, K. W., and P. G. Mahlberg. 2004. "A Systematic Analysis of Cannabinoid Variation in *Cannabis* (Cannabaceae)." *American Journal of Botany* 91:966–75.

Hirochika, N. 1988. "Divine Symbols in Japanese Festivals: The Ogi Festival in Kurokawa." In *Matsuri: Festival and Rite in Japanese Life*, edited by K. Ueda and M. Inoue, 78–104. Translated by N. Havens. Tokyo: Institute for Japanese Culture and Classics, Kokugakuin University.

———. 2003. *Japanese Religions at Home and Abroad: Anthropological Perspectives*. London: RoutledgeCurzon.

Hiroe, M. 1973. *The Plants of Basho's and Buson's Hokku Literature: Supplement: (the Plants of) Sora's Diary of a Journey with Basho*. The plants of Basho's Oku-no-hosomichi. Vol. 8 of The Plants of Classical Literature: Classical Taxonomy and Asiatic Plants. Tokyo: Ariake.

Hirth, F. 1966. China and the Roman Orient: Researches into their Ancient and Mediaeval Relations as Represented in Old Chinese Records. New York: Paragon Book Reprint Corporation. First published 1885, Shanghai and Hong Kong.

Hittell, T. H. 1897. History of California. Vol. 4. San Francisco, CA: N. J. Stone.

Ho, P. T. 1969. "The Loess and the Origin of Chinese Agriculture." American Historical Review 75:1–36.

Hochman, J. S. 1972. Marijuana and Social Evolution. Englewood Cliffs, NJ: Prentice-Hall Inc.

Hofstetter, S., W. Tinner, V. Valsecchi, G. Carraro, and M. Conedera. 2006. "Lateglacial and Holocene Vegetation History in the Insubrian Southern Alps—New Indications from a Amall-Scale Site." Vegetation History and Archaeobotany 15 (2): 1617–6278.

Holmboe, J. 1927. "Nytteplanter og ugræs i Osebergfundet." In Obsebergfundet, vol. V, edited by A. W. Brøggerand and H. Shetelig, 3–80. Oslo, Norway: Universitetets Oldsaksamling.

Holmstedt, B. 1973. "Introduction to Moreau de Tours." In Hashish and Mental Ilness, edited by H. Peters and G. G. Nahas, ix–xxii. New York: Raven.

Hölzer, A., and A. Hölzer. 1995. "Studies on the Vegetation History of the Hornisgrinde Area in the Northern Black Forest (SW Germany) by Means of Pollen, Macrofossils and Geochemistry." Carolinea 53:199–238 [in German].

———. 1998. "Silicon and Titanium in Peat Profiles as Indicators of Human Impact." The Holocene 8 (6): 685–96.

Honeychurch, W., and C. Amartuvshin. 2007. "Hinterlands, Urban Centers, and Mobile Settings: The 'New' Old World Archeology from the Eurasian Steppe." Asian Perspectives 46 (1): 36–64.

Hong, S., and R. C. Clarke. 1996. "Taxonomic Studies of Cannabis in China." Journal of the International Hemp Association 3 (2): 55–60.

Hong Kong Year Book. 2000. http://www.yearbook.gov.hk/2000/eng/.

Honkanen, S., and L.-S. Mäkelä. 2005. Traditional Net Making. Kuusankoski, Finland: Adult Education Centre.

Hopkins, J. F. 1951. A history of the Hemp Industry in Kentucky. Lexington, KY: University of Kentucky.

Hörnberg, G., E. Bohlin, E. Hellberg, I. Bergman, O. Zackrisson, A. Olofsson, J.-E. Wallin, and T. Påsse. 2005. "Effects of Mesolithic Hunter-Gatherers on Local Vegetation in a Non-Uniform Glacio-Isostatic Land Uplift Area, Northern Sweden." Vegetation History and Archaeobotany 15 (1): 13–26.

Hornsey, I. 2004. A History of Beer and Brewing. Cambridge: Royal Society of Chemistry.

House, J. D., J. Neufield, and D. Leson. 2010. "Evaluating the Quality of Protein from Hemp Seed (Cannabis Sativa L.) Products Through the Use of the Protein Digestibility-Corrected Amino Acid Score Method." Journal of Agricultural and Food Chemistry 58 (22): 11801–7.

Hozeski, B. W., trans. 2001. Hildegard's Healing Plants: From her Medieval Classic Physica. Boston: Beacon.

Hsü, C.-Y. 1977. Ancient China in Transition: An Analysis of Social Mobility 722–222 B.C. Palo Alto, CA: Stanford University.

———. 1978. "Agicultural Intensification and Marketing Agrarianism in the Han Dynasty." In Ancient China: Studies in Early Civilization, edited by D. T. Roy and T. Tsien, 263–64. Hong Kong: Chinese University.

———. 1980. Han Agriculture: The Formation of Early Chinese Agrarian Eeconomy (206 B.C.–A.D. 220). Han Dynasty China, Vol. 2. Edited by J. L. Dull. Seattle: University of Washington.

Hu, F. S., A. Hampe, and R. J. Petit. 2009. "Paleoecology Meets Genetics: Deciphering Past Vegetational Dynamics." Frontiers in Ecology and the Environment 7:371–79.

Huang, A. 1995. "Hanging, Cyanide Gas, and the Evolving Standards of Decency: The Ninth Circuit's Misapplication of the Cruel and Unusual Clause of the Eighth Amendment." Oregon Law Review 74:995–1030.

Huang, H. T. 2000. Science and Civilization in China. Volume 6: Biology and Biological Technology. Part V: Fermentations and Food Science. Cambridge: Cambridge University.

Hudson, M., and G. Barnes. 1991. Yoshinogari: A Yayoi Settlement in Northern Kyushu. Monumenta Nipponica 46 (Summer): 211–35.

Hughes, S. 1978. Washi, the World of Japanese Paper. New York: Kodansha International Tokyo.

Hulbert, H. B. 1969. The Passing of Korea. Seoul: Yonsei University.

Hulten, E. 1970. The Circumpolar Plants II Dicotyledons. Stockholm: Almqvist and Wiksell.

Hunan Sheng po-wu-kuan. 1973. Ch'ang-sha Ma-want-tui I-hao Han-mu. Vol. 1. Peking: Wen We Press.

Hunter, D. 1943. Papermaking: The History and Technique of an Ancient Craft. New York: Alfred A. Knopf.

———. 1978. Papermaking: The History and Technique of an Ancient Craft. Revised and enlarged, Mineola, NY: Dover Publications.

Huntley, J. P. 1987. "The Plant Remains." In "Excavations at Newcastle Quayside: the Crown Court site," edited by C. O'Brien, L. Bown, S. Dixon, L. Donel, L. J. Gidney, J. P. Huntley, R. Nicholson, and P. Walton. Archeologia Aeliana 17 (ser. 5): 180–82.

Husbands, J. D. 1909. USDA Bureau of Plant Industry Bulletin 153:42.

Hutchinson, H. W. 1975. "Patterns of Marijuana Use in Brazil." In Cannabis and Culture, edited by V. Rubin, 173–83. Paris: Mouton.

Hyams, E. 1971. Plants in the Service of Ma: 10,000 Years of Domestication. New York: J. B. Lippincot.

Hyland, D. C., I. S. Zhushchikhovskaya, V. E. Medvedev, A. P. Derevianko, and A. V. Tabarev. 2002. "Pleistocene Textiles in the Russian Far East: Impression from Some of the World's Oldest Pottery." Anthropologie (Brno, Czech Republic) 40 (1): 1–10.

Hyodo, M., H. Nakaya, A. Urabe, H. Saegusa, X. Shunrong, Y. Jiyun, and J. Xuepin. 2002. "Paleomagnetic Dates of Hominid Remains from Yuanmou, China, and Other Asian Sites." Journal of Human Evolution 43 (1): 27–41.

Hyo-Soon, C. 1995. "Korean Clothes and Fabrics." Koreana: A Quarterly on Korean Art & Culture 9 (3): 12–19.

IAASS. 1984. Recent Archeological Discoveries in the People's Republic of China. Institute of Archeology, Academy of Social Sciences, People's Republic of China. Tokyo: Centre for East Asian Cultural Studies. First published in Chinese by Z. An., C. Zhang, and P. Xu, edited by N. Xia.

IDPNI. 1995. International Dunhuang Project News iss. 3, July 1995. http://idp.bl.uk/downloads/newsletters/IDPNews03.pdf.

Iglésias, F. A. 1986. "Sobre o vício da diamba." In Diamba Sarabamba, edited by A. Henman and O. Pessoa Jr., 39–51 São Paulo, Brazil: Ground.

Ilijin, M. 1963. "Obredno ljuljanje u prolead" (Ritual swinging in the spring). RCE IX kongresa Saveza udruzenja folklorista Jugoslavije u Mostaru i Trebinju (Sarajevo), 1962, 282–83 [in Serbian].

Iltis, H. H. 1983. "From Teosinte to Maize: The Catastrophic Sexual Transmutation." Science 22:886–94.

Imamura, K. 1996a. Prehistoric Japan: New Perspectives on Insular East Asia. Honolulu: University of Hawaii.

———. 1996b. Jomon and Yayoi: The Transition to Agriculture in Japanese Prehistory. In The Origins and Spread of Agriculture and Pastoralism in Eurasia, edited by D. R. Harris, 442–64. Washington, DC: Smithsonian Institution.

INCSR. 2008. International Narcotics Control Strategy Report (Russia), March. Washington, DC: State Department's Bureau for International Narcotics and Law Enforcement Affairs.

Indreko, R. 1931. "Kiviaja võrgujäänuste leid Narvas." Eesti Rahva Muuseumi aastaraamat 7, 48–67.

International Association of Agricultural Economists. 1973. World Atlas of Agriculture Volume II. Southeast Asia and Oceania. Novara, Italy: Istituto Geografico De Agostini.

Ionică, I. 1996. Dealu Mohului: Ceremonia agrară în Țara Oltului. București, Romania: Editura Minerva.

Irniger, M., and M. Kühn. 1997. "Hanf und Flachs. Ein traditioneller Rohstoff in der Wirtschaft des Spätmittelalters und der frühen Neuzeit." Traverse 2: 100–115.

Isnard, S., and W. K. Silk. 2009. "Moving with Climbing Plants from Charles Darwin's Time into the 21st Century." American Journal of Botany 96 (7): 1205–21.

Ispas, S. 1993. Folclorul—proces viu de creație. Vol. 3. No. 8. Bucharest: Academica Ianuarie.

———. 2001. "Landmarks of Christianity-Related Balkan Folk Traditions." In Politics and Culture in Southeastern Europe: Studies on Science and Culture, edited by R. Theodorescu and L. C. Barrows, 171–228. Bucharest: UNESCO.

Ivanișević, F. 1905. "Podgajac, Vjerovanja" [Folk beliefs in Podgajac, Croatia]. [In Croatian.] Zbornik za narodni zivot i obicaje Juznih Slavena, Jugoslavenska akademija znanosti i umjetnosti, knj 10:46, 266, 289–90.

Ivanova, M. G. 1998. *Idnakar: Drevneudmurtskoe Gorodishche IX–XIII vv.* [In Russian.] Izhevsk, Russia: Udmurtskii institut istorii, iazyka i literatury.

Izzo, A. A., F. Borrelli, R. Capasso, V. Di Marzo, and R. Mechoulam. 2009. "Non-Psychotropic Plant Cannabinoids: New Therapeutic Opportunities from an Ancient Herb." *Trends in Pharmacological Sciences* 30 (10): 515–27.

Jaanits, L. 1992. "Põllumajanduse eelduste kujunemine" [Formation of agricultural presumptions]. [In Estonian.] In *Eesti talurahva ajalugu* [History of the rural population in Estonia], edited by J. Kahk, 42–56. Tallinn, Estonia: Olion.

Jacobs, J. 1894. *The Fables of Aesop.* London: Macmillan and Co.

Jacomet, S. 2002. "Switzerland Biesheim-Kunheim, Oedenburg Roman Iron Age. Les investigations archéobotaniques." In *Rapport triennal (2000–2002) sur les fouilles Franco-Germano-Suisses à Oedenburg (Haut-Rhin)*, edited by M. Reddé, 283–307. Paris: EPHE.

———. 2003. "Und zum Dessert Granatapfel—Ergebnisse der archäobotanischen Untersuchungen." In "Zur Frühzeit von Vindonissa. Auswertung der Holzbauten der Grabung Windsich-Breite 1996–1998," edited by A. Hagendorn, H. W. Doppler, A. Huber, H. Hüster-Plogmann, S. Jacomet, C. Meyer-Freuler, B. Pfäffli, and J. Schibler. *Veröffentlichungen der Gesellschaft Pro Vindonissa*, 18 (1): 53–71, 173–229, 483–92. Brugg, Switzerland: Aargauische Kantonsarchäologie.

Jagadish, V., J. Robertson, and A. Gibbs. 1996. "RAPD Analysis Distinguishes *Cannabis Sativa* Samples from Different Sources." *Forensic Science International* 79 (2): 113–21.

Jahns, S. 2000. "Late-Glacial and Holocene Woodland Dynamics and Land-Use History of the Lower Oder Valley, North-Eastern German, Based on Two, AMS ^{14}C-Dated, Pollen Profiles." *Vegetation History and Archaeobotany* 9:111–23.

Jain, S. K. 1999. *Dictionary of Ethnoveterinary Plants of India.* With assistance from S. Srinivasa. New Delhi: Deep Publications.

James, J. K., and R. J. Abbott. 2005. "Recent Allopatric, Homoploid Hybrid Speciation: The Origin of *Senecio squalidus* (Asteraceae) in the British Isles from a Hybrid Zone on Mount Etna, Sicily." *Evolution* 59:2533–47.

Jančerova, K. 1901. "*Trebarjevo: Narodni život i običaje* (Nastavak)" [Trebarjevo: Folk life and culture (continued)]. *ZNŽO* VI (2): 187–248 [in Croatian].

Janelli, R. L., and D. Y. Janelli. 1982. *Ancestor Worship and Korean Society.* Stanford, CA: Stanford University.

Janischevsky, D. E. 1924. *Cannabis Ruderalis.* Proceedings Saratov 2 (2): 14–15.

Jankovská, V. 1996. "Pylová analýza uloženin pozdně stedovkého vodovodu z Prachatic" (Pollen analysis of deposits from the late medieval water supply at Prachatice). *Zlatá Stezka* 3:182–88.

Jansen, T., P. Forster, M. A. Levine, H. Oelke, M. Hurles, C. Renfrew, J. Weber, and K. Olek. 2002. "Mitochondrial DNA and the Origins of the Domestic Horse." *Proceedings of the National Academy of Sciences* 99 (16): 10905–10.

Japan Foundation. 1993. *Breeze: The Japan Foundation Newsletter.* Los Angeles: Japan Foundation 22 (1): 15.

Jaques, D., and A. Hall. 2003. Technical Report: Biological Remains from Excavations at Grange Rath, Colp West, County Meath, Republic of Ireland (site codes: 03E0641 and 03E0660). Palaeoecology Research Services Report 2003/62: Shildon, County Durham, United Kingdom.

Jarvis, D. I., and T. Hodgkin. 1999. "Wild Relatives and Crop Cultivars: Detecting Natural Introgression and Farmer Selection of New Genetic Combinations in Agroecosystems." *Molecular Ecology* 8: S159–S173.

Jenkins, R. W., and D. A. Patterson. 1973. "The Rrelationship between Chemical Composition and Geographical Origin of *Cannabis.*" *Forensic Science* 2:59–66.

Jensen, G. 1996. "Discover the Belgian Hemp Museum." *Journal of the International Hemp Association* 3 (2): 80.

Jia, W. 2005. *Transition from Foraging to Farming in Northeast China.* PhD diss., University of Sydney, Department of Archeology.

Jiang, H.-E., X. Li, Y. X. Zhao, D. K. Ferguson, F. Hueber, S. Bera, Y. F. Wang, L. C. Zhao, C. J. Liu, and C. S. Li. 2006. "A New Insight into *Cannabis sativa* L. (Cannabaceae) Utilization from 2500-Year-Old Yanghai Tombs, Xinjiang, China." *Journal of Ethnopharmacology* 108 (3): 414–22.

Jianying, H. 2004. "Ambrosia of the Ancients." *China Today.* http://219.230.159.254/bjzs/chinatoday/chinatoday2004/200408/p70.asp.

Jinfen, Y. 2002. "A Feminine Expression of Mysticim, Romanticism and Syncretism. In *A Plaint of Lady Wang.*" *Inter-Religio* 42 (Winter): 3–20.

Jixu, Z. 2003. "Correspondences of Cultural Words between Old Chinese and Proto-Indo-European." *Sino-Platonic Papers* 125 (September): 1–17.

———. 2006. "The Rise of Agricultural Civilization in China: The Disparity between Archeological Discovery and the Documentary Record and its Explanation." *Sino-Platonic Papers* 175:1–38.

Jobling, M. A., and C. Tyler-Smith. 2003. "The Human Y Chromosome: An Evolutionary Marker Comes of Age." *Nature Reviews Genetics* 4 (8): 598–612.

Jochim, M. A. 1979. "Catches and Caches: Ethnographic Alternatives for Prehistory." In *Ethnoarcheology*, edited by C. Kramer, 219–46. New York: Columbia University.

Johnson, C., ed. 1985. *Dab Neeg Hmoob: Myths, Legends and Folk Tales from the Hmong of Laos.* St. Paul, MN: Macalester College.

Johnson, K. 1999. "The Roots of Renaissance Gardens." *Suite101.com: An Online Publishing Community.* http://web.archive.org/web/20051119112401/ http://www.harborside.com/~rayj/varro.html.

Johnson, M. D., W. A. Richards, and R. R. Griffiths. 2008. "Human Hallucinogen Research: Guidelines for Safety." *Journal of Psychopharmacology* 22 (6): 603–20.

Jones, K. 1995. *Nutritional and Medicinal Guide to Hemp Seed.* Gibsons, BC: Rainforest Botanical Laboratory.

Jones, G. E. M., V. Straker, and A. Davis. 1990. "Early Medieval Plant Use and Ecology in London." In "Aspects of Saxo-Norman London: II Finds and Environmental Evidence," edited by A. Vince. *Transactions of the London and Middlesex Archeological Society*, Special Paper 12.

Jones-Bley, K. 2007. *Pot in the Pots.* Paper presented at *40th Annual Chacmool Conference: Eat, Drink and Be Merry: The Archeology of Foodways*, November 11, University of Calgary, Canada.

Joosten, J. H. J. 1985. "A 130 Year Micro- and Macrofossil Record from Regeneration Peat in the Peel, The Netherlands: A Palaeoecological Study with Agricultural and Climatological Implications." *Palaeogeography, Palaeoclimatology, Palaeoecology* 49:277–312.

Joshi, S. G. 2006. *Medicinal Plants.* New Delhi: Oxford & IBH.

Jovanović, B. 1993. *Magija srpskih obreda* [Magic of the Serb rituals]. [In Serbian.] Novi Sad, Serbia: Svetovi.

Jovanović, M., and J. Bjeladinović. 1964. "Ženska domacá radinost" [Traditional women's textile craft (northwest Serbia)]. [In Serbian.] *Glasnik Etnografskog muzeja u Beogradu* 27:345–58.

Joy, J. E., S. J. Watson Jr., and J. A. Benson Jr., ed. 1999. *Marijuana and Medicine: Assessing the Science Base.* Division of Neuroscience and Behavioral Research, Institute of Medicine. Washington, DC: National Academy.

Joya, M. 1951. *Quaint Customs and Manners of Japan.* Tokyo: Tokyo News Service.

Julien, M. S. 1894. "Chirurgie Chinoise-Substance anesthétique employée en Chine, dans le commencement du IIIe siècle de notre ere, pour paralyser momentanement la sensibilite." *Comptes Rendus de l'Académie de Sciences* 28:195–98.

Julien, R. M., C. D. Advokat, and J. Comaty. 2011. *A Primer of Drug Action: A Concise, Nontechnical Guide to the Effects of Psychoactive Drugs*, 12th ed. London: Worth.

Justinová, Z., S. Yasar, G. H. Redhi, and S. R. Goldberg. 2011. "The Endogenous Cannabinoid 2-Arachidonoylglycerol is Intravenously Self-Administered by Squirrel Monkeys." *Journal of Neuroscience* 31 (19): 7043–48.

Kabelik, J., Z. Krejci, and F. Santavy. 1960. "Cannabis as a Medicament." *Bulletin on Narcotics* 12 (3): 5–23.

Kalant, H. 2001. "Medicinal Use of Cannabis: History and Current Status." *Pain Research and Management* 6 (2): 80–91.

Kalnina, L. 2006a. "Paleovegetation Changes since the Last Glaciation Recorded by Pollen Data from the Lubans Plain, Eastern Latvia." Paper presented at *7th European Paleobotany-Palynology Conference, Program and Abstracts*, June 6–11, Prague.

———. 2006b. "Paleovegetation and Human Impact in the Surrounding of the Ancient Burtnicks Lake as Reconstructed from Pollen Analysis." In *Back to the Origin: New research in the Mesoithic-Neolithic Zvejnieki Cemetery and Environment, Northern Latvia*, Acta Archeologica Lundensia ser. in 8° (52): 53–74, edited by Larsson, L. and I. Zagorska.

Kaltenrieder, P., C. A. Belis, S. Hofstetter, B. Ammann, C. Ravazzi, and W. Tinner. 2009. "Environmental and Climatic Conditions at a Potential Glacial Refugial Site of Tree Species Near the Southern Alpine Glaciers. New Insights from Multiproxy Sedimentary Studies at Lago della Costa (Euganean Hills, Northeastern Italy)." *Quaternary Science Reviews* 28:2647–62.

Kaltenrieder, P., G. Procacci, B. Vannière, and W. Tinner. 2010. "Vegetation and Fire History of the Euganean Hills (Colli Euganei) as Recorded by Lateglacial and Holocenesedimentary Series from Lago della Costa (Northeastern Italy)." *The Holocene* 20 (5): 679–95.

Kamruzzahan, S. 2003. *Is Alnus viridis 'a' Glacial Relict in the Black Forest?* Inaugural-Dissertation Zur Erlangung der Doktorwürde der Fakultät für Biologie der Albert-Ludwigs-Universität Freiburg im Breisgau. Freiburg, Germany: Freiburg University.

Kansu Museum. 1972. "Preliminary Report on the Excavation of Three Han Tombs at Ma-tsu-tze, Wu-wei." *Wen-wu* 12:9–21.

Kao, S. 1978. "The Clay Sculptures at the Ch'ing-lien-ssu Monastery, Chin-ch'eng, Shansi." *Wen-wu* 10 (1963): 7–12, pl. 7–10. In *Chinese Archeological Abstracts*, edited by R. C. Rudolph, 6th vol. of *Monumenta Archeologica*. Los Angeles: UCLA Institute of Archeology, 404.

Kaplan, J., ed. 1969. *Marijuana: Report of the Indian Hemp Drugs Commission 1893–1894.* Silver Spring, MD: Thomas Jefferson.

Kaplan, J. O., K. M. Krumhardt, and N. Zimmermann. 2009. "The Prehistoric and Preindustrial Deforestation of Europe." *Quaternary Science Reviews* 28 (27–28): 3016–34.

Karadžić, V. S. 1867. *Život i običaji naroda srpskog* [Life and customs of Serbian people]. [In Serbian.] U Beču, Vienna: A. Karacić.

Karafet, T. M., S. L. Zegura, O. Posukh, L. Osipova, A. Bergen, J. Long, D. Goldman, W. Klitz, S. Hrihara, P. de Knijff, V. Wiebe, R. C. Griffiths, A. R. Templeton, and M. F. Hammer. 1999. "Ancestral Asian Sources of New World Y-Chromosome Founder Haplotypes." *American Journal of Human Genetics* 64:817–31.

Karg, S. 1996. "Bizarre Früchte aus dem Wasser. Am Federsee wurde eine vergessene Nutzpflanze wiederentdeckt." *Schönes Schwaben* 7:8–11.

———. 2012. "Oil-Rich Seeds from Prehistoric Contexts in Southern Scandinavia—Reflections on Archaeobotanical Records of Flax, Hemp, Gold of Pleasure, and Corn Spurrey." *Acta Palaeobotanica* 52 (1): 17–24.

Karpowicz, A., and S. Selby. 2010. "Scythian Bow from Xinjang." *Journal of the Society of Archer-Antiquaries.* Vol. 53. See also http://www.atarn.org/chinese/Yanghai/Scythian_bow_ATARN.pdf

Karstien, C. 1936. "Indogermanisch und Germanisch." In *Germanen und Indogermanen. Festschrift furHermann Hirt*, vol. II, edited by Helmut Arntz, 297–327. Heidelberg, Germany: Carl Winters.

Karus, M. 2004. "European Hemp Industry 2002: Cultivation, Processing and Product Lines." *Journal of Industrial Hemp* 9 (2): 93–101.

Kasahara, Y. 1981. "Detection of Plant Seeds from the Torihama Shell Midden: Perilla Seeds and Tar Agglomerates." [In Japanese.] *Torihama Kaizuka*, 40: 65–87.

———. 1982. "Nabatake iseki no maizo shushi no bunseki, dōtei kenkyū" [Analysis and identification of ancient seeds from the Nabatake site]. [In Japanese.] In *Nabatake*, edited by T.-S. K. Iinkai, 354–79. Tosu-shi, Japan: Tosu-shi Kyōiku Iinkai.

———.1984. "Tohihama Kaizuka (dai-6-ji hakkutsu) no Shokubutsu Shushi no Kenshutsu to Dōtei ni tsuite" [The search for and indentification of plant seeds from the Torihama Shell Mound, 6th excavation]. [In Japanese.] In *Torihama Kaizuka* [The Torihama Shell Mound], edited by I. Okamoto, 47–64. Fukui, Japan: Fukui-Ken Kyōiku Iinkai.

Kasper, L. R. 1992. *The Splendid Table: Recipes from Emilia-Romagna, the Heartland of Northern Italian Food.* New York: William Morrow.

Kasuba, S. M. 1974. "Običaji i obredi u prolecnom kalendarskom ciklusu" [Spring customs and rites in annual custom calendar]. [In Serbian.] *Glasnik Etnografskog muzeja u Beogradu* 37 (Beograd, Serbia): 173.

Kaukonen, T.-I. 1946. "Pellavan ja hampun viljely ja muokkaus Suomessa: Kansatieteellinen tutkimus" [Flax and hemp cultivation and processing of Finland]. [In Finnish.] In *Kansatieteellinen arkisto*["Finnish scientific archive"] *VII.* Helsinki: Väitösk [University of Helsinki].

Kawamura, K. 1994. "The Life of a Shamaness: Scenes from the Shamanism of Northeastern Japan." Translated by N. Havens. In *Folk Beliefs in Modern Japan*, edited by I. Nobutaka, 94–124. Toyko: Kokugakuin University, Institute for Japanese Culture and Classics. Paper first published 1984 in Japanese as "Fusha no seikatsushi: Tôhoku chihô shamanizumu no ichidanmen" in *Nihon minzokugaku* 15:16–35.

KCNA. 2007. "Paper-Making Technology with Long History." *The Korean Central News Agency of the DPRK* (Democratic People's Republic of Korea [North Korea]), September 14. Provided by Korea News Service (KNS) in Tokyo. http://www.kcna.co.jp/item/2007/200709/news09/15.htm#8.

Kendall, L. 1985. "Death and Taxes: A Korean Approach to Hell." *Transactions of the Royal Asiatic Society* 60:1–14.

Kêng, H. 1974. "Economic Plants of Ancient North China as Mentioned in *Shih Ching* (Book of Poetry)." *Economic Botany* 28:391–410.

Kenward, H. K., and A. R. Hall. 1995. "Biological Evidence from Anglo-Scandinavian Deposits at 16–22 Coppergate." *The Archeology of York* 14 (7): 435–797.

Kenward, H. K., A. R. Hall, and A. K. G. Jones. 1986. "Environmental Evidence from a Roman Well and Anglian Pits in the Legionary Fortress." *The Archeology of York* 14 (5): 241–88 and fiche 2.

Kerr, H. C. 1877. "Report of the Cultivation of, and Trade in, Ganja in Bengal." In *Papers Relating to the Consumption of Ganja and Other Drugs in India*, 94–154. Sdimla, India: Government Central Printing Office.

Khalifa, A. 1975. "Traditional Patterns of Hashish Use in Egypt." In *Cannabis and Culture*, edited by V. Rubin, 195–206. Paris: Mouton.

Khamsi, R. 2007. "Neanderthals Roamed as Far as Siberia." *New Scientist News*, September 30, 2007. newscientist.com /article/dn12711-neanderthals-roamed-as-far-as-siberia.html.

Khan, R. U., F. R. Durrani, N. Chand, and H. Anwar. 2010. "Influence of Feed Supplementation with *Cannabis Sativa* on Quality of Broilers Carcass." *Pakistan Veterinary Journal* 30 (1): 34–38.

Khorikhin, V. V. 2001. "The Late Seventeenth-Century Tsar's Copy of Domostroi: A Problem of Origins." *Russian Studies in History* 40 (1): 75–93.

Kierman, F. A., Jr., and J. K. Fairbank, eds. 1974. *Chinese Ways in Warfare.* Cambridge, MA: Harvard University.

Kihno, K., and S. Hiie. 2008. "Evidence of Pollen and Plant Macro-Remains from the Sediments of Suburban Areas of Medieval Tartu." *Estonian Journal of Archeology* 12 (1): 30–50.

Kikvidze, Z., and M. Ohsawa. 2001. "Richness of Colchic Vegetation: Comparison between Refugia of South-Western and East Asia." *BMC Ecology* 1:6.

Kim, K. I. 1973. "Shamanist Healing Ceremonies in Korea." *Korea Journal* 13 (4): 41–48.

Kim, Y. G. 1979. "Evolutionary Characteristics of Korean Neolithic Transitional Clay Vessels." [In Korean.] *Kogo Minsok* 7:45–108.

Kimmens, A. C., ed. 1977. *Tales of Hashish.* New York: William Morrow.

King, R. S., and M. Mauer. 2006. "The War on Marijuana: The Transformation of the War on Drugs in the 1990s." *Harm Reduction Journal* 3:6, published online February 9. http://www.ncbi.nlm.nih.gov/ pmc/articles/PMC1420279/.

Kirby, D. 1998. *Northern Europe in the Early Modern Period: The Baltic World 1492–1772*, 3rd ed. New York: Longman.

Kirby, R. H. 1963. *Vegetable Fibers—Botany, Cultivation and Utilization.* New York: Interscience.

Kisgeci, J. 1994. *Konoplji hvala* [Gratitude to hemp]. [In Serbian.] Beograd, Serbia: Nolit.

Kisielienė, D., M. Stančikaitė, P. Blaževičius, J. Mažeika. 2007. "Reconstruction of Environmental Variations in the Area of Vilnius Royal Castle since the 5th Century AD, on the Basis of Palaeobotanical Data." Paper presented at *14th Symposium of the International Work Group for Palaeoethnobotany, Program and Abstracts*, June 17–23, Krakow, Poland.

Kister, D. A. 1997. *Korean Shamanist Ritual: Symbols and Dramas of Transformation.* Budapest: Bibliotheca Shamanistica, Akadémiai Kiadó.

Kitagawa, J. M. 1987. *On Understanding Japanese Religion.* Princeton, NJ: Princeton University.

Kittler, J. T., E. V. Gricorenko, C. Clayton, S.-Y. Zhuang, S. C. Bundey, M. M. Trower, D. Wallace, R. Hampson, and S. Deadwyer. 2000. "Large-Scale Analysis of Gene Expression Changes during Acute and Chronic Exposure to [Delta] 9-THC Exposure in Rats." *Physiological Genomics* 3:175–85.

Klages, K. H. W. 1942. *Ecological Crop Geography*. New York: Macmillan.

Klee, M., and C. Brombacher. 1996. "Botanische Makroreste aus 37 Proben von Develier/Courtetelle." In *Le haut Moyen Age à Develier, La Pran et à Courtetelle, Tivila (JU, Suisse)*, edited by M. Federici-Schenardi and R. Fellner, Archéologie et Transjurane, document 47:41–79. Porrentruy, Switzerland: Office du patrimoine historique.

Klein, R. G., and B. Edgar. 2002. *The Dawn of Human Culture*. New York: John Wiley and Sons.

Klein, S. 1908. *Tod und Begrabnis in Palistina*. Berlin: H. Itzkowski.

Kligman, G. 1988. *The Wedding of the Dead: Ritual, Poetics and Popular Culture in Transylvania*. Berkeley, CA: University of California.

Klinglehöfer, E. 1991. "Settlement and Landuse in Micheldever Hundred, Hampshire, 700–1100." *Transactions of the American Philosophical Society* 83 (3): 1–156.

Knab, S. H. 1995. *Polish Herbs, Flowers and Folk Medicine*. New York: Hippocrene Books.

Knapp, S. and J. Mallet. 2003. Refuting Refugia? *Science* 300 (5616): 71–72.

Knežević, S. 1964. "Zdravstvena kultura i problemi narodne medicine u Jadru" [Sanitary culture and traditional folk medicine problems in Jadru, North Serbia]. [In Serbian.] *Glasnik Etnografskog muzeja u Beogradu* 27:459–502.

Knežević, S., and M. Jovanović. 1958. "Jarmenovci." [In Serbian.] Rasprave i grada 4 [Dissertations and scientific material series 4]. *Srpski etnografski zbornik* 73. Beograd, Serbia: Srpska akademija nauka.

Knipping, M., M. Müllenhoff, and H. Brückner. 2008. "Human Induced Landscape Changes around Bafa Gölü (Western Turkey)." *Vegetation History and Archaeobotany* 17:365–80.

Knoblock, J., and J. Riegel. 2000. *The Annals of Lü Buwei: A Complete Translation and Study*. Palo Alto, CA: Stanford University.

Knörzer, K.-H. 2000. "3000 Years of Agriculture in a Valley of the High Himalayas." *Vegetation History and Archaeobotany* 9 (4): 219–22.

Kocsis, K., and E. Kocsis-Hodosi. 1998. *Hungarian Minorities in the Carpathian Basin*. Budapest: Simon Publications.

Kohn, L. 1993. *The Taoist Experience. An Anthology*. Ithica, NY: State University of New York.

Koizumi, A. 1934. *Rkuro Saikyu-bo: The Tomb of Painted Basket of Lo-Lang*. With English summary written by Kosaku Hamada. Seoul: Chosen-Koseki-Kengyu-Kwai (Society for the Study of Korean Antiquities).

Kokassaar, U. 2003. "Kanepiseemnetest tehti vanasti jurssi, piima ja putru" [Hemp seeds were used for making hemp butter, milk and porridge]. [In Estonian.] *Eesti Looduse* 10. http://www .loodusajakiri.ee/eesti_loodus/index.php?artikkel=485.

Kokugakuin University. 1997. *Basic Terms of Shinto*. Tokyo: Kokugakuin University, Institute for Japanese Culture and Classics. http://www2.kokugakuin.ac.jp/ijcc/wp/bts/bts_j .html#jingu_taima.

Kolander, C. 1995. *Hemp for Textile Artists*. Portland, OR: MAMA D. O. C.

Kolberg, O. 1899. "Mazowsze Lud." *Towarzystwo Ludoznawcze* 5:206.

Kolesov, V. V. 2001. *Domostroi* as a Work of Medieval Culture. *Russian Studies in History* 40 (1): 6–74.

Körber-Grohne, U. 1985. "Die biologischen Reste aus dem hallstattzeitlichen Fürstengrab von Hochdorf, Gemeinde Eberdingen (Kreis Ludwigsburg)." In "Hochdorf I," edited by H. Küster and U. Körber-Grohne. *Forschungen und Berichte zur Vor- und Frühgeschichte in Baden-Württemberg* 19:85–265.

———. 1987. *Nutzpflanzen in Deutschland*. Stuttgart, Germany: Konrad Theiss Verlag.

Korean National Heritage Private Museum. 1996. "Sinjungdo, Mok-A Buddhist Museum." http://web.archive.org/web/20090325075439/ http://www. heritage.go.kr/eng/mus/prv_03.jsp.

Korea Overseas Information Service. 2003. *Hanbok—Korean Dress*. http://www.koreaaward.com/korea.htm.

Korona, O. M. 2006. "Seed Flora from Cultural Layers of Archeological Sties in West Siberia Forest-Tundra." [Abstract.] *7th European Conference, Program and Abstracts*, Prague, September 6–11.

Kostić, P. 1975. "Godišnji Običaji" [Annual customs]. [In Serbian.] *Glasnik Etnografskog muzeja u Beogradu* 38:172, 323.

———. 1989. "Godišnji običaji. Sjeničko-Pešterska visoravan" (Calendaric customs of the Sjenica-Pešter plateau).*Glasnik Etnografskog muzeja u Beogradu* 52–53:73–123.

Krause, J., L. Orlando, D. B. Viola, K. Prüfer, M. P. Richards, J.-J. Hublin, C. Hänni, A. P. Derevianko, and S. Pääbo. 2007. "Neanderthals in Central Asia and Siberia." *Nature* 449:902–4.

Krebs, P., M. Conedera, M. Pradella, D. Torriani, M. Felber, and W. Tinner. 2004. "Quaternary Refugia of the Sweet Chestnut (*Castanea sativa* Mill.): An Extended Palynological Approach." *Vegetation History and Archaeobotany* 13:145–60.

Kreuz, A. 1999. "Becoming a Roman Farmer: Preliminary Report on the Environmental Evidence from the Romanization Project." In "Roman Germany. Studies in Cultural Interaction," edited by J. D. Creighton and R. J. A. Wilson. *Journal of Roman Archeology*, supplemental series, 32:71–98.

Kriiska, A., M. Lavento, and J. Peets. 2005. "New AMS Dates for the Neolithic and Bronze Age Ceramics in Estonia: Preliminary Results and Interpretations." *Estonian Journal of Archeology* 9 (1): 3–31.

Kroll, H. 1997. "Literature on Archeological Remains of Cultivated Plants 1996/1997." *Vegetation History and Archaeobotany* 6 (1): 25–67.

———. 1999. "Literature on Archeological Remains of Cultivated Plants 1997/1998." *Vegetation History and Archaeobotany* 8 (1–2): 129–63.

———. 2000. "Literature on Archeological Remains of Cultivated Plants 1998/1999." *Vegetation History and Archaeobotany* 9 (1): 31–68.

Kropotkin, P. P. A. 1911. "Orel or Orlov." In *Encyclopaedia Britannica*, vol. 20, edited by H. Chisholm, 250–51. Cambridge: Cambridge University.

Krstanova, K. 1999. "Domašno tkačestvo" [Domestic weaving]. [In Bulgarian.] In *Loveški kraj—Materialna i duhovna kultura* [The Lovec Region of Bulgaria: Material and spiritual culture], 163–80. Etnografski i ezikovi proučvanija: Sofija, Bulgaria.

Kudo, Y., M. Kobayashi, A. Momohara, T. Nakamura, S. Okitsu, S. Yanagisawa, and T. Okamoto. 2009. "Radiocarbon Dating of the Fossil Hemp Fruits in the Earliest Jomon Period from the Okinoshima Site, Chiba, Japan." [In Japanese with English abstract.] *Japanese Journal of Historical Botany* 17, 27–32.

Kuhaulua, J. 1973. *Takamiyama the World of Sumo*. Tokyo: Kodansha International Ltd.

Kühn, C. G., ed. 1965. *Galeni Opera Omnia*. 20 vols. Reprint, Hildesheim, Germany: Georg Olms Verlag AG. First published 1821–33, Leipzig, Germany: Knobloch.

Kühn, D. 1987. *Textile Technology: Spinning and Reeling*. In *Science and Civilization in China*, part 9, vol. 5, edited by J. Needham. Cambridge: Cambridge University.

Kuneš, P. 2006. "Impact of Early Holocene Hunter-Gatherers on Vegetation Derived from the Pollen Diagrams and Numerical Methods: An Example from the Czech Republic." In *7th European Paleobotany-Palynology Conference, Program and Abstracts*, September 6–11, 2006, Prague.

———. 2008. "Human-Driven and Natural Vegetation Changes of the Last Glacial and Early Holocene." PhD diss., Charles University—Prague, Faculty of Science, Department of Botany.

Kuneš, P., P. Pokorný, and P. Šída. 2008. "Detection of the Impact of Early Holocene Hunter-gatherers on Vegetation in the Czech Republic, Using Multivariate Analysis of Pollen Data." *Vegetation History and Archaeobotany* 17:269–87.

Kunwar, R. M., K. P. Shrestha, and R. W. Bussmann. 2010. "Traditional Herbal Medicine in Far-West Nepal: A Pharmacological Appraisal." *Journal of Ethnobiology and Ethnomedicine* 6:35. http:// www.ethnobiomed.com/content/6/1/35.

Kunzig, R. 2004. "The History of Men." *Discover* 25 (12): 32–39.

Kutcher, N. A. 1999. *Mourning in Late Imperial China: Filial Piety and the State*. Cambridge: Cambridge University.

Kuzmin, Y. V. 1995. "People and Environment in the Russian Far East from Paleolithic to Middle Ages: Chronology, Palaeogeography, Interaction." *Geojournal* 35:79–83.

Kuzmin, Y. V., and S. G. Keates. 2002. "Comment on 'Colonization of Northern Eurasia by Modern Humans: Radiocarbon Chronology and Environment' by Dolukhanov, P. M., A. M. Shukurov, P. E. Tarasov and G. I. Zaitseva." *Journal of Archeological Science* 29:593–606.

Kuzminova, N. N., and V. G. Petrenko. 1989. "Kulturnie rastenia na zapade Stepnogo Prichernomoria v seredine 3–2 tis. do n.e. (po dannim paleobotaniki)" [Cultivated plants in the West of Steppe Black Sea Coast in the middle of 3–2 millennium BC (according to paleobotany)]. [In Russian.] In *Problemu drevneyi istorii i arheologii Ukrainskoi SSR*, 119–20. Kiev, Ukraine: n.p.

Kvavadze, E., O. Bar-Yosef, A. Belfer-Cohen, E. Boaretto, N. Jakeli, Z. Matskevich, and T. Meshveliani. 2009. "30,000-Year-Old Wild Flax Fibers." *Science* 325 (5946): 1359.

Kynett, H., ed. 1895. "Cannabis Indica." *Medical and Surgical Reporter* (New York) 72 (1): 569.

Kyu, L. K. 1984. "The Concept of Ancestors and Ancestor Worship in Korea." *Asian Folklore Studies* 43 (2): 199–214.

La Barre, W. 1970. "Old and New World Narcotics: A Statistical Wuestion and an Ethnological Reply." *Economic Botany* 24:73–80.

———. 1977. "Anthropological Views of *Cannabis*." *Reviews in Anthropology* 4 (3): 237–50.

———. 1980. "History and Ethnography of *Cannabis*." In *Culture and Context: Selected Writings of Weston LaBarre*, edited by W. La Barre. Durham, NC: Duke University, 93–107.

Labrousse, A., and L. Laniel. 2001. *The World Geopolitics of Drugs, 1998/1999*. Dordrecht, The Netherlands: Kluwer Academic.

Lacey, R., and D. Danziger. 1999. *The Year 1000: What Life was Like at the Turn of the First Millennium—An Englishman's World*. Boston: Little, Brown.

Lagerås, P. 1996. "Long-Term History of Land-Use and Vegetation at Femtingagölen—A Small Lake in the Småland Uplands, Southern Sweden." *Vegetation History and Archeobotany* 5:215–28.

Lahanas, M. 2004. *The Syracusia Ship*. http://www.mlahanas.de/Greeks/Syracusia.htm.

Laine, A., E. Gauthier, J.-P. Garcia, C. Petit, F. Cruz, and H. Richard. 2010. "A Three-Thousand-Year History of Vegetation and Human Impact in Burgundy (France) Reconstructed from Pollen and Non-Pollen Palynomophs Analysis." *Comptes Rendus Biologies* 333 (11): 850–57.

Laitinen, E. 1996. "History of Hemp in Finland." Translated to English by Anita Hemmilä. *Journal of the International Hemp Association* 3 (1): 34–37. Originally published 1995 in Finnish as "Hampun Historia Soumessa" in *Hankasalmen hamppuseminaari, Hankasalmen kunnan monistamo*, edited by U. Kolehmainen, J. C. Callaway, and A. M. Hemmilä.

Lane, F. C., 1932. "The Rope Factory and Hemp Trade of Venice in the Fifteenth and Sixteenth Centuries." *Journal of Economic and Business History* 4:830–47.

Lang, V. 1999. "The Introduction. Early History of Farming in Estonia, as Revealed by Archeological Material." In *Environmental and Cultural History of the Eastern Baltic Region*, edited by U. Miller, T. Hackens, V. Lang, A. Raukas, and S. Hicks. *PACT* 57:325–38.

Langenheim, J. H. 2003. *Plant Resins: Chemistry, Evolution, Ecology and Ethnobotany*. Portland, OR: Timber.

Langerman. D. 2009. "The Tinderbox Revisited." *The Backwoodsman Magazine*, March/April, 2009. http://www.backwoodsmanmag.com/images/Feature%20Article%20PDFs/featarticle_marapr09.pdf.

Latalowa, M. 1992. "Man and Vegetation in the Pollen Diagrams from Wolin Island (NW Poland)." *Acta Palaeobotanica* (Krakow, Poland) 32 (1): 123–249.

———. 1999. "Palaeoecological Reconstruction of the Environmental Conditions and Economy in Early Medieval Wolin against a Background of the Holocene History of the Landscape." *Acta Palaeobotanica* (Krakow, Poland) 39 (2): 271.

Lattimore, O. 1950. *The Pivot of Asia*. Boston: Little, Brown.

Laufer, B. 1919. "Sino-Iranica." *Field Museum of Natural History, Publication 201, Anthropological Series* (Chicago) 15 (3): 293–94.

Laurence, J. 1960. *A History of Capital Punishment*. New York: The Citadel

Lavergne, S., and J. Molofsky. 2007. "Increased Genetic Variation and Evolutionary Potential Drive the Success of an Invasive Grass." *Proceedings of the Academy of Science* 104 (10): 3883–88.

Lawson, I. T., S. Al-Omari, P. C. Tzedakis, C. Bryant, and K. Christanis. 2005. "Lateglacial and Holocene Vegetation History at Nisi Fen and the Boras Mountains, Northern Greece." *The Holocene* 15 (6): 873–87.

le Bas, J. 1914. "Jersey Folklore Notes." *Folklore* 25:242–51.

Lechner, Z. 1954. "Obrada kudelje u baranjskim selima" [Hemp processing in villages of the region Baranja]. [In Serbian.] *Osjecki zbornik* IV:91–104.

Lee, G.-A., G. W. Crawford, L. Li, and X. Chen. 2007. "Plants and People from the Early Neolithic to Shang Periods in North China." *Proceedings of the National Academy of Science [PNAS]* 104 (3): 1087–92.

Lee, H. K. 1998. "A Lifetime with the Song of the Loom: Kim Chomsun." *Koreana* 12 (2): 64–69.

Lee, J. Y. 1981. *Korean Shamanistic Rituals*. Paris: Mouton.

———. 1973. "The Seasonal Rituals of Korean Shamanism." *History of Religions* 12 (3): 271–85.

Lee, K.-K. [Li, Kwang-kyu]. 1984. "The Concept of Ancestors and Ancestor Worship in Korea." *Asian Folklore Studies* 43: 199–214.

Legge, J. 1885a. *The Sacred Books of China: The Li Ki*. Part 3. Vol. 27. Oxford: Clarendon (Reprinted in 1966, Delhi: Motilal Banarsidass).

———. 1885b. *The Sacred Books of China: The Li Ki*. Part 4. Vol. 28. Oxford: Clarendon (Reprinted in 1966, Delhi: Motilal Banarsidass).

———, trans. 1967. *Li Chi Book of Rites, an Encyclopedia of Ancient Ceremonial Usages, Religious Creeds, and Social Institutions*, 2 vols. Edited by C. Chai and W.Chai. New Hyde Park, NY: University Books.

Lehmann, W. P., and J. Slocum. 2005. "Classical Greek Online." *Linguistics Research Center, University of Texas at Austin*. http://www.utexas.edu/cola/centers/lrc/eieol/grkol-5-R.html.

Lemoine, J. 1986. "Shamanism in the Context of Hmong Resettlement." In *The Hmong in Transition*, edited by G. L. Hendricks, B. T. Downing, and A. S. Deinard, 337–48. Staten Island, NY: Center for Migration Studies of New York.

Lempiäinen, T. 1999a. "On the History of Hemp (*Cannabis sativa* L.) in Finland and the Archaeobotanical Evidence." In *Museumslandskap. Artikkelsamling til Kerstin Griffin på 60-årsdagen*, edited by L. Selsing and G. Lillehammer, AmS-Rapport 12A:71–78. Stavanger: Arkeologiske Museum i Stavanger.

———. 1999b. "Hiiltyneitä perunoita ja kadonneita rikkaruohoja—kasvijäänteet kertovat Lahden kylän historista." In *Nuoren aupungin pitkä istoria*, edited by H. Takala and H. Poutiainen, 169–92. Lahti, Finland: Tutkimuksia Lahden menneisyydestä.

León Pinelo, A., and J. Solórzano Pereira. 1756. *Recopilación de Leyes de los Reynos de las Indias*. Madrid: Antonio Balba.

Lepetz, S., V. Matterne, M.-P. Ruas, and J.-H. Yvinec. 2002. "Culture et élévage en France septentrionale de l'âge du Fer à l'an Mil." In "Autour d'Olivier de Serres. Pratiques agricoles et pensées agronomique du Neolithique aux enjeux actuels. Actes du Colloque du Pradel (27–29 septembre 2000)," edited by A. Belmont. *Bibliothèque d'histoire rurale* (Caen, France) 6: 77–108.

Leroi-Gourhan, A. 1985. "Les Pollens en embaumement." In *La momie de Ramsès—Contribution scientifique à l'Egyptologie*, edited by Museum National d'Historie Naturelle, 162–65. Paris: Editions Recherches sur les Civilisations.

Leson, G., and P. Pless. 2002. "Hemp Seed and Hemp Oil." In *Cannabis and Cannabinoids: Pharmacology, Toxicology and Therapeutic Potential*, edited by F. Grotenhermen and E. Russo, 411–25 Binghampton, NY: Haworth.

Le Strange, R. 1977. *A History of Herbal Plants*. London: Angus and Robertson.

Le Thierry d'Ennequin, M., O. Panaud, B. Toupance, and A. Sarr. 2000. "Assessment of Genetic Relationships between *Setaria italica* and Its Wild Relative *S. Viridis* using AFLP Markers." *TAG Theoretical and Applied Genetics* 100 (7): 1061–66.

Lev, E., and Z. Amar. 2006. "Reconstruction of the Inventory of *Materia Medica* Used by Members of the Jewish Community of Medieval Cairo according to Prescriptions Found in the Taylor-Schechter Genizah Collection, Cambridge." *Journal of Ethnopharmacology* 108 (3): 428–44.

Lévesque, L. 1969. *Hakka Beliefs and Customs*. Tranlated by J. Maynard Murphy. Taichung, China: Kuang Chi.

Levey, M., and N. Al-Khaledy. 1967. *The Medical Formulary of Al-Samarqandi and the Relation of Early Arabic Simples to Those Found in the Indigenous Medicine of the Near East and India*. Philadelphia: University of Pennsylvania.

Levin, D. A. 2004. "Ecological Speciation: The Role of Disturbance." *Systematic Botany* 29 (2): 225–33.

Levine, J. 1944. "Origin of Cannabinol." *Journal of the American Chemical Society* 66:1886.

Levine, M. A. 1999a. "Botai and the Origins of Horse Domestication." *Journal of Anthropological Archeology* 18:29–78.

———. 1999b. "The Origins of Horse Husbandry on the Eurasian Steppe." In *Late Prehistoric Exploitation of the Eurasian Steppe*, edited by M. A. Levine, Y. Rassamakin, A. M. Kislenko, and N. S.

Tatarintseva, 5–58. Cambridge: McDonald Institute for Archeological Research.

———. 2002. "Domestication, Breed Diversification and Early History of the Horse." Paper presented at the Dorothy Russell Havemeyer Foundation Workshop on Horse Behavior and Welfare, June 13–16, Hólar, Iceland.

Lewin, L. 1924. *Phantastica*. [In German.] Berlin: Verlag Von Gorg Stilke. First English edition published 1931, London, by Kegan Paul, Trench Trubner. Reprinted 1964, New York, by E. P. Dutton.

Lewin, M. 2006. *Handbook of Fiber Chemistry*, 3rd ed. Boca Raton, FL: CRC, Taylor and Francis.

Lewis, P., and E. Lewis. 1984. *Peoples of the Golden Triangle—Six Tribes in Thailand*. London: Thames and Hudson.

Lewis, W. H., P. Vinay, and V. E. Zenger. 1983. *Airborne and Allergenic Pollen of North America*. Baltimore, MD: Johns Hopkins University.

Lexer, C., and A. Widmer. 2008. "The Genic View of Plant Speciation: Recent Progress and Emerging Questions." *Philosophical Transactions of the Royal Society B* 363:3023–36.

Leys, S. 2005. *The Wreck of the Batavia: A True Story*. Melbourne: Black Inc.

Li, C. 1977. *Anyang*. Seattle: University of Washington.

Li, H. L. 1966. *Tung-nan-ya Tsai-p'ei chih-wu chih ch'i-yuan* [Origins of the cultivated plants in South and East Asia]. [In Chinese.] Hong Kong: Chinese University of Hong Kong.

———. 1974a. "The Origin and Use of *Cannabis* in East Asia: Linguistic and Cultural Implications." *Economic Botany* 28 (2): 293–301.

———. 1974b. "An Archeological and Historical Account of *Cannabis* in China." *Economic Botany* 28 (4): 437–48.

———. 1975. "The Origin and Use of *Cannabis* in Eastern Asia." In *Cannabis and Culture*, edited by V. Rubin, 51–62. Paris: Mouton.

———. 1977. "Hallucinogenic Plants in Chinese Herbals." *Harvard University Botanical and Museum Leaflets* 25 (6): 161–81.

Li, J., X. J. Ge, H. L. Cao, and W. H. Ye. 2007. "Chloroplast DNA diversity in *Castanopsis Hystrix* Populations in South China." *Forest Ecology and Management* 243 (1): 94–101.

Li, S. 1984. "The 'Boat Coffins' of the Wuyi Mountains." In *Recent Discoveries in Chinese Archeology*, edited by F. Stockwell and T. Bowen, 79–80. Beijing: Foreign Language.

Li, X., and D. Walker. 1986. "The Plant Geography of Yunnan Province, Southwest China." *Journal of Biogeography* 13 (5): 367–97.

Liebenberg, L. 1990. *The Art of Tracking: The Origin of Science*. Claremont, South Africa: David Philip.

Lilek, E. 1894. "Vjerske starine iz Bosne i Hercegovine" [From the ancient religion in Bosnia and Herzegovina]. [In Serbian.] *Glasnik zemaljskog muzeja u Bosni i Hercegovini* 6:141–66.

Lip, E. 1993. *Out of China—Culture and Traditions*. New York: Addison-Wesley.

Lipson, E. 1931. *The Economic History of England*. London: A. C. Black.

Livingstone, D. 1858. *Missionary Travels and Researches in South Africa: Including a Sketch of Sixteen Years' Residence in the Interior of Africa, and a Journey from the Cape of Good Hope to Loanda on the West Coast: Thence Across the Continent, Down the River Zambesi, to the Eastern Ocean*. New York: Harper & Brothers.

Lomas-Clarke, S. H., and K. E. Barber. 2004. "Palaeoecology of Human Impact during the Historic Period: Palynology and Geochemistry of a Peat Deposit at Abbeyknockmoy, Co. Gaiway, Ireland." *The Holocene* 14 (7): 721–31.

López García, P. 1988. "Estudio polínico de seis yacimientos del Sureste Español." *Trabajos de Prehistoria*, 45:335–45.

———. 1991. "Estudios palinológicos." *El cambio cultural del IV al II milenios a.C. en la comarca noroeste de Murcia (Madrid) I*, 213–37, edited by P. López García.

Lorquin, A., and C. Moulherat. 2002. "Étude des vestiges textiles de la sépulture gallo-romaine de Fontvielle à Vareilles (Creuse)." *Aquitania* 18:171–85.

Lotter, A. F. 1999. "Late-Glacial and Holocene Vegetation History and Dynamics as Shown by Pollen and Plant Macrofossil Analyses in Annually Laminated Sediments from Soppensee, Cenral Switzerland." *Vegetation History and Archeobotany* 8:165–84.

Lotter, A. F. 2001. "The Palaeolimnology of Soppensee (Central Switzerland), as Evidenced by Diatoms, Pollen, and Fossil Pigment Analyses." *Journal of Paleolimnology* 25:65–79.

Lovett, R. A. 2006. "Ancient Manure may be Earliest Proof of Horse Domestication." *National Geographic News*, October 26.

Lovretić, J. 1902. "Otok, Vjerovanja" [Folk beliefs in Otok, Slavonia-Croatia]. [In Croatian.] *Zbornik za narodni zivot i obicaje Juznih Slavena, Jugoslavenska akademija znanosti i umjetnosti* 7:57–206.

Löw, I. 1924–34. *Die Flora Der Juden*. [In German.] 4 vols. Vienna: R. Löwit. Reprinted 1967, Hildesheim, Germany, by G. Olms.

Lowe, J. J., C. A. Accorsi, M. Bandini Mazzanti, A. Bishop, S. van Der Kaars, L. Forlani, A. M. Mercuri, et al. 1996. "Pollen Stratigraphy of Sediment Sequences from Lakes Albano and Nemi (near Rome) and from the Central Adriatic, Spanning the Interval from Oxygen Isotope Stage 2 to the Present Day." In *Palaeoenvironmental Analysis of Italian Crater Lake and Adriatic Sediments. Memorie dell'Istituto Italiano di Idrobiologia* 55:71–98, edited by P. Guilizzoni and F. Oldfield.

Lowell, P. 1895. *Occult Japan or the Way of the Gods*. New York: Houghton Mifflin.

Lozano, I. 2006. "The Therapeutic Use of *Cannabis Sativa* (L.) in Arabic Medicine." In *Handbook of Cannabis Therapeutics: From Bench to Bedside*, edited by E. B. Russo and F. Grotenhermen, 5–12. New York: Haworth.

Lozano Cámara, I. 1990. *Tres tratados árabes sobre el Cannabis indica: Textos para la historia del hachis en las sociedades islámicas s. XIII-XVI*. Madrid: Agencia Española de Cooperación Internacional.

Loze, I. 1997. "Indo-Europeans in the Eastern Baltic in the View of an Archeologist." *Humanities and Social Sciences. Latvia* 3 (16): 19–35.

Lu, X., and R. C. Clarke. 1995. "The Cultivation and Use of Hemp (*Cannabis Sativa* L.) in Ancient China." *Journal of the International Hemp Association* 2 (1): 26–30.

Lucas, R. A. 2005. "Industrial Milling in the Ancient and Medieval Worlds: A Survey of the Evidence for an Industrial Revolution in Medieval Europe." *Technology and Culture* 46 (1): 1–30.

Luo, Y. and Q. Zhong. 1999. *The Material Culture of Yunnan—Spinning and Weaving*. Kunming, China: Yunnan Education.

Lydon, J., A. H. Teramura, and C. B. Coffman. 1987. "UV-B Radiation Effects on Photosynthesis, Growth and Cannabinoid Production of Two *Cannabis Sativa* Chemotypes." *Photochemistry and Photobiology* 46 (2): 201–6.

Lyman, T. A. 1968. "Green Miao (Meo) Spirit Ceremonies." *Ethnologica* (Neue Folge) 4:1–28.

Ma, Y., and Y. Sun. 1994. "The Western Regions under the Hsiung-Nu and the Han." In *History of Civilizations of Central Asia, Vol. II. The Development of Sedentary and Nomadic Civilizations: 700 BC to AD 250*, edited by J. Harmatta, B. N. Puri, and G. F Etamadi, 227–46. Delhi: Motilal Banarsidass.

Macaj, S. 1966. "Običaji Rumuna" [Custom of "Romanian" Vlachs people in Eastern Serbia]. [In Serbian.] *Razvitak* VI/2: 58–69. MacDonald, G. M., A. A. Velichko, C. V. Kremenetski, O. K. Borisova, A. A. Goleva, A. A. Andreev, L. C. Cwynar, R. T. Riding, S. L. Forman, T. W. D. Edwards, R. Aravena, D. Hammarlund, J. M. Szeicz, and V. N. Gattaulin. 2000. "Holocene Treeline History and Climate Change across Northern Eurasia." *Quaternary Research* 53 (3): 302–11.

MacDonell, A., and A. Keith. 1958. *Vedic Index*. Vol. 2. Delhi: Mote Lal Banarsi Dass.

MacFadden, B. 1994. *Fossil Horses: Systematics, Paleobiology, and Evolution of the Family Equidae*. New York: Cambridge University.

Mack, G. R. and A. Surina. 2005. *Food Culture in Russia and Central Asia*. Westport, CT: Greenwood.

Maclagan, D. 1977. *Creation Myths—Man's Introduction to the World*. New York: Thames and Hudson.

Madella, M. 2003. "Investigating Agriculture and Environment in South Asia: Present and Future Contributions from Opal Phytoliths." In *Indus Ethnobiology: New Perspectives from the Field*, edited by S. A. Weber and W. R. Belcher, 199–249. Latham, MD: Lexington Books.

Madeyska, T. 1990. "The Distribution of Settlement in the Extra-Tropical Old World: 24,000–15,000 BP." In *The World at 18,000 BP*, vol. 2, edited by O. Soffer and C. Gamble, 24–40. London: Unwin Hyman.

Magny, M., and J.-N. Haas. 2004. "A Major Widespread Climatic Change around 5300 Cal. Yr BP at the Time of the Alpine Iceman." *Journal of Quaternary Science* 19 (5): 423–30.

Magri, D., G. G. Vendramin, B. Comps, I. Dupanloup, T. Geburek, D. Gömöry, M. Latałowa, T. Litt, L. Paule, J.-M. Roure, I. Tantau, W. O. van der Knaap, R. J. Petit, and J.-L. de Beaulieu. 2006. "A New Scenario for the Quaternary History of European Beech

Populations: Palaeobotanical Evidence and Genetic Conse-quences." *New Phytologist* 171:199–221.

Maher, B. A. 2008. "Holocene Variability of the East Asian Summer Monsoon from Cave Records: A Re-Assessment." *The Holocene* 18:861–66.

Majumdar, R. C., ed. 1952. *The Vedic Age*. London: George Allen and Unwin Ltd.

Majupuria, T. C., and I. Majupuria. 1978. *Sacred and Useful Plants and Trees of Nepal*. Kathmandu, Nepal: Sahayogi Prakashan.

Makino, H., and T. S. Melhem. 1973. "O polen de *Cannabis Sativa* L." *Ciência e Cultura* 25:535–38.

Mallinson, J., N. Donnelly, and L. Hang. (1988) 1996. *H'mong Batik—A Textile Technique from Laos*. Thailand ed. Qiang Mai, Thailand: Silkworm Books.

Mallory, J. P., and V. H. Mair. 2000. *The Tarim Mummies: Ancient China and the Mystery of the Earliest Peoples from the West*. New York: Thames and Hudson.

Malrain F., V. Matterne, and P. Méniel. 2002. *Les Paysans Gaulois III e siècle—52 av. J.-C.* Paris: éditions Errance.

Manandhar, N. P. 2002. *Plants and People of Nepal*. Portland, OR: Timber.

Mandal, S. K., and A. Mukherjee. 2003. "An Ethnobotanical Envision into Santhali Festivals in Purulia District, West Bengal." *Ethnobotany* 15:118–24.

Mandel, W. 1944. *The Soviet Far East and Central Asia*. New York: Dial.

Mann, J. G. 1957. "Armes et Armures." In *La Tapisserie de Bayeux*, edited by Frank Merry Stenton, 88–97. Paris: Flammarion.

Manniche, L. 1989. *An Ancient Egyptian Herbal*. Austin, TX: University of Texas.

Marcandier, M. A. (1758) 1766. *A Treatise on Hemp*. Boston, MA: Edes & Gill.

Marchesini, M. 1997. *Il paesaggio vegetale nella pianura bolognese in età romana sulla base di analisi archeopalinologiche ed archeocarpologiche*. PhD thesis, Università degli Studi di Firenze.

Margolis, M. 1908. *The Holy Scriptures*. Philadelphia: Jewish Publication Society of America.

Margreth, D. 1993. *Skythische Schamanen? Die Nachrichten über Enarees-Anarieis bei Herodot und Hippokrates [Scythian shamans? Information about Enarees-Anarieis in Herodotus and Hippocrates]*. Abhandlung zur Erlangung der Doktorwürde der Philosophischen Universität Zürich [Paper to obtain the Doctorate of Philosophy University of Zurich]. [In German.] Schaffhausen, Switzerland: Meier & Cie.

Markham, G. (1615) 1986. *The English Housewife, Containing the Inward and Outward Virtues which Ought to Be in a Complete Woman*. 2nd ed. Edited by M. R. Best. Reprinted in 1986, Montreal, QC: McGill-Queen's University.

Marshack, A. 1979. "Upper Paleolithic Symbol Systems of the Russian Plain: Cognitive and Comparative Analysis." *Current Anthropology* 20:271–311.

Marshall, J. A. 1992. *The Identification of Flax, Hemp, Jute and Ramie in Textile Artefacts*. Master's thesis, University of Alberta.

Martin, M. A. 1975. "Ethnobotanical Aspects of *Cannabis* in Southeast Asia." In *Cannabis and Culture*, edited by V. Rubin, 63–76. Paris: Mouton.

Marutani, A., J. Hayano, and M. Yamamoto. 2001. "Grandma Haru and Hemp Cloth: A Record of Life in Harihata, Kutsuki Village." *Journal of Kyoto Seika University* 4 (20): 240. Documentary film, 55 min., 16mm color, video transcription summary discussion in English.

Mathers, E. P. 1923. *The Book of the Thousand Nights and One Night*. Vol. 3. London: The Casanova Society. Rendered into English from the literal and complete French translation of Dr. J. C. Mardrus by Powys Mathers.

Mathieu, J., and R. Maneville. 1952. *Les Accoucheuses Musulmanes Traditionnelles de Casablanca*. Paris: IAC Paris.

Matsui, A. 1992. "Wetlands in Japan." In *The Wetlands Revolution in Prehistory*, edited by B. Coles, 5–14. Exeter, UK: The Prehistoric Society and WARP, University of Exeter.

———. 1996. "Postglacial Hunter-Gatherers in the Japanese Archipelago: Maritime Adaptations." In *Man and Sea in the Mesolithic: Coastal Settlement above and below Present Sea Level*, edited by A. Fischer, 327–34. Oxford: Oxbow Books.

Matsui, A., and M. Kanehara. 2006. "The Question of Prehistoric Plant Husbandry during the Jomon Period in Japan." *World Archeology* 38 (2): 259–73.

Matsui, A., M. Kanehara, and M. Kanehara. 2003. "Palaeoparasitology in Japan—Discovery of Toilet Features." *Memórias do Instituto Oswaldo Cruz, Rio de Janeiro* 98 (January), supplement 1, 127–36.

Matsuoka, Y., S. Takumi, and T. Kawahara. 2008. "Flowering Time Diversification and Dispersal in Central Eurasian Wild Wheat *Aegilops Tauschii* Coss.: Genealogical and Ecological Framework." *PLoS ONE* 3 (9): e3138.

Matterne, V. 2001. *Agriculture et alimentation végétale durant l'âge du Fer et l'époque gallo-romaine en France septentrionale Archéol Plantes Animaux 1*. Montagnac, France: M. Mergoil.

Matterne, V., J.-H. Yvinec, D. Gemehl, and C. Riquier. 1998. "Stockage de plantes alimentaires et infestation par les insectes dans un grenier incendié de la fin du 2e siècle après J.-C. à Amiens (Somme)." *Rev Archéol Picardie* 3 (4): 93–122.

Matyushin, G. 1986. "Mesolithic and Neolithic in South Urals and Central Asia." In *Hunters in Transition: Mesolithic Societies of Temperate Eursia and their Tranistion to Farming*, edited by M. Zvelebil, 133–50. Cambridge: Cambridge University.

Mavor, W., ed. (1573) 1812. *Five Hundred Points of Good Husbandry, as Well for the Champion, or Open Country, as for the Woodland or Several: Together with a Book of Huswifery by Tusser, Thomas*. London: Lackington, Allen.

Mayr, E. 1942. *Systematics and the Origin of Species*. New York: Columbia University.

Mažeika, J., P. Blaževičius, M. Stančikaitė, and D. Kisielienė. 2009. "Dating of the Cultural Layers from Vilnius Lower Castel, East Lithuania: Implications for Chronological Attribution and Environmental History." *Radiocarbon* 51 (2): 515–28.

McClure, H. E. 1974. *Migration and Survival of the Birds of Asia*, Bangkok: United States Army Medical Component, SEATO Medical Project.

McDougall, I., F. H. Brown, and J. G. Fleagle. 2005. "Stratigraphic Placement and Age of Modern Humans from Kibish, Ethiopia." *Nature* 433 (17 February): 733–36.

McGovern, P. E. 2003. *Ancient Wine: The Search for the Origins of Viniculture*. Princeton, NJ: Princeton University.

McGovern, W. M. 1939. *The Early Empires of Central Asia: A Study of the Scythians and the Huns and the Part they Played in World History with Special Reference to the Chinese Sources*. Chapel Hill, NC: University of North Carolina.

McKenna, T. 1992. *Food of the Gods: The Search for the Original Tree of Knowledge*. New York: Bantam Books.

McKenna, W. J. B. 1987. "The Environmental Evidence." In *Excavations in High Street and Blackfriargate: Hull Old Town Report, Series 5*, edited by P. Armstrong and B. Ayers. Hull, UK: East Riding Achaeology Society, 255–65.

———. 1992. "The Environmental Evidence." In "Excavations at 33-35 Eastgate, Beverley, 1983–86," edited by D. H. Evans and D. G. Tomlinson. *Sheffield Excutive Report* 3:227–33.

McKenny, M. 1939. *Birds in the Garden*. New York: Reynal and Hitchcock.

McNeill, W. H. 1963. *The Rise of the West*. Chicago: University of Chicago.

McPartland, J. M. 2004. "Phylogenetic and Chemotaxonomic Analysis of the Endocannabinoid System." *Brain Research Reviews* 45:18–29.

———. 2008. "The Endocannabinoid System: An Osteopathic Perspective." *Journal of the American Osteopathic Association* 108 (10): 586–600.

McPartland, J. M., and G. Guy. 2004a. "The Evolution of *Cannabis* and Coevolution with the Cannabinoid Receptor—A Hypothesis." In *The Medicinal Uses of Cannabis and Cannabinoids*, edited by G. W. Guy, B. A. Whittle, and P. J. Robson, 71–101. London: Pharmaceutical.

———. 2004b. "Random Queries Concerning the Evolution of *Cannabis* and Coevolution with the Cannabinoid Receptor." In *The Medicinal Use of Cannabis*, edited by G. Guy, R. Robson, K. Strong, and B. Whittle. London: Royal Society of Pharmacists, 71–102.

McPartland, J. M., and E. B. Russo. 2001. "Cannabis and Cannabis Extracts: Greater than the Sum of their Parts." *Journal of Cannabis Therapeutics* 1:103–32.

McPartland, J. M., J. Agraval, D. Gleeson, K. Heasman, and M. Glass. 2006. "Cannabinoid Receptors in Invertebrates." *Journal of Evolutionary Biology* 19 (2): 366–73.

McPartland, J. M., and J. Nicholson. 2003. "Using Parasite Databases to Identify Potential Nontarget Hosts of Biological Control Organisms." *New Zealand Journal of Botany* 41 (4): 699–706.

McPartland, J. M., and P. L. Pruit. 2002. "Sourcing the Code: Searching for the Evolutionary Origins of Cannabinoid Receptors, Vanilloid Receptors, and Anandamide." *Journal of Cannabis Therapeutics* 2 (1): 73 103.

McPartland, J. M., R. C. Clarke, and D. P. Watson. 2000. *Hemp Diseases and Pests: Management and Biological Control, an Advanced Treatise*. New York: CABI.

McPartland, J. M., R. W. Norris, and C. W. Kilpatrick. 2007a. "Tempo and Mode in the Endocannabinoid System." *Journal of Molecular Evolution* 65 (3): 267–76.

McPartland, J. M., R. W. Norris, and C. W. Kilpatrick. 2007b. "Coevolution between Cannabinoid Receptors and Endocannabinoid Ligands." *Gene* 397 (1–2): 126–35.

Mechoulam, R. 1986. "The Pharmacohistory of *Cannabis sativa*." In *Cannabinoids as Therapeutic Agents*, edited by R. Mechoulam, 1–19. Boca Raton, FL: CRC.

Mechoulam, R., and L. Hanus. 2000. "A Historical Overview of Chemical Research on Cannabinoids." *Cheistry and Physics of Lipids* 108:1–13.

Mechoulam, R., P. Maximilian, E. Murillo-Rodriguez, and L. O. Hanus. 2007. "Cannabidiol—Recent Advances." *Chemistry and Biodiversity* 4:1678–92.

Médail, F. 2009. "The Mediterranean Basin, a Hotspot for Plant Evolution." In *Abstracts of the 45th International SISV & FIP Congress on Biodiversity Hotspots in the Mediterranean Area: Species, Communities and Landscape Level*, June 22–24, 2009, University of Cagliari, Sardinia, Italy.

Médail, F., and K. Diadema. 2009. "Glacial Refugia Influence Plant Diversity Patterns in the Mediterranean Basin." *Journal of Biogeography* 36:1333–45.

Medjesi, L. 1978. "Konoplja u običajima, verovan-jima i praznover-cama vojvodjanskih rusina" [Hemp in the customs, beliefs, and superstitions of the Russians of the Vojvodina]. [In Serbian.] In *Rad XX kongresa Saveza udruzenja folklorista Jugoslavije u Novom Sadu 1973*, 347–53. Beograd, Serbia: Slavez udruzenja folklorista Jugoslavije.

Mee, A., ed. 1909. *The Children's Encyclopædia*. London: Educational Book Company.

Meersschaert, L., H. De Wolf, B. Klinck, D. Van Damme, and C. Verbruggen. 2007. "Archaeobtoanical Research on the 17th Century Ditch at Damme (Westflanders, Belgium)." *Program and Abstracts in the 14th Symposium of the International Work Group for Palaeoethnobotany in Krakow Poland 2007, 143*.

Meissner, B. 1925. *Babylonien und Assyrien*, 2 vols. Heidelberg, Germany: Carl Winters.

Mellars, P. 2006. "Going East: New Genetic and Archeological Perspectives on the Modern Colonization of Eurasia." *Science* 313 (5788): 796–800.

———. 2007. *Rethinking the Human Revolution: New Behavioural and Biological Perspectives on the Origin and Dispersal of Modern Humans*. Cambridge: McDonald Institute for Archeological Research.

Meltzer, D. 2009. *First Peoples in a New World: Colonizing Ice Age America*. Berkeley, CA: University of California.

Mendelbaum, I. 1982. *A History of the Mishnaic Law of Agriculture: Kilayim—Translation and Exegesis*. Chico, CA: Scholars.

Mercuri, A. M., C. A. Accorsi, and M. B. Bandini Mazzanti. 2002. "The Long History of *Cannabis* and Its Cultivation by the Romans in Central Italy, Shown by Pollen Records from Lago Albano and Lago di Nemi." *Vegetation history and Archaeobotany* 11 (4): 263–76.

Mercuri, A. M., C. A. Accorsi, M. B. Bandini Mazzanti, A. Cardarelli, D. Labate, M. Marchesini, and G. T. Grandi. 2006. "Economy and Environment of Bronze Age Settlements—Terramaras—on the Po Plain (Northern Italy): First results from the Archaeobotanical Research at the Terramara di Montale." *Vegetation History and Archaeobotany* 16 (1): 43–60.

Mercuri, A. M., E. Gasparini, G. Bosi, C. Guarnieri, and M. B. Bandini Mazzanti. 1999. "Seeds and Fruits from the Town of Ferrara (Emilia Romagna—Northern Italy) in the Middle Age (10–12 Century A.D.)." In *Archeologia e Ambiente*, edited by F. Lenzi, 231–36. Forli, Italy: ABACO Edizioni.

Mercuri, A. M., L. Sadori, and P. U. Ollero. 2011. "Mediterranean and North-African Cultural Adaptations to Mid-Holocene Environmental and Climatic Changes." *The Holocene* 21 (1): 189–206.

Merlin, M. D. 1972. *Man and Marijuana: Some Aspects of Their Ancient Relationship*. Rutherford NJ: Fairleigh Dickinson University.

———. 1984. *On the Trail of the Ancient Opium Poppy*. East Brunswick, NJ: Associated University.

———. 2003. "Archeological Evidence for the Tradition of Psychoactive Plant Use in the Old World." *Economic Botany* 57 (3): 295–323.

Merrillees, R. S. 1962. "Opium Trade in the Bronze Age Levant." *Antiquity* 36:287–92.

———. 1999. "How the Ancients Got High." *Odyssey* (Winter): 21–29.

Merzouki, A., and J. Molero Mesa. 1999. "Le chanvre (*Cannabis sativa* L.) dans la phamacopée traditionnelle du Rif (Nord du Maroc)." *Ars Pharmacology* 4:233–40.

Merzouki, A., F. Ed-derfoufi, and J. Molero Mesa. 1999. "A Polyherbal Remedy Used for Respiratory Affections in Moroccan Traditional Medicine." *Ars Pharmacology* 40:31–38.

———. 2000. "Hemp (*Cannabis Sativa* L.) and Abortion." *Journal of Ethnopharmacology* 73:501–3.

Meuli, K. 1935. Scythica. *Hermes* 70:121–76.

Meyer-Melikyan, N. R., and N. A. Avetov. 1998. "Analysis of Floral Remains in the Ceramic Vessel from the Gonur Temenos." In *Margiana and Protozoroastrism*, edited by V. Sarianidi, Appendix I, 176–77. Athens: Kapon Editions.

Meyerson, M. D. 1991. *The Muslims of Valencia: In the Age of Fernando and Isabel: Between Coexistence and Crusade*. Berkeley, CA: University of California.

Mićović, L. 1952. "Život i običaji Popovaca" [Folk life and customs among people of Popovo Polje in Montenegro]. [In Serbian.]. *Srpska akademija nauka, Srpski etnografski zbornik* 65, Beograd, Serbia.

Miettinen, J., E. Grönlund, H. Simola, and P. Huttunen. 2002. "Palaeolimnology of Lake Pieni-Kuuppalanlampi (Kurkijoki, Karelian Republic, Russia): Isolation History, Lake Ecosystem Development and Long-Term Agricultural Impact." *Journal of Paleolimnology* 27:29–44.

Migal, N. D. 1969. "On the Morphology of Hemp (*Cannabis* L.) Pollen." *Botanicheskii Zhurnal* (St. Petersburg, Russia) 54:274–76.

Mijatović, M. S. 1907. "Običaji naroda srpskoga" [Traditional customs among Serbs]. [In Serbian.] *Srpska kraljevska akademija, Srpski etnografski zbornik* 7:107–390.

———. 1909. "Narodna medicina" [Folk medicine]. [In Serbian.] *Srpska kraljevska akademija, Srpskietnografski zbornik* 13:279.

Mijatović, M. S., and M. T. Bušetić. 1925. "Tehnički radovi Srba seljaka u Levču i Temniću—Zivot i običaji narodni" [Traditional technical doings among Serb people peasants: Folk life and customs]. [In Serbian.] *Srpski etnografski zbornik* 14:127–29.

Mikuriya, T. 1969. "Historical Aspects of Cannabis Sativa in Western Medicine." *The New Physician* 18:902–8.

———. 1973. *Marijuana: Medical Papers 1839–1972*. Oakland, CA: Medi-Comp.

———. 1994. *India Hemp Drugs Report Centennial Volume 1. Policy, Social and Religious customs*. San Francisco: Last Gasp.

Milićević, M. Đ. 1894. "Zivot Srba seljaka." [In Serbian.] *Srpski etnografski zbornik*. Vol. I. Beograd, Serbia: Državna štamparija Kraljevine Srbije.

Mills, J. H. 2003. *Cannabis Britannica: Empire, Trade and Prohibition 1800–1928*. Oxford: Oxford University.

Milošević, M. J. 1936. "Narodne praznoverice u kopaonickim selima u Ibru" [Folk superstitions in the villages on Kopaonik Mountain]. [In Serbian.] *Glasnik Etnografskog Muzeja u Beogradu* 11: 51–52. Beograd, Serbia.

Min, K. J. 1985. "Study on the Weaving Technique of Textiles of Korea—Focus on the Hemp and Ramie in Ancient Three Kingdom and Goryeo Periods." *Educational Paper Collection* (Seoul: Kookmin University) 4:123–32.

Ming, R., J. Wang, P. H. Moore, and A. H. Paterson. 2007. "Sex Chromosomes in Flowering Plants." *American Journal of Botany* 94 (2): 141–50.

Miotik-Szpiganowicz, G. 1992. "The History of the Vegetation of Bory Tucholskie and the Role of Man in the Light of Palynological Investigations." *Acta Palaeobotanica* 32 (1): 39–122.

Mitchell, E. A. D., W. O. van der Knaap, J. F. N. van Leeuwen, A. Buttler, B. G. Warner, and J. M. Gobat. 2001. "The Palaeoecological History of the Praz-Rodet Bog (Swiss Jura) based on Pollen, Plant Macrofossils and Testate Amoebae (Protozoa)." *The Holocene* 11 (1): 65–80.

Mitchell, G. F., and C. A. Dickson. 1985. "Plant Remains and Other Items from Medieval Drogheda." *Circaea* 3 (1): 31–37.

Mitchell, G. F., C. A. Dickson, and J. H. Dickson. 1987. "Archeology and Environment in Early Dublin." *Medieval Dublin Excavations 1962–81*. Ser. C. Vol. 1. Dublin: Royal Irish Academy and National Museum of Ireland.

Mithen, S. 2004. *After the Ice: A Global Human History 20,000–5,000 B.C.* Cambridge, MA: Harvard University.

Moe, D., and W. O. van der Knaap. 1990. "Transhumance in Mountain Areas: Additional Interpretation of Three Pollen Diagrams from Norway, Portugal and Switzerland." In "Impact of Prehistoric and Medieval Man on the Vegetation: Man at the Forest Limit," by D. Moe and S. Hicks. *PACT* 31:91–105.

Moffat, B., and J. Fulton, eds. 1989. *Sharpe Practice 3: The Third Report on Researches into the Medieval Hospital Hospital at Soutra, Louthian/ Borders Region, Scotland*. Edinburgh: Soutra Hospital Archaeoethnopharmacological Research Project.

Moffett, L. 1992. "Fruits, Vegetables, Herbs and Other Plants from the Latrine at Dudley Castle in Central England, Used by the Royalist Garrison during the Civil War." In "Festrschrift for Professor van Zeist," edited by J. P. Pals, J. Buurman, and M. van der Veen. *Review of Palaeobotany and Palynology* 73:271–86.

Moldenke, H., and A. Moldenke. 1952. *Plants of the Bible*. Waltham, MA: Cronica Botanica.

Mölleken, H., and R. R. Theimer. 1997. "Survey of Minor Fatty Acids in *Cannabis Sativa* L. Fruits of Various Origins." *Journal of the International Hemp Association* 4 (1): 13–17.

Möller-Wiering, S. 2005. "Textiles for Transport." In *Northern Archeological Textiles Nesat VII: Textile Symposium in Edinburgh*, 5th-7th *May 1999*, edited by F. Pritchard and J. P. Peter Wild, 75–79. Oxford: Oxbow Books.

Molodkov, A. N., and N. S. Bolikhovskaya. 2006. "Long-Term Palaeoenvironmental Changes Recorded in Palynologically Studied Loess-Palaeosol and ESR-Dated Marine Deposits of Northern Eurasia: Implications for Sea-Land Correlation." *Quarternary International* 152–53:48–58.

Montgomery, D. H. (1887) 1912. *The Leading Facts of English History*. Boston: Ginn.

Moora, T., M. Ilomets, and L. Jaanits. 1988. "Muistsetest loodusoludest Akali kiviaja asulakoha lähiümbruses" [On ancient natural conditions in the vicinity of the Akali Neolithic settlement]. [In Estonian.] In *Loodusteaduslikke meetodeid Eesti arheoloogias*, edited by A.-M Rõuk and J. Seliran, 26–38. Tallinn, Estonia: NSV TA Ajaloo Instituut.

Moore, B. 1905. *The Hemp Industry in Kentucky*. Lexington, KY: James E. Hughes.

Moore, G. F. 1913. *History of Religions: Volume I, China, Japan, Egypt, Babylonia, Assyria, India, Persia, Greece, Rome*. New York: Charles Scribner's Sons.

Moreno, G. 1986. "Aspectos do Maconhismo em Sergipe." In *Diamba Sarabamba*, edited by A. Henman and O. Pessoa Jr., 53–68. São Paulo, Brazil: Ground.

Morgan, S., and K. Culhane-Pera. 1993. *Threads of Life: Hemp and Gender in a Hmong Village*. Watertown, MA: Documentary Educational Resources. Videocassette (VHS), 28 min., color, sound.

Morikawa, M., and S. Hashimoto. 1994. *The Torihama Shellmound: A Jomon Time Capsule*. [In Japanese.] Tokyo: Yomiuri Shimbunsha.

Morison, S. E. 1921. *The Maritime History of Massachusetts, 1783–1860*. Boston: Houghton Mifflin.

Morningstar, P. J. 1985. "Thandai and Chilam: Traditional Hindu Beliefs about the Proper Uses of *Cannabis*." *Journal of Psychoactive Drugs* 17 (3): 141–65.

Morris, J. 1980. *The Venetian Empire: A Sea Voyage*. London: Faber and Faber.

Morrison, K. 1994. "The Intensification of Production: Archeological Approaches." *Journal ofArcheological Method and Theory* 1 (2): 111–59.

Mosk, S. A. 1939. "Subsidized Hemp Production in Spanish California." *Agricultural History* 13 (4): 171–75.

Moulton, G. 2004. *The Complete Idiot's Guide to Biology*. East Rutherford, NJ: Alpha Books.

Mueggler, E. 1998. "The Poetics of Grief and the Price of Hemp in Southwest China." *Journal of Asian Studies* 57 (4): 979–1008.

Mukherjee, A., S. C. Roy, S. De Bera, H. Jiang, X. Li, C. Li, and S. Bera. 2008. "Results of Molecular Analysis of an Archeological Hemp (*Cannabis Sativa* L.) DNA Sample from North West China." *Genetic Resources and Crop Evolution* 55 (4): 481–85.

Muller, F. M. 2004. *Pahlavi Texts Part I: The Sacred Books of the East Part Five* (Sacred Books of the East). Translated by E. W. West. New Delhi, India: Motilal Banarsidass Publishers Pvt. Ltd.

Murad, W., A. Ahman, S. A. Gilani, and M. A. Khan. 2011. "Indigenous Knowledge and Folk Use of Medicinal Plants by the Tribal Communities of Hazar Nao Forest, Malakand District, North Pakistan." *Journal of Medicinal Plants Research* 5 (7): 1072–86.

Murakami, A., P. Darby, B. Javornik, M. S. S. Pais, E. Seigner, A. Lutz, and P. Svoboda. 2006. "Molecular Phylogeny of Wild Hops, *Humulus Lupulus* L." *Heredity* 97:66–74.

Murphy, P. 1983. "Plant Macrofossils." In "A waterfront Excavation at Whitefriears Street Car Park Norwich 1979," by B. Ayers and P. Murphy. *East Anglian Archeological Report* 17:40–44 and fiche.

———. 1984. "Plant Macrofossils from Site 1092." In "Excavations in Thetford 1948–59 and 1973–80," by A. Rogerson and C. Dallas. *East Anglian Archeological Report* 22: 194–95.

———. 1988. "VII. Plant macrofossils." In "Excavations at St. Martin-at-Palace Plain, Norwich. 1981," edited by B. Ayers. *East Anglian Archeology* 37:118–26 and fiche 2:A.12–14.

Murphy, T. M., N. Ben-Yehuda, R. E. Taylor, and J. R. Southon. 2011. "Hemp in Ancient Rope and Fabric from the Christmas Cave in Israel: Talmudic Background and DNA Sequence Identification." *Journal of Archeological Science* 38:2579–88

Musselman, L. J. 2007. *Figs, Dates, Laurel, and Myrrh: Plants of the Bible and the Quran*. Portland, OR: Timber.

Musty, R. E. 2004. "Natural Cannabinoids: Interactions and Effects." In *The Medicinal Uses of Cannabis and Cannabinoids*, edited by G. W. Guy, B. A. Whittle, and P. J. Robson, 165–204. London: Pharmaceutical.

Myles, S., A. R. Boyko, C. L. Owens, P. J. Brown, F. Grassi, M. K. Aradhya, B. Prins, A. Reynolds, J. Chia, D. Ware, C. D. Bustamante, and E. S. Buckler. 2011. "Genetic Structure and Domestication History of the Grape." *PNAS* 108 (9): 3530–35.

Nadkarni, K. M. 1954. *Indian Materia Medica*. 3rd ed. Vol. 1. Bombay: Popular Prakashan.

Nakagami, K. 1995. *Nakagami kenji zenshū*. Vol. 6. Tokyo: Shōeisha, 109.

Naquin, S. 1988. "Funerals in North China: Uniformity and Variation." In *Death Ritual in Late Imperial and Modern China*, edited by J. L. Watson and E. S. Rawski, 37–70. Berkeley, CA: University of California.

Naranjo, C. 1990. "A posthumous 'encounter' with R. Gordon Wasson." In *The Sacred Mushroom Seeker: Essays for R. Gordon Wasson: Ethnomycological Studies 11*, edited by T. J. Reidlinger, 177–81. Portland, OR: Dioscorides.

Naydenov, K., S. Senneville, J. Beaulieu, F. Tremblay, and J. Bousquet. 2007. "Glacial Vicariance in Eurasia: Mitochondrial DNA Evidence from Scots Pine for a Complex Heritage involving Genetically Distinct Refugia at Mid-Northern Latitudes and in Asia Minor." *BMC Evolutionary Biology* 7:233.

NCGR. 2010. *National Clonal Germplasm Repository, Corvallis, Oregon*. http://www.ars-grin.gov/cor/humulus/huminfo .html#plant.

Needham, J. 1954. *Science and Civilisation in China, Vol. I: Introductory Orientations*. With collaboration from W. Ling. Cambridge: Cambridge University.

———. 1956. *Science and Civilisation in China, Vol. II. History of Scientific Thought*. With research assitance from Wang Ling. Cambridge: Cambridge University.

———. 1974. "Spagyrical Discovery and Invention: Magisteries of Gold and Immortality." With collaboration from G.-D. Lu. In *Science and Civilisation in China, Vol. 5: Chemistry and Chemical Technology, Pt. II*, edited by J. Needham. Cambridge: Cambridge University.

———. 1980. *Science and Civilisation in China, Vol. 5: Chemistry and Chemical Technology*. Cambridge: Cambridge University.

———. 1984. *Science and Civilisation in China, Vol. 5: Chemistry and Chemical Technology*. Cambridge: Cambridge University.

———. 1996. *Agroindustries and Forestry*. With collaboration from L. Gwei-Djen, H.-T. Huang, H. Hsing-Tsung, C. Daniels, F. Bray, N. K. Menzies, and N. Sivin. In *Science and Civilisation in China, Vol 6: Biology and Biological Technology Pt. III*, edited by J. Needham. Cambridge: Cambridge University.

Needham, J., and G. Lu. 1986. "Botany." In *Science and Civilisation in China, Vol. VI: Biology and Biological Technology, Pt. 1*, edited by J. Needham, 174–75. Cambridge: Cambridge University.

Needham, J., G. Lu, and W. Ling. 1971. *Science and Civilisation in China, Vol. 4: Physics And Physical Technology, Part 3: Civil Engineering And Nautics*. Cambridge: Cambridge University.

Needham, J., and R. D. S. Yates. 1994. *Science and Civilization in China, Vol. V: Chemistry and Chemical Technology Pt. VI: Military Technology*. Cambridge: Cambridge University.

Needham, J., and T. Tsien. 1985. *Science and Civilisation in China, Vol. V: Chemistry and Chemical Technology, Part 1: Paper and Printing*. Cambridge: Cambridge University.

Negrutiu, I., B. Vyskot, N. Barbacar, S. Georgiev, and F. Moneger. 2001. "Dioecious Plants. A Key to the Early Events of Sex Chromosome Evolution." *Plant Physiology* 127:1418–24.

Nelson, R. 1996. *Hemp & History*. http://www.rexresearch.com/hhist/hhicon.htm. Original, unabridged manuscript of *The Great Book of Hemp*, Inner Traditions International, 1996, edited by Rowan Robinson.

Nelson, S. M. 1973. *Chulmun Period Villages on the Han River in Korea: Subsistence and Settlement*. PhD diss., University of Michigan.

———. 1993. *The Archeology of Korea*. Cambridge: Cambridge University.

Neményi, G. von. 1988. *Heidnische Naturreligion*. Bergen, Germany: Bohmeier Verlag.

Neusner, J. 1995. *The Talmud of Babylonia—An Academic Commentary: 25, Bavli tractate Abodah Zarah*. Atlanta, GA: Scholars.

"News Digest: Ancient Papermaking Still a Thriving Business." 2010. *The China*, December 27. http://thechina.biz/china-economy/ancient-papermaking-still-a-thriving-business.

Nicolas, M., V. Hykelova, B. Janousek, V. Laporte, B. Vyskot, D. Mouchiroud, I. Negrutiu, D. Charlesworth, and F. Monéger. 2005. "A Gradual Process of Recombination Restriction in the Evolutionary History of the Sex Chromosomes in Dioecious Plants." *PLoS Biology* 3 (1): e4. doi:10.1371/journal.pbio.0030004.

Nicholson, R., and A. R. Hall. 1988. "The Plant Remains." In *The Origins of the Newcastle Quayside. Excavation at Queen Street and Dog Bank.*, The Society of Antiquaries of Newcastle upon Tyne, Mongraph Series 3, edited by C. O'Brien, L. Brown, S. Dixon, and R. Nicholson, 112–29.

Niculiță-Voronca, E. 1903. *Datinele și credințele poporului român adunate șiașezate în ordine mitologică*. [In Romanian.] Cernțuăi, Romania: Tipografia Isidor Wiegler.

Nigro, D. 2012. *French Hemp Winery Is No Joke. Wine Spectator*, December 31.

Nihon Kogeika [Japan Art Crafts Association]. 2005. *Ningenkokuho Gallery (Photo Images) of Works by the Living National Treasures of Japan with Lists of Exhibits for the Annual Exhibitions of Japanese Traditional Art Crafts from 1983 to 2005*. Museum of Japanese Traditional Art Crafts in cooperation with Tokyo National Museum. http://www.nihon-kogeikai.com.

Niinimets, E., and L. Saarse. 2007. "Fine-Resolution Pollen-Based Evidences of Farming and Forest Development, South-Eastern Estonia." *Polish Journal of Ecology* 55 (2): 283–96.

Nikodinovski, T. 1984. "Elementarnite nepogodi i bolestite vo narodnoto tvorestvo vo Deborca-Ohridsko" [Weather storms and diseases in the folklore near the lake of Ohrid]. [In Serbian.] *Makedonski folklor* 17 (33): 231.

Nishida, M. 2002. "Another Neolithic in Holocene Japan." *Documenta Praehistorica* 29:20–28.

Nordal, A. 1970. "Microscopic Detection of *Cannabis* in the Pure State and in Semi-Combusted Residues." In *The Botany and Chemistry of* Cannabis, edited by C. R. B. Joyce and S. H. Curry, 61–68. London: J. & A. Churchill.

NORML. 2010. *United States National Organization for the Reform of Marijuana Laws Website*. http://blog.norml.org/2010/09/15/incarceration-nation-marijuana-arrests-for-year-2009-near-record-high.

———. 2012. *United States National Organization for the Reform of Marijuana Laws Website*. http://blog.norml.org/2012/10/29/marijuana-arrests-decline-in-2011-but-still-total-half-of-all-illicit-drug-violations/.

Nowak, M. 2000. "The Second Phase of Neolithization in East-Central Europe." *Antiquity* 75:582–92.

Now Lebanon. 2011. "Police Raze Bekaa Hashish Fields." *Now Lebanon*, June 27, 2011. https://now.mmedia.me/lb/en/latestnews/police_raze_bekaa_hashish_fields.

Nuñez, M., and T. Lempiäinen.1992. "A Late Iron Age Farming Complex from Kastelholms Kungsgård, Sund, Åland Islands." *Pact* 36:125–42.

Nunn, J. F. 1996. *Ancient Egyptian Medicine*. Norman, OK: University of Oklahoma.

Nunome, J. 1992. *The Archeology of Fiber*. Kyoto, Japan: Senshoku to Seikatsusha.

O'Callaghan, E. B., Brodhead, J. R., and B. Fernow, eds. 1860. *Documents Relative to the Colonial History of New York*. Vol. 5, no. 63. Albany, NY: Weed, Parsons Printers.

O'Connell, T., M. Levine, and R. Hedges. 2003. "The Importance of Fish in the Diet of Central Eurasian People from the Mesolithic to the Early Iron Age." In *Prehistoric Steppe Adaptation and the Horse*, edited by M. Levine, C. Renfrew and K. Boyle, 253–68. Cambridge: McDonald Institute for Archeological Research.

Ohr Somayach International. 1998. "Ask the Rabbi" http://www.ohr.edu.

Okamoto, I., ed. 1979. *Torihama Kaizuka* [The Torihama shell mound]. [In Japanese.] Fukui: Fukui-Ken Kyōiku iinkai.

Okazaki. H., M. Kobayashi, A. Momohara, S. Eguchi, T. Okamoto, S. Yanagisawa, S. Okubo, and J. Kiyonaga. 2011. "Early Holocene Coastal Environment Change Inferred from Deposits at Okinoshima Archeological Site, Boso Peninsula, Central Japan." *Quaternary International* 230:87–94.

Okura, N. 1974. "Seiyu Roku or 'On Oil Manufacturing.'" English translation reprint, New Brunswick, NJ: Olearius Editions. First published 1836 in Japanese.

Olsen, S., and D. Harding. 2005. "Fiber Technology in the Copper Age Botai Culture of Northern Kazakhstan." Paper presented at the Second University of Chicago Eurasian Archeology Conference, Social Orders and Social Landscapes: Interdisciplinary Approaches to Eurasian Archeology, April 15–16, Chicago.

Olsen, S. L., ed. 1996. *Horses through Time*. Boulder, CO: Roberts Rinehart.

———. 2003. "The Exploitation of Horses at Botai, Kazakhstan." In *Prehistoric Steppe Adaptation and the Horse*, edited by M. Levine, C. Renfrew, and K. Boyle, 83–104. McDonald Institute Monographs. Cambridge: McDonald Institute for Archeological Research.

Olsen, S. L. 2006a. "Early Horse Domestication on the Eurasian Steppe." In *Documenting Domestication: New Genetic and Archeological Paradigms*, edited by M. A. Zeder, D. G. Bradley, E. Emshwiller, and B. D. Smith, 245–69. Berkeley, CA: University of California.

———. 2006b. "Introduction." In *Horses and Humans: The Evolution of Human-Equine Relationships*, edited by S. L. Olsen, S. Grant, A. M. Choyke, and L. Bartosiewicz, British Archaeogical Reports International Series 1560:1–10. Oxford: Archaeopress.

———. 2006c. "Early Horse Domestication: Weighing the Evidence." In *Horses and Humans: The Evolution of Human-Equine Relationships*, edited by S. L. Olsen, S. Grant, A. M. Choyke, and L. Bartosiewicz, British Archaeogical Reports International Series 1560:81–113. Oxford: Archaeopress.

Olson, D. 1997. "Hemp Culture in Japan." *Journal of the International Hemp Association* 4 (1): 40–50.

———. 2002a. *Hempen Culture in Japan*. http://www.japanhemp.org/uncleweed/history.htm.

Olson, S. 2002b. *Mapping Human History: Discovering the Past through our Genes*. Boston: Houghton Mifflin.

Opravil, E. 1979. "*Hedera helix* L. aus der mittelalterlichen Stadt Most (Tschechoslowakei)." In *Veröffentlich mit Mitteln des Landschaftsverbandes Rheinland*, edited by M. Ludwig. Bonn, Germany, 209–15: Rheinisches Landesmuseum.

———. 1983. "Z historie šíření konopě seté (*Cannabis sativa* L.)" [From the history of hemp cultivation]. [In Czech.] *Archeologické rozhledy* 35:206–13.

———. 1997. "Vegetační poměry Sezimova ústí a jeho okolí středověku" [The vegetation conditions of Sezimovo Ústí and its surroundings in the Middle Ages]. [In Czech.] In *Život v archeologii středověku* [Archeological interpretations of life in the Middle Ages]: Sborník příspěvků věnovaných Miroslavu Richterovi a Zdeňku Smetánkovi [Papers in honor of M. Richter and Z. Smetánka], 498–506. Hrsg. Kubková, Jana u.a.: Praha, Czech Republic.

———. 1998a. "Gegenwärtiger Stand archäobotanischer Forschungen in der Siedlungsagglomeration von Staré Město in der

Burgwallzeit." In *Studien zum Burgwall von Mikulčice 3. Spisy Archeol ústav Akad Brno 11*, edited by L. Poláček, 327–56. Brno, Slovakia.

———. 1998b. "Makrozbytky rostlinného původu z Uherského Hradiště a Starého Města" [The botanical macrorelics from Uherské Hradiště and Staré Město]. *Slovácko* 40:115–21.

O'Shaughnessy, W. B. 1939. *On the Preparations of the Indian Hemp, or Gunjah (Cannabis Indica): Their Effects on the Animal System in Health, and their Utility in the Treatment of Tetanus and Other Convulsive Diseases*. Delivered to the Medical and Physical Society of Calcutta.

Otto-Bliesner, B. L., E. C. Brady, G. Clauzet, R. Tomas, S. Levis, and Z. Kothavala. 2006. "Last Glacial Maximum and Holocene Climate in CCSM3." *Journal of Climate* 19:2526–44.

Outram, A. K., N. A. Stear, R. Bendrey, S. Olsen, A. Kasparov, V. Zaibert, N. Thorpe, and R. P. Evershed. 2009. "The Earliest Horse Harnessing and Milking." *Science* 323 (5919): 1332–35.

Owen, L. A., M. W. Caffee, R. C. Finkel, and Y. B. Seong. 2008. "Quaternary Glaciation of the Himalayan-Tibetan Orogen." *Journal of Quaternary Science* 23:513–31.

Ozola, I., A. Cerina, and L. Kalnina. 2010. "Reconstruction of Palaeovegetation and Sedimentation Conditions in the Area of Ancient Lake Burtnieks, Northern Latvia." *Estonian Journal of Earth Science* 58 (2): 164–79.

Påhlsson, I. 1982. "*Cannabis Sativa* in Dalarna." *Striae* 14:79–82.

Painne, T., ed. 1766. *An Abstract of Most Useful Parts of a Late Treatise on Hemp Translated from the French by M. Marcandier*. Boston: Edes & Gill.

Pals, J. P. 1983. "Plant Remains from Aartswoud, a Neolithic Site in the Coastal Area." In *Plants and Ancient Man: Studies in Palaeoethnobotany*, edited by W. van Zeist and W. A. Casparie. Rotterdam, The Netherlands: A. A. Balkema, 313–21.

Pälsi, S. 1920. "Ein steinzeitlicher Moorfund bei Korpilahti im Kirchspiel Antrea. Län Viborg." *Suomen muinaismuistoyhdistyksen aikakauskirja* 28 (2): 1–19.

Pan, J. 1983. "The Invention and Development of Papermaking." In *Ancient China's Technology and Science*, edited by Institute of the History of Natural Sciences and Chinese Academy of Sciences, 176–83. Beijing: Foreign Languages.

———. *Zhongguo gu dai si da fa ming: Yuan liu, wai chuan ji shi jie ying xiang* [The four great inventions of ancient China: Their origin, development, spread and influence in the world]. [In Chinese.] Hefei shi, China: Zhongguo ke xue ji shu da xue chu ban she.

Pan, Z. 1979. "China." In *Handmade Papers of the World*, edited by Takeo Company, 29–41. Tokyo: Takeo

Panareda, J. M., and M. Boccio. 2009. "Mediterranean Mountains as a Refuge and Dispersal Area of Plants. The Study Case of Montseny Massif (Catalonia, Spain)." In *Abstracts of the the 45th International SISV & FIP Congress, Biodiversity Hotspots in the Mediterranean Area: Species, Communities and Landscape Level*, June 22–24, University of Cagliari, Sardinia, Italy.

Pandey, B. P. 1989. *Sacred Plants of India*. New Delhi: Shree.

Pantelić, N. 1974. "Etnološka građa iz Budžaka" [Ethnological scientific materials from Budzak]. [In Serbian.] *Glasnik Etnografskog muzeja u Beogradu* 37:215–17.

Paphitis, N. 2007. "2,700-Year-Old Fabric Found in Greece." *Associated Press Writer*, May 10.

Par Pharmaceutical. 2008. "Par Pharmaceutical Receives Final Approval to Market Generic Marinol® C-III (Dronabinol) Capsules." http://investors. parpharm.com/phoenix .zhtml?c=81806&p=irol-newsArticle&ID=1355051.

Parker, A. 1913. "Oxfordshire Village Folklore (1840–1900)." *Folklore* 24:74–91.

Parker, E. H. 1890. "On Race Struggles in Korea." *Transactions of the Asiatic Society of Japan* 23:137–228.

Parker, R. C., and Lux. 2008. "Psychoactive Plants in Tantric Buddhism: Cannabis and Datura Use in Indo-Tibetan Esoteric Buddhism." *Erowid Extracts* 14 (June): 6–11. http://www.erowid.org/ general/newsletter/erowid_newsletter14.pdf.

Parkinson, J. 1640. "Theatrum Botanicum." *The Theater of Plantes: Or, an Herball of a Large Extent*. London: Thomas Cotes.

Parpola, A. 1998. "Aryan Languages, Archeological Cultures, and Sinkiang: Where Did Proto-Iranian Come into Being and How Did it Spread?" In *The Bronze Age and Early Iron Age Peoples of Eastern Central Asia*, edited by V. Mair, 114–47. (*Journal of Indo-European Studies Monograph* 26 (1).) Washington, DC: Institute for the Study of Man.

Partridge, W. L. 1975. "Cannabis *and Cultural Groups in a Columbian Municipio*." In *Cannabis and Culture*, edited by V. Rubin, 147–72. The Hague: Mouton.

Pashkevich, G. A. 1997. "Early Farming in the Ukraine." In *Landscapes in Flux. Central and eastern Europe in antiquity. Colloquia Pontica* 3, edited by J. C. Chapman and P. M Dulokhanov, 267–74. Oxford: Oxbow Books.

———. 1998a. "Paleobotanichni doslidzhennya to deyaki pignnya vigotovlennya keramiki skifs'kogo chasu z Daiprovs'kogo lisostepovogo Livoberezhzhya." [In Ukrainian.] *Archeologiunnii Litopis Livoberezhnoyi Ukrayiini* 1–2:38–40.

———. 1998b. "Палеоботанічні дослідження матеріалів Пастирського городища" [Paleobotanical Researches of the Materials of Pastyrske Site]. [In Ukrainian.] *Археологія* 3:40–51.

———. 1999. "New Evidence for Plant Exploitation by the Scythian Tribes During the Early Iron Age in the Ukraine." In *Acta Palaeobotanica Supplement 2*, 597–601. July 2–19, 1999. Krákow, Poland.

———. 2003. "Palaeoethnobotanical Evidence of Agriculture in the Steppe and the Forest-Steppe of East Europe in the Late Neolithic and Bronze Age." In *Prehistoric Steppe Adaptation and the Horse*, edited by M. Levine, C. Renfrew, and K. Boyle, 287–97. Cambridge: McDonald Institute Monographs.

Pashkevych, G. 2012 [in press]. "Environment and Economic Activities of Neolithic and Bronze Age Populations of the Northern Pontic Area." *Quaternary International* 261: 176–82. doi:10.1016/ j.quaint.2011.01.024

Pate, D. W. 1983. "Possible Role of Ultraviolet Radiation in Evolution of *Cannabis* Chemotypes." *Economic Botany* 37 (4): 396–405.

Patil, D. A. 2000. "Sanskrit Plant Names in an Ethnobotanical Perspective." *Ethnobotany* 12:60–64.

Păun, M., E. Turenschi, L. Ifteni, V. Ciocarlan, and I. Moldovan. 1980. *Botanica*. Bucharest, Romania: Ed. Didactică și Pedagogică.

Payne, R. 1838. *Blo'Norton Tithe Map*. Norfolk, England: Norfolk Record Office.

Pearsall, D. M. 1989. *Paleoethnobotany: A Handbook of Procedures*, 2nd ed. (2000). New York: Academic.

Pearson, R. 2006. "Jomon Hot Spot: Increasing Sedentismin South-Western Japan in the Incipient Jomon (14,000–9250 cal. BC) and Earliest Jomon (9250–5300 cal. BC) Periods." *World Archeology* 38 (2): 239–58.

Peco, L. 1925. "Običaji i verovanja iz Bosne. Zivot i običaji narodni, knjiga 14" [Customs and beliefs in Bosnia. Folk life and customs, vol. 14]). [In Serbo-Croatian.] *Srpska kraljevska akademija, Srpski etnografski zbornik* 32:375.

Peglar, S. M. 1993a. "The Development of the Cultural Landscape around Diss Mere, Norfolk, UK during the Past 7000 Years." *Review of Palaeobotany and Palynology* 76 (1): 1–43.

———. 1993b. "Mid- and Late-Holocene Vegetation History of Quidenham Mere, Norfolk, UK Interpreted Using Recurrent Groups of Taxa." *Vegetation History and Archaeobotany* 2 (1): 15–28.

People's Daily Online. 2005. "2,000-Year-Old Periwig Unearthed in SW China." November 10. http://english.people.com.cn/200511/10/ eng20051110_220266.html.

Perry, L. M., and J. Metzger. 1980. *Medicinal Plants of East and Southeast Asia: Attributed Properties and Uses*. Cambridge, MA: MIT.

Petrović, P. Z. 1948. *Život i običaji narodni u Gruži* [Folk life and customs in the Gruza region]. [In Serbian.] Beograd, Serbia: Srpski etnografski zbornik, Srpska akademija nauka.

Petrović, S. 1992. *Mitologija, magija i običaji, Kulturna istorija Svrljiga I* [Mythology, magic and customs, Cultural history of Svrljig I, Southeast Serbia]. [In Serbian.] Niš, Serbia: Prosveta.

———. 2000. *Srpska mitologija: Mitologija, magija i običaji, istraživanje svrljiške oblasti, oblasti* [Serb mythology: Mythology, magic and customs, researches of Svrljig, Southeast Serbia]. [In Serbian.] Vol. 5. Niš, Serbia: Prosveta.

Pezzin, M. 1986. *Daily Life in Gaume in the 19th Century: Working with Hemp*. Translated by J. Genon. Virton, Belgium: Musée Gaumais.

Phillipson, D. W. 1965. "Early Smoking Pipes from Sebanzi, Zambia." *Arnoldia* (Rhodesia) 1, no. 40.

Pickersgill, B., and C. B. Heiser. 1976. "Cytogenics and Evolutionary Change under Domestication." *Philosophical Transactions of the Royal Society of London* B275:55–69.

Pillay, M., and S. T. Kenny. 2006. "Structural Organization of the Nuclear Ribosomal RNA Genes in *Cannabis* and *Humulus* (Cannabaceae)." *Plant Systematics and Evolution* 258 (1–2): 97–105.

Pinarkara, E., S. A. Kayis, E. E. Hakki, and A. Sag. 2009. "RAPD Analysis of Seized Marijuana (*Cannabis Sativa* L.) in Turkey." *Electronic Journal of Biotechnology* 12 (1): 1–13.

Pitkänen, A., and P. Huttunen. 1999. "A 1300-Year Forest-Fire History at a Site in Eastern Finland based on Charcoal and Pollen Records in Laminated Lake Sediment." *The Holocene* 9 (3): 311–20.

Pitkänen, A., H. Lehtonen, and P. Huttunen. 1999. "Comparison of Sedimentary Microscopic Charcoal Particle Records in a Small Lake with Dendrochronological Data: Evidence for the Local Origin of Microscopic Charcoal Produced by Forest Fires of Low Intensity in Eastern Finland." *The Holocene* 9 (5): 559–67.

Pliny (The Elder). 1950. *Natural History: With a Translation.* Vol. 5. Books 17–19. Edited by H. Rackham. London: W. Heinemann.

Pokorný, P., P. Šída, P. Kuneš, and O. Chvojka. 2008. "Mezolitické osídlení bývalého jezera Švarcenberk (jižní Čechy) v kontextu vývoje přírodního prostředí" [Mesolithic settlement of the former Lake Švarcenberk (south Bohemia) in its environmental context.]. In *Bioarcheologie v České Republice* [Bioarcheology in the Czech Republic], edited by J. Beneš and P. Pokorný, 145–76. Praha, Czech Republic: Jihočeská Univerzita & Archeologický ústav AV ČR.

Pokorný, P. 2002. "A High-Resolution Record of Late-Glacial and Early-Holocene Climatic and Environmental Change in the Czech Republic." *Quarternary International* 91:101–22.

Pollan, M. 2001. *The Botany of Desire.* New York: Random House.

Pollington, S. 2000. *Early English Charms Plantlore and Healing.* Trowbridge, England: Redwood Books.

Polunin, N. 1960. *Introduction to Plant Geography.* New York: McGraw-Hill.

Popović, C. 1953. "Bosansko-hercegovacke preslice i vretena" [Distaffs and spindles in Bosnia and Herzegovina]. [In Serbo-Croatian.] *Glasnik Zemaljskog muzeja u Sarajevu, Etnologija, Nova serija* 8:182–85.

Popper, V., and C. Hastorf. 1988. "Introduction." In *Current Paleo-ethnobotany: Analytical Methods and Cultural Interpretations of Archeological Plant Remains,* edited by C. Hastorf and V. Popper, 1–16. Chicago: University of Chicago.

Porcher, F. P. 1863. *Resources of the Southern Fields and Forests. Medical, Economical and Agricultural: Being also a Medical Botany of the Southern States.* Charleston, NC: Walker, Evans & Cogswell.

Portal, J. 2000. *Korea—Art and Archeology.* London: The British Museum.

Poska, A., and L. Saarse. 2006. "New Evidence of Possible Crop Introduction to North-Eastern Europe during the Stone Age: Cerealia Pollen Finds in Connection with the Akali Neolithic Settlement, East Estonia." *Vegetation History and Archaeobotany* 15 (3): 169–79.

Poska, A., L. Saarse, and S. Veski. 2004. "Reflections of Pre- and Early-Agrarian Human Impact in the Pollen Diagrams of Estonia." *Palaeogeography, Palaeoclimatology, Palaeoecology* 209:37–50.

Pouncy, C. J., ed. 1994. *The Domostroi: Rules for Russian Households in the Time of Ivan the Terrible.* Ithaca, NY: Cornell University.

Poznanović, R. 1988. *Tradicionalno usmeno narodno stvaralaštvo užičkog kraja* [Traditional oral folklore in Uzice area, northwestern Serbia]. [In Serbian.] Beograd, Serbia: Etnografski institut Srpske akademije nauka i umetnost.

Prabhavanda, S. 1963. *The Spiritual Heritage of India.* Garden City, NY: Doubleday.

Prain, M. D. 1904. "On the Morphology, Teratology and Diclinism of the Flowers of *Cannabis.*" *Memoirs by Officers of the Medical and Sanitary Departments of Government of India* 12:51–92.

Prakash, O. 1961. *Food and Drinks in Ancient India.* Delhi: Munshi Ram Manohar Lal.

Pratt, C. 2007. *The Encyclopedia of Shamanism.* New York: Rosen.

Price, T. D., R. A. Bentley, J. Lüning, D. Gronenborn, and J. Wahl. 2001. "Prehistoric Human Migration in the Linearbandkeramik of Central Europe." *Antiquity* 75:593–603.

Pringle, H. 1997. "Ice Age Communities May Be Earliest Known Net Hunters." *Science* 277:1203–4.

Prins, A. H. J. 1975. "Development in Arctic Boat Design: Efflorescence or Involution?" In *Netherlands-Swedish Symposiumon Developments in Scandinavian Arctic Culture, February 1974,* 12–30. Groningen, The Netherlands: University of Groningen.

Prioreschi, P., and D. Babin. 1993. "Ancient Use of Cannabis." *Nature* 364 (August 19): 680.

Prvulović, B. 1982. "Običaj i reč: prilozi proučavanju narodnih običaja i tekstova uz narodne običaje u istočnoj Srbiji." [In Serbian.] *Biblioteka zbornika Filozofskog Fakulteta u Nisu* 102:19–20.

Punt, W., and M. Malotaux. 1984. "Cannabaceae, Moraceae and Urticaceae." *Review of Paleobotany and Palynology* 42 (1–4): 23–44.

Pursehouse, E. 1961. "Hemp: A Forgotten Norfolk Crop." *Eastern Daily Press,* February 25.

Querino, M. n.d. "A raca africana e seus costumes no Baia." *Revista da Academia de Letras,* no. 70.

———. 1938. *Costumes Africanos no Brasil.* Rio de Janeiro: Civilização Brasileira.

Quintana-Murci, L., O. Semino, H. J. Bandelt, G. Passarino, K. McElreavey, and A. S. Santachiara-Benerecetti. 1999. "Genetic Evidence of an Early Exit of *Homo Sapiens Sapiens* from Africa through Eastern Africa." *Nat Genet* 23:437–41.

Quintana-Murci, L., R. Chaix, R. S. Wells, D. M. Behar, H. Sayar, R. Scozzari, C. Rengo, N. Al-Zahery, O. Semino, A. S. Santachiara-Benerecetti, A. Coppa, Q. Ayub, A. Mohyuddin, C. Tyler-Smith, S. Qasim Mehdi, A. Torroni, and K. McElreavey. 2004. "Where West Meets East: The Complex mtDNA Landscape of the Southwest and Central Asian Corridor." *American Journal of Human Genetics* 74:827–45.

Radauš-Ribarić, J. 1988. "O tekstilnom rukorvorstvu na tlu Jugoslavije kroz vjekove" [On textile handicraft in Yugoslavia through the ages]. [In Croatian.] In *Carolija niti,* edited by J. Radauš-Ribarić and R. Rihtman-Auguštin, 13–24. Zagreb, Croatia: Muzejski proctor.

Radenković, L. 1981. "O značenju jednog sakralnog teksta o konoplji ili lanu kod slovenskih i balkanskih naroda." [In Serbian.] *Naučni sastanak slavista u Vukove dane* 11 (2): 207–17.

———. 1996. *Narodna bajanja kod Južnih Slovena* [The South Slavic Charms]. [In Serbian.] Beograd, Serbia: Prosveta, Balkanološki Institute SANU.

Rafel i Fontanals, N., and M. Blasco i Arasanz. 1995. "El taller tèxtil del Coll del Moro de Gandesa (Terra Alta)." *Tribuna d'arqueologia* 1993/1994:37–50.

Raičević, S. 1935. "Običaji o Martincima u Južnoj Srbiji" [Folk customs and rituals connected with Martins days in South Serbia]. [In Serbian.] *Glasnik Etnografskog muzeja u Beogradu* 10:54–61.

Ramqvist, P. H. 1998. *Arnäsbacken. En gård från yngre järnålder och medeltid* [Arnäsbacken. A farm from the later Iron Age and medieval period]. [In Swedish.] Umeå, Sweden: HB Prehistorica.

Rasmussen, P. 2005. "Mid- to Late-Holocene Land-Use Change and Lake Development at Dallund Sø, Denmark: Vegetation and Land-Use History Inferred from Pollen Data." *The Holocene* 15 (8): 1116–29.

Rasmussen, P., and N. J. Anderson. 2005. "Natural and Anthropogenic Forcing of Aquatic Macrophyte Development in a Shallow Danish Lake during the Last 7000 Years." *Journal of Biogeography,* 32 (11): 1993.

Rathburn, M. Y. 1993. *Beyond the Tanabata Bridge, Traditional Japanese Textiles: Guide for a Seattle Asian Arts Museum Exhibit on Japanese Textiles.* Seattle: Seattle Art Museum, Seattle Asian Arts.

Rätsch, C. 2001. *Marijuana Medicine: A World Tour of the Healing and Visionary Powers of Cannabis.* Rochester, VT: Healing Arts.

Ravazzi, C., M. Marchetti, M. Zanon, R. Perego, T. Quirino, M. Deaddis, M. De Amicis, and D. Margaritora. In Press. "Lake Evolution and Landscape History in the Lower Mincio River Valley, Unravelling Drainage Changes in the Central Po Plain (N-Italy) since the Bronze Age." *Quaternary International forthcoming:* 1–11. doi:10.1016/j.quaint.2011.11.031.

Raven, P. H., R. F. Evert, and S. E. Eichhorn. 2005. *Biology of Plants,* 7th ed. New York: W. H. Freeman & Co.

Rawski, E. S. 1988. "A Historian's Approach to Chinese Death Ritual." In *Death Ritual in Late Imperial and Modern China,* edited by J. L. Watson and E. S. Rawski, 20–34. Berkeley, CA: University of California.

Ray, J. C. 1939. "Soma Plant." *Indian Historical Quarterly* 15 (2): 197–207.

Read, B. E. 1982. *Chinese Medicinal Plants from the Pen Ts'so Kang Mu of a Botanical, Chemical and Pharmacological Reference List. Chinese Medicine Series* 5. Taipei, China: Southern Materials Center Inc.

Reddé, M., H. U. Nuber, S. Jacomet, J. Schibler, C. Schucany, P.-A. Schwarz, and G. Seitz. 2005. "Oedenburg. Une agglomération d'époque romaine sur le Rhin Supérieur. Fouilles francaises, allemandes et suisses sur les communes de Biesheim et Kunheim (Haut-Rhin)." *Gallia* 62:215–77.

Ree, J. H. 1966. "Hemp Growing in the Republic of Korea." *Economic Botany* 20 (2): 176–86.

Regnéll, J. 1989. "Vegetation and Land Use during 6000 Years. Paleo-ecology of the Cultural Landscape at Two Lake Sites in Southern Skåne, Sweden." LUNDQUA Thesis 27.

Reich, D., N. Patterson, M. Kircher, F. Delfin, M. R. Nandineni, I. Pugach, A. M. Ko, Y.-C. Ko, T. A. Jinam, M. E. Phipps, N. Saitou, A. Wollstein, M. Kayser, S. Pääbo, and M. Stoneking. "2011 Denisova Admixture and the First Modern Human Dispersals into Southeast Asia and Oceania." *American Journal of Human Genetics* 89 (4): 1–13.

Reid, W. 1976. *The Lore of Arms*. London: Mitchell Beazley.

Reilly, T. M. 1987. "The Miao of Southwest China and Beyond." In *Richly Woven Traditions: Costumes of the Miao of Southwest China and Beyond*, edited by China Institute in America. New York: China Institute in America, 19–30.

Reininger, W. 1941. Zur Geschichte des Haschischgenusses. *Ciba Zeitschrift* 7 (80): 2765–88.

———. 1967. "Remnants from Prehistoric Times." In *The Book of Grass*, edited by G. Andrews and J. Vinkenoog, 14. New York: Grove.

Reljić, L. 1989. "Samrtni običaji i kult pokojnika—Sjenicko-pesterska visoravan" [Funeral customs and mortuary cult on Sjenicko-Pesterska plateau, southwest Serbia]. [In Serbian.] *Glasnik Etnografskog muzeja u Beogradu* 52–53:129.

Ren, G. 2007. "Changes in Forest Cover in China during the Holocene." *Vegetation History and Archaeobotany* 16 (2–3): 19–126.

Ren, G., and H. J. Beug. 2002. "Mapping Holocene Pollen Data and Vegetation of China." *Quaternary Science Reviews* 21 (12–13): 1395–1422.

Renfrew, C., P. Forster, and M. Hurles. 2000. "The Past within Us." *Nature Genetics* 26:253–54.

Renfrew, J. 1973. *Palaeoethnobotany: The Prehistoric Food Plants of the Near East and Europe*. New York: Columbia University.

"Report on the First (1956) and Second (1958) Excavation of the Chhien-shan-yang Remains at Wu-hsing, Chekiang Province." 1960. *Khao Ku Hsüeh Pao* (Acta Archeologia Sinica) 2:72–93.

Rexen, F., and U. Blicher-Mathiesen. 1998. *Report from the State of Denmark: Forming Part of the IENICA Project, the Interactive European Network for Industrial Crops and their Applications*. Copenhagen: Ministry of Food, Agriculture and Fisheries, Danish Directorate for Development, Non-food Secretariat, Denmark.

Reynolds, J. R., ed. 1879. *A System of Medicine*. London: Macmillan.

Richards, M. P., P. B. Pettitt, M. C. Stiner, and E. Trinkaus. 2001. "Stable Isotope Evidence from Increasing Dietary Breadth in the European Mid-Upper Paleolithic." *Proceedings of the National Academy of Sciences of the USA* 98 (11): 6528–32.

Richardson, J., and A. Woods. 1976. *Sinsemilla Marijuana Flowers*. Berkeley, CA: And/Or.

Richardson, T. H. 2010. "Cannabis Use and Mental Health: A Review of Recent Epidemiological Research." *International Journal of Pharmacology* 6:796–807.

Ridley, H. N. 1930. *The Dispersal of Plants throughout the World*. Ashford, KY: L. Reeve.

Riehl, S., and K. Pustovoytov. 2006. "Comment on van Geel et al., *Journal of Archeological Science* 31 (2004)." *Journal of Archeological Science* 33:143–44.

Riera, S., G. Wansard, and R. Julià. 2004. "2000-Year Environmental History of a Karstic Lake in the Mediterranean Pre-Pyrenees: the Estanya Lakes (Spain)." *Catena* 55:293–324.

Riera, S., J. A. López-Sáez, and R. Julià. 2006. "Lake Responses to Historical Land Use Changes in Northern Spain: The Contribution of Non-Pollen Palynomorphs in a Multiproxy Study." *Review of Paleobotany and Palynology* 141:127–37.

Rieseberg, L. H., and J. H. Willis. 2007. Plant Speciation. *Science* 317 (5840): 910–14.

Rieseberg, L. H., O. Raymond, D. M. Rosenthal, Z. Lai, K. Livingstone, T. Nakazato, J. L. Durphy, A. E. Schwarzbach, L. A. Donovan, and C. Lexer. 2003. "Major Ecological Transitions in Wild Sunflowers Facilitated by Hybridization." *Science* 301:1211–16.

Rimantiene, R. 1992a. "The Neolithic of the Eastern Baltic." *Journal of World Prehistory* 6 (1): 97–143.

———. 1992b. "Neolithic Hunter-Gatherers at Šventoji in Lithuania." *Antiquity* 66 (251): 367–76.

———. 1994. "Die Steinzeil in Litauen." *Berlicht der romischengermanischen Kommission* 75:23–146.

———. (1984) 1996. *Akmens amžaius Lietuvoje* [Stone Age in Lithuania]. 2nd ed. Vilnius, Lithuania: Mintis.

Rimantiene, R., U. Miller, T. Hackens, V. Lang, A. Raukas, and S. Hicks. 1999. "Traces of Agricultural Activity in the Stone Age Settlements of Lithuania." *PACT* 57:275–90.

Ripinsky-Naxon, M. 1993. *The Nature of Shamanism: Substance and Function of a Religious Metaphor*. Albany, NY: State University of New York.

Risberg, J., S. Karlsson, A.-M. Hansson, Λ. Hedenstrofm, J. Heimdahl, U. Miller, and C. Tingvall. 2002. "Environmental Changes and Human Impact as Recorded in a Sediment Sequence Offshore from a Viking Age Town, Birka, Southeastern Sweden." *The Holocene* 12 (4): 445–58.

Roberts, N. 1998. *The Holocene: An Environmental History*, 2nd revised edition. Oxford: Blackwell.

Robinson, D. E., and S. Karg. 2002. "Arkæobotaniske analyser, Nationalmuseet 2001." [In Danish.] In *Arkæologiske udgravninger I Danmark 2001. Archeological excavations in Denmark 2001*, 323–32. Copenhagen: Det Arkæologiske Naevn.

Robinson, D. E., J. A. Harild, and L. H. Pedersen. 2001. *Arkæobotaniske analyser af materiale fra to brønde ved Kragehavegård, Høje Taastrup*. [In Danish.] NNU Rapport 10. Copenhagen: Nationalmuseets Naturvidenskabelige Undersøgelser.

Robinson, R. 1996. *The Great Book of Hemp*. Rochester, VT: Park Street.

Robinson, R., and R. Nelson. 1995. *The Great Book of Hemp: The Complete Guide to the Commercial, Medicinal and Psychotropic Uses of the World's Most Extraordinary Plant*. Rochester, VT: Park Street.

Rogers, P. W., L. B. Jorgensen, and A. Rast-Eicher. 2001. *The Roman Textile Industry and Its Influence: A Birthday Tribute to Pohn Peter Wild*. Oxford: Oxbow Books.

Ronan, C. A. 1978. *The Shorter Science and Civilization in China: An Abridgement of Joseph Needhams's Original Text*. Vol. 3. Cambridge: Cambridge University.

Ronikier, M., and Intrabiodiv Consortium. 2008. "Phylogeographical Structure of Alpine Plants in the Carpathians: A Comparative Study." In *Book of Abstracts*, vol. 4, from Xth Symposium of the International Organization of Plant Biosystematists, 2–4 July, 2008, Vysoké Tatry, Slovakia.

Rosado, P. 1958. *O cicio da Liamba no Estudo do Pará-Una Toxicose que ressurge entre nós*. Cited in Hutchinson, 1975.

Rösch, M. 1998a. "The History of Crops and Crop Weeds in South-Western Germany from the Neolithic Period to Modern Times, as Shown by Archaeobotanical Evidence." *Vegetation History and Archaeobotany* 9 (2): 109–25.

———. 1998b. "Sieben Jahrtausende Ackerbau im Strohgäu- Bestand und Wandel anhand botanischer Untersuchungen." In *Dorfsterben . . . Vöhingen und was davon blieb. Archäologie eines mittelalterlichen Dorfes bei Schwieberdingen. Beigleith Ausst Landesdenkmalamt Baden-Württemberg Rathaus Schwieberdingen* [The dying village of Vöhingen and what remains of it. Archeology of a medieval village in Schwieberdingen. Accompanying the exhibition of Landesdenkmalamt Baden-Württemberg in the town hall Schwieberdingen]. 4:61–72. Stuttgart, Germany: Ges. für Vor- und Frühgeschichte in Württemberg und Hohenzollern.

———. 1998c. "Pflanzenreste als historische Quellen spätmittelalterlicher Alltagskultur-Neue Untersuchungen in Südwestdeutschland." [Plant remains as historical sources of late medieval popular culture—New studies in southwestern Germany.] In *Haus und Kultur im Spätmittelalter, Quellen und Materialien zur Hausforschung in Bayern 10/Schriften und Kataloge des Fränkischen Freilandmuseums, Bad Windsheim* 30, 1998, edited by K. Bedal, S. Fencer, and H. Heidrich, 56–74.

———. 1998d. "Die Pflanzenreste aus dem Haus Hochbrücktorstr. 27." [The plant remains from the house Hochbrücktorstr. 27]. In *Landesdenkmalamt Baden-Württemberg/Stadtarchiv Rottweil (Hg.), " . . . von anfang biss zu unsern zeiten. Das mittelalterliche Rottweil im Spiegel archäologischer Quellen . . . ,"* edited by Stadtarchiv Rottweil [Rottweil City Archives] *Archäologische Informationen aus Baden-Württemberg* 38: 113–19. Stuttgart, Germany.——— . 2007. "Waterlogged Material from Migration Time Sites in Southwestern Germany." In *Program and Abstracts from 14th Symposium of the International Work Group for Palaeoethnobotany*, June 13–23, Krakow, Poland.

Rösch, M., E. Fischer, and T. Märkle. 2005. "Human Diet and Land Use in the Time of the Khans: Archaeobotanical Research in the Capital of the Mongolian Empire, Qara Qorum, Mongolia." *Vegetation History and Archaeobotany* 14 (4): 485–92.

Rosen, A. 2000. "Phytolith Report for Results of Phytolith Analyses from Tseganka 8 (1998–1999)." In *Kazakh American Talgar Project, Tseganka 8 2000: Preliminary Report*. http://web.archive .org/web/20070609030403/http://www.faculty.sbc.edu/cchang/ Kaz%20web/TS8Phyto.htm.

———. 2001. "Phytolith Evidence for Agro-Pastoral Economies in the Scythian Period of Southern Kazakhstan." In *The Phytoliths: Applications in Earth Science and Human History*, edited by J. D. Meunierand and F. Colin. Aix en Provence, France: CEREGE.

Rosen, A., C. Chang, and F. Grigoriev. 2000. "Palaeoenvironments and Economy of Iron Age Saka-Wusun Agro-Pastoralists in South-eastern Kazakhstan." *Antiquity* 74 (285): 611–23.

Rosenthal, E. 2001. *The Big Book of Buds: Marijuana Varieties from the World's Great Seed Breeders*. Berkeley, CA: Quick Trading Company.

Rosenthal, F. 1971. *The Herb: Hashish versus Medieval Muslim Society*. Leiden, The Netherlands: E. J. Brill.

Royle, F. J. 1855. *Fibrous Plants of India Fitted for Cordage, Clothing and Paper with an Account of the Cultivation and Preparation of Flax, Hemp and their Substitutes*. London: Smith, Elder.

Ruas, M.-P. 1988. "Agriculture." In *Un village au temps de Charlemagne*, 203–13. Paris: Éditions de musées nationaux.

———. 1998. "Les plantes consommées au Moyen Âge en France méridionale d'après les semences archéologiques." *Archéologie Midi Médiéval* 15–16:179–204.

———. 2000. "Productions agricoles en Auvergne carolingienne d'après un dépotoir découvert à Saint-Germain-des-Fossés (Allier)." *Revue archéologique* du *Centre* de la *France* 39:137–60.

Rubim de Pinho, A. 1975. "Social and Medical Aspects of the Use of Cannabis in Brazil." In *Cannabis and Culture*, edited by Vera Rubin, 293–302. Paris: Mouton.

Rubin, V., ed. 1975. *Cannabis and Culture*. Paris: Mouton.

Rubin, V., and L. Comitas. 1975. *Ganja in Jamaica: A Medical Anthropological Study of Chronic Marijuana Use*. The Hague, The Netherlands: New Babylon Studies in the Social Sciences.

Rudenko, S. I. 1970. *Frozen Tombs of Siberia—the Pazyryk burials of Iron Age horsemen*. Los Angeles: University of California.

Rudgley, R. 1995. *Essential Substances: A Cultural History of Intoxicants in Society*. New York: Kodansha International.

———. 1998. *The Encyclopedia of Psychoactive Substances*. Boston: Little, Brown and Company.

———. 1999. *The Lost Civilizations of the Stone Age*. New York: Touchstone.

Rudolph, R. C. 1978. *Chinese Archeological Abstracts* 6:252. Los Angeles: University of California, Institute of Archeology.

Rui, M. 1983. "The Woollen Industry in Catalonia in the Later Middle Ages." In *Cloth and Clothing in Medieval Europe*, edited by N. B. Harte and K. G. Ponting, 205–29. London: Heinemann.

Rumpf, G. E., and E. M. Beekman. 1981. *The Poison Tree: Selected Writings of Rumphius on the Natural History of the Indies, Library of the Indies*. Amherst, MA: University of Massachusetts.

———. 1999. *The Ambonese Curiosity Cabinet—Georgius Everhardus Rumphius*. New Haven, CT: Yale University.

Russian Nature, ed. 2010. "Biomes and Regions of Northern Eurasia: The Caucasus." http://www.rusnature.info/reg/15_5.htm.

Russell, P. J., P. E. Hertz, and B. McMillan. 2011. *Biology: The Dynamic Science*. 2nd ed. Pacific Grove, CA: Brooks Cole.

Russo, E. B. 1998. "*Cannabis* for Migraine Treatment: The Once and Future Prescription? An Historical and Scientific Review." *Pain* 76:3–8.

———. 2001. "Hemp for Headache: An In-Depth Historical and Scientific Review of Cannabis in Migraine Treatment." *Journal of Cannabis Therapeutics* 1 (2): 21–92.

———. 2002a. "Cannabis Treatments in Obstetrics and Gynecology: A Historical Review. Haworth." *Journal of Cannabis Therapeutics* 2 (3–4): 5–35.

———. 2002b. "The Role of Cannabis and Cannabinoids in Pain Management." In *Pain Management: A Practical Guide for Clinicians*, edited by R. S. Weiner, 357–75. Boca Raton, FL: CRC.

———. 2004. "History of Cannabis as a Medicine." In *The Medicinal Uses of Cannabis and Cannabinoids*, edited by G. W. Guy, B. A. Whittle, and P. J. Robson. London: Pharmaceutical, 1–16.

———. 2005. "*Cannabis* in India: Indian Lore and Modern Medicine." In *Cannabinoids as Therapeutics: Milestones in Drug Therapy*, edited by R. Mechoulam, 1–22. Basel, Switzerland: Birkäuser Verlag.

———. 2006. "Cannabis Treatments in Obstetrics and Gynecology: A Historical Review." In *Handbook of Cannabis Therapeutics: From Bench to Bedside*, edited by E. Russo and F. Grotenhermen, 5–34. Binghamton, NY: Haworth.

———. 2007. "History of Cannabis and Its Preparations in Saga, Science, and Sobriquet." *Chemistry & Biodiversity* 4:614–48.

———. 2011. "Taming THC: Potential Cannabis Synergy and Phytocannabinoid-Terpenoid Entourage Effects." *British Journal of Pharmacology* 63 (7): 1344–64.

Russo, E. B., A. Merzouki, J. M. Mesa, K. A. Frey, and P. J. Bach. 2004. "*Cannabis* Improves Night Vision: A Case Study of Dark Adaptometry and Scotopic Sensitivity in Kif Smokers of the Rif Mountains of Northern Morocco." *Journal of Ethnopharmacology* 93 (1): 99–104.

Russo, E. B., and F. Grotenhermen, eds. 2006. *Handbook of Cannabis Therapeutics: From Bench to Bedside*. Binghamton, NY: Haworth.

Russo, E. B., H. Jiang, X. Li, A. Sutton, A. Carboni, F. Del Bianco, G. Mandolino, D. J. Potter, Y. Zhao, S. Bera, Y. Zhang, E. Lu, D. K. Ferguson, F. Hueber, L. Zhao, C. Liu, Y. Wang, and C. Li. 2008. "Phytochemical and Genetic Analyses of Ancient Cannabis from Central Asia." *Journal of Experimental Botany*, 59 (15): 4171–82.

Russo, E. B., M. Dreher, and M. L. Mathre, eds. 2003. *Women and Cannabis: Medicine, Science, and Sociology*. Binghamton, NY: Haworth.

Rybníček, K., J. Dickson, and E. Rybníčková. 1998. "Flora and Vegetation at about A. D. 1100 in the Vicinity of Brno, Czech Republic." *Vegetation History and Archaeobotany* 7:155–65.

Ryder, M. L. 1993. "Probable Hemp Fibre in Bronze Age Scotland." *Archeological Textiles Newsletter* 17:10–13.

———. 1999. "Probable Fibres from Hemp (*Cannabis sativa* L.) in Bronze Age Scotland." *Environmental Archeology*, 4:93–95.

Sagan, C. 1977. *The Dragons of Eden*. New York: Random House.

Saini, D. C., K. Kulshreshtha, S. Kumar, D. K. Gond, and G. K. Mishra. 2011. "Conserving Biodiversity based on Cultural and Religious Values." In *National Conference on Forest Biodiversity: Earth's Living Treasure*, edited by the Uttar Pradesh State Biodiversity Board, 145–52. Montreal, QC: Secretariat of the Convention on Biological Diversity. http://www.upsbdb.org/pdf/ Souvenir2011/22.pdf.

Salmon, L. 1902. "Folklore in the Kennet Valley." *Folklore* 13:418–29.

Salzberger, G. 1912. *Salomons Tempelbau und Thron*. Berlin: Mayer and Muller.

Santos, F. R., A. Pandya, C. Tyler-Smith, S. D. J. Pena, M. Schanfield, W. R. Leonard, L. Osipova, M. H. Crawford, and R. J. Mitchell. 1999. "Genetic Relationships of Asians and Northern Europeans, Revealed by Y-Chromosomal DNA Analysis." *American Journal of Human Genetics* 64:619–28.

Sanz, V. 1995. *D'artesans a Proletaris: La Manufactura del Cànem a Castelló, 1732–1843*. Castelló, Spain: Servei de Publicacions, Diputació de Castelló.

Saraswat, K. S. 2004. "Plant Economy of Early Farming Communities at Senuwar." In *Early Farming Communities of the Kaimur (Senuwar Excavations)*, vol. 2, edited by B. P. Singh, 416–535. Jaipur, India: Scheme.

Saraswat, K. S., and S. Chanchala. 1995. "Palaeobotanical and Pollen Analytical Investigations." In *Indian Archeology 1990–1991—A Review*, edited by the Archeological Survey of India, 103–4. New Delhi: Government of India.

Sarin, Y. K. 1990. "Some Less Known but Effective Folk Remedies in North-West Himalayan Region." *Ethnobotany* 2:39–43.

Sarianidi, V. 1994. "Temples of Bronze Age Margiana: Traditions of Ritual Architecture." *Antiquity* 8:88–397.

———. 1998. *Margiana and Protozoroastrism*. Athens: Kapon Editions.

———. 2003. "Margiana and Some-Hoama." *Electronic Journal of Vedic Studies* 9, iss. 1d (May 5). http://www.ejvs.laurasianacademy.com/ ejvs0901/ejvs0901d.txt.

Sastri, B. N., ed. 1950. *The Wealth of India: A Dictionary of Indian Raw Materials and Industrial Products*, vol. 2. Delhi: Council of Scientific and Industrial Research.

Sauer, C. O. 1952. *Agricultural Origins and Dispersals*. New York: American Geographical Society.

———. 1958. "Man in the Ecology of Tropical America." *Proceeding of the Ninth Pacific Science Congress, 1957*, 20:104–10.

———. 1967. "Environment and Culture During the Last Deglaciation." In *Land and Life*, edited by J. Leighly, 246–70. Berkeley, CA: University of California.

———. 1969. *Agricultural Origins and Dispersals: The Domestication of Animals and Foodstuffs*. Cambridge, MA: MIT.

Saul, N. E. 1969. "The Beginnings of the American-Russian Trade," 1763-1766. *The William and Mary Quarterly* 26 (4): 596–600.

Saxena, A., V. Prasad, I. B. Singh, M. S. Chauhan, and R. Hasa. 2006. "On the Holocene Record of Phytoliths of Wild and Cultivated Rice from Ganga Plain: Evidence for Rice-Based Agriculture." *Current Science* 90 (11): 1547–51.

Schaefer, F. 1945. "Hemp." *CIBA Review* 49: 1779–94.

Schafer, E. H. 1963. *The Golden Peaches of Samarkand; a Study of T`ang Exotics*. Berkeley, CA: University of California.

Schafer, R. J. 1958. *The Economic Societies in the Spanish World*. Syracuse, NY: Syracuse University.

Schilling, M. 1994. *Sumo a Fan's Guide*. Tokyo: The Japan Times.

Schleiffer, H., ed. 1979. *Narcotic Plants of the Old World—Used in Rituals and Everyday Life*. Monticello, NY: Lubrecht & Cramer.

Schlumbaum, A., M. Tensen, and V. Jaenicke-Després. 2008. "Ancient Plant DNA in Archaeobotany." *Vegetation History and Archaeobotany* 17:233–44.

Schmidt-Glintzer, H., trans. 1975. *Mo Ti. Solidarität und allgemeine Menschenliebe. Gegen den Krieg*. Vol. 2. Düsseldorf, Germany: Eugen Diederichs Verlag.

Schofield, J. E., and M. P. Waller. 2005. "A Pollen Analytical Record for Hemp Retting from Dungeness Foreland, UK." *Journal of Archeological Science* 32 (5): 715–26.

Schubert, G. 1984. "Konac, vrpce i tkanina kao magijska sredstva narodne medicine u Jugoistočnoj Evropi" [Thread, lace, and fabric tools of traditional folk medicine in the East Europe]. [In Serbian.] *Makedonski folklor* 17 (33): 135–51.

Schultes, R. E. 1969a. "Hallucinogens of Plant Origin." *Science* 163 (3864): 245–54.

———. 1969b. "The Plant Kingdom and Hallucinogens (Part I)." *Bulletin on Narcotics* 21 (3): 3–16.

———. 1970. "Random Thoughts and Queries on the Botany of Cannabis." In *The Botany and Chemistry of Cannabis*, edited by C. R. Joyce and S. H. Curry, 11–38. London: J. and A. Churchill.

Schultes, R. E., and A. Hofmann. (1979) 1992. *Plants of the Gods: Their Sacred, Healing and Hallucinogenic Powers*. Rochester, VT: Healing Arts.

Schultes, R. E., A. Hofmann, and C. Rätch. 2001. *Plants of the Gods: Their Sacred, Healing and Hallucinogenic Powers*, 2nd ed. Rochester, VT: Healing Arts.

Schultes, R. E., W. M. Klein, T. Plowman, and T. E. Lockwood. 1974. "*Cannabis*: An Example of Taxonomic Neglect." *Botanical Museum Leaflets—Harvard University* 23 (9): 337–64. Reprinted 1975, Paris, by Mouton, in *Cannabis and Culture*, edited by V. Rubin, 21–38.

Schwab, U. S., J. C. Callaway, A. T. Erkkilä, J. Gynther, M. I. J. Uusitupa, and T. Järvinen. 2006. "Effects of Hempseed and Flaxseed Oils on the Profile of Serum Lipids, Serum Total and Lipoprotein Lipid Concentrations and Haemostatic Factors." *European Journal of Nutrition* 45 (8): 470–77.

Schweinitz, L. D. de. 1836. "Remarks on the Plants of Europe which have become Naturalized in a More or Less Degree, in the United States." *Annals of the Lyceum of Natural History of New York* 3:148–55.

Schworer-Kohl, G. 1984. "Sprachgebundene Mundorgelmusik zum Totenritual bei den Hmong in Nord Thailand und Laos" [Language of the Mouth Organ Music during the Funeral Rituals of the Hmong in Northern Thailand and Laos]. In *Bericht Über den Internationalen Musikwissenschaftlichen Kongress, Bayreuth 1981* [Report on the International Musicological Congress, Bayreuth 1981], edited by Christoph-Hellmut Mahling and Sigrid Weismann, 609–17. Kassel and Basel: Bärenreiter. Sciolino, E. 2007. "Following in Lafayette's Footstep, or Rather, his Wake." *New York Times*, August 1, Page A4.

Scott, E. M., A. Alekseev, and G. Zaitseva, eds. 2004. *Impact of the Environment on Human Migration in Eurasia: Proceeding of the Nato Advanced Research Workshop, Held in St. Petersburg, 15–18 November 2003*. Nato Science Series, Earth and Environmnetal Sciences vol. 42. Dordrecht, The Netherlands: Kluwer Academic.

Seaman, G. 1994. "Central Asian Origins in Chinese Shamanism." In *Ancient Traditions—Shamanism in Central Asia and the Americas*, edited by G. Seaman and J. S. Day, 227–43. Boulder, CO: University of Colorado.

Seefelder, S., H. Ehrmaier, G. Schweizer, and E. Seigner. 2000. "Genetic Diversity and Phylogenetic Relationships among Accessions of Hop *Humulus Lupulus* as Determined by Amplified Fragment Length Polymorphism Fingerprinting Compared with Pedigree Data." *Plant Breeding* 119:257–63.

Sélincourt, A. de, trans. 1965. *Herodotus: The Histories*. Harmondsworth, England: Penguin Books Ltd. Reprint of 1954 original.

Sengbusch, R. von. 1956. "Le chanvre 'Fibrimon' et 'Fibridla' Hanf und Lein." In *Proceedings of the Second International Flax and Hemp Congress*, June 5–9. Wageningen, Holland: Berichte des Instituts für Bastfaserforschung.

Seoul Metropolitan Government. 2006. *Attractions & Cultural Heritages: Chaehwachilji*. http://web.archive.org/web/20041011060151/http://www.visitseoul.net/english_new/disappear_culture/chewha.htm.

Serebriakova, T. I. 1940. "Fiber Plants Volume 5, Part 1." In *Flora of Cultivated Plants*, edited by E. V. Wulff. Moscow: State Printing Office [in Russian].

Sethi, V. K., M. P. Jain, and R. S. Thakur. 1977. "Chemical Investigation of Wild *Cannabis Sativa* L. Roots." *Planta Medica* 32 (4): 378–79.

Shah, N. C. 1997. "Ethnobotany of *Cannabis Sativa* in Kumaon Region, India." *Ethnobotany* 9:117–21.

———. 2002. "Some Experience in the Field of Ethnobotany." *Ethnobotany* 14:63–72.

———. 2003. "Indigenous Uses and Ethnobotany of *Cannabis Sativa* L. (Hemp) in Uttaranchal (India)." *Journal of Industrial Hemp* 9 (1): 69–77.

Shamir, R., and D. Hacker. 2001. "Colonialism's Civilizing Mission: The Case of the Indian Hemp Drug Commission." *Law and Social Inquiry* 26 (2): 435–61.

Shao, H., and R. C. Clarke. 1996. "Taxonomic Studies of *Cannabis* in China." *Journal of the International Hemp Association* 3 (2): 55–60.

Sharma, G. K. 1977a. "*Cannabis* Folklore in the Himalayas." *Harvard Botanical Museum Leaflets* 25 (7): 203–15.

———. 1977b. "Ethnobotany and Its Significance for Cannabis Studies in the Himalayas." *Journal of Psychedelic Drugs* 9 (4): 337–39.

———. 1979. "Significance of Eco-Chemical Studies of *Cannabis*." *Science and Culture* 45 (8): 303–7.

———. 1980. "A Botanical Survey of *Cannabis* in the Himalayas." *Journal of the Bombay Natural Historical Society* 76:17–20.

Sharma, H., and C. Van Sumere. 1992a. "Enzyme Treatment of Flax." *Genetics, Engineering and Biotechnology* 12:19–23.

———, eds. 1992b. *The Biology and Processing of Flax*. Belfast, Northern Ireland: M Publications.

Shen, J., R. T. Jones, X. Yang, J. A. Dearing, and S. Wang. 2006. "The Holocene Vegetation History of Erhai Lake, Yunnan Province Southwestern China: The Role of Climate and Human Forcings." *The Holocene* 16 (2): 265–76.

Shen, L., X. Y. Chen, X. Zhang, Y. Y. Li, C. X. Fu, and Y. X. Qiu. 2005 "Genetic Variation of *Ginkgo Biloba* L. (Ginkgoaceae) based on cpDNA PCR-RFLPs: Inference of Glacial Refugia." *Heredity* 94:396–401.

Sherratt, A. G. 1981. "Plough and Pastoralism: Aspects of the Secondary Products Revolution." In *Pattern of the Past: Studies in Honour of David Clarke*, edited by I. Hodder, G. Isaac, and N. Hammond, 261–306. Cambridge: Cambridge University.

———. 1983. "The Secondary Exploitation of Animals in the Old World." *World Archeology* 15:90–104.

———. 1987. "Cups that Cheered: The Introduction of Alcohol to Prehistoric Europe." In *Bell Beakers of the Western Mediterranean*, edited by W. H. Waldren and R. C. Kennard, British Archaeogical Reports International Series 331, vol. 1:81–114. Oxford: British Archeological Reports.

———. 1991. "Sacred and Profane Substances: The Ritual Use of Narcotics in Later Neolithic Europe." In *Sacred and Profane: Proceedings of a Conference on Archeology, Ritual and Religion*, Oxford University Committee for Archeology Monographs 32:50–64, edited by P. Garwood, D. Jennings, R. Skeates, and J. Toms. Reprinted 1997, Princeton, NJ, by Princeton University, in *Economy and Society in Prehistoric Europe: Changing Perspectives*, 403–30.

———. 1995a. "Introduction: Peculiar Substances." In *Consuming Habits: Drugs in History and Anthropology*, edited by J. Goodman, P. E. Lovejoy, and A. Sherratt, 1–10. London: Routledge.

———. 1995b. "Alcohol and Its Alternatives: Symbol and Substance in Pre-Industrial Cultures." In *Consuming Habits: Drugs in History and Anthropology*, edited by J. Goodman, P. E. Lovejoy, and A. Sherratt. London: Routledge, 11–46

———. 1997. *Economy and Society in Prehistoric Europe: Changing Perspectives*. Princeton, NJ: Princeton University.

———. 2003. "The Horse and the Wheel: The Dialectics of Change in the Circum-Pontic and Adjacent Areas, 4500–1500 BC." In *Prehistoric Steppe Adaptation and the Horse*, McDonald Institute Monographs, edited by M. Levine, C. Renfrew, and K. Boyle, 233–52. Cambridge: University of Cambridge.

Shideler, J. C. 1983. *A Medieval Catalan Noble Family: The Montcadas, 1000–1230.* Berkeley, CA: The University of California.

Shih, S. 1959. *On "Fan Sheng-chih Shu": An Agriculturalist's Book of China Written by Fan Sheng-chih in the First Century B.C.* Peking, China: Science.

———. 1974. *A Preliminary Survey of the Book of Odes: An Agricultural Encyclopedia of the 6th Century.* 2nd ed. Beijing: Science. First written ca. 544–44 AD by Chia Ssu-hsieh as *Essential Ways for Living of the Common People.*

Shimamura, I., ed. 1991. *Kōjien* (Japanese Dictionary), 4th revised ed. Tokyo: Iwanami Shoten.

Shimura, A. 1980. *Early Chinese Papermaking.* Tokyo: Bunseido.

Shimwell, S. E. 2005. *An Analysis of Medical Use and Prohibition of the Cannabis Plant in Modern England.* Master's Thesis in History, University of East Anglia.

Shishlina, N. I., J. van der Plicht, R. E. M. Hedges, E. P. Zazovskaya, V. S. Sevastyanov, and O. A. Chichagova. 2007. "The Catacomb Culture of the North-West Caspian Steppe: 14C Chronology, Reservoir Effect and Paleodiet." *Radiocarbon* 49 (2): 713–26.

Shoemaker, J. V. 1899. "The Therapeutic Value of *Cannabis indica.*" *Texas Medical News* 8 (10): 477–88.

Shou-zhong, Y. 1997. *The Divine Farmer's Material Medica: A Translation of the Shen Nong Ben Cao Jing.* Boulder, CO: Blue Poppy.

Shoyama, Y., F. Taura, and S. Morimoto. 2001. "Expression of Tetrahydrocannabinolic Acid Synthase in Tobacco." *Proceedings, 2001 Symposium on the Cannabinoids.* Burlington, VT: International Cannabinoid Research Society.

Shuang, T., L. Luo, S. Ge, and Z. Zhang. 2008. "Clear Genetic Structure of *Pinus Kwangtungensis* (Pinaceae) Revealed by a Plastid DNA Fragment with a Novel Minisatellite." *Annals of Botany* 102: 69–78.

Shushan, E. R. 1990. *Grave Matters.* New York: Ballantine Books.

The Sikh Encyclopedia. 2012. http://www.thesikhencyclopedia.com.

Sillasoo, Ü. 1995. "Tartu 14. ja 15. sajandi jäätmekastide taimeleidudest." [In Estonian.] *Tartu Ülikooli ArheoloogiaKabineti Toimetised* 8:115–27.

———. 1997. "Eesti keskaegsete linnade ja nende lähiümbruse arheobotaanilisest uurimisest 1989–1996." [In Estonian.] *Tartu Ülikooli Arheoloogia Kabineti Toimetised* 9:109–19.

———. 2002. "Gardens and Garden Products in Medieval Turku, Estonia." In *Nordic Archaeobotany—NAG 2000 in Umeå. Archeology and Environment 15*, edited by K. Viklund, 181–92. Umeå, Sweden: Umeå University, Environmental Archeology Laboratory, Department of Archeology and Sami Studies.

Sim, Y. 2002. *5000 Years of Korean Textiles.* Seoul: Institute for Studies of Ancient Textiles. English text by Sun-ah Kim.

Sima, Q. 1993. *Records of the Grand Historian: Han Dynasty.* Translated by Burton Watson. Hong Kong and New York: Research Center for Translation, The Chinese University of Hong Kong and Columbia University Press.

Simić, S. 1964. "Narodna medicina u Kratovu" [Folk medicine in Kratovo, Macedonia]. [In Serbian.] *Zbornik za narodni život i običaje južnih Slavena* [Anthology for the life and customs of the Southern Slavs] 42:309–443. Zagreb, Croatia: Jugoslavenska akademija znanosti i umjetnosti.

Simmonds, N. W. 1976. "Hemp." In *Evolution of Crop Plants*, edited by N. W. Simmonds. London: Longman.

Simons, M. 1995. "Signs in Wind of Morocco Drug Crop." *The New York Times*, June 18.

Simoons, F. J. 1991. *Food in China: A Cultural and Historical Inquiry.* Boca Raton, FL: CRC.

Singh, N., and S. V. S. Chauhan. 2004. "Studies on Some Leaves (*Patra*) Used for Worship in Brij Mandal." *Ethnobotany* 16 (1–2): 69–71.

Singh, V., and R. P. Pandey. 1998. *Ethnobotany of Rajasthan, India.* Jodhpur, India: Scientific Publications.

Sinskaja, E. N. 1925. "Field Crops of the Altai." *Bulletin of Applied Botany and Plant Breeding* 14 (1): 367–70.

Sjögren, P. 2006. "The Development of Pasture Woodland in the Southwest Swiss Jura Mountains over 2000 Years, Based on Three Adjacent Peat Profiles." *The Holocene* 16 (2): 210–23.

Škarić, M. 1939. "Život i obiéaji planinaca pod Fruškom Gorom" [Way of living and beliefs among "Highlanders" in Fruska Gora foothill, Srem, Voivodina]. [In Serbian.] *Srpski etnografski zbornik* 23:104–12.

Skinner, C. M. 1911. *Myths and Legends of Flowers, Trees, Fruits and Plants: In All Ages and in All Climes.* London: Lippincott.

Slovak Heritage Live Newsletter. 1996. Vol. 4, no. 1, Spring. Burnaby, BC: Canada.

Small, E. 1978. "A Numerical and Nomenclatural Analysis of Morpho-Geographic Taxa of *Humulus.*" *Systematic Botany* 3 (1): 37–76.

———. 1979. *Cannabis: The Species Problem in Science and Semantics.* Toronto, ON: Corpus Information Services Ltd.

Small, E., and A. Cronquist. 1976. "A Practical and Natural Taxonomy for *Cannabis.*" *Taxon* 25 (4): 405–35.

Small, E., and D. Marcus. 2002. "Hemp: A New Crop with New Uses for North America." In *Trends in New Crops and New Uses*, edited by J. Janick and A. Whipkey, 284–326. Alexandria, VA: ASHS.

Small, E., and H. D. Beckstead. 1973a. "Cannabinoid Phenotypes in *Cannabis Sativa.*" *Nature* 245:147–48.

———. 1973b. "Common Cannabinoid Phenotypes in 350 Stocks of *Cannabis.*" *Lloydia* 36 (2): 144–65.

Small, E., H. D. Beckstead, and A. Chan. 1975. "The Evolution of Cannabinoid Phenotypes in *Cannabis.*" *Economic Botany* 29 (3): 219–32.

Small, E., and T. Antle. 2003. "A Preliminary Study of Pollen Dispersal in *Cannabis Sativa* in Relation to Wind Direction." *Journal of Industrial Hemp* 8 (2): 37–50.

Smartt, J., and N. W. Simmonds. 1995. *Evolution of Crop Plants.* Essex, England: Harlow.

Smekalova, T. N. 2008. "Major Crops and Crop Wild Relatives of Russia." *Crop Wild Relative* 6:18–19.

Smetlan, H. W. 1989. "Der Cannabis/Humulus pollen typ und seine auswertung im pollendiagram. Archäeobotanik." *Dissertationes Botanicae* 133:25–40.

Smith, H. 2000. *Cleansing the Doors of Perception: The Religious Significance of Entheogenic Plants and Chemicals.* New York: Tarcher/Putnam.

Smith, L., V. Harris, and T. Clark. 1990. *Japanese Art: Masterpieces in the British Museum.* London: British Museum.

Smith, R. E. F., and D. Christian. 1984. *Bread and Salt: A Social and Economic History of Food and Drink in Russia.* New York: Cambridge University.

So, J. F., and E. C. Bunker. 1995. *Traders and Raiders on China's Northern Frontier.* Arthur M. Sackler Gallery, Smithsonian Institution. Seattle: University of Washington.

Soderstrom, K., and F. Johnson. 2001. "Zebra Finch CB1 Cannabinoid Receptor: Pharmacology and in Vivo and in Vitro Effects of Activation." *The Journal of Pharmacology and Experimental Therapeutics* 297: 189–97.

Soffer, O. 1985. *The Upper Paleolithic of the Central Russion Plain.* Orlando, FL: Academic.

Soffer, O., and C. Gamble, eds. 1990. *The World at 18,000 BP*, 2 vols. London: Unwin Hyman.

Soltis, D. E., C. D. Bell, S. Kim, and P. S. Soltis. 2008. "Origin and Evolution of Angiosperms." *Annals of the New York Academy of Science* 1133:3–25.

Song, B., X. Wang, F. Li, and D. Hong. 2001. "Further Evidence for Paraphyly of the Celtidaceae from the Chloroplast Gene matK." *Plant Systematics and Evolution.* 228:107–15.

Soueif, M. I. 1972. "The Social Psychology of Cannabis Consumption: Myth, Mystery and Fact." *Bulletin on Narcotics* 24 (2): 1–10.

Speranza, A., J. Hanke, B. van Geel, and J. Fanta. 2000. "Late-Holocene Human Impact and Peat Development in the Černá Hora Bog, Krkonoše Mountains, Czech Republic." *The Holocene* 10 (5): 575–85.

Stafford, C. 1995. *The Roads of Chinese Childhood: Learning and Identification in Angang.* Cambridge: Cambridge University.

Stähli, M., W. Finsinger, W. Tinner, and B. Allgöwer. 2006. "Wildfire History and Fire Ecology of the Swiss National Park (Central Alps): New Evidence from Charcoal, Pollen and Plant Macrofossils." *The Holocene* 16 (6): 805–17.

Stančikaitė, M. 2000. *Natural and Human Initiated Environmental Changes throughout the Late Glacial and Holocene in Lithuania Territory.* PhD diss., Vilnius University.

———. 2004. "Neolithic Men and Environment in Šventoji Area, West Lithuania." Abstract of paper presented at *Rapid and*

Catastrophic Environmental Changes in the Holocene and Human Response, First Joint Meeting of IGCP 490 and ICSU Environmental Catastrophes in Mauritania, the Desert and the Coast, January 4–18. Atar, Mauritania.

Stančikaitė, M., P. Šinkūunas, J. Risberg, V. Šeirienė, N. Blažauskas, R. Jarockis, S. Karlsson, and U. Miller. 2009. "Human Activity and the Environment during the Late Iron Age and Middle Ages at the Impiltis Archeological Site, NW Lithuania." *Quaternary International* 203 (1–2): 74–90.

Stanley, H. M. 1879. *Through the Dark Continent.* Vol. I. New York: Harper & Brothers.

Stearns, P. N., M. Adas, and S. B. Schwartz. 1992. *World Civilizations: The Global Experience.* New York: Harpercollins College Division.

Stebbins, G. 1976. "Seeds, Seedlings, and the Origin of Angiosperm." In *Origin and Early Evolution of Angiosperms,* edited by C. Beck, 300–311. New York: Columbia University.

Steele, J. C. 1917. *The Yi- li: Book of Etiquette and Ceremonial.* London: Probsthain.

Stefanis, C., C. Ballas, and D. Madianou. 1975. "Sociocultural and Epidemiological Aspects of Hashish Use in Greece." In *Cannabis and Culture,* edited by V. Rubin, 303–26. Paris: Mouton.

Stefanova, I. and B. Ammann. 2003. "Lateglacial and Holocene Vegetation Belts in the Pirin Mountains (Southwestern Bulgaria)." *The Holocene* 13 (1): 97–107.

Stein, M. A. 1921. *Serindia: Detailed Report of Explorations in Central Asia and Westernmost China,* 5 vols. Oxford: Oxford University,

Stevens, P. F. 2008. *Angiosperm Phylogeny Website.* Version 9, June 2008. http://www.mobot.org/MOBOT/research/APweb.

Stewart, J. R., and A. M. Lister. 2001. "Cryptic Northern Refugia and the Origins of the Modern Biota." *Trends in Ecology & Evolution* 16 (11): 608–13.

Stewart, J. R., A. M. Lister, I. Barnes, and L. Dalén. 2010. "Refugia Revisited: Individualistic Responses of Species in Space and Time." *Proceedings of the Royal Society B* 277: 661–71.

Steyermark, J. A. 1963. *Flora of Missouri.* Ames, IA: Iowa State University.

Stix, G. 2008. "Traces of a Distant Past." *Scientific American* 299 (1): 56–63.

Stockberger, W. W. 1915. "Drug Plants under Cultivation." *USDA Farmer's Bulletin* 633:1–20.

Stojanović, D. 1968. "Posmrtni običaji u Negotinskoj Krajinia" [Funeral customs in Border area of Negotin]. [In Serbian.] *Razvitak* 7 (2): 60–62.

St. Penyak, P. 2005. "Cultivated Plants of Early Medieval Europe (according to Written and Archeological Data)." In *V International Conference: "Anthropization and environment of rural settlements. Flora and vegetation." Proceedings of the Conference,* edited by S. L. Mosyakinand and M. V. Shevera, 180–84. Kyiv, Ukraine: M. G. Kholodny Institute of Botany, NAS of Ukraine.

Strehlow, W., and G. Hertzka. 1988. *Hildegard of Bingen's Medicine.* Santa Fe, NM: Bear.

Stringer, C. B., and J. Hublin. 1999. "New Age Estimates for the Swanscombe Hominid, and their Significance for Human Evolution." *Journal of Human Evolution* 37:873–77.

Stuart, C. A., and F. P. Smith. 1911. *Chinese Materia Medica, Part 1, Vegetable Kingdom.* Shanghai: Presbyterian Mission.

Subotić, S. 1904. "O našim naro nim tkaninama i rukotvorinama, nešto o narodnim običajima u životu" [About our traditional folk linen and textile decorations in fabrics and handicrafts, with reference to continuing folk traditions]. [In Serbian.] *Letopis Matice srpske, Novi Sad* 3:223–28.

Suchá, R., and P. Kočár. 1996. "Výsledky archeobotanické *makrozbytkové analýzy středověkého vodovodu v* Prachaticích" (Results of archaeobotanical macrofossil analyses from medieval water pipes in Prachatice). *Zlatá Stezka* 3:189–203.

Suh, M. Y., and K. S. Park. 2008. "Traditional Textile Materials of Paekje Kingdom." *Textile Society of America Symposium Proceedings.* Paper 135. http://digitalcommons.unl.edu/tsaconf/135.

Sullivan, R. J., E. H. Hagen, and P. Hammerstein. 2008. "Revealing the Paradox of Drug Reward in Human Evolution." *Proceeding of the Royal Society Biological Sciences,* 275 (1640): 1231–41.

Sumach, A. 1976. *A Treasury of Hashish.* Toronto, ON: Stoneworks.

Sun, W., Q. Pan, Z. Liu, Y. Meng, T. Zhang, H. Wang, and X. Zeng. 2004. "Genetic Resources of Oilseed *Brassica* and Related Species in Gansu Province, China." *Plant Genetic Resources* 2:167–73.

Svensson, S. 1965. "Wasas segel och nàgot om älder segelmakeri." [In Swedish.] *Sjöhistorisk Arsbok* 1963/64:39–82.

Symonds, P. V. 1991. "Cosmology and the Cycle of Life: Hmong Views of Birth, Death and Gender in a Mountain Village in Northern Thailand." PhD diss., Brown University, Department of Anthropology.

Syrenius (Syreński), S. Z. 1613. *Zielnik* [Herbal or medicinal plants]. [In Polish.] Krakow, Poland: Typographia Basilii Skalski. Sytsma, K. J., J. Morawetz, J. C. Pires, M. Nepokroeff, E. Conti, M. Zjhra, J. C. Hall, and M. W. Chase. 2002. "Urticalean rosids: Circumscription, rosid ancestry, and phylogenetics based on *rbcL, tmL-F,* and *ndhF* sequences." *American Journal of Botany* 89: 1531–46.

Takahara, H., S. Sugita, S. P. Harrison, N. Miyoshi, Y. Morita, and T. Uchiyama. 2000. "Pollen-Based Reconstructions of Japanese Biomes at 6,000 and 18,000 C^{14} yr BP." *Journal of Biogeography* 27 (3): 665–83.

Takase, K. 2009. *Prehistoric and Protohistoric Plant Use in the Japanese Archipelago.* Lecture presented for Project for Premodern Japan Studies at University of Southern California, Meiji University–USC exchange program, December 7.

Takhtajan, A. 1987. "Flowering Plant Origin and Dispersal: The Cradle of the Angiosperms Revisited." In *Biogeographical Evolution of the Malay Archipelago,* edited by T. C. Whitmore. Oxford: Clarendon, 26–31.

"Tales of Some Witches." n.d. *The Fife Post.* http://www.thefifepost.com/?page_id=25.

Tan, S. 1984. *A Chinese Testament: The Autobiography of Tan Shih-Hua.* Translated by S. Tretiakov. New York: Simon and Schuster.

Tapp, N. 1989. *Sovereignty and Rebellion: The White Hmong of Northern Thailand.* Singapore: Oxford University.

———. 2001. *The Hmong of China—Context, Agency and the Imaginary.* Leiden, The Netherlands: Brill.

Tarasov, P. E., T. Webb III, A. A. Andreev, N. B. Afanas'eva, N. A. Berezina, L. G. Bezusko, T. A. Blyakharchuk, et al. 1998. "Present-Day and Mid-Holocene Biomes Reconstructed from Pollen and Plant Macrofossil Data from the Former Soviet Union and Mongolia." *Journal of Biogeography* 25 (6): 1029–53.

Tarasov, P. E., V. S. Volkova, T. Webb III, J. Guiot, A. A. Andreev, L. G. Bezusko, T. V. Bezusko, G. V. Bykova, N. I. Dorofeyuk, E. V. Kvavadze, I. M. Osipova, N. K. Panova, and D. V. Sevastyanov. 2000. "Last Glacial Maximum Biomes Reconstructed from Pollen and Plant Macrofossil Data from Northern Eurasia." *Journal of Biogeography* 27 (3): 609–20.

Tarasova, P., E. Bezrukovab, E. Karabanovc, T. Nakagawae, M. Wagnerf, N. Kulaginag, P. Letunovab, A. Abzaevab, W. Granoszewskih, and F. Riedela. 2007. "Vegetation and Climate Dynamics during the Holocene and Eemian Interglacials Derived from Lake Baikal Pollen Records." *Palaeogeography, Palaeoclimatology, Palaeoecology* 252:440–57.

Tarasov, P., J. Guiyun, and M. Wagner. 2006. "Mid-Holocene Environmental and Human Dynamics in Northeastern China Reconstructed from Pollen and Archeological Data." *Palaeogeography, Palaeoclimatology, Palaeoecology* 241 (2): 284–300.

Tarasov, P., W. Granoszewski, E. Bezrukova, S. Brewer, M. Nita, A. Abzaeva, and H. Oberhänsli. 2005. "Quantitative Reconstruction of the Last Interglacial Vegetation and Climate based on the Pollen Record from Lake Baikal, Russia." *Climate Dynamics* 25:625–37.

Tarling, N. 1999. *The Cambridge History of Southeast Asia: Part 1, from Early Times to c.1500.* Cambridge: Cambridge University.

Tarvel, E. 1992. "Asustus ja rahvastik" [Settlements and population]. [In Estonian.] In *Eesti talurahva ajalugu* [History of Estonian rural population], edited by J. Kahk, 130–46. Tallinn, Estonia: Olion.

Taura, F., S. Morimoto, and Y. Shoyama. 1996. "Purification and Characterization of Cannabidiolic-Acid Synthase from *Cannabis sativa* L." *Journal of Biological Chemistry* 271 (29): 17411–16.

Taura, F., S. Morimoto, Y. Shoyama, and R. Mechoulam. 1995. "First Direct Evidence for the Mechanism of Δ9-Tetrahydrocannabinol Acid Biosynthesis." *Journal of the American Chemical Society* 117 (38): 9766–67.

Tavernier, J. B. 1676–77. *Les six voyages qu'il a fait en Turquie, en Perse, et aux Indes.* Vol. 1. Paris: chez Gervais Clouzier et Claude Barbin.

Taylor, J. 1620. *The Praise of Hemp-Seed.* Transcribed by J. Gates from the 1869 reprint (facsimile reissued in 1967) of the 1630 folio edition of Taylor's works. First published 1620, London. https://

scholarsbank.uoregon.edu/xmlui/bitstream/handle/1794/830/praise.pdf.

Taylor, N. 1963. *Narcotics: Nature's Dangerous Gifts*. New York: Dell.

Teacher's College of Northwest China, Institute for Plant Research and the Gansu Provincial Museum. 1984. "Hemp and Millet from Majiayao Culture Cite of Linjia at Dongxiang, Gansu." [In Chinese.] *Kaogu* 7:654–55, 663.

Te-k'un, C. 1964. "New Light on Ancient China." *Antiquity* 38:179–83.

Temple, R. K. 1986. "The Genius of China: 3,000 Years of Science, Discovery, and Invention." New York: Simon and Schuster.

Tewari, R., R. K. Srivastava, K. K. Singh, K. S. Saraswat, I. B. Singh, M. S. Chauhan, A. K. Pokharia, A. Saxena, V. Prasad, and M. Sharma. 2006. "Second Preliminary Report of the Excavations at Lahuradewa District Sant Kabir Nagar, U.P.: 2002–4, 2005–6." *Pragdhara* (Journal of the Uttar Pradesh State Department of Archeology) 16:35–68.

Theobald, U. 2000. *Chinaknowledge—A Universal Guide for China Studies*. http://www.chinaknowledge.de/History/Song/song-tech.html.

This, P., T. Lacombe, and M. R. Thomas. 2006. "Historical Origins and Genetic Diversity of Wine Grapes." *Trends in Genetics* 22 (9): 511–19.

Thistle, J., and J. P. Cook, eds. 1972. *Seventeenth Century Documents*. Oxford: Clarendon.

Thomas, J. 1996. "The Cultural Context of the First Use of Domesticates in Continental Central and Northwest Europe." In *The Origins and Spread of Agriculture and Pastoralism in Eurasia*, edited by D. Harris, 310–22. Washington, DC: Smithsonian Institution.

Thompson, J. D. and J. Brunet. 1990. "Hypothesis for the Evolution of Dioecy in Seed Plants." *Trends in Ecology and Evolution* 5:11–16.

Thompson, R. A. 1951. *The Russian Settlement in California. Fort Ross. Founded 1812, Abandoned 1841. Why the Russians Came and Why They Left*. Oakland, CA: Biobooks.

Thompson, R. C. 1924. *The Assyrian Herbal*. London: Luzac.

———. 1949. *A Dictionary of Assyrian Botany*. London: British Academy.

Throop, P., trans. 1998. *Hildegarde von Bingen's Physica*. Rochester, VT: Healing Arts Press.

Tishkoff, S. A., E. Dietzsch, W. Speed, A. J. Pakstis, J. R. Kidd, K. Cheung, B. Bonné-Tamir, A. S. Santachiara-Benerecetti, P. Moral, and M. Krings. 1996. "Global Patterns of Linkage Disequilibrium at the CD4 Locus and Modern Human Origins." *Science* 271:1380–87.

Tishkoff, S. A., and K. K. Kidd. 2004. "Implications of Biogeography of Human Populations for 'Race' and Medicine." *Nature Genetics* 36:S21–S27.

Tolonen, M. 1978. "Paleoecology of Annually Laminated Sediments in Lake Ahvenainen, S. Finland. I. Pollen and Charcoal Analyses and their Relation to Human Impact." *Annales Botanici Fennici* 15:177–208.

Tomlinson, P. R. 1989. *Plant Remains from 118-26 Walmgate, York*. EAU Report 89/91. Prepared for York Archeological Trust and Ancient Monuments Laboratory.

Tomlinson, P., and A. R. Hall. 1996. "A Review of the Archeological Evidence for Food Plants from the British Isles: An Example of the Use of the Archaeobotanical Computer Database (ABCD)." *Internet Archeology*, iss. 1 (September). http://intarch.ac.uk/journal/issue1/tomlinson/index.html.

Tonkov, S., H. Panovska, G. Possnert, and E. Bozilova. 2002. "The Holocene Vegetation History of Northern Pirin Mountain, Southwestern Bulgaria: Pollen Analysis and Radiocarbon Dating of a Core from Lake Ribno Banderishko." *The Holocene* 12 (2): 201–10.

Török, G. 1954. "Graveyard of Halimba-Cseres Dating from the X–XIIth Centuries." [In Hungarian.] *Folia Archaelogica* 4:95–105, 207–8.

Touw, M. 1981. "The Religious and Medicinal Use of *Cannabis* in China, India and Tibet." *Journal of Psychoactive Drugs* 13 (1): 23–34.

Toyota, M., T. Shimamura, H. Ishii, M. Renner, J. Braggins, and Y. Asakawa. 2002. "New Bibenzyl Cannabinoid from the New Zealand Liverwort *Radula Marginata*." *Chemical and Pharmaceutical Bulletin* (Tokyo) 50(10):1390–2.

Tracz, O. P. 1999. "The Things We Do: Beyond the Smoke." *The Ukrainian Weekly*, 67 (33): 7, 12.

Traveset, A. 1999. "Ecology of Plant Reproduction: Mating Systems and Pollination." In *Handbook of Functional Plant Ecology*, edited by F. I. Pugnaire and F. Valladares, 545–88. New York: Marcel Dekker, Inc.

Trivedi, A., and M. S. Chauhan. 2009. "Holocene Vegetation and Climate Fluctuations in Northwest Himalaya, based on Pollen Evidence from Surinsar Lake, Jammu Region, India." *Journal of the Geoplogical Society of India* 74:402–12.

Trojanović, S. 1990. *Vatra u običajima i životu srpskog naroda naroda* [Fire in the life and customs among Serbs]. [In Serbian.] Beograd, Serbia: Prosveta.

———. 1911. "Glavni srpski žrtveni običaji: Etnološka gradja i rasprave" [Traditional customs of making sacrifice among Serbs: Ethnological scientific material and dissertations]. [In Serbian.] *Srpska kraljevska akademija, Srpski etnografski zbornik* (Beograd, Serbia) 17:114f.

Truţă, E., Z. Olteanu, S. Surdu, M.-M. Zamfirache, and L. Oprică. 2007. "Some Aspects of Sex Derterminism in Hemp." *Analele Ştiinţifice ale Universităţii "Alexandru Ioan Cuza," Secţiunea Genetică şi BiologieMoleculară* 8 (2): 31–39.

Tseng, K.-L. 1935. *Wu ching tsung yao* [Complete compendium of military classics]. [In Chinese.] 30 vols. Shanghai: Commercial. First written ca. 1040 CE.

Tsien, T. 2004. *Written on Bamboo and Silk: The Beginnings of Chinese Books and Inscriptions*. (Rev. and updated 2nd ed., originally published in 1962). Chicago: University of Chicago Press.

Tsunoda, R., W. T. de Bary, and D. Keene. 1958. *Sources of the Japanese Tradition*. New York: Columbia University.

Tuganaev, V., and A. Tuganaev. 2002. "Medieval Agroecosystems (9th–13th Centuries) in the Region of the Present-Day City of Glazov (Udmurt Republic)." *Russian Journal of Ecology* 33 (6): 388–91. Translated from *Ekologiya* 6 (2002) 412–15.

Turner, C. E., M. A. Elsohly, and E. G. Boeren. 1980. "Constituents of *Cannabis Sativa* L. XVII. A Review of the Natural Constituents." *Journal of Natural Products* 43 (2): 169–234.

Turner, C. E., M. A. Elsohly, P. C. Cheng, and G. Lewis. 1979. "Constituents of *Cannabis Sativa* L. XIV: Intrinsic Problems in Classifying *Cannabis* based on a Single Cannabinoid Analysis." *Lloydia: Journal of Natural Products* 42 (3): 317–19.

Tusser, T. 1984. *Five Hundred Points of Good Husbandry*. Oxford: Oxford University.

Ulenbrook, J. 1968–69. "Zum chinesischen Wort hüe für 'Blut.'" *Anthropos* 63–64:75–82.

Ulving, T. 1968–69. "Indo-European Elements in Chinese?" *Anthropos* 63–64:943–51.

Underhill, A. P. 1997. "Current Issues in Chinese Neolithic Archeology." *Journal of World Prehistory* 11 (2): 103–60.

Underhill, P. A., G. Passarino, A. A. Lin, P. Shen, M. Lahr, R. A. Foley, P. J. Oefner, and L. L. Cavalli-Sforza. 2001. "The Phylogeography of Y Chromosome Binary Haplotypes and the Origins of Modern Human Populations." *Annals of Human Genetics* 65:43–62.

Ungerer, M. C., S. J. E. Baird, J. Pan, and L. H. Rieseberg. 1998. "Rapid Hybrid Speciation in Wild Sunflowers." *Proceedings of the National Academy of Sciences of the United States of America* 95 (20): 11757–76.

United Nations Office on Drugs and Crime (UNODC). 1953. "The Surprising Extinction of the Charas Trade." *Bulletin on Narcotics* 5 (1): 1–7.

———. 2007. *World Drug Report 2005*. New York: United Nations.

———. 2011. *World Drug Report 2009*. New York: United Nations.

United States Department of Agriculture. 1914. *Farmer's Cyclopedia*. Vol. 5. Garden City, NY: Doubleday, Page.

University of Utah. 2005. "The Oldest *Homo Sapiens*: Fossils Push Human Emergence Back to 195,000 Years Ago." *Science Daily*, February 23, 2005. http://www.sciencedaily.com/releases/2005/02/050223122209.htm.

Unschuld, P. U. 1985. *Medicine in China: A History of Ideas*. Los Angeles: University of California.

Uustalu, E. 1952. *The History of Estonian People*. London: Boreas.

Vahtola, J. 2003. "Population and Settlement." In *The Cambridge History of Scandinavia*, vol. I, edited by K. Helle, 559–80. Cambridge: Cambridge University.

Valsecchi, V., W. Tinner, W. Finsinger, and B. Ammann. 2006. "Human Impact during the Bronze Age on the Vegetation at Lago Lucone (Northern Italy)." *Vegetation History and Archaeobotany* 15 (2): 99–113.

van Bakel, H., J. M. Stout, A. G. Cote1, C. M. Tallon, A. G. Sharpe, T. R. Hughes, and J. E. Page. 2011. "The Draft Genome and Transcriptome of *Cannabis Sativa*." *Genome Biology* 12:R102. http://genomebiology.com/2011/12/10/R102.

van der Knaap, W. O., J. F. N. van Leeuwen, A. Fankhauser, and B. Ammann. 2000. "Palynostratigraphy of the Last Centuries in Switzerland based on 23 Lake and Mire Deposits: Chronostratigraphic Pollen Markers, Regional Patterns, and Local Histories." *Review of Palaeobotany and Palynology* 108 (1–2): 85–142.

van der Merwe, N. J. 1975. "*Cannabis* Smoking in 13th–14th Century Ethiopia: Chemical Evidence." In *Cannabis and Culture*, edited by V. Rubin, 77–80. Paris: Mouton.

———. 2005. "Antiquity of the Smoking Habit in Africa." *Transactions of the Royal Society of South Africa* 60 (2): 147–50.

van der Zee, A., ed. 2002. *Batavia Yard Guide*. Translated by Ineke Touber. Leystad, The Netherlands: Information Department, Batavia yard, National Centre for Maritime History.

van der Zee, A., and A. Klein. 2005. "*De 7 Provinciën*: Building a Replica of a Famous Dutch Warship." *Nautical Research Journal* 50 (4): 1–8.

Vandorpe, P., L. Wick, A. Schlumbaum, and S. Jacomet. 2003. "Biesheim-Kunheim 2003: Analyses botaniques préliminaires des échantillons archéobiologiques." In *Oedenburg (Haut-Rhin). Rapport 2003 sur les fouilles Franco-Germano-Suisses*, edited by M. Reddé, 193–220. Paris: EPHE (L'Ecole pratique des hautes études).

van Geel, B., N. A. Bokovenko, N. D. Burova, K. V. Chugunov, V. A. Dergachev, V. G. Dirksen, M. Kulkova, et al. 2004a. "Climate Change and the Expansion of the Scythian Culture after 850 BC: A Hypothesis." *Journal of Archeological Science* 31:1735–42.

van Geel, B., N. A. Bokovenko, N. D. Burova, K. V. Chugunov, V. A. Dergachev, V. G. Dirksen, M. Kuldova, et al. 2004b. "The Sun, Climate Change and the Expansion of the Scythian Culture after 850 BC." In *Impact of the Environment on Human Migration in Eurasia*, edited by E. M. Scott, A. Y. Alekseev, and G. Zaitseva. New York: Springer-Verlag, 151–58.

van Geel, B., V. G. Dirksen, G. I. Zaitseva, N. A. Bokovenko, N. D. Burova, M. Kulkova, H. Parzinger, et al. 2006. "Reply to S. Riehl and K. Pustovoytov, 2006, in the *Journal of Archeological Science* 33:143–44." *Journal of Archeological Science* 33:145–48.

van Haaster, H. 2003. *Archeobotanica uit's-Hertogenbosch, illieuomstandigheden, bewoningsgeschiedenis en economische ontwikkelingen in en rond een (post)Middeleeuwse stad*. Amsterdam: Academisch proefschrift, Universiteit van Amsterdam.

van Haaster, H., and K. Hänninen. 1998. "Archeobotanisch onderzoek aan enkele afvalkuilen en beerputten van de locatie Korte Begijnestraat 10 te Haarlem" [Archaeobotanical research at several waste pits and cesspools at Khort Begijnestraat 10 in Haarlem]. [In Dutch.] *BIAXiaal* 57, Amsterdam.

van Haaster, H., and O. Brinkkemper. 1995. "RADAR, a Relational Archaeobotanical Database for Advanced Research." *Vegetation History & Archaeobotany* 4:117–25.

Vanniére, B., G. Bossuet, A.-V. Walter-Simonnet, E. Gauthier, P. Barral, C. Petit, M. Buatier, and A. Daubigney. 2003. "Land Use Change, Soil Erosion and Alluvial Dynamic in the Lower Doubs Valley over the 1st Millenium AD (Neublans, Jura, France)." *Journal of Archaeological Science* 30 (10): 1283–99.

van Riebeeck, J. A. 1900. "Letters Dispatched from the Cape, 1652–1662." In *Precis of the Archives of the Cape of Good Hope*, vol. 3, archives kept by H. C. V. Leibbrandt. Cape Town, South Africa: W. A. Richards & Sons.

Vantreese, V. L. 1998. "Industrial Hemp: What Can We Learn from the World Market?" *Foresight* 5 (4): 1, 6–10.

van Zeist, W. 1964. "A Paleobotanical Study of Some Bogs in Western Brittany (Finistère), France." *Palaeohistoria* 10:157–80.

Varvel, S. A., R. J. Hamm, B. R. Martin, and A. H. Lichtman. 2001. "Differential Effects of Delta9-THC on Spatial Reference and Working Memory in Mice." *Psychopharmacology* 157:142–50.

Vasks, A., B. Vaska, and R. Grāvere. 1997. *Latvijas aizvēsture: 8500. g. pr. Kr.-1200. g. pāc Kr.: eksperimentāls metodisks līdzeklis*. [In Latvian.] Riga, Latvia: Zvaigzne ABC.

Vasks, A., L. Kalnina, and R. Ritums. 1999. "The Introduction and Pre-Christian History of Farming in Latvia." In *Environmental and Cultural History of the Eastern Baltic Region*, edited by U. Miller, T. Hackens, V. Lang, A. Raukas, and S. Hicks. *PACT* 57:291–304.

Vasquez de Espinosa, A. (ca. 1620) 1960. *Description of the Indies*. Translated by C. Upson Clark. Smithsonian Miscellaneous Collections. Vol. 102. Reprint of 1942 translation by Clark, Washington, DC: Smithsonian Institute.

Vavilov, N. I. 1922. *Polevye kultury Yugo-Vostoka* [Field Crops of southeastern Russia]. [In Russian.] Works of Applied Botany and Plant Breeding. Supplement no. 23. Leningrad, Russia: VIR.

———. 1926. "Tzentry proiskhozhdeniya kulturnykh rastenii" [The centers of origin of cultivated plants]. [In Russian and English.] *Bulletin of Applied Botany, Genetics, and Plant Breeding* 16 (2): 1–248.

———. 1931. "Rol Tzentralnoi Azii v proiskhozhdenii kulturnykh rastenii" [The role of Central Asia in the origin of cultivated plants]. [In Russian and English.] *Bulletin of Applied Botany, Genetics, and Plant Breeding* 26 (3): 3–44.

———. 1949–51. *The Origin, Variation, Immunity and Breeding of Cultivated Plants*. Translated from Russian by K. Starr Chester. Waltham, MA: *Chronica Botanica*.

———. 1957. *Agroecological Survey of the Main Field Crops*. Moscow: Academy of Sciences of the USSR.

Vavilov, N. I., and D. D. Bukinich. 1929. "Zemledel'cheskii Afganistan" [Agricultural Afghanistan]. [In Russian and English.] *The Bulletin of Applied Botany, Genetics, and Plant Breeding*. Supplement no. 33. Leningrad, Russia: VIR.

Veljić, M. M. 1925. "Strižba. u: Život i običaji narodni" [Strižba: The life and customs of the people]. *Različita graĉa za narodni život i običaje* [Material culture in folk life and customs]. [In Serbian.] 14: 391–95.

Vereshchagina, I. V. 2000. "Periodizatsija i khronologija neolita krajnego evropeiskogo severo-vostoka Rossii." [In Russian.] In *Khronologiya Neolita Vostochnoi Evropy*, edited by E. N. Nosov, 10–13. St. Petersburg: IIMK.

Veski, S. 1996. "Vegetational History of Lake Maardu Sediments based on the Pollen Stratigraphy." In "Coastal Estonia. Recent advances in environmental and cultural history," edited by T. Hackens, S. Hicks, U. Miller, V. Lang, and L. Saarse. *PACT* 51:141–50.

———. 1998. "Vegetation History, Human Impact and Paleogeography of West Estonia: Pollen Analytical Studies of Lake and Bog Sediments." *Striae* 38:1–119.

Veski, S., K. Koppel, and A. Poska. 2005. "Integrated Palaeoecological and Historical Data in the Service of Fine-Resolution Land Use and Ecological Change Assessment during the Last 1000 Years in Rõuge, Southern Estonia." *Journal of Biogeography* 32:1473–88.

Vickery, R. 1995. *A Dictionary of Plant Lore*. Oxford: Oxford University.

Vilà, C., J. A. Leonard, A. Götherström, S. Marklund, K. Sandberg, K. Lidén, R. K. Wayne, and H. Ellegren. 2001. "Widespread Origins of Domestic Horse Lineages." *Science* 291:474–77.

Vindheim, J. B. 2002. "The History of Hemp in Norway." *Journal of Industrial Hemp* 7 (1): 89–103.

Vinogradov, A. V. 1981. "*Drevnie okhotniki i rybolovy Sredneaziatskogo Mezhdorechya*." *Trudy Khrezmskoi Arkheolog-Etnograficheskoi Ekspeditsii*, 10. [Proceedings of the Khorezm Archeological and Ethnographic Expedition.] [In Russian.] Moscow: Nauka.

Viski, K. 1937. *Hungarian Peasant Customs*. Budapest: George Vajna.

Vladić-Krstić, B. 1993. "Promene u tekstilnoj radinosti Djerdapskog podunavlja od poslednjih decenija XIX veka do savremenog doba" [Recent changes in traditional textile handwork in Djerdap (Danube Basin)]. [In Serbian.] *Glasnik Etnografskog muzeja u Beogradu* 57:103–57.

———. 1997. "Tekstilna radinost u Knjazevcu i okolini u proslosti i danas" [Traditional textile handwork in Knjazevac and surroundings in the past and present times (Eastern Serbia)]. [In Serbian.] *Glasnik Etnografskog muzeja u Beogradu* 61:263.

Voight, B. F., S. Kudaravalli, X. Wen, and J. K. Pritchard. 2006. "A Map of Recent Positive Selection in the Human Genome." *PLoS Biology* 4 (3): 72–154.

von Bingen, H. 2002. *Hildegard's Healing Plants: from her Medieval Classic Physica*. Boston: Beacon.

von Gernet, A. 1995. "Nicotine Dreams: The Prehistory and Early History of Tobacco in Eastern North America." In *Consuming Habits: Drugs in History and Anthropology*, edited by J. Goodman, P. E. Lovejoy, and A. Sherratt, 47–66. London: Routledge.

Von Wissmann, H. 1891. *Im Innern Afrikas*, 3rd ed. Leipzig, Germany: F. A. Brockhaus

Vukanović, T. 1938. "Preslice i predenje u Drenici" [Distaffs of the region of Drenica in Kosovo]. [In Serbian.] *Godisnjak skopskog filozofskog fakulteta* 2 (1931–33): 230–65.

———. 1986. *Srbi na Kosovu II* [Serbs in Kosovo II]. [In Serbian.] Vranje, Serbia: Nova Jugoslavija.

———. 1989. "Witchcraft in the Central Blakans II: Protection against Witches." *Folklore* 100 (2): 221–36.

Vukmanović, J. 1935. *O bozicnim običajima u Batinjanima* [Customs for Christmas in Batinjani (Slavonia, Croatia)]. [In Serbian.] Beograd, Serbia: Glasnik Etnografskog muzeja u Beogradu.

Vuorela, I. 1995. *Keski-Suomen asutushistoriaa Hankasalmen ja Saarijärven järvikerrostumien arkistoissa* [On the inhabited history of Central Finland, lake sediment archives of Hankasalmi and Saarijävi]. [In Finnish.] Raportti [report] P 34.4.114. Espoo, Finland: Geologian Tutkimuskeskus Maaperäosasto [Soil Department for the Center for Geologic Research].

Vuorela, I., and S. Hicks. 1995. "Human Impact on the Natural Landscape in Finland. A Review of the Pollen Evidence." In *PACT 50: Landscapes and Life*, edited by A. M. Robertsson, S. Hicks, A. Åkerlund, J. Risberg, and T. Hackens, 245–57. Rixensart, Belgium: Conseil de L'europe-Council of Europe, Division of Scientific Cooperation.

Vuorela, I., T. Lempiäinen, and M. Saarnisto. 2001. "Land Use Pollen Record from the Island of Valamo, Russian Karelia." *Annales Botanici Fennici* 38:139–65.

Wagner, W. 1926. *Die chinesische Landwirtschaft*. Berlin: Parey.

Walker, D. 1955. "Studies in the Post-Glacial History of British Vegetation. XIV: Skelsmergh Tarn and Kentmore, Westmoreland." *New Phytologist* 54 (2): 222–54.

———. 1986. "Late Pleistocene–Early Holocene Vegetational and Climatic Changes in Yunnan Province, Southwest China." *Journal of Biogeography* 13 (5): 477–86.

Walton, R. P. 1938. *Marijuana: Americas's New Drug Problem: A Sociologic Question with Its Basic Explanation Dependent on Biologic and Medical Principles*. Philadelphia: J. B. Lippincott.

Wang, C., and Y. Mou. 1980. "Concerning the Excavation of the Chhien-Shan-Yang Site at Wu-Hsing, in Chekiang Province." *Khao Ku* 4:353–58.

Wang, F.-Y., X. Gong, C.-M. Hu, and G. Hao. 2008a. Phylogeography of an alpine species *Primula secundiflora* inferred from the chloroplast DNA sequence variation. *Journal of Systematics and Evolution* 46(1):13–22.

Wang, F.-Y., X. J. Ge, X. Gong, C-M. Hu and G. Hao. 2008b. "Strong Genetic Differentiation of *Primula sikkimensis* in the East Himalaya-Hengduan Mountains." *Biochemical Genetics* 46 (1–2): 75–87.

Wang, J., P. Gao, M. Kang, A. J. Lowe, and H. Huang. 2009. "Refugia within Refugia: The Case Study of a Canopy Tree (*Eurycorymbus Cavaleriei*) in subtropical China." *Journal of Biogeography* 36 (11): 2156–64.

Wang, P., and X. Sun. 1994. "Last Glacial Maximum in China: Comparison between Land and Sea." *Catena* 23 (3–4): 341–53.

Wang, S. W., and D. Y. Gong. 2000. "Climate in China during the Four Special Periods in Holocene." *Progress in Nature Science* 10 (5): 379–86.

Wang, X.-Q. 2009. "Phylogeography of Plants in the Qinghai-Tibetan Plateau." Abstract from the Darwin-China 200 Conference, October 24–26, 2009, Peking University, Beijing, China.

Wasson, R. G. 1968. *Soma: Divine Mushroom of Immortality*. New York: Harcourt, Brace and Janovitch.

Watanabe, T., K. R. Rajbhandari, K. J. Malla, and S. Yahara. 2005. *A Handbook of Medicinal Plants of Nepal*. Bangkok: Kobfai Project.

Watkins, J. 1824. *The Sermons of the Right Reverend Father in God, and Constant Martyr of Jesus Christ, Hugh Latimer, Some Time Bishop of Worcester*. Vol. 2. London: James Duncan.

Watson, E., P. Forster, M. Richards, and H. J. Bandelt. 1997. "Mitochondrial Footprints of Human Expansions in Africa." *American Journal of Human Genetics* 61:691–704.

Watson, J. L. 1988. "The Structure of Chinese Funerary Rites: Elementary Forms, Ritual Sequence, and the Primacy of Performance." In *Death Ritual in Late Imperial and Modern China*, edited by J. L. Watson and E. S. Rawski. Berkeley: University of California.

Watt, G. 1889. *Dictionary of the Economic Products of India*. Vol. 2. Calcutta, India: Government Printing.

———. 1908. *Commercial Products of India*. Calcutta, India: E. P. Dutton.

Watt, J. M. 1961. "Dagga in Africa." *Bulletin on Narcotics* 13:9–14.

Watt, J. M., and M. G. Breyer-Brandwijk. 1932. *The Medicinal and Poisonous Plants of Southern Africa*. Edinburgh: E. & S. Livingstone.

Wear, A. 2000. *Knowledge and Practice in English Medicine, 1550–1680*. Cambridge: Cambridge University.

Weber, F. C. 1968. *The Present State of Russia*. 2 vols. London: Frank Cass. First published in Frankfurt by Nicolaus Förster as *Das veraenderte Russland* 3 volumes published in sequence (1721, 1739, 1740).

Wehrli, M., A. D. Mitchell, W. O. van der Knaap, B. Ammann, and W. Tinner. 2010. "Effects of Climatic Change and Bog Development on Holocene Tufa Formation in the Lorze Valley (Central Switzerland)." *The Holocene* 20 (3): 325–36.

Wei, H. Y., and S. Coutanceau. 1976. *Wine for the Gods—An Account of the Religious Traditions of Taiwan*. Taipei: Ch'eng Wen.

Wei, X.-X., and X.-Q. Wang. 2004. "Recolonization and Radiation in *Larix* (Pinaceae): Evidence from Nuclear Ribosomal DNA Paralogues." *Molecular Ecology* 13:3115–23.

Weightman, R., and D. Kindred. 2005. *Final Report for the Department for Environment Food and Rural Affairs Review and Analysis of Breeding and Regulation of Hemp and Flax Varieties Available for Growing in the UK*. Wolverhampton, England: ADAS UK Ltd.

Weigreffe, S. J., K. J. Sytsma, and R. P. Guries. 1998. "The *Ulmaceae*, one Family or Two? Evidence from Chloroplast DNA Restriction Site Mapping." *Plant Systematics and Evolution* 210 (3–4): 249–70.

Weiss, E., W. Wetterstrom, D. Nadel, and O. Bar-Yosef. 2004. "The Broad Spectrum Revisted: Evidence from Plant Remains." *Proceedings of the National Academy Science* 101 (26): 9551–55.

Wells, R. S. 2002. *The Journey of Man: A Genetic Odyssey*. Princeton, NJ: Princeton University.

Wells, R. S., N. Yuldasheva, R. Ruzibakiev, P. A. Underhill, I. Evseeva, J. Blue-Smith, L. Jin, B. Su, R. Pitchappan, S. Shanmugalakshmi, K. Balakrishnan, M. Read, N. M. Pearson, T. Zerjal, M. T. Webster, I. Zholoshvili, E. Jamarjashvili, S. Gambarov, B. Nikbin, A. Dostiev, O. Aknazarov, P. Zalloua, I. Tsoy, M. Kitaev, M. Mirrakhimov, A. Chariev, and W. F. Bodmer. 2001. "The Eurasian Heartland: A Continental Perspective on Y-Chromosome Diversity." *PNAS* 98 (18): 10244–49.

Welten, M. 1952. "Uber die spät-und postglaciale Vegetationsgeschichte des Simmentals sowie die fruhgeschichtliche and historische Wald- and Weiderodung auf Grund pollenanalytischer Untersuchungen." *Veröffentlichung des Geobotanischen Institutes* 26:135.

Werner, E. T. C. 1961. *A Dictionary of Chinese Mythology*. New York: Julian.

Werner, J. 1964. "Frankish Royal Tombs in the Cathedrals of Cologne and Saint-Denis." *Antiquity* 38:201–16.

West, D. P. 2005. *Hemp Archives*. http://www.newheadnews.com/hemp/archives.html.

West, E. W. 1880. *Pahlavi Texts, Sacred Books of the East*. Vol. 5. Oxford: Clarendon.

Westerdahl, C. 1985a. "Sewn Boats of the North: A Preliminary Catalogue with Introductory Comments. Part 1." *The International Journal of Nautical Archeology and Underwater Exploration* 14 (1): 33–62

———. 1985b. "Sewn Boats of the North: A Preliminary Catalogue with Introductory Comments. Part 2." *The International Journal of Nautical Archeology and Underwater Exploration* 14 (2): 119–42

Westheden-Olausson, C. 1988. "Vasa's Sails." In *Textiles in European Archeology*, edited by R. B. Jurgensen and C. Rinaldo, 301–15. Göteborg, Sweden: Göteborg University.

Wheildon, T. 2000. *Complete Guide to Fishing Skills*. London: Brockhampton.

White, R. K. 1980. "The Upper Paleolithic Occupation of the Périgord: A Topographic Approach to Subsistence and Settlement." PhD diss., University of Toronto, Department of Anthropology.

Whitehead, D. R. 1983. "Wind Pollination: Some Ecological and Evolutionary Perspectives." In *Pollination Biology*, edited by L. Real, 97–108. New York: Academic.

Whitfield, R. 1982–85. *Art of Central Asia: The Stein Collection in the British Museum*. Vol. 2. Tokyo: Kodansha International Ltd.

———. 1995. *Dunhuang: Caves of the Singing Sands: Buddhist Art for the Silk Road*, 2 vols. London: Textile and Art Publications.

Whitfield, R., and A. Farrer. 1990. *Caves of the Thousand Buddhas: Chinese Art from the Silk Route*. London: The British Museum.

Whitfeld, R., S. Whitfield, and N. Agnew. 2000. *Cave Temples of Mogao: Art and History on the Silk Road*. Los Angeles: Getty Conservation Institute.

Whitfield, S. 1999a. *Life along the Silk Road*. Berkeley: University of California.

———. 1999b. *The International Dunhuang Project: Addressing the Problem of Access and Conservation of a Scattered Manuscript Collection*.

Paper presented at the 4th International Conference (Preservation of Dunhuang and Central Asian Collections) Workshop, September, 1999, St. Petersburg, Russia.

Whittier, J. C. 1882. *The Complete Poetical Works of John Greenleaf Whittier*. Boston: Houghton Mifflin.

Whittington, G., and A. D. Gordon. 1987. "The Differentiation of the Pollen of *Cannabis sativa* L. from That of *Humulus lupulus* L." *Pollen et Spores* 29 (1): 111–20.

Whittington, G., and K. J. Edwards. 1989. "Problems in the Interpretation of Cannabaceae Pollen in the Stratigraphic Record." *Pollen et Spores* 31 (1–2): 79–96.

———. 1990. "The Cultivation and Utilisation of Hemp in Scotland." *Scottish Geographical Magazine* 160:167–73.

Wieger, L. 1981. *Moral Tenets and Customs in China*. Translated and annotated by L. Davrout. New York: Garland Press. Reprint of the 1913 edition published by Catholic Mission Press, Ho-Kien-fu, China, under title *Dr. Wieger's Moral tenets and customs in China*.

Wiegrefe, S. J., K. J. Sytsma, and R. D. Guries. 1998. "The *Ulmaceae*, one Family or Two? Evidence from Chloroplast DNA Restriction Site Mapping." *Plant Systematics and Evolution* 210:249–70.

Wiermann, R. 1965. "Moorkundliche and vegetations geschichtliche Betrachtungen zum Aussendeichsmoor bei Sehestedt (Jadebusen)." *Berichte der Deutschen Botanischen Gesellschaft* 78:269–78.

Wiethold, J. 1999. "Pflanzenreste des Mittelalters und der frühen Neuzeit aus zwei Kloaken in der Hansestadt Rostock. Die Ausgrabungen Kröpeliner Straße 34–36 / Kleiner Katthagen 4." *Bodendenkmalpflege Mecklenburg-Vorpommen Jahrbuch* 46 (1998): 409–32.

———. 2000. "Ernährung und Umwelt im spätmittelalterlichen Rostock. Archäobotanische Ergebnisse der Analyse zweier Kloaken in der Kröpeliner Straße 55–56/Kuhstraße." *Bodendenkmalpflege Mecklenburg-Vorpommern Jahrbuch* 47 (1999): 351–78.

———. 2001. "Recherches archéobotaniques en France Centre-Est. Campagne 2001." In *Rapport annuel d'activité 2001 du Centre archéologique européen du Mont Beuvray.*, edited by Guichard, V, 245–56. Glux-en-Glenne, France: Centre archéologique européen du Mont Beuvray.

———. 2002a. "Giff in de schottele. Strowe dar peper up . . . Botanische Funde als Quellen zur mittel-alterlichen Ernährungs-und Umweltgeschichte in Einbeck." In *Einbeck im Mittelalter. Eine archäologische Spurensuche. Studien zur Einbecker Geschichte 17*, edited by A. Heege, 240–46. Oldenburg, Germany: Isensee Verlag.

———. 2002b. "Pflanzenreste aus einem spätlatènezeitlichen Brunnen vom oppidum Fossé de Pandours, Col de Saverne (Bas-Rhin)—Vorbericht zu den archäobotanischen Analysen." In *L'oppidum médiomatrique du Fossé des Pandours au Col de Saverne (Bas-Rhin). Rapport triennal 2000–2002*, edited by S. Fichtl and A.-M. Adam, 177–86. Straßbourg, France: Université Marc Bloch.

———. 2003a. "Archäobotanische Untersuchungen zur Ernährungs-und Wirtschaftsgeschichte des Mittelalters und der frühen Neuzeit." In *Au-delà de l'écrit. Les hommes et leurs vécus matériels au Moyen Âge à la lumière des sciences et des techniques. Nouvelles perspectives*, edited by R. Noël, I. Paquay, and J.-P. Sosson, 461–99. Louvain-la-Neuve, Belgium: Brepols.

———. 2003b. "Archäobotanische Ergebnisse." In "Bemerkenswertes aus Stralsunds Altstadt—die Grabung Apollonienmarkt 6 und ihre Ergebnisse," edited by H. Fries and J. Wiethold. *Archäologische Berichte aus Mecklenburg-Vorpommern* 10:220–47.

———. 2003c. "'Nonnenstaub'—Pflanzenreste des späten Mittelalters und der frühen Neuzeit aus dem Fußbodenhohlraum unter dem Nonnengestühl des Klarissenklosters von Ribnitz." In *Klöster und monastische Kultur in Hansestädten*, edited by C. Kimminus-Schneider and M. Schneider, 277–88. Beiträge des 4. wissenschaftlichen Kolloquiums Stralsund, 12.–15. Dezember 2001. Stralsunder Beiträge zur Archaologie, Kunst und Volkskunde in Vorpommern 4, Rhaden, Westphalia.

———. 2003d. "How to trace 'romanisation' of central Gaule by archaeobotanical analysis?—Some Considerationes on New Archaeobotanical Results from France Centre-Est." In "Actualité de la recherche en histoire et archéologie agraires," edited by F. Favory and A. Vignot, 269–82. *Actes du colloque international AGER 5, septembre 2000. Annales Litteraires 764; Environnement, sociétés et archéologie 5*. Besançon, France.

———. 2012. "Hirse, Hanf und Hohldotter—Pflanzenfunde aus einem römischen Brunnen in Otterbach, Kr. Kaiserslautern." In *Verzweigungen: Eine Würdigung für A. J. Kalis und J. Meurers-Balke,*

edited by A. Stobbe and U. Tegtmeier, 311–24. Bonn, Germany: Verlag Dr. Rudolf Habelt GmbH.

Wikstrom, N., Savolainen, V., and M. W. Chase. 2001. "Evolution of the Angiosperms: Calibrating the Family Tree." *Proceedings of Biological Science* 268 (1482): 2211–20.

Wild, J. P. 2002. "The Textile Industries of Roman Britain." *Britannia: Journal of Romano-British and Kindred Spirits* 33:1–42.

Wilhelm, H. 1944. "[Review of] Wolfram Eberhard: Untersuchungen über den Aufbau der chinesischen Kultur II. Teil 2: Die Lokalkulturen des Südens und Ostens." *Monumenta Serica* 9:209–30.

Willan, T. S. 1953. *The Muscovy Merchants of 1555*. Manchester, England: Manchester University.

———. 1956. *The Early History of the Russia Company, 1555–1603*. Manchester, England: Manchester University.

Willerding, U. 1970. "Vor-und fruhgeschichtliche kulturpflanzenfunde in Mitteleuropa." *Neue Ausgrabungen und Forchungen in Niedersachen 5*. Hildesheim, Germany: Verlag August Lax.

Williams, D. 1977. "The Plant Macrofossil Contents of Medieval Pits at Sewer Lane, Hull." In "Excavations in Sewer Lane, Hull, 1974," edited by P. Armstrong. *East Riding Archeologist 3, Old Town Report Series* (Humberside, England) 1:18–32.

Williams-Garcia, R. 1975. "The Ritual Use of Cannabis in Mexico." In *Cannabis and Culture*, edited by V. Rubin, 133–47. Paris: Mouton.

Willis, K. J. 1992. "The Late Quaternary Vegetational History of Northwest Greece. I. Lake Gramousti." *New Phytologist* 121:101–17.

———. 1997. "The Impact of Early Agriculture upon the Hungarian Landscape." In *Landscapes in Flux: Central and Eastern Europe in Antiquity*, edited by J. Chapman and P. Dolukhanov, 193–209. Oxford: Oxbow Books Ltd.

Willis, K. J., P. Sumegi, M. Braun, K. D. Bennett, and A. Toth. 1998. "Prehistoric Land Degradation in Hungary: Who, How and Why?" *Antiquity* 72:101–13.

Wilson, K. 1979. *History of Textiles*. Boulder, CO: Westview, Inc.

Wimble, G., C. E. Wells, and D. Hodgkinson. 2000. "Human Impact on Mid- and Late Holocene Vegetation in South Cumbria, UK." *Vegetation History and Archaeobotany* 9 (1): 17–30.

Windler, R., and A. Rast-Eicher. 1999–2000. "Spätmittelalterliche Weberwerkstätten in der Winterthurer Altstadt." *Zeitschrift für Archäoogie des Mittelalters* 27/28:1–82

Wirtshafter, D. 1994. "The Schlichten Papers." In *Hemp Today*, edited by E. Rosenthal, 47–62. Oakland, CA: Quick American Archives. First published 1945, Athens, OH, by Ohio Hempery, as *The Schlichten Papers*.

———. 1995. "Nutrition of Hemp Seeds and Hemp Seed Oil." In *Bioresource Hemp: Proceedings of the Symposium Frankfurt am Main, Germany March 2–5, 1995*, edited by Nova Institute, 546–55. Ojai, California: Hemptech.

Withington, E. T. 1894. *Medical History from the Earliest Times: A Popular History of the Healing Art*. London: The Scientific.

Wodehouse, R. P. 1935. *Pollen Grains*. New York: McGraw-Hill Book.

Wolf, A. P. 1970. "Chinese Kinship and Mourning Dress." In *Family and Kinship in Chinese Society*, edited by M. Freedman, 189–207. Palo Alto, CA: Stanford University.

Wong, K. C., and W. Lien-Teh. 1936. *History of Chinese Medicine*. Shanghai: National Quarantine Service.

Wood, J. 1988. "The Old Man of the Mountain in Medieval Folklore." *Folklore* 99 (1): 78–87.

Wright, A. R. 1938. *British Calendar Customs England*. London: William Glaisher Ltd.

Wu, C., and P. H. Raven, eds. 1994. *Flora of China*. Vol. 5. Beijing: Science.

Xinjiang Institute of Cultural Relics and Archeology. 2004. "Tu lu fan kao gu xin shou huo: Shanshan Xian Yanghai mu di fa jue jian bao" [New results of archeological work in Turpan: Excavation of the Yanghai Graveyard]. [In Chinese.] *Tu lu fan Xue yan jiu* [Turfanological Research] 1:1–66.

Xue, Y., T. Zerjal, W. Bao, S. Zhu, Q. Shu, J. Xu, R. Du, S. Fu, P. Li, M. E. Hurles, H. Yang, and C. Tyler-Smith. 2006. "Male Demography in East Asia: A North-South Contrast in Human Population Expansion Times." *Genetics* 172:2431–39.

Yamada, G. 1993. "Plant Remains Unearthed from Sites in Hokkaido." [In Japanese.] *Kodai Bunka* 45: 13–22.

Yamada, K. 1995. "Liberation from Occupation! An Interview with Pon (Yamada Kaiya)." *Tokyo Observer* 15. First published December 1995 in Japanese in *Jiyu Ishi*, December issue.

Yan, H., C. Peng, C. Hu, and G. Hao. 2007. "Phylogeographic Structure of *Primula Obconica* (Primulaceae) Inferred from Chloroplast Microsatellites (cpSSRs) Markers." *Acta Phytotaxonomica Sinica* 45 (4): 488–96.

Yang, F., Y. Li, X. Ding, and X. Wang. 2008. "Extensive Population Expansion of *Pedicularis Longiflora* (Orobanchaceae) on the Qinghai-Tibetan Plateau and Its Correlation with the Quaternary Climate Change." *Molecular Ecology* 17 (23): 5135–45.

Yang, M. C. 1998. *A Chinese Village—Taitou, Shantung Province*. London: Taylor and Francis. First published 1945, New York, by Columbia University.

Yang, X., S. Ji, R. T. Jones, S. Wang, G. Tong, and Z. Zhang. 2005. "Pollen Evidence of Early Human Activities in Erhai Basin, Yunnan Province." *Chinese Science Bulletin* 50 (6): 569–77.

Yanushevich, Z. V. 1989 "Agricultural Evolution North of the Black Sea from the Neolithic to the Iron Age." In *Foraging and Farming—The Evolution of Plant Exploitation*, edited by D. R. Harris and G. C. Hillman, 607–19. London: Unwin Hyman.

Yates, R. D. S. 1982. "Siege Engines and Late Zhou Military Technology." In *Exploration in the History of Science and Technology in China, Collections of Essays on Chinese Literature and History*, edited by L. Guohao, Z. Mengwen, and C. Tianqin, 409–52. Shanghai: Shanghai Chinese Classics.

Yellowpageschina.com. 2006. Internet Business Directory. http://web.archive.org/web/20071008213639/ http://www.yellowpageschina.com/144/1.

Yeloff, D., and B. van Geel. 2007. "Abandonment of Farmland and Vegetation Succession Following the Eurasian Plague Pandemic of AD 1347–52." *Journal of Biogeography* 34 (4): 575–82.

Yeloff, D., P. Broekens, J. Innes, and B. van Geel. 2007. "Late Holocene Vegetation and Land-Use History in Denmark: A Multi-Decadally Resolved Record from Lille Vildmose, Northeast Jutland." *Review of Palaeobotany and Palynology* 146 (1–4): 182–92.

Yi, D. H. 1988. "Role Playing through Trance Possession in Korean Shamanism." Cited in *Shamanism: The spirit world of Korea*, edited by R. W. I. Guisso and Chai-shin Yu, 181–92. Berkeley, CA: Asian Humanities.

Yi, S. H., Y. Saito, H. Oshima, Y. Zhou, and H. Wei. 2002. "Holocene Environmental History Inferred from Pollen Assemblages in the Huanghe (Yellow River) Delta, China: Climatic Change and Human Impact." *Quaternary Science Reviews* 22 (5–7): 609–28.

Yi-jian, S. 1936. *Wenfang sipu: Collection Studies of the Four Articles for Writing in a Scholar's Studio*. CSJC ed. Shanghai: Commerical.

Yin, P., J. Kang, F. He, L. Qu, and H. Gu. 2010. "The Origin of Populations of *Arabidopsis Thaliana* in China, based on the Chloroplast DNA Sequences." *BMC Plant Biology* 10:22.

Yin, T. 1978. "The Bronzes Unearthed at Shu-ch'eng, Anhwei." *Khao Ku* 10 (1964): 490–503. In *Chinese Archeological Abstracts 116*, edited by R. C. Rudolph. Los Angeles: UCLA Institute of Archeology.

Ying-Hsing, S. 1966. *Chinese Technology in the Seventeenth Century*. Translated from Chinese and annotated by E.-T. Zen Sun and S.-C. Sun. University Park, PA: The Pennsylvania State University.

Yoneda, M., R. Suzuki, Y. Shibata, M. Morita, T. Sukegawa, N. Shigehara, and T. Akazawa. 2004. "Isotopic Evidence of Inland-Water Fishing by a Jomon Population Excavated from the Boji Site, Nagano, Japan." *Journal of Archeological Science* 31 (1): 97–107.

Yonhap News Agency. 2010. "South Korea Continues Rescue Operations on Sunken Ship." March 27, http://english.yonhapnews .co.kr.

Yoshida, S. I., and D. Williams. 1994. *Riches from Rags: Saki-ori & Other Recycling Traditions in Japanese Rural Clothing*. San Francisco, CA: Craft and Folk Art Museum.

Youngs, F. A. 1976. *The Proclamations of the Tudor Queens*. Cambridge: Cambridge University.

Yu, G., X. Chen, J. Ni, R. Cheddadi, J. Guiot, H. Han, S. P. Harrison, C. Huang, et al. 2000. "Palaeovegetation of China: A Pollen Data-Based Synthesis for the Mid-Holocene and Last Glacial Maximum." *Journal of Biogeography* 27:635–64.

Yu, Y. 1977. "Han China." In *Food in Chinese Culture: Anthropological and Historical Perspectives*, edited by K. C. Chang, 53–83. New Haven, CT: Yale University.

Yu, Y. 1987. "Agricultural History over Seven Thousand Years in China." In *Feeding a Billion: Frontiers of Chinese Agriculture*, edited by S. Wittwer, Y. Youtai, S. Han, and W. Lianzheng, 19–33. East Lansing, MI: Michigan State University.

Yuan, Q., Z. Zhang, H. Peng, and S. Ge. 2008. "Chloroplast Phylogeography of *Dipentodon* (Dipentodontaceae) in Southwest China and Northern Vietnam." *Molecular Ecology* 17:1054–65.

Yum, H. 2003. "Traditional Korean Papermaking: Analytical Examination of Historic Korean Papers and Research into History, Materials and Techniques of Traditional Papermaking of Korea." Research paper of the Getty Postgraduate Fellow, Cornell University Library, Department of Preservation and Collection Maintenance. http://www.library.cornell.edu/preservation/publications/koreanpapermaking.html.

———. Traditional Korean Papermaking. In *Scientific Research on the Pictorial Arts of Asia* (containing papers from the 2nd Forbes Symposium, 2003), edited by P. Jett, J. Winter, and B. McCarthy. London: Archetype Publications.

———. 2010. "A Brief Account of Traditional Korean Papermaking." *Papermaking* 14 (2): 8–11.

Zagorskis, F. 1987. *Zvejnieku Akmens Laikmeta Kapulauks*. Riga, Latvia: Zinatne.

Zaimeche, S. 2005. *Damascus*. Manchester, England: Foundation for Science Technology and Civilisation. Publication ID: 4079. http://www.muslimheritage.com/uploads/damascus.pdf.

Zaitseva, G. I., and B. van Geel. 2004. "The Occupation History of the Southern Eurasian Steppe during the Holocene: Chronology, the Calibration Curve and Methodological Problems of the Scythian Chronology." In *Impact of the Environment on Human Migration in Eurasia*, edited by E. M. Scott, A. Y. Alekseev, and G. Zaitsva, 63–82. Dordrecht, The Netherlands: Kluwer Academic.

Zaitseva, G. I., B. van Geel, N. A. Bokovenko, K. V. Chugunov, V. A. Dergachev, V. G. Dirksen, M. A. Koulkova, A. Nagler, G. Parzinger, J. van der Plicht, N. D. Bourova, and L. M. Lebedeva. 2004. "Chronology and Possible Links Between Climatic and Cultural Change During the First Millennium BC in Southern Siberia and Central Asia." *Radiocarbon* 46 (1): 259–76.

Zaitseva, G. I., K. V. Chugunov, V. I. Bokovenko, V. G. Dergachev, V. Dirksen, B. van Geel, M. A. Koulkova, L. M. Lebedeva, A. A. Sementsov, J. van der Plicht, E. M. Scott, S. S. Vasiliev, K. I. Lokhov, and N. Bokovenko. 2005. "Chronological Study of Archeological Sites and Environmental Change around 2600 BP in the Eurasian Steppe Belt (Uyuk Valley, Tuva Republic)." *Geochronometria* 24:97–107.

Zajaczkowa, J. 2002. "Hemp and Nettle: Two Food/Fiber/Medical Plants in Use in Eastern Europe." *Slovo, the Newsletter of the Slavic Interest Group*. http://www.gallowglass.org/jadwiga/SCA/hempnettle.html.

Zaninović, A. 1964. "Kad se zaprede kudelja" [When the hemp is spinning]. [In Croatian.] *Etnografski prilozi s Pelješca i Lastova. Zbornik za narodni život i običaje Južnih Slavena* 42:529–34. Zagreb: Jugoslavenska akademija znanosti i umjetnosti.

Zečević, S. 1968–69. "Praznik (slava) u severoistočnoj Srbiji" [Holiday celebrations in northeast Serbia]. [In Serbian.] *Glasnik Etnografskog muzeja u Beogradu* Vols. 31–32.

———. 1975. "Kult mrtvih i samrtni običaji u okolini Bora" [Mortuary cult and concomitant funeral ritual]. [In Serbian.] *Glasnik Etnografskog muzeja u Beogradu* 38:147–69.

Zerega, N. J. C., W. L. Clement, L. S. L. Datwyler, and G. D. Weiblen. 2005. "Biogeography and Divergence Times in the Mulberry Family (Moraceae)." *Molecular Phylogenetics and Evolution* 37: 402–16.

Zernitskaya, V., and N. Mikhailov. 2009. "Evidence of Early Farming in the Holocene Pollen Spectra of Belarus." *Quaternary International* 203 (1): 91–104.

Zhang, D., D. E. Boufford, R. H. Ree, and H. Sun. 2009. "The 29°N Latitudinal Line: An Important Division in the Hengduan Mountains, a Biodiversity Hotspot in Southwest China." *Nordic Journal of Botany* 27:405–12.

Zhang, D., and H. Sun. 2010. "The Floristic Division of the South and North Parts of Hengduan Mountains, a Biodiversity Hotspot in SW China." Abstract. Presented at the 2nd International GMBA-DIVERSITAS Conference, July 27–30, Chandolin (Valais), Switzerland.

Zhang, Y., Z. Kong, J. Ni, S. Yan, and Z. Yang. 2007. "Late Holocene Palaeoenvironment Change in Central Tianshan of Xinjiang, Northwest China." *Grana* 46 (3): 197–213.

Zhilin, S. G. 2006. "Zhaman-Kaindy-I-an unusual flora of the end of Late Eocene in Central Kazakhstan" Abstract. Presented at the 7th European Paleobotany-Palynology Conference, September 6–11,

Prague. http://www.mi.uni-hamburg.de/fileadmin/files/
forschung/theomet/docs/pdf/2006-eppc_programme.pdf.

Zhimin, A. 1989. "Prehistoric Agriculture in China." In *Foraging and Farming—The Evolution of Plant Exploitation.*, edited by D. R. Harris and G. C. Hillman, 641–49.London: Unwin Hyman.

———. 1999. "Neolithic Communities in Eastern Parts of Central Asia." In *History of Civilizations of Central Asia: A Publication of the UNESCO Central Asian Project*, vol. 1, edited by A. H. Dani and V. M. Masson, 153–68. Delhi: Motilal Banarsidass.

Zhmud, L. 1997. *Wissenschaft, Philosophie und Religion im frühen Pythagoreismus.* Berlin: Akademie Verlag.

Zhukovskii, P. M. 1950. *Cultivated Plants and their Wild Relatives.* Translated by P. S. Hudson. Abridged, London: Commonwealth Agricultural Bureaux. First published 1950 in Russian, as *Kul'turnye rasteniia i ikh sorodichi.*

Zhuo, Z., Y. Baoyin, and N. Petit-Maire. 1998. "Paleoenvironments in China during the Last Glacial Maximum and the Holocene Optimum." *Episodes* 21 (3): 152–58.

Zias, J., H. Stark, J. Sellgman, R. Levy, E. Werker, A. Breuer, and R. Mechoulam. 1993. "Early Medical Use of *Cannabis*." *Nature* 363:215.

Zimmer, A., A. M. Zimmer, A. G. Hohmann, M. Herkenham, and T. I. Bonner. 1999. "Increased Mortality, Hypoactivity, and Hypoalgesia in Cannabinoid CB1 Receptor Knockout Mice." *Proceedings National Academy of Sciences* 96:5780–85.

Zohary, D. 1984. "Modes of Evolution in Plants under Domestication." In *Plant Biosystematics*, edited by W. F. Grant. Toronto, ON: Academic, 579–86.

———. 1989. "Domestication of the Southwest Asian Neolithic Crop Assemblage of Cereals, Pulses, and Flax: The Evidence from the Living Plants." In *Foraging and Farming—The Evolution of Plant Exploitation*, edited by D. R. Harris and G. C. Hillman, 359–73. London: Unwin Hyman.

Zohary, D., and M. Hopf. 1988. *Domestication of the Plants in the Old World.* London: Clarendon.

Zuardi, A. W. 2006. "History of Cannabis as a Medicine: A Review." *Revista Brasileira de Psiquiatria* 28 (2): 153–57.

Zuener, F. E. 1954. "The Cultivation of Plants." In *A History of Technology*, edited by Charles Singer. Oxford: Clarendon, 520–57.

Zvelebil, M. 1980. "The Rise of the Nomads in Central Asia." In *The Cambridge Encyclopedia of Archeology.*, edited by A. Sherratt, 252–56. New York: Crown.

———. 1994. "Plant Use in the Mesolithic and Its Role in the Transition to Farming." *Proceedings of the Prehistoric Society* 60:35–74.

———. 1996. "The Transition to Farming in the Circum-Baltic Region." In *The Origins and Spread of Agriculture and Pastoralism in Eurasia*, edited by D. Harris, 323–45. Washington, DC: Smithsonian Institution.

———. 2006. "Mobility, Contact, and Exchange in the Baltic Sea Basin 6000–2000 BC." *Journal of Archeological Science* 25 (2): 178–92.

Zvelebil, M., and P. Dolukhanov. 1991. "The Transition to Farming in Eastern and Northern Europe." *Journal of World Prehistory* 5 (3): 233–78.

Zwalf, W., ed. 1985. *Buddhism: Art and Faith.* London: The British Museum.

Zygulski, Z., Jr. 1999. "Knightly Arms—Plebian Arms." http://web.archive.org/web/20110805101747/http://www.deremilitari.org/resources/articles/zygulski.htm. First published 1999, Warsaw, in *Quaestiones medii aevi novae, Peace and War.*, vol. 4, by Instytut Historyczny Uniwersytetu.

INDEX